ABEL SYMPOSIA

Edited by the Norwegian Mathematical Society

Participants to the Abel Symposium 2005

Organizers of the event

Photo credits: Heiko Junge, Scanpix

Fred Espen Benth · Giulia Di Nunno
Tom Lindstrøm · Bernt Øksendal
Tusheng Zhang
Editors

Stochastic Analysis and Applications

The Abel Symposium 2005

Proceedings of the Second Abel Symposium, Oslo,
July 29 – August 4, 2005, held in honor of Kiyosi Itô

ABEL PRISEN
Springer

Editors

Fred Espen Benth
Centre of Mathematics for Applications (CMA)
Department of Mathematics
University of Oslo
P. O. Box 1053 Blindern
0316 Oslo, Norway
e-mail: fredb@math.uio.no

Giulia Di Nunno
Centre of Mathematics for Applications (CMA)
Department of Mathematics
University of Oslo
P. O. Box 1053 Blindern
0316 Oslo, Norway
e-mail: giulian@math.uio.no

Tom Lindstrøm
Centre of Mathematics for Applications (CMA)
Department of Mathematics
University of Oslo
P. O. Box 1053 Blindern
0316 Oslo, Norway
e-mail: lindstro@math.uio.no

Bernt Øksendal
Centre of Mathematics for Applications (CMA)
Department of Mathematics
University of Oslo
P. O. Box 1053 Blindern
0316 Oslo, Norway
e-mail: oksendal@math.uio.no

Tusheng Zhang
Department of Mathematics
University of Manchester
Oxford Road
Manchester M13 9PL
United Kingdom
e-mail: tzhang@maths.man.ac.uk

Library of Congress Control Number: 2007921524

Mathematics Subject Classification (2000): 00B20, 28BXX, 28CXX–28C20, 35CXX–35C20, 35KXX–35C15, 35C55, 35QXX–35Q40, 46FXX, 46NXX–46N10, 49LXX–49L25, 60A10, 60BXX, 60EXX, 60GXX–60G07, 60G10, 60G15, 60G17, 60G18, 60G20, 60G35, 60G40, 60G46, 60G51, 60G57, 60G60, 60G70, 60HXX–60H05, 60H07, 60H10, 60H15, 60H30, 60H35, 60H40, 60JXX–60J25, 60J35, 60J60, 60J65, 81PXX–81P15, 81SXX–81S25, 81S40, 81TXX–81T10, 81T45, 91BXX–91B16, 91B24, 91B28, 91B70, 91B72, 91B84, 93EXX–93E20

ISBN-10 3-540-70846-4 Springer Berlin Heidelberg New York
ISBN-13 978-3-540-70846-9 Springer Berlin Heidelberg New York

Springer is a part of Springer Science+Business Media

springer.com

Typesetting by the editors and SPi using a Springer LaTeX macro package

Cover design: WMX Design GmbH, Heidelberg

Printed on acid-free paper SPIN: 12019052 VA46/3100/SPi 5 4 3 2 1 0

Preface to the Series

The Niels Henrik Abel Memorial Fund was established by the Norwegian government on January 1, 2002. The main objective is to honor the great Norwegian mathematician Niels Henrik Abel by awarding an international prize for outstanding scientific work in the field of mathematics. The prize shall contribute towards raising the status of mathematics in society and stimulate the interest for science among school children and students. In keeping with this objective the board of the Abel fund has decided to finance one or two Abel Symposia each year. The topic may be selected broadly in the area of pure and applied mathematics. The Symposia should be at the highest international level, and serve to build bridges between the national and international research communities. The Norwegian Mathematical Society is responsible for the events. It has also been decided that the contributions from these Symposia should be presented in a series of proceedings, and Springer Verlag has enthusiastically agreed to publish the series. The board of the Niels Henrik Abel Memorial Fund is confident that the series will be a valuable contribution to the mathematical literature.

Ragnar Winther
Chairman of the board of the Niels Henrik Abel Memorial Fund

Kiyosi Itô, at his residence in Kyoto, Japan, April 2005

Photo Credit: Søren-Aksel Sørensen

Preface

Kiyosi Itô, the founder of stochastic calculus, is one of the few central figures of the twentieth century mathematics who reshaped the mathematical world. Today stochastic calculus, also called Itô calculus, is a central research field in mathematics, with branches to several other mathematical disciplines and with many areas of application, for example physics, engineering and biology. Perhaps the most spectacular field of applications at present is economics and finance. Indeed, the Nobel Prize in Economics in 1997 was awarded to Robert Merton and Myron Scholes for their derivation of the celebrated Black–Scholes option pricing formula using stochastic calculus.

The Abel Symposium 2005 took place in Oslo, July 29th – August 4th 2005, and was organized as a tribute to Kiyosi Itô and his works on the occasion of his 90th birthday.

Distinguished researchers from all over the world were invited to present the newest developments within the exciting and fast growing field of stochastic calculus. We were happy that so many took part to this event. They were, in alphabetical order,

- Luigi Accardi, Universita' di Roma "Tor Vergata", Italy
- Sergio Albeverio, University of Bonn, Germany
- Ole E. Barndorff-Nielsen, University of Aarhus, Denmark
- Giuseppe Da Prato, Scuola Normale Superiore di Pisa, Italy
- Eugene B. Dynkin, Cornell University, USA
- David Elworthy, University of Warwick, UK
- Hans Föllmer, Humboldt University, Germany
- Masatoshi Fukushima, Kansai University, Japan
- Takeyuki Hida, Meijo University, Nagoya, Japan
- Yaozhong Hu, University of Kansas, USA
- Ioannis Karatzas, Columbia University, USA
- Claudia Klüppelberg, Technical University of Munich, Germany
- Torbjörn Kolsrud, KTH, Sweden
- Paul Malliavin, University of Paris VI, France

- Henry P. McKean, Courant Institute of Mathematical Science, New York, USA
- Shige Peng, Shandong University, China
- Yuri A. Rozanov, CNR, Milano, Italy
- Paavo Salminen, Åbo Akademy University, Finland
- Marta Sanz-Solé, Univeristy of Barcelona, Spain
- Martin Schweizer, ETH Zürich, Switzerland
- Michael Sørensen, Univeristy of Copenhagen, Denmark
- Esko Valkeila, Helsinki University of Technology, Finland
- Srinivasa Varadhan, Courant Institute of Mathematical Science, New York, USA
- Shinzo Watanabe, Ritsumeikan University, Japan
- Tusheng Zhang, University of Manchester, England
- Xianyin Zhou, Chinese University of Hong Kong

In addition there were many other international experts both attending and presenting valuable contributions to the conference. We are grateful to all for making this symposium so successful.

The present volume combines both papers from the invited speakers and contributions by the presenting lecturers. We are happy that so many sent their papers for publication in this proceedings, making it a valuable account of the research frontiers in stochastic analysis. Our gratitude is also directed to all the referees that put time and effort in reading the manuscripts.

A special feature of this volume is given by the Memoirs that Kiyosi Itô himself wrote for this occasion. We all thank him for these valuable pages which mean so much to both young and established researchers in the field.

We also thank the Abel Foundation, through the Norwegian Mathematical Society, and the Centre of Mathematics for Applications (CMA) at the University of Oslo, for their financial support and for their help with the preparation and organization of the symposium. Our special thanks go to Inga Bårdshaug Eide and Helge Galdal for their help with all practical matters before and during the conference.

Last but not least we are indebted to Sergio Albeverio for his scientific advice in the organization of the program.

Oslo, November 2006

Fred Espen Benth
Giulia Di Nunno
Tom Lindstrøm
Bernt Øksendal
Tusheng Zhang

Contents

Memoirs of My Research on Stochastic Analysis

Kiyosi Itô

Professor Emeritus, Kyoto University, Kyoto, 606-8501 Japan

It is with great honor that I learned of the 2005 Oslo Symposium on Stochastic Analysis and Applications, which is devoted to my work and its further developments. I would like to thank the symposium organizers for their tireless efforts in organizing this successful symposium and for providing me with the opportunity to present some memoirs of my research on stochastic analysis, which, I hope, will be of some interest to the participants.

My doctoral thesis published in 1942 [1] was on a decomposition of the sample path of the continuous time stochastic process with independent increments, now called the Lévy–Itô decomposition of the Lévy process. In the 1942 article written in Japanese [2] and the extended 1951 version that appeared in the Memoirs of the American Mathematical Society [3], I succeeded in unifying Lévy's view on stochastic processes and Kolmogorov's approach to Markov processes and created the theory of stochastic differential equations and the related stochastic calculus. As beautifully presented in a recent book by Daniel Stroock [11], *Markov Processes from K. Itô's Perspective,* my conception behind those works was to take, in a certain sense, a Lévy process as a tangent to the Markov process. The above mentioned papers are reprinted in *Kiyosi Itô Selected Papers* edited by Stroock and Varadhan [8], where the editors' introduction and my own foreword explain in some detail the circumstances leading to their development.

From 1954 to 1956, I was a Fellow at the Institute for Advanced Study at Princeton University, where Salomon Bochner and William Feller, both great mathematicians, were among the faculty members. In the preceding year, while still at Kyoto University, I had written a paper on stationary random distributions [4], using a Laurent Schwartz's extension of Bochner's theorem to a positive definite distribution representing it by a slowly increasing measure. As I learned from Bochner in Princeton, this had essentially already been obtained by Bochner himself by other means.

Feller had just finished his works on the most general one-dimensional diffusion process, especially representing its local generator as

$$\mathcal{G} = \frac{d}{dm}\frac{d}{ds}$$

by means of a canonical scale function s and a speed measure m. I learned about these from Henry McKean, a graduate student of Feller, while I explained my previous work to McKean. There was once an occasion when McKean tried to explain to Feller my work on the stochastic differential equations along with the above mentioned idea of tangent. It seemed to me that Feller did not fully understand its significance, but when I explained Lévy's local time to Feller, he immediately appreciated its relevance to the study of the one-dimensional diffusion. Indeed, Feller later gave us a conjecture that the Brownian motion on $[0, \infty)$ with an elastic boundary condition could be constructed from the reflecting barrier Brownian motion by killing its local time $\mathbf{t}(t, 0)$ at the origin by an independent exponentially distributed random time, which was eventually substantiated in my joint paper with McKean [9] published in 1963 in the Illinois Journal of Mathematics.

After my return to Kyoto from Princeton, McKean visited Kyoto in 1957–1958, and our intensive collaboration continued until our joint book *Diffusion Processes and Their Sample Paths* appeared from Springer in 1965 [10]. This coincides with the period when Dynkin and Hunt formulated the general theory of strong Markov processes along with their transformations by additive functionals and the associated probabilistic potential theory. The Kyoto probability seminars attracted many young probabilists in Japan; S. Watanabe, H. Kunita and M. Fukushima were among my graduate students. The primary concern of the seminar participants including myself was to fully understand the success of the study of one-dimensional diffusions and to look for its significant extensions to more general Markov processes. Let me mention some of the later developments of a different character that grew out of this exciting seminar atmosphere.

A popular saying by Feller goes as follows: A one-dimensional diffusion traveler X_t makes a trip in accordance with the road map indicated by the scale function s and with the speed indicated by the measure m appearing in the generator $\mathcal{G}$ of X_t. This was substantiated in my joint book with McKean in the following fashion. Given a one-dimensional standard Brownian motion X_t which corresponds to $ds = dx$, $dm = 2dx$, consider its local time $\mathbf{t}(t, x)$ at $x \in R^1$ and the additive functional defined by

$$A_t = \int_{R^1} \mathbf{t}(t, x) m(dx).$$

Then the time changed process X_{τ_t} by means of the inverse τ_t of A_t turns out to be the diffusion governed by the generator $\dfrac{d^2}{dmdx}$.

Observe that the transition function of the one-dimensional diffusion is symmetric with respect to the speed measure m and the associated Dirichlet form

$$\mathcal{E}(u,v) = -\int_{R^1} u \cdot \mathcal{G}v(x)dm(x) = \int_{R^1} \frac{du}{ds}\frac{dv}{ds}ds$$

is expressed only by the scale s, being separated from the symmetrizing measure m. Hence we are tempted to conjecture that the 0-order Dirichlet form $\mathcal{E}$ indicates the road map for the associated Markov process X_t and is invariant under the change of the symmetrizing measures m corresponding to the random time changes by means of the positive continuous additive functionals of X_t. The notion of the Dirichlet form was introduced by Beurling and Deny as a function space framework of an axiomatic potential theory in 1959, where already the road map was clearly indicated in analytical terms (the Beurling–Deny formula of the form) but the role of the symmetrizing measure m was much less clear. Being led by the above-mentioned picture of the one-dimensional diffusion path, the conjecture has been affirmatively resolved in later works by Fukushima and others (see the 1994 book by Fukushima, Takeda and Oshima *Dirichlet Forms and Symmetric Markov Processes,* [12]).

In 1965, M. Motoo and S. Watanabe wrote a paper [13] in which they made a profound analysis of the structure of the space of square integrable martingale additive functionals of a Hunt Markov process. In the meantime, the Doob–Meyer decomposition theorem of submartingales was completed by P.A. Meyer. These two works merged into a paper by H. Kunita and S. Watanabe which appeared in the Nagoya Mathematical Journal in 1967 [14] and a series of papers by P.A. Meyer in the Strasbourg Seminar Notes in 1967 [15], where the stochastic integral was defined for a general semi-martingale, and the stochastic calculus I initiated in 1942 and 1951 was revived in a new general context. Since then, various researchers including myself also became more concerned about the stochastic calculus and stochastic differential equations.

My joint paper [9] with McKean in 1963 gave a probabilistic construction of the Brownian motion on $[0, \infty)$ subjected to the most general boundary condition whose analytic study had been established by Feller under some restrictions. Our methods involved the probabilistic idea originated in Lévy about the local time and excursions away from 0. In 1970, the idea was extended in my paper in the Proceedings of the Sixth Berkeley Symposium [7], where I considered a general standard Markov process X_t for which a specific one point a is regular for itself. A Poisson point process taking values in the space U of excursions around point a was then associated, and its characteristic measure (a σ-finite measure on U) together with the stopped process obtained from X_t by the hitting time of a was shown to uniquely determine the law of the given process X_t. This approach may be considered as an infinite dimensional analogue to a part of the decomposition of the Lévy process I studied in 1942, and may have revealed a new aspect in the study of Markov processes.

The one-dimensional diffusion theory is still important as a basic prototype of Markov processes. Besides my joint book [10] with McKean, I also gave a comprehensive account of the Feller generator as a generalized second order differential operator in Section 6 of my *Lectures on Stochastic Processes* at the Tata Institute of Fundamental Research, Bombay, 1960 [6]. The second part of my book *Stochastic Processes* [5] written in Japanese and published in 1957 contains a detailed description of the Feller generator and, in addition, of the boundary behaviors of the solutions of the associated homogeneous equation

$$(\lambda - \mathcal{G})u = 0, \qquad \lambda > 0,$$

in an analytical way together with their probabilistic implications. I had sent the Japanese original of [5] at the time of its publication to Eugene B. Dynkin and it was translated into Russian by A.D. Wentzell in 1960 (Part I) and in 1963 (Part II). In 1959 Shizuo Kakutani at Yale University, noting the importance of my description of the one dimensional diffusions, advised Yuji Itô, at that time one of his graduate students, to produce a translation of the second part into English, which was then distributed among a limited circle of mathematicians around Yale University as a typewritten mimeograph. I am very glad to hear that a full English translation of the book [5] by Yuji Itô is now being prepared for publication by the American Mathematical Society under the title *Essentials of Stochastic Processes.*

Finally, let me extend my deepest gratitude to the symposium organizers and participants for honoring my 90th birthday with your work on stochastic analysis. I also wish to thank you again for allowing me to present these memoirs to you here, and I very much look forward to studying all the papers presented at this symposium.

References

1. K. Itô, On stochastic processes (infinitely divisible laws of probability) (Doctoral thesis), *Japan. Journ. Math.* **XVIII**, 261–301 (1942)
2. K. Itô, Differential equations determining a Markoff process (in Japanese), *Journ. Pan-Japan Math. Coll.* No. 1077 (1942); (in English) *Kiyosi Itô Selected Papers,* Springer-Verlag, 1986
3. K. Itô, On stochastic differential equations, *Mem. Amer. Math. Soc.* **4**, 1–51 (1951)
4. K. Itô, Stationary random distributions, *Mem. Coll. Science. Univ. Kyoto, Ser. A,* **28** 209–223 (1953)
5. K. Itô, *Stochastic Processes* **I, II** (in Japanese), Iwanami-Shoten, Tokyo, 1957
6. K. Itô, *Lectures on Stochastic Processes,* Tata Institute of Fundamental Research, Bombay, 1960
7. K. Itô, Poisson point processes attached to Markov processes, in: *Proc. Sixth Berkeley Symp. Math. Statist. Prob.* **III**, 225–239 (1970)
8. *Kiyosi Itô Selected Papers*, edited by D.W. Stroock and S.R.S. Varadhan, Springer-Verlag, 1986
9. K. Itô and H.P. McKean, Jr., Brownian motions on a half line, *Illinois Journ. Math.* **7**, 181–231 (1963)

10. K. Itô and H.P. McKean, Jr., *Diffusion Processes and Their Sample Paths*, Springer-Verlag, 1965; in Classics in Mathematics, Springer-Verlag, 1996
11. D. Stroock, *Markov Processes from K. Itô's Perspective*, Princeton University Press, 2003
12. M. Fukushima, Y. Oshima and M. Takeda, *Dirichlet Forms and Symmetric Markov Processes,* Walter de Gruyter, 1994
13. M. Motoo and S. Watanabe, On a class of additive functionals of Markov processes, *J. Math. Kyoto Univ.* **4**, 429–469 (1965)
14. H. Kunita and S. Watanabe, On square integrable martingales, *Nagoya Math. J.* **30**, 209–245 (1967)
15. P.A. Meyer, Intégrales stochastiques (4 exposés), in: *Séminaire de Probabilités I*, Lecture Notes in Math. **39**, Springer-Verlag, 72–162 (1967)

Itô Calculus and Quantum White Noise Calculus

Luigi Accardi[1] and Andreas Boukas[2]

[1] Centro Vito Volterra, Università di Roma Tor Vergata via Columbia, 2–00133 Roma, Italy. accardi@Volterra.mat.uniroma2.it, http://volterra.mat.uniroma2.it

[2] Department of Mathematics and Natural Sciences, American College of Greece, Aghia Paraskevi, Athens 15342, Greece, andreasboukas@acgmail.gr

Summary. Itô calculus has been generalized in white noise analysis and in quantum stochastic calculus. Quantum white noise calculus is a third generalization, unifying the two above mentioned ones and bringing some unexpected insight into some old problems studied in different fields, such as the renormalization problem in physics and the representation theory of Lie algebras. The present paper is an attempt to explain the motivations of these extensions with emphasis on open challenges.

The last section includes a result obtained after the Abel Symposium. Namely that, after introducing a new renormalization technique, the RHPWN Lie algebra includes (in fact we will prove elsewhere that this inclusion is an identification) a second quantized version of the extended Virasoro algebra, i.e. the Virasoro–Zamolodchikov $*$–Lie algebra w_∞, which has been widely studied in string theory and in conformal field theory.

1 Introduction

The year 2005 marks Kiyosi Itô's 90th birthday and, with it, the 63th birthday of stochastic calculus. The present Abel Symposium, devoted to the celebration of these events, offers to all mathematicians an important occasion to meditate on this important development in their discipline whose influence is going to follow the times of history, even in a period when the pace of scientific development has reached a level in which most papers have a life time of less than one year.

The applications of Itô's work have been so many, ranging from physics to biology, from logistics and operation research to engineering, from meteorology to mathematical finance, ..., that an exhaustive list is impossible.

From the mathematical point of view it is someteimes underestimated the fact that Itô calculus, with its radical innovation of the two basic operations of calculus – differentiation and integration – has been one of the few real conceptual breakthroughs in the development of classical analysis after Newton.

Itô laid down the foundations of stochastic calculus in his 1942 thesis [Itô42a, Itô42b] and the first systematic exposition of these ideas in English language appeared almost ten years later in [Itô51] and preceeded of about 15 years the now classical monograph [ItôMcKn65]. This gave rise to an impetuous development which has seen as protagonists several of the participants to the present conference and which will be reviewed by them.

My talk will take the move from one of the basic achievements of this development, completed in the late 1960's, and which led to the mathematical substantiation of a limpid and intuitive picture of the structure of a classical stochastic process indexed by the real line (interpreted as time) and with values in $\mathbb{R}^d$ (interpreted as a generalized phase space).

The sample space of a *generic process* of this type is identified to a space of $\mathbb{R}^d$–valued functions, interpreted as trajectories of a dynamical system, and each trajectory is canonically decomposed into a sum of two parts: a *regular* (bounded variation) part, corresponding to the drift in the stochastic equation and a *pure fluctuation* term, corresponding to the martingale part in the stochastic equation. The former part is handled with classical, *Newtonian*, calculus; the latter with Itô calculus. The picture is completed by the Kunita–Watanabe *martingale representation theorem* [KunWat67], which characterizes the *generic martingales* as stochastic integrals with respect to some stationary, independent increment process and by the *Lévy–Itô decomposition* of a stationary, independent increment process (Z_t):

$$Z_t = mt + \sigma B_t + X_t$$

where m is a constant, B_t is a Brownian motion and X_t is a compound Poisson process, i.e. an integral

$$X_T = \int_0^T dt \int P_{u,t} d\beta(u)$$

of independent Poisson processes $P_{u,t}$ with intensity of jumps equal to u, with respect to a measure $d\beta(u)$, called the **Levy measure** and with support in $\mathbb{R} \setminus \{0\}$.

The early generalizations of Itô calculus had gone in the direction of extending it to more general *state spaces* thus passing from $\mathbb{R}^d$ to manifolds or to infinite dimensions or both. Another, less developed extension was from vector valued to operator valued classical stochastic processes [Skor84]. However these extensions did not change the basic conceptual framework of the theory.

The situation changed in the past 30 years when *three qualitative innovations* appeared. This drastically enlarged not only the conceptual status of Itô calculus, and more generally of stochastic analysis, but also its technical apparatus. The traditional bridges between probability, classical analysis and combinatorics became an intricate network including practically every field of mathematics, from operator theory to graph theory, from Hopf algebras to group representations, . . .

The traditional applications to the classical world (physics, information, communications, engineering, finance,...) have now been expanded to the corresponding sectors in the quantum world thus bringing a remedy to the historical paradox according to which the mathematical discipline, dealing with the laws of chance, was not powerful enough to include into its framework the most advanced physical theory, quantum mechanics, in which chance enters in a much more intrinsic way than in any other physical theory.

These innovations begun with two, initially quite separated and independent, lines of research: *white noise analysis*, (1975) and *quantum stochastic calculus* (1982) and found their unification, starting from 1993, in *quantum white noise calculus.*

The rate of progression of these events, as well as the merging of different generalizations into a single, unified picture, has been so swift that, even for those who actively participated in the construction of these developments, it is quite hard to follow all the new ideas and to embrace the whole landscape in a single eyesight.

It is precisely on this broad picture that the present paper will be focused. Not only details, but also several important achievements, will be omitted from the exposition, in the attempt to convey an idea of some of the exciting new perspectives of *quantum stochastic analysis*.

The first attempts to go beyond the Itô calculus framework and to include processes which, although much more singular, were frequently used in the physics and engineering literature, was Hida *white noise theory*, first proposed in his Carleton lectures of 1975 [Hida75, Hida92].

The second conceptual generalization of Itô calculus took place in 1982 when Hudson and Parthasarathy developed their *quantum stochastic calculus* [HuPa82a, HuPa84c]. In it for the first time, the noises themselves (i.e. the martingales driving the stochastic differential equations) were no longer classical additive independent increment processes but *quantum independent increment processes.* This was the first quantum generalization of Itô calculus and opened the way to all subsequent ones. The culmination of the theory is the determination of the structure of those stochastic equations which admit a unitary solution. The reason why this result has fundamental implications both for quantum mechanics and for classical probability, will be explained starting from Section (7).

The Hudson–Parthasarathy theory inspired, directly or indirectly, most of the developments of quantum probability for the decade after its appearance. Its importance can be compared to the original Itô paper and the multiplicity of investigations it motivated was surveyed in [Partha92].

But the story does not end here: a third conceptual generalization, motivated by the stochastic limit of quantum theory, was developed between 1993 and 1995 and can be described as the unification of the white noise and the quantum stochastic approach: the non triviality of this unification will be clear starting from Section (12) of the present exposition. In particular this third step threw a new and unexpected light on the *microscopic structure of quantum, hence in particular classical, stochastic equations* as a consequence of:

(i) the discovery of the Hamiltonian structure of the (classical and quantum) stochastic differential equations
(ii) the discovery of the *translation code* between white noise and stochastic differential equations. This required the development of the *theory of distributions on the standard simplex* [AcLuVo99] which is the mathematical counterpart of the *time consecutive principle* of the stochastic limit of quantum theory.

However the main point of the new development was not so much the deeper understanding of the structure of classical and quantum stochastic calculus, but the possibilities it opened of further extensions, which cannot be obtained with the traditional tools of stochastic analysis. In fact the *white noise extension of the Itô table* opened the way to the nonlinear generalizations of Itô calculus to which is devoted the second part of the present report.

The beautiful landscape emerging from the simplest of these extensions, i.e. the one dealing with the second power of white noise, and the subsequent, totally unexpected unification of the five Meixner classes as classical subprocesses (algebraically: Cartan $*$–sub algebras) respectively of the first and second order white noise, rose strong hopes that this hierarchy could be extended from the second powers of white noise to its higher powers. This would lead to a new, interesting class of infinitely divisible processes (for a short while there was even the hope to obtain a new parametrization of all these processes).

This hope however collided with the wall of the *no go theorems* described in the last part of the present paper. Although negative results, these theorems are very interesting because they have revealed an hitherto unknown phenomenon relating stochastic analysis to two different fields, each of which has been the object of a huge literature outside probability theory namely:

(i) the representation theory of infinite dimensional Lie algebras
(ii) renormalization theory.

These two theories are at the core of contemporary theoretical physics and the fact that some developments, motivated by quantum white noise analysis, could bring new insight and new results in such a fundamental issue, which resisted decades of efforts from the best minds of theoretical physics, is an indication that this direction is deep and worth being pursued. For this reason while the first part of the present paper consists in an exposition of already established results, in the second part emphasis has been laid on the formulation of the problems facing the construction of a satisfactory theory of the higher powers of white noise. This has led to the introduction of some new notions, such as *Fock representation of a Lie algebra*, which are going to play an essential role in the development of the theory.

All these developments show that Itô calculus shares, with the richest and deepest mathematical theories, the germs of its radical innovation. Historical experience shows that these innovations often occur in directions which are quite unexpected for the experts of the field and this sometimes generates a feeling of extraneousness.

An instructive example is given by the theory of elliptic functions, originated from a deep intuition of Abel and initially developed within a purely analytical context, but now stably settled in a purely algebraic and geometrical framework.

The story we are going to tell shows that Itô calculus gives another important example in this direction.

2 Plan of the Present Paper

The goal of the present section is twofold: (i) to give a more analytical outline of the content of the present paper; (ii) to catch this occasion to say a few words about the motivations and the inner logic underlying the developments described here as well as about their connections with other sectors of quantum probability which could not be dealt with for reasons of space.

Section (3) defines the notion of quantum (Boson Fock) white noise and illustrates, in this basic particular case, one of the main ideas of quantum probability, i.e. the idea that *algebra implies statistics*. Let me just mention here that also the converse statement, i.e. that statistics implies algebra (e.g. commutation or anti commutation relations), is true and it lies at a deeper level. The first result in this direction was proved by von Waldenfels in the Bose and Fermi case [voWaGi78, voWa78] and about 20 years later, with the introduction of the notion of interacting Fock space [AcLuVo97b], this principle became a quite universal principle of probability theory and opened the way to the program of a full algebraic classification of probability measures. This is a quite interesting direction, and is also deeply related to the main topic of the present paper, stochastic and white noise calculus, but we will not discuss this connection and we refer the interested reader to [AcBo98, AcKuSt02, AcKuSt05a].

Section (4) describes another important new idea of quantum probability, i.e. the notion of *quantum decomposition of a classical random variable* (or stochastic process). This idea is illustrated in the important particular case of classical white noise and extended, in Section (6), to the Poisson noise.

The two above mentioned decompositions are at the root of Hudson–Parthasarathy's quantum extension of classical Itô calculus, briefly outlined in Section (6).

Section (7) briefly describes the classical Schrödinger and Heisenberg equations as a preparation to their stochastic and white noise versions.

The algebraic form of a classical stochastic process is described in Section (8). This leads to a reformulation, explained in Section (10), of classical stochastic differential equations, that makes quite transparent their equivalence to stochastic versions of the classical Schrödinger or of Heisenberg equations.

In Sections (9), (10) it is briefly outlined how this reformulation is nothing but a stochastic analogue of Koopman's algebraization of the theory of classical, deterministic dynamical systems.

Combining the content of Section (8) with the quantum decomposition of classical white and Poisson noise, described in Sections (4), (6), one arrives, in Section (11), to the full quantum versions of the stochastic Schrödinger and Heisenberg equations, which are the main object of study of the Hudson–Parthasarathy theory.

These equations are not of Hamiltonian type and they were developed by Hudson and Parthasarathy on the basis of a purely mathematical analogy with the classical Itô calculus. Hence their connection with the Hamiltonian equations of quantum physics was obscure and the early applications of these equations to physical problems, proposed by Barchielli [Barc88], Belavkin [Bela86a], Gardiner and Collet [GaCo85], ... , were built on a purely phenomenological basis. This led to some misgivings among physicists on the meaning of these models and their relations to the fundamental laws of quantum mechanics.

On the other hand, combining the main results of Hudson and Parthasarathy (construction of unitary Markovian cocycles) with the quantum Feynman–Kac formula of [Ac78b] we see that, by quantum conditioning of a stochastic Heisenberg evolution $X_0 \mapsto U_t X_0 U_t^*$ on the time zero algebra, one obtains a quantum Markov semigroup (P^t):

$$E_{0]}\left(U_t X_0 U_t^*\right) = P^t(X_0) \tag{2.1}$$

just as the analogue classical conditioning leads to a classical Markov semigroup. It was also known, since the early results of Pauli and van Hove, that quantum Markov semigroups (P^t) (and the associated master equations, which are the quantum analogue of the Chapman–Kolmogorov equations) can arise as appropriate time–scaling limits of reduced Heisenberg evolutions. The time–scaling being the same one used in classical stochastic homogenization (i.e. $t \to t/\lambda^2$), and known in the physical literature as *van Hove* or $1/\lambda^2$–scaling, and the limit being taken for $\lambda \to 0$. Since (in this particular context) the physical operation of *reducing an Heisenberg evolution to a subsystem*, used in these papers, is mathematically equivalent to conditioning on the time zero algebra, the above statement can be rewritten as:

$$\lim_{\lambda \to 0} E_{0]}\left(U^{(\lambda)}_{t/\lambda^2} X_0 U^{(\lambda)*}_{t/\lambda^2}\right) = P^t(X_0) \tag{2.2}$$

Comparing (2.1) with (2.2) it was therefore natural to conjecture that also the unconditioned limits,

$$\lim_{\lambda \to 0} U^{(\lambda)}_{t/\lambda^2} X_0 U^{(\lambda)*}_{t/\lambda^2} = U_t X_0 U_t^* \tag{2.3}$$

$$\lim_{\lambda \to 0} U^{(\lambda)}_{t/\lambda^2} = U_t \tag{2.4}$$

of the original Schrödinger and Heisenberg evolutions should exist, for some (at those times unspecified) topology, and satisfy some quantum stochastic Schrödinger and Heisenberg equations of Hudson–Parthasarathy type.

This conjecture was formulated by Frigerio and Gorini immediately after the development of quantum stochastic calculus [FrGo82a] and was proved a few years later by Accardi, Frigerio and Lu [AcFrLu87].

This result marked the beginning of the stochastic limit of quantum theory. It proved that *quantum stochastic differential equations arise as physically meaningful scaling and limiting procedures from the fundamental laws of quantum mechanics, expressed in terms of Hamiltonian equations.* This produced, among other things, a microscopic interpretation not only of the coefficients of the stochastic equations, but also of the fine structure of the driving martingales (quantum noises).

Several years later Accardi, Lu and Volovich [AcLuVo93] realized that in fact *stochastic differential equations (both classical and quantum) are themselves Hamiltonian equations* but not of usual type: they are *white noise Hamiltonian equations.* The identification of these two classes of equations required the development of new mathematical techniques such as the notion of *causal normal order* and the strictly related *time consecutive principle* and *theory of distributions on the standard simplex* (cf. [AcLuVo02] for a discussion of these notions).

The inclusion: *classical and quantum SDE* $\subseteq$ *WN Hamiltonian equations* is a consequence of this development and is described in Sections (12), (13). These few pages condensate a series of developments which took place in several years and in several papers. The interested reader is referred to [AcLuVo99] (the first attempt to systematize the impetuous development of the previous years) and to the more recent expositions [Ayed05] (thesis of Wided Ayed) and the papers [AcAyOu03, AcAyOu05a, AcAyOu05b]. The last of this papers deals with another one of the several interesting developments born from the stochastic limit of quantum theory which, for lack of space, are not discussed in the present paper, namely the module generalization of white noise calculus and the qualitatively new structure of the quantum noises emerging from it (the reader, interested in the first and main physical example of this new structure, is referred to [AcLuVo97b]).

Even more condensed is the description, in Sections (14), (15), (16), (17), of the renormalized square of WN. This is because the survey paper [AcBou04c] is specifically devoted to this subject and the interested reader can find there the necessary information.

On the contrary, since most of the material in Sections from (18) to (22) has not yet been published, we tried to give all the necessary definitions even if proofs had to be omitted for reasons of space.

The general problem, concerning the renormalized higher powers of WN, is formulated in Section (21) with the related no–go theorems. As explained in

Section (22), this problem is also related with an old open problem of classical probability, i.e. the infinite divisibility of the odd powers of a standard Gaussian random variable.

Further investigations are needed to understand the effective impact of these no–go theorems. Do they really close the hope of a general theory of higher powers of white noise? Our feeling is that the answer to this question is *no*! This hope is supported by the following considerations. The no–go theorems heavily depend on:

(i) the choice of a renormalization procedure;
(ii) the fact that we restrict our attention to a very special representation, i.e. the Fock one.

A way out of this conundrum has to be looked for in the relaxation of one of these assumption, i.e. one has to look for either new renormalization procedures or different representations. Both ways are now under investigation and raise challenging but fascinating mathematical problems.

The last section of the present paper refers to a development that took place after the end of the Abel Symposium and which shows that the idea to look for different types of renormalization procedures turned out particularly fruitful and brought to the fore a connection between the renormalized higher powers of white noise and the Virasoro algebra which promises to be as rich of developments as the connection between the renormalized square of white noise and the Meixner classes.

3 Fock Scalar White Noise (WN)

Definition 1. *The standard d–dimensional Fock scalar White Noise (WN) is defined by a quadruple*

$$\{\mathcal{H}, b_t, b_t^+, \Phi\}; \quad t \in \mathbb{R}^d$$

where $\mathcal{H}$ is a Hilbert space, $\Phi \in \mathcal{H}$ a unit vector called the (Fock) vacuum, and b_t, b_t^+ are operator valued distributions (for an explanation of this notion see the comment at the end of the present section and the discussion in [AcLuVo02], Section (2.1)) with the following properties.

The vectors of the form

$$b_{t_n}^+ \cdots b_{t_1}^+ \Phi \tag{3.1}$$

called the number vectors are well defined in the distribution sense and total in $\mathcal{H}$.

b_t is the adjoint of b_t^+ on the linear span of the number vectors

$$(b_t^+)^+ = b_t \tag{3.2}$$

Weakly on the same domain and in the distribution sense:

$$[b_s, b_t^+] := b_s b_t^+ - b_t^+ b_s = \delta(t - s) \tag{3.3}$$

where, here and in the following, the symbol $[\,\cdot\,,\,\cdot\,]$ *will denote the commutator:*

$$[A,B] := AB - BA$$

Finally b_t *and* Φ *are related by the Fock property (always meant in the distribution sense):*

$$b_t\Phi = 0 \tag{3.4}$$

The unit vector Φ *determines the expectation value*

$$\langle \Phi, X\Phi\rangle =: \langle X\rangle \tag{3.5}$$

which is well defined for any operator X *acting on* $\mathcal{H}$ *and with* Φ *in its domain.*

Remark. In the Fock case *algebra implies statistics* in the sense that the algebraic rules (3.3), (3.2), (3.4) uniquely determine the restriction of the expectation value (3.5) on the polynomial algebra generated by b_t and b_t^+. This is because, with the notation

$$X^\varepsilon = \begin{cases} X\,, & \varepsilon = -1 \\ X^*\,, & \varepsilon = +1 \end{cases} \tag{3.6}$$

the Fock prescription (3.4) implies that the expectation value

$$\langle b_{t_n}^{\varepsilon_n} \cdots b_{t_1}^{\varepsilon_1}\rangle \tag{3.7}$$

of any monomial in b_t and b_t^+ is zero whenever either n is odd or $b_{t_1}^{\varepsilon_1} = b_{t_1}$ or $b_{t_n}^{\varepsilon_n} = b_{t_n}^+$. If neither of these conditions is satisfied, then there is a $k \in \{2,\ldots,n\}$ such that the expectation value (3.7) is equal to

$$\langle b_{t_n}^{\varepsilon_n} \cdots b_{t_1}^{\varepsilon_1}\rangle = \langle b_{t_n}^{\varepsilon_n} \cdots b_{t_{k+1}}^{\varepsilon_{k+1}} [b_{t_k}, b_{t_{k-1}}^+ \cdots b_{t_1}^+]\rangle \tag{3.8}$$

Using the derivation property of the commutator $[b_{t_k}\,,\,\cdot\,]$ (i.e. (7.4)) one then reduces the expectation value (3.8) to a linear combination of expectation values of monomials of order less or equal than $n-2$. Iterating one sees that only the scalar term can give a nonzero contribution.

Remark. The **practical rule to deal with operator valued distributions** is the following: products of the form (3.7) are meant in the sense that, after multiplication by $\varphi(t_n)\cdot\ldots\cdot\varphi(t_1)$, where $\varphi_1,\ldots,\varphi_n$ are elements of an appropriate test function space (typically one chooses the space of smooth functions decreasing at infinity faster than any polynomial), and integration with respect to all variables $dt_1\cdot\ldots\cdot dt_n$ (each of which runs over $\mathbb{R}^d$) one obtains a product of well defined operators whose products contain the vector Φ in their domains. Here and in the following we will not repeat each time when an identity has to be meant in the distribution sense.

4 Classical Real Valued White Noise

Lemma. Let b_t, b_t^+ be a Fock scalar white noise. Then

$$w_t := b_t + b_t^+ \tag{4.1}$$

is a classical real random variable valued distribution satisfying:

$$w_t = w_t^+ \tag{4.2}$$

$$[w_s, w_t] = 0; \qquad \forall s, t \tag{4.3}$$

$$\langle w_t \rangle = 0 \tag{4.4}$$

$$\langle w_s w_t \rangle = \delta(t-s) \tag{4.5}$$

$$\langle w_{t_{2n}} \cdots w_{t_1} \rangle = \sum_{\{l_\alpha, r_\alpha\} \in p.p.\{1,\ldots,2n\}} \prod_{\alpha=1}^{n} \langle w_{t_{l_\alpha}} w_{t_{r_\alpha}} \rangle \tag{4.6}$$

moreover all odd moments vanish and $p.p.\{1,\ldots,2n\}$ denotes the set of all pair partitions of $\{1,\ldots,2n\}$.

Remark. The self–adjointness condition (4.2) and the commutativity condition (4.3) mean that (w_t) is (isomorphic to) a classical real valued process. Conditions (4.4) and (4.5) mean respectively that (w_t) is mean zero and δ–correlated. Finally (4.6), which follows from (3.4) and from the same arguments used to deduce the explicit form of (3.7), shows that the classical process (w_t) is Gaussian.

Definition 2. *The process (w_t) satisfying (4.2),…,(4.5) (one can prove its uniqueness up to stochastic equivalence) is called the standard d–dimensional classical real valued White Noise (WN). The identity (4.1) is called the quantum decomposition of the classical d–dimensional white noise.*

Remark. Notice that, for the classical process (w_t), it is not true that algebra implies statistics: this becomes true only using the quantum decomposition (4.1) combined with the Fock prescription (3.4).

Remark. In the case $d = 1$, integrating the classical WN one obtains the classical Brownian motion with zero initial condition:

$$W_t = B_t + B_t^+ = \int_0^t ds(b_s^+ + b_s) \tag{4.7}$$

Notice that (4.7) gives the q–decomposition of the classical BM just as (4.1) gives the q–decomposition of the classical WN.

From now on we will only consider the case $d = 1$.

5 Classical Subprocesses Associated to the First Order White Noise

An important generalization of the quantum decomposition (4.1) of the classical white noise is the identity:

$$p_t(\lambda) = b_t + b_t^+ + \lambda b_t^+ b_t; \quad \lambda \geq 0 \tag{5.1}$$

which can be shown to define (in the sense of vacuum distribution) a 1–parameter family of classical real valued distribution processes (i.e. $p_t(\lambda) = p_t(\lambda)^+$ and $[p_s(\lambda), p_t(\lambda)] = 0$). In fact this classical process can be identified, up to a time rescaling, to the compensated scalar valued standard classical Poisson noise with intensity $1/\lambda$ and the identity (5.1) gives a q–decomposition of this process.

Integrating (5.1), in analogy with (4.7), one obtains the standard compensated Poisson processes. Notice that the critical value

$$\lambda = 0$$

corresponds to the classical WN while any other value

$$\lambda \neq 0$$

gives a Poisson noise. As a preparation to the discussion of Section (17) notice that $\lambda = 0$ is the only critical point, i.e. a point where the vacuum distribution changes and that these two classes of stochastic processes exactly coincide with the first two Meixner classes.

6 The Hudson–Parthasarathy Quantum Stochastic Calculus

In the previous sections we have seen that, integrating the densities

$$w_t = b_t + b_t^+$$

$$p(\lambda)_t = b_t + b_t^+ + \lambda b_t^+ b_t$$

one obtains the stochastic differentials (random measures) as WN integrals

$$dW_t = \int_t^{t+dt} w_s ds = \int_t^{t+dt} (b_s + b_s^+) ds =: dB_t^+ + dB_t$$

$$dP_t(\lambda) = \int_t^{t+dt} p_s(\lambda) ds = \int_t^{t+dt} (b_s + b_s^+ + \lambda b_s^+ b_s) ds = dB_t^+ + dB_t + \lambda dN_t$$

Starting from these one defines the classical stochastic integrals with the usual constructions.

$$\int_0^t F_s dW_s; \qquad \int_0^t F_s dP_s(\lambda)$$

The passage to q–stochastic integrals consists in separating the stochastic integrals corresponding to the different pieces. In other words, the quantum decomposition (5.1) suggests to introduce separately the stochastic integrals

$$\int_0^t F_s dB_s; \quad \int_0^t F_s dB_s^+; \quad \int_0^t F_s dN_s$$

This important development was due to Hudson and Parthasarathy and we refer to the monograph [Partha92] for an exposition of the whole theory.

7 Schrödinger and Heisenberg Equations

A *Schrödinger equation* (also called an *operator Hamiltonian equation*) is an equation of the form:

$$\partial_t U_t = -iH_t U_t; \quad U_0 = 1; \quad t \in \mathbb{R} \tag{7.1}$$

where the 1–parameter family of symmetric operators on a Hilbert space $\mathcal{H}$

$$H_t = H_t^*$$

is called the *Hamiltonian*. In the pyhsics literature one often requires the positivity of H_t. We do not follow this convenction in order to give a unified treatment of the usual Schrödinger equation and of its so–called *interaction representation form*. This approach is essential to underline the analogy with the white noise Hamiltonian equations, to be discussed in Section (12).

When H_t is a self–adjoint operator independent of t, the solution of equation (7.1) exists and is a 1–parameter group of unitary operators:

$$U_t \in Un(\mathcal{H}); \; U_s U_t = U_{s+t}; \; U_0 = 1; \; U_t^* = U_t^{-1} = U_{-t}; \; s, t \in \mathbb{R}$$

Conversely every 1–parameter group of unitary operators is the solution of equation (7.1) for some self–adjoint operator $H_t = H$ independent of t.

An *Heisenberg equation*, associated to equation (7.1), is

$$\partial_t X_t = \delta_t(X_t); \qquad X_0 = X \in \mathcal{B}(\mathcal{H}) \tag{7.2}$$

where δ_t has the form

$$\delta_t(X_t) := -i[H_t, X_t]; \qquad X_0 = X \in \mathcal{B}(\mathcal{H}) \tag{7.3}$$

One can prove that δ_t is a $*$–derivation, i.e. a linear operator on an appropriate subspace of the algebra $\mathcal{B}(\mathcal{H})$ of all the bounded operators on $\mathcal{H}$, also called *the algebra of observables*, satisfying (on this subspace):

$$\delta_t(ab) = \delta_t(a)b + a\delta_t(b) \tag{7.4}$$

$$\delta_t^*(a) := \delta_t(a^*)^* = \delta_t(a)$$

Not all $*$–derivations δ_t on subspaces (or sub algebras) of $\mathcal{B}(\mathcal{H})$ have the form (7.3). If this happens, then the $*$–derivation, δ_t, and sometimes also the Heisenberg equation, is called *inner* and its solution has the form

$$X_t = U_t X_t U_t^* \tag{7.5}$$

where U_t is the solution of the corresponding Schrödinger equation (7.1). Conversely, every solution U_t of the Schrödinger equation (7.1) defines, through (7.5), a solution of the Heisenberg equation (7.2) with δ_t given by (7.3).

Thus every Schrödinger equation is canonically associated to an Heisenberg equation. The converse is in general false, i.e. there are Heisenberg equations with no associated Schrödinger equation (equivalently: not always a derivation is inner). The simplest physically relevant examples of this situation are given by the quantum generalization of the so called *interacting particle systems* [AcKo00b] which have been widely studied in classical probability.

8 Algebraic Form of a Classical Stochastic Process

Let (X_t) be a real valued stochastic process. Define

$$j_t(f) := f(X_t)$$

In the spirit of quantum probability, we realize f as a multiplication operator on $L^2(\mathbb{R})$ and $f(X_t)$ as a multiplication operator on

$$L^2(\mathbb{R} \times \Omega, \mathcal{B}_{\mathbb{R}} \times \mathcal{F}, dx \otimes P) \equiv L^2(\mathbb{R}) \otimes L^2(\Omega, \mathcal{F}, P)$$

where $(\Omega, \mathcal{F}, P)$ is the probability space of the process (X_t) and $\mathcal{B}_{\mathbb{R}}$ denotes the Borel σ–algebra on $\mathbb{R}$. Sometimes we use the notation:

$$M_f \varphi(x) := f(x)\varphi(x); \qquad \varphi \in L^2(\mathbb{R})$$

The same notation will be used if $x \in \mathbb{R}$ is replaced by $(x, \omega) \in \mathbb{R} \times \Omega$.

Thus $f(X_t)$ is realized as multiplication operator on $L^2(\mathbb{R}) \otimes L^2(\Omega, \mathcal{F}, P)$. With these notations, for each $t \geq 0$, j_t is a $*$–homomorphism

$$j_t : \mathcal{C}^2(\mathbb{R}) \subseteq \mathcal{B}(L^2(\mathbb{R})) \to \mathcal{B}(L^2(\mathbb{R}) \otimes L^2(\Omega, \mathcal{F}, P))$$

9 Koopman's Argument and Quantum Extensions of Classical Deterministic Dynamical Systems

The following considerations, due to Koopman, constitute the basis of the algebraic approach to dynamical systems which reduces the study of such systems to the study of 1–parameter groups of unitary operators or of

∗–automorphisms of appropriate commutative ∗–algebras or, at infinitesimal level, to the study of appropriate Schrödinger or Heisenberg equations.

To every ordinary differential equation in $\mathbb{R}^d$

$$dx_t = b(x_t)dt; \qquad x(0) = x_0 \in \mathbb{R}^d$$

such that the initial value problem admits a unique solution for every initial data x_0 and for every $t \geq 0$: one associates the 1–parameter family of maps

$$T_t : \mathbb{R}^d \to \mathbb{R}^d$$

characterized by the property that the image of x_0 under T_t is the value of the solution at time t:

$$x_t(x_0) =: T_t x_0; \qquad T_0 = id$$

Uniqueness then implies the semigroup property:

$$T_t T_s = T_{t+s}$$

If the above properties hold not only for every $t \geq 0$, but for every $t \in \mathbb{R}$, then the system is called reversible. In this case each T_t is invertible and

$$T_t^{-1} = T_{-t}$$

Typical examples of these systems are the classical Hamiltonian systems. They have the additional property that the maps T_t preserve the Lebesgue measure (Liouville's theorem).

Abstracting the above notion to an arbitrary measure space leads to the notion of (deterministic) dynamical system:

Definition 3. *Let (S, μ) be a measure space. A classical, reversible, deterministic dynamical system is a pair:*

$$\{(S, \mu); \ (T_t) \ t \in \mathbb{R}\}$$

where $T_t : S \to S$ $(t \in \mathbb{R})$ is a 1–parameter group of invertible bi–measurable maps of (S, μ) admitting μ as a quasi–invariant measure:

$$\mu \circ T_t \sim \mu$$

The quasi–invariance of (S, μ) is equivalent to the existence of a μ–almost everywhere invertible Radon–Nikodym derivative:

$$\frac{d(\mu \circ T_t)}{d\mu} =: p_{\mu,t} \in L^1(S, \mu)$$

$$p_{\mu,t} > 0; \ \mu - \text{a.e.}; \quad \int_S p_{\mu,t}(s) d\mu(s) = 1$$

Therefore for any $t \in \mathbb{R}$ the linear map

$$\varphi \mapsto U_t\varphi(s) := p_{\mu,t}(s)^{-1/2}\varphi(T_t^{-1}s)$$

is well defined for any measurable function φ. Moreover

$$\int_S (p_{\mu,t}(s))^{-1/2}\overline{f}(T_t^{-1}(s))(p_{\mu,t}(s))^{-1/2}g(T_t^{-1}(s))d\mu(s)$$

$$= \int_S \overline{f}(T_t^{-1}s)g(T_t^{-1}s)p_{\mu,t}(s)^{-1}d\mu(s)$$

$$= \int_S \overline{f}(T_t^{-1}s)g(T_t^{-1}s)p_{\mu,t}(T_tT_t^{-1}s)^{-1}d\mu(T_tT_t^{-1}s)$$

and since

$$d\mu(T_tT_t^{-1}s) = \frac{d\mu T_t}{d\mu}(T^{-1}s)d\mu(T^{-1}s)$$

the change of variables

$$T_t^{-1}s = s'$$

gives

$$\langle U_tf, U_tg\rangle = \int_S \overline{f}(s')g(s')d\mu(s') = \langle f, g\rangle$$

i.e. the map

$$U_t : f \in L^2(S,\mu) \to p_{\mu,t}^{-1/2} f \circ T_t^{-1} \in L^2(S,\mu)$$

defines a unitary operator in $L^2(S,\mu)$. Similarly one proves that the family (U_t) is a 1–parameter unitary group:

$$U_tU_s = U_{t+s}; \qquad U_t^* = U_{-t}; \qquad U_0 = id$$

By Stone's theorem there exists a self–adjoint operator H, on $L^2(S,\mu)$, which is the infinitesimal generator of this 1–parameter group, i.e.

$$U_t = e^{-itH}$$

and this is equivalent to the Schrödinger equation

$$\partial_t U_t = -iHU_t; \qquad U_0 = 1$$

The unitarity implies that:

$$U_t^*\varphi = U_t^{-1}\varphi = p_{\mu,t}(s)^{1/2}\varphi \circ T_t$$

therefore, recalling that M_f denotes the multiplication operator by f ($M_f\varphi = f\varphi$), one has

$$(U_t M_f U_t^* \varphi)(s) = (p_{\mu,t})(s)^{-1/2}(M_f U_t^* \varphi)(T_t^{-1}s)$$
$$= p_{\mu,t}(s)^{-1/2} f(T_t^{-1}s) p_{\mu,t}(s)^{1/2}\varphi(s) = f(T_t^{-1}s)\varphi(s)$$

In conclusion: the Heisenberg evolution on $\mathcal{B}(L^2(S,\mu))$:

$$x \mapsto U_t x U_t^* =: j_t(x)$$

canonically associated to the family (U_t), satisfies the identity:

$$M_{f\circ T_t^{-1}} = U_t M_f U_t^* =: j_t(M_f)$$

This implies that the Abelian algebra $L^\infty(S,\mu)$ (considered as a sub–algebra of on $\mathcal{B}(L^2(S,\mu))$) is left invariant by each j_t:

$$j_t(L^\infty(S,\mu)) = U_t L^\infty(S,\mu) U_t^* \subseteq L^\infty(S,\mu)$$

Replacing the multiplication operator M_f by an arbitrary bounded operator x, acting on $L^2(S,\mu)$ one obtains a quantum extension of a classical deterministic system. Abstraction from this situation suggests the following definition:

Definition 4. *A deterministic, reversible, dynamical system is a pair:*

$$\{\mathcal{A};\ (j_t)\ t \in \mathbb{R}\}$$

where $\mathcal{A}$ is a $*$–algebra and (j_t) a 1–parameter group of automorphisms of $\mathcal{A}$. If $\mathcal{A}$ is a C^* (W^*) algebra, then one speaks of a C^* (W^*) dynamical system. If $\mathcal{A}$ is Abelian the system is called classical; otherwise it is called quantum. Finally, if $\mathcal{A}$ is non Abelian but it contains an Abelian sub–algebra $\mathcal{A}_{cl}$ left invariant by j_t ($j_t(\mathcal{A}_{cl}) \subseteq \mathcal{A}$) then it is called a quantum extension of a classical system.

10 Stochastic Extension of Koopman's Approach: Emergence of Schrödinger and Heisenberg Equations in Classical Stochastic Analysis

In the present section we will replace, in the above Koopman's argument, the deterministic trajectory $(x_t(x_0))$ by a stochastic process (X_t) and show how the general algebraization procedure described in Section (8), when applied to the simple and important example of a classical diffusion flow (X_t), naturally leads to a classical stochastic generalization of the Heisenberg equation.

Let (X_t) denote the real valued solution of the classical stochastic differential equation

$$dX_t \; = \; ldt \; + \; adW_t; \; X(0) = X_0 \tag{10.1}$$

driven by classical Brownian motion (W_t) and with adapted coefficients l, a which guarantee the existence and uniqueness of a strong solution for all initial data X_0 in $L^2(\mathbb{R})$ and for all times. The initial value X_0 is a random variable independent of (W_t). By Itô's formula equation (10.1) is equivalent to

$$df(X_t) = \left(l\partial_x f + \frac{1}{2}a^2\partial_x^2 f \right) dt + \; a\partial_x f dW_t \tag{10.2}$$

where $f\colon \mathbb{R} \to \mathbb{R}$ varies in a space of sufficiently smooth functions.

Since X_t depends also on the initial condition $x \in \mathbb{R}$, $f(X_t)$ is realized as multiplication operator on

$$L^2(\mathbb{R}) \otimes L^2(\Omega, \mathcal{F}, P)$$

where $(\Omega, \mathcal{F}, P)$ is the probability space of the increment process of the Brownian motion. In the following we shall simply write f $(or\, f(X_t))$ to mean the multiplication operator by $f(or\ f(X_t))$. When confusion can arise we shall write M_f or $M_{f(X_t)}$. With these notations one has:

$$[\partial_x, f] = [\partial_x, M_f] = \partial_x \cdot f - f \cdot \partial_x = \partial_x \; f = M_{\partial_x f}$$

Therefore

$$[\partial_x, [\partial_x, f]] = [\partial_x, [\partial_x, M_f]] = M_{\partial_x^2 f} = \Delta \; f = M_{\Delta f}$$

Introducing the *momentum operator* on $L^2(\mathbb{R})$:

$$p := \frac{1}{i}\partial_x$$

defined on those functions in $L^2(\mathbb{R})$ with a derivative also in $L^2(\mathbb{R})$, we can write

$$\partial_x f = i[p, f]; \qquad \partial_x^2 f = -[p, [p, f]]$$

More generally, interpreting both f and l as multiplication operators and using the fact that f commutes with l, one finds:

$$\begin{aligned} lf' = l\partial_x f = li[p, f] = \frac{i}{2} l[p, f] + \frac{i}{2}[p, f]l = \frac{i}{2} lpf \\ -\frac{i}{2} lfp + \frac{i}{2} pfl - \frac{i}{2} fpl = i\left[\frac{1}{2} lp + \frac{1}{2} pl, f\right] =: i[p(l), f] \end{aligned} \tag{10.3}$$

and therefore:

$$a^2\partial_x^2 f = a\partial_x a\partial_x f - a(\partial_x a)\partial_x f = -[p(a), [p(a), f] - i[p(a\partial_x a), f]$$

Thus, with $p(a)$ and $p(l)$ defined by (10.3), equation (10.2) can be written in the form

$$df(X_t) = i[p(a), f(X_t)]dW_t$$
$$+\left(i\ [p(l),\ f(X_t)] - \frac{1}{2}[p(a),\ [p(a), f(X_t)]] - i\left[p\left(\frac{1}{4}\partial_x a^2\right), f(X_t)\right]\right)dt$$

In conclusion, denoting

$$l + \frac{1}{2}\partial_x a^2 =: l_1 \tag{10.4}$$

we write equation (10.2) in the form

$$df(X_t) = i[p(a), f(X_t)]dW_t - \frac{1}{2}\,[p(a), [p(a), f(X_t)]]dt + i[p(l_1), f(X_t)]dt$$

and, recalling the notation $f(X_t) = j_t(f)$, this is equivalent to

$$dj_t(f) = i[p(a), j_t(f)]dW_t - \frac{1}{2}\,[p(a), [p(a), j_t(f)]]dt + i[p(l_1), j_t(f)]dt \tag{10.5}$$

Since any operator acting on $L^2(\mathbb{R})$ can be identified to the operator $T\otimes 1$, acting on $L^2(\mathbb{R})\otimes L^2(\Omega, \mathcal{F}, P)$, also the multiplication operators by functions f in $\mathcal{C}^2(\mathbb{R})$, can be realized as pre–closed operators acting on $L^2(\mathbb{R})\otimes L^2(\Omega, \mathcal{F}, P)$. With this identification, the homomorphisms j_t can be considered as maps

$$j_t : \mathcal{C}^2(\mathbb{R}) \otimes 1 \subseteq \mathcal{B}(L^2(\mathbb{R})) \otimes 1 \subseteq \mathcal{B}(L^2(\mathbb{R})) \otimes \mathcal{B}(L^2(\Omega, \mathcal{F}, P))$$
$$\equiv \mathcal{B}\big(L^2(\mathbb{R}) \otimes (L^2(\Omega, \mathcal{F}, P)\big)$$

i.e. as homomorphisms from the Abelian sub–algebra $\mathcal{C}^2(\mathbb{R}) \otimes 1$ of $\mathcal{B}(L^2(\mathbb{R}) \otimes L^2(\Omega, \mathcal{F}, P))$ into $\mathcal{B}(L^2(\mathbb{R}) \otimes L^2(\Omega, \mathcal{F}, P))$. These homomorphisms j_t are characterized by the property of being the unique solution of the stochastic equation (10.5) with initial condition

$$j_0(f) = f$$

where we use the same symbol f to denote the function f and the multiplication operator by the function f acting on $L^2(\mathbb{R})$.

Now notice that the SDE (10.5) continues to have a meaning even if the multiplication operator by the function f is replaced by an arbitrary operator x acting on $L^2(\mathbb{R})$:

$$dj_t(x) = i[p(a), j_t(x)]dW_t - \frac{1}{2}\,[p,(a), [p(a), j_t(x)]]dt + i[p(l_1), j_t(x)]dt$$

and the initial condition by its natural generalization

$$j_0(x) = x$$

Under general conditions one can extend the existence and uniqueness theorem for the classical equation to the quantum equation, for example by proving the convergence of the iterated series in an appropriate topology.

Then, by a standard application of the classical Itô table, we can conclude that

$$j_t(x)j_t(y) = j_t(xy)$$

Similarly, using again uniqueness and the identity

$$[a, b]^* = [b^*, a^*]$$

one proves that

$$j_t(x)^* = j_t(x^*)$$

and therefore j_t is a $*$–homomorphism from $\mathcal{B}(L^2(\mathbb{R})) \equiv \mathcal{B}(L^2(\mathbb{R})) \otimes 1$ to $\mathcal{B}(L^2(\mathbb{R})) \otimes \mathcal{B}(L^2(\Omega, \mathcal{F}, P))$.

To complete the analogy with Koopman's argument we ask ourselves the following question: *does there exist a 1–parameter family of linear operators U_t acting on $L^2(\mathbb{R}) \otimes L^2$(Wiener space) such that, for any random variable X_0, independent of (W_t) and with initial distribution absolutely continuous with respect to the Lebesgue measure on $\mathbb{R}$, and for any $f \in L^\infty(\mathbb{R})$, one has*

$$U_t M_{f(X_0)} U_t^* = M_{f(X_t)} \tag{10.6}$$

Notice that, if such an U_t exist, then it must be unitary because:

(i) $U_t U_t^* = 1 \Leftrightarrow j_t(1) = 1$
(ii) $U_t^* U_t - 1 \Leftrightarrow$ the map $f \mapsto U_t M_{f(X_0)} U_t^*$ is a $*$–homomorphism

To answer the above question one can argue as follows: if such U_t exists it must be a functional of the $(W_s)_{s \leq t}$ because of (10.6). Thus we postulate an equation for U_t of the form

$$dU_t = (\alpha dt + \beta dW_t) U_t \tag{10.7}$$

Then, differentiating the unitarity conditions for U_t, i.e. $U_t U_t^* = U_t^* U_t = 1$ and using the classical Itô table, we deduce a relation between α and β. After that we differentiate, again using the classical Itô table, both sides of the identity

$$f(X_t) = j_t(M_f) = U_t M_f(X_0) U_t^*$$

and identify the coefficients of the differentials dW_t and dt. The result is a classical, i.e. driven by classical BM, stochastic Schrödinger equation (SSE):

$$dU_t = \left(-iK dW_t - \left[\frac{1}{2} K^2 + iH \right] dt \right) U_t \tag{10.8}$$

with (in the notations (10.3), (10.4)):

$$H = H^* = -p(l_1); \qquad K = K^* = -p(a)$$

In the absence of noise, i.e. $K = 0$ ($\Leftrightarrow a = 0$ in (10.1)), equation (10.8) becomes an usual Schrödinger equation (7.1). Notice that both H and K are symmetric hence, without the dissipation term $K^2/2$, equation (10.8) would be formally Hamiltonian. However the dissipation term is essential for the unitarity of the solution. Thus we see that, within the context of stochastic differential equations, the requirement of unitarity of the solution is in contradiction with the Hamiltonian character of the equation. This problem will persist in the context of quantum stochastic differential equations: only the white noise approach will be able to overcome this problem.

11 Quantum Stochastic Schrödinger and Heisenberg Equations

The transition from classical to quantum stochastic Schrödinger and Heisenberg equations is now accomplished by using the quantum decomposition of the classical Brownian motion $dW_t = dB_t^+ + dB_t$ and allowing for different coefficients of the quantum stochastic differentials dB_t^+ and dB_t.

Differentiating the unitarity conditions for U_t and using the Hudson–Parthasarathy Itô table, we deduce a relation between the coefficients of dB_t^+, dB_t and dt. The final form of the equation is then:

$$dU_t = \left(D dB_t^+ - D^+ dB_t - \left[\frac{1}{2} D^+ D + iH \right] dt \right) U_t \tag{11.1}$$

where D and H are arbitrary, say bounded, operators and $H = H^*$. The same argument, applied to a more general equation, including also the number differential dN_t leads to the most general Hudson–Parthasarathy stochastic Schrödinger equation:

$$dU_t = \left(SD dB_t^+ - D^* dB_t + (S-1) dN_t + \left(-\frac{1}{2} D^+ D + iH \right) dt \right) U_t \tag{11.2}$$

where D and H are as above and S must be a unitary operator. Notice that, contrary to the diffusion case (11.1), here the Hamiltonian nature of the equation is lost even at the level of the martingale term: the non Hamiltonian nature of equation (11.2) is not due only to the presence of the dissipative term $D^+D/2$ but also of the unitary operator S. The deep meaning of this apparently strange structure can only be understood in terms of quantum white noise calculus (see Section (13) below).

12 White Noise Schrödinger and Heisenberg Equations

The white noise equations live on spaces of the form

$$\mathcal{H} = \mathcal{H}_S \otimes \Gamma$$

where the Hilbert space $\mathcal{H}_S$ is called the initial (or system) space, and the Hilbert space Γ is called the noise space.

For 1–st order white noise equations the typical Γ is the same as for Hudson–Parthasarathy equations, i.e. a Fock space over a 1–particle space of the form $L^2(\mathbb{R};\mathcal{K})$ where $\mathcal{K}$ is another Hilbert space, called the *multiplicity space* (in mathematics) or *polarization space* (in physics). A WN Schrödinger (or Hamiltonian) equation is an equation of the form

$$\partial_t U_t = -iH_t U_t; \qquad U_0 = 1$$

where $H_t = H_t^*$ is a symmetric functional of white noise and the associated Heisenberg equation (from now on we will consider only the inner case)

$$\partial_t X_t = -i[H_t, X_t]; \qquad X_0 = X \in \mathcal{B}(\mathcal{H})$$

Since in the inner case, as explained in Section (7), the solution of the Heisenberg equation has the form

$$X_t = U_t X_t U_t^*$$

it will be sufficient to consider the Schrödinger equation.

13 Stochastic Equations Associated to 1–st Order WN Schrödinger Equations

The simplest WN equations are the 1–st order WN Schrödinger equations, for which H_t has the form:

$$H_t = Db_t^+ + D^+ b_t + Tb_t^+ b_t + C = D \otimes b_t^+ + \cdots$$

Notice that the right hand side is formally symmetric if

$$T^+ = T; \qquad C^+ = C$$

Diffusion WN equations are characterized by the condition:

$$T = 0$$

Example.

$$\partial_t U_t = -iH_t U_t = -i(Db_t^+ + D^+ b_t)U_t \tag{13.1}$$

if $D = D^+$ this becomes

$$\partial_t U_t = -iH_t U_t = -iD(b_t^+ + b_t)U_t = -iDw_t U_t$$

in terms of Brownian motion

$$\frac{d}{dt}U_t = -iD\frac{dW_t}{dt}U_t \tag{13.2}$$

Warning: in Section (4) one might be tempted to use the naive relation

$$\frac{d}{dt}W_t = w_t \Leftrightarrow dW_t = w_t dt \tag{13.3}$$

and to conclude that the classical WN equation (13.2) is equivalent to the classical stochastic differential equation

$$dU_t = -iDdW_tU_t \tag{13.4}$$

but this would lead to a contradiction because it can be proved that equation (13.4), does not admit any unitary solution while WN Hamiltonian equations of the form (13.1) can be shown to admit unitary solutions.

In fact it is true that WNH equations of the form (13.1) are canonically associated to stochastic differential equations but, for the determination of this stochastic equation, the naive prescription (13.3) is not sufficient and a much subtler rule must be used. The correct answer is given by the following theorem.

Theorem 5. *Let A, C and $T = T^*$ be bounded operators on the initial space $\mathcal{H}_S$. Then the white noise Schrödinger equation*

$$\partial_t U_t = -i(Ab_t + A^*b_t^+ + b_t^+Tb_t + C)U_t; \; U_0 = 1 \tag{13.5}$$

$(T = T^*; C = C^*)$ is equivalent to the following stochastic differential equation

$$dU_t = \Bigg(SDdB_t^+ - D^*dB_t + \frac{1}{2\text{Re}(\gamma_-)}(S-1)dN_t$$
$$+(-\gamma_-D^+D + i|\gamma_-|^2D^+TD - iC)dt\Bigg)U_t \tag{13.6}$$

where the unitary operator

$$S := \frac{1-iT}{1+iT} \tag{13.7}$$

is the Cayley transform of T and:

$$D^+ := iA\frac{1}{1+iT} \tag{13.8}$$

Remark. The two equations can be interpreted in the weak sense on the total domain of extended number vectors with continuous test functions (the vectors of the form $\xi_S \otimes n$ with $\xi_S \in \mathcal{H}_S$ and n a number vector with continuous test functions).

Once taken the matrix elements of both sides in a pair of such vectors both equations become (numerical) ordinary differential equations. Equivalence here means that the solutions of these two equations coincide.

The notion of weak solution of an equation is well established both in classical and stochastic analysis. Much stronger notions can be introduced in the present situation but, for the purpose of comparing the meaning and the solutions of the equations (13.5), (13.6), this is sufficient.

The transition from (13.5) to (13.6) is achieved through the operation of *causal normal order* which is also responsible for the emergence of the (complex) constant γ_-. This constant (and its generalizations to more complex systems) have a fundamental importance in physics (where they are interpreted as *generalized susceptivities*) but, in a purely mathematical treatment, they can be absorbed in the notations.

The proof of the equivalence is based on one of the deepest analytical principles emerged from the stochastic limit of quantum theory, *the time consecutive principle*, whose proof is based on another mathematical development motivated by the same theory, *the theory of distributions on the standard simplex*. These topics will not be discussed here. For additional information the reader is referred to the references mentioned in Section (2).

The most general unitary stochastic differential equation (11.2), in the sense of Hudson–Parthasarathy, is obtained from (13.6) by putting $\gamma_- = \frac{1}{2}$ and by the change of notation $|\gamma_-|^2 D^+ TD - C =: H$.

From formulae (13.7), (13.8), we can now understand in what sense the white noise approach unveils the microscopic structure of the coefficients of the unitary equations in the Hudson–Parthasarathy sense (i.e. (11.2)): the fact that only the special combination of coefficients of equation (11.2) can give a unitary solution is the result of a calculation based on the quantum Itô table, but the meaning of these conditions is quite obscure. On the contrary, looking at (13.5) we see that the unitarity condition is equivalent to the formal symmetry of the WH Hamiltonian. The analogue equivalence between white noise Heisenberg and stochastic Heisenberg equation has been recently established in [AcAyOu05] and, in its Hilbert module formulation [AcAyOu05], it leads to a purely algebraic stochastic generalization of the derivation condition (7.4) (an analytical generalization of this condition, based on operator valued measures and mutual quadratic variations was discussed in [AcHud89]).

14 The Renormalized Square of Classical WN

We have seen that the quantum decomposition of the 1–st order classical WN is:

$$w_t = b_t^+ + b_t$$

If one tries to do the square of w_t naively, one obtains:

$$w_t^2 = (b_t^+ + b_t)^2 = b_t^{+2} + b_t^2 + b_t^+ b_t + b_t b_t^+ = b_t^{+2} + b_t^2 + 2b_t^+ b_t + \delta(0) \quad (14.1)$$

where in the last identity we have applied the commutation relations (3.3) to the case $t = s$. This application is purely formal because $\delta(t-s)$ is a distribution and expressions like $\delta(0)$ are meaningless. The standard procedure to overcome this problem is to subtract the diverging quantity $\delta(0)$ (additive renormalization) and to conjecture that the result i.e.:

$$:w_t^2: = b_t^{+2} + b_t^2 + 2b_t^+ b_t \tag{14.2}$$

is, up to a constant, the quantum decomposition of the square of the classical white noise.

However, even after this renormalization the right hand side of (14.2) is ill defined. The problem is that, as will be shown in the following session, expressions like b_t^{+2}, b_t^2 are not well defined even as operator valued distributions!

15 Basic New Idea: Renormalize the Commutation Relations

The problem of giving a meaning to expressions like b_t^2, b_t^{+2} has its origins in the fact that the commutation relations

$$[b_s, b_t^+] = \delta(t-s)$$

imply that

$$[b_s^2, b_t^{+2}] = 4\delta(t-s)b_s^+ b_t + 2\delta(t-s)^2 \tag{15.1}$$

But what does it mean $\delta(t-s)^2$? We found in the literature [Ivanov79] (see also [BogLogTod69, Vlad66]) the following prescription: On an appropriate test function space the following identity holds

$$\delta(t)^2 = c\delta(t)$$

where the constant $c \in \mathbb{C}$ is arbitrary. (A poof of this statement and the description of the test function space can be found in [AcLuVo99].)

Using this prescription in (15.1) we obtain the renormalized commutation relations:

$$[b_s^2, b_t^{+2}] = 4\delta(t-s)b_s^+ b_t + 2c\delta(t-s) \tag{15.2}$$

Moreover (without any renormalization!)

$$[b_s^2, b_t^+ b_t] = 2\delta(t-s)b_t^2 \tag{15.3}$$

From (15.2) and (15.3) it follows that, after renormalization, the self–adjoint set of operators

$$b_s^2,\ b_s^{+2},\ b_t^+ b_t,\ c = \text{(central element)}$$

is closed under commutators, i.e. the linear span of these operators is a $*$–Lie algebra.

In test function language the renormalized square of white noise $*$–Lie algebra can be defined as follows

$$b_\varphi^+ = \int dt\varphi(t)b_t^{+2}; \quad b_\varphi = (b_\varphi^+)^+$$

$$n_\varphi = \int dt\varphi(t)b_t^+ b_t$$

where $\varphi \in \mathcal{S}$ and $\mathcal{S}$ is a space of test functions (for example one can choose the finitely valued step functions on $\mathbb{R}$). This leads to the commutation relations:

$$[b_\varphi, b_\psi^+] = \gamma\langle\varphi, \psi\rangle + n_{\overline{\varphi}\psi} \tag{15.4}$$

$$[n_\varphi, b_\psi] = -2b_{\overline{\varphi}\psi} \tag{15.5}$$

$$[n_\varphi, b_\psi^+] = 2b_{\varphi\psi}^+ \tag{15.6}$$

$$(b_\varphi^+)^+ = b_\varphi; \qquad n_\varphi^+ = n_{\overline{\varphi}} \tag{15.7}$$

Definition. *A Fock representation of the $*$–Lie algebra of the RSWN is a representation of this $*$–Lie algebra as operators on a Hilbert space $\mathcal{H}$ with a unit vector Φ satisfying the following conditions (for notational simplicity we will use the same symbol for an element of the $*$–Lie algebra of the RSWN and for its image acting on $\mathcal{H}$).*

(i) *The set of vectors (called the quadratic number vectors)*

$$\{(b_\varphi^+)^n\Phi \ : \ n \in \mathbb{N}, \ \varphi \in \mathcal{S}\}$$

is well defined and total in $\mathcal{H}$.

(ii) *The algebraic linear span of the quadratic number vectors is invariant under the action of the operators $\{b_\varphi^+, b_\varphi, n_\varphi \ : \ \varphi \in \mathcal{S}\}$.*

(iii) *The commutation relations (15.4), (15.5), (15.6), (15.7) take place on the quadratic number vectors.*

(iv) *The Fock property holds:*

$$b_\varphi\Phi = n_\varphi\Phi = 0 \tag{15.8}$$

16 Existence of Fock Representations

Having defined the Fock representation the first problem is its existence. In case of the first order white noise this is a well known result since the early days of quantum theory.

Theorem (Fock 1930). The Fock representation of the first order white noise (i.e. the current algebra over $\mathbb{R}$ of the CCR Lie algebra $[a, a^+] = 1$, for the notion of current algebra see Section (18)) exists and is unique up to unitary isomorphism.

The analogue for the RSWN Lie algebra was established more recently.

Theorem (Accardi, Lu, Volovich 1999). The Fock representation of the second order white noise (current algebra over $\mathbb{R}$ of the Lie algebra $sl(2,\mathbb{R})$) exists and is unique up to unitary isomorphism.

A direct proof of this result is a nontrivial application of the principle that *algebra implies statistics*, described in its simplest form in Section (3): one proves that, if the required Fock representation exists, then the scalar product of two number vectors is uniquely determined by the commutation relations (15.4), (15.5), (15.6), (15.7) and the Fock property (15.8). Then, and this is the difficult part, one has to prove that this is indeed a scalar product, i.e. that it is positive definite (cf. [AcLuVo99]).

In Section (21) we will come back to this point. Before that let us analyze some consequences of the above theorem. More precisely let us apply to this case the basic general principle of QP discussed in Section (3): **algebra implies statistics**. In Section (3) we have seen that the application of this principle to the first order white noise shows that the corresponding algebra implies Gaussian and Poisson statistics. It is therefore natural to rise the following question:

Which statistics is implied by the algebra of the renormalized Square of WN?

The answer to this question was given by Accardi, Franz and Skeide in the paper [AcFrSk00].

17 Classical Subprocesses Associated to the Second Order White Noise

To understand this answer it is convenient to take as starting point the analogy with the q-decomposition of the compensated classical Poisson process with intensity β^{-1}

$$\dot{p}_t = b_t^+ + b_t + \beta b_t^+ b_t$$

At the end of Section (5) we have seen that $\beta = 0$ is the only critical case and corresponds to the transition from classical scalar valued standard compensated Poisson process with intensity β^{-1}.

This analysis is extended in the paper [AcFrSk00] to the renormalized square of white noise by considering the classical subprocesses

$$X_\beta(t) := b_t^{+2} + b_t^2 + \beta b_t^+ b_t \tag{17.1}$$

where β is a real number. It is then proved that now there are 2 critical values of β, namely:

$$\beta = \pm 2$$

The value $+2$ corresponding to the renormalized square of the position (classical) white noise, i.e.

$$w_t^2 = | b_t^+ + b_t |^2 = b_t^{+2} + b_t^2 + b_t^+ b_t + b_t b_t^+$$
$$= b_t^{+2} + b_t^2 + 2b_t^+ b_t + \delta(0) \equiv b_t^{+2} + b_t^2 + 2b_t^+ b_t$$

and the value -2 to the renormalized square of the momentum white noise, i.e.

$$(b_t^+ - b_t)/i$$

The vacuum distribution of both processes is the Gamma–process

$$\mu(dx) = \frac{|x|^{m_0 - 1}}{\Gamma(m_0)} e^{-\beta x} \chi_{\beta \mathbb{R}_+}$$

whose parameter $m_0 > 0$ is uniquely determined by the choice of the unitary representation of $SL(2, \mathbb{R})$ corresponding to the representation of the SWN algebra (cf. [ACFRSK00]).

In this functional realization the number vectors become the Laguerre polynomials which are orthogonal for the Gamma distribution.

Since the Gamma–distributions are precisely the distributions of the χ^2–random variables, this result confirms the naive intuition that the distribution of the [renormalized] square of white noise should be a Gamma–distributions.

For $|\beta| < 2$ the intensity of the jumps is not strong enough and each of the classical random variables

$$X_\beta(t) := b_t^{+2} + b_t^2 + \beta b_t^+ b_t$$

still has a density whose explicit form, in terms of the Γ–function is:

$$\mu(dx) = C \exp\left(- \frac{(2 \arccos \beta + \pi) x}{2\sqrt{1 - \beta^2}} \right) \left| \Gamma \left(\frac{m_0}{2} + \frac{ix}{2\sqrt{1 - \beta^2}} \right) \right|^2$$

(C is a normalization constant). The corresponding orthogonal polynomials are the Meixner-Pollaczek polynomials. For m_0 integer more explicit formulae for these densities were found by Grigelionis [Grig00c, Grig01, Grig99]:

$$|\Gamma (n + ix)|^2 = \frac{\pi x (1 + x^2) \cdots ((n-1)^2 + x^2)}{\sinh(\pi x)}; \ n = 1, 2, \ldots, \ , \ x \in \mathbb{R}$$

$$\left| \Gamma \left(\frac{1}{2} + ix \right) \right|^2 = \frac{\pi}{\cosh(\pi x)}, \quad x \in \mathbb{R}$$

$$\left| \Gamma \left(n + \frac{1}{2} + ix \right) \right|^2 = \frac{\pi \left(\frac{1}{4} + x^2 \right) \cdots \left((n-1)n + \frac{1}{4} + x^2 \right)}{\cosh(\pi x)},$$
$$n = 1, 2, \ldots; \ x \in \mathbb{R}$$

Finally, for $|\beta| > 2$ the jumps dominate and the classical random variable

$$X_\beta(t) := b_t^{+2} + b_t^2 + \beta b_t^+ b_t$$

has a discrete vacuum distribution i.e. the negative binomial (Pascal) distribution which plays, for the square of white noise the a role analogue to the one played by the geometric (or Gibbs) distribution for the first order white noise:

$$\mu = C \sum_{n=0}^{\infty} \frac{c^{2n}(m_0)_n}{n!} \delta_{sgn(\beta((c-1/c)(n+m_0/2))}$$

where $(m_0)_n$ denotes the Pochammer symbol,

$$(m_0)_n = m_0(m_0+1)\cdots(m_0+n-1)$$

and

$$C^{-1} = \sum_{n=0}^{\infty} \frac{c^{2n}(m_0)_n}{n!} = (1-c^2)^{-m_0}$$

More precisely, if $\beta > 0$

$$P_n(x) = (-1)^n \prod_{k=1}^{n} \frac{n+m_0-1}{n} M_n\left(\frac{x}{c-1/c} - \frac{m_0}{2}; m_0; c^2\right)$$

$$\prod_{k=1}^{n} \frac{n+m_0-1}{n} M_n\left(-\frac{x}{c-1/c} + \frac{m_0}{2}; m_0; c^2\right)$$

where the M_n are the Meixner polynomials and

$$c = \beta - \sqrt{\beta^2 - 4}$$

if $\beta > +2$, while if $\beta < -2$ then:

$$c = -\beta - \sqrt{\beta^2 - 4}$$

In conclusion: the 1–parameter family (17.1) of vacuum operator processes gives rise, according to the values of the parameter, to three classes of processes:

$|\lambda| = 2$ classical Gamma processes (Laguerre polynomials)
$|\lambda| < 2$ Meixner processes (Meixner-Pollaczek polynomials)
$|\lambda| > 2$ negative binomial (Pascal) processes (Meixner polynomials)

These three classes of stochastic processes exactly coincide with the remaïning three Meixner classes! Comparing this result with the remark at the end of Section (5) we see that the first and second powers of white noise account exactly for the five Meixner classes.

These 5 classes were known since 1934 (date of publication of Meixner's paper), but only through their quantum decomposition was their common structure made clear. Moreover the infinite divisibility of these distributions, which is not easy to prove analytically, now became an easy corollary of the general phenomenon described in Section (19) below.

Since the 1–st and second powers of white noise gave rise, in the sense just explained, to the five Meixner classes, it is quite natural to ask oneself what happens with the higher powers of WN. In order to answer this question we have to recall the connection between additive independent increment processes and current algebras over Lie algebras.

18 Current Representations of Lie Algebras

Intuitively, if $\{\mathcal{L}, [\,\cdot\,,\,\cdot\,], *\}$ is a $*$–Lie algebra, a *current algebra of* $\mathcal{L}$ *over* $\mathbb{R}^d$ is a vector space $\mathcal{T}$ of $\mathcal{L}$–valued functions defined on $\mathbb{R}^d$ and closed under the pointwise operations:

$$[\varphi, \psi](t) := [\varphi(t), \psi(t)]; \quad \varphi^*(t) := \varphi(t)^*; \quad t \in \mathbb{R},\ \varphi \in \mathcal{T}$$

For example, if $X_1, \ldots, X_k$ are generators of $\mathcal{L}$ one can fix a space $\mathcal{S}$, of complex valued test functions on $\mathbb{R}$ and to each $\varphi \in \mathcal{S}$ and $j \in \{1, \ldots, k\}$ one can associate the $\mathcal{L}$–valued function on $\mathbb{R}X_j(\varphi)$ defined by:

$$X_j(\varphi)(t) := \varphi(t)X_j; \quad t \in \mathbb{R}$$

Definition 6. *Let $\mathcal{G}$ be a complex $*$–Lie algebra. A (canonical) set of generators of $\mathcal{G}$ is a linear basis of $\mathcal{G}$*

$$l^+_\alpha, l^-_\alpha, l^0_\beta, \alpha \in I, \quad \beta \in I_0$$

where I_0, I are sets, satisfying the following conditions:

$$(l^0_\beta)^* = l^0_\beta; \qquad \forall \beta \in I_0$$

$$(l^+_\alpha)^* = l^-_\alpha; \qquad \forall \alpha \in I$$

and all the central elements among the generators are of l^0–type (i.e. self-adjoint).

We will denote $c^\gamma_{\alpha\beta}(\varepsilon, \varepsilon', \delta)$ the structure constants of $\mathcal{G}$ with respect to the generators (l^ε_α), i.e., with $\alpha, \beta \in I \cup I_0$, $\varepsilon, \varepsilon', \delta \in \{+, -, 0\}$, and, assuming summation over repeated indices:

$$[l^\varepsilon_\alpha, l^{\varepsilon'}_\beta] = c^\gamma_{\alpha\beta}(\varepsilon, \varepsilon', \delta) l^\delta_\gamma =$$

$$:= \sum_{\gamma \in I_0} c^\gamma_{\alpha\beta}(\varepsilon, \varepsilon', 0) l^0_\gamma + \sum_{\gamma \in I} c^\gamma_{\alpha\beta}(\varepsilon, \varepsilon', +) l^+_\gamma + \sum_{\gamma \in I} c^\gamma_{\alpha\beta}(\varepsilon, \varepsilon', -) l^-_\gamma$$

In the following we will consider only *locally finite* Lie algebras, i.e. those such that, for any pair $\alpha, \beta \in I \cup I_0$ only a finite number of the structure constants $c^\gamma_{\alpha\beta}(\varepsilon, \varepsilon', \delta)$ is different from zero.

Definition 7. *Let be given:*

– *a $*$–Lie algebra $\mathcal{G}$*
– *a measurable space $(S, \mathcal{B})$*
– *a $*$–sub–algebra $\mathcal{C} \subseteq L^{\infty}_{\mathbb{C}}(S, \mathcal{B})$ for the pointwise operations.*

The current algebra of $\mathcal{G}$ over $\mathcal{C}$ is the $*$–Lie algebra

$$\mathcal{G}(\mathcal{C}) := \{\mathcal{C} \otimes \mathcal{G}, [\,\cdot\,,\,\cdot\,]\}$$

where $\mathcal{C} \otimes \mathcal{G}$ is the algebraic tensor product, the Lie brackets $[\,\cdot\,,\,\cdot\,]$ are given by

$$[f \otimes l, g \otimes l'] := fg \otimes [l, l']; \quad f, g \in \mathcal{C},\ l, l' \in \mathcal{G} \tag{18.1}$$

and the involution $*$ is given by

$$(f \otimes l)^* := \overline{f} \otimes l^*; \quad f \in \mathcal{C},\ l \in \mathcal{G}$$

where $\overline{f}$ denotes complex conjugate. In the following we shall use the notation:

$$l(f) := f \otimes l; \quad f, g \in \mathcal{C},\ l, \in \mathcal{G},\ f \in \mathcal{C} \tag{18.2}$$

Remark. If $(l^{\varepsilon}_{\gamma})$ is a (canonical) set of generators of $\mathcal{G}$ then the set

$$\{f \otimes l^{+}_{\alpha},\ f^* \otimes l^{-}_{\alpha},\ Re(f) \otimes l^{0}_{\gamma}\ :\ \alpha \in I,\ \gamma \in I_0,\ f \in \mathcal{C}\}$$

$(Re(f) := (f + \overline{f})/2)$ is a (canonical) set of generators of $\mathcal{G}(\mathcal{C})$.

In the following we will use the notations:

$$l^{+}_{\alpha}(f) := f \otimes l^{+}_{\alpha}, \qquad l^{-}_{\alpha}(f) := f^* \otimes l^{-}_{\alpha}, \qquad l^{0}_{\alpha}(f) := f \otimes l^{0}_{\gamma}$$

and, when no confusion can arise, we will often speak of *the current algebra* $(l^{\varepsilon}_{\alpha}(f))$.

Definition 8. *A representation of a $*$–Lie algebra $\mathcal{G}$ is a triple:*

$$\{\mathcal{H}, \mathcal{D}, \pi\}$$

where

– *$\mathcal{H}$ is an Hilbert space*
– *$\mathcal{D}$ is a total subset of $\mathcal{H}$*
– *$\pi : \mathcal{D} \to \mathcal{H}$ is a map such that:*

(i) for any $l \in \mathcal{G}$, $\pi(l)$ is a pre–closed operator on $\mathcal{D}$ with adjoint $\pi(l^*)$
(ii) for any $l, l' \in \mathcal{G}$,

$$\pi([l, l']) = [\pi(l), \pi(l')]$$

where the commutator on the right hand side is meant weakly on $\mathcal{D}$.

Remark. At the algebraic level the existence of representations of current algebras of an arbitrary pair $\{\mathcal{G}, (l^{\varepsilon}_{\alpha})\}$ over an arbitrary measure space $(S, \mathcal{B}, \mu)$ and sub–$*$–algebra

$$\mathcal{C} \subseteq L^{\infty}(S, \mathcal{B}, \mu)$$

is easily established. In fact, if $\{\pi, \mathcal{K}\}$ is any representation of $\mathcal{G}$ one can define a structure of $*$–Lie algebra on

$$\mathcal{C} \otimes \pi(\mathcal{G}) \in \mathcal{L}(L^2(S, \mathcal{B}, \mu) \otimes \mathcal{K})$$

in terms of the brackets

$$[f \otimes \pi(e), g \otimes \pi(l')] := fg \otimes \pi([l, l'])$$

Therefore, defining:

$$l^{+}_{\alpha}(f) := f \otimes \pi(l^{+}_{\alpha})$$
$$l^{\circ}_{\alpha}(f) := f \otimes \pi(l^{\circ}_{\alpha})$$

one has

$$l^{+}_{\alpha}(f)^{*} = \overline{f} \otimes \pi(l^{+}_{\alpha})^{*} = \overline{f} \otimes \pi((l^{+}_{\alpha})^{*}) = \overline{f} \otimes \pi(l^{-}_{\alpha})$$
$$(l^{0}_{\beta}(f))^{*} = \overline{f} \otimes \pi(l^{0}_{\beta})^{*} = \overline{f} \otimes \pi(l^{0}_{\beta}) = l^{0}_{\beta}(\overline{f})$$
$$\begin{aligned}[l^{\varepsilon}_{\alpha}(f), l^{\varepsilon'}_{\beta}(g)] &= [f^{\varepsilon} \otimes \pi(l^{\varepsilon}_{\alpha}), g^{\varepsilon'} \otimes \pi(l^{\varepsilon'}_{\beta})] = f^{\varepsilon} g^{\varepsilon'} \otimes [\pi(l^{\varepsilon}_{\alpha}), \pi(l^{\varepsilon'}_{\beta})] \\ &= f^{\varepsilon} g^{\varepsilon'} \otimes \pi[l^{\varepsilon}_{\alpha}, l^{\varepsilon'}_{\beta}] = c^{\gamma}_{\alpha\beta}(\varepsilon, \varepsilon', \delta) f^{\varepsilon} g^{\varepsilon'} \otimes l^{\delta}_{\gamma} \\ &= c^{\gamma}_{\alpha\beta}(\varepsilon, \varepsilon', \delta) l^{\delta}_{\gamma}(f^{\varepsilon} g^{\varepsilon'})\end{aligned}$$

Thus the current algebra relations are verified. This proves that any representation of $\mathcal{G}$ can be lifted to a representation of the current algebra $\mathcal{G} \otimes \mathcal{G}$.

Definition 9. *Let $\mathcal{G}$ be a $*$–Lie algebra with a (canonical) set of generators $(l^{\varepsilon}_{\alpha})$. A representation $\{\mathcal{K}, \mathcal{D}, \pi\}$ of $\mathcal{G}$ is called weakly irreducible if the images of the central elements are multiples of the identity.*

A representation $\{\mathcal{K}, \mathcal{D}, \pi\}$ of $\mathcal{G}$ on a Hilbert space $\mathcal{K}$ is called a Fock representation if it is weakly irreducible and:

(i) there exists a unit vector $\Phi \in \mathcal{K}$ such that $\forall\, \alpha \in I$ and $\forall\, \beta \in I_0$, with the exception of those $\beta \in I_0$ which correspond to central elements, one has

$$\pi(l^{-}_{\alpha})\Phi = \pi(l^{0}_{\beta})\Phi = 0$$

(ii) the set

$$\{\pi(l^{+}_{\alpha})^{n}\Phi \;:\; \alpha \in I,\; n \in \mathbb{N}\}$$

is total in $\mathcal{K}$.

All the difficulties found in the construction of a coherent theory of higher powers of white noise can be summarized in the following problem.

Problem. Let $\{\mathcal{G}, (l_\alpha^\varepsilon)\}$ be a $*$–Lie algebra with a set of generators. Suppose that $\{\mathcal{G}, (l_\alpha^\varepsilon)\}$ admits a Fock representation. Under which conditions on the measure space $(S, \mathcal{B}, \mu)$ and on the $*$–sub–algebra

$$\mathcal{C} \subseteq L^\infty(S, \mathcal{B}, \mu)$$

does the current algebra

$$\{l_\alpha^\varepsilon(f) : \varepsilon \in \{+, -, 0\}, \quad \alpha \in I \text{ or } \alpha \in I_0, \quad f \in \mathcal{C}\}$$

admit a Fock representation?

Example (1). The Heisenberg algebra is the $*$–Lie algebra with set of generators $\{a^+, a, 1\}$ (1 here and in the following denotes the central element) and relations

$$[a, a^+] = 1 \tag{18.3}$$

The associated current algebra over $\mathbb{R}^d$ with Lebesgue measure admits a Fock representation which is the standard d–dimensional white noise or the free boson Fock field over $\mathbb{R}^d$.

Example (2). The oscillator algebra is the $*$–Lie algebra with set of generators $\{a^+, a, a^+a, 1\}$ and relations deduced from (18.3). The associated current algebra over $\mathbb{R}^d$ with Lebesgue measure admits a Fock representation which can be called the Hudson–Parthasarathy algebra over $\mathbb{R}^d$.

Example (3). The square–oscillator algebra is the $*$–Lie algebra with set of generators $\{a^{+2}, a^2, a^+a, 1\}$ and relations deduced from (18.3). The associated current algebra over $\mathbb{R}^d$ with Lebesgue measure admits a Fock representation which is the renormalized square of white noise algebra over $\mathbb{R}^d$.

The square oscillator algebra is canonically isomorphic to (a central extension of) the $*$–Lie algebra $sl(2, \mathbb{R})$. This is easily seen because $sl(2, \mathbb{R})$ is the $*$–Lie algebra with 3 generators B^-, B^+, M and relations

$$(B^-)^* = B^+; \quad M^* = M \tag{18.4}$$

$$[B^-, B^+] = M \tag{18.5}$$

$$[M, B^\pm] = \pm 2B^\pm \tag{18.6}$$

From this it follows that the algebra of the RSWN is isomorphic to The Fock representation of a current algebra over a central extension of $sl(2, \mathbb{R})$.

Example (4). The Schrödinger algebra is the $*$–Lie algebra with generators $\{a^+, a\, a^{+2}, a^2, a^+a, 1\}$ and relations deduced from (18.3).

Example (5). The full oscillator algebra is the $*$–Lie algebra with set of generators $\{a^{+h}a^k \; : \; \forall h, k \in \mathbb{N}\}$ and relations deduced from (18.3).

Notice that all the $*$–Lie algebras listed in the above examples admit a Fock representation. In Section (21) we will see that this is not true for the associated current algebra over $\mathbb{R}^d$ with Lebesgue measure (for any $d > 0$).

19 Connections with Classical Independent Increment Processes

In this section we look for some necessary conditions for the solution of the problem stated in the previous section. This will naturally lead to an interesting connection with the theory of classical independent increment processes which was first noticed in Araki's thesis [Arak60]. We refer to the monographs of K.R. Parthasarathy and K. Schmidt [PaSch72] and of Guichardet [Gui72] for a systematic exposition. In the notations of Section (18) we consider:

- a pair $\{\mathcal{G}, (l_\alpha^\varepsilon)\}$ of a $*$–Lie algebra and a set of generators which admits a Fock representation.
- a measure space (S, μ)
- a $*$–sub–algebra $\mathcal{C} \subseteq L^\infty_{\mathbb{C}}(S, \mathcal{B}, \mu)$

such that the current algebra

$$\{l_\alpha^\varepsilon(f) : \varepsilon \in \{+, -, 0\},\ \alpha \in I \text{ or } \alpha \in I_0,\ f \in \mathcal{C}\}$$

admits a Fock representation on some Hilbert space $\mathcal{H}$ with cyclic vector Φ. We identify the elements of this current algebra with their images in this representation and we omit from the notation the symbol π of the representation. Moreover we add the following assumptions:

(i) among the generators (l_α^ε) there is exactly one (self–adjoint) central element, denoted l_0^0.
(ii) for any $f \in \mathcal{C}$ one has:

$$l_0^0(f) = \int_S f d\mu \tag{19.1}$$

where the scalar on the right hand side is identified to the corresponding multiple of the identity operator on $\mathcal{H}$. In particular the representation is weakly irreducible.

Under these conditions it is not difficult to see that the general principle that algebra implies statistics can be applied and that the vacuum mixed moments of the operators $l_\alpha^\varepsilon(f)$ are uniquely determined by the structure constants of the Lie algebra. Another important property is that, by fixing a measurable subset $I \subseteq S$ such that

$$\mu(I) = 1 \tag{19.2}$$

and denoting χ_I the corresponding characteristic function, the $*$–Lie algebra generated by the operators $l_\alpha^\varepsilon(\chi_I)$ is isomorphic to $\mathcal{G}$ and therefore it has the same vacuum statistics.

Finally the commutation relations (18.1) imply that the maps $f \mapsto l_\alpha^\varepsilon(f)$ define an independent increment process of boson type, i.e. the restriction of the vacuum state on the polynomial algebra generated by two families

$(l_\alpha^\varepsilon(f))_{\varepsilon,\alpha}$ and $(l_\alpha^\varepsilon(g))_{\varepsilon,\alpha}$ with f and g having disjoint supports, coincides with the tensor product of the restrictions on the single algebras.

In particular, if $X(I)$ is any self–adjoint linear combination of operators of the form $l_\alpha^\varepsilon(\chi_I)$, then the map $I \subseteq S \mapsto X(I)$ defines an additive independent increment process on $(S, \mathcal{B}, \mu)$. Thus the law of every random variable of the form $X(I)$ will be an infinitely divisible law on $\mathbb{R}$ whenever the set I can be written as a countable union of subsets of nonzero μ–measure.

If $S = \mathbb{R}^d$ and μ is the Lebesgue measure, then any such process $X(I)$ $(I \subseteq \mathbb{R}^d)$ will also be translation invariant.

Combining together all the above remarks one obtains a necessary condition for the existence of the Fock representation of the current algebra of a $*$–Lie algebra $\mathcal{G}$ and a set of generators namely: *the pair* $\{\mathcal{G}, (l_\alpha^\varepsilon)\}$ *must admit a Fock representation and the vacuum distribution of any self–adjoint linear combination* X *of generators must be infinitely divisible.*

Since there is no reason to expect that any pair $\{\mathcal{G}, (l_\alpha^\varepsilon)\}$ will have this property, this gives a probabilistic intuition of the reason why it might happen that a $*$–Lie algebra and a set of generators $\{\mathcal{G}, (l_\alpha^\varepsilon)\}$ might admit a Fock representation without this being true for the associated current algebra.

In the following section we review some progresses made in the past few years in one important special case: the full oscillator algebra.

20 Current Algebras over the Full Oscillator Algebra

We have seen how the developments reviewed in the previous sections naturally lead to the following problem: can we extend to the renormalized higher powers of quantum white noise what has been achieved for the second powers? To answer this question we start with the Heisenberg algebra

$$[a, a^+] = 1 \tag{20.1}$$

Its universally enveloping algebra is generated by the products of monomials of the form

$$a^n, \quad a^{+m}$$

and their commutation relations are deduced from (20.1) and the derivation property of the commutator. The problem we want to study is the following: **does there exist a current representation of this algebra over $\mathbb{R}^d$ for some $d>0$?**

In order to define the current algebra of the full oscillator algebra, we have first to overcome the **renormalization problem**, illustrated in Section (14) in the case of the second powers of white noise. In fact, dealing with higher powers of white noise we meet higher powers of the δ–function. A natural way out is to write

$$\delta^n = \delta^2(\delta^{n-2}); \qquad n \geq 2; \qquad \delta^0 := 1$$

and to apply iteratively the renormalization prescription used in Section (14). This leads to the following:

Definition. *The boson Fock white noise, renormalized with the prescription:*

$$\delta(t)^l = c^{l-1}\,\delta(t),\ c>0,\ l=2,3,\dots. \tag{20.2}$$

simply called RBFWN in the following, over a Hilbert space $\mathcal{H}$ with vacuum (unit) vector Φ is the locally finite $*$–Lie algebra canonically associated to the associative unital $*$–algebra of operator–valued distributions on $\mathcal{H}$ with generators

$$b_t^{+n} b_t^k, \qquad k,n\in\mathbb{N}, \qquad t\in\mathbb{R}^d$$

and relations deduced from:

$$[b_t, b_s^+] = \delta(t-s)$$

$$[b_t^+, b_s^+] = [b_t, b_s] = 0$$

$$(b_s)^* = b_s^+$$

$$b_t\,\Phi = 0$$

Here locally finite means that the commutator of any pair of generators is a finite linear combination of generators.

Lemma. The $*$–Lie algebra, associated to the RBFWN (renormalized boson Fock white noise), is the Lie algebra with generators

$$b_t^{+k} b_t^n =: b_n^k(t)$$

central element $b_t^0 b_t^{+0} =: 1$ and relations

$$(b_s^{+k} b_t^n)^+ = (b_t^{+n}) b_s^k$$

$$[b_t^n, b_s^{+k}] = \epsilon_{n,0}\epsilon_{k,0} \sum_{l\geq 1} \binom{n}{l}\, k^{(l)}\, c^{l-1}\, b_s^{+k-l}\, b_t^{n-l}\, \delta(t-s)$$

where:

$$k = 0,1,2,\dots$$

$$\epsilon_{n,k} := 1 - \delta_{n,k}$$

$$k^{(l)} = k(k-1)(k-2)\cdots(k-l+1)$$

and, if $l > k$, by definition

$$k^{(l)} = 0$$

$$\binom{n}{l} = 0$$

In our case we fix the test function space to be the algebra of finite step functions. In terms of test functions we have the identifications:

$$B_k^n(f) = \int_{\mathbb{R}^d} f(t)\, b_t^{+n}\, b_t^k\, dt$$

$$(B_k^n(f))^* = B_n^k(\bar{f})$$

21 No–go Theorems

The first no–go theorem, showing that it is not true that, if a Lie algebra admits a Fock representation, then any associated current algebra also admits one was proved by Śniady [Śnia99]. In the terminology introduced in the present paper Śniady's result can be rephrased as follows:

Theorem 10. *The Schrödinger algebra admits a Fock representation but its associated current algebra over* $\mathbb{R}$ *with Lebesgue measure doesn't.*

Since the Schrödinger algebra is contained in the full oscillator algebra, which clearly admits a Fock representation, Śniady's theorem also rules out the possibility of a Fock representation for the current algebra of the full oscillator algebra over $\mathbb{R}$ with Lebesgue measure.

Recalling, from the examples at the end of Section (18), that the Schrödinger algebra is the smallest $*$–Lie algebra containing the oscillator algebra (with generators $\{a^+, a, a^+a, 1\}$) and the square–oscillator algebra, i.e. $sl(2, \mathbb{R})$ (with generators $\{a^{+2}, a^2, a^+a, 1\}$), we see that the difficulty comes from the combination of two closed Lie algebras. More precisely: consider the two sets of generators

$$\{a^+, a, a^+a, 1\}$$

$$\{a^{+2}, a^2, a^+a, 1\}$$

We know that the current algebra over $\mathbb{R}^d$ associated to each of them has a Fock representation. However the union of the two sets, i.e.

$$\{a^+, a, a^{+2}, a^2, a^+a, 1\}$$

is also a set of generators of a $*$–Lie algebra whose associated current algebra over $\mathbb{R}^d$ does not admit a Fock representation.

Notice that the first of the two algebras is generated by the first powers of the white noise and the number operator while the second one is generated by the second powers of the white noise and the number operator. An extrapolation of this argument suggested the hope that a similar thing could happen also for the higher powers, i.e. that, denoting $\mathcal{G}_3$ the $*$–Lie algebra generated by the cube of the white noise b_t^3 and the number operator; and, for $n \geq 4$, $\mathcal{G}_n$ the $*$–Lie algebra generated by the number operator and the smallest power of the white noise not included in $\bigcup_{1\leq k\leq n-1} \mathcal{G}_k$, the current algebra of $\mathcal{G}_n$ over $\mathbb{R}^d$ admits a Fock representation.

This hope was ruled out by the following generalization of Śniady's theorem, due to Accardi, Boukas and Franz [AcBouFr05] and by its corollary reported below.

Theorem 11. *Let $\mathcal{L}$ be a Lie $*$–algebra with the following properties:*

(i) $\mathcal{L}$ contains B_0^n, and B_0^{2n} where by definition:

$$B_k^n := \int_I b_t^{+n} b_t^k \, dt; \qquad I \subseteq \mathbb{R}^d \tag{21.1}$$

(ii) the B_K^N satisfy the commutation relations of the higher powers of white noise.

Then, in the notation (21.1), $\mathcal{L}$ has not a Fock representation if the interval I is such that

$$\mu(I) \leq \frac{1}{c} \tag{21.2}$$

where c is the renormalization constant.

Corollary 12. *The current algebra over $\mathbb{R}^d$ of the $*$–Lie algebra $\mathcal{G}_3$, generated by the cube of the quantum white noise b_t^3 and the number operator, does not admit a Fock representation if, in the notation (21.1), the interval I is such that*

$$\mu(I) \leq \frac{1}{c} \tag{21.3}$$

where c is the renormalization constant.

Idea of the Proof. One proves that, if a Lie algebra contains b_t^3 and b_t^{+3}, then it contains b_t^6 hence b_t^{+6}. The thesis then follows from Theorem (11).

Theorem 13. *Let $\mathcal{L}$ be a Lie $*$–subalgebra of the RPWN Lie algebra which contains B_0^n for some $n \geq 3$. Then $\mathcal{L}$ does not admit a Fock space representation if, in the notation (21.1), the interval I is such that*

$$\mu(I) \leq \frac{1}{c} \tag{21.4}$$

where c is the renormalization constant.

Another hope was that may be the root of the difficulty is in the insistence on the Fock representation and that maybe the analogue of the finite temperature (equilibrium) representations might exist.

The problem here is that, at the moment there is no general structure theory of free dynamics and of the associated KMS (Kubo–Martin–Schwinger) states on current subalgebras of the full oscillator algebra.

In the case of the square of white noise a classification of free dynamics was obtained and a first class of examples of KMS states was built in the paper [AcAmFr02] and A more general class of KMS functionals was built in the paper [AcPeRo05] but a proof of the positivity of these functionals was missing. The proof of the positivity of these functionals was obtained for an interesting and physically meaningful class of functionals by Prohorenko [Prohor05].

The following theorem, also proved in [AcBouFr05], concerns the non compatibility between first and second order white noise dropping the Fock assumption and replacing it by the use of the natural commutation relations between first and second order noise.

Theorem 14. *Let $\mathcal{A}$ be a $*$–algebra of operator valued distributions on $\mathbb{R}^d$ whose test function space includes the characteristic functions of intervals in $\mathbb{R}^d$ defined as follows:*

$$(a,b) = \begin{cases} \{x = (x_j) : a_j < x_j < b_j; \quad j = 1, \dots, d\}, \text{ if } a_j < b_j, \ \forall j \\ \phi, \text{ if } a_j > b_j \text{ for some } j \end{cases} \tag{21.5}$$

and let

$$b_k^2, b_k^{+2}, b_h, b_h^+ \tag{21.6}$$

be elements of $\mathcal{A}$. There exists no state $\langle \cdot \rangle$ on $\mathcal{A}$ with the following properties:

$$\langle b_k b_h^+ \rangle = \mu(k)\, \delta(k-h) \tag{21.7}$$

$$\langle b_{k'} b_k b_{h'}^+ b_h^+ \rangle = 2\mu(k)\, \mu(k')\, \delta(k'-h')\, \delta(k-h) \tag{21.8}$$

$$\langle b_k^2 b_{h'}^+ b_h^+ \rangle = 2\mu(k)^2\, \delta(k-h')\, \delta(k-h) \tag{21.9}$$

$$\langle b_k^2 b_h^{+2} \rangle = \sigma(k)\, \delta(k-h) \tag{21.10}$$

where $\sigma \in L^1_{loc}(\mathbb{R}^d)$ and $\mu \in L^1_{loc} \cap L^2_{loc}(\mathbb{R})$ are such that there exist an interval $I \subseteq \mathbb{R}^d$ and constants M_I, $\varepsilon_I > 0$ such that

$$+\infty > M_I > \sigma(k), \quad \mu(k); \ \mu(k) \geq \varepsilon_I > 0; \quad \forall k \in I \tag{21.11}$$

A third possible loophole is to consider commutation relations not of Boson type. The simplest choice is provided by the q–deformed commutation relations:

$$a\, a^+ - q\, a^+\, a = 1 \tag{21.12}$$

Bozeiko, Kümmerer and Speicher [BoKüSp96b] proved that this algebra and the associated current algebra over $\mathbb{R}^d$, admits a Fock representation if $q \in (-1,1)$ and Bozeiko and Speicher [BozSpe96a] proved that this algebra and the associated current algebra over $\mathbb{R}^d$, does not admits a Fock representation if $|q| > 1$.

Theorem 15. *([AcBou05a]) Let $q \in (-1,1), q \neq 0$ and for a fixed interval $I \subset \mathbb{R}$ and $n, k \geq 0$ let $\mathcal{L}_q$ denote the $*$–Lie algebra generated by the operators*

$$B_k^n := B_k^n(\chi_I)$$

defined as in (21.1) but for the q–deformed white noise. Then $\mathcal{L}_q$ does not admit a Fock representation if, the interval I is such that

$$\mu(I) \le \frac{1}{c} \tag{21.13}$$

where c is the renormalization constant.

Remark. It is interesting to notice that the lower bound $1/c$ seems to be universal, i.e. independent of the type of noise considered.

22 Connection with an Old Open Problem in Classical Probability

Since the vacuum distribution of the first order classical white noise is a Gaussian, any reasonable renormalization should lead to the conclusion that the n–th power of the first order classical white noise is still the n–th power of a Gaussian. But the δ–correlation implies that the corresponding integrated process will be a stationary additive independent increment process on $\mathbb{R}$.

These heuristic ideas, which can be put in a satisfactory mathematical form with some additional work, lead to the conjecture that a necessary condition for the existence of the n–th power of white noise, renormalized as in [AcBouFr05], is that the n–th power of a classical Gaussian random variable is infinitely divisible.

The n–th powers of the standard Gaussian random variable γ and their distributions have been widely studied. It is known that, $\forall k \ge 1$ γ^{2k} is infinitely divisible, but it is not known if, $\forall k \ge 1$ γ^{2k+1} is infinitely divisible (and the experts conjecture that, at least for γ^3, the answer is negative).

23 Renormalized Powers of White Noise and the Virasoro–Zamolodchikov Algebra

In the present section we will use the notations of Section (20) and the results of the papers [AcBou06a, AcBou06b, AcBou06c] which contain the proofs of all the results discussed here.

The formal extension of the white noise commutation relations to the associative $*$–algebra generated by $b_t, b_s^\dagger, 1$, called from now on the renormalized higher powers of (Boson) white noise (RHPWN) algebra, leads to the identities:

$$\begin{aligned}[b_t^{\dagger^n} b_t^k, b_s^{\dagger^N} b_s^K] &= \epsilon_{k,0}\epsilon_{N,0} \sum_{L\ge 1} \binom{k}{L} N^{(L)}\, b_t^{\dagger^n}\, b_s^{\dagger^{N-L}}\, b_t^{k-L}\, b_s^K\, \delta^L(t-s) \\ &\quad -\epsilon_{K,0}\epsilon_{n,0} \sum_{L\ge 1} \binom{K}{L} n^{(L)}\, b_s^{\dagger^N}\, b_t^{\dagger^{n-L}}\, b_s^{K-L}\, b_t^k\, \delta^L(t-s)\end{aligned} \tag{23.1}$$

In Section (20) we have given a meaning to these formal commutation relations, i.e. to the ill defined powers of the δ–function, through the renormalization prescription (20.2).

In the present note we will use a different renormalization rule, introduced in [AcBou06a] and whose motivations are discussed in [AcBou06b, AcBou06c], namely:

$$\delta^l(t-s) = \delta(s)\,\delta(t-s), \quad l = 2,3,4,\dots \tag{23.2}$$

where the right hand side is defined as a convolution of distributions. Using this (23.1) can be rewritten in the form:

$$\begin{aligned}
[b_t^{\dagger^n} b_t^k, b_s^{\dagger^N} b_s^K] &= \epsilon_{k,0}\epsilon_{N,0}\Bigg(k\,N\, b_t^{\dagger^n}\, b_s^{\dagger^{N-1}}\, b_t^{k-1}\, b_s^K\,\delta(t-s) \\
&\quad + \sum_{L\geq 2}\binom{k}{L} N^{(L)}\, b_t^{\dagger^n}\, b_s^{\dagger^{N-L}}\, b_t^{k-L}\, b_s^K\,\delta(s)\,\delta(t-s)\Bigg) \\
&\quad - \epsilon_{K,0}\epsilon_{n,0}\Bigg(K\,n\, b_s^{\dagger^N}\, b_t^{\dagger^{n-1}}\, b_s^{K-1}\, b_t^k\,\delta(t-s) \\
&\quad + \sum_{L\geq 2}\binom{K}{L} n^{(L)}\, b_s^{\dagger^N}\, b_t^{\dagger^{n-L}}\, b_s^{K-L}\, b_t^k\,\delta(s)\,\delta(t-s)\Bigg)
\end{aligned} \tag{23.3}$$

Introducing test functions and the associated smeared fields

$$B_k^n(f) := \int_{\mathbb{R}} f(t)\, b_t^{\dagger^n}\, b_t^k\, dt$$

The commutation relations (23.2) become:

$$\begin{aligned}
[B_k^n(\bar g), B_K^N(f)] &= (\epsilon_{k,0}\epsilon_{N,0}\, k\,N - \epsilon_{K,0}\epsilon_{n,0}\, K\,n)\; B_{K+k-1}^{N+n-1}(\bar g f) \\
&\quad + \sum\nolimits_{L=2}^{(K\wedge n)\vee(k\wedge N)} \theta_L(n,k;N,K)\,\bar g(0)\, f(0)\, b_0^{\dagger^{N+n-l}}\, b_0^{K+k-l}
\end{aligned} \tag{23.4}$$

$$\theta_L(N,K;n,k) := \epsilon_{K,0}\,\epsilon_{n,0}\binom{K}{L} n^{(L)} - \epsilon_{k,0}\,\epsilon_{N,0}\binom{k}{L} N^{(L)} \tag{23.5}$$

The commutation relations (23.4) still contain the ill defined symbol $b_0^{\dagger^{N+n-l}}\, b^{K+k-l}$. However, if the test function space is chosen so that

$$f(0) = g(0) = 0 \tag{23.6}$$

then the singular term in (23.4) vanishes and the commutation relations (23.4) become:

$$[B_k^n(\bar g), B_K^N(f)]_R := (k\,N - K\,n)\; B_{k+K-1}^{n+N-1}(\bar g f) \tag{23.7}$$

which no longer include ill defined objects. In the following, the symbol $[\,\cdot\,,\,\cdot\,]_R$ denotes these renormalized commutation relations.

A direct calculation shows that the commutation relations (23.7) define, on the family of symbols $B_k^n(f)$, a structure of $*$–Lie algebra with involution

$$B_k^n(f)^* := B_n^k(\overline{f})$$

The commutation relations (23.7) imply that, fixing a sub–set $I \subseteq \mathbb{R}^d$, not containing 0, and the test function

$$\chi_I(s) = \begin{cases} 1, & s \in I \\ 0, & s \notin I \end{cases} \tag{23.8}$$

the (self–adjoint) family

$$\{B_k^n := B_k^n(\chi_I) \;:\; n, k \in \mathbb{N},\; n, k \geq 1,\; n + k \geq 3\} \tag{23.9}$$

satisfies the commutation relations

$$[B_k^n, B_K^N]_R := (k\,N - K\,n)\; B_{k+K-1}^{n+N-1} \tag{23.10}$$

The comments to condition (19.2) then suggest the natural interpretation of the $*$–Lie–algebra, defined by the relations (23.9), (23.10), as the 1–*mode algebra* of the RHPWN and, conversely, the interpretation of the RHPWN $*$–Lie–algebra as a current algebra of its 1–mode version.

Now recall the following definition (see, for example, [Ketov95, Pope91]):

Definition 16. *The w_∞–$*$–Lie–algebra is the infinite dimensional Lie algebra spanned by the generators $\hat{B}_k^n$, where $n, N \in \mathbb{N}$, $n, N \geq 2$ and $k, K \in \mathbb{Z}$, with commutation relations:*

$$[\hat{B}_k^n, \hat{B}_K^N]_{w_\infty} = (k\,(N-1) \;-\; K\,(n-1))\; \hat{B}_{k+K}^{n+N-2} \tag{23.11}$$

and involution

$$\left(\hat{B}_k^n\right)^* = \hat{B}_{-k}^n \tag{23.12}$$

Remark 17. The w_∞ – $*$–Lie–algebra, whose elements are interpreted as area preserving diffeomorphisms of 2–manifolds, contains as a sub–Lie–algebra (not as $*$–sub–Lie–algebra) the (centerless) Virasoro (or Witt) algebra with commutations relations

$$[\hat{B}_k^2(\bar{g}), \hat{B}_K^2(f)]_V := (k-K)\,\hat{B}_{k+k}^2(\bar{g}f)$$

Both w_∞ and a quantum deformation of it, denoted W_∞, have been studied extensively in connection to two-dimensional Conformal Field Theory and Quantum Gravity.

The striking similarity between the commutation relations (23.11) and (23.10) suggests that the two algebras are deeply related. The following theorem shows that the current algebra, over $\mathbb{R}$, of the $w_\infty - *$–Lie–algebra can be realized in terms of the renormalized powers of white noise.

Theorem 18. *Let $\mathcal{S}_0$ be the test function space of complex valued (right-continuous) step functions on $\mathbb{R}$ assuming a finite number of values and vanishing at zero, and let the powers of the δ–function be renormalized by the prescription*

$$\delta^l(t-s) = \delta(s)\,\delta(t-s), \quad l = 2, 3, \ldots \tag{23.13}$$

Then the white noise operators

$$\hat{B}^n_k(f) := \int_{\mathbb{R}} f(t)\, e^{\frac{k}{2}(b_t - b_t^\dagger)} \left(\frac{b_t + b_t^\dagger}{2}\right)^{n-1} e^{\frac{k}{2}(b_t - b_t^\dagger)}\, dt; \; n \in \mathbb{N},\; n \geq 2,\; k \in \mathbb{Z} \tag{23.14}$$

satisfy the relations (23.11) and (23.12) of the w_∞–Lie algebra.

Remark 19. The integral on the right hand side of (23.14) is meant in the sense that one expands the exponential series, applies the commutation relations (23.1) to bring the resulting expression to normal order, introduces the renormalization prescription (23.13) and integrates the resulting expressions after multiplication by a test function. The proof shows in particular that the result is an element of the RHPWN $*$–Lie–algebra.

References

[Itô42b] K. Itô: Differential equations determining a Markoff process (in Japanese), Journ. Pan–Japan Math. Coll. 1077 (1942); (in English)

[Itô51] K. Itô: On stochastic differential equations, Mem. Amer. Math. Soc. 4, 1–51 (1951)

[ItôMcKn65] K. Itô and H.P. McKean, Jr.: Diffusion Processes and Their Sample Paths, Springer–Verlag (1965); in Classics in Mathematics, Springer–Verlag (1996)

[AcLuVo02] Accardi L., Lu Y.G., Volovich I.: Quantum Theory and its Stochastic Limit, Springer Verlag (2002)

[AcBou06a] Accardi L., Boukas A., Renormalized Higher Powers of White Noise (RHPWN) and Conformal Field Theory: IDA-QP (Infinite Dimensional Analysis, Quantum Probability and Related Topics), 9 (3) (2006) 353–360

[AcBou06b] Accardi L., Boukas A., The emergence of the Virasoro and w_∞ algebras through the renormalized powers of quantum white noise, International Journal of Mathematics and Computer Science 1 (3) (2006) 315–342

[AcBou06c] Accardi L., Boukas A., Lie algebras associated with the RHPWN, submitted to: Communications on Stochastic Analysis (COSA), 1 (2007)

[AcBouFr06] Accardi L., Boukas A., Franz U.: Renormalized powers of quantum white noise, IDA–QP (Infinite Dimensional Anal. Quantum Probab. Related Topics) 9 (1) (2006) 129–147 Preprint Volterra n. 597 (2006)

[AcKuSt05a] Accardi L., Kuo H.–H., Stan A.: Moments and commutators of probability measures Preprint Volterra (2005)

[AcKuSt02] Accardi L., Kuo H.–H., Stan A.: Orthogonal polynomials and interacting Fock spaces, IDA-QP (Infinite Dimensional Anal. Quantum Probab. Related Topics) 7 (4) (2004) 485–505

[AcBou05a] Accardi L., Boukas A., Higher Powers of q-deformed White Noise, submitted to: Methods of Functional Analysis and Topology (2005)

[AcBou04c] L. Accardi, Andreas Boukas: White noise calculus and stocastic calculus Talk given at: International Conference on "Stochastic analysis: classical and quantum, Perspectives of white noise theory", Meijo University, Nagoya, November 1-5, 2003 in: Quantum Information, T. Hida, K. SaItô (eds.) World Scientific (2005) 260–300 Preprint Volterra n. 579 (2005)

[AcBou03] Accardi L., Boukas A.: Unitarity conditions for the renormalized square of white noise, Infinite Dimensional Anal. Quantum Probab. Related Topics 6 (2) (2003) 197–222

[AcRosc05] L. Accardi, R. Roschin: Renormalized squares of Boson fields, IDA–QP (Infinite Dimensional Analysis, Quantum Probability and Related Topics) 8 (2) (2005) 307–326

[AcAmFr02] Luigi Accardi, Grigori Amosov, Uwe Franz: Second quantized automorphisms of the renormalized square of white noise (RSWN) algebra IDA-QP (Infinite Dimensional Analysis, Quantum Probability and Related Topics) 7 (2) (2004) 183–194 Preprint Volterra (2002)

[AcBou01a] Accardi L., Boukas A.: Square of white noise unitary evolutions on Boson Fock space, in: Proceedings International conference on stochastic analysis in honor of Paul Kree, Hammamet, Tunisie, October 22-27, 2001, S. Albeverio, A. Boutet de Monvel, H. Oueridiane (eds.) Kluwer (2004) 267–302

[AcLuVo99] Accardi L., Lu Y.G., Volovich I.V.: White noise approach to classical and quantum stochastic calculi, Lecture Notes of the Volterra–CIRM International School with the same title, Trento, Italy, 1999, Volterra Preprint n. 375 July (1999)

[AcAyOu05a] Accardi L., Ayed W., Ouerdiane H.: White noise Heisenberg evolution and Evans–Hudson flows, 10 (1) 2007 IDA-QP (Infinite Dimensional Analysis, Quantum Probability and Related Topics), Preprint Volterra n. 590 (2005)

[AcAyOu05b] Accardi L., Ayed W.: Module Form of The White Noise Stochastic Calculus, to appear in: Random Operators and Stochastic Equations (ROSE) 2007

[AcAyOu03] Luigi Accardi, Wided Ayed, Habib Ouerdiane: White Noise Approach to Stochastic Integration, Random Operators and Stochastic Equations 13 (4) (2005) 369–398 Preprint Volterra n. 573 (2004)

[AcBo98] Accardi L., Bożejko M.: Interacting Fock spaces and Gaussianization of probability measures, Infinite dimensional analysis, quantum probability and related topics, 1, N. 4 (1998) 663–670

[AcLuVo97b] Accardi L., Lu Y.G., Volovich I.: The QED Hilbert module and Interacting Fock spaces, Publications of IIAS (Kyoto) (1997)

[AcLuVo93] Accardi L., Lu Y.G., Volovich I.: The Stochastic Sector of Quantum Field Theory, Matematicheskie Zametki (1994) Volterra Preprint N. 138 (1993)

[AcHud89] Accardi L., Hudson R.: Quantum stochastic flows and non abelian cohomology, in: Quantum Probability and Applications V, Springer LNM 1442 (1990) Volterra preprint N. 16 (1989)

[AcFrLu87] Accardi L., Frigerio A., Lu Y.G.: On the weak coupling limit problem, in: [QP–PQ IV], Quantum Probability and Applications IV Springer LNM N. 1396 (1987) 20–58

[Ac78b] Accardi L.: On the quantum Feynmann-Kac formula, Rendiconti del seminario Matematico e Fisico, Milano 48 (1978) 135–179

[Arak60] Araki H.: Princeton Thesis (1960)

[Ayed05] Ayed Wided: White Noise Approach to Quantum Stochastic Calculus, Thesis, University Tunis El Manar dan University, Tor Vergata (2005)

[Barc88] Barchielli A.: Input and output channels in quantum systems and quantum stochastic differential equations, in: [QP–PQ III], SLNM 1303 (ed) Accardi L. von Waldenfels W, Springer (1988)

[Bela86a] Belavkin V.P.: Optimal linear filtering and control in quantum linear systems, In: Proc of I F A C Conference on Statistical Control Theory, Vilnius (1986)

[BogLogTod69] Bogolyubov, N.N., Logunov, A.A., Todorov, I.T.: Fundamentals of the axiomatic approach in quantum field theory (Russian) [Izdat. "Nauka", Moscow (1969) English translation: Bogolubov, N.N.; Logunov, A.A.; Todorov, I.T.: Introduction to axiomatic quantum field theory, Mathematical Physics Monograph Series, N. 18. W.A. Benjamin Inc. (1975)

[BoKüSp96b] Bozejko M., Kümmerer B., Speicher R.: q–Gaussian processes: non commutative and classical aspects, Comm. Math. Phys. 185 (1997) 129–154 Preprint (1996)

[BozSpe96a] Bozejko M., Speicher R.: Interpolations between bosonic and Fermionic relations given by generalized Brownian motions, Math. Zeit. 322 (1996) 135–160

[FrGo82a] Frigerio A., Gorini V.: On stationary Markov dilations of quantum dynamical semigroups, In: [QP–PQ II], Quantum Probability and Applications to the Quantum Theory of Irreversible Processes (a cura di L. Accardi, A. Frigerio, V. Gorini), Proceedings, Villa Mondragone (1982), Springer LNM 1055 (1984) 119–125

[GaCo85] Gardiner C.W., Collet M.J.: Input and output in damped quantum systems: quantum stochastic differential equation and master equation, Phys. Rev. A 31 (1985) 3761–3774

[Grig01] B. Grigelionis: Generalized z–distributions and related stochastic processes, Lithuanian Math. J. 41 (2001) 303–319

[Grig99] B. Grigelionis: Processes of Meixner type, Lithuanian Math. J. 39 (1999) 33–41

[Grig00c] B. Grigelionis: On generalized z–diffusions, Preprint Vilnius (2000)
[Gui72] Guichardet A.: Symmetric Hilbert spaces and related topics, Lect. Notes Math.: 261, Springer, Berlin (1972)
[Hida92] Hida T.: Selected papers, World Scientific (2001)
[Hida75] Hida T.: Analysis of Brownian Functionals, Carleton Mathematical Lecture notes 13 (1975); 2nd ed., 1978
[HuPa84c] Hudson R.L., Parthasarathy K.R.: Quantum Itô's formula and stochastic evolutions, Commun. Math. Phys. 93 (1984) 301–323
[HuPa82a] Hudson R.L., Parthasarathy K.R.: Construction of quantum diffusions, in: Quantum probability and applications to the quantum theory of irreversible processes, Proc. 2–d Conference: Quantum Probability and applications to the quantum theory of irreversible processes, 6–11, 9 (1982) Villa Mondragone (Rome), Springer LNM N. 1055 (1984)
[Ketov95] Ketov S.V., Conformal field theory, World Scientific (1995)
[KunWat67] H. Kunita and S. Watanabe: On square integrable martingales, Nagoya Math. J. 30, 209–245 (1967)
[Ivanov79] Ivanov, V.K.: The algebra of elementary generalized functions. (Russian) Dokl. Akad. Nauk SSSR 246 (1979), no. 4, 805–808. English translation: Soviet Math. Dokl. 20 (1979), no. 3, 553–556
[Partha92] Parthasarathy K.R.: An introduction to quantum stochastic calculus, Birkhäuser (1992)
[PaSch72] Parthasarathy K.R., Schmidt K.: Positive definite kernels continuous tensor products and central limit theorems of probability theory, Springer Lecture Notes in Mathematics no. 272 (1972)
[Pope91] Pope C. N., Lectures on W algebras and W gravity, Lectures given at the Trieste Summer School in High-Energy Physics, August 1991
[Prohor05] D.B. Prohorenko: Squares of white noise, $SL(2,\mathbb{C})$ and Kubo-Martin-Schwinger states, IDA-QP (Infinite Dimensional Analysis, Quantum Probability and Related Topics) 9 (4) (2006) 491–511 Preprint 25-4-2005
[Skor84] Skorodod A.V.: Random linear operators, Reidel (1984)
[Śnia99] P. Śniady: Quadratic bosonic and free white noises, Commun. Math. Phys. 211 (3) (2000) 615–628 Preprint (1999)
[Vlad66] Vladimirov, V. S. : (Russian) [Methods in the theory of functions of several complex variables, Izdat. "Nauka", Moscow (1964), English translation: Vladimirov, V. S.: Methods of the theory of functions of many complex variables, Scripta Technica, Inc. Translation edited by Leon Ehrenpreis The M.I.T. Press (1966)
[voWaGi78] von Waldenfels W., Giri N.: An Algebraic Version of the Central Limit Theorem, Z. Wahrscheinlichkeitstheorie verw. Gebiete 42 (1978) 129–134
[voWa78] von Waldenfels W.: An Algebraic Central Limit Theorem in the Anticommuting Case. Z. Wahrscheinlichkeitstheorie verw. Gebiete 42 (1978) 135–140

Homogenization of Diffusions on the Lattice $\mathbf{Z}^d$ with Periodic Drift Coefficients, Applying a Logarithmic Sobolev Inequality or a Weak Poincaré Inequality

Sergio Albeverio[1], M. Simonetta Bernabei[2], and Michael Röckner[3], and Minoru W. Yoshida[4]

[1] Inst. Angewandte Mathematik, Universität Bonn, Wegelerstr. 6, D-53115, Bonn, Germany. SFB611; BiBoS; CERFIM, Locarno; Acc. Architettura USI, Mendrisio, albeverio@uni-bonn.de

[2] Dipartimento di Matematica e Informatica, Università di Camerino, Via Madonna delle Carceri, 9 I-62032 Camerino, Italy, simone-bernabei@unicam.it

[3] Fakultät für Mathematik, Universität Bielefeld, D-33615 Bielefeld, Germany, roeckner@math.uni-bielefeld.de
Department of Mathematics, Purdue University, Math. Sci. Building 150N. University Street, West Lafayette, IN 47907-2067, USA; BiBoS, roeckner@math.purdue.edu

[4] Kansai Univ. Dept. Math. 564-8680 Yamate-Tyou, Suita-shi, Osaka, Japan wyoshida@ipcku.kansai-u.ac.jp

1 Introduction

In this paper we treat limit theorems for diffusions on the lattice $\mathbf{Z}^d$ of the form of those constituting the solution of the homogenization problem of diffusions. For finite dimensional diffusion processes, various models of homogenization (generalized in several directions) have been studied in detail (cf. eg. [F2, FNT, FunU, O, PapV, Par] and references therein). On the other hand, for corresponding problems of infinite dimensional diffusions only few results are known (cf. [FunU, ABRY1,2,3]). In this paper we consider a homogenization problem of infinite dimensional diffusion processes indexed by $\mathbf{Z}^d$ having periodic drift coefficients with the period 2π (cf. (2.1)), by applying an L^2 type ergodic theorem for the corresponding quotient processes taking values in $[0, 2\pi)^{\mathbf{Z}^d}$ (cf. Prop. 1). The ergodic theorem which is based on a (weak) Poincaré inequality.

In [ABRY3] the same problem has been discussed by applying the uniform ergodic theorem for the corresponding quotient process, that is available by assuming that the Markov semi-group of the quotient process of the original process satisfies a logarithmic Sobolev inequality. In the same paper it has also been shown that a homogenization property of the processes starting from an almost every arbitrary point in the state space with respect to an invariant measure of the quotient process holds (cf. also [ABRY1, ABRY2]). In this occasion, the main purpose of the present paper is the comparison between the results derived under the assumption of logarithmic Sobolev inequality and the corresponding results proven by assuming L^2 ergodic theorem based on (weak) Poincaré inequality, which is strictly weaker than the one for logarithmic Sobolev inequality (cf. [AKR, G]). This paper is a series of works on the considerations of several types of homogenization models for infinite dimensional diffusion processes.

For an adequate understanding of crucial differences between homogenization problems in finite and infinite dimensional situations, we first briefly review a simple case of the homogenization problem for finite dimensional diffusions.

On some complete probability space, suppose that we are given a one dimensional standard Brownian motion process $\{B_t\}_{t\in\mathbf{R}_+}$ and consider the stochastic differential equation for each initial state $x \in \mathbf{R}$ and each scaling parameter $\epsilon > 0$ given by

$$\begin{aligned} X^\epsilon(t,x) = x + \frac{1}{\epsilon}\int_0^t b\left(\frac{X^\epsilon(s,x)}{\epsilon}\right) ds \\ +\sqrt{2}\int_0^t a\left(\frac{X^\epsilon(s,x)}{\epsilon}\right) dB_s, \quad t\in\mathbf{R}_+, \end{aligned} \tag{1.1}$$

where $a \in C^\infty(\mathbf{R}\to\mathbf{R})$ is a periodic function with period 2π which satisfies

$$\lambda \le a(x) \le \lambda^{-1}, \qquad \forall x\in\mathbf{R},$$

for some constant $\lambda > 0$ and $b(x) \equiv \frac{d}{dx}a^2(x)$.

Let $p_t^\epsilon(x,y)$ be the transition density function corresponding to the diffusion process defined through (1.1). Then by Nash's inequality (cf. eg. [S]) we have that there exist constants $c_1, c_2 > 0$ such that

$$p_t^\epsilon(x,y) \le c_1 t^{-\frac{1}{2}} \exp\{-c_2|x-y|^2 t^{-1}\}, \qquad \forall t>0, \quad \forall \epsilon\in(0,1]. \tag{1.2}$$

Also, there exists a periodic function $\chi \in C^2(\mathbf{R})$ with period 2π such that

$$a^2(x)\chi''(x) + b(x)\chi'(x) = b(x), \quad x\in\mathbf{R}, \qquad \chi(0)=0. \tag{1.3}$$

Then by Itô's formula and using (3) we see that

$$\begin{aligned} X^\epsilon(t,x) = x - \epsilon\chi\left(\frac{x}{\epsilon}\right) + \epsilon\chi\left(\frac{X^\epsilon(t,x)}{\epsilon}\right) \\ + \sqrt{2}\int_0^t \left(1-\chi'\left(\frac{X^\epsilon(s,x)}{\epsilon}\right)\right) a\left(\frac{X^\epsilon(s,x)}{\epsilon}\right) dB_s, \quad t\in\mathbf{R}_+. \end{aligned} \tag{1.4}$$

The (probabilistic) homogenization problem consists of proving weak convergence of the process $\{\{X^\epsilon(t,x)\}_{t\in\mathbf{R}_+}\}_{\epsilon>0}$. In this simple situation and also in various generalized models of finite dimensional diffusions, it is shown that

$$\lim_{\epsilon\downarrow 0} E^{P_x^\epsilon}[\varphi(\cdot)] = E^{P_x}[\varphi(\cdot)], \tag{1.5}$$

$$\forall x\in\mathbf{R}, \qquad \forall\varphi\in C_b(C(\mathbf{R}_+\to\mathbf{R})\to\mathbf{R}),$$

where P_x^ϵ is the probability law of the process $\{X^\epsilon(t,x)\}_{t\in\mathbf{R}_+}$, as a $C(\mathbf{R}_+\to\mathbf{R})$ valued random variable, P_x is the probability law of the continuous Gaussian process starting at $x\in\mathbf{R}$ with constant diffusion coefficient given by

$$\sigma \equiv \left\{2\int_0^{2\pi}\{(1-\chi'(y))a(y)\}^2 dy\right\}^{\frac{1}{2}}, \tag{1.6}$$

$E^{P_x^\epsilon}[\cdot]$, $E^{P_x}[\cdot]$ are respectively the expectations with respect to the corresponding probability measures, and $C_b(\cdot)$ is the space of bounded continuous functions.

Here, our problem is a homogenization problem of a diffusion

$$\{\mathbf{X}_\mathbf{k}^\epsilon(t,\mathbf{x})\}_{\mathbf{k}\in\mathbf{Z}^d}$$

with index set $\mathbf{Z}^d$ involving the periodic drift coefficients of period 2π defined by (2.1) in the next section. Since the constants in Nash's inequality depend on the dimensions, for our infinite dimensional diffusions we can not use the uniform bound (1.2) for the Markovian transition density functions which are based on Nash's inequality. By Lemma 3 below for our infinite dimensional diffusions we can define a family of functions $\chi_\mathbf{k}$, ($\mathbf{k}\in\mathbf{Z}^d$) (cf. (3.4)) that is an infinite dimensional version of χ defined by (1.3). But except for some trivial cases, we can not expect the regularity $\chi_\mathbf{k}\in C^2$. Thus the same strategy developed for the consideration of finite dimensional problems can not be applied directly in infinite dimension. These are the main difficulties which appear in the alter situations.

But in Theorem 9 and Theorem 12, by using the L^2 ergodic theorem (3.2), which is a consequence of a weak Poincaré inequality for the corresponding quotient process of $\{\mathbf{X}_\mathbf{k}^1(t,\mathbf{x})\}_{\mathbf{k}\in\mathbf{Z}^d}$, instead we can show that in the infinite dimensional situation a homogenization holds, roughly speaking, in the following weaker sense than (1.5): Let a Polish space W be a subspace of $C(\mathbf{R}_+\to\mathbf{R}^{\mathbf{Z}^d})$ equipped with a topology which is sufficiently stronger than the product topology on $C(\mathbf{R}_+\to\mathbf{R}^{\mathbf{Z}^d})$ (cf. (2.4)), $C_b(W)$ be the space of bounded continuous functions on W, P_0 be the probability law of an infinite dimensional diffusion $\{\mathbb{Y}_t\}_{t\in\mathbf{R}_+}$ with a constant covariance matrix defined through $\chi_\mathbf{k}$, ($\mathbf{k}\in\mathbf{Z}^d$) starting from the initial state 0 (cf. (3.7) and Definition 5), also let $\tilde{P}_\mathbf{x}^\epsilon$ be the probability law of the process

$$\left\{\mathbf{X}_\mathbf{k}^\epsilon(t,\mathbf{x}) - \epsilon\chi_\mathbf{k}\left(\frac{\mathbf{X}_\mathbf{k}^\epsilon(t,\mathbf{x})}{\epsilon}\right) + \epsilon\chi_\mathbf{k}\left(\frac{\mathbf{x}}{\epsilon}\right)\right\}_{\mathbf{k}\in\mathbf{Z}^d},$$

then it holds that

$$\lim_{\epsilon \downarrow 0} E^{\tilde{P}^{\epsilon}_{\nu_\epsilon}}[\varphi(\cdot)] = E^{P_0}[\varphi(\cdot)], \quad \forall \varphi \in C_b(W \to \mathbf{R}), \tag{1.7}$$

where for each $\epsilon \in [0,1)$, the probability measure $\tilde{P}^{\epsilon}_{\nu_\epsilon}$ on $(W, \mathcal{B}(W))$ is defined by

$$\tilde{P}^{\epsilon}_{\nu_\epsilon}(B) \equiv \int_{[0,2\pi)^{\mathbf{Z}^d}} \tilde{P}^{\epsilon}_{\epsilon \mathbf{y}}(B)\nu(d\mathbf{y}), \quad \forall B \in \mathcal{B}(W), \tag{1.8}$$

for a probability measure ν on $([0,2\pi)^{\mathbf{Z}^d}, \mathcal{B}([0,2\pi)^{\mathbf{Z}^d}))$ such that

$$\left\| \frac{d\nu}{d\mu} \right\|_{L^\infty([0,2\pi)^{\mathbf{Z}^d})} < \infty, \tag{1.9}$$

and μ is the unique invariant measure of the quotient diffusion process on the infinite dimensional torus (identified with $[0,2\pi)^{\mathbf{Z}^d}$) of the original diffusion $\{\mathbf{X}^1_{\mathbf{k}}(t,\mathbf{x})\}_{\mathbf{k}\in\mathbf{Z}^d}$ (cf. Proposition 1).

2 Fundamental Notations

Let $\mathbf{N}$ and $\mathbf{Z}$ be the set of natural numbers and integers respectively. For $d \in \mathbf{N}$ let $\mathbf{Z}^d$ be the d-dimensional lattice. We consider the problem for the diffusions taking values in $\mathbf{R}^{\mathbf{Z}^d}$. We use the following notions and notations:

By $\mathbf{k}$ we denote $\mathbf{k} = (k^1, \ldots, k^d) \in \mathbf{Z}^d$. For a subset $\Lambda \subseteq \mathbf{Z}^d$, we define $|\Lambda| \equiv \mathrm{card}\,\Lambda$. For $\mathbf{k} \in \mathbf{Z}^d$ and $\Lambda \subseteq \mathbf{Z}^d$ let

$$\Lambda + \mathbf{k} \equiv \{\mathbf{l} + \mathbf{k} \,|\, \mathbf{l} \in \Lambda\}.$$

For any non-empty $\Lambda \subseteq \mathbf{Z}^d$, we assume that $\mathbf{R}^\Lambda$ is the topological space equipped with the direct product topology. For each non-empty $\Lambda \subseteq \mathbf{Z}^d$, by $\mathbf{x}_\Lambda$ we denote the image of the projection onto $\mathbf{R}^\Lambda$:

$$\mathbf{R}^{\mathbf{Z}^d} \ni \mathbf{x} \longmapsto \mathbf{x}_\Lambda \in \mathbf{R}^\Lambda.$$

For each $p \in \mathbf{N} \cup \{0\} \cup \{\infty\}$ we define the set of p-times continuously differentiable functions with support Λ: $C^p_\Lambda(\mathbf{R}^{\mathbf{Z}^d}) \equiv \{\varphi(\mathbf{x}_\Lambda) \,|\, \varphi \in C^p(\mathbf{R}^\Lambda)\}$, where $C^p(\mathbf{R}^\Lambda)$ is the set of real valued p-times continuously differentiable functions on $\mathbf{R}^\Lambda$. For $p = 0$, we simply denote $C^0_\Lambda(\mathbf{R}^{\mathbf{Z}^d})$ by $C_\Lambda(\mathbf{R}^{\mathbf{Z}^d})$. Also we set

$$C^p_0\left(\mathbf{R}^{\mathbf{Z}^d}\right) \equiv \left\{\varphi \in C^p_\Lambda\left(\mathbf{R}^{\mathbf{Z}^d}\right) \,|\, |\Lambda| < \infty\right\}.$$

$\mathcal{B}(\mathbf{R}^{\mathbf{Z}^d})$ is the Borel σ-field of $\mathbf{R}^{\mathbf{Z}^d}$ and $\mathcal{B}_\Lambda(\mathbf{R}^{\mathbf{Z}^d})$ is the sub σ-field of $\mathcal{B}(\mathbf{R}^{\mathbf{Z}^d})$ that is generated by the family $C_\Lambda(\mathbf{R}^{\mathbf{Z}^d})$. For each $\mathbf{k} \in \mathbf{Z}^d$, let $\vartheta^{\mathbf{k}}$ be the shift operator on $\mathbf{R}^{\mathbf{Z}^d}$ such that

$$(\vartheta^{\mathbf{k}}\mathbf{x})_{\{\mathbf{j}\}} \equiv \mathbf{x}_{\{\mathbf{k}+\mathbf{j}\}}, \ \mathbf{x} \in \mathbf{R}^{\mathbf{Z}^d}, \ \mathbf{j} \in \mathbf{Z}^d,$$

where $\mathbf{x}_{\{\mathbf{k}+\mathbf{j}\}}$ is the $\mathbf{k}+\mathbf{j}$-th component of the vector $\mathbf{x}$.

We shall define the infinite dimensional diffusions we are interested in through a stochastic differential equation (SDE). On a complete probability space $(\Omega, \mathcal{F}, P; \mathcal{F}_t)$ with an increasing family of sub σ-field $\{\mathcal{F}_t\}_{t\in\mathbf{R}_+}$ we are given a family of independent 1-dimensional $\mathcal{F}_t$-standard Brownian motion processes $\{B_{\mathbf{k}}(t)\}_{t\in\mathbf{R}_+}$, $\mathbf{k} \in \mathbf{Z}^d$. For each $\epsilon \in (0,1]$ and each $\mathbf{x} = \{x_{\mathbf{k}}\}_{\mathbf{k}\in\mathbf{Z}^d} \in \mathbf{R}^{\mathbf{Z}^d}$, consider the following system of SDE's:

$$X^{\epsilon}_{\mathbf{k}}(t,\mathbf{x}) = x_{\mathbf{k}} + \sqrt{2}B_{\mathbf{k}}(t) + \frac{1}{\epsilon}\int_0^t b_{\mathbf{k}}\left(\frac{\mathbb{X}^{\epsilon}(s,\mathbf{x})}{\epsilon}\right) ds, \quad t \in \mathbf{R}_+, \quad \mathbf{k} \in \mathbf{Z}^d, \tag{2.1}$$

where we set

$$\mathbb{X}^{\epsilon}(s,\mathbf{x}) \equiv \{X^{\epsilon}_{\mathbf{k}}(s,\mathbf{x})\}_{\mathbf{k}\in\mathbf{Z}^d}, \quad \text{and define} \quad b_{\mathbf{k}}(\mathbf{x}) \equiv \sum_{\Lambda \ni \mathbf{k}} \left(-\frac{\partial}{\partial x_{\mathbf{k}}} J_{\Lambda}(\mathbf{x})\right),$$

for a given family of potentials $\mathcal{J} \equiv \{J_{\Lambda} \,|\, \Lambda \subset \mathbf{Z}^d, |\Lambda| < \infty\}$ such that

J-1) (Periodicity) for each $\Lambda \subset \mathbf{Z}^d$ such that $|\Lambda| < \infty$,

$$J_{\Lambda} \in C^{\infty}_{\Lambda}(\mathbf{R}^{\mathbf{Z}^d}),$$

and it is a periodic function with respect to each variable with the period 2π;

J-2) (Shift invariance)

$$J_{\Lambda+\mathbf{k}} = J_{\Lambda} \circ \vartheta^{\mathbf{k}}, \qquad \forall \mathbf{k} \in \mathbf{Z}^d;$$

J-3) (Finite range) there exists an $L < \infty$ and $J_{\Lambda} = 0$ holds for any Λ such that $\Lambda \ni 0$ and $\Lambda \not\subset [-L, +L]^d$.

By Lemma 1.2 of [HS] we have the following: Under the assumption J-1), J-2) and J-3), for each $\epsilon > 0$, SDE (10) has a strong unique solution. Also, for any $T < \infty$ and $\epsilon > 0$ there exists a constant $A^{\epsilon}_T < \infty$ and for any $\mathbf{x} = \{x_{\mathbf{k}}\}_{\mathbf{k}\in\mathbf{Z}^d}$, $\mathbf{x}' = \{x'_{\mathbf{k}}\}_{\mathbf{k}\in\mathbf{Z}^d}$ one has that

$$E\left[\sum_{\mathbf{k}\in\mathbf{Z}^d} \frac{1}{2^{|\mathbf{k}|}} \sup_{0\le t\le T} \left|X^{\epsilon}_{\mathbf{k}}(t,\mathbf{x}) - x_{\mathbf{k}}\right|^2\right] \le A^{\epsilon}_T, \tag{2.2}$$

$$E\left[\sum_{\mathbf{k}\in\mathbf{Z}^d} \frac{1}{2^{|\mathbf{k}|}} \sup_{0\le t\le T} \left|X^{\epsilon}_{\mathbf{k}}(t,\mathbf{x}) - X^{\epsilon}_{\mathbf{k}}(t,\mathbf{x}')\right|^2\right] \le A^{\epsilon}_T \left(\sum_{\mathbf{k}\in\mathbf{Z}^d} \frac{1}{2^{|\mathbf{k}|}} \left|x_{\mathbf{k}} - x'_{\mathbf{k}}\right|^2\right). \tag{2.3}$$

By (2.2) and (2.3) we can define a metric ρ on a linear subspace W of $C(\mathbf{R}_+ \to \mathbf{R}^{\mathbf{Z}^d})$, on which the trajectories of the diffusions $\mathbb{X}^\epsilon(\cdot, \mathbf{x})$ exist if their initial states satisfy $\mathbf{x} \in \mathcal{H}$. Namely, let

$$\mathcal{H} \equiv \left\{ \mathbf{x} = \{x_{\mathbf{k}}\}_{\mathbf{k}\in\mathbf{Z}^d} \in \mathbf{R}^{\mathbf{Z}^d} \,\Big|\, \sum_{\mathbf{k}\in\mathbf{Z}^d} \frac{1}{2^{|\mathbf{k}|}} x_{\mathbf{k}}^2 < \infty \right\},$$

denoting $\mathbf{x}(\cdot) \equiv \{x_{\mathbf{k}}(\cdot)\}_{\mathbf{k}\in\mathbf{Z}^d}$ and define

$$W \equiv \left\{ \mathbf{x}(\cdot) \in C\left(\mathbf{R}_+ \to \mathbf{R}^{\mathbf{Z}^d}\right) \Big|\, \sum_{\mathbf{k}\in\mathbf{Z}^d} \frac{1}{2^{|\mathbf{k}|}} \sup_{0\le t\le T} |x_{\mathbf{k}}(t)|^2 < \infty,\ \forall T < \infty \right\}.$$

We define a metric ρ on W, and denote the Polish space equipped with this metric by the same symbol W:

$$\rho(\mathbf{x}(\cdot), \mathbf{x}'(\cdot)) \equiv \sum_{n\in\mathbf{N}} \frac{1}{2^n} \left\{ \left\{ \sum_{\mathbf{k}\in\mathbf{Z}^d} \frac{1}{2^{|\mathbf{k}|}} \sup_{0\le t\le n} |x_{\mathbf{k}}(t) - x'_{\mathbf{k}}(t)|^2 \right\}^{\frac{1}{2}} \wedge 1 \right\}, \tag{2.4}$$

$$\text{for} \qquad \mathbf{x}(\cdot) \equiv \{x_{\mathbf{k}}(\cdot)\}_{\mathbf{k}\in\mathbf{Z}^d},\ \mathbf{x}'(\cdot) \equiv \{x'_{\mathbf{k}}(\cdot)\}_{\mathbf{k}\in\mathbf{Z}^d} \in W.$$

The metric ρ gives a stronger topology than the product topology and keeps the Borel structure unchanged. We note that the metric ρ is also stronger than the following metric (cf. [AKR])

$$\sum_{n\in\mathbf{N}} \frac{1}{2^n} \left\{ \sup_{0\le t\le n} \left\{ \sum_{\mathbf{k}\in\mathbf{Z}^d} \frac{1}{2^{|\mathbf{k}|}} |x_{\mathbf{k}}(t) - x'_{\mathbf{k}}(t)|^2 \right\}^{\frac{1}{2}} \bigwedge 1 \right\}.$$

Let $\mathcal{B}(W)$ be the Borel σ-field of W and $\mathcal{B}_t(W)$, $t \in \mathbf{R}_+$, be the sub σ-field of $\mathcal{B}(W)$ generated by the cylinder sets of $(C([0,t] \to \mathbf{R}))^{\mathbf{Z}^d}$.

For each $t \ge 0$, let $\boldsymbol{\xi}_t$ be the measurable map given by

$$\boldsymbol{\xi}_t : W \ni \mathbf{x}(\cdot) \longmapsto \mathbf{x}(t) \in \mathbf{R}^{\mathbf{Z}^d},$$

then $\mathcal{B}_t(W)$ is the σ-field generated by $\boldsymbol{\xi}_s$, $s \in [0,t]$.

For each $\mathbf{x} \in \mathcal{H}$ and $\epsilon > 0$, let $P^\epsilon_{\mathbf{x}}$ be the probability measure on $(W, \mathcal{B}(W))$ which is the probability law of the process $\{\mathbb{X}^\epsilon(t, \mathbf{x})\}_{t\in\mathbf{R}_+}$:

$$P(\{\omega \,|\, \mathbb{X}^\epsilon(\cdot, \mathbf{x}) \in B\}) = P^\epsilon_{\mathbf{x}}(\boldsymbol{\xi}_\cdot \in B), \qquad \forall B \in \mathcal{B}(W).$$

Let $T = \{y \in \mathbf{R}^2 : |y| = 1\}$ be the unit circle equipped with the natural Riemannian metric. Let $T^{\mathbf{Z}^d}$ be the product space of T endowed with the direct product topology, so that $T^{\mathbf{Z}^d}$ is a Polish space. Let $(W_T, \mathcal{B}(W_T);$

$\mathcal{B}_t(W_T))$ be the measurable space of the Polish space $W_T \equiv C(\mathbf{R}_+ \to T^{\mathbf{Z}^d})$, such that $\mathcal{B}(W_T)$ is the Borel σ-field of W_T and $\mathcal{B}_t(W_T)$, $(t \in \mathbf{R}_+)$, is the sub σ-field of $\mathcal{B}(W_T)$ generated by the cylinder sets of $C([0,t] \to T^{\mathbf{Z}^d})$.

Corresponding to the previously defined notations $\mathbf{x}_\Lambda$, resp. $C^p_\Lambda(\mathbf{R}^{\mathbf{Z}^d})$ and $C^p_0(\mathbf{R}^{\mathbf{Z}^d})$, we define the following: For each non-empty $\Lambda \subseteq \mathbf{Z}^d$, by $\mathbf{y}_\Lambda$ we denote the image of the projection onto T^Λ:

$$T^{\mathbf{Z}^d} \ni \mathbf{y} \longmapsto \mathbf{y}_\Lambda \in T^\Lambda.$$

Also, $C^p_\Lambda(T^{\mathbf{Z}^d})$, $C^p_0(T^{\mathbf{Z}^d})$, and $C_\Lambda(T^{\mathbf{Z}^d})$ are defined correspondigly.

We use the notation $\mathbf{y} = \{y_\mathbf{k}\}_{\mathbf{k}\in\mathbf{Z}^d}$ to denote a point in $T^{\mathbf{Z}^d}$.

In order to give a correspondence between the points in $\mathbf{R}^{\mathbf{Z}^d}$ and the points in $T^{\mathbf{Z}^d}$, we introduce the function

$$\Theta : T^{\mathbf{Z}^d} \ni \{y_\mathbf{k}\}_{\mathbf{k}\in\mathbf{Z}^d} \longmapsto \{\theta_\mathbf{k}\}_{\mathbf{k}\in\mathbf{Z}^d} \in [0, 2\pi)^{\mathbf{Z}^d}$$

where $\theta_\mathbf{k} = \theta(y_\mathbf{k})$ and the function $\theta : T \longrightarrow [0, 2\pi)$ is defined by

$$y = \begin{pmatrix} \cos\theta(y) \\ \sin\theta(y) \end{pmatrix} \in T \subset \mathbf{R}^2.$$

Let $\widehat{C}([0,2\pi)^{\mathbf{Z}^d})$ be the linear subspace of $C([0,2\pi)^{\mathbf{Z}^d})$ such that

$$\widehat{C}\left([0,2\pi)^{\mathbf{Z}^d}\right) \equiv \Big\{\phi \in C\left([0,2\pi)^{\mathbf{Z}^d}\right) \Big| \lim_{\theta_\mathbf{k}\uparrow 2\pi} \phi(\boldsymbol{\theta}) = \phi(\boldsymbol{\theta}|_{\theta_\mathbf{k}=0}),$$
$$\forall \boldsymbol{\theta} \equiv \{\theta_\mathbf{l}\}_{\mathbf{l}\in\mathbf{Z}^d} \in [0,2\pi)^{\mathbf{Z}^d}, \forall \mathbf{k} \in \mathbf{Z}^d\Big\},$$

where $\boldsymbol{\theta}|_{\theta_\mathbf{k}=0}$ is the vector defined by changing the $\mathbf{k}$-th component $\theta_\mathbf{k}$ of $\boldsymbol{\theta}$ to 0. Then, each $\varphi \in C(T^{\mathbf{Z}^d})$ has a corresponding element $\phi \in \widehat{C}([0,2\pi)^{\mathbf{Z}^d})$ such that

$$\varphi(\mathbf{y}) = \phi \circ \Theta(\mathbf{y}), \qquad \forall \mathbf{y} \in T^{\mathbf{Z}^d}$$

By this, we will identify the elements of $C(T^{\mathbf{Z}^d})$ with the corresponding elements in $\widehat{C}([0,2\pi)^{\mathbf{Z}^d})$.

In addition we define $\Phi(x_\mathbf{k}) = \theta_\mathbf{k} \in [0,2\pi)$ if $x_\mathbf{k} = \theta_\mathbf{k} \, mod \, 2\pi$. Then we can define a surjection from $\mathbf{R}^{\mathbf{Z}^d}$ to $T^{\mathbf{Z}^d}$ such that

$$\Theta^{-1} \circ \boldsymbol{\Phi} : \mathbf{R}^{\mathbf{Z}^d} \ni \mathbf{x} = \{x_\mathbf{k}\}_{\mathbf{k}\in\mathbf{Z}^d} \longmapsto \{\theta^{-1} \circ \Phi(x_\mathbf{k})\}_{\mathbf{k}\in\mathbf{Z}^d} \in T^{\mathbf{Z}^d}. \tag{2.5}$$

In the sequel, if there is no ambiguity, to denote such interpretation $\varphi(\boldsymbol{\Theta}^{-1} \circ \boldsymbol{\Phi}(\cdot)) \in C(\mathbf{R}^{\mathbf{Z}^d})$ of $\varphi(\cdot) \in C(T^{\mathbf{Z}^d})$ we will use the same notation φ, i.e., we will not always write the corresponding **periodic function** by $\varphi(\boldsymbol{\Theta}^{-1} \circ \boldsymbol{\Phi}(\mathbf{x}))$ but simply $\varphi(\mathbf{x})$.

The following Proposition 1-i), ii) resp. and iii) are results of Theorem 2.23 of [HS] resp. and Proposition 1.2 of [S] (cf. also [BoRW]):

Proposition 1. ***(Quotient process of*** $\{\mathbb{X}^1(t,\mathbf{x})\}_{t\geq 0}$***)***
Let $\mathcal{J}$ be a potential that satisfies the conditions J-1), J-2) and J-3).
i) For each $t \geq 0$, let $\boldsymbol{\eta}_t$ be the measurable function defined by

$$\boldsymbol{\eta}_t : W_T \ni \mathbf{y}(\cdot) \longmapsto \mathbf{y}(t) \in T^{\mathbf{Z}^d}.$$

Let $\mathbf{y} \in T^{\mathbf{Z}^d}$ and take $\mathbf{x} \in \mathcal{H}$ such that $\mathbf{y} = \boldsymbol{\Theta}^{-1} \circ \boldsymbol{\Phi}(x)$.
On $(W_T, \mathcal{B}(W_T))$ define the probability measure

$$Q_{\mathbf{y}} \equiv P^1_{\mathbf{x}} \circ \Theta^{-1} \circ \boldsymbol{\Phi},$$

i.e.

$$Q_{\mathbf{y}}(B) \equiv P^1_{\mathbf{x}}\Big(\{\mathbf{x}(\cdot) \in W \,|\, \Theta^{-1} \circ \boldsymbol{\Phi}(\mathbf{x}(\cdot)) \in B\} \Big), \ \forall B \in \mathcal{B}(W_T),$$

where the probability measure $P^1_{\mathbf{x}}$ on $(W, \mathcal{B}(W))$ is the probability law of the process $\{\mathbb{X}^1(t,\mathbf{x})\}_{t\in\mathbf{R}_+}$. Then, $Q_{\mathbf{y}}$ satisfies the following:

$$Q_{\mathbf{y}}(\boldsymbol{\eta}_0 = \mathbf{y}) = 1 \quad \text{and} \quad \left(f(\boldsymbol{\eta}_t) - \int_0^t (Lf)(\boldsymbol{\eta}_s)ds, \ \mathcal{B}_t(W_T), \ Q_{\mathbf{y}} \right)$$

is a martingale for each $f \in C_0^\infty(T^{\mathbf{Z}^d})$, where

$$(L\,f)(\mathbf{y}) = \sum_{k \in \mathbf{Z}^d} \left\{ \frac{\partial^2 f}{\partial y_{\mathbf{k}}^2}(\mathbf{y}) + b_{\mathbf{k}}(\Theta(\mathbf{y})) \frac{\partial}{\partial y_{\mathbf{k}}} f(\mathbf{y}) \right\}.$$

Furthermore, $Q_{\mathbf{y}}$ is the unique solution of the above martingale problem.
ii) Let $p(t, \mathbf{y}, \cdot)$ be the transition function associated with the diffusion process $(\boldsymbol{\eta}_t, Q_{\mathbf{y}} : \mathbf{y} \in T^{\mathbf{Z}^d})$. For $N \in \mathbf{N}$, $\mathbf{y} \in T^{\mathbf{Z}^d}$, let $p^{(N)}(t, \mathbf{y}, \cdot)$ be such that

$$p^{(N)}(t, \mathbf{y}, \Gamma) = p(t, \mathbf{y}, \tilde{\Gamma}) \qquad \text{for} \quad \Gamma \in \mathcal{B}(T^{[-N,+N]^d}),$$

where $\tilde{\Gamma} = \{\mathbf{y} \in T^{\mathbf{Z}^d} \,|\, \mathbf{y}_{(N)} \equiv \mathbf{y}_{[-N,N]^d} \in \Gamma\}$. Then $p^{(N)}(t, \mathbf{y}, d\mathbf{y}_{(N)})$ has a density $p^{(N)}(t, \mathbf{y}, \mathbf{y}_{(N)})$ with respect to Lebesgue measure on $T^{[-N,+N]^d}$ whose partial derivatives in the variable $\mathbf{y}_{(N)}$ of all order exist and are continuous functions of $(t, \mathbf{y}, \mathbf{y}_{(N)})$ in $(0, \infty) \times T^{\mathbf{Z}^d} \times T^{[-N,+N]^d}$.
iii) There exists at least one Gibbs probability measure μ on $(T^{\mathbf{Z}^d}, \mathcal{B}(T^{\mathbf{Z}^d}))$ such that

$$\langle \mathbb{E}^\Lambda \varphi, \mu \rangle = \langle \varphi, \mu \rangle, \quad \forall \Lambda \subset \mathbf{Z}^d \ s.t. \ |\Lambda| < \infty, \quad \forall \varphi \in C_0\left(T^{\mathbf{Z}^d}\right), \qquad (2.6)$$

where

$$[\mathbb{E}^{\Lambda}\varphi](\mathbf{y}) = \frac{1}{Z_{\Lambda}(\mathbf{y}_{\Lambda^c})}\int_{T^{\mathbf{Z}^d}} \varphi(\mathbf{y}'_{\Lambda}\cdot\mathbf{y}_{\Lambda^c})e^{-U^{\Lambda}(\Theta(\mathbf{y}'_{\Lambda}\cdot\mathbf{y}_{\Lambda^c}))}d\mathbf{y}',$$

with

$$U^{\Lambda}(\mathbf{x}) \equiv \sum_{\Lambda'\cap\Lambda\neq\emptyset} J_{\Lambda'}(\mathbf{x}), \quad \mathbf{x}\in\mathbf{R}^{\mathbf{Z}^d},$$

$$Z_{\Lambda}(\mathbf{y}_{\Lambda^c}) = \int_{T^{\mathbf{Z}^d}} e^{-U^{\Lambda}(\Theta(\mathbf{y}'_{\Lambda}\cdot\mathbf{y}_{\Lambda^c}))}d\mathbf{y}'.$$

Here we use the notation $\mathbf{y}'_{\Lambda}\cdot\mathbf{y}_{\Lambda^c}\equiv\mathbf{y}''\in T^{\mathbf{Z}^d}$, *so that* $\mathbf{y}''_{\Lambda}=\mathbf{y}'_{\Lambda}$ *and* $\mathbf{y}''_{\Lambda^c}=\mathbf{y}_{\Lambda^c}$.

□

Remark 2. (Q_y and Dirichlet forms)
Let μ be some Gibbs measure on $(T^{\mathbf{Z}^d},\mathcal{B}(T^{\mathbf{Z}^d}))$, and consider the Dirichlet space $(\mathcal{E}^{\mu},\mathcal{D}(\mathcal{E}^{\mu}))$ on $L^2(\mu)$ that is a quasi-regular Markovian extension of the form

$$\sum_{\mathbf{k}\in\mathbf{Z}^d}\int_{T^{\mathbf{Z}^d}}\frac{\partial\varphi}{\partial y_{\mathbf{k}}}\cdot\frac{\partial\psi}{\partial y_{\mathbf{k}}}\mu(d\mathbf{y}), \qquad \varphi,\psi\in C_0^{\infty}\left(T^{\mathbf{Z}^d}\right), \quad \text{on } L^2(\mu).$$

(closability holds according to [AKR].) Let $\mathbf{M}$ be the (strong) Markov process properly associated with the Dirichlet space $(\mathcal{E}^{\mu},\mathcal{D}(\mathcal{E}^{\mu}))$, which exists by [MR]. Denote the corresponding Markovian transition function and the probability law of the process $\mathbf{M}$ starting at $\mathbf{y}\in T^{\mathbf{Z}^d}$ by $p^{\mathbf{M}}(t,\mathbf{y},\cdot)$ ($t\in\mathbf{R}_+$, $\mathbf{y}\in T^{\mathbf{Z}^d}$) and $Q^{\mathbf{M}}_{\mathbf{y}}(\cdot)$ on $(W_T,\mathcal{B}(W_T))$ respectively.

By the uniqueness statement given in Prop. 1 we see that the Markov process $\left(\{\boldsymbol{\eta}_t\}_{t\geq 0},Q_{\mathbf{y}}:\mathbf{y}\in T^{\mathbf{Z}^d}\right)$ defined by Prop. 1 is equivalent to $\mathbf{M}$ above, hence properly associated to the Dirichlet space $(\mathcal{E}^{\mu},\mathcal{D}(\mathcal{E}^{\mu}))$ (for precise arguments cf. Remark 1.1 of [ABRY3], also cf. [F1], [MR]).

□

3 Theorems

In [ABRY3] we have considered the homogenization problem of the sequence of the diffusions $\{\{\mathbb{X}^{\epsilon}(t,\mathbf{x})\}_{t\in\mathbf{R}_+}\}_{\epsilon>0}$ in the case where the the following uniform ergodicity (3.1) holds for the quotient process $(\{\boldsymbol{\eta}_t\}_{t\geq 0},Q_{\mathbf{y}}:\mathbf{y}\in T^{\mathbf{Z}^d})$. Here we consider the same problem for $\{\{\mathbb{X}^{\epsilon}(t,\mathbf{x})\}_{t\in\mathbf{R}_+}\}_{\epsilon>0}$ in the case where the L^2-type ergodicity holds for $(\boldsymbol{\eta}_t,Q_{\mathbf{y}}:\mathbf{y}\in T^{\mathbf{Z}^d})$, and compare the results available under these two different assumptions of (3.1) and (3.2). Each comparison will be given as a Remark following each Theorem resp. Lemma.

In the sequel we denote the uniform ergodicity (3.1) as (LS) and the L^2-type ergodicity (3.2) as (WP) respectively. We have to remark that if the potential $\mathcal{J}$, that satisfies J-1), J-2) and J-3), satisfies in addition *Dobrushin-Shlosman mixing condition*, then (3.1) holds, more precisely in this case the *logarithmic Sobolev inequality (LS)* holds for the Dirichlet form $\mathcal{E}(u(\cdot), v(\cdot))$ defined in Remark 2, then the stronger inequality such that the term $(c+t)^{-\alpha}$ in (3.1) is replaced by $e^{-\alpha t}$ for some $\alpha > 0$ holds (cf. [S]).

Correspondingly, if $\mathcal{E}(u(\cdot), v(\cdot))$ satisfies the *weak Poincaré (WP) inequality*, then (3.2) holds. We remark that the *logarithmic Sobolev inequality* is strictly stronger than the *the weak Poincaré inequality* (cf. [RWang]).

Precisely, we define the ergodicities (LS) and (WP) as follows:

(LS) For some Gibbs state μ, there exists a $c = c(\mathcal{J}) > 0$ and an $\alpha = \alpha(\mathcal{J}) > 1$ which depend only on $\mathcal{J}$, such that for each $\Lambda \in \mathbf{Z}^d$ with $|\Lambda| < \infty$ there exists $K(\Lambda) \in (0, \infty)$ and for $\forall t > 0$, $\forall \varphi \in C^{\infty}_{\Lambda}(T^{\mathbf{Z}^d})$ the following holds

$$\left\| \int_{T^{\mathbf{Z}}} \varphi(\mathbf{y}_\Lambda) p(t, \cdot, d\mathbf{y}) - \langle \varphi, \mu \rangle \right\|_{L^\infty} \leq K(\Lambda)(c+t)^{-\alpha}(\|\nabla\varphi\|_{L^\infty} + \|\varphi\|_{L^\infty}), \quad (3.1)$$

(WP) There exist $c = c(\mathcal{J}) > 0$, $\alpha = \alpha(\mathcal{J}) > 1$ and $K > 0$, that depends only on $\mathcal{J}$, and the following holds

$$\|\mathcal{P}_t\varphi - < \varphi, \mu > \|_{L^2(\mu)} \leq K(c+t)^{-\alpha}\|\varphi\|_{L^2(\mu)}, \; \forall t > 0, \; \forall \varphi \in C\left(T^{\mathbf{Z}^d}\right). \quad (3.2)$$

We also remark that (3.1) or (3.2) gives the uniqueness of the Gibbs state, since by (3.1) or (3.2) we see that a Gibbs state μ that satisfies (3.1) or (3.2) is the only invariant measure for $p(t, \cdot, d\mathbf{y})$, but every Gibbs state is an invariant measure. From now on we denote the unique Gibbs measure by μ (cf. [ABRY3, AKR]).

Lemma 3. *Assume that J-1), J-2), J-3) and the L^2 ergodicity (WP) is satisfied. Then, for any $\mathbf{k} \in \mathbf{Z}^d$,*

$$\chi_{\mathbf{k}}(\mathbf{y}) \equiv E^{Q_{\mathbf{y}}} \left[\int_0^\infty \{b_{\mathbf{k}}(\Theta(\eta_s(\cdot))\} ds \right],$$

is well defined as a measurable function of $\mathbf{y} \in T^{\mathbf{Z}^d}$.

Let for $u, v \in \mathcal{D}(\mathcal{E})$

$$\mathcal{E}(u(\cdot), v(\cdot)) \equiv \sum_{\mathbf{j} \in \mathbf{Z}^d} \int_{T^{\mathbf{Z}^d}} \left(\frac{\partial}{\partial y_{\mathbf{j}}} u(\mathbf{y})\right) \left(\frac{\partial}{\partial y_{\mathbf{j}}} v(\mathbf{y})\right) \mu(d\mathbf{y}),$$

then for any $\mathbf{k} \in \mathbf{Z}^d$

$$\chi_{\mathbf{k}}(\cdot) \in \mathcal{D}(\mathcal{E}), \qquad \chi_{\mathbf{k}} \text{ is quasi-continuous} \quad (3.3)$$

$$\mathcal{E}(\chi_{\mathbf{k}}, \chi_{\mathbf{k}}) \leq \frac{5}{4},$$

$$\mathcal{E}(\chi_{\mathbf{k}}(\cdot), v(\cdot)) = \int_{T^{\mathbf{Z}^d}} b_{\mathbf{k}}(\Theta(\mathbf{y}))v(\mathbf{y})\mu(d\mathbf{y}) \quad \forall v \in C_0^\infty\left(T^{\mathbf{Z}^d}\right). \quad (3.4)$$

□

Remark 4. Under the assumption (LS), we have a stronger result than (3.3) (cf. Lemma 2.1 of [ABRY3]):

$$\chi_{\mathbf{k}}(\cdot) \in \mathcal{D}(\mathcal{E}) \quad \textit{and} \quad \chi_{\mathbf{k}}(\cdot) \in C(T^{\mathbf{Z}^d}).$$

□

By Lemma 3 we define

$$\chi'_{\mathbf{k},\mathbf{j}}(\mathbf{y}) = \sqrt{2}\frac{\partial}{\partial y_{\mathbf{j}}}\chi_{\mathbf{k}}(\mathbf{y}) \quad \text{if } \mathbf{j} \neq \mathbf{k}$$

and

$$\chi'_{\mathbf{k},\mathbf{j}}(\mathbf{y}) = \sqrt{2}\left(1 - \frac{\partial}{\partial y_{\mathbf{k}}}\chi_{\mathbf{k}}(\mathbf{y})\right) \quad \text{if } \mathbf{j} = \mathbf{k}.$$

Let χ' and $\mathbb{A}$ be the matrices whose components are functions such that respectively

$$\chi'(\mathbf{y}) \equiv \left(\chi'_{\mathbf{k},\mathbf{j}}(\mathbf{y})\right)_{\mathbf{k},\mathbf{j}\in\mathbf{Z}^d}, \quad (3.5)$$

$$\mathbb{A}(\mathbf{y}) \equiv (a_{\mathbf{k},\mathbf{j}}(\mathbf{y}))_{\mathbf{k},\mathbf{j}\in\mathbf{Z}^d}, \quad \text{with} \quad a_{\mathbf{k},\mathbf{l}}(\mathbf{y}) \equiv \sum_{\mathbf{j}\in\mathbf{Z}^d} \chi'_{\mathbf{k},\mathbf{j}}(\mathbf{y}) \cdot \chi'_{\mathbf{l},\mathbf{j}}(\mathbf{y}). \quad (3.6)$$

By (3.4), we can define a matrix $\bar{\mathbb{A}}$ whose components are constants as follows:

$$\bar{\mathbb{A}} \equiv (\bar{a}_{\mathbf{k},\mathbf{j}})_{\mathbf{k},\mathbf{j}\in\mathbf{Z}^d}, \quad \text{with} \quad \bar{a}_{\mathbf{k},\mathbf{l}} \equiv \sum_{\mathbf{j}\in\mathbf{Z}^d} \int_{T^{\mathbf{Z}^d}} \chi'_{\mathbf{k},\mathbf{j}}(\mathbf{y}) \cdot \chi'_{\mathbf{l},\mathbf{j}}(\mathbf{y})\mu(d\mathbf{y}). \quad (3.7)$$

For each $M \in \mathbf{N}$ let $\bar{\mathbb{A}}|_M$ be the submatrix of $\bar{\mathbb{A}}$ such that

$$\bar{\mathbb{A}}|_M = \left\{\bar{a}_{\mathbf{k},\mathbf{l}}\right\}_{|\mathbf{k}|,|\mathbf{l}|\leq M},$$

then by (3.4) and Fubini's Lemma, for $\mathbf{z} = \{z_{\mathbf{k}}\}_{|\mathbf{k}|\leq M}$

$$\begin{aligned} 0 \leq\ & \mathbf{z} \cdot \bar{\mathbb{A}}|_M \cdot {}^t\mathbf{z} \\ =\ & \int_{T^{\mathbf{Z}^d}} \sum_{\mathbf{j}\in\mathbf{Z}^d} \left(\sum_{|\mathbf{k}|\leq M} \left(\chi'_{\mathbf{k},\mathbf{j}}(\mathbf{y})\right) z_{\mathbf{k}} \right)^2 \mu(d\mathbf{y}) < +\infty. \end{aligned}$$

Hence, by the martingale representation theorem by means of the Brownian motion processes (cf. for e.g. Section II-6 of [IW]) the finite dimensional quadratic variation matrix $\bar{\mathbb{A}}|_M$ determines uniquely an $M' \equiv \sharp\{\mathbf{k}\big| |\mathbf{k}| \le M\}$-dimensional continuous Gaussian process on some adequate probability space. Since the sequence of the probability laws of such M'-dimensional processes, that is a sequence of Borel probability measures on $C(\mathbf{R}_+ \to \mathbf{R}^{M'})$, is consistent, by the Kolmogorov's extention theorem there exists a unique probability measure on $\big(C(\mathbf{R}_+ \to \mathbf{R}^{\mathbf{Z}^d}), \mathcal{B}(C(\mathbf{R}_+ \to \mathbf{R}^{\mathbf{Z}^d}))\big)$, such that any of its M'-dimensional marginals is identical to the probability law of the continuous Gaussian process characterized by $\bar{\mathbb{A}}|_M$.

By this construction, we denote by $\{\mathbb{Y}_t\}_{t\in\mathbf{R}_+}$ with $\mathbb{Y}_0 = 0$ as the unique continuous Gaussian process taking values in $\mathbf{R}^{\mathbf{Z}^d}$ (namely $\mathbb{Y}_\cdot$ is a $C(\mathbf{R}_+ \to \mathbf{R}^{\mathbf{Z}^d})$ valued random variable) with covariance matrix $t \cdot \bar{\mathbb{A}}$ $(t \in \mathbf{R}_+)$ defined on a complete probability space.

If $\mathbf{x} \equiv \{x_{\mathbf{k}}\}_{\mathbf{k}\in\mathbf{Z}^d} \in \mathcal{H}$, then by (3.4), (3.7) and above mentioned construction of $\{\mathbb{Y}_t\}_{t\in\mathbf{R}_+}$ with $\mathbb{Y}_0 = 0$ by means of $\bar{\mathbb{A}}|_M$, by using the martingale inequality we see that the trajectories of $\{\mathbb{Y}_t + \mathbf{x}\}_{t\in\mathbf{R}_+}$ stay in W with probability 1, and $\mathcal{B}(C(\mathbf{R}_+ \to \mathbf{R}^{\mathbf{Z}^d})) \cap W = \mathcal{B}(W)$ is identical with $(W, \mathcal{B}(W))$. We can then set the following definition:

Definition 5. *Let $\{\mathbb{Y}_t\}_{t\in\mathbf{R}_+}$ with $\mathbb{Y}_0 = 0$ be the unique continuous Gaussian process defined above, with a law which is a Borel probability measure on $(W, \mathcal{B}(W))$. For each $\mathbf{x} \in \mathcal{H}$, let $P_{\mathbf{x}}$ be the probability measure on $(W, \mathcal{B}(W))$ that is the probability law of the process $\{\mathbf{x} + \mathbb{Y}_t\}_{t\in\mathbf{R}_+}$.* □

Heuristically, $\{\mathbb{Y}_t\}_{t\in\mathbf{R}_+}$ can be expressed by

$$\mathbb{Y}_t = \int_0^t \bar{\mathbb{A}}^{\frac{1}{2}} d\mathbf{B}_t, \quad t \in \mathbf{R}_+,$$

where $\{\mathbf{B}_t\}_{t\in\mathbf{R}_+} \equiv \big\{\{B_{\mathbf{k},t}\}_{t\in\mathbf{R}_+}\big\}_{\mathbf{k}\in\mathbf{Z}^d}$ and $\{B_{\mathbf{k},t}\}_{t\in\mathbf{R}_+}$ $(\mathbf{k} \in \mathbf{Z}^d)$ are some independent sequences of one-dimensional standard Brownian motion processes.

Lemma 6. *Assume that J-1), J-2), J-3) and that the L^2 ergodicity (WP) is satisfied. Let $\chi_{\mathbf{k}}(\cdot) \in \mathcal{D}(\mathcal{E})$ $(\mathbf{k} \in \mathbf{Z}^d)$ be the functions defined by Lemma 3, denote $\chi_{\mathbf{k}}(\boldsymbol{\Theta}^{-1} \circ \boldsymbol{\Phi}(\mathbf{x}))$ simply by $\chi_{\mathbf{k}}(\mathbf{x})$. For each $\epsilon > 0$ let*

$$M_t^{\epsilon,\mathbf{k}}(\cdot) = (\xi_t^{\mathbf{k}}(\cdot) - \xi_0^{k}(\cdot)) - \left(\epsilon\chi_{\mathbf{k}}\left(\frac{\boldsymbol{\xi}_t(\cdot)}{\epsilon}\right) - \epsilon\chi_{\mathbf{k}}\left(\frac{\boldsymbol{\xi}_0(\cdot)}{\epsilon}\right)\right). \tag{3.8}$$

Set $\tilde{\mathbf{y}} = \boldsymbol{\Theta}(\mathbf{y})$, for $\boldsymbol{\Theta}$ the mapping from $T^{\mathbf{Z}^d}$ to $[0, 2\pi)^{\mathbf{Z}^d}$ defined in the previous section. Then, for each $\epsilon > 0$ and $\mathcal{E}$-q.e. the processes $\{M_t^{\epsilon,\mathbf{k}}\}_{t\in\mathbf{R}_+}$, $\mathbf{k} \in \mathbf{Z}^d$, on $(W, \mathcal{B}(W), P^{\epsilon}_{\epsilon\tilde{\mathbf{y}}})$ are $L^2(P^{\epsilon}_{\epsilon\tilde{\mathbf{y}}})$, continuous $\mathcal{B}_t(W)$-martingales whose quadratic variations are given by

$$< M^{\epsilon,\mathbf{k}}(\cdot), M^{\epsilon,\mathbf{l}}(\cdot) >_t = \int_0^t a_{\mathbf{k},\mathbf{l}}\left(\frac{\boldsymbol{\xi}_s(\cdot)}{\epsilon}\right) ds, \quad \mathbf{k}, \mathbf{l} \in \mathbf{Z}^d, \tag{3.9}$$

where

$$a_{\mathbf{k},\mathbf{l}}(\mathbf{y}) \equiv \sum_{\mathbf{j}\in\mathbf{Z}^d} {\chi'}_{\mathbf{k},\mathbf{j}}(\mathbf{y}){\chi'}_{\mathbf{l},\mathbf{j}}(\mathbf{y}),$$

with

$${\chi'}_{\mathbf{k},\mathbf{j}}(\mathbf{y}) = \begin{cases} \sqrt{2}\dfrac{\partial}{\partial y_{\mathbf{j}}}\chi_{\mathbf{k}}(\mathbf{y}) & \mathbf{j} \neq \mathbf{k} \\ \sqrt{2}\left(1 - \dfrac{\partial}{\partial y_{\mathbf{k}}}\chi_{\mathbf{k}}(\mathbf{y})\right) & \mathbf{j} = \mathbf{k}. \end{cases}$$

□

Remark 7. If we assume (LS), then for each $\epsilon > 0$ and each $\mathbf{x} \in \mathcal{H}$ the process $\{M_t^{\epsilon,\mathbf{k}}\}_{t\in\mathbf{R}_+}$, $\mathbf{k} \in \mathbf{Z}^d$, on $(W, \mathcal{B}(W), P_{\mathbf{x}}^\epsilon)$ is an $L^2(P_{\mathbf{x}}^\epsilon)$ continuous $\mathcal{B}_t(W)$-martingale, with quadratic variations given by (24). □

Let ν be a probability measure on $(T^{\mathbf{Z}^d}, \mathcal{B}(T^{\mathbf{Z}^d}))$ such that

$$\left\| \frac{d\nu}{d\mu} \right\|_{L^\infty(T^{\mathbf{Z}^d})} < \infty. \tag{3.10}$$

For each $\epsilon \in [0, 1)$, define a probability measure $P_{\nu_\epsilon}^\epsilon$ on $(W, \mathcal{B}(W))$ such that

$$P_{\nu_\epsilon}^\epsilon(B) \equiv \int_{T^{\mathbf{Z}^d}} P_{\epsilon\tilde{\mathbf{y}}}^\epsilon(B)\nu(d\mathbf{y}), \quad \forall B \in \mathcal{B}(W), \tag{3.11}$$

where as above (and in the sequel) $\tilde{\mathbf{y}} = \boldsymbol{\Theta}(\mathbf{y})$.

Remark 8. We remark that by Lemma 6, the processes $\{M_t^{\epsilon,\mathbf{k}}\}_{t\in\mathbf{R}_+}$, $\mathbf{k} \in \mathbf{Z}^d$, on $(W, \mathcal{B}(W), P_{\nu_\epsilon}^\epsilon)$ are $L^2(P_{\nu_\epsilon}^\epsilon)$, continuous $\mathcal{B}_t(W)$-martingales with quadratic variations given by (3.9). □

Theorem 9. *Assume that J-1), J-2), J-3) and (WP) are satisfied. Then, for each $\epsilon > 0$ and each probability measure ν on $(T^{\mathbf{Z}^d}, \mathcal{B}(T^{\mathbf{Z}^d}))$ satisfying (3.10), it is possible to construct a probability space $(\bar{W}, \mathcal{B}(\bar{W}), \bar{P}_{\nu_\epsilon}^\epsilon; \mathcal{B}_t(\bar{W}))$, which is a standard extension of $(W, \mathcal{B}(W), P_{\nu_\epsilon}^\epsilon; \mathcal{B}_t(W))$, and a $\mathcal{B}_t(W)$-adapted $\mathbf{R}^{\mathbf{Z}^d}$-valued continuous process $\{\boldsymbol{\zeta}_t^\epsilon\}_{t\in\mathbf{R}_+}$ (defined precisely in the next section) that satisfies the following: $\boldsymbol{\zeta}_\cdot^\epsilon$ is a W valued random variable whose probability law $\bar{P}_{\nu_\epsilon}^\epsilon \circ \boldsymbol{\zeta}_\cdot^\epsilon$ forms a relatively compact set $\{\bar{P}_{\nu_\epsilon}^\epsilon \circ \boldsymbol{\zeta}_\cdot^\epsilon\}_{\epsilon>0}$ in the space of probability measures on $(W, \mathcal{B}(W))$ equipped with the weak topology, and for any $\varphi \in C_b(W \to \mathbf{R})$, the following holds:*

$$\lim_{\epsilon\downarrow 0} E^{\bar{P}_{\nu_\epsilon}^\epsilon}\left[\varphi(\boldsymbol{\zeta}_\cdot^\epsilon(\cdot))\right] = E^{P_0}\left[\varphi\left(\hat{\boldsymbol{\xi}}_\cdot(\cdot)\right)\right], \tag{3.12}$$

$$\lim_{\epsilon\downarrow 0} E^{\bar{P}_{\nu_\epsilon}^\epsilon}\left[\rho\left(\hat{\boldsymbol{\xi}}_\cdot(\cdot), \boldsymbol{\zeta}_\cdot^\epsilon(\cdot)\right)\right] = 0, \tag{3.13}$$

where

$$\hat{\boldsymbol{\xi}}_t(\cdot) \equiv \boldsymbol{\xi}_t(\cdot) - \epsilon\boldsymbol{\chi}\left(\frac{\boldsymbol{\xi}_t(\cdot)}{\epsilon}\right) + \epsilon\boldsymbol{\chi}\left(\frac{\boldsymbol{\xi}_0(\cdot)}{\epsilon}\right).$$

□

Remark 10. Under the assumption (LS), we can take the initial states as Dirac point measures (cf. Theorem 2.1 of [ABRY3]):

$$\lim_{\epsilon\downarrow 0} E^{\bar{P}^\epsilon_{\mathbf{x}}}[\varphi(\boldsymbol{\zeta}^\epsilon_\cdot(\cdot))] = E^{P_{\mathbf{x}}}[\varphi(\boldsymbol{\xi}_\cdot(\cdot))]. \qquad \forall \mathbf{x} \in \mathcal{H}, \tag{3.14}$$

One then also have:

$$\lim_{\epsilon\downarrow 0} E^{\bar{P}^\epsilon_{\epsilon\mathbf{x}}}[\varphi(\boldsymbol{\zeta}^\epsilon_\cdot(\cdot))] = E^{P_0}[\varphi(\boldsymbol{\xi}_\cdot(\cdot))], \qquad \forall \mathbf{x} \in [0, 2\pi)^{\mathbf{Z}^d}, \tag{3.15}$$

where the approximation sequence $\left\{\{\boldsymbol{\zeta}^\epsilon_t\}_{t\in\mathbf{R}_+}\right\}_{\epsilon>0}$ satisfies

$$\lim_{\epsilon\downarrow 0} \int_{T^{\mathbf{Z}^d}} E^{\bar{P}^\epsilon_{\epsilon\tilde{\mathbf{y}}}}[\rho(\boldsymbol{\xi}_\cdot(\cdot), \boldsymbol{\zeta}^\epsilon_\cdot(\cdot))]\mu(d\mathbf{y}) = 0. \tag{3.16}$$

□

Remark 11. In order to show that $\chi_{\mathbf{k}} \in \mathcal{D}(\mathcal{E})$ satisfies $\chi_{\mathbf{k}} \in C(T^{\mathbf{Z}^d} \to \mathbf{R})$ we used crucially (LS) in [ABRY3]. Here, in Lemma 3 we assume (WP) and we can show $\chi_{\mathbf{k}} \in \mathcal{D}(\mathcal{E})$ only, and we can not see in general that $\chi_{\mathbf{k}}$ is bounded. By this we are not able to assert that the term

$$-\epsilon\boldsymbol{\chi}\left(\frac{\boldsymbol{\xi}_t(\cdot)}{\epsilon}\right) + \epsilon\boldsymbol{\chi}\left(\frac{\boldsymbol{\xi}_0(\cdot)}{\epsilon}\right)$$

vanishes as $\epsilon \downarrow 0$, and we have to modify $\boldsymbol{\xi}_t(\cdot)$ by $\hat{\boldsymbol{\xi}}_t(\cdot)$ in Theorem 9 above and 12 below (cf. Remarks 4 and 10).

□

Theorem 12. *Let P_0 be the probability law of the process $\{\mathbb{Y}_t\}_{t\in\mathbf{R}_+}$. Assume that the assumptions of Theorem 9 are satisfied, then the following hold:*

$$\lim_{\epsilon\downarrow 0} E^{P^\epsilon_{\tilde{\nu}_\epsilon}}[\varphi(\hat{\boldsymbol{\xi}}_\cdot(\cdot))] = E^{P_0}[\varphi(\boldsymbol{\xi}_\cdot(\cdot))], \quad \forall \varphi \in C_b(W \to \mathbf{R}). \tag{3.17}$$

□

Remark 13. Under the assumptions J-1), J-2), J-3) and (LS), in [ABRY3] we proved the following: For $\tilde{\mathbf{y}} = \Theta(\mathbf{y})$ with $\Theta : T^{\mathbf{Z}^d} \to [0, 2\pi)^{\mathbf{Z}^d}$,

$$\lim_{\epsilon\downarrow 0} \int_{T^{\mathbf{Z}^d}} \left| E^{P^\epsilon_{\epsilon\tilde{\mathbf{y}}}}[\varphi(\boldsymbol{\xi}_\cdot(\cdot))] - E^{P_0}[\varphi(\boldsymbol{\xi}_\cdot(\cdot))] \right| \mu(d\mathbf{y}) = 0, \quad \forall \varphi \in C_b(W \to \mathbf{R}). \tag{3.18}$$

Also there exists an $\mathcal{N} \in \mathcal{B}(T^{\mathbf{Z}^d})$ such that $\mu(\mathcal{N}) = 0$, and a subsequence

$$\{\epsilon_n\}_{n\in\mathbf{N}} \subset \{\epsilon \,|\, \epsilon \in (0,1]\},$$

and the following holds (cf. [PapV] for the finite dimensional case):

$$\lim_{\epsilon_n \downarrow 0} E^{P^{\epsilon_n}_{\epsilon_n \mathbf{x}}}[\varphi(\boldsymbol{\xi}.(\cdot))] = E^{P_0}[\varphi(\boldsymbol{\xi}.(\cdot))], \quad \forall \varphi \in C_b(W \to \mathbf{R}) \tag{3.19}$$

$$\forall \mathbf{x} \equiv \{x_{\mathbf{k}}\}_{\mathbf{k}\in\mathbf{Z}^d} \in \mathbf{R}^{\mathbf{Z}^d} \quad \text{such that} \quad \Theta^{-1}\boldsymbol{\Phi}(\mathbf{x}) \in T^{\mathbf{Z}^d} \setminus \mathcal{N}, \quad \sup_{\mathbf{k}\in\mathbf{Z}^d} |x_{\mathbf{k}}| < \infty.$$

□

Remark 14. For the present problem we use crucially the ergodicity of the corresponding quotient process $(\{\boldsymbol{\eta}_t\}_{t\geq 0}, Q_{\mathbf{y}} : \mathbf{y} \in T^{\mathbf{Z}^d})$ whose existence depends essentially on the periodicity of the coefficients of original process. If we consider the homogenization problems based on the diffusions processes taking values in $\mathbf{R}^{\mathbf{Z}^d}$ with the index set $\mathbf{Z}^d$ which are defined through Dirichlet forms with convex potential terms (cf. [AKR]), then there are no corresponding quotient processes and the present formulation is impossible.

□

4 Construction of $\{\{\zeta_t^\epsilon\}_{t\in\mathbf{R}_+}\}_{\epsilon>0}$ and an Outline of the Proofs

In order to get the results on the homogenization problem for the infinite dimensional diffusions $\{\mathbb{X}^\epsilon(t,\mathbf{x})\}_{t\in\mathbf{R}_+}$ $(\epsilon > 0)$, we firstly pass through the discussion of a sequence of approximating processes $\{\boldsymbol{\zeta}_t^\epsilon\}_{t\in\mathbf{R}_+}$ $(\epsilon > 0)$ of the original diffusions introduced in Theorem 9. $\{\{\boldsymbol{\zeta}_t^\epsilon\}_{t\in\mathbf{R}_+}\}_{\epsilon>0}$ is composed in order that the sequence of probability laws of $\{\{\boldsymbol{\zeta}_t^\epsilon\}_{t\in\mathbf{R}_+}\}_{\epsilon>0}$ forms a relatively compact set in the space of Borel probability measures on $(W, \mathcal{B}(W))$ equipped with the relative topology. For each $\epsilon > 0$ the dimension of $\{\boldsymbol{\zeta}_t^\epsilon\}_{t\in\mathbf{R}_+}$ is essentially finite, that is controlled by the parameter $\epsilon > 0$ with a tricky way composed by using the **uniform ergodic theorem** (LS) or **L^2 ergodic theorem** (WP). In [ABRY3] under the assumption (LS) this subsidiary sequence of processes $\{\{\boldsymbol{\zeta}_t^\epsilon\}_{t\in\mathbf{R}_+}\}_{\epsilon>0}$ has been constructed in order that it satisfies the pointwise homogenization property given by (3.14).

Here, we show how the approximating processes $\{\{\boldsymbol{\zeta}_t^\epsilon\}_{t\in\mathbf{R}_+}\}_{\epsilon>0}$ that satisfy the homogenization property given by (3.12) are constructed by using the assumption (WP). Once $\{\{\boldsymbol{\zeta}_t^\epsilon\}_{t\in\mathbf{R}_+}\}_{\epsilon>0}$ is constructed, the proofs of Theorems 9 and 12 of the present paper are very similar to the ones of Theorems 2.1 and 2.2 in [ABRY3], we do not repeat them here. Also, since Lemmas 3 and 6 in this paper are included in Lemmas 2.1 and 3.1 of [ABRY3], therefore we also omit these proofs here.

For the (WP) case we construct $\left\{\{\zeta_t^\epsilon\}_{t\in\mathbf{R}_+}\right\}_{\epsilon>0}$ as follows. Let $\mathbb{A}(\mathbf{y})$ be the matrix valued function defined by (3.6), and for each $N \in \mathbf{N}$ let

$$N' \equiv \sharp\{\mathbf{k}|\ |\mathbf{k}| \le N\},$$

and define an $N' \times N'$ matrix that is a submatrix of $\mathbb{A}(\mathbf{y})$ such that

$$\mathbb{A}(\mathbf{y})|_N \equiv \left(a_{\mathbf{k},\mathbf{j}}(\mathbf{y})\right)_{|\mathbf{k}|,|\mathbf{j}|\le N'}, \qquad \mathbf{y} \in T^{\mathbf{Z}^d}.$$

Then by (3.4) and Fubini's Lemma, for any real vector $\mathbf{z} = \{z_\mathbf{k}\}_{|\mathbf{k}|\le N}$

$$\begin{aligned} 0 \le\ & \mathbf{z}\cdot\mathbb{A}(\mathbf{y})|_N \cdot {}^t\mathbf{z} \\ &= \sum_{\mathbf{j}\in\mathbf{Z}^d}\left(\sum_{|\mathbf{k}|\le N}(\chi'_{\mathbf{k},\mathbf{j}}(\mathbf{y}))z_\mathbf{k}\right)^2 < +\infty, \qquad \mu - a.s.\ \ \mathbf{y}\in T^{\mathbf{Z}^d}. \end{aligned}$$

By this for each $N \in \mathbf{N}$, there exists a matrix $\left(\sigma^N_{\mathbf{k},\mathbf{l}}(\mathbf{y})\right)_{|\mathbf{k}|,|\mathbf{j}|\le N}$ such that

$$a_{\mathbf{k},\mathbf{l}}(\mathbf{y}) = \sum_{|\mathbf{j}|\le N}\sigma^N_{\mathbf{k},\mathbf{j}}(\mathbf{y})\cdot\sigma^N_{\mathbf{l},\mathbf{j}}(\mathbf{y}), \quad |\mathbf{k}|,\ |\mathbf{l}| \le N, \quad \mu - a.s.\ \mathbf{y}\in T^{\mathbf{Z}^d}. \tag{4.1}$$

By (3.4), (3.5) and (3.6) since for any $\mathbf{k}\in\mathbf{Z}^d$

$$\int_{T^{\mathbf{Z}^d}} a_{\mathbf{k},\mathbf{k}}(\mathbf{y})\mu(d\mathbf{y}) \le \frac{5}{2},$$

we see that

$$\sum_{|\mathbf{l}|\le N}\left\|\sigma^N_{\mathbf{k},\mathbf{l}}\right\|^2_{L^2(\mu)} \le \frac{5}{2}, \qquad \text{for any } \mathbf{k} \text{ such that} \quad |\mathbf{k}| \le N. \tag{4.2}$$

By this, for each $N \in \mathbf{N}$, there exists a sequence of $N' \times N'$ matrices

$$\{\sigma^{N,n}_{\mathbf{k},\mathbf{l}}(\mathbf{y})\}_{|\mathbf{k}|,|\mathbf{l}|\le N} \qquad n = 1, 2, \cdots,$$

such that

$$\sigma^{N,n}_{\mathbf{k},\mathbf{l}} \in C^\infty_{\Lambda_{N,n}}(T^{\mathbf{Z}^d} \to \mathbf{R}), \quad \text{for some bounded } \Lambda_{N,n} \subset \mathbf{Z}^d, \quad n \in \mathbf{N},$$

$$\lim_{n\to\infty}\left\|\sigma^{N,n}_{\mathbf{k},\mathbf{l}}(\cdot) - \sigma^N_{\mathbf{k},\mathbf{l}}(\cdot)\right\|_{L^2(\mu)} = 0, \qquad |\mathbf{k}|, |\mathbf{l}| \le N.$$

Next, define a natural number valued function $n(\cdot)$ as follows:

$$n(N) \equiv \min\left\{ n\in\mathbf{N} \,\Big|\, \sum_{|\mathbf{l}|\le N}\left\|\sigma^{N,n}_{\mathbf{k},\mathbf{l}} - \sigma^N_{\mathbf{k},\mathbf{l}}\right\|_{L^2(\mu)} < \frac{1}{N},\ \forall |\mathbf{k}| \le N \right\}, \tag{4.3}$$

and then define

$$\tilde{\sigma}^N_{\mathbf{k},\mathbf{l}}(\mathbf{y}) \equiv \sigma^{N,n(N)}_{\mathbf{k},\mathbf{l}}(\mathbf{y}), \qquad \mathbf{y} \in T^{\mathbf{Z}^d}, \qquad |\mathbf{k}|, |\mathbf{l}| \le N. \tag{4.4}$$

By construction we see that

$$\tilde{\sigma}^N_{\mathbf{k},\mathbf{l}} \in C^\infty_{\Lambda_N}(T^{\mathbf{Z}^d} \to \mathbf{R}), \qquad \text{where} \quad \Lambda_N \equiv \bigcup_{n \le n(N)} \Lambda_{N,n}.$$

Let

$$\tilde{a}^N_{\mathbf{k},\mathbf{k}}(\mathbf{y}) \equiv \sum_{|\mathbf{j}|\le N} \tilde{\sigma}^N_{\mathbf{k},\mathbf{j}}(\mathbf{y}) \cdot \tilde{\sigma}^N_{\mathbf{k},\mathbf{j}}(\mathbf{y}) \ge 0, \qquad \mathbf{y} \in T^{\mathbf{Z}^d}. \tag{4.5}$$

Finally, by using the constants $c(\mathcal{J}) > 0$, $\alpha \equiv \alpha(\mathcal{J}) > 1$ and the constant $K > 0$ which appeared in (WP) we define

$$K_N = K \left\{ \max\left(\frac{2}{c^\alpha}, 1 \right) \right\} \cdot \left\{ \max_{|\mathbf{k}|\le N} \|\tilde{a}^N_{\mathbf{k},\mathbf{k}}\|_{L^2(\mu)} \right\}, \tag{4.6}$$

and then, for each $\epsilon > 0$, we define

$$N(\epsilon) \equiv \max\left\{ N \in \mathbf{N} \,\middle|\, \sqrt{\epsilon} K_N M_{N,\mathbf{k},\mathbf{l}} \le 1, \quad \forall |\mathbf{k}|, \forall |\mathbf{l}| \le N \right\}, \tag{4.7}$$

where (cf. (4.5))

$$M_{N,\mathbf{k},\mathbf{l}} \equiv \sup_{\mathbf{y}\in T^{\Lambda_N}} \left(\tilde{a}^N_{\mathbf{k},\mathbf{k}}(\mathbf{y}) \cdot \tilde{a}^N_{\mathbf{l},\mathbf{l}}(\mathbf{y}) \right)^{\frac{1}{2}}. \tag{4.8}$$

Now, we define the approximation sequence of the original process as follows. By Lemma 6, for each $\epsilon > 0$ (hence for $N(\epsilon)$ defined by (4.7)), since the quadratic variation of the $L^2(P^\epsilon_{\nu_\epsilon})$ continuous $\mathcal{B}_t(W)$-martingale $\{M^{\epsilon,\mathbf{k}}_t\}_{t\in\mathbf{R}_+}$, $\mathbf{k} \in \mathbf{Z}^d$, on $(W, \mathcal{B}(W), P^\epsilon_{\nu_\epsilon})$ is given by

$$< M^{\epsilon,\mathbf{k}}(\cdot), M^{\epsilon,\mathbf{l}}(\cdot) >_t = \int_0^t a_{\mathbf{k},\mathbf{l}}\left(\frac{\boldsymbol{\xi}_s(\cdot)}{\epsilon} \right) ds, \quad \mathbf{k}, \mathbf{l} \in \mathbf{Z}^d,$$

from the expression of $a_{\mathbf{k},\mathbf{l}}(\mathbf{y})$ given by (4.1), by applying the martingale representation theorem by means of the Brownian motion processes for the finite dimensional continuous L^2 martingales (cf., for e.g., Section II-7 of [IW]), we see that on a probability space $(\bar{W}, \mathcal{B}(\bar{W}), \bar{P}^\epsilon_{\nu_\epsilon}; \mathcal{B}_t(\bar{W}))$ there exists an $N'(\epsilon) \equiv \sharp\{\mathbf{k} \,|\, |\mathbf{k}| \le N(\epsilon)\}$ dimensional standard Brownian motion process

$$\{B^\epsilon_{\mathbf{k}}(t)\}_{t\in\mathbf{R}_+}, \qquad |\mathbf{k}| \le N(\epsilon),$$

and the following holds:

$$\xi^{\mathbf{k}}_t(\cdot) = \xi^{\mathbf{k}}_0(\cdot) - \epsilon\chi_{\mathbf{k}}\left(\frac{\boldsymbol{\xi}_0(\cdot)}{\epsilon} \right) + \epsilon\chi_{\mathbf{k}}\left(\frac{\boldsymbol{\xi}_t(\cdot)}{\epsilon} \right)$$
$$+ \sum_{|\mathbf{l}|\le N(\epsilon)} \int_0^t \sigma^{N(\epsilon)}_{\mathbf{k},\mathbf{l}}\left(\frac{\boldsymbol{\xi}_s(\cdot)}{\epsilon} \right) dB^\epsilon_l(s), \ \bar{P}^\epsilon_{\nu_\epsilon} - a.s., \ |\mathbf{k}| \le N(\epsilon). \tag{4.9}$$

Then, by using $\tilde{\sigma}^{N(\epsilon)}_{\mathbf{k},\mathbf{l}}$ defined by (4.4) and (4.7), we define the approximating process $\{\boldsymbol{\zeta}^{\epsilon}_t\}_{t\in\mathbf{R}_+} = \{\{\zeta^{\epsilon,\mathbf{k}}_t\}_{t\in\mathbf{R}_+}\}_{\mathbf{k}\in\mathbf{Z}^d}$ on $(\bar{W},\mathcal{B}(\bar{W}),\bar{P}^{\epsilon}_{\nu_\epsilon};\mathcal{B}_t(\bar{W}))$ as follows:

$$\zeta^{\epsilon,\mathbf{k}}_t(\cdot) = \xi^{\mathbf{k}}_0(\cdot) + \sum_{|\mathbf{l}|\le N(\epsilon)} \int_0^t \tilde{\sigma}^{N(\epsilon)}_{\mathbf{k},\mathbf{l}}\left(\frac{\boldsymbol{\xi}_s(\cdot)}{\epsilon}\right) dB^{\epsilon}_l(s),\ |\mathbf{k}| \le N(\epsilon); \tag{4.10}$$

$$\zeta^{\epsilon,\mathbf{k}}_t(\cdot) = \xi^{\mathbf{k}}_0(\cdot),\ |\mathbf{k}| > N(\epsilon),\ \forall t \in \mathbf{R}_+. \tag{4.11}$$

Let us explain the key point of the proof of the tightness of $\{\{\boldsymbol{\zeta}^{\epsilon}_t\}_{t\in\mathbf{R}_+}\}_{\epsilon>0}$. Let

$$\overline{\tilde{a}^{N(\epsilon)}_{\mathbf{k},\mathbf{k}}} = \left\|\tilde{a}^{N(\epsilon)}_{\mathbf{k},\mathbf{k}}\right\|_{L^1(\mu)}.$$

By (4.5) using

$$\begin{aligned} &p_{u_1}\left(\tilde{a}^{N(\epsilon)}_{\mathbf{k},\mathbf{k}}(\cdot)\cdot p_{u_2-u_1}\left(\tilde{a}^{N(\epsilon)}_{\mathbf{k},\mathbf{k}}(\cdot)\right)\right)(\mathbf{y}) \\ &\le \left(\overline{\tilde{a}^{N(\epsilon)}_{\mathbf{k},\mathbf{k}}}\right)^2 + \overline{\tilde{a}^{N(\epsilon)}_{\mathbf{k},\mathbf{k}}}\left|p_{u_1}\left(\tilde{a}^{N(\epsilon)}_{\mathbf{k},\mathbf{k}}(\cdot)\right)(\mathbf{y}) - \overline{\tilde{a}^{N(\epsilon)}_{\mathbf{k},\mathbf{k}}}\right| \\ &\quad + \left\|\tilde{a}^{N(\epsilon)}_{\mathbf{k},\mathbf{k}}\right\|_{L^\infty} p_{u_1}\left(\left|p_{u_2-u_1}\left(\tilde{a}^{N(\epsilon)}_{\mathbf{k},\mathbf{k}}(\cdot)\right)(\mathbf{y}) - \overline{\tilde{a}^{N(\epsilon)}_{\mathbf{k},\mathbf{k}}}\right|\right), \end{aligned}$$

by (4.10), (4.11), Fubini's Lemma and (WP) we see that

$$\begin{aligned} &E^{\bar{P}^{\epsilon}_{\nu_\epsilon}}\left[\left|\zeta^{\epsilon,\mathbf{k}}_t(\cdot) - \zeta^{\epsilon,\mathbf{k}}_0(\cdot)\right|^4\right] \\ &= E^{\bar{P}^{\epsilon}_{\nu_\epsilon}}\left[\left|\sum_{|\mathbf{l}|<N(\epsilon)}\int_0^t \tilde{\sigma}^{N(\epsilon)}_{\mathbf{k},\mathbf{l}}\left(\frac{\boldsymbol{\xi}_s(\cdot)}{\epsilon}\right) dB^{\epsilon}_{\mathbf{l}}(s)\right|^4\right] \\ &\le \epsilon^4 \int_{T^{\mathbf{Z}^d}}\left\{\int_0^{\frac{t}{\epsilon^2}}\int_{u_1}^{\frac{t}{\epsilon^2}} p_{u_1}\left(\tilde{a}^{N(\epsilon)}_{\mathbf{k},\mathbf{k}}(\cdot)\cdot p_{u_2-u_1}\left(\tilde{a}^{N(\epsilon)}_{\mathbf{k},\mathbf{k}}(\cdot)\right)\right)(\mathbf{y})du_1du_2\right\}\nu(d\mathbf{y}) \\ &\le \frac{1}{2}\left(\overline{\tilde{a}^{N(\epsilon)}_{\mathbf{k},\mathbf{k}}}\right)^2 t^2 + \epsilon^4\overline{\tilde{a}^{N(\epsilon)}_{\mathbf{k},\mathbf{k}}}\left\|\frac{d\nu}{d\mu}\right\|_{L^\infty}\left\|\tilde{a}^{N(\epsilon)}_{\mathbf{k},\mathbf{k}}\right\|_{L^2(\mu)}\int_0^{\frac{t}{\epsilon^2}} K\left(\frac{t}{\epsilon^2} - u_1\right)(c+u_1)^{-\alpha}du_1 \\ &\quad + \epsilon^4\left\|\frac{d\nu}{d\mu}\right\|_{L^\infty}\left\|\tilde{a}^{N(\epsilon)}_{\mathbf{k},\mathbf{k}}\right\|_{L^2(\mu)}\left\|\tilde{a}^{N(\epsilon)}_{\mathbf{k},\mathbf{k}}\right\|_{L^\infty}\int_0^{\frac{t}{\epsilon^2}}\int_{u_1}^{\frac{t}{\epsilon^2}} K(c+(u_2-u_1))^{-\alpha}du_1du_2. \end{aligned} \tag{4.12}$$

But, for $\alpha > 1$, using

$$\epsilon^2\int_0^{\frac{t}{\epsilon^2}}\frac{1}{(c+s)^\alpha}ds \le \epsilon\max\left(\frac{2}{c^\alpha},1\right)t^{\frac{1}{2}},\quad \forall\epsilon\in(0,1],$$

togetherwith (4.6), (4.7), (4.8) and using the bounds (4.2) and (4.3) we see that the RHS of (4.12) is dominated by

$$c't^2 + c''t^{\frac{3}{2}},$$

for some constants c', $c'' > 0$. Thus we have

$$E^{\bar{P}^{\epsilon}_{\nu_\epsilon}}\left[\left|\zeta^{\epsilon,\mathbf{k}}_t(\cdot) - \zeta^{\epsilon,\mathbf{k}}_0(\cdot)\right|^4\right] \leq c't^2 + c''t^{\frac{3}{2}}, \qquad \forall t > 0, \quad \forall \mathbf{k} \in \mathbf{Z}^d, \quad \forall \epsilon > 0. \tag{4.13}$$

From (4.13) through a similar discussion as for the proof of Theorem 2.1 of [ABRY3] we can complete the proof of Theorem 9 in the present paper.

□

Remark 15. In [ABRY3] by using the ergodicity (LS), we constructed the corresponding approximation sequence $\left\{\{\zeta^\epsilon_t\}_{t\in\mathbf{R}_+}\right\}_{\epsilon>0}$ which satisfies

$$E^{\bar{P}^{\epsilon}_{\mathbf{x}}}\left[\left|\zeta^{\epsilon,\mathbf{k}}_t(\cdot) - \zeta^{\epsilon,\mathbf{k}}_0(\cdot)\right|^4\right] \leq c't^2 + c''t^{\frac{3}{2}}, \forall t > 0, \quad \forall \mathbf{k} \in \mathbf{Z}^d, \quad \forall \epsilon > 0, \quad \forall \mathbf{x} \in \mathcal{H}. \tag{4.14}$$

□

Acknowledgements

We have to express our deep acknowledgments to the organizers in Oslo, Prof. F.E. Benth, Prof. G. Di Nunno, Prof. T. Lindstrøm, Prof. B. Øksendal and Prof. T. Zhang, of the Abel Symposium 2005 where the last named author could get a chance to present this result. We are also very much grateful to Prof. P. Malliavin and Prof. S.R.S. Varadhan for very stimulating and interesting discussions. Financial supports by SFB 611 (Bonn), ZiF (Bielefeld) and DAAD (Vigoni Program) is also gratefully acknowledged.

References

[ABRY1] S. Albeverio, M.S. Bernabei, M. Röckner, M.W. Yoshida: *Homogenization of infinite dimensional diffusion processes with periodic drift coefficients.* "Proceedings of Quantum Information and Complexity", Meijo Univ., 2003 Jan.", World Sci. Publishing, River Edge, NJ, 2004.

[ABRY2] S. Albeverio, M.S. Bernabei, M. Röckner, M.W. Yoshida: *Homogenization with respect to Gibbs measures for periodic drift diffusions on lattices. C. R. Acad. Sci. Paris. Ser. I* in press (2005).

[ABRY3] S. Albeverio, M.S. Bernabei, M. Röckner, M.W. Yoshida: *Homogenization of diffusions on the lattice* $\mathbf{Z}^d$ *with periodic drift coefficients; Application of logarithmic Sobolev inequality.* SFB pre-print 2005.

[AKR] S. Albeverio, Y.G. Kondratiev and M. Röckner: *Ergodicity of* L^2*-semigroups and extremality of Gibbs states, J. Funct. Anal.* 144 (1997), 394–423.

[BoRW] V.I. Bogachev, M. Röckner and F-Y. Wang: *Elliptic equations for invariant measures on finite and infinite dimensional manifolds,* *J. Math. Pures Appl.* 80, 2 (2001) 177–221.

[F1] M. Fukushima: *Dirichlet forms and Markov processes,* North-Holland, 1980.

[F2] M. Fukushima: *A generalized Stochastic Calculus in Homogenization,* in Proc. Sympos., Univ. Bielefeld, 1978, pp. 41–51, Springer, Vienna, 1980.

[FNT] M. Fukushima, S. Nakao and M. Takeda: *On Dirichlet forms with random data- recurrence and homogenization,* in Lecture Notes in Mathematics, 1250, Springer-Verlag, Berlin (1987).

[FunU] N. Funaki and K. Uchiyama: *From the Micro to the Macro, 1, 2, (in Japanese)* Series of Springer Contemporary Mathematics, (2003) Springer-Verlag, Tokyo.

[G] L. Gross: *Logarithmic Sobolev inequalities and contractive properties of semigroups,* in Lecture Notes in Mathematics 1563, Springer-Verlag, Berlin (1993).

[HS] R. Holley and D. Stroock: *Diffusions on an infinite dimensional torus,* *J. Funct. Anal.* 42 (1981), 29–63.

[IW] N. Ikeda and S. Watanabe: *Stochastic Differential Equations and Diffusion Processes, second edition,* North-Holland, 1989.

[MR] Z.M. Ma and M. Röckner: *Introduction to the theory of (Non-Symmetric) Dirichlet Forms,* Springer-Verlag, Berlin, 1992.

[O] H. Osada: *Homogenization of diffusion processes with random stationary coefficients,* Probability theory and mathematical statistics (Tbilisi, 1982), 507–517, Lecture Notes in Math., 1021, Springer, Berlin, 1983

[PapV] G. Papanicolaou, S. Varadhan: *Boundary value problems with rapidly oscillating random coefficients,* Seria Coll. Math. Soc. Janos Bolyai 27 (1979) North-Holland Publ.

[Par] E. Pardoux: *Homogenization of linear and semilinear second order parabolic PDEs with periodic coefficients: a probabilistic approach,* *J. Funct. Anal.* 167 (1999), 498–520.

[RWan] M. Röckner, F-Y. Wang: *Weak Poincaré inequalities and L^2-Convergence rates of Markov Semigroups,* *J. Funct. Anal.* 185 (2001), 564–603.

[S] D. Stroock: *Logarithmic Sobolev inequalities for Gibbs states,* in Lecture Notes in Mathematics 1563, Springer-Verlag, Berlin (1993).

[SZ] D. Stroock, B. Zegarlinski: *The equivalence of the logarithmic Sobolev inequality and the Dobrushin-Shlosman mixing condition,* *Comm. Math. Phys.* 144 (1992), no. 2, 303–323.

Theory and Applications of Infinite Dimensional Oscillatory Integrals

Sergio Albeverio[1] and Sonia Mazzucchi[2]

[1] Institut für Angewandte Mathematik, Wegelerstr. 6, 53115 Bonn, Germany, Dip. Matematica, Università di Trento, 38050 Povo (I), BiBoS; IZKS; SFB611; CERFIM (Locarno); Acc. Arch. (Mendrisio), albeverio@uni-bonn.de

[2] Institut für Angewandte Mathematik, Wegelerstr. 6, 53115 Bonn, Germany, Dip. Matematica, Università di Trento, 38050 Povo (I), mazzucch@science.unitn.it

Summary. Theory and main applications of infinite dimensional oscillatory integrals are discussed, with special attention to the relations with the original work of K. Itô in this area. New developments related to polynomial interactions are also presented.

Mathematics Subject Classification: 28C20, 35Q40, 46F20, 60G60, 81S40, 81T10, 81T45, 81P15, 35C20

Keywords: oscillatory integrals, Feynman path integrals, stationary phase, K. Itô, polynomial potentials, semiclassical expansions, trace formula, Chern-Simons model

1 Introduction

Professor K. Itô's work on the topic of infinite dimensional oscillatory integrals has been very germinal and stimulated much of the subsequent research in this area. It is therefore a special honour and pleasure to be able to dedicate the present pages to him. We shall give a short exposition of the theory of a particular class of functionals, the oscillatory integrals:

$$I^{\frac{\Phi}{\epsilon}}(f) = \text{``}\int_{\Gamma} e^{i\frac{\Phi}{\epsilon}(\gamma)} f(\gamma) d\gamma\text{''} \tag{1.1}$$

where Γ denotes either a finite dimensional space (e.g. $\mathbb{R}^s$, or an s-dimensional differential manifold M^s), or an infinite dimensional space (e.g. a "path space"). $\Phi : \Gamma \to \mathbb{R}$ is called phase function, while $f : \Gamma \to \mathbb{C}$ is the function to be integrated and $\epsilon \in \mathbb{R}\setminus\{0\}$ is a parameter. The symbol $d\gamma$ denotes a "flat" measure. In particular, if $dim(\Gamma) < \infty$ then $d\gamma$ is the Riemann-Lebesgue volume measure, while if $dim(\Gamma) = \infty$ an analogue of Riemann-Lebesgue measure is not mathematically defined and $d\gamma$ is just a heuristic expression.

1.1 Finite Dimensional Oscillatory Integrals

In the case where Γ is a finite dimensional vector space, i.e. $\Gamma = \mathbb{R}^s$, $s \in \mathbb{N}$, the expression (1.1)

$$\text{“}\int_{\mathbb{R}^s} e^{i\frac{\Phi}{\epsilon}(\gamma)} f(\gamma) d\gamma\text{”} \tag{1.2}$$

can be defined as an improper Riemann integral. The study of finite dimensional oscillatory integrals of the type (1.2) is a classical topic, largely developed in connection with several applications in mathematics (such as the theory of Fourier integral operators [48]) and physics. Interesting examples of integrals of the form (1.2) in the case $s = 1$, $\epsilon = 1$, $f = \chi_{[0,w]}$, $w > 0$, and $\Phi(x) = \frac{\pi}{2}x^2$, are the Fresnel integrals, that are applied in optics and in the theory of wave diffraction. If $\Phi(x) = x^3 + ax$, $a \in \mathbb{R}$ we obtain the Airy integrals, introduced in 1838 in connection with the theory of the rainbow.

Particular interest has been devoted to the study of the asymptotic behavior of integrals (1.2) when ϵ is regarded as a small parameter converging to 0. Originally introduced by Stokes and Kelvin and successively developed by several mathematicians, in particular van der Corput, the “stationary phase method” provides a powerful tool to handle the asymptotics of (1.2) as $\epsilon \downarrow 0$. According to it, the main contribution to the asymptotic behavior of the integral should come from those points $\gamma \in \mathbb{R}^s$ which belong to the critical manifold:

$$\Gamma_c^\Phi := \{\gamma \in \mathbb{R}^s, \mid \Phi'(\gamma) = 0\},$$

that is the points which make stationary the phase function Φ. Beautiful mathematical work on oscillatory integrals and the method of stationary phase is connected with the mathematical classification of singularities of algebraic and geometric structures (Coxeter indices, catastrophe theory), see, e.g. [31].

1.2 Infinite Dimensional Oscillatory Integrals

The extension of the results valid for $\Gamma = \mathbb{R}^s$ to the case where Γ is an infinite dimensional space is not trivial. The main motivation is the study of the “Feynman path integrals”, a class of (heuristic) functional integrals introduced by R.P. Feynman in 1942[1] in order to propose an alternative, Lagrangian, formulation of quantum mechanics. According to Feynman, the solution of the Schrödinger equation describing the time evolution of the state $\psi \in L^2(\mathbb{R}^d)$ of a quantum particle moving in a potential V

$$\begin{cases} i\hbar\frac{\partial}{\partial t}\psi = -\frac{\hbar^2}{2m}\Delta\psi + V\psi \\ \psi(0,x) = \psi_0(x) \end{cases} \tag{1.3}$$

[1] The first proposal going in the direction of Feynman's formulation can be found in work by P. Dirac in 1935, which inspired Feynman's own work.

(where $m > 0$ is the mass of the particle, $\hbar$ is the reduced Planck constant, $t \geq 0$, $x \in \mathbb{R}^d$) can be represented by a "sum over all possible histories", that is an integral over the space of paths γ with fixed end point

$$\psi(t,x) = \text{“}\int_{\{\gamma|\gamma(t)=x\}} e^{\frac{i}{\hbar}S_t(\gamma)}\psi_0(\gamma(0))d\gamma\text{”} \tag{1.4}$$

$S_t(\gamma) = S^0(\gamma) - \int_0^t V(s,\gamma(s))ds$, $S^0(\gamma) = \frac{m}{2}\int_0^t |\dot{\gamma}(s)|^2 ds$, is the classical action of the system evaluated along the path γ and $d\gamma$ a heuristic "flat" measure on the space of paths (see e.g. [40] for a physical discussion of Feynman's approach and its applications). The Feynman path integrals (1.4) can be regarded as oscillatory integrals of the form (1.1), where

$$\Gamma = \{ \text{ paths } \gamma : [0,t] \to \mathbb{R}^s,\ \gamma(t) = x \in \mathbb{R}^s\},$$

the phase function Φ is the classical action functional S_t, $f(\gamma) = \psi_0(\gamma(0))$, the parameter ϵ is the reduced Planck constant $\hbar$ and $d\gamma$ denotes heuristically

$$d\gamma = \text{“}C \prod_{s\in[0,t]} d\gamma(s)\text{”}, \tag{1.5}$$

$C :=$ "$(\int_{\{\gamma|\gamma(t)=x\}} e^{\frac{i}{\hbar}S_0(\gamma)}d\gamma)^{-1}$" being a normalization constant.

The Feynman's path integral representation (1.4) for the solution of the Schrödinger equation is particularly suggestive. Indeed it creates a connection between the classical (Lagrangian) description of the physical world and the quantum one and makes intuitive the study of the semiclassical limit of quantum mechanics, that is the study of the detailed behavior of the wave function ψ in the case where the Planck constant $\hbar$ is regarded as a small parameter. According to an (heuristic) application of the stationary phase method, in the limit $\hbar \downarrow 0$ the main contribution to the integral (1.4) should come from those paths γ which make stationary the action functional S_t. These, by Hamilton's least action principle, are exactly the classical orbits of the system.

Despite its powerful physical applications, formula (1.4) lacks mathematical rigour, in particular the "flat" measure $d\gamma$ given by (1.5) has no mathematical meaning.

In 1949 Kac [54, 55] observed that, by considering the heat equation (with $m = \hbar = 1$ for simplicity)

$$\begin{cases} \frac{\partial}{\partial t}u = \frac{1}{2}\Delta u - Vu \\ u(0,x) = u_0(x) \end{cases} \tag{1.6}$$

instead of the Schrödinger equation and by replacing the oscillatory factor $e^{iS_t(\gamma)}d\gamma$ by the non oscillatory $e^{-S_t(\gamma)}d\gamma$, one can give (for "good" V) a mathematical meaning to Feynman's formula in terms of a well defined Gaussian integral on the space of continuous paths: an integral with respect to the well known Wiener measure

$$u(t,x) = \text{“}\int e^{-S_t(\omega)} u_0(\omega(t)) d\omega\text{”} = \mathbb{E}\Big[e^{-\int_0^t V(\omega(s)+x)ds} u_0(\omega(t)+x)\Big] \quad (1.7)$$

(with $\mathbb{E}$ standing for expectation with respect to the standard Wiener process (mathematical Brownian motion) ω started at time 0 at the origin). Equation (1.7) is called Feynman-Kac formula.

In 1956 I.M. Gelfand and A.M. Yaglom [44] tried to realize Feynman's heuristic complex measure $e^{\frac{i}{\hbar}\Phi(\gamma)}d\gamma$ by means of a limiting procedure:

$$e^{\frac{i}{\hbar}\Phi(\gamma)}d\gamma := \lim_{\sigma\downarrow 0} e^{\frac{i}{\hbar - i\sigma}\Phi(\gamma)}d\gamma$$

In 1960 Cameron [34] proved however that the resulting measure cannot be $\sigma-$ additive and of bounded variation, even on very "nice" subsets of paths' space, and it is not possible to implement an integration in the Lebesgue's traditional sense (not even locally in space). As a consequence mathematicians tried to realize the integral (1.4) as a linear continuous functional on a suitable Banach algebra of integrable functions.

A particularly interesting approach can be found in the two pioneering papers by K. Itô [51, 52]. Itô was aware of the interest of Feynman's formula, as well as of the mathematical problems involved in it:

"It is easy to see that (1.4) *solves* (1.3) *unless we require mathematical rigour."* [51]

In the first paper in 1961 the author starts to study the problem by assuming that the potential V has a simple form, postponing the study of a more general case:

"It is our purpose to define the generalized measure $d\gamma$ (that, in our terms, is the integral $I^{\frac{\Phi}{\epsilon}}(f)$) *rigorously and prove* (1.4) *solves* (1.3) *in the case* $V \equiv 0$ *(case of no force) or* $V(x) = x$ *(case of constant force). We hope that this fact will be proven for a general* V *with some appropriate regularity conditions."*

Very shortly, what Itô does is to define rigorously the "generalized measure" (1.5), hence the heuristic integral (1.4), for V of above form and ψ_0 having a Fourier transform of compact support as a linear functional, taken to be the limit for $n \to \infty$ of finite dimensional approximations $I_n(\psi_0) = C_n \int_{L_x} e^{\frac{i}{2\hbar}\int_0^t \dot\gamma(s)^2 ds} \psi_0(\gamma(t)) P_n^{(x)}(d\gamma)$, with L_x the "translate by x of Cameron-Martin space", $P_n^{(x)}$ a suitable Gaussian measure associated with a certain compact operator T concentrated on L_x and $C_n \equiv \prod_j (1 - in\nu_j\hbar)^{\frac{1}{2}}$, $\{\nu_j\}$ being the eigenvalues of T. In the second paper [52] on the subject in 1967 K. Itô extended the class of potentials which can be handled and covers the case where the function $V : \mathbb{R}^d \to \mathbb{C}$ is the Fourier transform of a complex bounded variation measure on $\mathbb{R}^d$.

Itô's definition (in [62]) of the heuristic integral (1.4) is of the form

$$\lim_V \prod_{j=1}^{\infty} (1 - i\mu_j)^{\frac{1}{2}} E\left(e^{\frac{i}{2\hbar}\int_0^t \dot\gamma(s)^2 ds} \psi_0(\gamma(t)); a; V\right),$$

with E meaning expectation with respect to the Gaussian measure with mean a in L_x and a nuclear covariance operator V with eigenvalues μ_j (lim being taken along the directed system of all such V's, being independent of a). Itô's method for the definition of the Feynman's functional applies also to the Wiener integral and to the path integral representation (1.7) of the solution of the heat equation: *"Our definition is also applicable to the Wiener integral; namely, using it, we shall prove that the solution of the heat equation* (1.6) *is given by*

$$u(t,x) = \int_\Gamma e^{-\int_0^t \left(\frac{\dot\gamma^2(s)}{2} + V(\gamma(s))\right) ds} u_0(\gamma(t)) d\gamma$$

for any bounded continuous function $V(x)$.... This should be called the Feynman's version of Kac's theorem".

"Now that Kac's theorem is well known to probabilists, no one bothers with its Feynman version. However it is interesting that Kac had the Feynman version ... in mind and formulated it as ... to make it rigorous".

1.3 Other Examples of "Feynman Type Formulae"

The path integral representation (1.4) has been extended to more general dynamical systems. As we have already seen, its probabilistic version, i.e. the Feynman-Kac formula (1.7), is a representation of the solution of the heat equation. More generally probabilistic type integrals, which can be heuristically represented by expressions of the following form

$$\text{“}\int_\Gamma e^{-\frac{\Phi}{\epsilon}} f(\gamma) d\gamma\text{”} \tag{1.8}$$

(with the function $\Phi : \Gamma \to \mathbb{R}$ lower bounded and $\epsilon > 0$) have several applications, e.g. in stochastic analysis, statistical mechanics, hydrodynamics and in the theory of acoustic and electromagnetic waves.

The original Feynman path integral representation (1.4) for the solution of the Schrödinger equation and, more generally, heuristic oscillatory integrals of the type

$$\text{“}\int_\Gamma e^{i\frac{\Phi}{\epsilon}} f(\gamma) d\gamma\text{”} \tag{1.9}$$

can also be extended to the study of more general quantum systems. Feynman himself generalized formula (1.4) to a corresponding formula describing (relativistic) quantum fields. Recent applications of heuristic path integrals can be found in gauge theory (Yang-Mills fields), quantum gravity and in string theory.

Particularly interesting is the application to topological field theory, e.g. Chern-Simons' model. In this case the integration is performed on a space Γ

of geometric objects, i.e. on the space of connection 1-forms on the principal fiber bundle over a 3-dimensional manifold M. The phase function Φ is the Chern-Simons action functional:

$$\Phi(\gamma) \equiv \frac{k}{4\pi} \int_M \Big(\langle \gamma \wedge d\gamma \rangle + \frac{1}{3} \langle \gamma \wedge [\gamma \wedge \gamma] \rangle \Big), \tag{1.10}$$

where γ denotes a $g-$valued connection 1-form, g being the Lie algebra of a compact Lie group G (the "gauge group"). Φ is metric independent. The function f to be integrated is given by

$$f(\gamma) := \prod_{i=1}^{n} \mathrm{Tr}(Hol(\gamma, l_i)) \in \mathbb{C}, \tag{1.11}$$

where $(l_1, ..., l_n)$, $n \in \mathbb{N}$, are loops in M whose arcs are pairwise disjoint and $Hol(\gamma, l)$ denotes the holonomy of γ around l. According to a conjecture by Witten [70] and Schwartz the integral $I^\Phi(f)$ should represent a topological invariant. In particular, if $M = S^3$ and $G = SU(2)$ resp. $G = SU(N)$ resp. $G = SO(N)$, $I^\Phi(f)$ gives the Jones polynomials, resp. the Homfly polynomials resp. the Kauffmann polynomials. In the next section we shall see how a good part of these statements can be rigorously implemented using an adequate mathematical definition of Feynman path integrals.

2 Mathematical Definition of Infinite Dimensional Oscillatory Integrals

The heuristic Feynman integrals given by formula (1.4) and its generalization (1.9) have lead to fascinating and fruitful applications in physics and mathematics, even though as as they stand they do not have a well defined mathematical meaning. The present section is devoted to the description of the mathematical definition of the Feynman functional, and more generally of the infinite dimensional oscillatory integrals. In order to mirror the features of the heuristic Feynman measure, the Feynman functional should have some basic properties:

1. It should behave in a simple way under "translations and rotations in Γ", reflecting the fact that $d\gamma$ is a "flat" measure.
2. It should satisfy a Fubini type theorem, concerning iterated integrations along subspaces of Γ (allowing the construction, in physical applications, of a one-parameter group of unitary operators).
3. It should be approximable by finite dimensional oscillatory integrals, allowing a sequential approach in the spirit of Feynman's original work.
4. It should be related to probabilistic integrals with respect to the Wiener measure, allowing an "analytic continuation approach to Feynman path integrals from Wiener type integrals".

5. It should be sufficiently flexible to yield a rigorous mathematical implementation of an infinite dimensional version of the stationary phase method and the corresponding study of the semiclassical limit of quantum mechanics.

2.1 Finite Dimensional Case

The first step is the definition of the oscillatory integrals on a finite dimensional space $\Gamma := \mathbb{R}^n$, whose elements will be denoted by $x \in \mathbb{R}^n$:

$$\text{“}\int_{\mathbb{R}^n} e^{i\frac{\Phi}{\epsilon}(x)} f(x)dx\text{”} \tag{2.1}$$

If Φ is e.g. continuous and the function $f : \mathbb{R}^n \to \mathbb{C}$ is Lebesgue integrable, then the integral (2.1) is well defined in Lebesgue's sense. However, for suitable non integrable functions f, e.g. $f \equiv 1$, it is still possible to define expression (2.1) by exploiting the cancellations due to the oscillatory term $e^{i\frac{\Phi}{\epsilon}(x)}$. The following definition was proposed in [38] and is a modification of the one introduced by Hörmander [48].

Definition 1. *The oscillatory integral of a Borel function $f : \mathbb{R}^n \to \mathbb{C}$ with respect to a (continuous) phase function $\Phi : \mathbb{R}^n \to \mathbb{R}$ is well defined if and only if for each test function $\phi \in \mathcal{S}(\mathbb{R}^n)$ such that $\phi(0) = 1$ the integral*

$$I_\delta(f,\phi) := \int_{\mathbb{R}^n} (2\pi i\epsilon)^{-n/2} e^{i\frac{\Phi(x)}{\epsilon}} f(x)\phi(\delta x)dx$$

exists for all $\delta > 0$ and the limit $\lim_{\delta\to 0} I_\delta(f,\phi)$ exists and is independent of ϕ. In this case the limit is called the oscillatory integral of f with respect to Φ and denoted by

$$I^\Phi(f) \equiv \widetilde{\int_{\mathbb{R}^n}} e^{i\frac{\Phi(x)}{\epsilon}} f(x)dx. \tag{2.2}$$

The symbol $\tilde{\int}$ recalls the normalization factor $(2\pi i\epsilon)^{-n/2}$ which makes the integral "normalized" in the case $\Phi(x) = \frac{|x|^2}{2}$, in the sense that $I^\Phi(1) = 1$ for such a Φ.

A "complete direct characterization" of the class of functions f and phases Φ for which the integral (2.2) is well defined is still an open problem. However, for suitable Φ, it is possible to find an interesting set of "integrable functions" f, for which the oscillatory integral $I^\Phi(f)$ is well defined and can be explicitly computed in terms of an absolutely convergent integral thanks to a Parseval-type equality.

We shall shortly introduce a setting first presented in [15]. Given a (finite or infinite dimensional) real separable Hilbert space $(\mathcal{H}, \langle\ ,\ \rangle)$, let us denote by $\mathcal{M}(\mathcal{H})$ the Banach space of the complex bounded variation measures on $\mathcal{H}$, endowed with the total variation norm, that is:

$$\mu \in \mathcal{M}(\mathcal{H}), \qquad \|\mu\| = \sup \sum_i |\mu(E_i)|,$$

where the supremum is taken over all sequences $\{E_i\}$ of pairwise disjoint Borel subsets of $\mathcal{H}$, such that $\cup_i E_i = \mathcal{H}$. $\mathcal{M}(\mathcal{H})$ is a Banach algebra, where the product of two measures $\mu * \nu$ is by definition their convolution:

$$\mu * \nu(E) = \int_{\mathcal{H}} \mu(E-x)\nu(dx), \qquad \mu, \nu \in \mathcal{M}(\mathcal{H}).$$

and the unit element is the Dirac point measure δ_0 (with support at the origin).

Let $\mathcal{F}(\mathcal{H})$ be the space of complex functions on $\mathcal{H}$ which are Fourier transforms of measures belonging to $\mathcal{M}(\mathcal{H})$, that is:

$$f : \mathcal{H} \to \mathbb{C} \qquad f(x) = \int_{\mathcal{H}} e^{i\langle x,\beta\rangle} \mu_f(d\beta) \equiv \hat{\mu}_f(x).$$

$\mathcal{F}(\mathcal{H})$ is a Banach algebra of functions, where the product is the pointwise one; the unit element is the function 1, i.e. $1(x) = 1 \; \forall x \in \mathcal{H}$ and the norm is given by $\|f\| = \|\mu_f\|$.

It is possible to prove [19] that if $\mathcal{H} \in \mathbb{R}^n$ and $f \in \mathcal{F}(\mathbb{R}^n)$, $f = \hat{\mu}_f$, and if the phase function Φ is such that $F^\Phi \equiv \frac{e^{i\frac{\Phi}{\epsilon}}}{(2\pi i\epsilon)^{n/2}}$ has a Fourier transform $\hat{F}^\Phi$ having the property that the integral

$$\int_{\mathbb{R}^n} \hat{F}^\Phi(\alpha) d\mu_f(\alpha)$$

exists, then the oscillatory integral $I^\Phi(f)$ exists (in the sense of definition 1) and it is given by the following "Parseval formula":

$$I^\Phi(f) = \int_{\mathbb{R}^n} \hat{F}^\Phi(\alpha) d\mu_f(\alpha) \tag{2.3}$$

Equation (2.3) holds for smooth phase functions Φ of at most even polynomial growth at infinity (see [19] for more details). It is worthwhile to recall that $I^\Phi(f)$ can be defined for more general f as proved in [48], but in this case formula (2.3) is no longer valid in general.

It is interesting to analyze two particular cases, which we in Sect. 3 shall extend to the infinite dimensional case. Let us assume that $\varepsilon > 0$ and Φ has one of the following forms

$$\Phi(x) := \frac{1}{2}\langle x, Qx\rangle - V(x), \tag{2.4}$$

$$\Phi(x) := \frac{1}{2}\langle x, Qx\rangle - V(x) - \lambda P(x), \tag{2.5}$$

where $V \in \mathcal{F}(\mathbb{R}^n)$, $Q : \mathbb{R}^N \to \mathbb{R}^n$ is a linear, symmetric invertible operator, $\lambda < 0$ and P is an homogeneous 4-degree polynomial. Since the function $e^{iV} f$ belongs to $\mathcal{F}(\mathbb{R}^n)$ for $V, f \in \mathcal{F}(\mathbb{R}^n)$, we do not loose generality in setting

$V \equiv 0$ (which simplifies notations). If Φ is of the type (2.4), then formula (2.3) assumes the following form [6, 38]:

$$I^{\Phi}(f) = \det Q^{-1/2} \int_{\mathbb{R}^n} e^{-i\frac{\epsilon}{2}\langle \alpha, Q^{-1}\alpha\rangle} d\mu_f(\alpha) \tag{2.6}$$

while in the case where Φ is of the type (2.5), with $Q > 0$, then formula (2.3) is still valid with $\hat{F}^{\Phi}$ given by

$$\hat{F}^{\Phi}(\alpha) = (2\pi\epsilon)^{-n/2} \int_{\mathbb{R}^n} e^{ie^{i\frac{\pi}{4}}\langle \alpha, x\rangle} e^{-\frac{1}{2\epsilon}\langle x, Qx\rangle} e^{i\frac{\lambda}{\epsilon}P(x)} dx$$
$$= \mathbb{E}(e^{ie^{i\frac{\pi}{4}}\langle \alpha, x\rangle} e^{\frac{1}{2\epsilon}\langle x, (I-Q)x\rangle} e^{i\frac{\lambda}{\epsilon}P(x)}), \tag{2.7}$$

$\alpha \in \mathbb{R}^n$, where the expectation is taken with respect to the standard Gaussian measure $N(0, \epsilon I_{\mathbb{R}^n})$. Moreover under some analyticity assumptions on the function f, the integral $I^{\Phi}(f)$ can be computed by means of the following formula:

$$I^{\Phi}(f) = \mathbb{E}(f(e^{i\frac{\pi}{4}}x)e^{\frac{1}{2\epsilon}\langle x, (I-Q)x\rangle} e^{i\frac{\lambda}{\epsilon}P(x)}). \tag{2.8}$$

The r.h.s. of formula (2.8) extends to an analytic function of the variable λ, for $Im(\lambda) > 0$, and is still continuous for $Im(\lambda) = 0$ [20, 21].

The leading idea of the proof is the computation of the Fourier transform F^{Φ} by means of a rotation of $\pi/4$ in counterclock direction of the integration contour. This operation maps the quadratic part $e^{i\frac{|x|^2}{2\epsilon}}$ of $e^{i\frac{\Phi}{\epsilon}}$ into the Gaussian density $e^{-\frac{|x|^2}{2\epsilon}}$ while the quartic part $e^{-i\frac{\lambda P(x)}{\epsilon}}$ of $e^{i\frac{\Phi}{\epsilon}}$ remains bounded, going over to $e^{i\frac{\lambda P(x)}{\epsilon}}$. For more details see [20].

2.2 Infinite Dimensional Case

The results of the previous section can be partially extended to the case where Γ is an infinite dimensional real separable Hilbert space $(\mathcal{H}, \langle\ ,\ \rangle)$. An infinite dimensional oscillatory integral can be defined as the limit of a sequence of finite dimensional approximations, as proposed in [6, 38].

Definition 2. *A function $f : \mathcal{H} \to \mathbb{C}$ is said to be integrable with respect to the phase function $\Phi : \mathcal{H} \to \mathbb{R}$ if for any sequence P_n of projectors onto n-dimensional subspaces of $\mathcal{H}$, such that $P_n \leq P_{n+1}$ and $P_n \to 1$ strongly as $n \to \infty$ (1 being the identity operator in $\mathcal{H}$), the finite dimensional approximations*

$$\widetilde{\int_{P_n\mathcal{H}}} e^{i\frac{\Phi(P_n x)}{\epsilon}} f(P_n x) d(P_n x),$$

are well defined (in the sense of definition 1) and the limit

$$\lim_{n\to\infty} \widetilde{\int_{P_n\mathcal{H}}} e^{i\frac{\Phi(P_n x)}{\epsilon}} f(P_n x) d(P_n x) \tag{2.9}$$

exists and is independent of the sequence $\{P_n\}$.

In this case the limit is called oscillatory integral of f with respect to the phase function Φ and is denoted by

$$I^{\Phi}(f) \equiv \widetilde{\int_{\mathcal{H}}} e^{i\frac{\Phi(x)}{\epsilon}} f(x)dx.$$

Again it is important to find classes of functions f and phases Φ, for which $I^{\Phi}(f)$ is well defined. All basic properties valid in the finite dimensional case remain valid in the infinite dimensional case. Indeed the fundamental space is again $\mathcal{F}(\mathcal{H})$, the space of functions which are Fourier transforms of complex bounded variation measures on $\mathcal{H}$. K. Itô was first in understanding the important role of this space in connection with the mathematical definition of Feynman path integrals. He introduced $\mathcal{F}(\mathcal{H})$ in his second paper on the topic [52], where he generalized the results of [51] to the case where the potential V belongs to $\mathcal{F}(\mathbb{R}^d)$ (as we briefly discussed in Sect. 1). Itô's results were extensively developed by S. Albeverio and R. Høegh-Krohn [15, 16] and later by D. Elworthy and A. Truman [38]. In the case where the phase function Φ is of the form

$$\Phi(x) = \frac{1}{2}\langle x, Qx\rangle + \langle a, x\rangle + \Phi_{int}(x), \tag{2.10}$$

where $Q : \mathcal{H} \to \mathcal{H}$ is a linear invertible self-adjoint operator, $I - Q$ is of trace class, $a \in \mathcal{H}$ and $\Phi_{int} \in \mathcal{F}(\mathcal{H})$, and, moreover, $f \in \mathcal{F}(\mathcal{H})$, these authors prove that $I^{\Phi}(f)$ is well defined and can be explicitly computed in terms of a well defined absolutely convergent integral with respect to a bounded variation measure by means of a Parseval-type equality similar to (2.6). Some time later the definition of $I^{\Phi}(f)$ was generalized to unbounded functions Φ_{int} that are Laplace transforms of complex bounded variation measures on $\mathcal{H}$ [8,17,59]. More recently a breakthrough in handling the case where Φ_{int} is a fourth-order polynomial has been achieved [20, 21]. In fact formula (2.8) valid in the finite dimensional case has been generalized to the infinite dimensional case. Let us describe in more details this newer development, because of its relevance for applications (the quartic potential model is one of the most discussed ones in the physical literature).

Given a real separable infinite dimensional Hilbert space $(\mathcal{H}, \langle\ ,\ \rangle)$ with norm $|\ |$, let ν be the finitely additive cylinder measure on $\mathcal{H}$, defined by its characteristic functional $\hat{\nu}(x) = e^{-\frac{1}{2}|x|^2}$. Let $\|\ \|$ be a "measurable" norm on $\mathcal{H}$, that is $\|\ \|$ is such that for every $\delta > 0$ there exist a finite-dimensional projection $P_\delta : \mathcal{H} \to \mathcal{H}$, such that for all $P \perp P_\delta$ one has $\nu(\{x \in \mathcal{H} |\ \|P(x)\| > \delta\}) < \delta$, where P and P_δ are called orthogonal ($P \perp P_\delta$) if their ranges are orthogonal in $(\mathcal{H}, \langle , \rangle)$. One can easily verify that $\|\ \|$ is weaker than $|\ |$. Denoting by $\mathcal{B}$ the completion of $\mathcal{H}$ in the $\|\ \|$-norm and by i the continuous inclusion of $\mathcal{H}$ in $\mathcal{B}$, one proves that $\mu \equiv \nu \circ i^{-1}$ is a countably additive Gaussian measure on the Borel subsets of $\mathcal{B}$. The triple $(i, \mathcal{H}, \mathcal{B})$ is called an *abstract Wiener space* (in the sense of L. Gross). Let us consider a phase function of the following form:

$$\Phi(x) = \frac{1}{2}\langle x, Qx\rangle - \lambda P(x), \tag{2.11}$$

with $Q : \mathcal{H} \to \mathcal{H}$ a self-adjoint strictly positive operator such that $I - Q$ is trace class, and $P : \mathcal{H} \to \mathbb{R}$ is given by $P(x) = B(x, x, x, x)$, with $B : \mathcal{H}\times\mathcal{H}\times\mathcal{H}\times\mathcal{H} \to \mathbb{R}$ a completely symmetric positive covariant tensor operator on $\mathcal{H}$ such that the map $V : \mathcal{H} \to \mathbb{R}^+$, $x \mapsto V(x) \equiv B(x, x, x, x)$ is continuous in the $\|\ \|$ norm. Under these assumptions, it is possible to prove that the functions on $\mathcal{H}$ defined by

$$x \in \mathcal{H} \mapsto \langle k, x\rangle \text{ resp. } \langle x, (I - Q)x\rangle \text{ resp. } P(x),$$

can be lifted to random variables on $\mathcal{B}$, denoted by

$$\omega \in \mathcal{B} \mapsto \langle n(k)(\omega)\rangle \text{ resp. } \langle \omega, (I - Q)\omega\rangle \text{ resp. } P(\omega) \text{ (with } k \in \mathcal{H})$$

Moreover the following holds [20]:

Theorem 1. *Let $f : \mathcal{H} \to \mathbb{C}$ be the Fourier transform of a measure $\mu_f \in \mathcal{M}(\mathcal{H})$, $f \equiv \hat{\mu}_f$, satisfying the following assumption*

$$\int_{\mathcal{H}} e^{\frac{\epsilon}{4}\langle k, Q^{-1}k\rangle} |\mu_f|(dk) < +\infty. \tag{2.12}$$

Then the infinite dimensional oscillatory integral

$$\widetilde{\int_{\mathcal{H}}} e^{\frac{i}{2\epsilon}\langle x, Qx\rangle} e^{-i\frac{\lambda}{\epsilon}P(x)} f(x)dx \tag{2.13}$$

exists and is given by:

$$\int_{\mathcal{H}} \mathbb{E}[e^{in(k)(\omega)e^{i\pi/4}} e^{\frac{1}{2\epsilon}\langle \omega, (I-Q)\omega\rangle} e^{i\frac{\lambda}{\epsilon}P(\omega)}]\mu_f(dk) \tag{2.14}$$

It is also equal to:

$$\mathbb{E}[e^{\frac{1}{2\epsilon}\langle \omega, (I-Q)\omega\rangle} e^{i\frac{\lambda}{\epsilon}P(\omega)} f(e^{i\pi/4}\omega)] \tag{2.15}$$

$\mathbb{E}$ denotes the expectation value with respect to the Gaussian measure μ on $\mathcal{B}$ (described before the statement of the theorem).

2.3 Properties and Comparison with Other Approaches

The infinite dimensional oscillatory integral $I^{\Phi}(f)$, with $\Phi \equiv \frac{|x|^2}{2}$ and $f \in \mathcal{F}(\mathcal{H})$, was originally defined [15] by "duality" by means of the Parseval type equality (2.6). The more recent definition of $I^{\Phi}(f)$ (see definition 2, based on [38]) by means of finite dimensional approximations maintains this property: indeed for suitable Φ the application $f \mapsto I^{\Phi}(f)$ is a linear continuous functional on $\mathcal{F}(\mathcal{H})$.

The realization of the integral $I^{\Phi}(f)$ by means of a duality relation is typical of several approaches to the definition of the Feynman path integral. In other words one tries to define the Feynman density $e^{i\Phi(\gamma)}$ as an "infinite dimensional distribution". Besides [15] origins of this idea can be found in work by C. DeWitt-Morette (see, e.g. [35], see also e.g. [57]). It was systematically developed in the framework of white noise calculus by T. Hida and L. Streit [47,59]. In the latter setting the integral $I^{\Phi}(f)$ is realized as the pairing $\langle T_{\Phi}, f\rangle$ with respect to the standard Gaussian measure $N(0, I_{L^2(\mathbb{R}^n)})$ of a white noise distribution $T_{\phi} \in (S')$ (which, heuristically, can be interpreted as $e^{\frac{i}{2}\Phi(\gamma)+\frac{1}{2}\langle\gamma,\gamma\rangle)}$ and a regular $f \in (S)$, where $(S), (S')$ are elements of the Gelfand triple $(S) \subset L^2(N(0, I_{L^2(\mathbb{R}^n)})) \subset (S')$ (see [47] for details).

It is interesting to note that formula (2.15) shows a deep connection between infinite dimensional oscillatory integrals and probabilistic Gaussian integrals. Indeed, under suitable assumptions on the function f that is integrated and on the phase function Φ, the oscillatory integral of f with respect to Φ is equal to a Gaussian integral. On the other hand one of the first approaches to the rigorous mathematical definition of Feynman path integrals was by means of analytic continuation of Gaussian Wiener integrals [34,37,51,53,56,62,64,67,69]. The leading idea of this approach is the analogy between Schrödinger and heat equation on one hand, and between the rigorous Feynman-Kac formula (1.7) and the heuristic Feynman representation (1.4) on the other hand. By introducing in the heat equation (1.6) and in the corresponding path integral solution (1.7) a suitable parameter λ, proportional for instance to the time, or to the mass, or to the Planck constant, and by allowing λ to assume complex values, then one gets, at least heuristically, the Schrödinger equation and its solution. This procedure can be made completely rigorous under suitable conditions on the potential V and initial datum ψ_0.

Another approach to the mathematical definition of Feynman path integrals, which is very close to Feynman's original derivation, is the "sequential approach". It was originally proposed by A. Truman [68] and further extensively developed by D. Fujiwara and N. Kumano-go [41–43]. In this approach the paths γ in formula (1.4) are approximated by piecewise linear paths and the Feynman path integral is correspondingly approximated by a finite dimensional integral.

Two other alternative approaches to the mathematical definition of Feynman path integrals are based on Poisson measures respectively on nonstandard analysis. The first one was originally proposed by A.M. Chebotarev and V.P. Maslov [63] and further developed by several authors as S. Albeverio, Ph. Blanchard, Ph. Combe, R. Høegh-Krohn, M. Sirugue [2, 3] and V. Kolokol′tsov [58]. The second was proposed in the 80's by S. Albeverio, J.E. Fenstad, R. Høegh-Krohn and T. Lindstrøm [13], but it has not been much further developed yet.

2.4 The Method of Stationary Phase

One of the main motivations for the rigorous mathematical definition of the infinite dimensional oscillatory integrals is the implementation of a corresponding infinite dimensional version of the method of the stationary phase and its application to the study of the asymptotic behavior of the expressions in formula (1.4) in the limit $\hbar \to 0$. The first results were obtained in [16] and further developed in [6,65] (see also e.g. [3–5]). Up to now, only the case where the phase function Φ is of the form (2.10) has been handled rigorously. In this case the detailed asymptotic expansion of the infinite dimensional oscillatory integral has been computed, and, in the case where the phase function has a unique stationary point, the Borel summability of the expansion has been proved [65].

For results on the study of the asymptotic behavior of infinite dimensional probabilistic integrals and its connection with the semiclassical limit of Schrödinger equation see e.g. [18, 26, 27, 32, 33, 50, 61, 66].

3 Applications

3.1 The Schrödinger Equation

Infinite dimensional oscillatory integrals, as defined in Section 1.2, provided a rigorous mathematical realization of the heuristic Feynman path integral representation (1.4) for the solution of the following Schrödinger equation

$$\begin{cases} i\hbar\frac{\partial}{\partial t}\psi = H\psi \\ \psi(0,x) = \psi_0(x) \end{cases} \tag{3.1}$$

where

$$H = -\frac{\hbar^2}{2m}\Delta + \frac{1}{2}xA^2x + V(x) + \lambda P(x),$$

$V, \psi_0 \in \mathcal{F}(\mathbb{R}^d)$, $A : \mathbb{R}^d \to \mathbb{R}^d$ is a symmetric positive operator, $\lambda \geq 0$, P is an homogeneous fourth order polynomial. In other words, under the assumptions above, the heuristic path integral (1.4) can be realized as a well defined infinite dimensional oscillatory integral on a suitable Hilbert space $\mathcal{H}$ with parameter $\epsilon \equiv \hbar$. We describe here the result in the case $\lambda = 0$ [6, 15, 16, 38], recalling that the general case with $\lambda \neq 0$ has been recently handled in [20].

Let us consider the Cameron-Martin space $(\mathcal{H}_t, \langle\ ,\ \rangle)$, i.e. the Hilbert space of absolutely continuous paths $\gamma : [0, t] \to \mathbb{R}^d$ such that $\gamma(t) = 0$ and $\dot\gamma \in L_2([0, t]; \mathbb{R}^d)$, endowed with the inner product

$$\langle\gamma_1, \gamma_2\rangle = \int_0^t \dot\gamma_1(s)\dot\gamma_2(s)ds.$$

From now on we shall assume for notational simplicity that $m = 1$. Let us consider the operator L on $\mathcal{H}_t$ given by

$$\langle \gamma, L\gamma \rangle \equiv \int_0^t \gamma(s) A^2 \gamma(s) ds,$$

and the function $v : \mathcal{H}_t \to \mathbb{C}$

$$v(\gamma) \equiv \int_0^t V(\gamma(s) + x) ds + 2xA^2 \int_0^t \gamma(s) ds \qquad \gamma \in \mathcal{H}_t.$$

By analyzing the spectrum of the operator L (see [38]) one can easily verify that L is trace class and $I - L$ is invertible. The following holds:

Theorem 2. *Under the assumptions above, the function $f : \mathcal{H}_t \to \mathbb{C}$ given by*

$$f(\gamma) := e^{-\frac{i}{\hbar} v(\gamma)} \psi_0(\gamma(0) + x)$$

is the Fourier transform of a complex bounded variation measure μ_f on $\mathcal{H}_t$ and the infinite dimensional oscillatory integral of the function $g(\gamma) = e^{-\frac{i}{2\hbar}\langle \gamma, L\gamma \rangle} f(\gamma)$

$$\widetilde{\int_{\mathcal{H}_t}} e^{\frac{i}{2\hbar}\langle \gamma, (I-L)\gamma \rangle} e^{-\frac{i}{\hbar} v(\gamma)} \psi_0(\gamma(0) + x) d\gamma. \tag{3.2}$$

is well defined (in the sense of definition (2)) and it is equal to

$$\det(I - L)^{-1/2} \int_{\mathcal{H}_t} e^{-\frac{i\hbar}{2}\langle \gamma, (-L)^{-1}\gamma \rangle} d\mu_f(\gamma),$$

$\det(I - L)$ being the Fredholm determinant of the operator $(I - L)$.

Moreover it is a representation of the solution of the Schrödinger equation (3.1) evaluated at $x \in \mathbb{R}^d$ at time t.

For a proof see [38]. An extension of this result to the case of the presence of a polynomial potential (i.e. $\lambda \neq 0$ in the expression for it) has been obtained in [20] (on the basis of theorem 1) In this case the Borel summability of the asymptotic expansion of $I^{\Phi}(f)$ in powers of the coupling constant λ has also been proven.

The method of the stationary phase in infinite dimensions has been applied to the study of the asymptotic behavior of the integral (3.2) in the limit $\hbar \to 0$, in the case $\lambda = 0$ [6, 16, 65] (for other methods leading to similar results, see e.g. [32]).

The result of theorem 2 has been recently generalized to the case where the potential V, the matrix A and the coupling constant λ are explicitly time dependent [23, 25, 30]. The case $\lambda \neq 0$ requires special attention because of the superquadratic growth of the term $\lambda P(x)$ at infinity which excludes the possibility of applying the usual methods of the theory of hyperbolic evolution equations.

Let us also mention that infinite dimensional oscillatory integrals are a flexible tool and provide a rigorous mathematical realization for other large classes of Feynman path integral representations, such as the "phase space Feynman path integrals" [10] and the "Feynman path integrals with complex phase functions" that are applied to the solution of a stochastic Schrödinger equation [11,12]. Other interesting applications are the solution of the Schrödinger equation with a magnetic field [7], the trace formula for the Schrödinger group [4,5] (which includes a rigorous proof of "Gutzwiller's trace formula", of basic importance in the study of quantum chaos, and with interesting connections with number theory), the dynamics of Dirac systems [49] and of quantum open systems [9,39].

3.2 The Chern-Simons Model

The application of the infinite dimensional oscillatory integrals to the mathematical definition of the Chern-Simons functional integral described in section 1.3 has been realized in [28] in the case where the gauge group G is abelian. It has been proven in particular that if $H^1(M) = 0$ then $I^\Phi(f)$ gives the linking numbers. The same results were obtained in [60] in the framework of white noise calculus. These result have been extended to the case where G is not abelian and $M = \mathbb{R}^3$ in [29,45] by means of white noise analysis (see also [14] for a detailed exposition of this topic). The case $M = S^1 \times S^2$ has been recently handled in [46]. Rigorous asymptotic expansions have been discussed in [24]. There is certainly still a large gap between the extensive and productive heuristic use of Feynman type integrals in this area (and in related areas connected with quantum gravity and string theory) and what can be achieved rigorously. This is a great challenge for the future.

Acknowledgements

The first author would like to express his gratitude to the organizers, in particular G. Di Nunno, for the kind invitation to speak at a conference in honour of Professor K. Itô. He met Professor K. Itô at a Katata workshop in 1981 and on several other occasions, including a visit to his department at Gakushuin University. He always appreciated his very pleasant human side, besides his extraordinary scientific qualities. This contribution is dedicated to him with great admiration and gratitude.

References

1. S. Albeverio. Wiener and Feynman Path Integrals and Their Applications. Proceedings of the Norbert Wiener Centenary Congress, 163–194, Proceedings of Symposia in Applied Mathematics **52**, Amer. Math. Soc., Providence, RI, 1997.

2. S. Albeverio, Ph. Blanchard, Ph. Combe, R. Høegh-Krohn, M. Sirugue, Local relativistic invariant flows for quantum fields., Comm. Math. Phys. 90(3), 329–351 (1983).
3. S. Albeverio, Ph. Blanchard, R. Høegh-Krohn, Feynman path integrals, the Poisson formula and the theta function for the Schrödinger operators, in: Trends in Applications of Pure Mathematics to Mechanics, Vol III. Pitman, Boston, 1–21 (1981).
4. S. Albeverio, Ph. Blanchard, R. Høegh-Krohn, Feynman path integrals and the trace formula for the Schrödinger operators, Comm. Math. Phys. 83 n. 1, 49–76 (1982).
5. S. Albeverio, A.M. Boutet de Monvel-Berthier, Z. Brzeźniak, The trace formula for Schrödinger operators from infinite dimensional oscillatory integrals, Math. Nachr. 182, 21–65 (1996).
6. S. Albeverio and Z. Brzeźniak. Finite-dimensional approximation approach to oscillatory integrals and stationary phase in infinite dimensions. J. Funct. Anal., 113(1): 177–244, 1993.
7. S. Albeverio, Z. Brzeźniak, Oscillatory integrals on Hilbert spaces and Schrödinger equation with magnetic fields, J. Math. Phys. 36(5), 2135–2156 (1995).
8. S. Albeverio, Z. Brzeźniak, Z. Haba. On the Schrödinger equation with potentials which are Laplace transform of measures. Potential Anal., **9** n. 1, 65–82, 1998.
9. S. Albeverio, L. Cattaneo, L. DiPersio, S. Mazzucchi, An infinite dimensional oscillatory integral approach to the Feynman-Vernon influence functional I, II, in preparation.
10. S. Albeverio, G. Guatteri, S. Mazzucchi, Phase space Feynman path integrals, J. Math. Phys. 43, 2847–2857 (2002).
11. S. Albeverio, G. Guatteri, S. Mazzucchi, Representation of the Belavkin equation via Feynman path integrals, Probab. Theory Relat. Fields 125, 365–380 (2003).
12. S. Albeverio, G. Guatteri, S. Mazzucchi, Representation of the Belavkin equation via phase space Feynman path integrals, Infin. Dimens. Anal. Quantum Probab. Relat. Top. 7, no. 4, 507–526 (2004).
13. S. Albeverio, J.E. Fenstad, R. Høegh-Krohn, T. Lindstrøm, Non Standard Methods in Stochastic Analysis and Mathematical Physics, Pure and Applied Mathematics **122**, Academic Press, Inc., Orlando, FL, (1986).
14. S. Albeverio, A. Hahn, A. Sengupta, Rigorous Feynman path integrals, with applications to quantum theory, gauge fields, and topological invariants, Stochastic analysis and mathematical physics (SAMP/ANESTOC 2002), 1–60, World Sci. Publishing, River Edge, NJ, (2004).
15. S. Albeverio and R. Høegh-Krohn. Mathematical theory of Feynman path integrals. Springer-Verlag, Berlin, 1976. Lecture Notes in Mathematics, Vol. 523. 2nd edn with S. Mazzucchi, (2007).
16. S. Albeverio and R. Høegh-Krohn. Oscillatory integrals and the method of stationary phase in infinitely many dimensions, with applications to the classical limit of quantum mechanics. Invent. Math., 40(1): 59–106, 1977.
17. S. Albeverio, A. Khrennikov, O. Smolyanov. The probabilistic Feynman-Kac formula for an infinite-dimensional Schrödinger equation with exponential and singular potentials. Potential Anal. 11, no. 2, 157–181, (1999).

18. S. Albeverio, S. Liang, Asymptotic expansions for the Laplace approximations of sums of Banach space-valued random variables. Ann. Probab. 33, no. 1, 300–336, (2005).
19. S. Albeverio and S. Mazzucchi. Generalized Fresnel Integrals. Bull. Sci. Math. 129 (2005), no. 1, 1–23.
20. S. Albeverio and S. Mazzucchi. Feynman path integrals for polynomially growing potentials. J. Funct. Anal. **221** no. 1 (2005), 83–121.
21. S. Albeverio and S. Mazzucchi. Generalized infinite-dimensional Fresnel Integrals. C. R. Acad. Sci. Paris **338** n. 3, 255–259, 2004.
22. S. Albeverio and S. Mazzucchi. Some New Developments in the Theory of Path Integrals, with Applications to Quantum Theory. J. Stat. Phys., **115** n.112, 191–215, 2004.
23. S. Albeverio and S. Mazzucchi. Feynman path integrals for time-dependent potentials. "Stochastic Partial Differential Equations and Applications-VII", G. Da Prato and L. Tubaro eds, Lecture Notes in Pure and Applied Mathematics, vol. 245, Taylor & Francis, 2005, pp 7–20.
24. S. Albeverio and I. Mitoma, in preparation
25. S. Albeverio and S. Mazzucchi. Feynman path integrals for the time dependent quartic oscillator. C. R. Math. Acad. Sci. Paris 341 (2005), no. 10, 647–650.
26. S. Albeverio, H. Röckle, V. Steblovskaya, Asymptotic expansions for Ornstein-Uhlenbeck semigroups perturbed by potentials over Banach spaces. Stochastics Stochastics Rep. 69, no. 3–4, 195–238, (2000).
27. S. Albeverio, V. Steblovskaya, Asymptotics of infinite-dimensional integrals with respect to smooth measures. I. Infin. Dimens. Anal. Quantum Probab. Relat. Top. 2, no. 4, 529–556 (1999).
28. S. Albeverio, J. Schäfer, Abelian Chern-Simons theory and linking numbers via oscillatory integrals, J. Math. Phys. 36, 2157–2169 (1995).
29. S. Albeverio, A. Sengupta, A mathematical construction of the non-Abelian Chern-Simons functional integral, Commun. Math. Phys. 186, 563–579 (1997).
30. S. Albeverio and S. Mazzucchi. The time dependent quartic oscillator - a Feynman path integral approach. J. Funct. Anal. 238, 42, 471–488 (2006).
31. V.I. Arnold, S.N. Gusein-Zade, A.N. Varchenko, Singularities of differentiable maps, Vol. II Birkhäuser, Basel (1988).
32. R. Azencott, H. Doss, L'équation de Schrödinger quand h tend vers zéro: une approche probabiliste. (French) [The Schrödinger equation as h tends to zero: a probabilistic approach], Stochastic aspects of classical and quantum systems, Eds. S. Albeverio et al. (Marseille, 1983), 1–17, Lecture Notes in Math., 1109, Springer, Berlin, 1985.
33. G. Ben Arous, F. Castell, A Probabilistic Approach to Semi-classical Approximations, J. Funct. Anal. **137**, 243–280 (1996)
34. R.H. Cameron. A family of integrals serving to connect the Wiener and Feynman integrals. J. Math. and Phys. **39**, 126–140, 1960.
35. P. Cartier and C. DeWitt-Morette. Functional integration. J. Math. Phys. **41**, no. 6, 4154–4187, 2000.
36. P.A.M. Dirac, The Lagrangian in quantum mechanics, Phys. Zeitschr. d. Sowjetunion, 3, No 1, 64–72, 1933.
37. H. Doss. Sur une Résolution Stochastique de l'Equation de Schrödinger à Coefficients Analytiques. Commun. Math. Phys., 73, 247–264, 1980.

38. D. Elworthy and A. Truman. Feynman maps, Cameron-Martin formulae and anharmonic oscillators. Ann. Inst. H. Poincaré Phys. Théor., 41(2): 115–142, 1984.
39. P. Exner, Open quantum systems and Feynman integrals, D. Reidel, Dordrecht, 1985.
40. R.P. Feynman, A.R. Hibbs. Quantum mechanics and path integrals. Macgraw Hill, New York, 1965.
41. D. Fujiwara, Remarks on convergence of Feynman path integrals, Duke Math. J. 47, 559–600 (1980).
42. D. Fujiwara, The Feynman path integral as an improper integral over the Sobolev space. Journées "Équations aux Dérivées Partielles" (Saint Jean de Monts, 1990), Exp. No. XIV, 15 pp., École Polytech., Palaiseau, (1990).
43. D. Fujiwara, N. Kumano-go Smooth functional derivatives in Feynman path integrals by time slicing approximation, Bull. Sci. Math. 129, 57–79 (2005).
44. I. M. Gel′fand, A. M. Yaglom. Integration in functional spaces, J. Math. Phys. 1, 48–69 (1960) (transl. from Usp. Mat. Nauk. 11, Pt. 1; 77–114 (1956)).
45. A. Hahn, The Wilson loop observables of Chern-Simons theory on $\mathbb{R}^3$ in axial gauge, Commun. Math. Phys. 248 (3), 467–499 (2004).
46. A. Hahn, Chern-Simons models on $S^2 \times S^1$, torus gauge fixing, and link invariants I, J. Geom. Phys. 53 (3), 275–314 (2005).
47. T. Hida, H.H. Kuo, J. Potthoff, L. Streit, *White Noise* Kluwer, Dordrecht (1995).
48. L. Hörmander, Fourier integral operators I. Acta Math., 127(1): 79–183, 1971.
49. T. Ichinose, Path integrals for the Dirac equation, Sugaku Exp. **6**, 15–31, 1993
50. N. Ikeda, S. Manabe, Asymptotic formulae for stochastic oscillatory integrals. Asymptotic problems in probability theory: Wiener functionals and asymptotics (Sanda/Kyoto, 1990), 136–155, Pitman Res. Notes Math. Ser., 284, Longman Sci. Tech., Harlow, 1993.
51. K. Itô. Wiener integral and Feynman integral. Proc. Fourth Berkeley Symposium on Mathematical Statistics and Probability. Vol. 2, pp. 227–238, California Univ. Press, Berkeley, 1961.
52. K. Itô. Generalized uniform complex measures in the hilbertian metric space with their applications to the Feynman path integral. Proc. Fifth Berkeley Symposium on Mathematical Statistics and Probability. Vol. 2, part 1, pp. 145–161, California Univ. Press, Berkeley, 1967.
53. G.W. Johnson, M.L. Lapidus, The Feynman integral and Feynman's operational calculus. Oxford University Press, New York, 2000.
54. M. Kac. On distributions of certain Wiener functionals. Trans. Amer. Math. Soc. 1–13, 65 (1949).
55. M. Kac. On some connections between probability theory and differential and integral equations. Proceedings of the Second Berkeley Symposium on Mathematical Statistics and Probability, 1950, pp. 189–215. University of California Press, Berkeley and Los Angeles, 1951.
56. G. Kallianpur, D. Kannan, R.L. Karandikar. Analytic and sequential Feynman integrals on abstract Wiener and Hilbert spaces, and a Cameron Martin Formula. Ann. Inst. H. Poincaré, Prob. Th. **21**, 323–361, 1985.
57. O.G. Smolyanov, A.Yu. Khrennikov, The central limit theorem for generalized measures on infinite-dimensional spaces. (Russian) Dokl. Akad. Nauk SSSR 281 (1985), no. 2, 279–283.
58. V. N. Kolokoltsov, Semiclassical analysis for diffusions and stochastic processes, Lecture Notes in Mathematics, 1724. Springer-Verlag, Berlin, (2000).

59. T. Kuna, L. Streit, W. Westerkamp. Feynman integrals for a class of exponentially growing potentials. J. Math. Phys., 39 (9): 4476–4491, 1998.
60. S. Leukert, J. Schäfer, A Rigorous Construction of Abelian Chern-Simons Path Integral Using White Noise Analysis, Reviews in Math. Phys. 8, 445–456 (1996).
61. P. Malliavin, S. Taniguchi, Analytic functions, Cauchy formula, and stationary phase on a real abstract Wiener space. J. Funct. Anal. 143 (1997), no. 2, 470–528.
62. V. Mandrekar. Some remarks on various definitions of Feynman integrals, in Lectures Notes Math., Eds K. Jacob Beck, pp. 170–177, 1983.
63. V.P. Maslov, A.M. Chebotarev, Processus à sauts et leur application dans la mécanique quantique, In: S. Albeverio et al. (ed) Feynman path integrals. Springer Lecture Notes in Physics 106, 58–72 (1979).
64. E. Nelson. Feynman integrals and the Schrödinger equation. J. Math. Phys. **5**, 332–343, 1964.
65. J. Rezende, The method of stationary phase for oscillatory integrals on Hilbert spaces, Comm. Math. Phys. 101, 187–206 (1985).
66. H. Sugita, S. Taniguchi, Oscillatory integrals with quadratic phase function on a real abstract Wiener space. J. Funct. Anal. 155 (1998), no. 1, 229–262.
67. H. Thaler. Solution of Schrödinger equations on compact Lie groups via probabilistic methods. Potential Anal. **18**, n.2, 119–140, 2003.
68. A. Truman, The polygonal path formulation of the Feynman path integral, In: S. Albeverio et al. (ed) Feynman path integrals. Springer Lecture Notes in Physics 106, (1979).
69. A. Truman, The Feynman maps and the Wiener integral. J. Math. Phys. **19**, 1742–1750, 1978.
70. E. Witten, Quantum field theory and the Jones polynomial, Commun. Math. Phys. 121, 353–389 (1989).
71. T. Zawstawniak, Path integrals for the Dirac-equation - some recent developments in mathematical theory, pp. 243–264 in K. D. Elworthy, J. C. Zambrini, eds., Stochastic analysis, path integration and dynamics, Pitman Res. Notes **200**, 1989.

Ambit Processes; with Applications to Turbulence and Tumour Growth

Ole E. Barndorff-Nielsen and Jürgen Schmiegel

Thiele Centre for Applied Mathematics in Natural Science, Department of Mathematical Sciences, University of Aarhus, DK-8000 Aarhus, Denmark

Summary. The concept of ambit processes is outlined. Such stochastic processes are of interest in spatio-temporal modelling, and they play a central role in recent studies of velocity fields in turbulence and of the growth of cancer tumours. These studies are reviewed, and some open problems are outlined.

Keywords: Brownian sheet, Lévy basis, normal inverse Gaussian, quadratic variation, refined similarity hypotheses, spatio-temporal modelling, stochastic differentials, stochastic integration, stochastic intermittency, tumour growth, turbulence

1 Introduction

The concept of ambit processes discussed in this paper arose out of a current study (Barndorff-Nielsen and Schmiegel (2005), Schmiegel et al. (2006), Barndorff-Nielsen and Schmiegel (2004), Schmiegel et al. (2004) and Schmiegel (2005)) the ultimate aim of which is to build a realistic stochastic process model of 3-dimensional turbulent velocity fields, in the spirit of Kolmogorov's phenomenological theory (Frisch (1995)) – and beyond. Besides applications to turbulence, the concept has also been used in modelling the growth of cancer tumours (Schmiegel (2006)), and it should be of interest to other fields as well.

Section 2 outlines the idea of ambit processes and lists a number of basic questions that need to be resolved in order to have a fullfledged stochastic analysis theory for such processes. In some important special settings, relevant for the turbulence context, the questions can be answered positively. Section 3 provides some background on the physics of turbulence while Section 4 discusses the phenomenology of turbulence. We then, in Section 5, turn to the formulation of a stochastic modelling framework for the velocity and intermittency fields, using the idea of ambit sets, and we outline how it is possible within this framework to capture main features of the phenomenological theory. Applications to cancer growth are briefly indicated in Section 6. Section 7 concludes.

2 Ambit Processes

In this Section we consider a rather general type of spatio-temporal processes that we shall refer to as *ambit processes*. We do not, at present, have a strict mathematical specification of what should be called an ambit process, but processes of the kind we have in mind would seem to be of interest in a variety of situations, and have in fact been applied not only in turbulence (cf. Section 5) but also for modelling cancer growth (see Section 6).

2.1 On Spatio-Temporal Processes

Let t denote time and σ a point in some space $\mathcal{S}$. To each point $(t, \sigma) \in \mathbb{R} \times \mathcal{S}$ let there be associated a random variable $Y_t(\sigma)$. Let $\omega(w) = (t(w), \sigma(w))$, where $-\infty < w < \infty$, be a smooth curve in space-time, such that $w \to t(w)$ is nondecreasing, and let $X_w = Y_{t(w)}(\sigma(w))$. We assume that $X = \{X_w\}_{w \in \mathbb{R}}$ is welldefined as a stochastic process. Unless otherwise specified we let $\mathcal{S} = \mathbb{R}$.

A key question is when the quadratic variation $[X]$ of X is well-defined in the sense of being a stochastic process such that

$$[X]_w = \mathrm{p} - \lim \sum \left(X_{w_j} - X_{w_{j-1}}\right)^2 \tag{2.1}$$

for any sequence of subdivisions $0 = w_0 < w_1 < \cdots < w_j = w$ with $\max(w_j - w_{j-1}) \to 0$. Intimately connected to this is the question of whether it is possible to define stochastic differentials $\mathrm{d}X_w$ and an associated symbolic calculus under which $(\mathrm{d}X_w)^2 = \mathrm{d}[X]_w$.

In settings such that $X_w = Y_{t(w)}(\sigma(w))$ is a semimartingale or a linear combination of semimartingales the existence of $[X]$ and of such differentials is of course ensured.

2.2 Ambit Sets and Lévy Bases

Turning now to a more specific setting, suppose that to each point (t, σ) is associated a set $A_t(\sigma)$, which we refer to as an *ambit set*. We take $A_t(\sigma)$ to be of the form

$$A_t(\sigma) = \left\{(s, \rho) : s \le t, \sigma - c_t^-(s; \sigma) \le \rho \le \sigma + c_t^+(s; \sigma)\right\} \tag{2.2}$$

for some nonnegative functions $c_t^-(s; \sigma)$ and $c_t^+(s; \sigma)$.

A particularly simple case is that of a *homogeneous* family of ambit sets where

$$A_t(\sigma) = \{(s, \rho) : (s - t, \rho - \sigma) \in A_0(0)\} \tag{2.3}$$

in which case $c_t^-(s; \sigma)$ and $c_t^+(s; \sigma)$ are independent of σ and of the form

$$c_t^\pm(s; \sigma) = c_{t-s}^\pm. \tag{2.4}$$

We write the cumulant function of an arbitrary random variable X as

$$\mathrm{C}\{\zeta \ddagger X\} = \log \mathrm{E}\left\{e^{i\zeta X}\right\} \tag{2.5}$$

and denote the m-th order cumulant of X by $c_m(X)$, i.e.

$$c_m(X) = i^m \frac{\mathrm{d}^m \mathrm{C}\{\zeta \ddagger X\}}{\mathrm{d}\zeta^m}. \tag{2.6}$$

Let L be a *Lévy basis*, i.e. an independently scattered random measure whose values are infinitely divisible. Then L has a Lévy-Khintchine representation

$$\mathrm{C}\{\zeta \ddagger L(B)\} = i\zeta a(B) - \frac{1}{2}\zeta^2 b(B) + \int_{\mathbb{R}} \{e^{i\zeta x} - 1 - i\zeta x \mathbf{1}_{[-1,1]}(x)\}\mu(\mathrm{d}x, B) \tag{2.7}$$

where a is a signed measure, b is a measure, and $\mu(\mathrm{d}x, B)$ is (for fixed B) a Lévy measure on $\mathbb{R}$ and a measure for fixed $\mathrm{d}x$. Heuristically it is useful to express (2.7) in infinitesimal form as

$$\mathrm{C}\{\zeta \ddagger L(\mathrm{d}z)\} = ia(\mathrm{d}z) - \frac{1}{2}\zeta^2 b(\mathrm{d}z) + \int_{\mathbb{R}} \left\{e^{i\zeta x} - 1 - i\zeta x \mathbf{1}_{[-1,1]}(x)\right\} \mu(\mathrm{d}x, \mathrm{d}z). \tag{2.8}$$

If the Lévy basis L is such that $L(B)$ is Poisson distributed for all B then L is a *Poisson basis*. In this case the generalised Lévy measure is of the form $\mu(\mathrm{d}x, B) = \mathrm{Leb}(B)\delta_1(\mathrm{d}x)$ where Leb denotes Lebesgue measure and δ_1 is the Dirac measure at 1.

The Lévy basis is said to be *factorisable* provided μ factorises as

$$\mu(\mathrm{d}x, \mathrm{d}z) = \nu(\mathrm{d}x)\, c(\mathrm{d}z) \tag{2.9}$$

for some σ-finite measure $c(\mathrm{d}z)$ and where ν is a Lévy measure on $\mathbb{R}$. If, moreover, a, b and c are proportional to Lebesgue measure then L is called *homogeneous*.

The *Brownian sheet* is the homogeneous Lévy basis on $\mathbb{R}^2$ with $a(\mathrm{d}z) = 0$, $b(\mathrm{d}z) = \mathrm{Leb}(\mathrm{d}z)$ and $\mu(\mathrm{d}x, \mathrm{d}z) = 0$ in (2.8).

Remark 1. Integration of deterministic functions with respect to Lévy bases is discussed in detail in Rajput and Rosinski (1989). Here we shall need more general types of integration. However, for the time being we shall argue under the presumption that all the integrals and differentials, and the manipulations with these, are rigourously justifiable, taking up the questions of rigour briefly in Section 2.5.

2.3 Ambit Processes

Let $\{Y_t(\sigma)\}_{t\in\mathbb{R}}$ be a spatio-temporal stochastic process of the form

$$Y_t(\sigma) = \mu + \int_{A_t(\sigma)} g(t-s,\rho-\sigma) I_s(\rho) L(\mathrm{d}s\mathrm{d}\rho) + \int_{D_t(\sigma)} h(t-s,\rho-\sigma) J_s(\rho)\, \mathrm{d}s\mathrm{d}\rho \tag{2.10}$$

where μ is a constant, $\{A_t(\sigma) : (t,\sigma) \in \mathbb{R}^2\}$ and $\{D_t(\sigma) : (t,\sigma) \in \mathbb{R}^2\}$ are families of ambit sets, g and h are damping functions (ensuring the convergence of the integrals), $I_s(\sigma)$ and $J_s(\sigma)$ are random fields on $\mathbb{R}^2$, and L is a Lévy basis on $\mathbb{R}^2$. A related type of process $\{Y_t(\sigma)\}_{t\geq 0}$ defines

$$Y_t(\sigma) = Y_0(\sigma) + \int_{A_t^+(\sigma)} g(t-s,\rho-\sigma) I_s(\rho) L(\mathrm{d}s\mathrm{d}\rho) + \int_{D_t^+(\sigma)} h(t-s,\rho-\sigma) J_s(\rho)\, \mathrm{d}s\mathrm{d}\rho \tag{2.11}$$

where $A_t^+(\sigma) = \{(s,\rho) \in A_t(\sigma) : s \geq 0\}$ and $D_t^+(\sigma) = \{(s,\rho) \in D_t(\sigma) : s \geq 0\}$. We refer to processes of these kinds as *ambit processes*, and we say that such a process is of *Brownian type* if L is the Brownian sheet BS and of *shot noise type* in case L is a pure jump basis (i.e. $a = b = 0$ in (2.7)).

Now suppose that L is Brownian sheet and, for simplicity, that the ambit sets are homogeneous (cf. (2.3)–(2.4)) and $A_t(\sigma) = D_t(\sigma)$. Then (2.10) may be written

$$Y_t(\sigma) = \mu + \int_{-\infty}^{t} \int_{\sigma - c_{t-s}^-}^{\sigma + c_{t-s}^+} g(t-s,\rho-\sigma) I_s(\rho) BS(\mathrm{d}s\mathrm{d}\rho) + \int_{-\infty}^{t} \int_{\sigma - c_{t-s}^-}^{\sigma + c_{t-s}^+} h(t-s,\rho-\sigma) J_s(\rho)\, \mathrm{d}s\mathrm{d}\rho. \tag{2.12}$$

In particular, if

$$c_{t-s}^{\pm} = c^{\pm}(t-s) \tag{2.13}$$

for some nonnegative constants c^- and c^+ (a choice motivated in the turbulence context, see Section 5.1) then

$$Y_t(\sigma) = \mu + \int_{-\infty}^{t} \int_{\sigma - c^-(t-s)}^{\sigma + c^+(t-s)} g(t-s,\rho-\sigma) I_s(\rho) BS(\mathrm{d}s\mathrm{d}\rho) + \int_{-\infty}^{t} \int_{\sigma - c^-(t-s)}^{\sigma + c^+(t-s)} h(t-s,\rho-\sigma) J_s(\rho)\, \mathrm{d}s\mathrm{d}\rho. \tag{2.14}$$

Note that if $I_s(\rho)$ and $J_s(\rho)$ are stationary processes in s for fixed ρ then $Y_t(\sigma)$ is a stationary process in t for fixed σ.

2.4 Lagrangian Dynamics

We proceed to discuss associated questions of dynamics, for processes $X_w = Y_{t(w)}(\sigma(w))$, as introduced in Section 2.1.

Figure 1 illustrates the dynamics of X_w along the curve $(t(w), \sigma(w))$ for arbitrary ambit sets $A_{t(w)}(\sigma(w))$. We assume that $J = I^2$ (this setting is sufficient for the applications to turbulence that will be reviewed later). Further, for notational simplicity we let $\mu = 0$ and we restrict consideration to the specification (2.14).

The process X_w may be written as

$$\begin{aligned} X_w = & \int_{-\infty}^{t} \int_{-\infty}^{\sigma + c^+(t-s)} g(t-s, \rho - \sigma) I_s(\rho) BS(\mathrm{d}s\mathrm{d}\rho) \\ & + \int_{-\infty}^{t} \int_{-\infty}^{\sigma + c^+(t-s)} h(t-s, \rho - \sigma) J_s(\rho) \mathrm{d}s\mathrm{d}\rho \\ & - \int_{-\infty}^{t} \int_{-\infty}^{\sigma - c^-(t-s)} g(t-s, \rho - \sigma) I_s(\rho) BS(\mathrm{d}s\mathrm{d}\rho) \\ & - \int_{-\infty}^{t} \int_{-\infty}^{\sigma - c^-(t-s)} h(t-s, \rho - \sigma) J_s(\rho) \mathrm{d}s\mathrm{d}\rho \end{aligned} \tag{2.15}$$

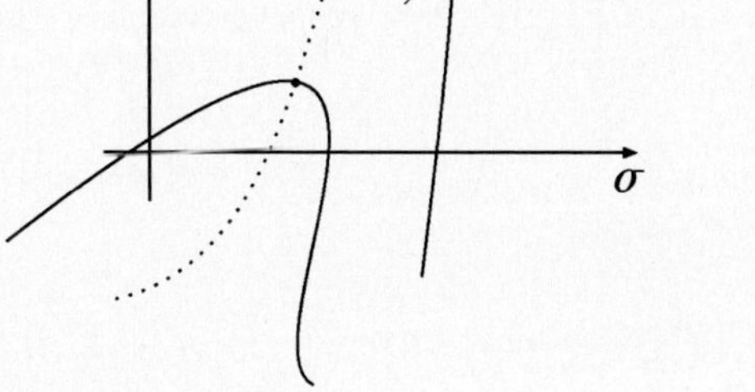

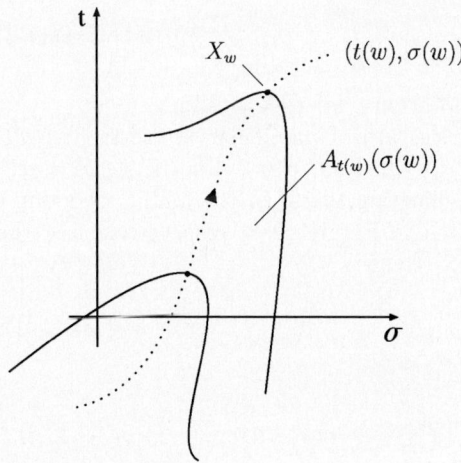

Fig. 1. Illustration of the dynamics of the process X_w (2.15) along the curve $(t(w), \sigma(w))$

with $(t,\sigma) = (t(w), \sigma(w))$. Continuing to argue formally, and suppressing in the notation the dependence of t and σ on w, we find

$$\begin{aligned} \mathrm{d}X_w = &\int_{-\infty}^{t} g\left(t-s, c^+(t-s)\right) I_s\left(\sigma + c^+(t-s)\right) BS\left(\mathrm{d}s\mathrm{d}_w\left(\sigma + c^+(t-s)\right)\right) \\ &- \int_{-\infty}^{t} g\left(t-s, -c^-(t-s)\right) I_s\left(\sigma - c^-(t-s)\right) BS\left(\mathrm{d}s\mathrm{d}_w\left(\sigma - c^-(t-s)\right)\right) \\ &+ \mathrm{d}R_w \end{aligned}$$

where

$$\begin{aligned} \frac{\mathrm{d}R_w}{\mathrm{d}w} = &\int_{-\infty}^{t}\int_{\sigma-c^-(t-s)}^{\sigma+c^+(t-s)} \mathrm{d}_w g\left(t-s, \rho-\sigma\right) I_s(\rho) BS(\mathrm{d}s\mathrm{d}\rho) \\ &+ \int_{-\infty}^{t}\int_{\sigma-c^-(t-s)}^{\sigma+c^+(t-s)} \mathrm{d}_w h\left(t-s, \rho-\sigma\right) J_s(\rho)\,\mathrm{d}s\mathrm{d}\rho \\ &+ \int_{-\infty}^{t} h\left(t-s, \sigma + c^+(t-s)\right) J_s\left(\sigma + c^+(t-s)\right) \mathrm{d}s\mathrm{d}_w\left(\sigma + c^+ t\right) \\ &- \int_{-\infty}^{t} h\left(t-s, \sigma - c^-(t-s)\right) J_s\left(\sigma - c^-(t-s)\right) \mathrm{d}s\mathrm{d}_w\left(\sigma - c^- t\right). \end{aligned}$$

Consequently,

$$\begin{aligned} \frac{(\mathrm{d}X_w)^2}{\mathrm{d}w} &= \left|\sigma' + c^+ t'\right| \int_{-\infty}^{t} g^2\left(t-s, c^+(t-s)\right) J_s\left(\sigma + c^+(t-s)\right) \mathrm{d}s \\ &\quad + \left|\sigma' - c^- t'\right| \int_{-\infty}^{t} g^2\left(t-s, -c^-(t-s)\right) J_s\left(\sigma - c^-(t-s)\right) \mathrm{d}s \\ &= \left|\sigma' + c^+ t'\right| \int_0^{\infty} g^2(s, c^+ s) J_s(\sigma + c^+ s)\mathrm{d}s \\ &\quad + \left|\sigma' - c^- t'\right| \int_0^{\infty} g^2(s, -c^- s) J_s(\sigma - c^- s)\mathrm{d}s. \end{aligned} \tag{2.16}$$

We adopt the notation ε for $(\mathrm{d}X_w)^2/\mathrm{d}w$.

Three special cases are of particular interest: (i) $t(w) = w$, $\sigma(w) = \sigma$ constant (ii) $t(w) = t$ constant, $\sigma(w) = w$ (iii) $t(w) = w$, $\sigma(w) = \sigma + c^- w$. The triangular specification (2.13) of the ambit set along the curves (i), (ii) and (iii) is illustrated in Figures 2–4, respectively. For these the expression (2.16) becomes respectively

$$\varepsilon_{time}(t,\sigma) = \int_0^{\infty} \left[c^- g^2(s, -c^- s) J_{t-s}\left(\sigma - c^- s\right) + c^+ g^2(s, c^+ s) J_{t-s}\left(\sigma + c^+ s\right)\right] \mathrm{d}s \tag{2.17}$$

$$\varepsilon_{space}(t,\sigma) = \int_0^{\infty} \left[g^2\left(s, -c^- s\right) J_{t-s}\left(\sigma - c^- s\right) + g^2\left(s, c^+ s\right) J_{t-s}\left(\sigma + c^+ s\right)\right] \mathrm{d}s \tag{2.18}$$

$$\varepsilon_{Lagr}(t,\sigma) = \left(c^- + c^+\right) \int_0^{\infty} g^2\left(s, c^+ s\right) J_{t-s}\left(\sigma + c^+ s\right) \mathrm{d}s. \tag{2.19}$$

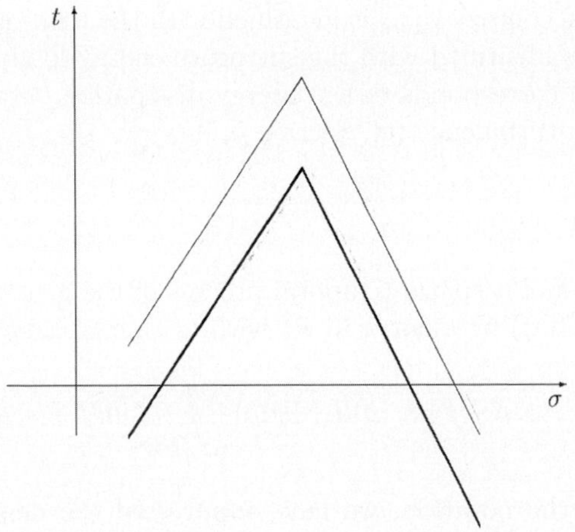

Fig. 2. Illustration of the dynamics of the process X_w (2.15) along the curve $t(w) = w$, $\sigma(w) = \sigma$ constant

Fig. 3. Illustration of the dynamics of the process X_w (2.15) along the curve $t(w) = t$ constant, $\sigma(w) = w$

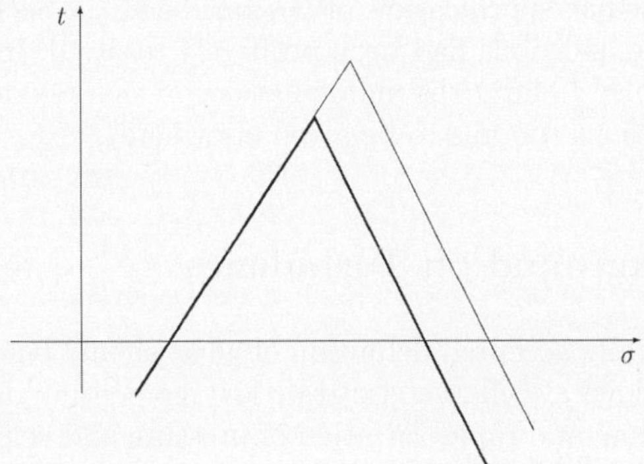

Fig. 4. Illustration of the dynamics of the process X_w (2.15) along the curve $(t(w), \sigma(w)) = (w, \sigma + c^{-}w)$

In the turbulence context ε_{time} is identified with the temporal energy dissipation and ε_{space} is identified with the surrogate energy dissipation (cf. Section 3). The case (iii) corresponds to the energy dissipation obtained by following the mean flow in turbulence (cf. Section 5.1).

2.5 Discussion

Let $Y_t(\sigma)$ be an ambit spatio-temporal process of the general form (2.10), let $\omega(w) = (t(w), \sigma(w))$ be a curve in $\mathbb{R}^2$ with $t(w)$ increasing in w, and let

$$X_w = \int_{A_t(\sigma)} g(t-s, \rho-\sigma) I_s(\rho) L(\mathrm{d}s\mathrm{d}\rho) + \int_{D_t(\sigma)} h(t-s, \rho-\sigma) J_s(\rho) \mathrm{d}s\mathrm{d}\rho \tag{2.20}$$

where again, in the notation, we have suppressed the dependence of t and σ on w. The following questions are of interest and do not seem answerable in any immediate fashion from existing results in the literature on spatio-temporal processes and on stochastic integration with respect to multiparameter martingales. (In the present context, key references to that literature are Cairoli and Walsh (1975), Khoshnevisan (2002), Klein and Giné (1975), Walsh (1986a,b) and Wong and Zakai (1974).)

(i) Under what conditions, especially on the ambit sets $A_t(\sigma)$ and $D_t(\sigma)$, does the quadratic variation $[X]$ exist (in the sense of (2.1)).
(ii) Under what conditions, especially on the ambit sets $A_t(\sigma)$ and $D_t(\sigma)$, is it possible meaningfully to define the differential $\mathrm{d}X$.
(iii) Related to (ii), what meaning should be given to an expression like $BS\left(\mathrm{d}s\mathrm{d}_w\left(\left(\sigma + c_t^+(s;\sigma)\right)\right)\right)$.
(iv) Supposing $L = BS$ (the Brownian sheet), when is $(\mathrm{d}X)^2 = \mathrm{d}[X]$ (or, otherwise put, when is $[X]_w = \int_0^w (\mathrm{d}X_s)^2$).
(v) When is X a linear combination of semimartingales.

For the particular specification of X considered in Section 2.4, we have argued as if these questions had been positively resolved. In fact, under mild assumptions on g, h, I and J the manipulations in that Section can be verified by direct calculations (details to be given elsewhere).

3 Some Background on Turbulence

There is no generally accepted definition of what should be called a turbulent flow. Turbulent flows are characterized by low momentum diffusion, high momentum convection, and rapid variation of pressure and velocity in space and time. Flow that is not turbulent is called laminar flow. The non-dimensional Reynolds number R characterizes whether flow conditions lead to laminar or turbulent flow. Increasing the Reynolds number increases the turbulent character and the limit of infinite Reynolds number is called the fully developed turbulent state.

Turbulence as part of hydrodynamics is governed by the Navier-Stokes equation which has been known since 1823. Its non-linear and non-local character does so far not allow to describe the wide range of turbulent phenomena from basic principles. Consequently, a great deal of phenomenological models have emerged that are based on and designed for certain aspects of turbulent dynamics. Most of these models can be classified according to the physical observable they address (see Section 4). The most prominent observables are the velocity field and the energy dissipation process.

In general, turbulence concerns the dynamics in a fluid flow of the three-dimensional velocity vector $\mathbf{u}(\mathbf{r},t) = (u_x(\mathbf{r},t), u_y(\mathbf{r},t), u_z(\mathbf{r},t))$ as a function of position $\mathbf{r} = (x,y,z)$ and time t. A derived quantity is the energy dissipation, defined as

$$\varepsilon(\mathbf{r},t) \equiv \frac{\nu}{2} \sum_{i,j=x,y,z} \left(\partial_i u_j(\mathbf{r},t) + \partial_j u_i(\mathbf{r},t)\right)^2 \tag{3.1}$$

describing the loss of kinetic energy due to friction forces characterized by the viscosity ν.

A pedagogical valuable illustration of a turbulent flow can be gained from the Kolmogorov cascade (Frisch (1995)). In this representation kinetic energy is injected into the flow at large scales through large scale forcing. Non-linear effects redistribute the kinetic energy towards smaller scales. This cascade of energy stops at small scales where dissipation transforms kinetic energy into heat. It is traditional to call the large scale L of energy input the integral scale and the small scale η of dissipation the dissipation scale or Kolmogorov scale. With increasing Reynolds number the fraction L/η increases, giving space for the so called inertial range $\eta \ll l \ll L$ where turbulent statistics are expected to have some universal character.

The resolution of all dynamically active scales in experiments is at present not achievable for the full three-dimensional velocity vector. Most experiments measure a time-series of one component u (in direction of the mean flow) of the velocity vector at a fixed single location $\mathbf{r}_0$ (in the stochastic framework we denote the spatial location by σ). Based on this restriction one defines the temporal energy dissipation

$$\varepsilon_{time}(\mathbf{r}_0,t) \equiv \frac{15\nu}{\overline{u}^2} \left(\frac{\mathrm{d}u(\mathbf{r_0},t)}{\mathrm{d}t}\right)^2, \tag{3.2}$$

where $\overline{u}$ denotes the mean velocity.

In going from (3.1) to (3.2) one assumes the flow to be stationary, homogeneous and isotropic. In this case (3.1) may be approximated as (Elsner and Elsner (1996))

$$\varepsilon_{space}(\mathbf{r},t) \equiv 15\nu \left(\frac{\partial u(\mathbf{r},t)}{\partial x}\right)^2 \tag{3.3}$$

which is believed to have similar statistical properties as the true energy dissipation at not too small scales. Discrepancies appear at small scales and are termed surrogacy effects. In particular, the autocorrelation function of the surrogate energy dissipation (3.3) shows an additional increase at small time scales (Cleve et al. (2003)).

The transformation of the spatial derivative in (3.3) to the temporal derivative in (3.2) is performed under the assumption of Taylor's Frozen Flow Hypothesis (Taylor (1938)) which states that spatial structures of the flow are predominantly swept by the mean velocity $\overline{\mathbf{u}}$ without relevant distortion. Under this hypothesis, widely used in analyzing turbulent time series, spatial increments along the direction of the mean flow (in direction x) are expressed in terms of temporal increments

$$u_{t+s}(\mathbf{r}) - u_t(\mathbf{r}) = u_t(\mathbf{r} - \overline{\mathbf{u}}s) - u_t(\mathbf{r}). \tag{3.4}$$

Remark 2. The temporal energy dissipation (3.2) is expected to approximate the true energy dissipation (3.1) for stationary, homogeneous and isotropic flows. Nevertheless, the temporal energy dissipation contains for all flow conditions important statistical information about the turbulent velocity field.

4 Turbulence Phenomenology

The statistical analysis of a great variety of time series has revealed a number of universal stylized facts of homogeneous and isotropic turbulent flows. Here we restrict the discussion to the so-called intermittency and to the statistics associated with the Kolmogorov variable, leaving aside, among others, the important characterization of turbulent statistics in terms of scaling relations (Meneveau and Sreenivasan (1991) and Sreenivasan and Antonia (1997) and references therein). Scaling relations are expected to hold for fully developed turbulent flows while being hard to detect for small and moderate Reynolds number flows. Intermittency and universality of the statistics associated to the Kolmogorov variable are found for a much wider range of Reynolds numbers (Castaing et al. (1990), Vincent and Meneguzzi (1991), Barndorff-Nielsen et al. (2004), Stolovitzky et al. (1992), Zhu et al. (1995) and Hosokawa et al. (1994)).

4.1 Intermittency

Since the pioneering work of Kolmogorov (1962) and Obukhov (1962), intermittency of the turbulent velocity field is of major interest in turbulence research. From a probabilistic point of view, intermittency refers, in particular, to the increase of the non-Gaussian behaviour of the probability density function (pdf) of velocity increments with decreasing scale. A typical scenario is characterized by an approximate Gaussian shape for the large scales, turning to exponential tails for the intermediate scales and stretched exponential tails for dissipation scales (Castaing et al. (1990) and Vincent and Meneguzzi (1991)).

It was reported in Barndorff-Nielsen et al. (2004) that the evolution of the pdf of velocity increments for all amplitudes and all scales can be described within one class of analytically tractable distributions, the normal inverse Gaussian (NIG) distributions. This class of distributions equals the family of possible distributions at time $t = 1$ of the NIG Lévy process, which is defined as Brownian motion with drift subordinated by the inverse Gaussian Lévy process, i.e. the Lévy process of first passage times to constant levels of (another) Brownian motion. The Appendix provides a brief summary of the definition and properties of NIG laws.

The NIG laws and associated processes have found widespread application, particularly in finance, see for instance Barndorff-Nielsen (1998a,b), Barndorff-Nielsen and Shephard (2001), Barndorff-Nielsen and Shephard (2007), Øigård et al. (2005), Corsi et al. (2005), Carr et al. (2003), Forsberg (2002), Lindberg (2005), Eberlein and Prause (2002) and further references there, cf. also Shiryaev (1999) and Cont and Tankov (2004).

Figure 5 shows, as an example, the log densities of velocity increments $\Delta u_s = u_{t+s} - u_t$ measured in the atmospheric boundary layer for various time scales s. The solid lines denote the approximation of these densities within the class of NIG distributions. NIG distributions fit the empirical densities equally well for all time scales s.

A subsequent analysis of the observed parameters of the NIG distributions from many, widely different data sets with Reynolds numbers ranging from $R_\lambda = 80$ up to $R_\lambda = 17000$ (where R_λ is the Taylor based Reynolds number, see below) led to the formulation of a key universality law (Barndorff-Nielsen et al. (2004)): The temporal development of a turbulent velocity field has an intrinsic clock which depends on the experimental conditions but in terms of which the one-dimensional marginal distributions of the velocity differences become independent of the experimental conditions. Figure 6 provides an empirical validation of this. As a consequence, the collapse of pdf's immediately resulted in a substantially wider and more general reformulation of the concept of Extended Self Similarity (Benzi et al. (1993)) in terms of a stochastic equivalence class. For details we refer to Barndorff-Nielsen et al. (2004).

4.2 Kolmogorov's Refined Hypotheses

In 1962, Kolmogorov published two hypotheses (usually refered to as K62) about a quantity V that combines velocity increments, being a large scale quantity, and the energy dissipation, being a small scale quantity. The first hypothesis states that the pdf of the stochastic variable

$$V_r = \frac{\Delta u_t(r)}{(r\varepsilon_r)^{1/3}} \tag{4.1}$$

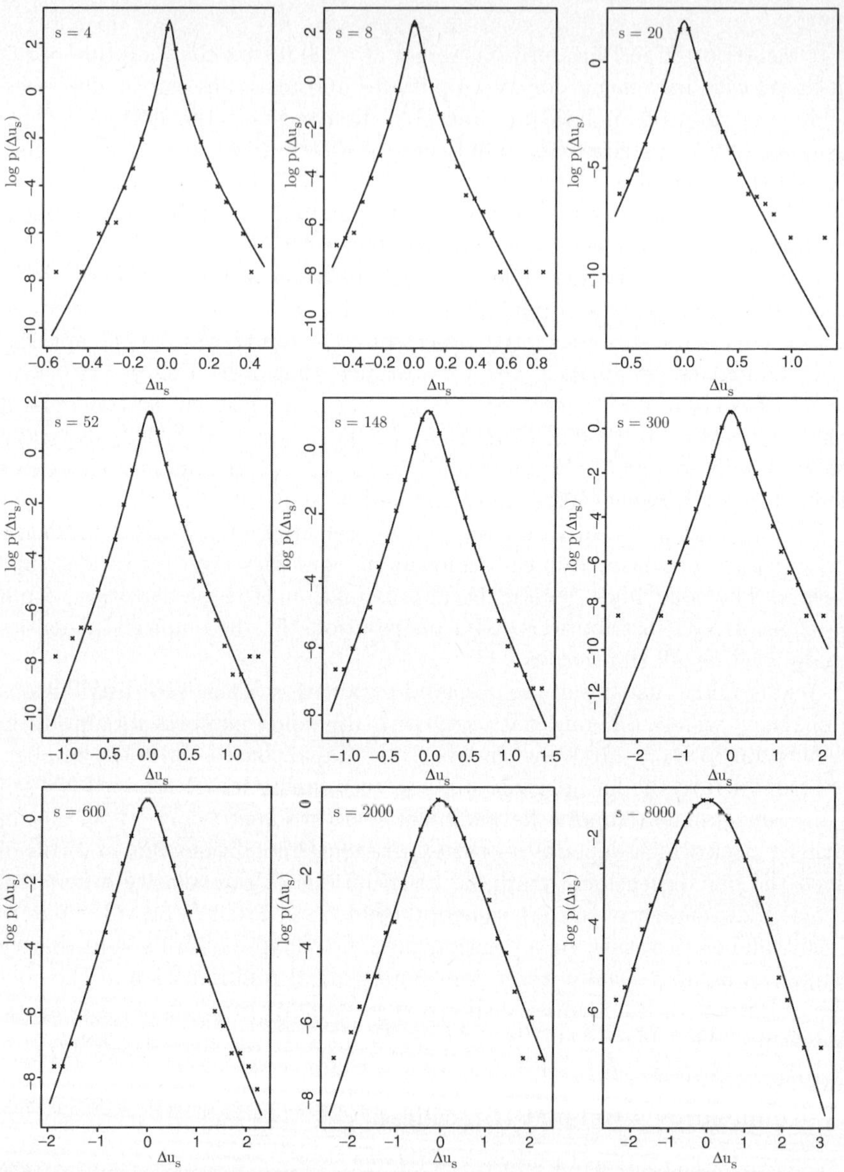

Fig. 5. Approximation of the pdf of velocity increments within the class of NIG distributions (solid lines, fitting by maximum likelihood) for data from the atmospheric boundary layer (kindly provided by K.R. Sreenivasan) with $R_\lambda = 17000$ and time scales $s = 4, 8, 20, 52, 148, 300, 600, 2000, 8000$ (in units of the finest resolution)

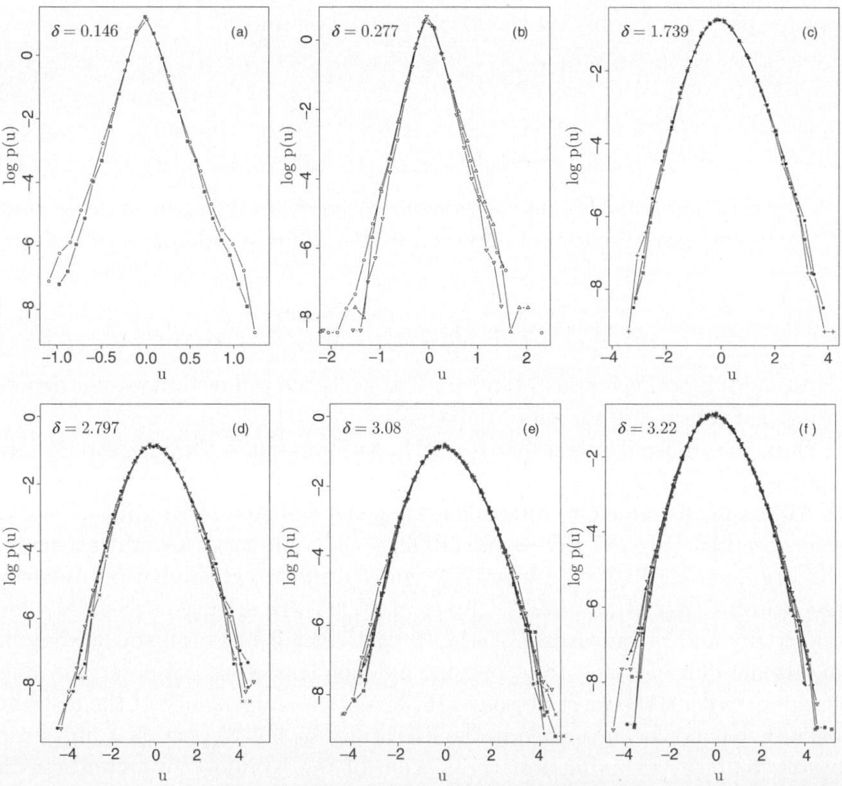

Fig. 6. Collapse of the densities of velocity increments at time scale s for various fixed values of the scale parameter $\delta(s)$ of the approximating NIG-distributions. The data are from the atmospheric boundary layer (data set (at) with $R_\lambda = 17000$, kindly provided by K.R. Sreenivasan), from a free jet experiment (data set (j) with $R_\lambda = 190$, kindly provided by J. Peinke), from a wind tunnel experiment (data set (w) with $R_\lambda = 80$, kindly provided by B.R. Pearson) and from a gaseous helium jet flow (data sets (h85), (h124), (h208), (h283), (h352), (h703), (h885), (h929), (h985) and (h1181) with $R_\lambda = 85, 124, 208, 283, 352, 703, 885, 929, 985, 1181$, respectively, kindly provided by B. Chabaud). The corresponding values of the time scales s (in units of the finest resolution of the corresponding data set) and the codes for the data sets are (a) $(s = 116, (\text{at}))$ (○), $(s = 4, (\text{h352}))$ (⊞), (b) $(s = 440, (\text{at}))$ (○), $(s = 8, (\text{j}))$ (△), $(s = 8, (\text{h929}))$ (▽), (c) $(s = 192, (\text{h885}))$ (■), $(s = 88, (\text{h352}))$ (⊞), $(s = 10, (\text{w}))$ (+), (d) $(s = 380, (\text{h885}))$ (■), $(s = 410, (\text{h929}))$ (▽), $(s = 350, (\text{h703}))$ (×), $(s = 340, (\text{h985}))$ (●), (e) $(s = 420, (\text{h703}))$ (×), $(s = 440, (\text{h929}))$ (▽), $(s = 180, (\text{h352}))$ (⊞), $(s = 270, (\text{h283}))$ (•), $(s = 108, (\text{h124}))$ (∗), $(s = 56, (\text{h85}))$ (⊠), (f) $(s = 470, (\text{h929}))$ (▽), $(s = 116, (\text{h124}))$ (∗), $(s = 60, (\text{h85}))$ (⊠), $(s = 188, (\text{h352}))$ (⊞), $(s = 470, (\text{h1181}))$ (▲), $(s = 140, (\text{h208}))$ (♦)

depends, for $r \ll L$, only on the local Reynolds number

$$\mathrm{Re}_r = r(r\varepsilon_r)^{1/3}/\nu. \tag{4.2}$$

Here,

$$\Delta u_t(r) = u_t(x + r, y, z) - u_t(x, y, z) \tag{4.3}$$

denotes the increment of one component of the velocity vector at scale r and $r\varepsilon_r$ is the integrated energy dissipation over a domain of linear size r

$$\varepsilon_r = \frac{1}{r}\int_{x_0-r/2}^{x_0+r/2} \varepsilon(\mathbf{r}, t)\mathrm{d}x. \tag{4.4}$$

The second hypothesis states that, for $\mathrm{Re}_r \gg 1$, the pdf of V_r does not depend on Re_r, either, and is therefore universal.

Note the unusual power 1/3 in (4.1). An immediate thinking would have expected power 1/2.

Although, for small r, an additional r dependence of the pdf of V_r has been observed (Stolovitzky et al. (1992)), the validity of several aspects of K62 has been verified experimentally and by numerical simulation of turbulence (Stolovitzky et al. (1992), Zhu et al. (1995), Hosokawa et al. (1994) and Stolovitzky and Sreenivasan (1994)). In particular it has been shown that the conditional densities $p(V_r|r\varepsilon_r)$ become independent of $r\varepsilon_r$ for a certain range of scales r within the inertial range. However, the universality of the distribution of V has not been verified in the literature. In this respect, it is important to note that the experimental verification of the Kolmogorov hypotheses is, with reasonable resolution of scales, restricted to temporal statistics and as such relies on the use of the temporal energy dissipation (3.2) instead of the true energy dissipation (3.1).

We take up the discussion of the Kolmogorov variable V in Section 5.2.

5 Stochastic Modelling of Turbulent Velocity Fields

The modelling framework we propose for the velocity field specifies this as an ambit process and incorporates the energy dissipation, also in the form of an ambit process, as a building block. As we shall discuss, basic stylized facts of turbulent statistics are captured by the model without specifying the degrees of freedom in all detail.

Remark 3. For the energy dissipation, discrete cascade processes are one of the most basic and successful models (Meneveau and Sreenivasan (1991), Jouault et al. (1999), Jouault et al. (2000) and Cleve and Greiner (2000)). However, these models lack translational invariance and moreover, they introduce an artifical and discrete hierarchy of scales. To overcome these drawbacks, ambit processes can be used as continuous and translation invariant generalisations of discrete cascade models (cf. Section 5.3).

5.1 A Spatio-Temporal Modelling Framework

We propose to model one component of the velocity vector in homogeneous and stationary turbulence as in (2.14)

$$u_t(\sigma) = \mu + \int_{-\infty}^{t} \int_{\sigma - c^-(t-s)}^{\sigma + c^+(t-s)} g(t-s, \rho - \sigma) I_s(\rho) BS(\mathrm{d}s\mathrm{d}\rho)$$
$$+\beta \int_{-\infty}^{t} \int_{\sigma - c^-(t-s)}^{\sigma + c^+(t-s)} h(t-s, \rho - \sigma) J_s(\rho) \mathrm{d}s\mathrm{d}\rho \qquad (5.1)$$

where μ and β are constants, c^+ and c^- are positive constants and we assume $J = I^2$, which is sufficiently general in the turbulence context. Here we adopt the notation u (instead of Y) for the velocity as is customary in the physics literature. The specific choice of a triangular ambit set corresponds to a constant maximum speed for information to arrive at a given site (σ, t). In this simple set-up the influence of an event sitting at $\rho < \sigma$ or $\sigma < \rho$ is experienced at σ with a delay of $(\sigma - \rho)/c^-$ or $(\rho - \sigma)/c^+$, respectively. The difference in the propagation velocities for $\sigma > \rho$ and $\sigma < \rho$ is due to the presence of a mean velocity. In general, interactions in the flow are due to pressure fluctuations traveling with the speed of sound c and interactions that are sweeping with the flow. Here we only deal with the simplest case where the sweeping velocity is assumed to be the mean velocity $\overline{u} > 0$. In this case

$$c^+ = c - \overline{u}, \quad c^- = c + \overline{u}. \qquad (5.2)$$

In this definition, density fluctuations are taken into account which corresponds to compressible flows. The ratio $\overline{u}/c$ is called the Mach number.

For incompressible flows, density fluctuations are neglected and this is encompassed by the model (5.1) in setting

$$c^+ = 0, \quad c^- = \overline{u}. \qquad (5.3)$$

The mean velocity $\overline{u}$ is a free parameter of the model related to μ by

$$\mu = \overline{u} - \beta c_1(J) \int_0^{\infty} \int_{-c^- s}^{c^+ s} h(s, \rho) \mathrm{d}s\mathrm{d}\rho. \qquad (5.4)$$

In the setting of stochastic differential equations of the Brownian semimartingale type (5.1) the quantity $[\mathrm{d}u_t(\sigma)]^2/\mathrm{d}t$ (2.17) is the natural analogue of the squared first order derivative of the velocity, which in the classical formulation is taken to express the temporal local energy dissipation (3.2) (up to a constant pre-factor). In a similar reasoning, $[\mathrm{d}u_t(\sigma)]^2/\mathrm{d}\sigma$ (2.18) may be identified with (3.3) (up to a constant pre-factor). In both cases, the local energy dissipation is independent of the second term in (5.1) which, importantly, allows to choose the function h and the constant β in (5.1) independently of the energy dissipation process.

The intermittency of the model, i.e. its non-Gaussian statistics, arises from both terms in (5.1). In particular the third order cumulant results in a polynomial of third order in β. Here we do not present the results for the full cumulant function of velocity increments. We rather specify the intermittent and turbulent character of the model in terms of the Taylor based Reynolds number (Frisch (1995)) defined as

$$R_\lambda = \frac{c_2(u)}{\nu\sqrt{\mathrm{E}\{\varepsilon_{space}\}}}. \tag{5.5}$$

Using (2.18) and (5.1), we calculate this most prominent characteristic of turbulence to be

$$R_\lambda = \frac{1}{\nu}\left(G_1 + G_2(\beta)\right) \tag{5.6}$$

where

$$G_1 = \sqrt{c_1(J)}\frac{\int_0^\infty \int_{-c^- s}^{c^+ s} g^2(s,\rho)\mathrm{d}s\mathrm{d}\rho}{\sqrt{\int_0^\infty \left(g^2(s,-c^- s) + g^2(s,c^+ s)\right)\mathrm{d}s}} \tag{5.7}$$

and $G_2(\beta) = \beta^2 G_2$ where

$$G_2 = \frac{\int_0^\infty \int_0^\infty \int_{-c^- s}^{c^+ s} \int_{-c^- s'}^{c^+ s'} h(s,\rho)h(s',\rho')\mathrm{Cov}\{J_s(\rho), J_{s'}(\rho')\}\mathrm{d}s\mathrm{d}s'\mathrm{d}\rho\mathrm{d}\rho'}{\sqrt{c_1(J)}\sqrt{\int_0^\infty \left(g^2(s,-c^- s) + g^2(s,c^+ s)\right)\mathrm{d}s}}, \tag{5.8}$$

Cov indicating covariance.

The first term (5.7) is independent of the weight function βh. Therefore we are able to increase R_λ by manipulating β and/or the function h without changing the statistics of the energy dissipation (see (2.17), (2.18) and (2.19)). In other words, the level of turbulence can be increased independently of the energy dissipation process. This type of behaviour has been observed for flows with strong shear where the intermittency of the velocity field (measured in terms of structure functions) shows an enhanced degree while the energy dissipation behaves in a universal fashion (Casciola et al. (2001)).

5.2 A Temporal Modelling Framework

The spatio-temporal (1 + 1)-dimensional model (5.1) and its generalization to higher dimensional modelling provides the general modelling framework for the turbulent velocity field. For a preliminary verification of the proposed modelling framework with experimental data we restrict ourselves to purely temporal statistics at a fixed spatial position σ, which are by now the type of data that are accessible with reasonable quality. For mathematical simplicity, we define a purely temporal version of (5.1) as

$$u_t = \mu + \int_{-\infty}^t g(t-s)I_s\mathrm{d}B_s + \beta\int_{-\infty}^t g(t-s)J_s\mathrm{d}s, \tag{5.9}$$

where B denotes Brownian motion. This model is in fact a limiting case of (5.1) with $h = g$, for $c^- = c^+ = c/2 \to \infty$ and $g(s,\rho) = c^{-1}s^{-1+c}g(s)$. The statistical properties of (5.9) are reported in more detail in Barndorff-Nielsen and Schmiegel (2005) where it was shown that a considerable part of its statistics are mediated by the structure of the model without specifying the intermittency J and the weight function g in all details. In the following we review the validation of the model (5.9) concerning the evolution of the density of velocity increments across time scales and the experimental verification of the statistics of the Kolmogorov variable.

In the setting of the model (5.9), the local energy dissipation can be identified with $[\mathrm{d}u_t]^2/\mathrm{d}t = J_t$ and consequently the quadratic variation $[u]_t$ is the stochastic analogue of the integrated energy dissipation.

As for the spatio-temporal model (5.1), the energy dissipation process does not depend on β. The constant β introduces a non-vanishing skewness in accordance with Kolmogorov's famous 4/5-the law (Kolmogorov (1941)). For the calculations below and for simulations we set $\beta = 0$, for convenience. This restriction does not essentially alter the results we derived by simulations.

Temporal Model and Shape Dynamics

As mentioned earlier, the density of empirical velocity increments evolves from heavy tails at small time scales s towards an approximate Gaussian shape at large scales s, in a manner that is well described within the class of NIG-distributions.

In comparing this to properties of the model (5.9) the first thing to note is that under (5.9) the asymptotic law of $u_t - u_0$ for $t \to \infty$ will not be Gaussian unless the intermittency field I is deterministic. This is in accordance with experimental findings, as illustrated by Figure 7 which shows the estimated (by maximum likelihood) asymmetry and steepness parameters χ and ξ of the fitted NIG laws, plotted in the NIG shape triangle (see Appendix). Note that the normal law occurs as a limiting case near $(\chi, \xi) = (0, 0)$. The data are from the atmospheric boundary layer (see also Figure 5).

To quantify the non-Gaussian character of the density of velocity increments in the model (5.9) we first focus on the standardized fourth order cumulant $\bar{c}_4$ which, in the absence of skewness $\beta = 0$, is the first order that distinguishes between a Gaussian shape and a heavytailed distribution. A specific result can be obtained by setting

$$g(t) = e^{-\gamma t} \tag{5.10}$$

and assuming J to be of Ornstein-Uhlenbeck-type, i.e.

$$J_t = \int_{-\infty}^{t} e^{-\lambda(t-s)} \mathrm{d}L_s \tag{5.11}$$

where L is the inverse Gaussian Lévy process. For brevity a process J of this form is refered to as an OU-IG process.

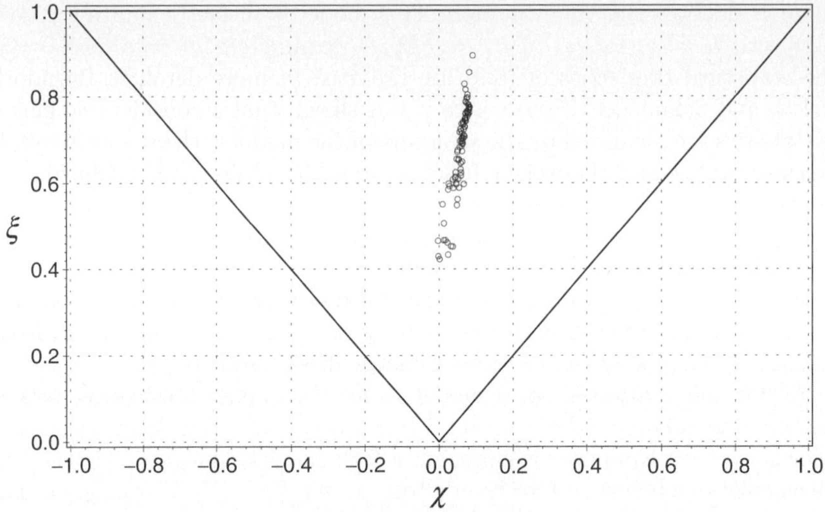

Fig. 7. Shape triangle for the evolution of the pdf of velocity increments across time scales (time scales increase from top to bottom) for data from the atmospheric boundary layer (kindly provided by K.R. Sreenivasan)

The parameters λ and γ control the autocorrelation functions of J and u, respectively. In this case we obtain

$$\lim_{t\to\infty} \overline{c}_4(u_t - u_0) = \frac{3c_2(L_1 - L_0)}{2c_1(L_1 - L_0)^2} \frac{\gamma\lambda}{2\gamma + \lambda}. \tag{5.12}$$

and

$$\lim_{t\to 0} \overline{c}_4(u_t - u_0) = \frac{3c_2(L_1 - L_0)}{2c_1(L_1 - L_0)^2}\lambda. \tag{5.13}$$

The heaviness of the tails of the pdf of velocity increments increases with increasing λ, i.e. with a faster decrease of correlations of the local energy dissipation. Qualitatively, the same behaviour is observed for turbulent flows where the heaviness of the tails of the pdf of velocity increments increases with increasing Reynolds number and with increasing intermittency exponent μ_2 (Cleve et al. (2004)), defined as $\mathrm{E}\{\varepsilon_0\varepsilon_t\} \sim t^{-\mu_2}$. (Due to this power-law behaviour, the assumption of ε $(= J)$ following an OU-IG process, for which $\mathrm{E}\{\varepsilon_0\varepsilon_t\} = c_2(L_1 - L_0)(2\lambda)^{-1}e^{-\lambda t} + c_1(L_1 - L_0)^2\lambda^{-2}$, is not a quite realistic approach for modelling the local energy dissipation. We come back to this point in Section 5.3.)

The corresponding results for moderate time scales are only accessible through numerical simulation. For the simulations we set $\beta = 0$ in (5.9) and model J as an OU-IG-process. Figure 8 shows the evolution of the probability densities of the simulated increments $u_t - u_0$ for various time

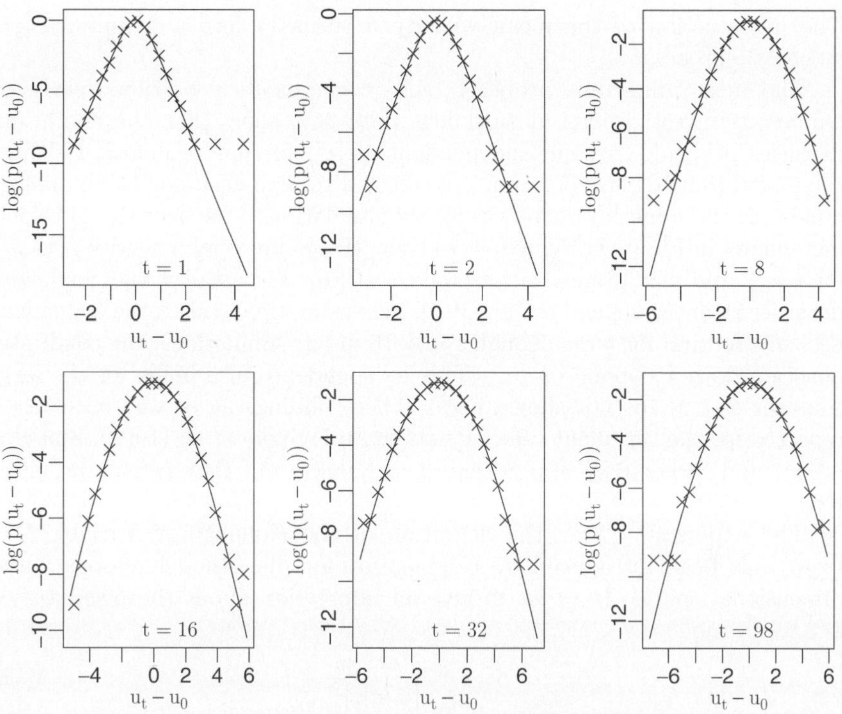

Fig. 8. Logarithm of the probability densities of the simulated increments $u_t - u_0$ (arbitrary units) under the model (5.9) with $t = 1, 2, 8, 16, 32, 98$ (in units of the finest resolution). The solid lines denote the approximation within the class of NIG distributions (fitting by maximum likelihood)

scales t. We clearly observe heavy tails for the small scales and an approximately Gaussian shape for the large scales. The solid lines denote the approximation of the densities within the class of NIG-distributions. The densities of $u_t - u_0$ qualitatively display the empirical findings about the evolution across scales of turbulent velocity increments shown in Figure 5.

Temporal Model and K62

As the second validation of the model (5.9) we briefly discuss K62 and its experimental verification. The original definition of V in (4.1) relates to spatial statistics which are not accessible in experiments. Therefore, the experimental verification of K62 has been performed in terms of temporal analysis. In the temporal model (5.9), the Kolmogorov variable may be defined as

$$V_t = \frac{u_t - u_0}{\{\bar{u}\,[u]_t\}^{1/3}}. \tag{5.14}$$

The introduction of the mean velocity $\bar{u}$ turns V_t into a non-dimensional stochastic process.

The most important property of V concerns its conditional statistics. Numerous investigations of turbulent data sets show that the conditional densities $p(V_t|[u]_t)$ become independent of $[u]_t$ for not too small t. Within the model (5.9) this observation is confirmed to high accuracy by simulations done with the same parameters as for the simulation of the densities of velocity increments in Figure 8. Figures 9–10 show the conditional densities $p(V_t|[u]_t)$ for $t = 2$ and $t = 16$ and various values of $[u]_t$. For small t, the conditional densities strongly depend on $[u]_t$. With increasing time scale t, the dependence gets smaller and for large enough t ($t \approx 16$ in our simulation), the conditional densities do not depend on $[u]_t$. This independence also holds for the larger time scales $t > 16$ (not shown here). These findings agree well with results reported for the turbulent velocity field in Stolovitzky et al. (1992), Zhu et al. (1995) and Stolovitzky and Sreenivasan (1994), and they reveal the gist of K62.

The exponent 1/3 in the definition of the Kolmogorov Variable V in (5.14) has been introduced by Kolmogorov for dimensional reasons (three-dimensional space). In order to give an impression about the peculiarity of 1/3 we define

$$V_{\alpha,t} = \frac{u_t - u_0}{\{\bar{u}\,[u]_t\}^{\alpha}}. \tag{5.15}$$

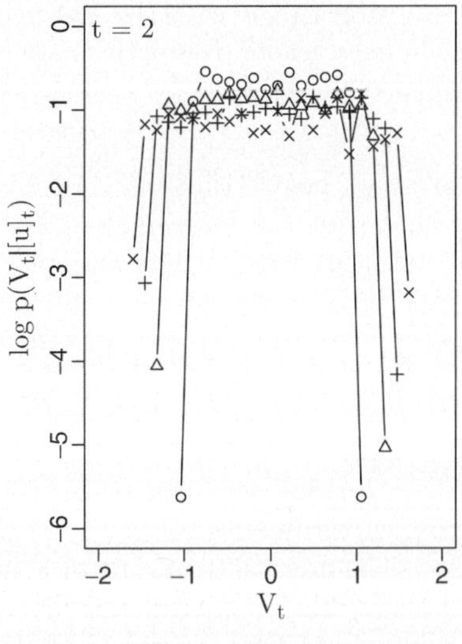

Fig. 9. Logarithm of the conditional densities $p(V_t|[u]_t)$ of the simulated Kolmogorov variable V_t under the model (5.9) for $t = 2$ (in units of the finest resolution) with $[u]_t^{1/3} = 0.45$ (∘), $[u]_t^{1/3} = 0.77$ (△), $[u]_t^{1/3} = 0.99$ (+) and $[u]_t^{1/3} = 1.20$ (×) (in arbitrary units)

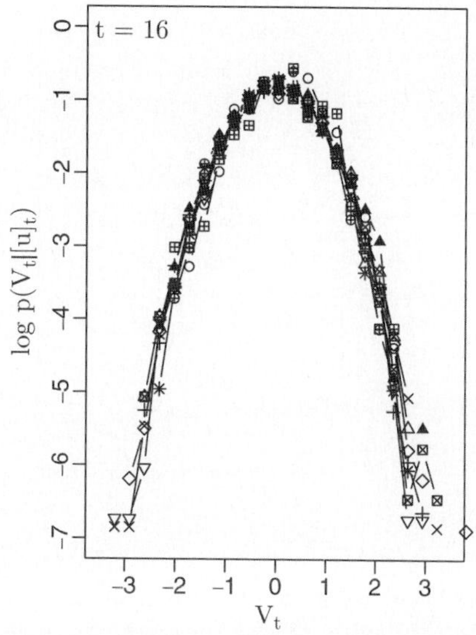

Fig. 10. Logarithm of the conditional density $p(V_t|[u]_t)$ of the simulated Kolmogorov variable V_t under the model (5.9) for $t = 16$ (in units of the finest resolution) with $[u]_t^{1/3} = 0.98$ (○), $[u]_t^{1/3} = 1.16$ (△), $[u]_t^{1/3} = 1.26$ (+), $[u]_t^{1/3} = 1.35$ (×), $[u]_t^{1/3} = 1.44$ (◇), $[u]_t^{1/3} = 1.53$ (▽), $[u]_t^{1/3} = 1.63$ (⊠), $[u]_t^{1/3} = 1.72$ (∗), $[u]_t^{1/3} = 1.81$ (▲), $[u]_t^{1/3} = 1.9$ (⊕) and $[u]_t^{1/3} = 2.0$ (⊞) (in arbitrary units)

which coincides with (5.14) for $\alpha = 1/3$. To assess the question of how much the independence of the conditional densities $p(V_t|[u]_t)$ on $[u]_t$ depends on the specific choice $\alpha = 1/3$ we analyse the dependence of the second-order conditional cumulants $c_2(V_{\alpha,t}|[u]_t)$ on $[u]_t$ for different values of α. Figure 11 compares $c_2(V_{\alpha,t}|[u]_t)$ for $\alpha = 1/2$ and $\alpha = 1/3$. The conditional cumulants are estimated from simulations with the same parameters as used for the simulation of velocity increments in Figure 8. For $\alpha = 1/2$ the conditional cumulants considerably decrease with increasing $[u]_t$. For $\alpha = 1/3$ the conditional cumulants stay roughly constant. For the moment, we have no explanation for why the model seems to be adapted to the exponent 1/3 (or at least to an exponent close to 1/3).

5.3 The Energy Dissipation Process

The basic ingredient of the model for the turbulent velocity is the intermittency process J. For the temporal model (5.9), J coincides with the temporal energy dissipation ε_{time}. For the more general spatio-temporal model (5.1), the energy dissipation is expressed as an integral over the weighted J process (see Section 2.4). In the following, we discuss a particular model for the energy

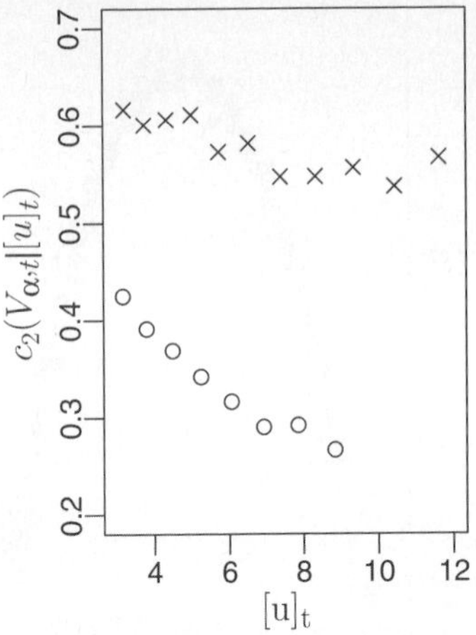

Fig. 11. Comparison of the simulated conditional variances $c_2(V_{\alpha,t}|[u]_t)$ for $\alpha = 1/3$ ($\times$) and $\alpha = 1/2$ ($\circ$) as a function of $[u]_t$ (in arbitrary units) with $t = 32$ (in units of the finest resolution)

dissipation process ε that is along the line of ambit processes (Schmiegel et al. (2006), Barndorff-Nielsen and Schmiegel (2004), Schmiegel et al. (2004) and Schmiegel (2005)).

We model the energy dissipation process as an ambit process of the exponential form

$$\varepsilon_t(\sigma) = \exp\left\{\int_{C_t(\sigma)} f(|t-s|, |\sigma-\rho|)L(\mathrm{d}s\mathrm{d}\rho)\right\}, \tag{5.16}$$

where L is a homogeneous and factorisable Lévy basis and f is an integrable deterministic function. Then we have the fundamental relation

$$\mathrm{E}\left\{\exp\left\{\int_C f(c)L(\mathrm{d}c)\right\}\right\} = \exp\left\{\int_C \mathrm{K}[f(c)]\mathrm{d}c\right\}, \tag{5.17}$$

where K denotes the cumulant function of $L(\mathrm{d}c)$, defined by

$$\ln \mathrm{E}\left\{\exp\left\{\xi L(\mathrm{d}c)\right\}\right\} = \mathrm{K}[\xi]. \tag{5.18}$$

The usefulness of (5.17) is obvious: it permits explicit calculation of the correlation functions of the integrated and f-weighted noise field $L(\mathrm{d}c)$ once the cumulant function K is known.

The generality of the model (5.16) is based on the possibility of choosing the constituents of the process $\varepsilon_t(\sigma)$ independently. The available degrees of freedom are an arbitrary infinitely divisible law for the Lévy basis L, the deterministic function f and the shape of the family C of ambit sets.

Despite its generality, the model is tractable enough to yield explicit expressions for arbitrary n-point correlations $\mathrm{E}\{\varepsilon_{t_1}(\sigma_1)\cdot\ldots\cdot\varepsilon_{t_n}(\sigma_n)\}$ in closed form.

Here, we focus on two-point correlators of order (n_1, n_2), defined as

$$c_{n_1,n_2}(\sigma_1, t_1; \sigma_2, t_2) \equiv \frac{\mathrm{E}\{\varepsilon_{t_1}(\sigma_1)^{n_1}\varepsilon_{t_2}(\sigma_2)^{n_2}\}}{\mathrm{E}\{\varepsilon_{t_1}(\sigma_1)^{n_1}\}\,\mathrm{E}\{\varepsilon_{t_2}(\sigma_2)^{n_2}\}}. \tag{5.19}$$

In the following we set $f \equiv 1$. This choice of the weight function f is motivated by the fact that two-point correlators obtained from a variety of turbulent data sets show the property of self-scaling (see below). Moreover, the freedom of choosing an arbitrary shape of the ambit set C is sufficient to model a wide range of two-point correlators of order $(1, 1)$ which are of primary interest in the present context.

Using (5.17), it is straightforward to show that

$$c_{n_1,n_2}(\sigma_1, t_1; \sigma_2, t_2) = \exp\left\{\overline{\mathrm{K}}[n_1, n_2]\int_{C_{t_1}(\sigma_1)\cap C_{t_2}(\sigma_2)} \mathrm{d}\sigma\mathrm{d}t\right\}, \tag{5.20}$$

with the abbreviation $\overline{\mathrm{K}}[n_1, n_2] = \mathrm{K}[n_1 + n_2] - \mathrm{K}[n_1] - \mathrm{K}[n_2] > 0$ (as follows from the Minkowski inequality). The important point here is the fact that the exponent in (5.20) factorizes into the Euclidean volume of the overlap of the two ambit sets times a factor depending only on the order (n_1, n_2). Thus we are able to rewrite (5.20) as a self-scaling relation of two point correlators of orders (n_1, n_2) and (m_1, m_2) (Schmiegel (2005))

$$c_{n_1,n_2}(\sigma_1, t_1; \sigma_2, t_2) = c_{m_1,m_2}(\sigma_1, t_1; \sigma_2, t_2)^{k[m_1,m_2;n_1,n_2]} \tag{5.21}$$

with the abbreviation

$$k[m_1, m_2; n_1, n_2] = \frac{\overline{\mathrm{K}}[n_1, n_2]}{\overline{\mathrm{K}}[m_1, m_2]}, \tag{5.22}$$

called the self-scaling exponent.

The self-scaling relation (5.21) implies that correlators of arbitrary order (n_1, n_2) are determined by the correlator of order $(1, 1)$ and the knowledge of the self-scaling exponents k of all orders. Note that the self-scaling exponents k only depend on the Lévy basis L.

For a given Lévy basis L it is possible to extract the shape of the ambit set directly from two-point correlators of order $(1, 1)$ which are accessible in experiments. For that we assume the ambit set $C_t(\sigma)$ to be of the form

$$C_t(\sigma) = \{(\rho, s) : \; t - T < s < t, \rho \in [\sigma - q(s - t + T), \sigma + q(s - t + T)]\} \tag{5.23}$$

where the function $q(s)$, defined on $[0,T]$, is nonnegative and decreasing. The constant T introduces a decorrelation time for the energy dissipation process. We further assume q and its inverse $q^{(-1)}$ to be differentiable. In this case it is easy to give necessary and sufficient conditions on spatial two-point correlators of order $(1,1)$ to be modelled by the Ansatz (5.16). From (5.20) it follows that

$$\begin{aligned}\frac{\partial}{\partial l}\ln c_{1,1}(\sigma,t;\sigma+l,t) &= \overline{\mathrm{K}}[1,1]\frac{\partial}{\partial l}\left(2\int_0^{q^{(-1)}(l/2)}(q(s)-l/2)\mathrm{d}s\right)\\ &= -\overline{\mathrm{K}}[1,1]q^{(-1)}(l/2) \qquad (5.24)\end{aligned}$$

and

$$\frac{\partial^2}{\partial l^2}\ln c_{1,1}(\sigma,t;\sigma+l,t) = -\frac{1}{2}\overline{\mathrm{K}}[1,1]\frac{\partial}{\partial l}q^{(-1)}(l/2). \qquad (5.25)$$

Thus, the Ansatz (5.16) together with a decreasing boundary $q(t)>0$ is able to model any twice differentiable spatial two point correlator that has the properties

$$\frac{\partial}{\partial l}\ln c_{1,1}(\sigma,t;\sigma+l,t) < 0 \qquad (5.26)$$

and

$$\frac{\partial^2}{\partial l^2}\ln c_{1,1}(\sigma,t;\sigma+l,t) > 0. \qquad (5.27)$$

Relation (5.24) has been applied to turbulent data in Schmiegel et al. (2004) where the shape of the ambit set has been extracted from scaling two-point correlators. As a consequence the higher order correlators are fixed and the three-point correlators have been successfully compared to experimental data.

In the temporal set-up (5.9) the intermittency process J is identified with the local energy-disspation ε and as such directly accessible to turbulent data analysis. For the more general spatio-temporal model (2.14) correlators of the energy-disspation can be expressed as weighted integrals over the correlators of the intermittency process J and as such can be modelled by suitably adapting the weight function g and the statistics of J.

6 Modelling Tumour Growth

The potential of processes of the type (5.16) for modelling a certain, well defined correlation structure may also be useful for modelling tumour dynamics (Schmiegel (2006) and Jensen et al. (2006)). The object of interest in that context is the star-shaped approximation of planar tumour tissue characterized by a radius function

$$R_t(\phi) = \max\{R : \mathbf{c}_0 + R\mathbf{e}_\phi \in S_t\} \qquad (6.1)$$

where S_t denotes the two-dimensional domain occupied by the tumour at time t, $\mathbf{c}_0$ denotes the centre of mass of the tumour at time $t = 0$ and $\mathbf{e}_\phi$ is the unit vector in direction $\phi \in [0, 2\pi)$.

Tumour profiles show structures at very different scales with strongly localized outbursts of different size. Due to the unrestricted growth of the tumour in in vitro experiments we can expect the profiles to be statistically isotropic. A comparison of these star-shaped profiles with the original profiles as observed in the experiment (Brú et al. (1998)) shows that (6.1) approximates the growing tumour to a high accuracy. For the star-shaped approximation, we neglect details of the tumour profiles where small regions of non-tumour tissue are surrounded by tumour cells.

For the stochastic modelling of profiles we normalize the radial function

$$r_t(\phi) \equiv \frac{R_t(\phi)}{\mathrm{E}\{R_t(\phi)\}}, \tag{6.2}$$

where $\mathrm{E}\{R_t(\phi)\}$ is the mean radius at time t, assumed to be independent of ϕ. Thus, $\mathrm{E}\{r_t(\phi)\} = 1$ for all times t. For the estimation of expectations, we perform spatial averaging.

Spatial correlators of star shaped tumour profiles have the form (Schmiegel (2006))

$$\ln\left(c_{n_1,n_2}(t,\phi;t,\phi+\Delta\phi)\right) = d_{n_1,n_2}(t) f_t(\Delta\phi) \mathbf{1}_{[0,\phi_0(t)]}(\Delta\phi) + b_{n_1,n_2}(t)\cos(\Delta\phi) \tag{6.3}$$

where the critical angle $\phi_0(t)$ confines the validity of the cosine behaviour for $\Delta\phi > \phi_0$. For $\Delta\phi < \phi_0$ deviations from the cosine behaviour occur and are denoted by f_t. The factors d and b are independent of $\Delta\phi$ and depend only on the order (n_1, n_2).

To account for the particular correlation structure (6.3) we propose an exponential ambit process of the type

$$\begin{aligned} r_t(\phi) = \exp\Bigg\{ a(t) \int_{t-T(t)}^{t-t_0(t)} \int_{\phi-\pi}^{\phi+\pi} \cos(\phi-\phi') BS(\mathrm{d}t'\mathrm{d}\phi') \\ + h(t) \int_{t-t_0(t)}^{t} \int_{\phi-q_t(t'-t+t_0(t))}^{\phi+q_t(t'-t+t_0(t))} BS(\mathrm{d}t'\mathrm{d}\phi') \Bigg\}, \end{aligned} \tag{6.4}$$

with cyclic definition in the angle and where BS is a Brownian sheet. The first term on the right hand side of (6.4) is responsible for the validity of the cosine law (second term in (6.3)) and the second term on the right hand side of (6.4) is associated with the deviations from the cosine law at small angular distances. We call the first term in the exponent on the right hand side of (6.4) the large scale term and the second term the small scale term.

The ambit set associated with the large scale term is a rectangle of the form $[t-T(t), t-t_0(t)] \times [\phi-\pi, \phi+\pi]$. The deterministic function $T(t)$ can be interpreted as the decorrelation time of the radius process and $t_0(t)$ expresses the decorrelation time of the small scale term.

The ambit set associated with the small scale term is assumed to be determined by a deterministic and monotonically decreasing function q_t defined on $[0, t_0(t)]$ and satisfying $q_t(t_0(t)) = 0$. These two parts of the ambit set are weighted differently according to the deterministic functions $a(t)\cos(\phi - \phi')$ and $h(t)$ for the large scale term and the small scale term, respectively.

Within the modelling framework (6.4) the two point correlators are of the specific form (6.3) where

$$b_{n_1,n_2}(t) = n_1 n_2 a(t)^2 \pi \left(T(t) - t_0(t)\right) \tag{6.5}$$

and the small scale amplitude $d_{n_1,n_2}(t)$ has the form

$$d_{n_1,n_2}(t) = n_1 n_2 h(t)^2 \tag{6.6}$$

and we identify f_t by

$$f_t(\Delta\phi) = V_t(\Delta\phi) = \int_0^{q_t^{(-1)}(\Delta\phi/2)} (2q_t(s) - \Delta\phi)\mathrm{d}s \tag{6.7}$$

where $V_t(\Delta\phi)$ is the Euclidean volume of the overlap of the ambit sets of the small scale terms separated by the angular distance $\Delta\phi$. The critical angle $\phi_0(t)$ is given by

$$\phi_0(t) = 2q_t(0), \tag{6.8}$$

and is independent of the order (n_1, n_2).

The modelling potential of the Ansatz (6.4) for the dynamics of tumour profiles lies in the fact that the cosine behaviour at large scales can be modelled independently of the deviations at the small scales. In particular, a suitable choice of the bounding function $q_t(s)$ allows to model any monotonically decreasing overlap $V_t(\phi)$ and, consequently, any monotonically decreasing deviation $d_{n_1,n_2}(t)f_t(\Delta\phi)$.

The assumption of a Brownian sheet in (6.4) is motivated by the implied order dependence of the amplitudes b and d in (6.5) and (6.6), respectively, and the fact that tumour profiles show self-scaling of spatial correlators (Schmiegel (2006))

$$c_{n_1,n_2}(t, \Delta\phi) = \left(c_{m_1,m_2}(t, \Delta\phi)\right)^{k_t[m_1,m_2;n_1,n_2]} \tag{6.9}$$

with self-scaling exponents k_t of the form

$$k_t[m_1, m_2; n_1, n_2] = \frac{n_1 n_2}{m_1 m_2}. \tag{6.10}$$

The self-scaling property (6.9) holds for all angular distances $\Delta\phi$ implying

$$\frac{d_{n_1,n_2}(t)}{d_{m_1,m_2}(t)} = \frac{b_{n_1,n_2}(t)}{b_{m_1,m_2}(t)} = \frac{n_1 n_2}{m_1 m_2} \tag{6.11}$$

in accordance with (6.5) and (6.6).

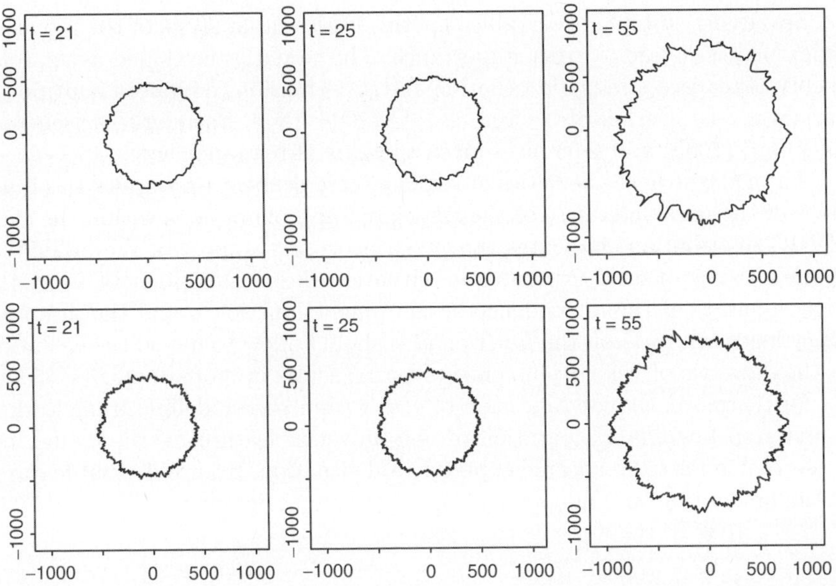

Fig. 12. Comparison of the simulated tumour profiles (bottom row) under the model (6.4) with the star-shaped tumour profiles (top row) at times $t = 21, 25, 55$ (arbitrary units)

Figure 12 shows a comparison of star-shaped brain tumour profiles with simulations of the model (6.4). For the estimation of the parameters used for the simulation we refer to Schmiegel (2006).

Remark 4. The modelling framework (6.4) has been defined for the normalized radius $r_t(\phi)$. However, it equally applies to the modelling of the non-normalized radius $R_t(\phi)$. The definition of correlators is invariant under rescaling with the mean radius. Going from $r_t(\phi)$ to $R_t(\phi)$ is equivalent to replacing $h(t)$ with $h(t) - \log(\mathrm{E}\{R_t(\phi)\})/f_t(0)$, keeping all other parameters of the model (6.4) unchanged.

7 Concluding Remarks

The modelling of the turbulent energy dissipation (5.16) and the turbulent velocity fields (5.1) and (5.9) within the class of ambit processes, as outlined in Section 5, poses various important questions, in addition to the purely mathematical problems listed in Section 2.5.

Of major interest is the identification of the parameters of the model with physical observables. For the temporal model (5.9) the intermittency process J is identified with the local energy dissipation and as such accessible to data analysis. For the spatio-temporal model (5.1), the energy dissipation is identified with a weighted integral of the intermittency process.

Specifying suitable observables for the statistical analysis of the intermittency process J are of great importance. The recently developed asymptotic theory of realised quadratic variation and its extension to realised multipower variation, see Barndorff-Nielsen et al. (2006) and Barndorff-Nielsen and Shephard (2005) and references given there, is of relevance here.

The collapse of the densities of velocity increments at time scales s as functions of the parameter $\delta(s)$ of the associated approximations within the class of NIG distributions indicates that $\delta(s)$ incorporates most of the individual characteristics of each experimental situation. From this point of view, the determination of the dependence of the weight function g and the intermittency field J in (5.1) on the function $\delta(s)$ should allow to model the evolution of the densities of velocity increments across scales in more detail.

Furthermore, the identification of $\delta(s)$ within the modelling framework is a first step towards a separation of non-universal features of the model, i.e. those that reflect the specific experimental situation, from universal features of the model that are independent of experimental details.

Appendix

A Normal Inverse Gaussian Distribution

The normal inverse Gaussian law, with parameters α, β, μ and δ, is the distribution on the real axis $\mathbf{R}$ having probability density function

$$p(x;\alpha,\beta,\mu,\delta) = a(\alpha,\beta,\mu,\delta) q\left(\frac{x-\mu}{\delta}\right)^{-1} K_1\left\{\delta\alpha q\left(\frac{x-\mu}{\delta}\right)\right\} e^{\beta x} \tag{A.1}$$

where $q(x) = \sqrt{1+x^2}$ and

$$a(\alpha,\beta,\mu,\delta) = \pi^{-1}\alpha \exp\left\{\delta\sqrt{\alpha^2-\beta^2} - \beta\mu\right\} \tag{A.2}$$

and where K_1 is the modified Bessel function of the third kind and index 1. The domain of variation of the parameters is given by $\mu \in \mathbf{R}$, $\delta \in \mathbf{R}_+$, and $0 \leq |\beta| < \alpha$. The distribution is denoted by $\mathrm{NIG}(\alpha,\beta,\mu,\delta)$.

If X is a random variable with distribution $\mathrm{NIG}(\alpha,\beta,\mu,\delta)$ then the cumulant generating function of X, i.e. $\mathrm{K}(\theta;\alpha,\beta,\mu,\delta) = \log \mathrm{E}\{e^{\theta X}\}$, has the simple explicit form

$$\mathrm{K}(\theta;\alpha,\beta,\mu,\delta) = \delta\{\sqrt{\alpha^2-\beta^2} - \sqrt{\alpha^2-(\beta+\theta)^2}\} + \mu\theta. \tag{A.3}$$

We note that the NIG distribution (A.1) has semiheavy tails; specifically,

$$p(x;\alpha,\beta,\mu,\delta) \sim \mathrm{const.}\, |x|^{-3/2} \exp\left(-\alpha|x| + \beta x\right), \ x \to \pm\infty. \tag{A.4}$$

The normal inverse Gaussian law can be characterized in terms of subordinated Brownian motion. For that, let B_t be a Brownian motion starting at the point μ and having constant drift β. Let Z_t be the inverse Gaussian

Lévy process, assumed independent of the process B_t. The inverse Gaussian Lévy process is defined as the Lévy process for which $Z_1 \overset{law}{=} Z$ and where the distribution of Z is the inverse Gaussian law whose probability density function is given by

$$(2\pi)^{-1/2}\delta e^{\delta\gamma}x^{-3/2}\exp\left\{-\left(\delta^2 x^{-1}+\gamma^2 x\right)/2\right\}.$$

This distribution is denoted $\mathrm{IG}(\delta,\gamma)$. Then, the process

$$X_t = B_{z_t} + \mu t$$

is also a Lévy process, termed the normal inverse Gaussian Lévy process, whose distribution at time $t=1$ is $\mathrm{NIG}(\alpha,\beta,\mu,\delta)$ where $\alpha=\sqrt{\beta^2+\gamma^2}$.

NIG shape triangle For some purposes it is useful, instead of the classical skewness and kurtosis quantities $\bar{c}_3$ and $\bar{c}_4$, to work with the alternative asymmetry and steepness parameters χ and ξ defined by

$$\chi = \rho\xi \tag{A.5}$$

and

$$\xi = [1+\bar{\gamma}]^{-1/2} \tag{A.6}$$

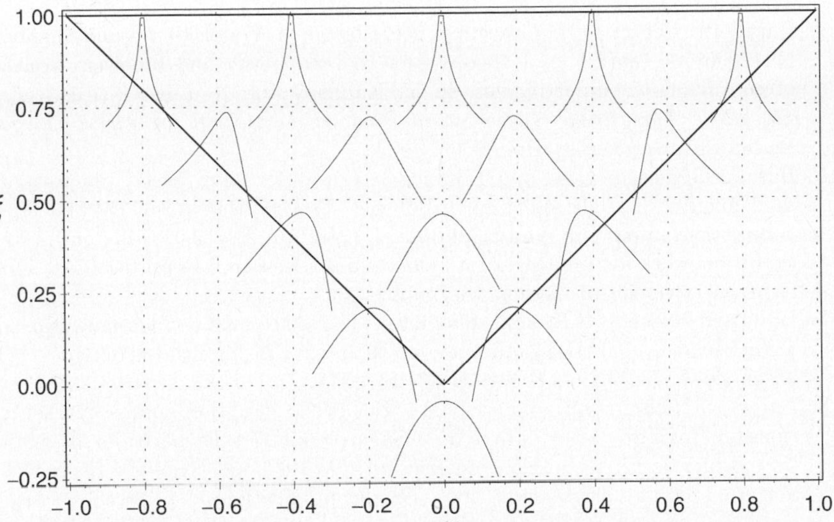

Fig. 13. The shape triangle of the NIG distributions with the log density functions of the standardized distributions, i.e. with mean 0 and variance 1, corresponding to the values $(\chi,\xi)=(\pm 0.8, 0.999)$, $(\pm 0.4, 0.999)$, $(0.0, 0.999)$, $(\pm 0.6, 0.75)$, $(\pm 0.2, 0.75)$, $(\pm 0.4, 0.5)$, $(0.0, 0.5)$, $(\pm 0.2, 0.25)$ and $(0.0, 0.0)$. The graphs of the log densities are placed at the corresponding values of (χ,ξ)

where $\bar{\gamma} = \delta\sqrt{\alpha^2 - \beta^2}$. Like $\bar{c}_3$ and $\bar{c}_4$, these parameters are invariant under location-scale changes and the domain of variation for (χ, ξ) is the *normal inverse Gaussian shape triangle*

$$\{(\chi, \xi) : -1 < \chi < 1, 0 < \xi < 1\}. \tag{A.7}$$

The distributions with $\chi = 0$ are symmetric, and the normal and Cauchy laws occur as limiting cases for (χ, ξ) near to $(0, 0)$ and $(0, 1)$, respectively. Figure 13 gives an impression of the shape of the NIG distributions for various values of (χ, ξ).

References

1. Barndorff-Nielsen, O.E. (1998a): Probability and Statistics; selfdecomposability, finance and turbulence. In L. Accardi and C.C. Heyde (Eds.): Proceedings of the Conference *"Probability towards 2000"*, held at Columbia University, New York, 2-6 October 1995. Berlin: Springer-Verlag. Pp. 47–57.
2. Barndorff-Nielsen, O.E. (1998b): Processes of normal inverse Gaussian type. *Finance and Stochastics* **2**, 41–68.
3. Barndorff-Nielsen, O.E., Blæsild, P. and Schmiegel, J. (2004): A parsimonious and universal description of turbulent velocity increments. *Eur. Phys. J. B* **41**, 345–363.
4. Barndorff-Nielsen, O.E., Graversen, S.E., Jacod, J., Podolskij, M. and Shephard, N. (2006): A central limit theorem for realised power and bipower variations of continuous semimartingales. In Y. Kabanov and R. Lipster (Eds.): *From Stochastic Analysis to Mathematical Finance, Festschrift for Albert Shiryaev.* Berlin: Springer-Verlag. Pp. 33–68.
5. Barndorff-Nielsen, O.E. and Schmiegel, J. (2004): Lévy based tempo-spatial modelling; with applications to turbulence. *Uspekhi Mat. Nauk* **159**, 63–90.
6. Barndorff-Nielsen, O.E. and Schmiegel, J. (2005): A stochastic differential equation framework for the turbulent velocity field. Research Report 2005-4. Thiele Centre, University of Aarhus. (Submitted.)
7. Barndorff-Nielsen, O.E. and Shephard, N. (2001): Non-Gaussian Ornstein-Uhlenbeck-based models and some of their uses in financial economics (with Discussion). *J. R. Statist. Soc.* **B 63**, 167–241.
8. Barndorff-Nielsen, O.E. and Shephard, N. (2005): Variation, jumps, market frictions and high frequency data in financial econometrics. Prepared for the invited symposium on Financial Econometrics, 9th World Congress of the Econometric Society, London, 20th August 2005. (Under publication.)
9. Barndorff-Nielsen, O.E. and Shephard, N. (2007): *Financial Volatility in Continuous Time.* Cambridge University Press. (To appear.)
10. Benzi, R., Ciliberto, S., Tripiccione, R., Baudet, C., Massaioli, F. and Succi, S. (1993): Extended self-similarity in turbulent flows. *Phys. Rev. E* **48**, R29-R32.
11. Brú, A., Pastor, J.M., Fernaud, I., Brú, I., Melle, S. and Berenguer, C. (1998): Super rough dynamics on tumor growth. *Phys. Rev. Lett.* **81** 4008–4011.
12. Cairoli, R. and Walsh, J.B. (1975): Stochastic integrals in the plane. *Acta Math.* **134**, 111–183.

13. Casciola, C.M., Benzi, R., Gualtieri, P., Jacob, B. and Piva, R. (2001): Double scaling and intermittency in shear dominated flows. *Phys. Rev. E* **65**, 015301(R).
14. Carr, P., Geman, H., Madan, D. and Yor, M. (2003): Stochastic volatility for Lévy processes. *Mathematical Finance* **13**, 345–382.
15. Castaing, B., Gagne, Y. and Hopfinger, E.J. (1990): Velocity probability density functions of high Reynolds number turbulence. *Physica D* **46**, 177–200.
16. Cleve, J. and Greiner, M. (2000): The markovian metamorphosis of a simple turbulent cascade model. *Phys. Lett. A* **273**, 104–108.
17. Cleve, J., Greiner, M. and Sreenivasan, K.R. (2003): On the effects of surrogacy of energy dissipation in determining the intermittency exponent in fully developed turbulence. *Europhys. Lett.* **61**, 756–761.
18. Cleve, J., Greiner, M., Pearson, B.R. and Sreenivasan, K.R. (2004): Intermittency exponent of the turbulent energy cascade. *Phys. Rev. E* **69**, 066316.
19. Cont, R. and Tankov, P. (2004): *Financial Modelling With Jump Processes.* London: Chapman & Hall/CRC.
20. Corsi, F., Kretschmer, U., Mittnik, S. and Pigorsch, C. (2005): The volatility of volatility. Unpublished working paper.
21. Eberlein, E. and Prause, K. (2002): The Generalized Hyperbolic Model: Financial Derivatives and Risk Measures. In H. Geman, D. Madan, S. Pliska and T. Vorst (Eds.): *Mathematical Finance – Bachelier Congress 2000.* Berlin: Springer-Verlag. Pp. 245–267.
22. Elsner, J.W. and Elsner, W. (1996): On the measurement of turbulence energy dissipation. *Meas. Sci. Technol.* **7**, 1334–1348.
23. Forsberg, L. (2002): *On the Normal Inverse Gaussian Distribution in Modelling Volatility in the Financial Markets.* Uppsala University Press.
24. Frisch, U. (1995): *Turbulence. The legacy of A.N. Kolmogorov.* Cambridge University Press.
25. Hosokawa, I., Van Atta, C.W. and Thoroddsen, S.T. (1994): Experimental study of the Kolmogorov refined similarity variable. *Fluid Dyn. Res.* **13**, 329–333.
26. Jensen, E.B.V, Jonsdottir, K.Y., Schmiegel, J. and Barndorff-Nielsen, O.E. (2006): Spatio-temporal modelling - with a view to biological growth. In B. Finkenstädt, L. Held and V. Isham (Eds.): Statistical Methods for Spatio-Temporal Systems, Monographs on Statistics and Applied Probability, Chapman & Hall/CRC.
27. Jouault, B., Lipa, P. and Greiner, M. (1999): Multiplier phenomenology in random multiplicative cascade processes. *Phys. Rev. E* **59**, 2451–2454.
28. Jouault, B., Greiner, M. and Lipa, P. (2000): Fix-point multiplier distributions in discrete turbulent cascade models. *Physica D* **136**, 125–144.
29. Khoshnevisan, D. (2002): *Multiparameter Processes. An Introduction to Random Fields.* Heidelberg: Springer.
30. Klein, R. and Giné, E. (1975): On quadratic variation of processes with Gaussian increments. *Ann. Prob.* **3**, 716–721.
31. Kolmogorov, A.N. (1941): Dissipation of energy in locally isotropic turbulence. *Dokl. Akad. Nauk. SSSR* **32**, 16–18.
32. Kolmogorov, A.N. (1962): A refinement of previous hypotheses concerning the local structure of turbulence in a viscous incompressible fluid at high Reynolds number, *J. Fluid Mech* **13**, 82–85.
33. Lindberg, C. (2005): Portfolio Optimization and Statistics in Stochastic Volatility Markets. Ph. D. Thesis, Chalmers University of Technology and Göteborg University.

34. Meneveau, C. and Sreenivasan, K.R. (1991): The multifractal nature of turbulent energy dissipation. *J. Fluid Mech* **224**, 429–484.
35. Obukhov, A.M. (1962): Some specific features of atmospheric turbulence. *J. Fluid Mech.* **13**, 77–81.
36. Rajput, B. and Rosinski, J. (1989): Spectral representations of infinitely divisible processes. *Probab. Th. Rel. Fields* **82**, 451–487.
37. Schmiegel, J. (2005): Self-scaling of turbulent energy dissipation correlators. *Phys. Lett. A* **337**, 342–353.
38. Schmiegel, J. (2006): Self-scaling tumor growth. Physica A 367, 509–524.
39. Schmiegel, J., Cleve, J., Eggers, H.C., Pearson, B.R. and Greiner, M. (2004): Stochastic energy-cascade model for (1+1)-dimensional fully developed turbulence. *Phys. Lett. A* **320**, 247–253.
40. Schmiegel, J., Barndorff-Nielsen, O.E. and Eggers, H.C. (2006): A class of spatio-temporal and causal stochastic processes, with application to multiscaling and multifractality. South African Journal of Science **101**, 513–519.
41. Shiryaev, A.N. (1999): *Essentials of Stochastic Finance.* Singapore: World Scientific.
42. Sreenivasan, K.R. and Antonia, R.A. (1997): The phenomenology of small-scale turbulence. *Ann. Rev. Fluid Mech* **29**, 435–472.
43. Stolovitzky, G., Kailasnath, P. and Sreenivasan, K.R. (1992): Kolmogorov's refined similarity hypothesis. *Phys. Rev. Lett.* **69**, 1178–1181.
44. Stolovitzky, G. and Sreenivasan, K.R. (1994): Kolmogorov's refined similarity hypotheses for turbulence and general stochastic processes. *Rev. Mod. Phys.* **66**, 229–239.
45. Taylor, G.I. (1938): The spectrum of turbulence. *Proc. R. Soc. Lond. A* **164**, 476–490.
46. Vincent, A. and Meneguzzi, M. (1991): The spatial structure and statistical properties of homogeneous turbulence. *J. Fluid Mech.* **225**, 1–25.
47. Walsh, J.B. (1986a): An introduction to stochastic partial differential equations. In *École d'Été de Probabilités de Saint-Flour* **XIV-1984**. Berlin: Springer. Pp. 265–439.
48. Walsh, J.B. (1986b): Martingales with a multidimensional parameter and stochastic integrals in the plane. In *Lectures in Probability and Statistics (Santiago de Chile 1986).* Berlin: Springer. Pp. 329–491.
49. Wong, E. and Zakai, M. (1974): Martingales and stochastic integrals for processes with a multi-dimensional parameter. *Z. Wahrscheinlichkeitstheorie verw. Geb.* **29**, 109–122.
50. Zhu, Y., Antonia, R.A. and Hosokawa, I. (1995): Refined similarity hypotheses for turbulent velocity and temperature fields. *Phys. Fluids* **7**, 1637–1648.
51. Øigård, T.A., Hanssen, A., Hansen, R.E. and Godtliebsen, F. (2005): EM-estimation and modelling of heavy-tailed processes with the multivariate normal inverse Gaussian distribution. *Signal Processing* **85**, 1655–1673.

A Stochastic Control Approach to a Robust Utility Maximization Problem

Giuliana Bordigoni[1], Anis Matoussi[2], and Martin Schweizer[3]

[1] Dipartimento di Matematica, Politecnico di Milano, Piazza Leonardo da Vinci 32, I – 20133 Milano, Italy, bordigoni@mate.polimi.it

[2] Laboratoire de Statistique et Processus, Université du Maine, Avenue Olivier Messiaen, F – 72085 Le Mans Cedex 9, France, Anis.Matoussi@univ-lemans.fr

[3] ETH Zürich, Departement Mathematik, ETH-Zentrum, HG G 51.2, CH – 8092 Zürich, Switzerland, martin.schweizer@math.ethz.ch

Summary. We study a stochastic control problem arising in the context of utility maximization under model uncertainty. The latter is formulated as a sup-inf problem over strategies π and models (measures) Q, and we treat the inner problem of minimizing over Q the sum of a Q-expected utility term and a penalty term based on the relative entropy of Q with respect to a reference measure P. We prove in general that there exists a unique optimal measure Q^* and show that Q^* is equivalent to P. For a continuous filtration, we characterize the dynamic value process of our stochastic control problem as the unique solution of a generalized backward stochastic differential equation with a quadratic driver. Our results extend earlier work in [21] and are based on a different approach.

Mathematics Subject Classification (2000): 93E20, 91B16, 60H10, 46N10

Keywords: robust control, model uncertainty, quadratic BSDE, stochastic control, relative entropy, entropic penalty, martingale optimality principle, utility maximization, multiple priors, robust utility

JEL Classification Numbers: C60, G10

1 Introduction

This paper takes one step in the problem of *utility maximization* under *model uncertainty*. At a very general level, the latter could be formulated as

$$\text{find } \sup_{\pi} \inf_{Q} \mathbf{U}(\pi, Q), \tag{1.1}$$

where π runs through a set of strategies (portfolios, investment decisions, ...) and $Q \in \mathcal{Q}$ through a set of models (measures, scenarios, ...). In the simplest case, there is one known model so that $\mathcal{Q} = \{P\}$ for a fixed probability measure P, and $\mathbf{U}(\pi, P)$ has the form of a P-expected utility from terminal wealth and/or consumption, both of which are determined by π. There is a vast literature on this by now classical problem; but there is always the drawback that one assumes the underlying model to be exactly known.

To address this issue, one recent line of research considers a non-singleton set $\mathcal{Q}$ of probability measures while keeping for $\mathbf{U}(\pi, Q)$ a Q-expected utility. Such a setting for $\mathcal{Q}$ is often called a *multiple priors model*, and the corresponding optimization problem (1.1) is known as *robust utility maximization*. Some results in this area have been obtained in [10,16,18], among others, and the overall approach relies a lot on convex duality ideas. The set $\mathcal{Q}$ of models under consideration is assumed to have certain properties, but is otherwise quite abstract and usually not specified in any detail.

Instead of working with a somehow given set of models, an alternative is to allow a priori all possible models Q, but to include in $\mathbf{U}(\pi, Q)$ a *penalization* term; this should only depend on Q (not on π) and quantify the decision maker's attitude towards the model Q. Such an approach has for instance been suggested in [1,11]; they take as $\mathbf{U}(\pi, Q)$ the sum of a Q-expected utility, like above, plus a penalty term based on the *relative entropy* of Q with respect to a reference model (measure) P. This is also the setting that we use here. For a very recent treatment of a closely related problem via duality methods, see [17].

The focus of the analysis in [1, 11] is on general insights about the qualitative behaviour of a solution to (1.1) in their framework. This is done via a mostly formal study of the corresponding Hamilton–Jacobi–Bellman (HJB) equations in a Markovian setting. Our goal in contrast is to obtain rigorous mathematical results, and the present paper achieves some progress in that direction for the partial (inner) problem of minimizing $\mathbf{U}(\pi, Q)$ over Q when π is kept fixed. This problem has also been studied by [21] who has obtained very similar results, but with a different approach; see Section 6 for a more detailed comparison.

The paper is structured as follows. Section 2 sets the stage by giving a precise definition of the functional $Q \mapsto \mathbf{U}(\pi, Q)$ with fixed π and of the corresponding optimization problem, and by introducing notations and key assumptions. Section 3 provides a number of auxiliary results for subsequent use. In Section 4, we show with the help of a standard Komlós-type argument that there exists a unique minimizing measure Q^*, and we prove that Q^* is even equivalent to P. This mainly functional analytic approach is complemented by Section 5. There we treat our optimization problem by *stochastic control* methods and show that for a continuous filtration, the corresponding dynamic value process is characterized as the unique solution of a *generalized backward stochastic differential equation (BSDE)* with a quadratic term in its driver. Our BSDE is a slight generalization of an equation studied in detail by [19], but our method of attack is rather different. Like in [19], however, our BSDE involves unbounded terms in the driver and the terminal value which cannot be handled by existing general techniques from the BSDE literature. Hence our approach has to exploit the precise structure of our equation. The final Section 6 contains a brief comparison with some of the most closely related literature.

2 The Basic Optimization Problem

This section gives a precise formulation of our optimization problem and introduces a number of notations for later use.

We start with a filtered probability space $(\Omega, \mathcal{F}, \mathbb{F}, P)$ over a finite time horizon $T \in (0, \infty)$. The filtration $\mathbb{F} = (\mathcal{F}_t)_{0 \le t \le T}$ satisfies the usual conditions of right-continuity and P-completeness. For any probability measure $Q \ll P$ on $\mathcal{F}_T$, the density process of Q with respect to P is the RCLL P-martingale $Z^Q = (Z_t^Q)_{0 \le t \le T}$ with

$$Z_t^Q = \left.\frac{\mathrm{d}Q}{\mathrm{d}P}\right|_{\mathcal{F}_t} = E_P\left[\frac{\mathrm{d}Q}{\mathrm{d}P}\,\middle|\,\mathcal{F}_t\right], \qquad 0 \le t \le T.$$

Since Z^Q is closed on the right by $Z_T^Q = \left.\frac{\mathrm{d}Q}{\mathrm{d}P}\right|_{\mathcal{F}_T}$, we can and do identify Z^Q with Q. (More precisely, Z^Q determines the restriction of Q to $\mathcal{F}_T$, but this will be enough for our purposes.)

The basic ingredients for our optimization problem are now

- parameters $\alpha, \alpha' \in [0, \infty)$ and $\beta \in (0, \infty)$;
- progressively measurable processes $\delta = (\delta_t)_{0 \le t \le T}$ and $U = (U_t)_{0 \le t \le T}$;
- an $\mathcal{F}_T$-measurable random variable U'_T.

Interpretations will follow presently. We define the discounting process

$$S_t^\delta := \exp\left(-\int_0^t \delta_s \,\mathrm{d}s\right), \qquad 0 \le t \le T,$$

the auxiliary quantities

$$\mathcal{U}_{t,T}^\delta := \alpha \int_t^T \frac{S_s^\delta}{S_t^\delta} U_s \,\mathrm{d}s + \alpha' \frac{S_T^\delta}{S_t^\delta} U'_T, \qquad 0 \le t \le T,$$

$$\mathcal{R}_{t,T}^\delta(Q) := \int_t^T \delta_s \frac{S_s^\delta}{S_t^\delta} \log \frac{Z_s^Q}{Z_t^Q} \,\mathrm{d}s + \frac{S_T^\delta}{S_t^\delta} \log \frac{Z_T^Q}{Z_t^Q}, \qquad 0 \le t \le T$$

for $Q \ll P$ on $\mathcal{F}_T$ and consider the cost functional

$$c(\omega, Q) := \mathcal{U}_{0,T}^\delta(\omega) + \beta \mathcal{R}_{0,T}^\delta(Q)(\omega).$$

The basic goal is to

$$\text{minimize the functional } Q \mapsto \Gamma(Q) := E_Q[c(\cdot, Q)] \tag{2.1}$$

over a suitable class of probability measures $Q \ll P$ on $\mathcal{F}_T$. Note that in the language of the introduction, $\Gamma(Q)$ represents $\mathbf{U}(\pi, Q)$ for fixed π.

A closer look at the cost functional $c(\omega, Q)$ shows that

$$\Gamma(Q) = E_P\left[Z_T^Q\left(\alpha\int_0^T S_s^\delta U_s\,\mathrm{d}s + \alpha' S_T^\delta U_T'\right)\right] \tag{2.2}$$
$$+\beta E_P\left[\int_0^T \delta_s S_s^\delta Z_s^Q \log Z_s^Q\,\mathrm{d}s + S_T^\delta Z_T^Q \log Z_T^Q\right]$$

consists of two terms. The first is a Q-expected discounted utility with discount rate δ, utility rate U_s at time s and terminal utility U_T' at time T. Usually, U_s comes from consumption and U_T' from final wealth. As explained above, we consider the strategy decisions π as being frozen for the moment; a maximization over some π determining $U_{\bullet}(\pi)$ and $U_T'(\pi)$ will only be done in a second step. The weights α and α' can be used to obtain as special cases the extreme situations of utility rate only or terminal utility only. The second summand is a sort of discounted relative entropy term with both an "entropy rate" as well as a "terminal entropy". The (constant) factor β determines the strength of this penalty term.

Definition 1. *$D_0^{\exp}$ is the space of all progressively measurable processes $y = (y_t)_{0\le t\le T}$ with*

$$E_P\left[\exp\left(\gamma \operatorname*{ess\,sup}_{0\le t\le T}|y_t|\right)\right] < \infty \qquad \textit{for all } \gamma > 0.$$

$D_1^{\exp}$ denotes the space of all progressively measurable processes $y = (y_t)_{0\le t\le T}$ such that

$$E_P\left[\exp\left(\gamma\int_0^T |y_s|\,\mathrm{d}s\right)\right] < \infty \qquad \textit{for all } \gamma > 0.$$

Definition 2. *For any probability measure Q on $(\Omega, \mathcal{F})$,*

$$H(Q|P) := \begin{cases} E_Q\left[\log \frac{\mathrm{d}Q}{\mathrm{d}P}\Big|_{\mathcal{F}_T}\right] & \textit{if } Q \ll P \textit{ on } \mathcal{F}_T \\ +\infty & \textit{otherwise} \end{cases}$$

denotes the relative entropy of Q with respect to P on $\mathcal{F}_T$. We denote by $\mathcal{Q}_f$ the space of all probability measures Q on $(\Omega, \mathcal{F})$ with $Q \ll P$ on $\mathcal{F}_T$, $Q = P$ on $\mathcal{F}_0$ and $H(Q|P) < \infty$. Clearly $P \in \mathcal{Q}_f^e := \{Q \in \mathcal{Q}_f \,|\, Q \approx P$ on $\mathcal{F}_T\}$.

For a precise formulation of (2.1), we now assume

(A1) $0 \le \delta \le \|\delta\|_\infty < \infty$ for some constant $\|\delta\|_\infty$;
(A2) the process U is in $D_1^{\exp}$;

(A3) $E_P\left[\exp\left(\gamma|U'_T|\right)\right] < \infty$ for all $\gamma > 0$.

We shall see below that $E_Q[c(\cdot, Q)]$ is then well-defined and finite for $Q \in \mathcal{Q}_f$. Due to (A1), a simple estimation gives

$$E_P\left[S_T^\delta Z_T^Q \log Z_T^Q\right] \geq -\mathrm{e}^{-1} + \mathrm{e}^{-\|\delta\|_\infty T} H(Q|P).$$

Hence the second term in $\Gamma(Q)$ explodes unless $H(Q|P) < \infty$. Because we want to minimize $\Gamma(Q)$, this explains why we only consider measures Q in $\mathcal{Q}_f$.

Remark 3. The special case $\delta \equiv 0$ is much simpler and already gives a flavour of the results we obtain for general δ. In fact, $\delta \equiv 0$ yields $S^\delta \equiv 1$ and allows us to rewrite $\Gamma(Q)$ as

$$\Gamma(Q) = E_Q[\mathcal{U}^0_{0,T}] + \beta H(Q|P) = \beta H(Q|P_\mathcal{U}) - \beta \log E_P\left[\exp\left(-\frac{1}{\beta}\mathcal{U}^0_{0,T}\right)\right]$$

if we define a new probability measure $P_\mathcal{U} \approx P$ by

$$\mathrm{d}P_\mathcal{U} := \text{const.} \exp\left(-\frac{1}{\beta}\mathcal{U}^0_{0,T}\right) \mathrm{d}P.$$

Hence (2.1) amounts to minimizing the relative entropy of Q with respect to $P_\mathcal{U}$, and it is well known from [4] that there exists a unique solution Q^* to this problem and that Q^* is equivalent to $P_\mathcal{U}$, hence also to P. In fact, the minimizer obviously is $Q^* = P_\mathcal{U}$.

For $\delta \not\equiv 0$, we shall also find that there exists a unique minimizer Q^* of $\Gamma(Q)$ and that $Q^* \approx P$. However, it does not seem possible to reduce the general δ case to $\delta \equiv 0$ in a simple way. The presence of a discounting term with positive δ is indispensable for an infinite horizon version of (2.1); see [11] and forthcoming work by G. Bordigoni for more on this issue. ◇

We later embed the minimization of $\Gamma(Q)$ in a stochastic control problem and to that end now introduce a few more notations. Let $\mathcal{S}$ denote the set of all $\mathbb{F}$-stopping times τ with values in $[0, T]$ and $\mathcal{D}$ the space of all density processes Z^Q with $Q \in \mathcal{Q}_f$. Recall that we can identify Q with Z^Q. We define

$$\mathcal{D}(Q, \tau) := \left\{ Z^{Q'} \in \mathcal{D} \,\middle|\, Q' = Q \text{ on } \mathcal{F}_\tau \right\},$$
$$\Gamma(\tau, Q) := E_Q[c(\cdot, Q) \,|\, \mathcal{F}_\tau]$$

and the minimal conditional cost at time τ,

$$J(\tau, Q) := Q\text{ - }\operatorname*{ess\,inf}_{Z^{Q'} \in \mathcal{D}(Q,\tau)} \Gamma(\tau, Q').$$

Then (2.1) can be reformulated to

$$\text{find } \inf_{Q \in \mathcal{Q}_f} \Gamma(Q) = \inf_{Q \in \mathcal{Q}_f} E_Q[c(\cdot, Q)] = E_P[J(0, Q)] \tag{2.3}$$

by using the dynamic programming equation and the fact that $Q = P$ on $\mathcal{F}_0$ for every $Q \in \mathcal{Q}_f$.

3 Auxiliary Estimates

In this section, we prove a number of auxiliary estimates that will help us later in establishing our main results. We frequently use the inequalities

$$x \log x \geq -\mathrm{e}^{-1} \qquad \text{for all } x \geq 0, \tag{3.1}$$
$$|x \log x| \leq x \log x + 2\mathrm{e}^{-1} \qquad \text{for all } x \geq 0$$

(where we set $0 \log 0 := 0$) and

$$xy \leq y \log y - y + \mathrm{e}^{x} \qquad \text{for all } x \in \mathbf{R}, y \geq 0. \tag{3.2}$$

The latter is simply the observation that the function $x \mapsto xy - \mathrm{e}^{x}$ on $\mathbf{R}$ takes its maximum for $y > 0$ in $x = \log y$. *Throughout this section, we assume that (A1) – (A3) hold.*

We first show that $\Gamma(Q)$ can be controlled by $H(Q|P)$.

Lemma 4. *There is a constant $C \in (0,\infty)$ depending only on α, α', β, δ, T, U, U'_T such that*

$$\Gamma(Q) \leq E_Q[|c(\cdot,Q)|] \leq C\big(1 + H(Q|P)\big) \qquad \textit{for all } Q \in \mathcal{Q}_f.$$

Proof. The first inequality is obvious. To prove the second, we introduce $R := \alpha \int_0^T |U_s|\,\mathrm{d}s + \alpha'|U'_T|$ and use first the definition of $c(\omega,Q)$, the Bayes formula, (A1) and $0 \leq S^\delta \leq 1$, and then (3.1) and (3.2) to obtain

$$\begin{aligned} E_Q[|c(\cdot,Q)|] &\leq E_P[Z_T^Q R] + \beta E_P\left[\|\delta\|_\infty \int_0^T |Z_s^Q \log Z_s^Q|\,\mathrm{d}s + |Z_T^Q \log Z_T^Q|\right] \\ &\leq E_P[Z_T^Q \log Z_T^Q - Z_T^Q + \mathrm{e}^R] + 2\mathrm{e}^{-1}\beta(\|\delta\|_\infty T + 1) \\ &\quad + \beta E_P\left[\|\delta\|_\infty \int_0^T Z_s^Q \log Z_s^Q\,\mathrm{d}s + Z_T^Q \log Z_T^Q\right]. \end{aligned}$$

By Jensen's inequality and conditioning on $\mathcal{F}_s$, we have

$$E_P[Z_s^Q \log Z_s^Q] \leq E_P[Z_T^Q \log Z_T^Q] = H(Q|P)$$

and therefore

$$E_Q[|c(\cdot,Q)|] \leq E_P[\mathrm{e}^R] + 2\mathrm{e}^{-1}\beta(\|\delta\|_\infty T + 1) + \big(1 + \beta(\|\delta\|_\infty T + 1)\big)H(Q|P).$$

Hence

$$C := \max\big(E_P[\mathrm{e}^R] + 2\mathrm{e}^{-1}\beta(\|\delta\|_\infty T + 1), 1 + \beta\|\delta\|_\infty T + \beta\big)$$

will do, and $C < \infty$ due to (A1) – (A3) and the definition of R. □

An immediate but very useful consequence is

Corollary 5. *Assume (A1) – (A3). Then*

$$c(\cdot, Q) \in L^1(Q) \qquad \textit{for every } Q \in \mathcal{Q}_f,$$

and in particular $\Gamma(Q)$ is well-defined and finite for every $Q \in \mathcal{Q}_f$.

Our next result now shows that conversely, $H(Q|P)$ can also be controlled by $\Gamma(Q)$. This is a bit more tricky and will be crucial later on. Note how the argument exploits almost the full strength of the integrability assumptions (A2) and (A3).

Proposition 6. *There is a constant $C \in (0, \infty)$ depending only on α, α', β, δ, T, U, U'_T with*

$$H(Q|P) \leq C\big(1 + \Gamma(Q)\big) \qquad \textit{for all } Q \in \mathcal{Q}_f. \tag{3.3}$$

In particular, $\inf_{Q \in \mathcal{Q}_f} \Gamma(Q) > -\infty$.

Proof. We first prove for later use an auxiliary inequality in somewhat greater generality. Fix a stopping time $\tau \in \mathcal{S}$. Using the Bayes formula, (A1), then $0 \leq S^\delta \leq 1$ and (3.1) gives

$$\begin{aligned} E_Q\left[\int_0^T \delta_s S_s^\delta \log Z_s^Q \,\mathrm{d}s \,\middle|\, \mathcal{F}_\tau\right] &= \int_0^\tau \delta_s S_s^\delta \log Z_s^Q \,\mathrm{d}s \\ &\quad + \frac{1}{Z_\tau^Q} E_P\left[\int_\tau^T \delta_s S_s^\delta Z_s^Q \log Z_s^Q \,\mathrm{d}s \,\middle|\, \mathcal{F}_\tau\right] \\ &\geq \int_0^\tau \delta_s S_s^\delta \log Z_s^Q \,\mathrm{d}s - \frac{1}{Z_\tau^Q} \|\delta\|_\infty T \mathrm{e}^{-1}. \end{aligned}$$

Similarly, using $1 \geq S_T^\delta \geq \mathrm{e}^{-\|\delta\|_\infty T}$ yields

$$\begin{aligned} E_Q[S_T^\delta \log Z_T^Q \,|\, \mathcal{F}_\tau] &= \frac{1}{Z_\tau^Q} E_P[S_T^\delta Z_T^Q \log Z_T^Q \,|\, \mathcal{F}_\tau] \\ &\geq \frac{1}{Z_\tau^Q}\left(-\mathrm{e}^{-1} + \mathrm{e}^{-\|\delta\|_\infty T}(\mathrm{e}^{-1} + E_P[Z_T^Q \log Z_T^Q \,|\, \mathcal{F}_\tau])\right) \\ &\geq \frac{1}{Z_\tau^Q}\left(-\mathrm{e}^{-1} + \mathrm{e}^{-\|\delta\|_\infty T} E_P[Z_T^Q \log Z_T^Q \,|\, \mathcal{F}_\tau]\right). \end{aligned}$$

Moreover, using $0 \leq S^\delta \leq 1$ and again setting $R := \alpha \int_0^T |U_s| \,\mathrm{d}s + \alpha' |U'_T|$ gives

$$E_Q[\mathcal{U}_{0,T}^\delta | \mathcal{F}_\tau] \geq -E_Q[R|\mathcal{F}_\tau] = -\frac{1}{Z_\tau^Q} E_P[Z_T^Q R \,|\, \mathcal{F}_\tau]$$

so that we get

$$\Gamma(\tau,Q) \geq -\frac{1}{Z_\tau^Q}\Bigg(E_P\left[Z_T^Q \left(\alpha \int_0^T |U_s|\,\mathrm{d}s + \alpha'|U_T'| \right) \,\middle|\, \mathcal{F}_\tau \right] \tag{3.4}$$
$$+\beta\left(-\|\delta\|_\infty T\mathrm{e}^{-1} - \mathrm{e}^{-1} + \mathrm{e}^{-\|\delta\|_\infty T} E_P[Z_T^Q \log Z_T^Q \,|\, \mathcal{F}_\tau]\right)\Bigg)$$
$$+\beta\int_0^\tau \delta_s S_s^\delta \log Z_s^Q\,\mathrm{d}s.$$

To estimate the first term in (3.4), we now use (3.2) with $x = \gamma R$, $y = \frac{1}{\gamma}Z_T^Q$ and $\gamma > 0$ to be chosen later. This yields

$$E_P[Z_T^Q R\,|\,\mathcal{F}_\tau] \leq E_P\left[\frac{1}{\gamma}Z_T^Q \log Z_T^Q - \frac{1}{\gamma}Z_T^Q \log\gamma - \frac{1}{\gamma}Z_T^Q \,\middle|\, \mathcal{F}_\tau\right] + E_P[\mathrm{e}^{\gamma R}\,|\,\mathcal{F}_\tau]$$
$$= \frac{1}{\gamma}E_P[Z_T^Q \log Z_T^Q\,|\,\mathcal{F}_\tau] - \frac{1}{\gamma}(\log\gamma + 1)Z_\tau^Q + E_P[\mathrm{e}^{\gamma R}\,|\,\mathcal{F}_\tau].$$

We plug this into (3.4) to obtain for later use

$$\Gamma(\tau,Q) \geq \frac{1}{\gamma}(\log\gamma+1) - \frac{1}{Z_\tau^Q}E_P\left[\exp\left(\gamma\alpha\int_0^T |U_s|\,\mathrm{d}s + \gamma\alpha'|U_T'|\right)\,\middle|\,\mathcal{F}_\tau\right] \tag{3.5}$$
$$-\frac{1}{Z_\tau^Q}\beta\mathrm{e}^{-1}(\|\delta\|_\infty T + 1) + \frac{1}{Z_\tau^Q}E_P[Z_T^Q \log Z_T^Q\,|\,\mathcal{F}_\tau]\left(\beta\mathrm{e}^{-\|\delta\|_\infty T} - \frac{1}{\gamma}\right)$$
$$+\beta\int_0^\tau \delta_s S_s^\delta \log Z_s^Q\,\mathrm{d}s.$$

If we choose $\tau = 0$ and take expectations, this gives in particular

$$\Gamma(Q) \geq \frac{1}{\gamma}(\log\gamma+1) - E_P[\mathrm{e}^{\gamma R}] - \beta\mathrm{e}^{-1}(\|\delta\|_\infty T + 1) + H(Q|P)\left(\beta\mathrm{e}^{-\|\delta\|_\infty T} - \frac{1}{\gamma}\right).$$

For any $\gamma > 0$ such that

$$\beta\mathrm{e}^{-\|\delta\|_\infty T} - \frac{1}{\gamma} \geq \eta > 0$$

we thus obtain (3.3) with

$$C := \frac{1}{\eta}\max\left(1, E_P[\mathrm{e}^{\gamma R}] + \frac{1}{\gamma}(|\log\gamma| + 1) + \beta\mathrm{e}^{-1}(\|\delta\|_\infty T + 1)\right),$$

and $C < \infty$ due to (A1) – (A3) and the definition of R. The final assertion is clear since $H(Q|P) \geq 0$. □

A slight modification in the proof of Proposition 6 also yields the following technical estimate.

Lemma 7. *For any $\gamma > 0$ and any set $A \in \mathcal{F}_T$, we have*

$$E_Q[|\mathcal{U}^\delta_{0,T}|I_A] \leq \frac{1}{\gamma}H(Q|P) + \frac{1}{\gamma}(\mathrm{e}^{-1} + |\log\gamma| + 1) \qquad (3.6)$$
$$+E_P\left[I_A \exp\left(\gamma\alpha\int_0^T |U_s|\,\mathrm{d}s + \gamma\alpha'|U'_T|\right)\right].$$

Proof. We again use (3.2) with $x = \gamma R := \gamma\left(\alpha\int_0^T S^\delta_s|U_s|\,\mathrm{d}s + \alpha' S^\delta_T|U'_T|\right)$, $y = \frac{1}{\gamma}Z^Q_T$ and then multiply by I_A to obtain

$$Z^Q_T|\mathcal{U}^\delta_{0,T}|I_A \leq Z^Q_T R I_A \leq I_A\left(\frac{1}{\gamma}Z^Q_T\log Z^Q_T - \frac{1}{\gamma}Z^Q_T(\log\gamma + 1) + \mathrm{e}^{\gamma R}\right).$$

Adding e^{-1} and using (3.1) then yields

$$Z^Q_T|\mathcal{U}^\delta_{0,T}|I_A \leq \frac{1}{\gamma}(Z^Q_T\log Z^Q_T + \mathrm{e}^{-1}) + Z^Q_T\frac{1}{\gamma}(|\log\gamma| + 1) + \mathrm{e}^{\gamma R}I_A,$$

and (3.6) follows by taking expectations under P. □

We later want to use the martingale optimality principle from stochastic control theory. Although we know from Corollary 5 that $c(\cdot, Q)$ is Q-integrable for every $Q \in \mathcal{Q}_f$, this is not enough since we have no uniformity in Q. Therefore we prove here directly that each $J(\tau, Q)$ is Q-integrable.

Lemma 8. *For each $\tau \in \mathcal{S}$ and $Q \in \mathcal{Q}_f$, the random variable $J(\tau, Q)$ is in $L^1(Q)$.*

Proof. By definition,

$$J(\tau, Q) \leq \Gamma(\tau, Q) \leq E_Q\left[|c(\cdot, Q)|\,\middle|\,\mathcal{F}_\tau\right]$$

so that

$$\left(J(\tau, Q)\right)^+ \leq E_Q\left[|c(\cdot, Q)|\,\middle|\,\mathcal{F}_\tau\right]$$

is Q-integrable by Corollary 5. Dealing with the negative part is a bit more delicate. We first fix $Z^{Q'} \in \mathcal{D}(Q, \tau)$ and consider $\Gamma(\tau, Q')$. Our goal is to find a Q-integrable lower bound for $\Gamma(\tau, Q')$ which does not depend on Q', because this will then also work for $J(\tau, Q) = \operatorname{ess\,inf}_{Z^{Q'}\in\mathcal{D}(Q,\tau)} \Gamma(\tau, Q')$. To that end, we use (3.5) with Q' instead of Q and observe that $Z^{Q'} = Z^Q$ on $[\![0, \tau]\!]$ because $Q' = Q$ on $\mathcal{F}_\tau$. Choosing $\gamma > 0$ to satisfy $\beta\mathrm{e}^{-\|\delta\|_\infty T} - \frac{1}{\gamma} = 0$ thus yields

$$\left(\Gamma(\tau, Q')\right)^- \leq B := \frac{1}{Z^Q_\tau}\Bigg(E_P\left[\exp\left(\gamma\alpha\int_0^T |U_s|\,\mathrm{d}s + \gamma\alpha'|U'_T|\right)\,\middle|\,\mathcal{F}_\tau\right]$$
$$+ \beta\mathrm{e}^{-1}(\|\delta\|_\infty T + 1)\Bigg)$$
$$+\frac{1}{\gamma}(|\log\gamma| + 1) + \beta\int_0^\tau \delta_s S^\delta_s \log Z^Q_s\,\mathrm{d}s.$$

But this nonnegative random variable does not depend on Q', and thus we conclude that

$$J(\tau,Q)=\operatorname*{ess\,inf}_{Z^{Q'}\in\mathcal{D}(Q,\tau)}\Gamma(\tau,Q')\geq\operatorname*{ess\,inf}_{Z^{Q'}\in\mathcal{D}(Q,\tau)}-\big(\Gamma(\tau,Q')\big)^{-}\geq -B$$

so that $\big(J(\tau,Q)\big)^{-}\leq B$. Finally, $B\in L^1(Q)$ because (A1) – (A3) yield that

$$E_Q[B]\leq E_P\left[\exp\left(\gamma\alpha\int_0^T|U_s|\,\mathrm{d}s+\gamma\alpha'|U'_T|\right)\right]+\beta\mathrm{e}^{-1}(\|\delta\|_\infty T+1)$$
$$+\frac{1}{\gamma}(|\log\gamma|+1)+\beta E_P\left[\int_0^T\delta_s S_s^\delta|\log Z_s^Q|\,\mathrm{d}s\right]<\infty;$$

in fact, the last summand is at most $\beta\|\delta\|_\infty T\big(H(Q|P)+2\mathrm{e}^{-1}\big)$ by the same computation as in the proof of Lemma 4. □

4 Existence of an Optimal Measure Q^*

The main result of this section is that the problem (2.1) of minimizing $\Gamma(Q)=E_Q[c(\cdot,Q)]$ over $Q\in\mathcal{Q}_f$ has a unique solution $Q^*\in\mathcal{Q}_f$, and that Q^* is even equivalent to P. This is proved for a general filtration $\mathbb{F}$.

Theorem 9. *Assume (A1) – (A3). Then there exists a unique $Q^*\in\mathcal{Q}_f$ which minimizes $Q\mapsto\Gamma(Q)$ over all $Q\in\mathcal{Q}_f$.*

Proof. 1) $x\mapsto x\log x$ is strictly convex and δ and S^δ are nonnegative; hence $Q\mapsto\Gamma(Q)$ is also strictly convex and Q^* must be unique if it exists.

2) Let $(Q^n)_{n\in\mathbf{N}}$ be a sequence in $\mathcal{Q}_f$ such that

$$\searrow\text{-}\lim_{n\to\infty}\Gamma(Q^n)=\inf_{Q\in\mathcal{Q}_f}\Gamma(Q)>-\infty$$

and denote by $Z^n=Z^{Q^n}$ the corresponding density processes. Since each $Z_T^n\geq 0$, it follows from Komlós' theorem that there exists a sequence $(\bar{Z}_T^n)_{n\in\mathbf{N}}$ with $\bar{Z}_T^n\in\operatorname{conv}(Z_T^n,Z_T^{n+1},\dots)$ for each $n\in\mathbf{N}$ and such that $(\bar{Z}_T^n)$ converges P-a.s. to some random variable $\bar{Z}_T^\infty$, which is then also nonnegative but may take the value $+\infty$. Because $\mathcal{Q}_f$ is convex, each $\bar{Z}_T^n$ is again associated to some $\bar{Q}^n\in\mathcal{Q}_f$. We claim that this also holds for $\bar{Z}_T^\infty$, i.e., that $\mathrm{d}\bar{Q}^\infty:=\bar{Z}_T^\infty\,\mathrm{d}P$ defines a probability measure $\bar{Q}^\infty\in\mathcal{Q}_f$. To see this, note first that we have

$$\Gamma(\bar{Q}^n)\leq\sup_{m\geq n}\Gamma(Q^m)=\Gamma(Q^n)\leq\Gamma(Q^1)\tag{4.1}$$

because $Q\mapsto\Gamma(Q)$ is convex and $n\mapsto\Gamma(Q^n)$ is decreasing. Hence Proposition 6 yields

$$\sup_{n\in\mathbf{N}} E_P[\bar{Z}_T^n \log \bar{Z}_T^n] = \sup_{n\in\mathbf{N}} H(\bar{Q}^n|P) \leq C\left(1 + \sup_{n\in\mathbf{N}} \Gamma(\bar{Q}^n)\right) \tag{4.2}$$
$$\leq C\left(1 + \Gamma(Q^1)\right) < \infty.$$

Thus $(\bar{Z}_T^n)_{n\in\mathbf{N}}$ is P-uniformly integrable by de la Vallée-Poussin's criterion and therefore converges in $L^1(P)$ as well. This implies that $E_P[\bar{Z}_T^\infty] = \lim_{n\to\infty} E_P[\bar{Z}_T^n] = 1$ so that $\bar{Q}^\infty$ is indeed a probability measure and $\bar{Q}^\infty \ll P$ on $\mathcal{F}_T$. Because $x \mapsto x\log x$ is bounded below by $-\mathrm{e}^{-1}$, Fatou's lemma and (4.2) yield

$$H(\bar{Q}^\infty|P) = E_P[\bar{Z}_T^\infty \log \bar{Z}_T^\infty] \leq \liminf_{n\to\infty} E_P[\bar{Z}_T^n \log \bar{Z}_T^n] < \infty. \tag{4.3}$$

Finally, we also have $\bar{Q}^\infty = P$ on $\mathcal{F}_0$; in fact, $(\bar{Z}_T^n)$ converges to $\bar{Z}_T^\infty$ strongly, hence also weakly in $L^1(P)$ and so we have for every $A \in \mathcal{F}_0$

$$\bar{Q}^\infty[A] = E_P[\bar{Z}_T^\infty I_A] = \lim_{n\to\infty} E_P[\bar{Z}_T^n I_A] = \lim_{n\to\infty} \bar{Q}^n[A] = P[A]$$

since all the $\bar{Q}^n$ are in $\mathcal{Q}_f$ and hence agree with P on $\mathcal{F}_0$. This shows that $\bar{Q}^\infty \in \mathcal{Q}_f$.

3) We now want to show that $Q^* := \bar{Q}^\infty$ attains the infimum of $Q \mapsto \Gamma(Q)$ in $\mathcal{Q}_f$ and therefore examine $\Gamma(\bar{Q}^\infty)$ more closely. Let $\bar{Z}^\infty$ be the density process of $\bar{Q}^\infty$ with respect to P. Because we know that $(\bar{Z}_T^n)$ converges to $\bar{Z}_T^\infty$ in $L^1(P)$, Doob's maximal inequality

$$P\left[\sup_{0\leq t\leq T} |\bar{Z}_t^\infty - \bar{Z}_t^n| \geq \epsilon\right] \leq \frac{1}{\epsilon} E_P\left[|\bar{Z}_T^\infty - \bar{Z}_T^n|\right]$$

implies that $\left(\sup_{0\leq t\leq T} |\bar{Z}_t^\infty - \bar{Z}_t^n|\right)_{n\in\mathbf{N}}$ converges to 0 in P-probability. By passing to a subsequence that we still denote by $(\bar{Z}^n)_{n\in\mathbf{N}}$, we may thus assume that $(\bar{Z}_\cdot^n)$ converges to $\bar{Z}_\cdot^\infty$ uniformly in t with P-probability 1. This implies that

$$\bar{Z}_T^n c(\,\cdot\,, \bar{Q}^n) \longrightarrow \bar{Z}_T^\infty c(\,\cdot\,, \bar{Q}^\infty) \qquad P\text{-a.s.}$$

and in more detail with

$$\bar{Y}_1^n := \bar{Z}_T^n \mathcal{U}_{0,T}^\delta,$$
$$\bar{Y}_2^n := \beta\left(\int_0^T \delta_s S_s^\delta \bar{Z}_s^n \log \bar{Z}_s^n \,\mathrm{d}s + S_T^\delta \bar{Z}_T^n \log \bar{Z}_T^n\right) = \beta \mathcal{R}_{0,T}^\delta(\bar{Q}^n)$$

for $n \in \mathbf{N} \cup \{\infty\}$ that

$$\lim_{n\to\infty} \bar{Y}_i^n = \bar{Y}_i^\infty \qquad P\text{-a.s. for } i = 1, 2.$$

Since $\bar{Y}_2^n$ is by (A1) like $x\log x$ bounded below, uniformly in n and ω, Fatou's lemma yields

$$E_P[\bar{Y}_2^\infty] \le \liminf_{n\to\infty} E_P[\bar{Y}_2^n]. \tag{4.4}$$

We prove below that we also have

$$E_P[\bar{Y}_1^\infty] \le \liminf_{n\to\infty} E_P[\bar{Y}_1^n]. \tag{4.5}$$

Adding (4.5) and (4.4) then yields by (4.1) that

$$\Gamma(\bar{Q}^\infty) = E_P[\bar{Y}_1^\infty + \bar{Y}_2^\infty] \le \liminf_{n\to\infty} \Gamma(\bar{Q}^n) \le \liminf_{n\to\infty} \Gamma(Q^n) = \inf_{Q\in\mathcal{Q}_f} \Gamma(Q)$$

which proves that $\bar{Q}^\infty$ is indeed optimal.

4) Although $\bar{Y}_1^n$ is linear in $\bar{Z}_T^n$, it is more difficult to handle than $\bar{Y}_2^n$ because the factor $\mathcal{U}_{0,T}^\delta$ is not bounded. However, the random variables $\mathcal{U}_{0,T}^\delta$ and $R := \alpha \int_0^T |U_s|\,\mathrm{d}s + \alpha'|U_T'|$ are still manageable thanks to the exponential integrability properties from (A2) and (A3); these imply that R is almost bounded in the sense that $e^{\gamma R} \in L^1(P)$ for all $\gamma > 0$. To exploit this, we set

$$\tilde{R}_m := \mathcal{U}_{0,T}^\delta I_{\{\mathcal{U}_{0,T}^\delta \ge -m\}} \ge -m \qquad \text{for } m \in \mathbf{N}$$

so that we have for each $n \in \mathbf{N} \cup \{\infty\}$

$$\bar{Y}_1^n = \bar{Z}_T^n \mathcal{U}_{0,T}^\delta = \bar{Z}_T^n \tilde{R}_m + \bar{Z}_T^n \mathcal{U}_{0,T}^\delta I_{\{\mathcal{U}_{0,T}^\delta < -m\}}.$$

Because $\tilde{R}_m \ge -m$ and each $\bar{Z}_T^n$ has P-expectation 1, Fatou's lemma yields

$$E_P[\bar{Z}_T^\infty \tilde{R}_m] = -m + E_P[\bar{Z}_T^\infty(\tilde{R}_m + m)] \le \liminf_{n\to\infty} E_P[\bar{Z}_T^n \tilde{R}_m]$$

and therefore adding and subtracting $E_P\left[\bar{Z}_T^n \mathcal{U}_{0,T}^\delta I_{\{\mathcal{U}_{0,T}^\delta < -m\}}\right]$ gives

$$\begin{aligned}
E_P[\bar{Y}_1^\infty] &\le \liminf_{n\to\infty} E_P[\bar{Z}_T^n \tilde{R}_m] + E_P\left[\bar{Z}_T^\infty \mathcal{U}_{0,T}^\delta I_{\{\mathcal{U}_{0,T}^\delta < -m\}}\right] \\
&\le \liminf_{n\to\infty} E_P[\bar{Y}_1^n] + 2 \sup_{n\in\mathbf{N}\cup\{\infty\}} E_P\left[\bar{Z}_T^n |\mathcal{U}_{0,T}^\delta| I_{\{\mathcal{U}_{0,T}^\delta < -m\}}\right].
\end{aligned}$$

Hence (4.5) will follow once we prove that

$$\lim_{m\to\infty} \sup_{n\in\mathbf{N}\cup\{\infty\}} E_P\left[\bar{Z}_T^n |\mathcal{U}_{0,T}^\delta| I_{\{\mathcal{U}_{0,T}^\delta < -m\}}\right] = 0. \tag{4.6}$$

However, Lemma 7 yields for each $n \in \mathbf{N} \cup \{\infty\}$

$$\begin{aligned}
E_P\left[\bar{Z}_T^n |\mathcal{U}_{0,T}^\delta| I_{\{\mathcal{U}_{0,T}^\delta < -m\}}\right] &= E_{\bar{Q}^n}\left[|\mathcal{U}_{0,T}^\delta| I_{\{\mathcal{U}_{0,T}^\delta < -m\}}\right] \\
&\le \frac{1}{\gamma} H(\bar{Q}^n|P) + \frac{1}{\gamma}(\mathrm{e}^{-1} + |\log\gamma| + 1) \\
&\quad + E_P\left[I_{\{\mathcal{U}_{0,T}^\delta < -m\}} \mathrm{e}^{\gamma R}\right]
\end{aligned}$$

and therefore by using (4.2) and (4.3)

$$\sup_{n\in\mathbf{N}\cup\{\infty\}} E_P\left[\bar{Z}^n_T|\mathcal{U}^\delta_{0,T}|I_{\{\mathcal{U}^\delta_{0,T}<-m\}}\right] \leq \frac{1}{\gamma}\left(C(1+\Gamma(Q^1)+\mathrm{e}^{-1}+|\log\gamma|+1\right)$$
$$+E_P\left[I_{\{\mathcal{U}^\delta_{0,T}<-m\}}\mathrm{e}^{\gamma R}\right]$$

for each $\gamma > 0$. The first term on the right-hand side becomes arbitrarily small for γ large enough, and the second converges for each fixed γ to 0 as $m \to \infty$ by dominated convergence, due to the exponential integrability of R from (A1) – (A3). This proves (4.6) and completes the proof. □

Remark 10. In abstract terms, the proof of Theorem 9 can morally be summarized as follows:

a) Use Komlós' theorem to produce a candidate $\bar{Q}^\infty$ for the optimal measure, where $\bar{Z}^\infty_T$ is a P-almost sure limit of convex combinations $\bar{Z}^n_T$ formed from a minimizing sequence $(Z^n_T)_{n\in\mathbf{N}}$.
b) View $\Gamma(Q)$ like in (2.2) as a function $g(Z^Q)$ defined on density processes Z^Q. Minimality of $\bar{Q}^\infty$ then follows by standard reasoning if g is convex and lower semicontinuous with respect to P-almost sure convergence of Z^Q_T.

While convexity of g is immediate, lower semicontinuity is not obvious at all. For the entropy term (the second summand in (2.2)), we can use Fatou's lemma, but we first need the convergence of the entire density process $Z^Q_\cdot$ and not only of its final value Z^Q_T. We have done this above by using $L^1(P)$-convergence of the final values, but this requires of course P-uniform integrability. Thanks to the linearity in Z^Q_T, there is no convergence problem for the integrand of the first summand in (2.2); but we cannot use Fatou's lemma there since we have no uniform lower bound. The arguments in steps 3) and 4) of the above proof show that while g is probably not lower semicontinuous on all of $\mathcal{D}$ with respect to P-almost sure convergence of Z^Q_T, it is so at least along any sequence $\left(Z^{Q^n}\right)_{n\in\mathbf{N}}$ which is bounded in entropy in the sense that $\sup_{n\in\mathbf{N}} H(Q^n|P) < \infty$. Note that we exploit here the full strength of the assumptions (A2) and (A3) because we need to let γ tend to ∞.

The above problems disappear if the utility terms U and U'_T are uniformly bounded below or if we have a uniform bound on $H(Q|P)$ for all measures Q we allow in the minimization problem. In [2], this is for instance achieved by minimizing over a set $\tilde{\mathcal{Q}} \subseteq \mathcal{Q}_f$ which is convex and satisfies $\sup_{Q\in\tilde{\mathcal{Q}}}\left\|\frac{\mathrm{d}Q}{\mathrm{d}P}\right\|_{L^p(P)} < \infty$ for some $p > 1$. One major achievement of the present work is that it avoids such restrictive assumptions on U, U'_T and Q. ◇

Having established existence and uniqueness of an optimal Q^*, our next goal is to prove that Q^* is equivalent to P. This uses an adaptation of an argument in [9], and we start with an auxiliary result.

Lemma 11. *Suppose for $i = 0, 1$ that $Q^i \in \mathcal{Q}_f$ with density processes $Z^i = Z^{Q^i}$. Then*

$$\sup_{0 \le t \le T} E_P\left[(Z_t^1 \log Z_t^0)^+\right] \le 2 + \mathrm{e}^{-1} + H(Q^1|P) < \infty. \tag{4.7}$$

Proof. This slightly sharpens a result obtained in the proof of Lemma 2.1 in [9]. For completeness we give details. If we set $Z^x := xZ^1 + (1-x)Z^0$, $\psi(x) := x \log x$ and

$$H(x;t) := \frac{1}{x}\left(\psi(Z_t^x) - \psi(Z_t^0)\right) \qquad \text{for } x \in (0,1] \text{ and } t \text{ fixed,} \tag{4.8}$$

the random function $x \mapsto H(x;t)$ is increasing because ψ is convex, and so

$$\begin{aligned} H(1;t) \ge \lim_{x \searrow 0} \frac{\psi(Z_t^x) - \psi(Z_t^0)}{x} = \frac{\mathrm{d}}{\mathrm{d}x}\psi(Z_t^x)\bigg|_{x=0} &= \psi'(Z_t^0)(Z_t^1 - Z_t^0) \\ &= (\log Z_t^0 + 1)(Z_t^1 - Z_t^0). \end{aligned}$$

Rearranging terms gives

$$Z_t^1 \log Z_t^0 \le \psi(Z_t^1) - \psi(Z_t^0) + Z_t^0 \log Z_t^0 + Z_t^0 - Z_t^1 \le \psi(Z_t^1) + \mathrm{e}^{-1} + Z_t^0 + Z_t^1, \tag{4.9}$$

and the right-hand side is by (3.1) nonnegative with

$$E_P[\psi(Z_t^1)] \le E_P[\psi(Z_T^1)] = H(Q^1|P)$$

by Jensen's inequality. Hence (4.7) follows from (4.9). □

Now we are ready to prove the second main result of this section.

Theorem 12. *Assume (A1) – (A3). Then the optimal measure Q^* from Theorem 9 is equivalent to P.*

Proof. 1) Like in the proof of Lemma 11, we take $Q^0, Q^1 \in \mathcal{Q}_f$, define $Q^x := xQ^1 + (1-x)Q^0$ for $x \in [0,1]$ and denote by Z^x the density process of Q^x with respect to P. With $\psi(x) = x \log x$ and H as in (4.8), we then obtain

$$\begin{aligned} &\frac{1}{x}\left(\Gamma(Q^x) - \Gamma(Q^0)\right) = E_P[(Z_T^1 - Z_T^0)\mathcal{U}_{0,T}^\delta] \\ &\qquad + \frac{1}{x}\beta E_P\left[\int_0^T \delta_s S_s^\delta \left(\psi(Z_s^x) - \psi(Z_s^0)\right) \mathrm{d}s + S_T^\delta\left(\psi(Z_T^x) - \psi(Z_T^0)\right)\right] \\ &\quad = E_P[(Z_T^1 - Z_T^0)\mathcal{U}_{0,T}^\delta] + \beta E_P\left[\int_0^T \delta_s S_s^\delta H(x;s)\,\mathrm{d}s + S_T^\delta H(x;T)\right]. \end{aligned}$$

For x decreasing to 0, $H(x;s)$ decreases like in the proof of Lemma 11 to $(\log Z_s^0 + 1)(Z_s^1 - Z_s^0)$, and

$$H(x;s) \le H(1;s) = \psi(Z_s^1) - \psi(Z_s^0) \le \psi(Z_s^1) + \mathrm{e}^{-1}$$

shows that we have an integrable upper bound. Hence we can use monotone convergence to conclude that

$$\begin{aligned} &\frac{\mathrm{d}}{\mathrm{d}x}\Gamma(Q^x)\bigg|_{x=0} \qquad (4.10)\\ &\quad = E_P[(Z_T^1 - Z_T^0)\mathcal{U}_{0,T}^\delta] \\ &\qquad + \beta E_P\left[\int_0^T \delta_s S_s^\delta(\log Z_s^0 + 1)(Z_s^1 - Z_s^0)\,\mathrm{d}s + S_T^\delta(\log Z_T^0 + 1)(Z_T^1 - Z_T^0)\right] \\ &\quad =: E_P[Y_1] + E_P[Y_2]. \end{aligned}$$

As in the proof of Lemma 4, (A1) – (A3) imply that $Y_1 \in L^1(P)$, and since $x \mapsto H(x;s)$ is increasing, Y_2 is majorized by

$$\begin{aligned} \int_0^T \delta_s S_s^\delta H(1;s)\,\mathrm{d}s + S_T^\delta H(1;T) \ \le\ &\int_0^T \delta_s S_s^\delta \left(\psi(Z_s^1) + \mathrm{e}^{-1}\right)\mathrm{d}s \\ &+ S_T^\delta \left(\psi(Z_T^1) + \mathrm{e}^{-1}\right) \end{aligned}$$

which is P-integrable because $Q^1 \in \mathcal{Q}_f$. Hence $Y_2^+ \in L^1(P)$ and so the right-hand side of (4.10) is well-defined in $[-\infty, +\infty)$.

2) Now take $Q^0 = Q^*$ and any $Q^1 \in \mathcal{Q}_f$ which is equivalent to P; this is possible since $\mathcal{Q}_f$ contains P. The optimality of Q^* yields $\Gamma(Q^x) - \Gamma(Q^*) \ge 0$ for all $x \in (0,1]$, hence also

$$\frac{\mathrm{d}}{\mathrm{d}x}\Gamma(Q^x)\bigg|_{x=0} \ge 0. \qquad (4.11)$$

Therefore the right-hand side of (4.10) is nonnegative which implies that Y_2 must be in $L^1(P)$. This allows us to rearrange terms and rewrite (4.11) by using (4.10) as

$$\begin{aligned} &\beta E_P\left[\int_0^T \delta_s S_s^\delta Z_s^1 \log Z_s^*\,\mathrm{d}s + S_T^\delta Z_T^1 \log Z_T^*\right] \qquad (4.12)\\ &\ge -E_P[(Z_T^1 - Z_T^*)\mathcal{U}_{0,T}^\delta] + \beta E_P\left[\int_0^T \delta_s S_s^\delta Z_s^* \log Z_s^*\,\mathrm{d}s + S_T^\delta Z_T^* \log Z_T^*\right] \\ &\quad - \beta E_P\left[(Z_T^1 - Z_T^*)\int_0^T \delta_s S_s^\delta\,\mathrm{d}s + S_T^\delta\right]. \end{aligned}$$

But the right-hand side of (4.12) is $> -\infty$ and the first term on the left-hand side is $< +\infty$ due to (A1) and Lemma 11. Moreover, (A1) implies that $S_T^\delta \ge \mathrm{e}^{-\|\delta\|_\infty T} > 0$. So if we have $Q^* \not\approx P$, we get $(\log Z_T^*)^- = \infty$ on the set $A := \{Z_T^* = 0\}$ and $P[A] > 0$. This gives $(Z_T^1 \log Z_T^*)^- = \infty$ on

A because $Z_T^1 > 0$ since $Q^1 \approx P$. But since we know from Lemma 11 that $(Z_T^1 \log Z_T^*)^+ \in L^1(P)$, we then conclude that $E_P[S_T^\delta Z_T^1 \log Z_T^*] = -\infty$, and this gives a contradiction to (4.12). Therefore $Q^* \approx P$. □

5 A BSDE Description for the Dynamic Value Process

In this section, we use stochastic control techniques to study the dynamic value process V associated to the optimization problem (2.1) or (2.3). We show that V is the unique solution of a backward stochastic differential equation (BSDE) with a quadratic driver, if the underlying filtration is continuous. This extends earlier work by [13, 19, 21].

We first recall from Section 2 the conditional cost $\Gamma(\tau, Q) = E_Q[c(\cdot, Q) \,|\, \mathcal{F}_\tau]$ and the minimal conditional cost

$$J(\tau, Q) = Q\text{ - }\operatorname*{ess\,inf}_{Z^{Q'} \in \mathcal{D}(Q,\tau)} \Gamma(\tau, Q') \qquad \text{for } \tau \in \mathcal{S} \text{ and } Q \in \mathcal{Q}_f.$$

A measure $\tilde{Q} \in \mathcal{Q}_f$ is called *optimal* if it minimizes $Q \mapsto \Gamma(Q) = E_Q[c(\cdot, Q)]$ over $Q \in \mathcal{Q}_f$. Then we have the following *martingale optimality principle* from stochastic control.

Proposition 13. *Assume (A1) – (A3). Then:*

1. *The family $\{J(\tau, Q) \,|\, \tau \in \mathcal{S}, Q \in \mathcal{Q}_f\}$ is a submartingale system; this implies that for any $Q \in \mathcal{Q}_f$, we have for any stopping times $\sigma \le \tau$ the Q-submartingale property*
$$E_Q[J(\tau, Q) \,|\, \mathcal{F}_\sigma] \ge J(\sigma, Q) \qquad Q\text{-a.s.} \tag{5.1}$$
2. *$\tilde{Q} \in \mathcal{Q}_f$ is optimal if and only if $\{J(\tau, \tilde{Q}) \,|\, \tau \in \mathcal{S}\}$ is a $\tilde{Q}$-martingale system; this means that instead of (5.1), we have for any stopping times $\sigma \le \tau$*
$$E_Q[J(\tau, \tilde{Q}) \,|\, \mathcal{F}_\sigma] = J(\sigma, \tilde{Q}) \qquad \tilde{Q}\text{-a.s.}$$
3. *For each $Q \in \mathcal{Q}_f$, there exists an adapted RCLL process $J^Q = (J_t^Q)_{0 \le t \le T}$ which is a right-closed Q-submartingale such that*
$$J_\tau^Q = J(\tau, Q) \qquad Q\text{-a.s. for each stopping time } \tau.$$

Proof. This is almost a direct consequence of Theorems 1.15 (for 1.), 1.17 (for 2.) and 1.21 (for 3.) in [6]. It is straightforward (but a little tedious; see [2] for details) to check that our control problem satisfies all the assumptions required for these results, with just one exception; we have neither $c \ge 0$ nor $\inf_{Z^{Q'} \in \mathcal{D}(Q,\tau)} E_{Q'}[|c(\,\cdot\,, Q')|] < \infty$ for all $\tau \in \mathcal{S}$ and $Q \in \mathcal{Q}_f$ as required in [6]. However, closer inspection of the proofs in [6] shows that all the required assertions from there still hold true if one can show that $E_Q[|J(\tau, Q)|] < \infty$ for each $Q \in \mathcal{Q}_f$ and $\tau \in \mathcal{S}$. Because we have proved this in Lemma 8, our assertion follows. □

We already know from Theorem 9 that there exists an optimal $Q^* \in \mathcal{Q}_f$, and we even have $Q^* \in \mathcal{Q}_f^e$ by Theorem 12. Hence we may equally well minimize $Q \mapsto \Gamma(Q)$ only over $Q \in \mathcal{Q}_f^e$ without losing any generality. For each $Q \in \mathcal{Q}_f^e$, $\tau \in \mathcal{S}$ and $Q' \in \mathcal{D}(Q,\tau)$, we now define

$$\tilde{V}(\tau, Q') := E_{Q'}[\mathcal{U}^\delta_{\tau,T}|\mathcal{F}_\tau] + \beta E_{Q'}[\mathcal{R}^\delta_{\tau,T}(Q')|\mathcal{F}_\tau]$$

and

$$V(\tau, Q) := Q\text{ - }\operatorname*{ess\,inf}_{Z^{Q'}\in\mathcal{D}(Q,\tau)} \tilde{V}(\tau, Q').$$

The latter is the *value* of the control problem started at time τ instead of 0 and assuming one has used the model Q up to time τ. By using the Bayes formula and the definition of $\mathcal{R}^\delta_{\tau,T}(Q')$, one easily sees that each $\tilde{V}(\tau, Q')$ depends only on the values of $Z^{Q'}$ on $]\!]\tau, T]\!]$ and therefore not on Q, since $Z^{Q'} \in \mathcal{D}(Q,\tau)$ only says that $Z^{Q'} = Z^Q$ on $[\![0,\tau]\!]$. So we can equally well take the ess inf under $P \approx Q$ and over all $Q' \in \mathcal{Q}_f$ and call the result $V(\tau)$ since it does not depend on $Q \in \mathcal{Q}_f^e$.

From the definition of $\mathcal{R}^\delta_{\tau,T}(Q')$, we have for Q' with $Z^{Q'} \in \mathcal{D}(Q,\tau)$ that

$$\begin{aligned}\mathcal{R}^\delta_{0,T}(Q') &= \int_0^\tau \delta_s S^\delta_s \log Z^{Q'}_s \,ds + S^\delta_\tau \mathcal{R}^\delta_{\tau,T}(Q') + \left(\int_\tau^T \delta_s S^\delta_s \,ds + S^\delta_T\right) \log Z^{Q'}_\tau \\ &= S^\delta_\tau \mathcal{R}^\delta_{\tau,T}(Q') + \int_0^\tau \delta_s S^\delta_s \log Z^Q_s \,ds + S^\delta_\tau \log Z^Q_\tau.\end{aligned}$$

Comparing the definitions of $V(\tau) = V(\tau, Q)$ and $J(\tau, Q)$ therefore yields for $Q \in \mathcal{Q}_f^e$

$$J(\tau, Q) = S^\delta_\tau V(\tau) + \alpha \int_0^\tau S^\delta_s U_s \,ds + \beta \int_0^\tau \delta_s S^\delta_s \log Z^Q_s \,ds + \beta S^\delta_\tau \log Z^Q_\tau,$$

because we can also take the ess inf for $J(\tau, Q)$ under $P \approx Q$. Since each $J(\,\cdot\,, Q)$ admits an RCLL version by Proposition 13, we can choose an adapted RCLL process $V = (V_t)_{0\le t\le T}$ such that

$$V_\tau = V(\tau) = V(\tau, Q) \qquad P\text{-a.s., for each } \tau \in \mathcal{S} \text{ and } Q \in \mathcal{Q}_f^e,$$

and then we have for each $Q \in \mathcal{Q}_f^e$

$$J^Q = S^\delta V + \alpha \int S^\delta_s U_s \,ds + \beta \int \delta_s S^\delta_s \log Z^Q_s \,ds + \beta S^\delta \log Z^Q. \tag{5.2}$$

As P is in $\mathcal{Q}_f^e$ and J^P is a P-submartingale by Proposition 13, (5.2) yields via $J^P = S^\delta V + \alpha \int S^\delta_s U_s \,ds$ that V is a P-special semimartingale. We write its canonical decomposition as

$$V = V_0 + M^V + A^V$$

and want to know more about M^V and A^V. Since S^δ is uniformly bounded from below and J^P is a P-submartingale, (A2) implies that M^V is a P-martingale. In a continuous filtration, we even obtain much stronger results a bit later.

Consider now the semimartingale backward equation

$$\begin{aligned} \mathrm{d}Y_t &= (\delta_t Y_t - \alpha U_t)\,\mathrm{d}t + \frac{1}{2\beta}\mathrm{d}\langle M\rangle_t + \mathrm{d}M_t, \\ Y_T &= \alpha' U'_T. \end{aligned} \tag{5.3}$$

A *solution* of (5.3) is a pair (Y, M) satisfying (5.3), where Y is a P-semimartingale and M is a locally square-integrable local P-martingale null at 0. Note that Y is then automatically P-special, and that if M is continuous, so is Y.

Remark 14. Like the optimization problem (2.1), the BSDE (5.3) becomes much simpler when $\delta \equiv 0$; in fact, one can explicitly write down its solution. This has already been observed in [19], and we come back to this point at the end of this section. ◇

Our main result in this section shows that (V, M^V) is the unique solution of (5.3) if the filtration $\mathbb{F}$ is continuous. As a preliminary, we first establish some auxiliary results about the structure and uniqueness of solutions to (5.3).

Lemma 15. *Assume (A1), (A2) and let (Y, M) be a solution of (5.3) with M continuous. Assume either $Y \in D_0^{\exp}$ or that $\mathcal{E}\left(-\frac{1}{\beta}M\right)$ is a true P-martingale. For any pair of stopping times $\sigma \le \tau$, we then have the recursive relation*

$$Y_\sigma = -\beta \log E_P\left[\exp\left(\frac{1}{\beta}\int_\sigma^\tau (\delta_s Y_s - \alpha U_s)\,\mathrm{d}s - \frac{1}{\beta}Y_\tau\right)\,\middle|\,\mathcal{F}_\sigma\right]. \tag{5.4}$$

Proof. From (5.3), we have

$$Y_\tau - Y_\sigma = \int_\sigma^\tau \mathrm{d}Y_s = \int_\sigma^\tau (\delta_s Y_s - \alpha U_s)\,\mathrm{d}s + M_\tau - M_\sigma + \frac{1}{2\beta}(\langle M\rangle_\tau - \langle M\rangle_\sigma).$$

Divide by $-\beta$, exponentiate and use continuity of M to obtain

$$\frac{\mathcal{E}\left(-\frac{1}{\beta}M\right)_\tau}{\mathcal{E}\left(-\frac{1}{\beta}M\right)_\sigma} = \exp\left(\frac{1}{\beta}Y_\sigma + \frac{1}{\beta}\int_\sigma^\tau (\delta_s Y_s - \alpha U_s)\,\mathrm{d}s - \frac{1}{\beta}Y_\tau\right). \tag{5.5}$$

If $\mathcal{E}\left(-\frac{1}{\beta}M\right)$ is a P-martingale, (5.4) follows directly by conditioning on $\mathcal{F}_\sigma$ and solving for Y_σ. In general, we stop $\mathcal{E}\left(-\frac{1}{\beta}M\right)$ after σ by τ_n to have the P-martingale property and thus obtain (5.5) and (5.4) with $\tau_n \wedge \tau$ instead of τ. Then (A1), (A2) and the assumption that $Y \in D_0^{\exp}$ yield a P-integrable majorant for the right-hand side of (5.5) and so we can use dominated convergence to let $n \to \infty$ and again get (5.4) for τ. □

The argument for the next result is a simple adaptation of the proof for Lemma A2 in [19].

Lemma 16. *1. For any semimartingale Y, there is at most one local P-martingale M such that (Y, M) solves* (5.3).
2. Assume (A1), (A2). Then (5.3) *has at most one solution (Y, M) with $Y \in D_0^{\exp}$ and M continuous.*

Proof. 1. For any solution (Y, M) of (5.3), Y is P-special, and its unique local P-martingale part is M by (5.3).

2. Let (Y, M) and $(\tilde{Y}, \tilde{M})$ be two solutions as stated. Suppose that for some $t \in [0, T]$, the event $A := \{Y_t > \tilde{Y}_t\}$ has $P[A] > 0$. Since $Y_T = \alpha' U_T' = \tilde{Y}_T$, the stopping time $\tau := \inf\{s \geq t \,|\, Y_s \leq \tilde{Y}_s\}$ has values in $[t, T]$, and since $Y, \tilde{Y}$ are both continuous, we have $Y_\tau = \tilde{Y}_\tau$ on A and $Y_s > \tilde{Y}_s$ on $A \cap \{t \leq s < \tau\}$. This implies that

$$\int_t^\tau (\delta_s Y_s - \alpha U_s)\,\mathrm{d}s - Y_\tau > \int_t^\tau (\delta_s \tilde{Y}_s - \alpha U_s)\,\mathrm{d}s - \tilde{Y}_\tau \qquad \text{on } A \in \mathcal{F}_t$$

so that Lemma 15 yields

$$\begin{aligned}\exp\left(-\frac{1}{\beta} Y_t\right) &= E_P\left[\exp\left(\frac{1}{\beta}\int_t^\tau (\delta_s Y_s - \alpha U_s)\,\mathrm{d}s - \frac{1}{\beta} Y_\tau\right) \,\middle|\, \mathcal{F}_t\right] \\ &> \exp\left(-\frac{1}{\beta}\tilde{Y}_t\right) \qquad \text{on } A.\end{aligned}$$

Hence $Y_t < \tilde{Y}_t$ on A, in contradiction to the definition of A, and so Y and $\tilde{Y}$ must be indistinguishable. By part 1., M and $\tilde{M}$ must then coincide as well.
□

Armed with the above results, we can now prove the announced characterization of (V, M^V) as the unique solution of the generalized BSDE (5.3).

Theorem 17. *Assume (A1) – (A3). If $\mathbb{F}$ is continuous, the pair (V, M^V) is the unique solution in $D_0^{\exp} \times \mathcal{M}_{0,\mathrm{loc}}(P)$ of the BSDE*

$$\begin{aligned}\mathrm{d}Y_t &= (\delta_t Y_t - \alpha U_t)\,\mathrm{d}t + \frac{1}{2\beta}\mathrm{d}\langle M\rangle_t + \mathrm{d}M_t, \\ Y_T &= \alpha' U_T'.\end{aligned}$$

Moreover, $\mathcal{E}\left(-\frac{1}{\beta}M^V\right)$ is a true P-martingale.

Proof. 1) We first use the martingale optimality principle to show that (V, M^V) is indeed a solution. For each $Q \in \mathcal{Q}_f^e$, we have $Z^Q = \mathcal{E}(L^Q)$ for some continuous local P-martingale L^Q null at 0 since $Q = P$ on $\mathcal{F}_0$. This implies that $\mathrm{d}(\log Z^Q) = \mathrm{d}L^Q - \frac{1}{2}\,\mathrm{d}\langle L^Q\rangle$, and combining this with Itô's formula applied to (5.2) yields

$$\mathrm{d}J^Q = S^\delta(\mathrm{d}M^V + \mathrm{d}A^V) - \delta S^\delta V\,\mathrm{d}t + \alpha S^\delta U\,\mathrm{d}t + \beta S^\delta\left(\mathrm{d}L^Q - \frac{1}{2}\mathrm{d}\langle L^Q\rangle\right). \tag{5.6}$$

By Girsanov's theorem,

$$N^Q := M^V + \beta L^Q - \langle M^V + \beta L^Q, L^Q\rangle$$

is a local Q-martingale. Together with (5.6), this gives the Q-canonical decomposition

$$\begin{aligned}\mathrm{d}J^Q = S^\delta\,\mathrm{d}N^Q + S^\delta\Big(&\mathrm{d}A^V - \delta V\,\mathrm{d}t + \alpha U\,\mathrm{d}t + \mathrm{d}\left\langle M^V, L^Q\right\rangle \\ &+ \beta\,\mathrm{d}\langle L^Q\rangle - \frac{\beta}{2}\mathrm{d}\langle L^Q\rangle\Big).\end{aligned} \tag{5.7}$$

Because J^Q is by Proposition 13 a Q-submartingale for any $Q \in \mathcal{Q}_f^e$ and a Q^*-martingale for the optimal Q^* (which exists and is in $\mathcal{Q}_f^e$ by Theorem 9 and Theorem 12), the second term in (5.7) is increasing for any $Q \in \mathcal{Q}_f^e$ and constant (at 0) for $Q = Q^*$. Thus we have

$$A^V = \int(\delta V - \alpha U)\,\mathrm{d}t - \operatorname*{ess\,inf}_{Q\in\mathcal{Q}_f^e}\left(\langle M^V, L^Q\rangle + \frac{\beta}{2}\langle L^Q\rangle\right),$$

where the ess inf is taken with respect to the strong order $\preceq$ (so that $A \preceq B$ means that $B - A$ is increasing). Step 2) shows that the ess inf term equals $-\frac{1}{2\beta}\langle M^V\rangle$ so that we get

$$\mathrm{d}V_t = \mathrm{d}M_t^V + \mathrm{d}A_t^V = (\delta_t V_t - \alpha U_t)\,\mathrm{d}t + \frac{1}{2\beta}\mathrm{d}\langle M^V\rangle_t + \mathrm{d}M_t^V.$$

Since clearly from the definitions $V_T = V(T, Q) = \alpha' U_T'$, (5.3) follows with $M = M^V$.

2) We claim that

$$\operatorname*{ess\,inf}_{Q\in\mathcal{Q}_f^e}\left(\langle M^V, L^Q\rangle + \frac{\beta}{2}\langle L^Q\rangle\right) = -\frac{1}{2\beta}\langle M^V\rangle, \tag{5.8}$$

and that the ess inf is attained for $L^{Q^*} = -\frac{1}{\beta}M^V$. To prove this, choose localizing stopping times $(\tau_n)_{n\in\mathbf{N}}$ such that for $L^n := -\frac{1}{\beta}(M^V)^{\tau_n}$, the process $Z^n := \mathcal{E}(L^n)$ is bounded. Then Z^n is a strictly positive P-martingale starting at 1 with $E_P[|Z_T^n \log Z_T^n|] < \infty$ so that $\mathrm{d}Q^n := Z_T^n\,\mathrm{d}P$ defines an element $Q^n \in \mathcal{Q}_f^e$. Moreover, the definition of L^n gives

$$\left\langle M^V, L^n\right\rangle_t + \frac{\beta}{2}\langle L^n\rangle_t = -\frac{1}{\beta}\langle M^V\rangle_{t\wedge\tau_n} + \frac{1}{2\beta}\langle M^V\rangle_{t\wedge\tau_n}$$

and so we get for $n \to \infty$ that

$$\operatorname*{ess\,inf}_{Q\in\mathcal{Q}_f^e}\left(\langle M^V,L^Q\rangle+\frac{\beta}{2}\langle L^Q\rangle\right)\le\lim_{n\to\infty}\left(\langle M^V,L^n\rangle+\frac{\beta}{2}\langle L^n\rangle\right)=-\frac{1}{2\beta}\langle M^V\rangle.$$

Because we also have

$$\langle M^V,L^Q\rangle+\frac{\beta}{2}\langle L^Q\rangle=\frac{\beta}{2}\left\langle L^Q+\frac{1}{\beta}M^V\right\rangle-\frac{1}{2\beta}\langle M^V\rangle\ge-\frac{1}{2\beta}\langle M^V\rangle, \quad (5.9)$$

(5.8) follows. Finally, since the ess inf in (5.8) is attained by Q^* due to Proposition 13, combining (5.8) with (5.9) for $Q=Q^*$ yields

$$-\frac{1}{2\beta}\langle M^V\rangle=\left(\langle M^V,L^{Q^*}\rangle+\frac{\beta}{2}\langle L^{Q^*}\rangle\right)=\frac{\beta}{2}\left\langle L^{Q^*}+\frac{1}{\beta}M^V\right\rangle-\frac{1}{2\beta}\langle M^V\rangle,$$

and this implies that $L^{Q^*}=-\frac{1}{\beta}M^V$.

3) By step 2), $\mathcal{E}\left(-\frac{1}{\beta}M^V\right)=\mathcal{E}\left(L^{Q^*}\right)=Z^{Q^*}$ is a true P-martingale.

4) Since $\mathbb{F}$ is continuous, so is M^V; hence uniqueness follows from Lemma 16 once we show that $V\in D_0^{\exp}$. This is done below in Proposition 20, and completes the proof. □

A closer look at the proof of Theorem 17 shows that we have the following additional integrability property for the P-martingale M^V.

Corollary 18. *Assume (A1) – (A3). If $\mathbb{F}$ is continuous, the optimal measure Q^* is given by $Z^{Q^*}=\mathcal{E}\left(-\frac{1}{\beta}M^V\right)$, and $\mathcal{E}\left(-\frac{1}{\beta}M^V\right)$ is a P-martingale whose supremum is in $L^1(P)$.*

Proof. The first assertion is just step 3) from the preceding proof. Because $Q^*\in\mathcal{Q}_f^e$, we have $E_P\left[Z_T^{Q^*}\log Z_T^{Q^*}\right]=H(Q^*|P)<\infty$, and this implies that $\sup_{0\le t\le T}Z_t^{Q^*}$ is in $L^1(P)$. □

To finish the proof of Theorem 17, it remains to show that $V\in D_0^{\exp}$. We begin with

Lemma 19. *Assume (A1) – (A3). Then the process $(J^P)^+$ is in $D_0^{\exp}$.*

Proof. We have seen in the proof of Lemma 8 that

$$\big(J(\tau,P)\big)^+\le E_P\big[|c(\,\cdot\,,P)|\,\big|\,\mathcal{F}_\tau\big].$$

Now fix $\gamma>0$ and choose for N an RCLL version of the P-martingale $N:=E_P\left[e^{\gamma|c(\,\cdot\,,P)|}\,\middle|\,\mathbb{F}\right]$. Then Proposition 13, right-continuity of J^P and Jensen's inequality imply that

$$\exp\left(\gamma\operatorname*{ess\,sup}_{0\le t\le T}\left(J_t^P\right)^+\right)=\exp\left(\gamma\sup_{0\le t\le T}\left(J_t^P\right)^+\right)\le\sup_{0\le t\le T}N_t. \quad (5.10)$$

Since $S^\delta \le 1$, we have $|c(\,\cdot\,,P)| = |\mathcal{U}^\delta_{0,T}| \le \alpha \int_0^T |U_s|\,\mathrm{d}s + \alpha'|U'_T| =: R$, and since $e^{\gamma R} \in L^p(P)$ for every $p \in (1,\infty)$ by (A2) and (A3), Doob's inequality implies that $\sup_{0\le t\le T} N_t$ is in $L^p(P)$ for every $p \in (1,\infty)$. Hence the assertion follows from (5.10). □

We have already shown that (V, M^V) is a solution of (5.3) and also that $\mathcal{E}\left(-\frac{1}{\beta}M^V\right)$ is a true P-martingale. This allows us now to use Lemma 15 and prove that V inherits the good integrability properties of U and U'_T.

Proposition 20. *Assume (A1) – (A3). If $\mathbb{F}$ is continuous, the process V is in $D_0^{\exp}$.*

Proof. Because $D_0^{\exp}$ is a vector space, it is enough to prove that V^+ and V^- lie both in it. Using (5.4) for V with $\sigma = t$, $\tau = T$ and Jensen's inequality gives

$$-V_t \ge E_P\left[\int_t^T (\delta_s V_s - \alpha U_s)\,\mathrm{d}s - \alpha' U'_T \,\middle|\, \mathcal{F}_t\right]$$

and therefore

$$V_t^+ = V_t + V_t^- \le V_t^- + E_P\left[\|\delta\|_\infty T \sup_{0\le s\le T} V_s^- + \alpha\int_0^T |U_s|\,\mathrm{d}s + \alpha'|U'_T| \,\middle|\, \mathcal{F}_t\right].$$

Due to (A2) and (A3), the same argument via Doob's inequality as in the proof of Lemma 19 shows that the last term is in $D_0^{\exp}$ as soon as V^- is, and this implies then in turn that V^+ is in $D_0^{\exp}$. Hence it only remains to prove that V^- is in $D_0^{\exp}$.

Now (5.2) for $Q = P$ gives

$$\begin{aligned}\delta_s V_s &= \delta_s\left(J_s^P - \alpha\int_0^s S_r^\delta U_r\,\mathrm{d}r\right)\Big/ S_s^\delta \\ &\le \|\delta\|_\infty\left(\sup_{0\le t\le T}\left(J_t^P\right)^+ + \alpha\int_0^T |U_r|\,\mathrm{d}r\right)\mathrm{e}^{\|\delta\|_\infty T},\end{aligned}$$

and combining this with (5.4) for V with $\sigma = t$, $\tau = T$ yields

$$\begin{aligned}-V_t &\le \beta\log\left(1 + E_P\left[\exp\left(\frac{1}{\beta}\int_t^T (\delta_s V_s - \alpha U_s)\,\mathrm{d}s - \frac{1}{\beta}\alpha' U'_T\right)\middle|\, \mathcal{F}_t\right]\right) \quad (5.11)\\ &\le \beta\log\left(1 + E_P\left[\exp\left(\frac{1}{\beta}\|\delta\|_\infty \mathrm{e}^{\|\delta\|_\infty T}\left(\sup_{0\le t\le T}\left(J_t^P\right)^+ + \alpha\int_0^T |U_r|\,\mathrm{d}r\right)\right.\right.\right.\\ &\qquad\left.\left.\left. + \frac{1}{\beta}\alpha\int_0^T |U_s|\,\mathrm{d}s + \frac{1}{\beta}\alpha'|U'_T|\right)\middle|\, \mathcal{F}_t\right]\right)\\ &=: \beta\log\left(1 + E_P[\mathrm{e}^B|\mathcal{F}_t]\right).\end{aligned}$$

Thanks to (A2), (A3) and Lemma 19, the above random variable B satisfies $E_P[\mathrm{e}^{\gamma|B|}] < \infty$ for all $\gamma > 0$. Hence the martingale $E_P[\mathrm{e}^B|\mathbb{F}]$ has its supremum in $L^p(P)$ for every $p \in (1,\infty)$ by Doob's inequality, and this implies by (5.11) that V^- is in $D_0^{\exp}$. □

Remark 21. 1. The above argument rests on continuity of $\mathbb{F}$ because we exploit via Lemma 15 the BSDE for V. However, one feels that the integrability of V should be a general result, and this raises the question if there is an alternative proof for Proposition 20 which works for general $\mathbb{F}$.
2. The BSDE (5.3) is very similar to an equation studied in detail in [15], but has a crucial difference: If the final value $Y_T = \alpha' U'_T$ is unbounded, there is no evident way in which the results from [15] could be used or adapted. ◇

By exploiting the BSDE for (V, M^V), we can show that the P-martingale M^V has very good integrability properties. This adapts an argument in the proof of Lemma A1 from [19].

Proposition 22. *Assume (A1) – (A3). If $\mathbb{F}$ is continuous, then M^V lies in the martingale space $\mathcal{M}_0^p(P)$ for every $p \in [1,\infty)$.*

Proof. Because $V \in D_0^{\exp}$ by Proposition 20, (A1) – (A3) imply via Doob's inequality that the (continuous) P-martingale

$$N := E_P\left[\exp\left(\frac{1}{\beta}\int_0^T (\delta_s V_s - \alpha U_s)\,\mathrm{d}s - \frac{1}{\beta}\alpha' U'_T\right)\middle|\,\mathbb{F}\right]$$

lies in every $\mathcal{M}_0^p(P)$, and so $\langle N\rangle_T \in L^p(P)$ for every p by the BDG inequalities. Moreover, Lemma 15 applied to (V, M^V) with $\sigma = t$, $\tau = T$ yields

$$V_t = -\beta \log N_t + \int_0^t (\delta_s V_s - \alpha U_s)\,\mathrm{d}s, \qquad 0 \le t \le T \tag{5.12}$$

which implies that

$$\frac{1}{N_t} = \exp\left(\frac{1}{\beta}V_t - \frac{1}{\beta}\int_0^t (\delta_s V_s - \alpha U_s)\,\mathrm{d}s\right), \qquad 0 \le t \le T. \tag{5.13}$$

Using (5.3) for (V, M^V) and comparing the local P-martingale parts in (5.12) gives via Itô's formula that $M^V = -\beta \int \frac{1}{N}\,dN$. Combining this with (5.13), we get

$$\begin{aligned}
\langle M^V\rangle_T &= \beta^2 \int_0^T \frac{1}{N_t^2}\,\mathrm{d}\langle N\rangle_t \\
&\le \beta^2 \langle N\rangle_T \sup_{0\le t\le T} \frac{1}{N_t^2} \\
&\le \beta^2 \langle N\rangle_T \exp\left(\frac{2}{\beta}\sup_{0\le t\le T}|V_t|(1 + \|\delta\|_\infty T) + \frac{2}{\beta}\alpha\int_0^T |U_s|\,\mathrm{d}s\right).
\end{aligned}$$

Due to (A1), (A2) and $V \in D_0^{\exp}$, all the terms on the right-hand side are in $L^p(P)$ for every $p \in [1, \infty)$, and hence so is $\langle M^V \rangle_T$ by Hölder's inequality. So the assertion follows by the BDG inequalities. □

We have formulated Theorem 17 as a result on the characterization of the dynamic value process V for the stochastic control problem (2.3). If we want to restate our results in pure BSDE terms, we have also shown

Theorem 23. *Let δ and ρ be progressively measurable processes and B an $\mathcal{F}_T$-measurable random variable. Assume that δ is nonnegative and uniformly bounded, that $\rho \in D_1^{\exp}$ and that $\exp(\gamma|B|) \in L^1(P)$ for every $\gamma > 0$. If the filtration $\mathbb{F}$ is continuous, there exists for every $\beta > 0$ a unique solution $(Y, M) \in D_0^{\exp} \times \mathcal{M}_{0,\mathrm{loc}}(P)$ to the BSDE*

$$\begin{aligned} \mathrm{d}Y_t &= (\delta_t Y_t + \rho_t)\,\mathrm{d}t + \frac{1}{2\beta}\,\mathrm{d}\langle M \rangle_t + \mathrm{d}M_t, \\ Y_T &= B. \end{aligned} \tag{5.14}$$

For this solution, we have $M \in \mathcal{M}_0^p(P)$ for every $p \in [1, \infty)$.

Remark 24. As mentioned above, the BSDE (5.3) or (5.14) can be explicitly solved for $\delta \equiv 0$. This has already been observed in [19], Appendix A; in fact, it follows immediately from Lemma 15 which gives for $\sigma = t$ and $\tau = T$ the representation

$$Y_t = -\beta \log E_P \left[\exp\left(-\frac{1}{\beta} \int_t^T \rho_s \,\mathrm{d}s - \frac{1}{\beta} B \right) \middle| \mathcal{F}_t \right]$$

for the solution of (5.14). Choosing $\rho = \alpha U$ and $B = \alpha' U'_T$ gives the solution to (5.3). ◇

6 A Comparison with Related Results

This section is an attempt to position the results of the present paper in relation to other work in the area. Such a comparison naturally cannot be complete, but we have made an effort to include at least some of the most relevant papers.

6.1 Skiadas (2003) and Schroder/Skiadas (1999)

Our primary inspiration clearly comes from the two papers [19, 21]. In [21], Skiadas studies essentially the same optimization problem as (2.1) or (2.3), and proves that its dynamic value process V can be described by the BSDE

$$\begin{aligned} \mathrm{d}V_t &= (\delta_t V_t - \alpha U_t)\,\mathrm{d}t + \frac{1}{2\beta}|z_t|^2\,\mathrm{d}t + z_t\,\mathrm{d}W_t, \\ V_T &= \alpha' U'_T. \end{aligned} \tag{6.1}$$

This is clearly our BSDE (5.3) specialized to the case of a filtration $\mathbb{F} = \mathbb{F}^W$ generated by a P-Brownian motion W. It is a minor point that [21] only treats the case $\alpha' = 0$. The important differences to our work lie in the interpretation and in the way that [21] derives its results. The main point Skiadas wants to make is that the BSDE (6.1) coincides with one describing a stochastic differential utility; hence working with a standard expected utility under (a particular form of) model uncertainty is observationally equivalent to working with a corresponding stochastic differential utility under one fixed model. For the derivation, Skiadas argues in a first step that (6.1) does have a solution (V^*, z^*), since this is proved in [19]. In a second step, he uses explicit computations to show that z^* induces an optimal measure Q^*: in our terminology, he proves for every $\tau \in \mathcal{S}$ that

$$V_\tau = \tilde{V}(\tau, Q^*) \leq \tilde{V}(\tau, Q') \qquad \text{for every } Q' \text{ with } Z^{Q'} \in \mathcal{D}(Q^*, \tau).$$

However, this approach has a disadvantage. The existence proof for (V^*, z^*) relies on a fixed point argument in [19], and thus from the beginning uses the assumption that $\mathbb{F} = \mathbb{F}^W$. (One could slightly generalize this fixed point method to a continuous filtration; see forthcoming work by G. Bordigoni.) In contrast, our method first shows for a general filtration the existence of an optimal measure Q^*. Only then do we assume and use continuity of $\mathbb{F}$ to deduce via the martingale optimality principle that V satisfies a BSDE. As a further minor point, the integrability of M^V in Proposition 22 is not given in [19].

An alternative proof for the result in [21] can be found in [13]. These authors also assume $\mathbb{F} = \mathbb{F}^W$ and in addition impose the severe condition that U and U'_T are bounded. The argument then uses a comparison result for BSDEs from [12].

6.2 Robustness, Control and Portfolio Choice

Our second important source of inspiration has been provided by the work of L.P. Hansen and T. Sargent with coauthors; see for instance the homepage of Hansen at the URL `http://home.uchicago.edu/~lhansen`. We explicitly mention here the two papers [1, 11] which also contain more references. They both introduce and discuss (in slightly different ways) the basic problem of robust utility maximization when model uncertainty is penalized by a relative entropy term. Both papers are cast in Markovian settings and use mainly formal manipulations of Hamilton–Jacobi–Bellman (HJB) equations to provide insights about the optimal investment behaviour in these situations. While the authors of [11] find that "One Hamilton–Jacobi–Bellman (HJB) equation is worth a thousand words", our (still partial) analysis here is driven by a desire to obtain rigorous results in a general setting by stochastic methods.

The related paper [14] studies (also via formal HJB analysis) a problem where the penalization parameter β is allowed to depend on V; this is also briefly discussed in [21]. And when the present paper was almost finished, we discovered that A. Schied has also been working on the problem (2.1) with a fairly general penalization term for Q; see [17]. However, his (static) results do not contain ours even without the dynamic parts in Section 5 — they only cover as one example the simple case $\delta \equiv 0$.

6.3 BSDEs with Quadratic Drivers

In the setting of a Brownian filtration $\mathbb{F} = \mathbb{F}^W$, the pure BSDE (5.14) takes the form

$$\begin{aligned} \mathrm{d}Y_t &= \left(\delta_t Y_t + \rho_t + \frac{1}{2\beta}|z_t|^2 \right) \mathrm{d}t + z_t \,\mathrm{d}W_t, \\ Y_T &= B. \end{aligned} \tag{6.2}$$

This is one particular BSDE with a driver (dt-term) which is quadratic in the z-variable. Such BSDEs have been much studied recently and typically appear in problems from mathematical finance; see [5] for probably the first appearance of such a BSDE (derived in the context of stochastic differential utility), and for instance [7, 8, 20] for some recent references. However, almost all (existence and comparison) results for these equations (with nonvanishing quadratic term) assume that the terminal value B is bounded. This condition is too restrictive for our purposes and seems very difficult to get rid of. A class of BSDEs with quadratic growth and unbounded terminal value has recently been studied in [3], but (6.2) does not satisfy the assumptions of that paper as soon as ρ is unbounded.

Acknowledgments

GB thanks Politecnico di Milano and in particular Marco Fuhrman for providing support. AM thanks the members of the financial and insurance mathematics group for their warm hospitality during his visit at ETH Zürich in December 2004. MS thanks Stefan Geiss for a number of very stimulating discussions around the topic of this paper.

References

1. E. Anderson, L. P. Hansen, and T. Sargent. A quartet of semigroups for model specification, robustness, prices of risk, and model detection. *Journal of the European Economic Association*, 1:68–123, 2003.
2. G. Bordigoni. Robust utility maximization with an entropic penalty term: Stochastic control and BSDE methods. Master's thesis, ETH Zürich and University of Zürich, February 2005. `http://www.msfinance.ch/pdfs/GiulianaBordigoni.pdf`.

3. P. Briand and Y. Hu. BSDE with quadratic growth and unbounded terminal value. *Probability Theory and Related Fields*, 136:604–618, 2006.
4. I. Csiszár. I-divergence geometry of probability distributions and minimization problems. *Annals of Probability*, 3:146–158, 1975.
5. D. Duffie and L. G. Epstein. Stochastic differential utility. *Econometrica*, 60:353–394, 1992.
6. N. El Karoui. Les aspects probabilistes du contrôle stochastique. In *École d'Été de Probabilités de Saint Flour IX*, Lecture Notes in Mathematics 876, pages 73–238. Springer, 1981.
7. N. El Karoui and S. Hamadène. BSDEs and risk-sensitive control, zero-sum and nonzero-sum game problems of stochastic functional differential equations. *Stochastic Processes and their Applications*, 107:145–169, 2003.
8. N. El Karoui, S. Peng, and M.-C. Quenez. A dynamic maximum principle for the optimization of recursive utilities under constraints. *Annals of Applied Probability*, 11:664–693, 2001.
9. M. Frittelli. The minimal entropy martingale measure and the valuation problem in incomplete markets. *Mathematical Finance*, 10:39–52, 2000.
10. A. Gundel. Robust utility maximization for complete and incomplete market models. *Finance and Stochastics*, 9:151–176, 2005.
11. L. P. Hansen, T. J. Sargent, G. A. Turmuhambetova, and N. Williams. Robust control and model misspecification. *Journal of Economic Theory*, 128:45–90, 2006.
12. M. Kobylanski. Backward stochastic differential equations and partial differential equations with quadratic growth. *Annals of Probability*, 28:558–602, 2000.
13. A. Lazrak and M.-C. Quenez. A generalized stochastic differential utility. *Mathematics of Operations Research*, 28:154–180, 2003.
14. P. Maenhout. Robust portfolio rules and asset pricing. *Review of Financial Studies*, 17:951–983, 2004.
15. M. Mania and M. Schweizer. Dynamic exponential utility indifference valuation. *Annals of Applied Probability*, 15:2113–2143, 2005.
16. M.-C. Quenez. Optimal portfolio in a multiple-priors model. In R. Dalang, M. Dozzi, and F. Russo, editors, *Seminar on Stochastic Analysis, Random Fields and Applications IV*, volume 58 of *Progress in Probability*, pages 292–321. Birkhäuser, 2004.
17. A. Schied. Optimal investments for risk- and ambiguity-averse preferences: a duality approach. *Finance and Stochastics*, 11:107–129, 2007.
18. A. Schied and C.-T. Wu. Duality theory for optimal investments under model uncertainty. *Statistics & Decisions*, 23:199–217, 2005.
19. M. Schroder and C. Skiadas. Optimal consumption and portfolio selection with stochastic differential utility. *Journal of Economic Theory*, 89:68–126, 1999.
20. M. Schroder and C. Skiadas. Lifetime consumption-portfolio choice under trading constraints, recursive preferences, and nontradeable income. *Stochastic Processes and their Applications*, 115:1–30, 2005.
21. C. Skiadas. Robust control and recursive utility. *Finance and Stochastics*, 7:475–489, 2003.

Extending Markov Processes in Weak Duality by Poisson Point Processes of Excursions

Zhen-Qing Chen[1], Masatoshi Fukushima[2], and Jiangang Ying[3]

[1] Department of Mathematics, University of Washington, Seattle, WA 98195, USA, zchen@math.washington.edu. The research of this author is supported in part by NSF Grant DMS-0303310

[2] Department of Mathematics, Kansai University, Suita, Osaka 564-8680, Japan, fuku2@mx5.canvas.ne.jp. The research of this author is supported in part by Grant-in-Aid for Scientific Research of MEXT No. 15540142

[3] Department of Mathematics, Fudan University, Shanghai, China, jgying@fudan.edu.cn. The research of this author is supported in part by NSFC No. 10271109

Dedicated to Professor Kiyosi Itô on the occasion of his 90th birthday

Summary. Let a be a non-isolated point of a topological space E. Suppose we are given standard processes X^0 and $\widehat{X}^0$ on $E_0 = E \setminus \{a\}$ in weak duality with respect to a σ-finite measure m on E_0 which are of no killings inside E_0 but approachable to a. We first show that their extensions X and $\widehat{X}$ to E admitting no sojourn at a and keeping the weak duality are uniquely determined by the approaching probabilities of X^0, $\widehat{X}^0$ and m up to a non-negative constant δ_0 representing the killing rate of X at a. We then construct, starting from X^0, such X by piecing together returning excursions around a and a possible non-returning excursion including the instant killing. This extends a recent result by M. Fukushima and H. Tanaka [16] which treats the case where X^0, X are m-symmetric diffusions and X admits no sojourn nor killing at a. Typical examples of jump type symmetric Markov processes and non-symmetric diffusions on Euclidean domains are given at the end of the paper.

1 Introduction

Let a be a non-isolated point of a topological space E and $X^0 = \{X_t^0, \zeta^0, \mathbf{P}_x^0\}$ be a strong Markov process on $E_0 = E \setminus \{a\}$ which admits no killings inside E_0 and satisfies

$$\varphi(x) := \mathbf{P}_x^0(\zeta^0 < \infty, X_{\zeta^0-}^0 = a) > 0 \qquad \text{for every } x \in E_0.$$

We are concerned with a strong Markovian extension X of X^0 from E_0 to E such that X admits no sojourn at the one-point set $\{a\}$. Natural questions arise: is X uniquely determined by X^0 and how can it be constructed from X^0?

When both X^0 and X are required to be diffusions that are symmetric with respect to a σ-finite measure m on E_0 with $m(\{a\}) = 0$, affirmative answers to these questions were given quite recently in M. Fukushima and H. Tanaka [16]. It is shown in [16] that the entrance law and the absorption rate for the absorbed Poisson point processes of excursions attached to X away from a (due to K. Itô [24] and P.A. Meyer [M]) are uniquely determined by the approaching probability φ to a for X^0 and the measure m, yielding the uniqueness of the extension X that admits no sojourn nor killing at the point a. Conversely such extension X can be constructed from X^0 by piecing together the associated returning excursions and possibly a non-returning one away from a.

The purpose of the present paper is to generalize the stated results of [16] to general standard processes X^0 and X which are not necessarily symmetric but admitting weak dual standard processes $\widehat{X}^0$ and $\widehat{X}$, respectively. We can no longer use the Dirichlet form theory which has played an important role in [16].

Nevertheless, the entrance law and the absorption rate for the absorbed Poisson point process of excursions of X at the point a can still be identified in §2 and §3 in terms of the approaching probabilities to a by X^0 and $\widehat{X}^0$ and m owing to the recent works on the exit system by P.J. Fitzsimmons and R.G. Getoor [12] and by the present authors [5]. It turns out that we must allow the killings of X and $\widehat{X}$ at the point a in order to preserve the duality of X^0 and $\widehat{X}^0$ so that the uniqueness of extensions holds only up to a parameter δ_0 that represents the killing rate of X at a (see Theorem 4.2.).

In §5, we shall construct such an extension X starting from X^0 by piecing together the returning excursions around a and possibly a non-returning excursion from a including a killing at a. X^0 and its dual $\widehat{X}^0$ are assumed to be of no killings inside E_0. The sample path of the constructed process X is cadlag and is continuous at the times t when $X_t = a$. If X^0 is of continuous sample path, then so is X. In this construction, we can proceed along essentially the same line laid in [16] although some natural additional conditions on X^0 and $\widehat{X}^0$ including an off-diagonal finiteness of jumping measures will be required due to the lack of the symmetry and the path continuity. But we shall see that an integrability condition of the α-order approaching probability being imposed on X^0 in [16] can be removed under a fairly general circumstance.

As a typical example of a jump type Markov process, we consider in §6 the case where X^0 is a censored symmetric α-stable process on an open set of $\mathbb{R}^n$ studied by K. Bogdan, K. Burdzy and Z.-Q. Chen [3]. An example is also given on extending non-symmetric diffusions in Euclidean domains. Finally we remark at the end of §6 that the present results on the one point extensions can be applied to obtaining an extension to infinitely many points.

2 Exit System and Point Process of Excursions Around a Point

Let E be a Lusin space (i.e. a space that is homeomorphic to a Borel sunset of a compact metric space), $\mathcal{B}(E)$ be the Borel σ-algebra on E and m be a σ-finite Borel measure on E. We consider a pair of Borel right processes $X = (X_t, \zeta, \mathbf{P}_x)$ and $\widehat{X} = (\widehat{X}_t, \widehat{\zeta}, \widehat{\mathbf{P}}_x)$ on E that are in weak duality with respect to m:

(C.1) $\int_E \widehat{G}_\alpha f(x) g(x) m(dx) = \int_E f(x) G_\alpha g(x) m(dx)$
for every $f, g \in \mathcal{B}^+(E)$ and $\alpha > 0$, where G_α, $\widehat{G}_\alpha$ denote the resolvents of X, $\widehat{X}$ respectively.

We fix a point $a \in E$ which is regular for $\{a\}$ with respect to X:

(C.2) $\mathbf{P}_a(\sigma_a = 0) = 1$.
Here $\sigma_a = \inf\{t > 0 : X_t = a\}$ with the convention of $\inf \varnothing := \infty$.

Under **(C.1)**, we may and do assume that both X and $\widehat{X}$ are of cadlag paths up to their lifetimes (c. [21, §9]).

Let $E_0 := E \setminus \{a\}$, $m_0 := m|_{E_0}$, and

$$\varphi(x) := \mathbf{P}_x(\sigma_a < \infty), \qquad u_\alpha(x) := \mathbf{E}_x\left[e^{-\alpha\sigma_a}\right] \quad \text{for every } x \in E.$$

The corresponding functions for $\widehat{X}$ will be denoted by $\widehat{\varphi}$ and $\widehat{u}_\alpha(x)$, respectively. For $u, v \in \mathcal{B}^+(E_0)$, (u, v) will denote the inner product of u and v in $L^2(E_0; m_0)$, that is, $(u, v) := \int_{E_0} u(x) v(x) m_0(dx)$.

Denote by $X^0 = (X_t^0, \zeta^0, \mathbf{P}_x^0)$ and $\widehat{X}^0 = (\widehat{X}_t^0, \widehat{\zeta}^0, \widehat{\mathbf{P}}_x^0)$ the subprocesses of X and $\widehat{X}$ killed upon leaving E_0, respectively. It is known that they are in weak duality with respect to m_0. The X^0-energy functional $L^{(0)}(\widehat{\varphi} \cdot m_0, v)$ of the X^0-excessive measure $\widehat{\varphi} \cdot m_0$ and an X^0-excessive function v is then well defined by

$$L^{(0)}(\widehat{\varphi} \cdot m_0, v) = \lim_{t \downarrow 0} \frac{1}{t}(\widehat{\varphi} - \widehat{P}_t^0 \widehat{\varphi}, v),$$

where $\widehat{P}_t^0$ is the transition semigroup of $\widehat{X}^0$ (see [16, Lemma 2.1]).

We shall now work with the exit system of X for the point a. To this end, it is convenient to take as the sample space Ω of the process X the space of all paths ω on $E_\Delta = E \cup \Delta$ which are cadlag up to the lifetime $\zeta(\omega)$ and stay at the cemetery Δ after ζ. Thus, $X_t(\omega)$ is just t-th coordinate of ω. Ω is equipped with the minimal completed admissible filtration $\{\mathcal{F}_t,\ t \geq 0\}$ for $\{X_t,\ t \geq 0\}$. The shift operator θ_t is defined by $X_s(\theta_t\omega) = X_{s+t}(\omega),\ s \geq 0$. We also introduce an operator k_t, $t \geq 0$, on Ω defined by

$$X_s(k_t\omega) = \begin{cases} X_s(\omega) & \text{if} \quad s < t \\ \Delta & \text{if} \quad s \geq t. \end{cases}$$

We adopt the usual convention that any numerical function of E is extended to E_Δ by setting its value at Δ to be zero.

Let us consider the random time set $M(\omega)$

$$M(\omega) := \overline{\{t \in [0,\infty) : X_t(\omega) = a\}}, \tag{2.1}$$

where $^-$ indicates the closure in $[0,\infty)$. The random set $M(\omega)$ is closed and homogeneous on $[0,\infty)$.

Define $R_t(\omega) := t + \sigma_a(\theta_t\omega)$ for every $t > 0$ and $L(\omega) := \sup\{s > 0 : s \in M(\omega)\}$, with the convention that $\sup\emptyset := 0$. The connected components of the open set $[0,\infty) \setminus M(\omega)$ are called the excursion intervals. The collection of the strictly positive left end points of excursion intervals will be denoted by $G(\omega)$. We can easily see that

$$t \in G(\omega) \quad \text{if and only if} \quad R_{t-}(\omega) < R_t(\omega),$$

and in this case $R_{t-}(\omega) = t$. In particular, $L(\omega) \in G(\omega)$ whenever $L(\omega) < \infty$. We further define the operator i_t, $t \geq 0$, on Ω by $i_t = k_{\sigma_a} \circ \theta_t$. Then

$$\{i_s\omega : s \in G\} \qquad \text{and} \qquad \{i_s\omega : s \in G, R_s < \infty\}$$

are by definition the collection of excursions and the collection of returning excursions respectively of the path ω away from F, while $i_{L(\omega)}(\omega) = \theta_{L(\omega)}(\omega)$ is the non-returning excursion whenever $L(\omega) < \infty$.

Note that those excursions belong to the excursion space W specified by

$$W = \{k_{\sigma_a}\omega : \omega \in \Omega, \sigma_a(\omega) > 0\}, \tag{2.2}$$

which can be decomposed as

$$W = W^+ \cup W^- \cup \{\partial\} \tag{2.3}$$

with

$$W^+ = \{w \in W : \sigma_a < \infty\} \quad \text{and} \quad W^- = \{w \in W : \sigma_a = \infty \text{ and } \zeta > 0\}.$$

Here ∂ denotes the path identically equal to Δ.

The unit mass $\delta_{\{a\}}$ concentrated at the point a is smooth in the sense of [11] because $\{a\}$ is not semipolar by the assumption **(C.2)**. Hence there is a unique positive continuous additive functional (PCAF in abbreviation) $\ell = \{\ell_t, t \geq 0\}$ of X with Revuz measure $\delta_{\{a\}}$. Clearly ℓ is supported by $\{a\}$ and any PCAF of X supported by $\{a\}$ is a constant multiple of ℓ. We call ℓ the local time of X at the point a.

Since the point a is assumed to be regular for $\{a\}$, $\{t \geq 0 : X_t = a\}$ has no isolated points, and the equilibrium 1-potential $\mathbf{E}_x[e^{-\sigma_a}]$ is regular in the sense of [2, Definition IV.3.2] because $\mathbf{E}^x[e^{-\sigma_a}] = c\,\mathbf{E}_x\left[\int_0^\infty e^{-t}d\ell_t\right]$ on E for some $c > 0$. Thus according to [26, §9] (see also [1,8,12] and [20]), there exists a unique σ-finite measure $\mathbf{P}^*$ on Ω carried by $\{\sigma_a > 0\}$ and satisfying

$$\mathbf{P}^* \left[1 - e^{-\sigma_a}\right] < \infty \tag{2.4}$$

such that

$$\mathbf{E}_x \left[\sum_{s \in G} Z_s \cdot \Gamma \circ \theta_s \right] = \mathbf{P}^*(\Gamma) \cdot \mathbf{E}_x \left[\int_0^\infty Z_s d\ell_s \right] \qquad \text{for } x \in E, \tag{2.5}$$

for every non-negative predictable process Z and every non-negative random variable Γ on Ω. Here $\mathbf{E}^*$ is the expectation under the law $\mathbf{P}^*$. The pair $(\mathbf{P}^*, \ell)$ is the predictable version of the exist system for a originated in Maisonneuve [26, §9]. The measure $\mathbf{P}^*$ is Markovian with respect to the transition semigroup of X. We are particularly concerned with the σ-finite measure $\mathbf{Q}^*$ on the space of excursions W induced from $\mathbf{P}^*$ by $\mathbf{Q}^*(\Gamma) = \mathbf{E}_*(\Gamma \circ k_{\sigma_a})$. The measure $\mathbf{Q}^*$ is Markovian with respect to the semigroup $\{P_t^0, t \geq 0\}$ of X^0 and satisfies

$$\mathbf{E}_x \left[\sum_{s \in G} Z_s \cdot \Gamma \circ i_s \right] = \mathbf{Q}^*[\Gamma] \cdot \mathbf{E}_x \left[\int_0^\infty Z_s d\ell_s \right], \qquad x \in E, \tag{2.6}$$

for every non-negative predictable process Z_s and every non-negative random variable Γ on W.

We define for $f \in \mathcal{B}^+(E)$

$$\nu_t(f) := \mathbf{Q}^*[f(X_t)] = \mathbf{E}^*[f(X_t); t < \sigma_a], \qquad t > 0.$$

By the Markov property of $\mathbf{Q}^*$, we readily see that $\{\nu_t : t > 0\}$ is an entrance law for X^0: $\nu_t P_s^0 = \nu_{t+s}$.

Proposition 2.1. (i) *$\{\nu_t\}_{t>0}$ is the unique X^0-entrance law characterized by*

$$\widehat{\varphi} \cdot m_0 = \int_0^\infty \nu_t \, dt. \tag{2.7}$$

Moreover $\nu_t(E_0)$ is finite for each $t > 0$.

(ii) $\mathbf{Q}^*[W^-] = L^{(0)}(\widehat{\varphi} \cdot m_0, 1 - \varphi)$.

Proof. (i). We put $\check{\nu}_\alpha(f) = \displaystyle\int_0^\infty e^{-\alpha t} \nu_t(f) dt$. Then, for $f \in \mathcal{B}_b^+(E)$ and for $v \in C_b(E)$ vanishing at a, we have, using **(C.1)**, (2.6) and the Revuz formula [21, (2.13)],

$$\begin{aligned}
&(\widehat{u}_\alpha, v)\widehat{G}_\alpha f(a) = (\widehat{G}_\alpha f - \widehat{G}_\alpha^0 f, v) = (f, G_\alpha v - G_\alpha^0 v) \\
&= \mathbf{E}_{f\cdot m} \left[\int_{\sigma_a}^\infty e^{-\alpha t} v(X_t) 1_{M^c}(t) dt \right] = \mathbf{E}_{f\cdot m} \left[\sum_{s \in M} \int_s^{s+\sigma_a \circ \theta_s} e^{-\alpha t} v(X_t) dt \right] \\
&= \mathbf{E}_{f\cdot m} \left[\sum_{s \in M} e^{-\alpha s} \int_0^{\sigma_a} e^{-\alpha t} v(X_t) dt \circ \theta_s \right] = \check{\nu}_\alpha(v) \mathbf{E}_{f\cdot m} \left[\int_0^\infty e^{-\alpha s} d\ell_s \right] \\
&= \check{\nu}_\alpha(v) \widehat{G}_\alpha f(a).
\end{aligned}$$

Hence

$$\widehat{u}_\alpha \cdot m_0 = \check{\nu}_\alpha, \tag{2.8}$$

from which (2.7) follows by letting $\alpha \downarrow 0$. Since $\widehat{\varphi} \cdot m_0$ is a purely excessive measure of X^0, the uniqueness follows (cf. [20]). The finiteness of ν_t follows from (2.4).

(ii) By (i) and [5, Lemma 3.1], $L^{(0)}(\widehat{\varphi} \cdot m_0, v) = \lim_{t \downarrow 0} \nu_t(v)$ for any X^0-excessive function v. Hence

$$\begin{aligned} L^{(0)}(\widehat{\varphi} \cdot m_0, 1-\varphi) &= \lim_{t\downarrow 0} \mathbf{Q}^*[(1-\varphi)(X_t)] = \lim_{t\downarrow 0} \mathbf{Q}^*[1_{\sigma_a=\infty} \circ \theta_t;\ t<\zeta \wedge \sigma_a] \\ &= \mathbf{Q}^*[W^-]. \end{aligned}$$

□

Remark 1. In the next section, we shall identify $\mathbf{Q}^*$ with the characteristic measure $\mathbf{n}$ of the absorbed Poisson point process of excursions associated with ℓ. Proposition 2.1. was first proved by Fukushima-Tanaka [16] for $\mathbf{n}$ in the case that X is an m-symmetric diffusion by making use of the Dirichlet form of X. In a recent paper of Fitzsimmons-Getoor [12], various properties of some basic quantities for the exit system of a one point set including those in the above proposition have been obtained in the most general setting that X is just a Borel right process with an excessive measure m, in which case $\widehat{X}$ can be taken to be a dual left continuous moderate Markov process. But the present proof, taken from a recent paper by Chen-Fukushima-Ying [5], is simpler under the condition **(C.1)** as far as Proposition 2.1. is concerned.

The next proposition is taken from Fitzsimmons-Getoor [12, (2.10) and (2.17)]. Recall that $L(\omega) := \sup M(\omega)$.

Proposition 2.2. *Put $\delta = \mathbf{P}^*(\sigma_a = \infty)$. Then the followings are true:*
(i) $\mathbf{P}_a(\ell_\infty > t) = \exp(-\delta t), \qquad t > 0.$

(ii) $\mathbf{P}_a(L < \infty) = 0$ *or* 1 *according to* $\delta = 0$ *or* $\delta > 0$.

Let $\{\tau_t, t \geq 0\}$ be the right continuous inverse of $\ell = \{\ell_t, t \geq 0\}$, that is,

$$\tau_t := \inf\{s \geq 0 : \ell_s > t\}, \tag{2.9}$$

with the convention that $\inf \emptyset = \infty$. Since ℓ is supported by a, we have (cf. [4, §5]) $\mathbf{P}_a$-a.s.

$$\tau_{\ell_t} = R_t \qquad \text{for every } t \geq 0.$$

We see from the above that, after removing from Ω a $\mathbf{P}_a$-negligible set,

$$L(\omega) < \infty \quad \text{ if and only if } \ell_\infty(\omega) < \infty,$$

and in this case,

$$\ell_\infty(\omega) = \ell_L(\omega), \quad \tau_{\ell_\infty -}(\omega) = L(\omega) \quad \text{and} \quad \tau_{\ell_\infty}(\omega) = \infty.$$

Hence, if we let

$$J_\ell(\omega) := \{s \in (0,\infty) : \tau_{s-}(\omega) < \tau_s(\omega)\},$$

then

$$J_\ell(\omega) := \{\ell_t : t \in G(\omega)\} \tag{2.10}$$

and $s \in J_\ell(\omega)$ implies that $s = \ell_t(\omega)$ for some $t \in G(\omega)$ with $\tau_{s-}(\omega) = R_{t-}(\omega) = t$ and $\tau_s(\omega) = R_t(\omega)$.

In particular, $\ell_\infty(\omega) \in J_\ell(\omega)$ whenever it is finite.

Finally the W-valued point process $\mathbf{p} = \mathbf{p}(\omega)$ associated with the local time ℓ is introduced by

$$\mathcal{D}_{\mathbf{p}(\omega)} = J_\ell(\omega) \qquad \text{and} \qquad \mathbf{p}_s(\omega) = i_{\tau_{s-}}\omega \ \text{ for } s \in \mathcal{D}_{\mathbf{p}(\omega)}. \tag{2.11}$$

Note that $\{\mathbf{p}_s(\omega) : s \in \mathcal{D}_{\mathbf{p}(\omega)}\} \subset W$ and $\{\mathbf{p}_s(\omega) : s \in \mathcal{D}_{\mathbf{p}(\omega)},\ \tau_s < \infty\} \subset W^+$ is the collections of excursions and of the returning excursions away from a, respectively, while $\mathbf{p}_{\ell_\infty}(\omega)(= \theta_L(\omega)) \in W^- \cup \{\partial\}$ is the non-returning excursion whenever $\ell_\infty(\omega) < \infty$ or, equivalently, $L(\omega) < \infty$.

The counting measure of $\mathbf{p}$ is defined by

$$n_{\mathbf{p}}((s,t],\Lambda) = \sum_{u \in \mathcal{D}_{\mathbf{p}} \cap (s,t]} 1_\Lambda(\mathbf{p}_u), \qquad \Lambda \in \mathcal{B}(W), \tag{2.12}$$

and $n_{\mathbf{p}}(t,\Lambda) = n_{\mathbf{p}}((0,t],\Lambda)$ is then $\mathcal{F}_{\tau_t}$-adapted as a process in $t \geq 0$.

Using (2.10), we now make the time substitute in the relation (2.6) to obtain

$$\mathbf{E}_a\left[\sum_{s \in J_\ell} Z_{\tau_{s-}} \cdot \Gamma \circ i_{\tau_{s-}}\right] = \mathbf{Q}^*[\Gamma] \cdot \mathbf{E}_a\left[\int_0^{\ell_\infty} Z_{\tau_s} ds\right]. \tag{2.13}$$

Inserting the predictable process $Z_u = 1_{(0,\tau_{t-}]}(u)$, we arrive at the formula holding for the counting measure of the point process $\mathbf{p}$ associated with ℓ:

$$\mathbf{E}_a[n_{\mathbf{p}}(t,\Lambda)] = \mathbf{Q}^*[\Lambda] \cdot \mathbf{E}_a[t \wedge \ell_\infty] \qquad \text{for every } t \geq 0 \text{ and } \Lambda \in \mathcal{B}(W). \tag{2.14}$$

This formula will be utilized in the next section.

3 Characteristic Measure of Absorbed Poisson Point Process

In this section, we continue to work with the setting in §2 and investigate properties of the point process $(\mathbf{p}_s, \mathcal{D}_{\mathbf{p}})$ defined by (2.11) for the local time

$\ell = \{\ell_t, t \geq 0\}$ at the point a. By the observation made after (2.11), it then holds that

$$\ell_\infty = T \quad \text{where} \quad T = \inf\{s > 0 : \mathbf{p}_s \in W^- \cup \{\partial\}\}. \tag{3.1}$$

In view of Proposition 2.2., T is exponentially distributed with parameter $\delta = \mathbf{P}^*(\sigma_a = \infty)$.

Lemma 3.1. *Under measure $\mathbf{P}_a$, $\mathbf{p}$ is an absorbed Poisson point process with absorption time T in Meyer's sense ([M]), that is,*

$$\begin{aligned}\mathbf{P}_a\Big(n_{\mathbf{p}}((r+s_1, r+t_1], \Lambda_1) \in H_1, \cdots, n_{\mathbf{p}}((r+s_n, r+t_n], \Lambda_n) \in H_n \,\Big|\, \mathcal{F}_{\tau_r}\Big) \\ = 1_{\{T>r\}}\mathbf{P}_a\left(n_{\mathbf{p}}((s_1, t_1], \Lambda_1) \in H_1, \cdots, n_{\mathbf{p}}((s_n, t_n], \Lambda_n) \in H_n\right) \\ +1_{\{T\leq r\}}1_{H_1}(0)\cdots 1_{H_n}(0),\end{aligned} \tag{3.2}$$

for any $s_1 < t_1, \cdots, s_n < t_n$, $H_1, \cdots, H_n \subset \mathbb{Z}_+$, $r > 0$, $\Lambda_1, \cdots, \Lambda_n \in \mathcal{B}(W)$.

Proof. The proof is the same as in [M, §2] although [M] considered only the conservative case. In fact, the identity $\tau_{r+u} = \tau_r + \tau_u \circ \theta_{\tau_r}$ implies $n_{\mathbf{p}}((r+s, r+t], \Lambda) = n_{\mathbf{p}}((s,t], \Lambda) \circ \theta_{\tau_r}$ and consequently we see from (3.1) and the strong Markov property of X that the left hand side of (3.2) (with $n = 1$) equals

$$\begin{aligned}\mathbf{P}_{X_{\tau_r}}(n_{\mathbf{p}}((s,t], \Lambda) \in H) = 1_{\{T>r\}}\mathbf{P}_a(n_{\mathbf{p}}((s,t], \Lambda) \in H) \\ +1_{\{T\leq r\}}\mathbf{P}_\Delta(n_{\mathbf{p}}((s,t], \Lambda) \in H),\end{aligned}$$

whose last factor is equal to $1_H(0)$. □

By virtue of [M, §1], there is on a certain probability space $(\widetilde{\Omega}, \widetilde{\mathbf{P}})$ a W-valued Poisson point process $\widetilde{\mathbf{p}} = \{\widetilde{\mathbf{p}}, s > 0\}$ with domain $\mathcal{D}_{\widetilde{\mathbf{p}}}$ satisfying the following property.

Let $\widetilde{T} = \inf\{s > 0 : \widetilde{\mathbf{p}}_s \in W^- \cup \{\partial\}\}$ and consider the stopped process $\{\overline{\mathbf{p}}_s, s > 0\}$:

$$\overline{\mathbf{p}}_s = \widetilde{\mathbf{p}}_s \quad \text{for} \quad s \in \mathcal{D}_{\overline{\mathbf{p}}} = \mathcal{D}_{\widetilde{\mathbf{p}}} \cap (0, \widetilde{T}]. \tag{3.3}$$

Then the point process $\{\mathbf{p}_s, s > 0\}$ under $\mathbf{P}_a$ and $\{\overline{\mathbf{p}}_s, s > 0\}$ under $\widetilde{\mathbf{P}}$ are equivalent in law.

Let us denote by $\mathbf{n}$ the characteristic measure of the W-valued Poisson point process $\{\widetilde{\mathbf{p}}, s > 0\}$.

Theorem 3.2. *It holds that*

$$\mathbf{n} = \mathbf{Q}^*. \tag{3.4}$$

Therefore $\mathbf{n}$ is a σ-finite measure on W with $\mathbf{n}(\sigma_a > t) < \infty$ for every $t > 0$, and $\mathbf{n}$ is Markovian with respect to the transition semigroup $\{P_t^0, t \geq 0\}$ of X^0. The X^0-entrance law $\{\nu_t, t > 0\}$ of $\mathbf{n}$ defined by

$$\nu_t(f) = \mathbf{n}(f(X_t); t < \sigma_a), \quad t > 0, \quad f \in \mathcal{B}^+(E)$$

is characterized by

$$\int_0^\infty \nu_t \, dt = \widehat{\varphi} \cdot m_0. \tag{3.5}$$

Define δ_0 *by*

$$\delta_0 = \mathbf{n}(\zeta = 0). \tag{3.6}$$

Then $\widetilde{T}$ *is exponentially distributed with parameter* $L^{(0)}(\widehat{\varphi} \cdot m_0, 1 - \varphi) + \delta_0$*:*

$$\widetilde{\mathbf{P}}(\widetilde{T} > t) = \exp\left(-t\left(L^{(0)}(\widehat{\varphi} \cdot m_0, 1 - \varphi) + \delta_0\right)\right) \qquad \text{for every } t > 0. \tag{3.7}$$

Moreover, $\nu_t(E_0) < \infty$ *for each* $t > 0$ *and* $L^{(0)}(\widehat{\varphi} \cdot m_0, 1 - \varphi) < \infty$.

Proof. Since $\{\widetilde{\mathbf{p}}_s : s \in \mathcal{D}_{\widetilde{\mathbf{p}}}, \ \widetilde{\mathbf{p}}_s \in W^+\}$ and $\widetilde{T}$ are independent, we have by (3.1)

$$\mathbf{E}_a[n_{\mathbf{p}}(t, \Lambda)] = \widetilde{\mathbf{E}}\left[\sum_{u \in \mathcal{D}_{\widetilde{\mathbf{p}}} \cap (0, t \wedge \widetilde{T}]} 1_\Lambda(\widetilde{\mathbf{p}}_u)\right] = \mathbf{n}(\Lambda) \cdot \widetilde{\mathbf{E}}[t \wedge \widetilde{T}] = \mathbf{n}(\Lambda) \cdot \mathbf{E}_a[t \wedge \ell_\infty],$$

which compared with (2.14) leads us to (3.4).

Identities (3.5) and (3.7) are the consequences of Proposition 2.1. as

$$\mathbf{Q}^*(W^- \cup \{\partial\}) = \mathbf{Q}^*(W^-) + \mathbf{Q}^*(\{\partial\}) = L^{(0)}(\widehat{\varphi} \cdot m_0, 1 - \varphi) + \delta_0.$$

Then σ-finiteness of $\mathbf{n}$ and the last statement follow from (2.4). □

4 Duality Preserving One-Point Extension

Let E be a locally compact separable metric space, a be a non-isolated point of E and m be a σ-finite measure on $E_0 := E \setminus \{a\}$. Contrarily to the preceding two sections, we shall start in this section with two given strong Markov processes X^0 and $\widehat{X}^0$ on E_0 that are in weak duality with respect to m_0 and have no killings inside E_0. We are concerned with their possible duality preserving extensions X and $\widehat{X}$ to E that admit no sojourn at a. It turns out that we need to allow X and $\widehat{X}$ have killings at a in order to guarantee their weak duality but they are unique up to a parameter δ_0 that represents the killing rate of X at a.

We shall assume that we are given two Borel standard processes $X^0 = (X_t^0, \mathbf{P}_x^0, \zeta^0)$ and $\widehat{X}^0 = (\widehat{X}_t^0, \widehat{\mathbf{P}}_x^0, \widehat{\zeta}^0)$ on E_0 satisfying the next three conditions.

(A.1) X^0 and $\widehat{X}^0$ are in weak duality with respect to m_0; that is, for every $\alpha > 0$ and $f, g \in \mathcal{B}^+(E_0)$,

$$\int_{E_0} \widehat{G}^0_\alpha f(x) g(x) m_0(dx) = \int_{E_0} f(x) G^0_\alpha g(x) m_0(dx),$$

where G^0_α and $\widehat{G}^0_\alpha$ are the resolvent of X^0 and $\widehat{X}^0$, respectively.

(A.2) X^0 and $\widehat{X}^0$ are approachable to $\{a\}$ but admit no killings inside E_0: for every $x \in E_0$,

$$\mathbf{P}^0_x\left(\zeta^0 < \infty,\, X^0_{\zeta^0-} = a\right) > 0 \quad \text{and} \quad \mathbf{P}^0_x\left(\zeta^0 < \infty,\, X^0_{\zeta^0-} \in E_0\right) = 0, \tag{4.1}$$

$$\widehat{\mathbf{P}}^0_x\left(\widehat{\zeta}^0 < \infty,\, \widehat{X}^0_{\widehat{\zeta}^0-} = a\right) > 0 \quad \text{and} \quad \widehat{\mathbf{P}}^0_x\left(\widehat{\zeta}^0 < \infty, \widehat{X}^0_{\widehat{\zeta}^0-} \in E_0\right) = 0. \tag{4.2}$$

Here for a Borel set $B \subset E$, the notation "$X^0_{\zeta^0-} \in B$" means that the left limit of X^0_t at $t = \zeta^0$ exists under the topology of E and takes values in $B \subset E$. We use the same convention for $\widehat{X}$.

We shall use the same notations as in [16]: for $x \in E_0$ and $\alpha > 0$,

$$\varphi(x) := \mathbf{P}^0_x\left(\zeta^0 < \infty, X^0_{\zeta^0-} = a\right) \text{ and } u_\alpha(x) := \mathbf{E}^0_x\left[e^{-\alpha\zeta^0} : X^0_{\zeta^0-} = a\right]. \tag{4.3}$$

As in §2, the X^0-energy functional of X^0-excessive measure μ and X^0-excessive function v is denoted by $L^{(0)}(\mu, v)$. The corresponding notations for $\widehat{X}^0$ will be designated by $\widehat{\varphi}$, $\widehat{u}_\alpha$, $\widehat{L}^{(0)}$. We use (u, v) to denote the inner product between u and v in $L^2(E_0, m_0)$, that is, $(u, v) = \int_{E_0} u(x)v(x)m_0(dx)$.

We say that a strong Markov process X (resp. $\widehat{X}$) on E is an extension of X^0 (resp. $\widehat{X}^0$) if the subprocess on E_0 of X (resp. $\widehat{X}$) killed upon hitting the point a is identical in law to X^0 (resp. $\widehat{X}^0$).

Let us now consider two Borel right processes $X = (X_t, \mathbf{P}_x, \zeta)$ and $\widehat{X} = (\widehat{X}_t, \widehat{\mathbf{P}}_x, \widehat{\zeta})$ on E satisfying the next four conditions.

(1) X and $\widehat{X}$ are in weak duality with respect to a σ-finite measure m on E with $m|_{E_0} = m_0$.

(2) X and $\widehat{X}$ are extensions of X^0 and $\widehat{X}^0$ respectively.

(3) The point a is regular for itself with respect to X:

$$\mathbf{P}_a(\sigma_a = 0) = 1,$$

where $\sigma_a = \inf\{t > 0 : X_t = a\}$ is the hitting time of a by X.

(4) X admits no sojourn at the point a, that is,

$$\mathbf{P}_x\left(\int_0^\infty 1_{\{a\}}(X_s)ds = 0\right) = 1 \qquad \text{for every } x \in E.$$

Under **(1)**, we can and do assume that both X and $\widehat{X}$ possess cadlag paths up to their lifetimes.

Proposition 4.1. *Assume that the above conditions* **(1)**, **(2)**, **(3)** *and* **(4)** *hold. Then*

(i) The measure m does not charging on $\{a\}$: $m(\{a\}) = 0$

(ii) X admits no jumping from E_0 to the point a: for every $x \in E_0$,

$$\mathbf{P}_x\left(X_{t-} \in E_0,\ X_t = a \ \text{ for some } t \in (0, \zeta)\right) = 0, \tag{4.4}$$

(iii) X admits no jump from the point a to E_0 in the following sense:

$$\mathbf{P}_x\left(X_{t-} = a,\ X_t \in E_0 \ \text{ for some } t \in (0, \zeta)\right) = 0 \quad \text{for q.e. } x \in E. \tag{4.5}$$

Here q.e. *means except on an m-polar set for X.*

(iv) The one point set $\{a\}$ is not m-polar for X. Let functions φ and u_α be defined as in (4.3). *Then*

$$\varphi(x) = \mathbf{P}_x(\sigma_a < \infty) \text{ and } u_\alpha(x) = \mathbf{E}_x\left[e^{-\alpha\sigma_a}\right] \qquad \text{for } x \in E_0. \tag{4.6}$$

(v) $u_\alpha,\ \widehat{u}_\alpha \in L^1(E_0, m_0)$ for every $\alpha > 0$.

Proof. (i). This is immediate from **(1)**, **(4)** for X as

$$\widehat{G}_\alpha f(a) m(\{a\}) = \int_E f(x) G_\alpha 1_{\{a\}}(x) m(dx) = 0 \qquad \text{for every } f \in \mathcal{B}^+(E).$$

(ii). It follows from (4.1) and **(2)** that

$$\mathbf{P}_x\left(X_{\sigma_a-} \in E_0,\ \sigma_a < \infty\right) = 0 \qquad \text{for every } x \in E_0. \tag{4.7}$$

For any open set O that has a positive distance from $\{a\}$, let $\{\sigma_a^n, n \geq 0\}$, $\{\eta^n, n \geq 0\}$ be the stopping times defined by

$$\eta^0 = 0,\ \sigma_a^0 = \sigma_a,\ \eta^n = \sigma_a^{n-1} + \sigma_O \circ \theta_{\sigma_a^{n-1}},\ \sigma_a^n = \eta^n + \sigma_a \circ \theta_{\eta^n} \tag{4.8}$$

with an obvious modification after one of them becomes infinity. Clearly the time set

$$\{t \in (0, \zeta(\omega)) : X_{t-}(\omega) \in O,\ X_t(\omega) = a\} \subset \{\sigma_a^n(\omega);\ n = 0, 1, 2, \cdots\}.$$

Thus it follows from the strong Markov property of X and (4.7) that for every $x \in E_0$,

$$\mathbf{P}_x\left(\text{there is some } t > 0 \text{ such that } X_{t-} \in O,\ X_t = a\right) = 0.$$

Letting O increase to E_0 establishes (4.4).

(iii). Clearly, property (ii) also holds for $\widehat{X}$:

$$\widehat{\mathbf{P}}_x\left(\widehat{X}_{t-} \in E_0,\ \widehat{X}_t = a \text{ for some } t \in (0, \widehat{\zeta})\right) = 0 \qquad \text{for every } x \in E_0. \quad (4.9)$$

We combine the above with a time reversal argument based on the stationary Kuznetsov process $(\mathbf{P}, Z_t, \alpha < t < \beta)$ associated with X and $\widehat{X}$ as was formulated in [21, §10] : the σ-finite measure $\mathbf{P}$ on a path space $D((-\infty,\infty), E_\Delta)$ with a random birth time α and a random death time β is stationary under the time shift of the path, and furthermore, if we put

$$\widehat{Z}_t = Z_{(-t)-} \text{ for } t \in \mathbb{R}, \qquad \widehat{\alpha} = -\beta \text{ and } \widehat{\beta} = -\alpha,$$

then $\{Z_t, 0 \le t < \beta\}$ (resp. $\{\widehat{Z}_t, 0 \le t < \widehat{\beta}\}$) on $\{Z_0 \in E\}$ (resp. $\{\widehat{Z}_0 \in E\}$) is a copy of $\{X_t, 0 \le t < \zeta\}$ (resp. $\{\widehat{X}_t, 0 \le t < \widehat{\zeta}\}$) under $\mathbf{P}_m$(resp. $\widehat{\mathbf{P}}_m$). We shall use the formula (10.5) of [21, §10] which express a precise meaning of this property.

Consider the set

$$\Lambda = \{Z_{t-} = a \text{ and } Z_t \in E_0, \text{ for some } t \in (\alpha, \beta)\}.$$

Then

$$\Lambda = \{\widehat{Z}_{t-} \in E_0 \text{ and } \widehat{Z}_t = a, \text{ for some } t \in (\widehat{\alpha}, \widehat{\beta})\},$$

and thus $\Lambda = \bigcup_{r \in \mathbb{Q}^+} \Lambda_r$ with

$$\Lambda_r = \{\widehat{\alpha} < r < \widehat{\beta},\ \widehat{Z}_{t-} \in E_0,\ \widehat{Z}_t = a \text{ for some } t > r\}.$$

According to (10.5) of [21, §10], $\mathbf{P}(\Lambda_r)$ is equal to the integral of the left hand side of (4.7) with respect to m for each rational r. Therefore $\mathbf{P}(\Lambda) = 0$.

Denote by $h(x)$ the function of $x \in E$ appearing in the left hand side of (4.5). By (10.5) of [21, §10] again, we have

$$\begin{aligned}\int_E h(x)m(dx) &= \mathbf{P}\left(Z_{t-} = a \text{ and } Z_t \in E_0, \text{ for some } t \in (0, \beta),\ \alpha < 0 < \beta\right) \\ &\le \mathbf{P}(\Lambda) = 0.\end{aligned}$$

Consequently, $h = 0$ m-a.e. and hence q.e. on E because h is X-excessive (cf. [5, §2]).

(iv). On account of [2, p. 59] (see also [21, Proposition 15.7] when E is a Lusin space),

$$\mathbf{P}_x(0 < \sigma'_a < \sigma_a) = 0, \quad \text{where } \sigma'_a = \inf\{t : X_{t-} = a\}, \quad x \in E.$$

On the other hand, **(A.2)** and **(2)** imply for $\zeta^0 = \sigma_a \wedge \zeta$ that for $x \in E_0$,

$$\mathbf{P}_x(\sigma_a < \sigma'_a) \le \mathbf{P}_x(\sigma_a < \infty, X_{\sigma_a-} \ne a) \le \mathbf{P}_x(\zeta^0 < \infty, X_{\zeta^0-} \in E_0) = 0.$$

Hence $\mathbf{P}_x(\sigma_a = \sigma'_a) = 1$ and

$$\varphi(x) = \mathbf{P}_x(\zeta^0 < \infty, X_{\zeta^0 -} = a) = \mathbf{P}_x(\sigma_a < \infty) \quad \text{for } x \in E_0.$$

In particular,

$$\mathbf{P}_m(\sigma_a < \infty) = \int_{E_0} \varphi(x) m(dx) > 0$$

by **(A.2)** and therefore $\{a\}$ is not m-polar for X.

(v). By the strong Markov property of $\widehat{X}$,

$$\widehat{G}_\alpha f(x) = \widehat{G}^0_\alpha f(x) + \widehat{u}_\alpha(x) \widehat{G}_\alpha f(a), \quad x \in E.$$

We can take a non-negative m-integrable function f on E such that $\widehat{G}_\alpha f(a) > 0$. Then

$$\widehat{G}_\alpha f(a)(\widehat{u}_\alpha, 1) \leq (\widehat{G}_\alpha f, 1) = (f, G_\alpha 1) \leq \frac{1}{\alpha}(f, 1) < \infty,$$

yielding the m_0-integrability of $\widehat{u}_\alpha$. Similar, we have $u_\alpha \in L^1(E_0, m_0)$. □

Theorem 4.2. *Assume that X and $\widehat{X}$ are two Borel right processes on E satisfying conditions* **(1)**, **(2)**, **(3)** *and* **(4)** *in this section. Let $\{G_\alpha, \alpha > 0\}$ and $\{\widehat{G}_\alpha, \alpha > 0\}$ denote the resolvents of X and $\widehat{X}$, respectively. Then there exist constants $\delta_0 \geq 0$, $\widehat{\delta}_0 \geq 0$ such that*

$$L^{(0)}(\widehat{\varphi} \cdot m_0, 1 - \varphi) + \delta_0 = \widehat{L}^{(0)}(\varphi \cdot m_0, 1 - \widehat{\varphi}) + \widehat{\delta}_0, \tag{4.10}$$

and for every $f \in \mathcal{B}^+(E)$ and $\alpha > 0$,

$$G_\alpha f(a) = \frac{(\widehat{u}_\alpha, f)}{\alpha(\widehat{u}_\alpha, \varphi) + L^{(0)}(\widehat{\varphi} \cdot m_0, 1 - \varphi) + \delta_0}, \tag{4.11}$$

$$G_\alpha f(x) = G^0_\alpha f(x) + u_\alpha(x) G_\alpha f(a) \qquad \text{for } x \in E_0, \tag{4.12}$$

$$\widehat{G}_\alpha f(a) = \frac{(u_\alpha, f)}{\alpha(u_\alpha, \widehat{\varphi}) + \widehat{L}^{(0)}(\varphi \cdot m_0, 1 - \widehat{\varphi}) + \widehat{\delta}_0}, \tag{4.13}$$

$$\widehat{G}_\alpha f(x) = \widehat{G}^0_\alpha f(x) + \widehat{u}_\alpha(x) \widehat{G}_\alpha f(a) \qquad \text{for } x \in E_0. \tag{4.14}$$

Corollary 4.3. *Borel right processes X and $\widehat{X}$ on E satisfying conditions* **(1)**-**(4)** *of this section are unique in law up to a parameter δ_0 satisfying*

$$\delta_0 \geq \max\left\{ \widehat{L}^{(0)}(\varphi \cdot m_0, 1 - \widehat{\varphi}) - L^{(0)}(\widehat{\varphi} \cdot m_0, 1 - \varphi), 0 \right\}.$$

Proof of Theorem 4.2. In view of conditions **(1)**-**(4)** of this section and Proposition 4.1., X satisfies the conditions **(C.1)**-**(C.2)** of §2 so that Theorem 3.2. is applicable to X.

The identity (4.12) is a simple consequence of the strong Markov property of X applied to the hitting time σ_a. In order to show (4.11), we consider the local time $\ell = \{\ell_t, t \geq 0\}$ of X with Revuz measure $\delta_{\{a\}}$ and the W-valued point process $\mathbf{p}$ associated with ℓ defined by (2.11). By Lemma 3.1., $\mathbf{p}$ under $\mathbf{P}_a$ is an absorbed Poisson point process and admits the representation (3.3) in terms of a W-valued Poisson point process $\widetilde{\mathbf{p}}$ defined on some probability space $(\widetilde{\Omega}, \widetilde{\mathbf{P}})$ together with its hitting time $\widetilde{T}$ of $W^- \cup \{\partial\}$.

Let $\mathbf{n}$ be the characteristic measure of $\widetilde{\mathbf{p}}$. Then, for any non-negative predictable process $\{a(t, w, \widetilde{\omega}),\ t \geq 0, w \in W, \widetilde{\omega} \in \widetilde{\Omega}\}$, we have

$$\widetilde{\mathbf{E}}\left[\sum_{s \leq t} a(s, \widetilde{\mathbf{p}}_s, \widetilde{\omega})\right] = \widetilde{\mathbf{E}}\left[\int_{W \times (0,t]} a(s, w, \widetilde{\omega}) \mathbf{n}(dw) ds\right], \tag{4.15}$$

because the compensator of $\widetilde{\mathbf{p}}$ equals $t\,\mathbf{n}(\cdot)$ (cf. [23, §II.3]).

We now proceed along the same line as in [16, Remark 4.2]. The terminal time of $w \in W$ is denoted by $\zeta(w)$: for $w = k_{\sigma_a}(\omega)$ with $\omega \in \Omega$, $\zeta(w) = \sigma_a(\omega)$. We put for $f \in \mathcal{B}^+(E_0)$

$$\check{f}_\alpha(w) = \int_0^{\zeta(w)} e^{-\alpha t} f(w(t)) dt, \quad w \in W, \quad \alpha > 0.$$

Note that $t \mapsto X_t(\omega)$ has only at most countably many discontinuous points. Thus by (2.2) and the condition **(4)**, $M(\omega)$ has zero Lebesgue measure almost surely. So we have $\mathbf{P}_a$-a.s.

$$\begin{aligned}\int_0^\infty e^{-\alpha t} f(X_t) dt &= \sum_{s < \ell_\infty} \int_{\tau_{s-}}^{\tau_s} e^{-\alpha t} f(X_t) dt + \int_{\tau_{\ell_\infty -}}^\infty e^{-\alpha t} f(X_t) dt \\ &= \sum_{s < \ell_\infty} e^{-\alpha \tau_{s-}} \check{f}_\alpha(\mathbf{p}_s) + e^{-\alpha \tau_{L_\infty} -} \check{f}_\alpha(\mathbf{p}_{\ell_\infty}),\end{aligned} \tag{4.16}$$

which is equivalent in law to

$$\sum_{s < \widetilde{T}} e^{-\alpha S(s-)} \check{f}_\alpha(\widetilde{\mathbf{p}}_s^+) + e^{-\alpha S(\widetilde{T}-)} \check{f}_\alpha(\widetilde{\mathbf{p}}_{\widetilde{T}}), \quad \text{under } \widetilde{\mathbf{P}}, \tag{4.17}$$

where $\{\widetilde{\mathbf{p}}_s^+, s > 0\}$ is a Poisson point process defined by $\widetilde{\mathbf{p}}_s^+ = \widetilde{\mathbf{p}}_s$ for $s \in \mathcal{D}_{\widetilde{\mathbf{p}}^+} = \{s \in \mathcal{D}_{\widetilde{\mathbf{p}}} : \widetilde{\mathbf{p}}_s \in W^+\}$ and $S(s) = \sum_{r \leq s} \zeta(\widetilde{\mathbf{p}}_r^+)$. The characteristic measure of $\{\widetilde{\mathbf{p}}_s^+, s > 0\}$ is the restriction $\mathbf{n}^+$ of $\mathbf{n}$ on W^+.

First we claim that

$$\widetilde{\mathbf{E}}\left[e^{-\alpha S(s)}\right] = \exp(-\alpha(\widehat{u}_\alpha, \varphi)s). \tag{4.18}$$

Since

$$e^{-\alpha S(s)} - 1 = \sum_{r \leq s} \left\{e^{-\alpha S(r)} - e^{-\alpha S(r-)}\right\} = \sum_{r \leq s} e^{-\alpha S(r-)} \left\{e^{-\alpha \zeta(\mathbf{p}_r^+)} - 1\right\},$$

it follows from (4.15) that

$$\widetilde{\mathbf{E}}\left[e^{-\alpha S(s)}\right] - 1 = -c\int_0^s \widetilde{\mathbf{E}}\left[e^{-\alpha S(r)}\right] dr,$$

with

$$c = \mathbf{n}^+(1 - e^{-\alpha\zeta}) = \mathbf{n}(1 - e^{-\alpha\zeta}; \zeta < \infty) = \mathbf{n}\left\{\alpha\int_0^\zeta e^{-\alpha t}dt; \zeta < \infty\right\}$$
$$= \alpha\int_0^\infty e^{-\alpha t}\mathbf{n}(t < \zeta < \infty)dt.$$

Due to (3.5) (see also (2.8)), we have accordingly

$$c = \alpha\int_0^\infty e^{-\alpha t}\nu_t(\varphi)dt = \alpha(\widehat{u}_\alpha, \varphi),$$

which is finite by Proposition 4.1.(v). The identity (4.18) then follows.

On the other hand, we have from Theorem 3.2. and the basic properties of Poisson point proccsses,

(i) $\widetilde{T}$ has an exponential distribution with exponent $L^{(0)}(\widehat{\varphi}\cdot m_0, 1-\varphi)+\delta_0$, where δ_0 is defined by (3.6).
(ii) The three objects $\{\widetilde{\mathbf{p}}_s^+, s > 0\}$, $\widetilde{T}$ and $\widetilde{\mathbf{p}}_{\widetilde{T}}$ are independent.
(iii) The law of $\widetilde{\mathbf{p}}_{\widetilde{T}}$ is $\bar{\mathbf{n}}^-(W^- \cup \{\partial\})^{-1}\bar{\mathbf{n}}^- = (L^{(0)}(\widehat{\varphi}\cdot m_0, 1-\varphi)+\delta_0)^{-1}\bar{\mathbf{n}}^-$, where $\bar{\mathbf{n}}^-$ is the restriction of $\mathbf{n}$ on $W^- \cup \{\partial\}$.

Taking these facts and formula (4.15) for $\widetilde{\mathbf{p}}^+$ into account, we get from (4.16), (4.17) and (4.18),

$$G_\alpha f(a) = \widetilde{\mathbf{E}}\left[\sum_{s<\widetilde{T}} e^{-\alpha S(s-)}\check{f}_\alpha(\widetilde{\mathbf{p}}_s^+) + e^{-S(\widetilde{T}-)}\check{f}_\alpha(\widetilde{\mathbf{p}}_{\widetilde{T}})\right]$$
$$= \widetilde{\mathbf{E}}\left[\int_0^{\widetilde{T}} e^{-\alpha(\widehat{u}_\alpha,\varphi)s}ds\right]\mathbf{n}^+(\check{f}_\alpha)$$
$$+ \widetilde{\mathbf{E}}\left(e^{-\alpha(\widehat{u}_\alpha,\varphi)\widetilde{T}}\right)(L^{(0)}(\widehat{\varphi}\cdot m_0, 1-\varphi)+\delta_0)^{-1}\mathbf{n}^-(\check{f}_\alpha)$$
$$= \frac{\mathbf{n}^+(\check{f}_\alpha)}{\alpha(\widehat{u}_\alpha,\varphi) + L^{(0)}(\widehat{\varphi}\cdot m_0, 1-\varphi)+\delta_0}$$
$$+ \frac{\mathbf{n}^-(\check{f}_\alpha)}{\alpha(\widehat{u}_\alpha,\varphi) + L^{(0)}(\widehat{\varphi}\cdot m_0, 1-\varphi)+\delta_0}$$
$$= \frac{\mathbf{n}(\check{f}_\alpha)}{\alpha(\widehat{u}_\alpha,\varphi) + L^{(0)}(\widehat{\varphi}\cdot m_0, 1-\varphi)+\delta_0},$$

which coincides with the right hand side of (4.11) because we have from Theorem 3.2.

$$\mathbf{n}(\check{f}_\alpha) = \int_0^\infty e^{-\alpha t}\nu_t(f)dt = (\widehat{u}_\alpha, f).$$

(4.13) can be obtained analogously.

Under the weak duality assumption **(1)**, the denominators of (4.11) and (4.13) must be equal. Since $(\widehat{u}_\alpha, \varphi) = (u_\alpha, \widehat{\varphi})$ (see the first two equations in the proof of Lemma 5.8.), we must have the identity (4.10). □

In the above proof, we did not use the property of X having no jumps from the point a to E_0, which is proved in Proposition 4.1.(iii). But this property reflects on the following property of the characteristic measure $\mathbf{n}$ of the absorbed Poisson point process $\mathbf{p}$ considered in the above proof.

Proposition 4.4. $\mathbf{n}\{w(0) \neq a\} = 0.$

Proof. By (4.5), we have $\mathbf{E}_a(\sum_{s \in G} 1_\Lambda \circ i_s) = 0$ for $\Lambda = \{w(0) \neq a\}$ and we get $\mathbf{n}(\Lambda) = \mathbf{Q}^*(\Lambda) = 0$ from (2.6) and (3.4). □

Remark 2. In this section, we have assumed that E is a locally compact separable metric space. But all assertions in this section remain valid for a general Lusin space E except that the identities (4.6), (4.12), (4.14) hold only for q.e. $x \in E_0$ rather than for every $x \in E_0$, because we need to replace the usage of [2, p. 59] by [21, (15.7)] in the proof of (4.6). The uniqueness statement in Corollary 4.3. should be modified accordingly in the Lusin space case.

We also note that the expression (4.11) of the resolvent has been obtained in [12] by a different method for a general right process X and its excessive measure m, in which case $\widehat{X}$ can be taken to be a dual moderate Markov process. But the present proof is more useful in the next section.

5 Extending Markov Process via Poisson Point Processes of Excursions

As in §4, let E be a locally compact separable metric space and a be a fixed non-isolated point of E and m_0 be a σ-finite measure on $E_0 := E \setminus \{a\}$ with $\mathrm{Supp}[m_0] = E$. We extend m_0 to a measure m on E by setting $m(\{a\}) = 0$. Note that m could be infinity on a compact neighborhood of a in E. Let $E_\Delta = E \cup \{\Delta\}$ be the one point compactification of E. When E is compact, Δ is added as an isolated point.

5.1 Excursion Laws in Duality

We shall assume that we are given two Borel standard processes $X^0 = \{X_t^0, \mathbf{P}_x^0, \zeta^0\}$ and $\widehat{X}^0 = \left\{\widehat{X}_t^0, \widehat{\mathbf{P}}_x^0, \widehat{\zeta}^0\right\}$ on E_0 satisfying the following conditions.

(A.1) X^0 and $\widehat{X}^0$ are in weak duality with respect to m_0, that is, for every $\alpha > 0$, and $f, g \in \mathcal{B}^+(E_0)$,

$$\int_{E_0} \widehat{G}^0_\alpha f(x) g(x) m_0(dx) = \int_{E_0} f(x) G^0_\alpha g(x) m_0(dx),$$

where G^0_α and $\widehat{G}^0_\alpha$ are the resolvents of X^0 and $\widehat{X}^0$, respectively.
(A.2) X^0 and $\widehat{X}^0$ satisfy, for every $x \in E_0$,

$$\begin{aligned}
&\mathbf{P}^0_x\left(\zeta^0 < \infty,\ X^0_{\zeta^0-} = a\right) > 0,\\
&\mathbf{P}^0_x\left(\zeta^0 < \infty,\ X^0_{\zeta^0-} \in \{a, \Delta\}\right) = \mathbf{P}^0_x(\zeta < \infty), \qquad (5.1)\\
&\widehat{\mathbf{P}}^0_x\left(\widehat{\zeta}^0 < \infty,\ \widehat{X}^0_{\widehat{\zeta}^0-} = a\right) > 0,\\
&\widehat{\mathbf{P}}^0_x\left(\widehat{\zeta}^0 < \infty,\ \widehat{X}^0_{\widehat{\zeta}^0-} \in \{a, \Delta\}\right) = \widehat{\mathbf{P}}^0_x(\widehat{\zeta} < \infty).x \qquad (5.2)
\end{aligned}$$

Here, as in §4, for a Borel set $B \subset \mathbf{E}_\Delta$, the notation "$X^0_{\zeta^0-} \in B$" means that the left limit of $t \mapsto X^0_t$ at $t = \zeta^0$ exists under the topology of E_Δ and takes values in B.

The first condition in (5.1) (resp. (5.2)) means that X^0 (resp. $\widehat{X}^0$) is approachable to the point a, while the second condition in (5.1) (resp. (5.2)) implies that X^0 (resp. $\widehat{X}^0$) admits no killings inside E_0.

As in §4, we put for $x \in E_0$ and $\alpha > 0$,

$$\varphi(x) := \mathbf{P}^0_x\left(\zeta^0 < \infty,\ X^0_{\zeta^0-} = a\right) \quad \text{and} \quad u_\alpha(x) := \mathbf{E}^0_x\left[e^{-\alpha\zeta^0};\ X^0_{\zeta^0-} = a\right]. \tag{5.3}$$

The corresponding notations for $\widehat{X}^0$ will be designated by $\widehat{\varphi}$ and $\widehat{u}_\alpha$. As in §2, the X^0-energy functional $L^{(0)}(\widehat{\varphi} \cdot m_0, v)$ of the X^0-excessive measure $\widehat{\varphi} \cdot m_0$ and an X^0-excessive function v is well defined. Similarly the $\widehat{X}^0$-energy functional $\widehat{L}^{(0)}(\varphi \cdot m_0, \widehat{v})$ is well defined. The inner product of u, v in $L^2(E_0, m_0)$ will be denoted by (u, v), that is, $(u, v) = \int_{E_0} u(x) v(x) m_0(dx)$. The space of all bounded continuous functions on E_0 will be denoted by $C_b(E_0)$.

We impose some more assumptions:
(A.3) $u_\alpha,\ \widehat{u}_\alpha \in L^1(E_0, m_0)$ for every $\alpha > 0$.

(A.4) $G^0 f(x),\ \widehat{G}^0 f(x),\ x \in E_0$, are lower semi-continuous for any Borel $f \geq 0$. Here G^0 denotes the 0-order resolvent of X^0:

$$G^0 f(x) := \mathbf{E}_x\left[\int_0^\infty f(X_t) dt\right] =\uparrow \lim_{\alpha \downarrow 0} G^0_\alpha f(x)$$

for $x \in E$ and Borel function $f \geq 0$ on E. The 0-order resolvent $\widehat{G}^0$ of $\widehat{X}^0$ is similarly defined.

We note that, if $G^0_\alpha(C_b(E_0)) \subset C_b(E_0),\ \widehat{G}^0_\alpha(C_b(E_0)) \subset C_b(E_0),\ \alpha > 0$, then **(A.4)** is satisfied by the monotone class lemma.

The next condition will be imposed only when X^0 is non-symmetric, namely, when $X^0 \neq \widehat{X}^0$.

(A.5) $\lim_{x\to a} u_\alpha(x) = \lim_{x\to a} \widehat{u}_\alpha(x) = 1$, for every $\alpha > 0$.

The next condition **(A.6)** will be imposed only when X^0 is not a diffusion, namely, when

$$\mathbf{P}^0_m\left(X^0_{t-} \neq X^0_t \quad \text{for some } t \in (0, \zeta^0)\right) > 0.$$

Note that $\widehat{X}^0$ then has the same property in view of [21, §10]. According to [31, (73.1), (47.10)], the standard process X^0 on E_0 has a Lévy system (N, H) on E_0. That is, $N(x, dy)$ is a kernel on $(E_0, \mathcal{B}(E_0))$ and H is a PCAF of X^0 in the strict sense with bounded 1-potential such that for any nonnegative Borel function f on $E_0 \times (E_0 \cup \{\Delta_0\})$ that vanishes on the diagonal and is extended to be zero outside $E_0 \times E_0$,

$$\mathbf{E}^0_x\left[\sum_{s\le t} f(X^0_{s-}, X^0_s)\right] = \mathbf{E}^0_x\left[\int_0^t \int_{E_0} f(X^0_s, y) N(X^0_s, dy) dH_s\right] \tag{5.4}$$

for every $x \in E_0$ and $t \geq 0$. Similarly, the standard process $\widehat{X}^0$ has a Lévy system $(\widehat{N}, \widehat{H})$. Let μ_H and $\mu_{\widehat{H}}$ be the Revuz measure of the PCAF H of X^0 and the PCAF $\widehat{H}$ of $\widehat{X}^0$ with respect to the measure m_0 on E_0, respectively. Define

$$J_0(dx, dy) := N(x, dy)\mu_H(dx) \quad \text{and} \quad \widehat{J}_0(dx, dy) := \widehat{N}(x, dy)\mu_{\widehat{H}}(dx). \tag{5.5}$$

The measures J_0 and $\widehat{J}_0$ are called the jumping measure of X^0 and $\widehat{X}^0$, respectively. It is known (see [18]) that

$$J_0(dx, dy) = \widehat{J}_0(dy, dx) \qquad \text{on } E_0 \times E_0. \tag{5.6}$$

We now state the condition **(A.6)**.

(A.6) Either $E \setminus U$ is compact for any neighborhood U of a in E, or for any open neighborhood U_1 of a in E, there exists an open neighborhood U_2 of a in E with $\overline{U}_2 \subset U_1$ such that

$$J_0(U_2 \setminus \{a\}, E_0 \setminus U_1) < \infty \quad \text{and} \quad \widehat{J}_0(U_2 \setminus \{a\}, E_0 \setminus U_1) < \infty.$$

Throughout this section, we assume that we are given a pair of Borel standard processes X^0 and $\widehat{X}^0$ on E_0 satisfying conditions **(A.1)**, **(A.2)**, **(A.3)**, **(A.4)**, and additionally **(A.5)** in non-symmetric case and **(A.6)** in non-diffusion case. We aim at constructing (see Theorem 5.15.) under these conditions their right process extensions X, $\widehat{X}$ to E with resolvents (4.11), (4.13) respectively. Theorem 5.16. will then be concerned with some stronger conditions **(A.1)'** and **(A.4)'** to ensure the quasi-left continuity of the constructed processes so that they become standard.

We note that, if X^0 is an m_0-symmetric diffusion on E_0, then the present conditions **(A.2), (A.3)** are the same as the conditions **(A.1), (A.2), (A.3)** assumed in [16, §4], while the present **(A.4)** is weaker than **(A.4)** of [16, §4] as is noted in the paragraph below **(A.4)**. Therefore the results of this paper extend the construction problem treated in [16, §4] to a more general case. However we shall proceed along the same line as was laid in [16, §4].

In Theorem 5.17. at the end of this section, we shall present a stronger variant **(A.2)'** of the condition **(A.2)** and prove using a time change argument that, under the conditions **(A.1), (A.2)', (A.4)** and additionally **(A.5)** in non-symmetric case and **(A.6)** in non-diffusion case, the integrability condition **(A.3)** holds automatically and therefore can be dropped.

As is shown in [5, Lemma 3.1], the measure $\widehat{\varphi} \cdot m_0$ is X^0-purely excessive and accordingly there exists a unique entrance law $\{\mu_t\}_{t>0}$ for X^0 characterized by

$$\widehat{\varphi} \cdot m_0 = \int_0^\infty \mu_t dt. \tag{5.7}$$

Analogously there exists a unique $\widehat{X}^0$-entrance law $\{\widehat{\mu}_t\}_{t>0}$ characterized by

$$\varphi \cdot m_0 = \int_0^\infty \widehat{\mu}_t dt. \tag{5.8}$$

Further by [5, Lemma 3.1], the Laplace transforms of μ_t, $\widehat{\mu}_t$ satisfy

$$\int_0^\infty e^{-\alpha t} \langle \mu_t, f \rangle dt = (\widehat{u}_\alpha, f) \quad \text{and} \quad \int_0^\infty e^{-\alpha t} \langle \widehat{\mu}_t, f \rangle dt = (u_\alpha, f) \tag{5.9}$$

for every $\alpha > 0$ and $f \in \mathcal{B}^+(E_0)$. On account of the assumption **(A.3)**, we then have that for every $t > 0$,

$$\mu_t(E_0) < \infty, \ \widehat{\mu}_t(E_0) < \infty, \quad \text{and} \quad \int_0^1 \mu_s(E_0) ds < \infty, \ \int_0^1 \widehat{\mu}_s(E_0) ds < \infty. \tag{5.10}$$

We now introduce the spaces W' and W of excursions by

$$W' = \{w : \text{a cadlag function from } \ (0, \zeta(w)) \text{ to } E_0 \text{ for some } \zeta(w) \in (0, \infty]\},$$
$$W = \left\{ w \in W' : \text{if } \zeta(w) < \infty \text{ then } w(\zeta(w)-) := \lim_{t \uparrow \zeta(\omega)} w(t) \in \{a, \Delta\} \right\}. \tag{5.11}$$

We call $\zeta(w)$ the *terminal time* of the excursion w.

We are concerned with a measure $\mathbf{n}$ on the space W specified in terms of the entrance law $\{\mu_t,\ t>0\}$ and the transition semigroup $\{P_t^0, t\ge 0\}$ of X^0 by

$$\int_W f_1(w(t_1))f_2(w(t_2))\cdots f_n(w(t_n))\mathbf{n}(dw) = \mathbf{E}_{\mu_{t_1}}\left[\prod_{k=1}^n f_k(X^0_{t_k-t_1})\right]$$
$$= \mu_{t_1} f_1 P^0_{t_2-t_1} f_2 \cdots P^0_{t_{n-1}-t_{n-2}} f_{n-1} P^0_{t_n-t_{n-1}} f_n, \qquad (5.12)$$

for any $0<t_1<t_2<\cdots<t_n$, $f_1, f_2, \cdots, f_n \in B_b(E_0)$. Here, we use the convention that $w\in W$ satisfies $w(t) := \Delta$ for $w \in W$ and $t \ge \zeta(w)$, and any function f on E_0 is extended to $E_0 \cup \Delta$ by setting $f(\Delta)=0$. Further, on the right hand side of (5.12), we employ an abbreviated notation for the repeated operations

$$\mu_{t_1}\left(f_1 P^0_{t_2-t_1}\left(f_2\cdots P^0_{t_{n-1}-t_{n-2}}\left(f_{n-1}P^0_{t_n-t_{n-1}}f_n\right)\cdots\right)\right).$$

Proposition 5.1. *There exists a unique measure* $\mathbf{n}$ *on the space* W *satisfying* (5.12).

Proof. Let $\mathbf{n}$ be the Kuznetsov measure on W' uniquely associated with the transition semigroup $\{P_t^0, t\ge 0\}$ and the entrance rule $\{\eta_u, u\in\mathbb{R}\}$ defined by

$$\eta_u = 0 \quad \text{for } u\le 0 \qquad \text{and} \qquad \eta_u=\mu_u \quad \text{for } u>0,$$

as is constructed in [8, Chap. XIX, §9] for a right semigroup. Because of the present choice of the entrance rule, it holds that the random birth time α for the Kuznetsov process is identically 0 (cf. [20, p. 54]).

On account of the assumption **(A.2)** for the standard process X^0 on E_0, the same method of the construction of the Kuznetsov measure as in [8, Chap. XIX, §9] works in proving that $\mathbf{n}$ is carried on the space W and satisfies (5.12). □

We call $\mathbf{n}$ the *excursion law* associated with the entrance law $\{\mu_t\}$ for X^0. It is strong Markov with respect to the transition semigroup $\{P_t^0, t\ge 0\}$ of X^0. Analogously we can introduce the *excursion law* $\widehat{\mathbf{n}}$ on the space W associated with the entrance law $\widehat{\mu}_t$ for $\widehat{X}^0$.

We split the space W of excursions into two parts:

$$W^+ := \{w\in W : \zeta(w)<\infty \text{ and } w(\zeta-)=a\} \text{ and } W^- := W\setminus W^+. \qquad (5.13)$$

For $w\in W^+$, we define time-reversed path $\widehat{w}\in W'$ by

$$\widehat{w}(t) := w((\zeta-t)-) = \lim_{t'\uparrow t} w(\zeta - t'), \qquad 0<t<\zeta. \qquad (5.14)$$

The next lemma asserts that the excursion laws $\mathbf{n}$ and $\widehat{\mathbf{n}}$ restricted to W^+ are interchangeable under this time reversion.

Lemma 5.2. *For any $t_k > 0$ and $f_k \in \mathcal{B}_b(S_0)$, $(1 \le k \le n)$,*

$$\mathbf{n}\left\{\prod_{k=1}^{n} f_k(w(t_1+\cdots+t_k));\ W^+\right\} = \mu_{t_1} f_1 P^0_{t_2} f_2 \cdots P^0_{t_{n-1}} f_{n-1} P^0_{t_n} f_n \varphi, \tag{5.15}$$

$$\mathbf{n}\left\{\prod_{k=1}^{n} f_k(w(t_1+\cdots+t_k));\ W^+\right\} = \widehat{\mathbf{n}}\left\{\prod_{k=1}^{n} f_k(\widehat{w}(t_1+\cdots+t_k));\ W^+\right\}. \tag{5.16}$$

Proof. (5.15) readily follows from (5.12) and the Markov property of $\mathbf{n}$. As for (5.16), we observe that, for $\alpha_1, \cdots, \alpha_n > 0$,

$$\int_0^\infty \cdots \int_0^\infty e^{-\sum_{k=1}^n \alpha_k t_k} \mathbf{n}\left\{\prod_{k=1}^{n} f_k(w(t_1+\cdots+t_k)); W^+\right\} dt_1 \cdots dt_n$$
$$= \mathbf{n}\{F(w); \zeta < \infty,\ w(\zeta-) = a\}, \tag{5.17}$$

where, with $t+0 := 0$,

$$F(w) = n! \int_{0<t_1<\cdots<t_n<\zeta} \prod_{k=1}^{n} \left\{ e^{-\alpha_k(t_k - t_{k-1})} f_k(w(t_k)) \right\} dt_1 \cdots dt_n.$$

Hence, for (5.16), it suffices to prove for $f_k \in C_b(E_0)$, $1 \le k \le n$,

$$\mathbf{n}\{F(w); \zeta < \infty,\ w(\zeta-) = a\} = \widehat{\mathbf{n}}\{F(\widehat{w}); \zeta < \infty,\ w(\zeta-) = a\}. \tag{5.18}$$

Changing of variables $\zeta - t_k = s_k$ for $0 \le k \le n$ in the following expression

$$F(\widehat{w}) = n! \int_{0<t_1<\cdots<t_n<\zeta} \prod_{k=1}^{n} \left\{ e^{-\alpha_k(t_k - t_{k-1})} f_k(w((\zeta - t_k)-)) \right\} dt_1 \cdots dt_n,$$

where $t_0 := 0$, and noting that

$$s_0 = \zeta \ \text{ and } \ 0 < t_1 < \cdots < t_n < \zeta \ \text{ if and only if } \ 0 < s_n < \cdots < s_1 < \zeta,$$

we obtain

$$F(\widehat{w}) = n! \int_{0<s_n<\cdots<s_1<\zeta} \prod_{k=1}^{n} \left\{ e^{-\alpha_k(s_{k-1} - s_k)} f_k(w(s_k)) \right\} ds_1 \cdots ds_n$$
$$= n! \int_{0<s_n<\cdots<s_1<\infty} \Gamma_{s_1\cdots s_n}(w) ds_1 \cdots ds_n,$$

where

$$\Gamma_{s_1\cdots s_n}(w) = \prod_{k=2}^{n} \left\{ e^{-\alpha_k(s_{k-1} - s_k)} f_k(w(s_k)) \right\} \cdot e^{-\alpha_1(\zeta - s_1)} f_1(w(s_1)) 1_{(0,\zeta)}(s_1).$$

On the other hand, we get from (5.10) and the Markov property of $\widehat{\mathbf{n}}$ that

$$\begin{aligned}
&\widehat{\mathbf{n}}\{\Gamma_{s_1 s_2\cdots s_n}(w);\zeta<\infty,\ w(\zeta-)=a\}\\
&=\widehat{\mathbf{n}}\Big\{f_n(w(s_n))e^{-\alpha_n(s_{n-1}-s_n)}\cdots f_2(w(s_2))e^{-\alpha_2(s_1-s_2)}\\
&\quad f_1(w(s_1))u_{\alpha_1}(w(s_1));s_1<\zeta\Big\}\\
&=e^{-\sum_{k=2}^n \alpha_k(s_{k-1}-s_k}\widehat{\mu}_{s_n}f_n\widehat{P}^0_{s_{n-1}-s_n}f_{n-1}\widehat{P}^0_{s_{n-2}-s_{n-1}}f_{n-1}\\
&\quad\cdots\widehat{P}^0_{s_2-s_3}f_2\widehat{P}^0_{s_1-s_2}f_1\widehat{u}_{\alpha_1}.
\end{aligned}$$

Therefore,

$$\begin{aligned}
&\widehat{\mathbf{n}}\{F(\widehat{w});\zeta<\infty,w(\zeta-)=a\}\\
&=\int_0^\infty ds_n\widehat{\mu}_{s_n}f_n\widehat{G}^0_{\alpha_n}f_{n-1}\widehat{G}^0_{\alpha_{n-1}}\cdots f_3\widehat{G}^0_{\alpha_3}f_2\widehat{G}^0_{\alpha_2}f_1\widehat{u}_{\alpha_1}.
\end{aligned}$$

In view of (5.8), the weak duality **(A.1)**, (5.15) and (5.17), we arrive at

$$\begin{aligned}
&\widehat{\mathbf{n}}\{F(\widehat{w});\ \zeta<\infty,w(\zeta-)=a\}\\
&=\Big\langle\varphi\cdot m_0,\ f_n\widehat{G}^0_{\alpha_n}f_{n-1}\widehat{G}^0_{\alpha_{n-1}}\cdots f_3\widehat{G}^0_{\alpha_3}f_2\widehat{G}^0_{\alpha_2}f_1\widehat{u}_{\alpha_1}\Big\rangle\\
&=\Big(f_n\varphi,\ \widehat{G}^0_{\alpha_n}f_{n-1}\widehat{G}^0_{\alpha_{n-1}}\cdots f_3\widehat{G}^0_{\alpha_3}f_2\widehat{G}^0_{\alpha_2}f_1\widehat{u}_{\alpha_1}\Big)\\
&=\big(f_1G^0_{\alpha_2}f_2G^0_{\alpha_3}f_3\cdots G^0_{\alpha_n}f_n\varphi,\ \widehat{u}_{\alpha_1}\big)\\
&=\int_0^\infty e^{-\alpha_1 t_1}\mu_{t_1}f_1G^0_{\alpha_2}f_2G^0_{\alpha_3}f_3\cdots G^0_{\alpha_n}f_n\varphi dt_1\\
&=\mathbf{n}\{F(w);\ \zeta<\infty\text{ and }w(\zeta-)=a\},
\end{aligned}$$

the desired identity (5.18). This establishes (5.16). □

Next we define

$$W_a:=\{w\in W:\ w(0+):=\lim_{t\downarrow 0}w(t)=a\}. \tag{5.19}$$

Lemma 5.3. $\mathbf{n}\{W\setminus W_a\}=0$ *and* $\widehat{\mathbf{n}}\{W\setminus W_a\}=0$.

Proof. The preceding lemma implies that

$$\begin{aligned}
\mathbf{n}\left\{W^+\setminus W_a\right\}&=\mathbf{n}\left\{W^+\cap(w(0+)=a)^c\right\}\\
&=\widehat{\mathbf{n}}\left\{W^+\cap(\widehat{w}(0+)=a)^c\right\}\\
&=\widehat{\mathbf{n}}\left\{W^+\cap(w(\zeta-)=a)^c\right\}\\
&=0.
\end{aligned}$$

We then have for each $t > 0$

$$\mathbf{n}\left\{\varphi(w(t)); (\zeta > t) \cap (w(0+) = a)^c\right\} = \mathbf{n}\left\{(W^+ \setminus W_a) \cap (\zeta > t)\right\} = 0.$$

As $\varphi(x) > 0$ for every $x \in E_0$ by the assumption **(A.2)**. we conclude that

$$\mathbf{n}\left\{(W \setminus W_a) \cap (\zeta > t)\right\} = 0 \qquad \text{for every } t > 0,$$

and therefore $\mathbf{n}\left\{(W \setminus W_a)\right\} = 0$ after letting $t \downarrow 0$. The same property of $\widehat{\mathbf{n}}$ can be shown analogously. □

Lemma 5.4. *For any neighborhood U of a in E, define*

$$\tau_U(w) = \inf\{t > 0 : w(t) \notin U\} \qquad \textit{for } w \in W.$$

Then

$$\mathbf{n}\left\{\tau_U < \zeta\right\} < \infty \qquad \textit{and} \qquad \widehat{\mathbf{n}}\left\{\tau_U < \zeta\right\} < \infty.$$

Proof. We only give a proof for $\mathbf{n}$. Let V be any neighborhood of a in E. It suffices to show

$$\mathbf{n}(\tau_U < \zeta) < \infty$$

for some neighborhood U of a with $U \subset V$. We choose such U as follows. Let us fix a relatively compact open neighborhood U_1 of a in E. When X^0 is a diffusion, we put $U = V \cap U_1$. When X^0 is not a diffusion and the second condition of **(A.5)** is fulfilled, we take U_2 in the condition for U_1 and put $U = V \cap U_2$.

By virtue of the relation

$$\varphi - u_1 = G_1^0 \varphi = G^0 u_1$$

and the assumption **(A.4)**, the function $G_1^0\varphi$ is lower semi-continuous on E_0. Furthermore, since φ is X^0-excessive and strictly positive by assumption **(A.2)**, $G_1^0\varphi$ is moreover strictly positive on E_0. As $\overline{U_1}$ is compact in E,

$$\delta := \frac{1}{2} \inf_{x \in \overline{U}_1 \setminus U} G_1^0\varphi(x) > 0. \tag{5.20}$$

Since $G_1^0\varphi(x) = \displaystyle\int_0^\infty e^{-t}\,\mathbf{P}_x\left(t < \zeta^0 < \infty, X^0_{\zeta^0-} = a\right) dt$, we have

$$\mathbf{P}_x\left(\delta < \zeta^0 < \infty, X^0_{\zeta^0-} = a\right) > \delta \qquad \text{for every } x \in \overline{U}_1 \setminus U. \tag{5.21}$$

We shall use the notation τ_U not only for $w \in W$ but also for the sample path of the Markov process X^0. Using the preceding lemma, we have

$$\mathbf{n}\left\{\tau_U < \zeta^0\right\} = \lim_{\epsilon \downarrow 0} \mathbf{n}\left\{\epsilon < \tau_U < \zeta^0\right\} = \lim_{\epsilon \downarrow 0} \int_U \mu_\epsilon(dx) \mathbf{P}_x^0\left\{\tau_U < \zeta^0\right\} = I + II,$$

where

$$I := \lim_{\epsilon \downarrow 0} \int_U \mu_\epsilon(dx) \mathbf{P}_x^0 \left(\tau_U < \zeta^0,\ X_{\tau_U}^0 \in \overline{U}_1 \setminus U\right),$$

$$II := \lim_{\epsilon \downarrow 0} \int_U \mu_\epsilon(dx) \mathbf{P}_x^0 \left(\tau_U < \zeta^0,\ X_{\tau_U}^0 \in E_0 \setminus U_1\right).$$

From (5.21) and (5.10), it follows that

$$\begin{aligned}
I &\leq \overline{\lim}_{\epsilon \downarrow 0} \int_U \mu_\epsilon(dx) \mathbf{E}_x^0 \Big[\delta^{-1} \mathbf{P}_{X_{\tau_U}^0} \left(\delta < \zeta^0 < \infty, X_{\zeta^0 -}^0 = a\right); \\
&\qquad\qquad \tau_U < \zeta^0,\ X_{\tau_U}^0 \in \overline{U}_1 \setminus U\Big] \\
&\leq \delta^{-1} \lim_{\epsilon \downarrow 0} \int_{E_0} \mu_\epsilon(dx) \mathbf{P}_x^0 (\delta < \zeta^0 < \infty, X_{\zeta^0 -}^0 = a) \\
&\leq \delta^{-1} \lim_{\epsilon \downarrow 0} \int_{E_0} \mu_\epsilon(dx) \mathbf{P}_x^0 (\delta < \zeta^0) \\
&= \delta^{-1} \lim_{\epsilon \downarrow 0} \mu_{\epsilon+\delta}(E_0) \\
&\leq \delta^{-1} \mu_\delta(E_0) < \infty.
\end{aligned}$$

II may not vanish when X^0 is not a diffusion. In this case, let $(N(x,dy),H)$ be the Lévy system of X^0 appearing in the condition **(A.5)**. Note that

$$\begin{aligned}
II &= \lim_{\epsilon \downarrow 0} \int_U \mu_\epsilon(dx) \mathbf{E}_x^0 \left[\int_0^{\tau_U} 1_U(X_s^0) N(X_s^0, E \setminus U_1) dH_s\right] \\
&\leq \lim_{\epsilon \downarrow 0} \int_{E_0} \mu_\epsilon(dx) \mathbf{E}_x^0 \left[\int_0^{\infty} 1_U(X_s^0) N(X_s^0, E \setminus U_1) dH_s\right] \\
&= \lim_{\epsilon \downarrow 0} \int_{E_0} \mu_\varepsilon(dx)\, G^0 \mu_K(x)
\end{aligned}$$

where $\mu_K(dx) := 1_U(x) N(x, E_0 \setminus U_1) \mu_H(dx)$ is the Revuz measure of the PCAF of X^0

$$K_t := \int_0^t 1_U(X_s^0) N(X_s^0, E \setminus U_1) dH_s, \qquad t \geq 0,$$

and $G^0 \mu_K(x) := \mathbf{E}_x[K_\infty]$. Note that μ_K is a finite measure on E_0 by assumption **(A.5)**. For $\alpha > 0$ and $x \in E_0$, we define

$$G_\alpha^0 \mu_K(x) := \mathbf{E}_x \left[\int_0^\infty e^{-\alpha t} dK_t\right].$$

Observe that $\alpha G^0_\alpha G^0 \mu_K$ increases to $G^0\mu_K$ as $\alpha \uparrow \infty$. We have, by (5.7), the identity $G^0_\alpha G^0 \mu_K = G^0 G^0_\alpha \mu_K$ and [21, (9.3)],

$$\begin{aligned}
\int_{E_0} \mu_\varepsilon(dx)\, G^0\mu_K(x) &= \lim_{\alpha\to\infty} \alpha \int_{E_0} \mu_\varepsilon(dx)\, G^0 G^0_\alpha \mu_K(x) \\
&= \lim_{\alpha\to\infty} \int_0^\infty \langle \mu_\varepsilon P^0_t,\, \alpha G^0_\alpha \mu_K\rangle dt \\
&\le \lim_{\alpha\to\infty} \int_0^\infty \langle \mu_t,\, \alpha G^0_\alpha \mu_K\rangle dt = \lim_{\alpha\to\infty} \langle \widehat{\varphi}\cdot m_0,\, \alpha G^0_\alpha \mu_K\rangle \\
&= \lim_{\alpha\to\infty} \langle \alpha \widehat{G}^0_\alpha \widehat{\varphi},\, \mu_K\rangle = \int_{E_0} \widehat{\varphi}(x)\mu_K(dx) \le \mu_K(E_0) < \infty.
\end{aligned}$$

Hence we get the desired finiteness of II.

When the first condition of **(A.5)** is fulfilled, the first half of the preceding proof is enough if we replace U, $\overline{U}_1$ with V, E_0 respectively. □

Lemma 5.5. $\mathbf{n}(W^-) = L^0(\widehat{\varphi}\cdot m_0, 1-\varphi) < \infty$ *and*
$\widehat{\mathbf{n}}(W^-) = \widehat{L}^0(\varphi\cdot m_0, 1-\widehat{\varphi}) < \infty.$

Proof. Since $\mathbf{n}\left(\zeta > t;\, W^-\right) = \langle \mu_t, 1-\varphi\rangle$, the first identity follows from [5, Lemma 3.1] by letting $t \downarrow 0$. Take a relatively compact neighborhood U of a in E. Since $a \in E$ and Δ is a one-point compactification of E, we have

$$\{\zeta < \infty \text{ and } w(\zeta-) = \Delta\} \subset \{\tau_U < \zeta\}. \tag{5.22}$$

Hence for any $t > 0$,

$$\begin{aligned}
\mathbf{n}\left(W^-\right) &= \mathbf{n}\{\zeta < \infty, w(\zeta-) = \Delta\} + \mathbf{n}\{\zeta = \infty\} \\
&\le \mathbf{n}\{\tau_U < \zeta\} + \mathbf{n}\{\zeta > t\} \\
&= \mathbf{n}\{\tau_U < \zeta\} + \mu_t(E_0),
\end{aligned}$$

which is finite by Lemma 5.4. and (5.10). The second assertion can be shown similarly. □

5.2 Poisson Point Processes on $W_a \cup \{\partial\}$ and a New Process X^a

By Lemma 5.3., the excursion law $\mathbf{n}$ is concentrated on the space W_a defined by (5.19). In correspondence to (5.13), we define

$$W^+_a := \left\{ w \in W^+ :\ \lim_{t\downarrow 0} w(t) = a \right\} \ \text{ and } \ W^-_a := \left\{ w \in W^- :\ \lim_{t\downarrow 0} w(t) = a \right\},$$

so that $W_a = W_a^+ + W_a^-$. In the sequel however, we shall employ slightly modified but equivalent definitions of those spaces by extending each w from an E_0-valued excursion to E-valued one as follows:

$$\begin{aligned} W_a = \{w : \ & \text{a cadlag function from } [0,\zeta(w)) \text{ to } E \text{ for some } \zeta(w) \in (0,\infty] \\ & \text{with } w(0) = a,\ w(t) \in E_0 \text{ for } t \in (0,\zeta(w)) \\ & \text{and } w(\zeta(w)-) \in \{a, \Delta\} \text{ if } \zeta(w) < \infty\}. \end{aligned} \tag{5.23}$$

Any $w \in W_a$ with the properties $\zeta(w) < \infty$ and $w(\zeta(w)-) = a$ will be regarded to be a cadlag function from $[0,\zeta(w)]$ to E by setting $w(\zeta(w)) = a$. We further define

$$\begin{aligned} W_a^+ := \{w : \ & \text{a cadlag function from } [0,\zeta(w)] \text{ to } E \text{ for some } \zeta(w) \in (0,\infty) \\ & \text{with } w(t) \in E_0 \text{ for } t \in (0,\zeta(w)) \text{ and } w(0) = w(\zeta(w)) = w(\zeta(w)-) = a\}, \\ W_a^- := W_a & \setminus W_a^+. \end{aligned}$$

The excursion law $\mathbf{n}$ will be considered to be a measure on W_a defined by (5.23). Let us add an extra point ∂ to W_a which represents a specific path constantly equal to Δ. Fix a non-negative constant δ_0 and we assign a point mass δ_0 to $\{\partial\}$ and extend the measure $\mathbf{n}$ on W_a to a measure $\bar{\mathbf{n}}$ on $W_a \cup \{\partial\}$ by

$$\bar{\mathbf{n}}(\Lambda) = \begin{cases} \mathbf{n}(\Lambda) & \text{if} \quad \Lambda \subset W_a \\ \mathbf{n}(\Lambda \cap W_a) + \delta_0 & \text{if} \quad \partial \in \Lambda \end{cases} \tag{5.24}$$

for $\Lambda \subset W_a \cup \{\partial\}$. The restrictions of $\bar{\mathbf{n}}$ to W_a^+ and $W_a^- \cup \{\partial\}$ are denoted by $\mathbf{n}^+$ and $\bar{\mathbf{n}}^-$, respectively.

Let $\mathbf{p} = \{\mathbf{p}_s : s \in \mathcal{D}_\mathbf{p}\}$ be a Poisson point process on $W_a \cup \{\partial\}$ with characteristic measure $\bar{\mathbf{n}}$ defined on an appropriate probability space $(\Omega_a, \mathbf{P})$. We then let $\mathbf{p}^+$ and $\mathbf{p}^-$ be the point processes obtained from $\mathbf{p}$ by restricting to W_a^+ and $W_a^- \cup \{\partial\}$ respectively, that is,

$$\mathcal{D}_{\mathbf{p}^+} = \{s \in \mathcal{D}_\mathbf{p} : \mathbf{p}_s \in W_a^+\} \qquad \text{and} \qquad \mathcal{D}_{\mathbf{p}^-} = \{s \in \mathcal{D}_\mathbf{p} : \mathbf{p}_s \in W_a^- \cup \{\partial\}\}. \tag{5.25}$$

Then $\{\mathbf{p}_s^+, s > 0\}$, $\{\mathbf{p}_s^-, s > 0\}$ are mutually independent Poisson point processes on W_a^+ and $W_a^- \cup \{\partial\}$ with characteristic measures $\mathbf{n}^+$ and $\bar{\mathbf{n}}^-$, respectively. Clearly,

$$\mathbf{p}_s = \mathbf{p}_s^+ + \mathbf{p}_s^-.$$

Recall that $\zeta(\mathbf{p}_r^+)$ denotes the terminal time of the excursion $\mathbf{p}_r^+$. We define

$$J(s) := \sum_{r \le s} \zeta(\mathbf{p}_r^+) \quad \text{for } s > 0 \qquad \text{and} \qquad J(0) := 0. \tag{5.26}$$

Lemma 5.6. (i) $J(s) < \infty$ a.s. *for* $s > 0$.
(ii) $\{J(s)\}_{s \ge 0}$ *is a subordinator with*

$$\mathbf{E}\left[e^{-\alpha J(s)}\right] = \exp\left(-\alpha(\widehat{u}_\alpha, \varphi)s\right). \tag{5.27}$$

Proof. (i) We write $J(s)$ as $J(s) = I + II$ with

$$I := \sum_{r \le s,\ \zeta(\mathbf{p}_r^+) \le 1} \zeta(\mathbf{p}_r^+) \qquad \text{and} \qquad II := \sum_{r \le s,\ \zeta(\mathbf{p}_r^+) > 1} \zeta(\mathbf{p}_r^+).$$

Since $\mathbf{n}^+(\zeta > 1) \le \mu_1(E_0) < \infty$ by (5.10), r in the sum II is finite a.s. and hence $II < \infty$ a.s. On the other hand,

$$\begin{aligned}\mathbf{E}(I) &= s\,\mathbf{n}^+(\zeta;\ \zeta \le 1) \le s\,\mathbf{n}^+(\zeta \wedge 1)\\ &= s\,\mathbf{n}^+\left\{\int_0^1 1_{(0,\zeta)}(t)dt\right\} = s\int_0^1 \mathbf{n}^+(\zeta > t)dt \le s\int_0^1 \mu_t(E_0)dt,\end{aligned}$$

which is finite by (5.10). Hence $I < \infty$ a.s.

(ii) This can be shown exactly in the same way as that for (4.18) in the proof of Theorem 4.2. by using the identity (5.9).

□

In view of Lemma 5.4. and Lemma 5.6., by subtracting a $\mathbf{P}$-negligible set from Ω_a if necessary, we may and do assume that the next three properties hold for every $\omega \in \Omega_a$:

$$J(s) < \infty \qquad \text{for every } s > 0, \tag{5.28}$$

$$\lim_{s \to \infty} J(s) = \infty, \tag{5.29}$$

and, for any finite interval $I \subset (0, \infty)$ and any neighborhood U of a in E,

$$\{s \in I : \tau_U(\mathbf{p}_s^+) < \zeta(\mathbf{p}_s^+)\} \text{ is a finite set.} \tag{5.30}$$

Let T be the first time of occurrence of the point process $\{\mathbf{p}_s^-, s > 0\}$, namely,

$$T = \inf\{s > 0 : s \in \mathcal{D}_{\mathbf{p}^-}\}. \tag{5.31}$$

Since by Lemma 5.5.

$$\bar{\mathbf{n}}^-(W_a^- \cup \{\partial\}) = \mathbf{n}(W_a^-) + \delta_0 = L^0(\widehat{\varphi} \cdot m_0, 1 - \varphi) + \delta_0 < \infty,$$

we see that T and $\mathbf{p}_T^-$ are independent and

$$\mathbf{P}(T > t) = e^{-(L(\widehat{\varphi} \cdot m_0, 1-\varphi) + \delta_0)t} \quad \text{and} \quad \mathbf{p}_T^- \overset{dist}{=} (L(\widehat{\varphi} \cdot m_0, 1 - \varphi) + \delta_0)^{-1}\,\bar{\mathbf{n}}^-. \tag{5.32}$$

We are now in a position to produce a new process $X = \{X_t, t \ge 0\}$ out of the point processes of excursions $\mathbf{p}^\pm$.

(i) For $0 \le t < J(T-)$, there is an $s \ge 0$ such that

$$J(s-) \le t \le J(s).$$

We define

$$X_t^a := \begin{cases} \mathbf{p}_s^+(t - J(s-)) & \text{if } J(s) - J(s-) > 0, \\ a & \text{if } J(s) - J(s-) = 0. \end{cases} \tag{5.33}$$

It is easy to see that X^a is well-defined.

(ii) If $\mathbf{p}_T^- \in W_a^-$, then we define

$$\zeta_\omega := J(T-) + \zeta(\mathbf{p}_T^-) \quad \text{and} \quad X_t^a := \mathbf{p}_T^-(t - J(T-)) \quad \text{for} \quad J(T-) \le t < \zeta_\omega. \tag{5.34}$$

(iii) If $\mathbf{p}_T^- = \partial$, then we define

$$\zeta_\omega := J(T-). \tag{5.35}$$

In this way, the E-valued path

$$\{X_t^a,\ 0 \le t < \zeta_\omega\}$$

is well-defined and enjoys the following properties:

$$\begin{aligned} &X_0^a = a, \quad \text{is cadlag in } t \in [0, \zeta_\omega) \text{ and continuous when } X_t^a = a, \\ &\text{and } X_{\zeta_\omega -}^a \in \{a, \Delta\} \quad \text{whenever} \quad \zeta_\omega < \infty. \end{aligned} \tag{5.36}$$

The second property is a consequence of (5.30). If $\mathbf{p}_T^- \in W_a^-$ and $\zeta_\omega < \infty$, then $X_{\zeta_\omega -}^a = \Delta$. If $T < \infty$, $\mathbf{p}_T^- = \partial$, then $T \notin \mathcal{D}_{\mathbf{p}^+}$ and hence by (5.35), we have $X_{\zeta_\omega -}^a = X_{J(T-)-}^a = a$. Thus the third property holds.

For this process $X^a = \{X_t^a, 0 \le t < \zeta_\omega, \mathbf{P}\}$, let us put

$$G_\alpha f(a) = \mathbf{E}\left[\int_0^{\zeta_\omega} e^{-\alpha t} f(X_t^a) dt\right], \quad \alpha > 0,\ f \in \mathcal{B}(E). \tag{5.37}$$

Similarly we assign a non-negative mass $\widehat{\delta}_0$ to the death path ∂ and extend the measure $\widehat{\mathbf{n}}$ on W_a to a measure $\overline{\widehat{\mathbf{n}}}$ on $W_a \cup \{\partial\}$. By making use of the Poisson point process $\widehat{\mathbf{p}}$ on $W_a \cup \{\partial\}$ with the characteristic measure $\overline{\widehat{\mathbf{n}}}$ on a certain probability space $(\widehat{\Omega}_a, \widehat{\mathbf{P}})$, we can construct a cadlag process $\{\widehat{X}_t^a, 0 \le t < \widehat{\zeta}_{\widehat{\omega}}, \widehat{\mathbf{P}})$ on E quite analogously. The corresponding quantity to (5.37) is denoted by $\widehat{G}_\alpha f(a)$. We can then obtain the first identity of the next proposition exactly in the same way as in the proof of Theorem 4.2. using (5.9), Lemma 5.6. and (5.32). An analogous consideration gives the second identity.

Proposition 5.7. *For $\alpha > 0$ and $f \in \mathcal{B}(E)$, it holds that*

$$G_\alpha f(a) = \frac{(\widehat{u}_\alpha, f)}{\alpha(\widehat{u}_\alpha, \varphi) + L^{(0)}(\widehat{\varphi} \cdot m_0, 1 - \varphi) + \delta_0}. \tag{5.38}$$

$$\widehat{G}_\alpha f(a) = \frac{(u_\alpha, f)}{\alpha(u_\alpha, \widehat{\varphi}) + \widehat{L}^0(\varphi \cdot m_0, 1 - \widehat{\varphi}) + \widehat{\delta}_0}. \tag{5.39}$$

For $\alpha > 0$ and $f \in \mathcal{B}(E)$, define

$$G_\alpha f(x) := G_\alpha^0 f(x) + G_\alpha f(a) u_\alpha(x) \qquad \text{for } x \in E_0, \tag{5.40}$$

$$\widehat{G}_\alpha f(x) := \widehat{G}_\alpha^0 f(x) + \widehat{G}_\alpha f(a) \widehat{u}_\alpha(x) \qquad \text{for } x \in E_0. \tag{5.41}$$

Lemma 5.8. *$\{G_\alpha, \alpha > 0\}$ and $\{\widehat{G}_\alpha, \alpha > 0\}$ are sub-Markovian resolvents on E. They are in weak duality with respect to m if and only if*

$$L^{(0)}(\widehat{\varphi} \cdot m_0, 1 - \varphi) + \delta_0 = \widehat{L}^{(0)}(\varphi \cdot m_0, 1 - \widehat{\varphi}) + \widehat{\delta}_0. \tag{5.42}$$

Proof. By making use of the resolvent equations for G_α^0, $\widehat{G}_\alpha^0$, their weak duality with respect to m_0 and the equations

$$u_\alpha(x) - u_\beta(x) + (\alpha - \beta) G_\alpha^0 u_\beta(x) = 0, \quad \alpha, \beta > 0, \ x \in E_0, \tag{5.43}$$

$$\widehat{u}_\alpha(x) - \widehat{u}_\beta(x) + (\alpha - \beta) \widehat{G}_\alpha^0 \widehat{u}_\beta(x) = 0, \quad \alpha, \beta > 0, \ x \in E_0, \tag{5.44}$$

we can easily check the resolvent equations

$$G_\alpha f(x) - G_\beta f(x) + (\alpha - \beta) G_\alpha G_\beta f(x) = 0, \quad x \in E,$$

$$\widehat{G}_\alpha f(x) - \widehat{G}_\beta f(x) + (\alpha - \beta) \widehat{G}_\alpha \widehat{G}_\beta f(x) = 0, \quad x \in E.$$

Moreover we get as in [16, Lemma 2.1] that

$$\begin{aligned} \alpha G_\alpha 1(x) &= \alpha G_\alpha^0 1(x) + u_\alpha(x) \frac{\alpha(\widehat{u}_\alpha, \varphi) + \alpha(\widehat{u}_\alpha, 1 - \varphi)}{\alpha(\widehat{u}_\alpha, \varphi) + L(\widehat{\varphi} \cdot m_0, 1 - \varphi) + \delta_0} \\ &\leq 1 - u_\alpha(x) + u_\alpha(x) = 1, \quad x \in E_0, \end{aligned}$$

and similarly, $\alpha G_\alpha 1(a) \leq 1$.

The m-weak duality

$$\int_E \widehat{G}_\alpha f(x) g(x) m(dx) = \int_E f(x) G_\alpha g(x) m(dx), \quad f, g \in \mathcal{B}^+(E),$$

holds if and only if the denominators of the right hand sides of (5.38) and (5.39) coincide. Since $(\widehat{u}_\alpha, \varphi) = (u_\alpha, \widehat{\varphi})$ by the above equations for u_α, $\widehat{u}_\alpha$, we get the last conclusion. □

5.3 Regularity of Resolvent Along the Path of X^a

Let $\{U_n\}$ be a decreasing sequence of open neighborhoods of the point a in E such that $U_n \supset \overline{U}_{n+1}$ and $\bigcap_{n=1}^{\infty} U_n = \{a\}$. For $\alpha > 0$ and $0 < \rho < 1$, let

$$A = A_{\alpha,\rho} := \{x \in E_0 : u_\alpha(x) < \rho\}.$$

We then define

$$\sigma_n := \inf\{t > 0 : X_t^0 \in U_n \cap E_0\}, \quad \tau_n := \inf\{t > 0 : X_t^0 \in U_n \cap A\},$$

and $\sigma := \lim_{n\to\infty} \sigma_n$, with the convention that $\inf \emptyset = \infty$. The stopping time σ may be called the approaching time to a of X^0.

The next lemma can be proved exactly in the same way as the proof of [16, Lemma 4.7].

Lemma 5.9. *For any $\alpha > 0$, $\rho \in (0,1)$ and $x \in E_0$,*

$$\lim_{n\to\infty} \mathbf{P}_x^0 \{\tau_n < \sigma < \infty\} = 0. \tag{5.45}$$

Lemma 5.10. *The following are ture.*

(i) *For any $x \in E_0$, $\mathbf{P}_x^0$-a.s. on $\{\sigma < \infty\}$,*

$$\lim_{t\uparrow\sigma} u_\alpha(X_t^0) = 1 \quad \textit{for every } \alpha > 0. \tag{5.46}$$

(ii) $\mathbf{n}(\Lambda \cap W_a^+) = 0$ *where*

$$\Lambda = \left\{ w \in W_a : \exists \alpha > 0,\ \liminf_{t\uparrow\zeta} u_\alpha(w(t)) < 1 \right\}.$$

(iii) $\mathbf{n}(\widehat{\Lambda}) = 0$ *where*

$$\widehat{\Lambda} = \left\{ w \in W_a : \exists \alpha > 0,\ \liminf_{t\downarrow 0} \widehat{u}_\alpha(w(t)) < 1 \right\}.$$

Proof. Let $0 < \rho < 1$. If $\sigma < \infty$ and if $\underline{\lim}_{t\uparrow\sigma} u_\alpha(X_t^0) < \rho$, then for any small $\epsilon > 0$ there exists $t \in (\sigma - \epsilon, \sigma)$ such that $u_\alpha(X_t^0) < \rho$, and so $\tau_n < \sigma$ for all n. Therefore by the preceding lemma

$$\mathbf{P}_x^0 \left(\liminf_{t\uparrow\sigma} u_\alpha(X_t^0) < \rho,\ \sigma < \infty \right) = 0.$$

Since u_α is decreasing in α and ρ can be taken arbitrarily close to 1, we obtain (5.46).

(ii) follows from (i) as

$$\mathbf{n}(\Lambda \cap W_a^+) = \lim_{\epsilon\downarrow 0} \mathbf{n}(\Lambda \cap W_a^+ \cap \{\epsilon < \zeta\})$$
$$= \lim_{\epsilon\downarrow 0} \int_{E_0} \mu_\epsilon(dx) \mathbf{P}_x^0 \left(\liminf_{t\uparrow\sigma} u_\alpha(X_t^0) < 1,\ \sigma < \infty \text{ for every } \alpha > 0 \right) = 0.$$

(iii) Part (ii) combined with Lemma 5.2. and a similar reasoning as in the proof of Lemma 5.3. leads us to

$$\mathbf{n}(\widehat{\Lambda} \cap W_a^+) = \widehat{\mathbf{n}}(\{\widehat{w} \in \widehat{\Lambda}\} \cap W_a^+) = 0,$$

and also $\mathbf{n}(\widehat{\Lambda}) = 0$. □

Denote by Q^+ the set of all positive rational number and by $C_b(E)$ the space of all bounded continuous functions on E. Let us fix an arbitrary countable subfamily $\mathbf{L}$ of $C_b(E)$. We extend functions $u_\alpha(x)$ and $G^0_\alpha f(x)$ for $f \in C_b(E)$ to be functions on E by setting $u_\alpha(a) = 1$ and $G^0_\alpha f(a) = 0$ respectively. Functions $\widehat{u}_\alpha$ and $\widehat{G}^0_\alpha f$ are similarly extended to E.

As u_α and $G^0_\alpha f$ for a non-negative $f \in C_b(E)$ are α-excessive with respect to the process X^0, it is well-known (cf. [2]) that

$$u_\alpha(X^0_t),\ G^0_\alpha f(X^0_t) \text{ are right continuous in } t \in [0,\zeta) \quad \mathbf{P}^0_x\text{–a.s.} \quad x \in E_0. \tag{5.47}$$

Suppose that X^0 is m_0-symmetric: $X^0 = \widehat{X}^0$. Then $u_\alpha = \widehat{u}_\alpha$ and hence by Lemma 5.10.

$$\mathbf{n}\left(\liminf_{t\downarrow 0} u_\alpha(w(t)) < 1\right) = 0.$$

On account of (5.47) and the inequality $aG^0_\alpha 1(x) \le 1 - u_\alpha(x),\ x \in E$, after subtracting a suitable $\mathbf{n}$-negligible set from W_a if necessary, we may and do assume that, for any $f \in \mathbf{L},\ \alpha \in Q^+$,

$$\begin{aligned} &u_\alpha(w(t)) \text{ and } G^0_\alpha f(w(t)) \text{ are right continuous in } t \in [0,\zeta) \text{ for } w \in W_a,\\ &u_\alpha(w(\zeta-)) = 1\,, G^0_\alpha f(w(\zeta-)) = 0, \text{ for } w \in W^+_a. \end{aligned} \tag{5.48}$$

When X^0 is non-symmetric, $u_\alpha \ne \widehat{u}_\alpha$ and the above argument does not work. However, since we have assumed in this non-symmetric case the condition **(A.5)**, the above property (5.48) holds by Lemma 5.3.

Lemma 5.11. *Let $0 < \rho < 1$ and set, for $\alpha > 0$,*

$$\widetilde{W}_\rho = \left\{ w \in W^+_a :\ \sup_{0\le t\le\zeta} \{1 - u_\alpha(w(t))\} > \rho \right\}.$$

Then $\mathbf{n}^+(\widetilde{W}_\rho) < \infty$.

Proof. Define $\delta := -\frac{1}{\alpha}\ \log(1 - \frac{\rho}{2}) > 0$. For any x with $1 - u_\alpha(x) \ge \rho$, we have

$$\begin{aligned} \mathbf{P}^0_x(\sigma > \delta) &\ge \mathbf{E}^0_x\left[1 - e^{-\alpha\sigma}; \sigma > \delta\right] = \mathbf{E}^0_x\left[1 - e^{-\alpha\sigma}\right] - \mathbf{E}^0_x\left[1 - e^{-\alpha\sigma}; \sigma \le \delta\right] \\ &\ge 1 - u_\alpha(x) - (1 - e^{-\alpha\delta}) \ge \rho - (1 - e^{-\alpha\delta}) = \frac{\rho}{2}. \end{aligned}$$

Therefore if we define

$$\tau := \inf\{t > 0 :\ 1 - u_\alpha(w(t)) > \rho\},$$

then for any neighborhood U of a,

$$
\begin{aligned}
\mathbf{n}^+(\widetilde{W}_\rho) &= \mathbf{n}^+(\tau < \zeta^0) = \lim_{\epsilon \downarrow 0} \mathbf{n}^+(\epsilon < \tau < \zeta^0) \\
&= \lim_{\epsilon \downarrow 0} \int_{E_0} \mu_\epsilon(dx) \mathbf{P}^0_x(\tau < \zeta^0 < \infty) \\
&\leq \liminf_{\epsilon \downarrow 0} \int_{E_0} \mu_\epsilon(dx) \mathbf{E}^0_x \left[\left(\frac{2}{\rho} \right) \mathbf{P}^0_{X^0_\tau}(\sigma > \delta); \tau < \zeta^0 \right] \\
&\leq \frac{2}{\rho} \liminf_{\epsilon \downarrow 0} \int_{E_0} \mu_\epsilon(dx) \mathbf{P}^0_x(\sigma > \delta,\ \zeta^0 < \infty) \\
&\leq \frac{2}{\rho} \lim_{\epsilon \downarrow 0} \int_{E_0} \mu_\epsilon(dx) \mathbf{P}^0_x(\zeta^0 > \delta) + \frac{2}{\rho} \lim_{\epsilon \downarrow 0} \int_{E_0} \mu_\epsilon(dx) \mathbf{P}^0_x(\zeta^0 < \infty, X_{\zeta^0 -} = \Delta) \\
&\leq \frac{2}{\rho} \lim_{\epsilon \downarrow 0} \mu_{\epsilon+\delta}(E_0) + \frac{2}{\rho} \mathbf{n}(\tau_U < \zeta),
\end{aligned}
$$

which is finite in view of (5.10) and Lemma 5.4. □

In last subsection, we have constructed a process $X^a = \{X^a_t, t \in [0, \zeta_\omega)\}$ starting from a out of the Poisson point processes $\mathbf{p}^+$ and $\mathbf{p}^-$ on W^+_a and $W^-_a \cup \{\partial\}$ defined on a probability space $(\Omega, \mathbf{P})$, respectively. A process $\{\widehat{X}^a_t, t \in [0, \zeta_{\widehat{\omega}})\}$ can be constructed similarly.

Proposition 5.12. *Let $v(x) = G_\alpha f$ with $f \in C_b(E)$ be defined by (5.38) and (5.40). Then $v(X^a_t)$ is right continuous in $t \in [0, \zeta_\omega)$ and is continuous when $X_t = a$ for every $f \in \mathbf{L}$ and every $\alpha \in Q^+$ $\mathbf{P}$-a.s. An analogous property holds for $\widehat{X}^a$.*

Proof. We already saw that the functions u_α and $G^0_\alpha f$ for $f \in \mathbf{L}$, $\alpha \in \mathbf{Q}^+$, have the property (5.48) along any sample point functions of $\mathbf{p}^+ = \{\mathbf{p}^+_s, s > 0\}$ and $\mathbf{p}^- = \{\mathbf{p}^-_s, s > 0\}$. Moreover, by Lemma 5.11., after subtracting a suitable $\mathbf{P}$-negligible set from Ω if necessary, we can assume that, in addition to the properties (5.28), (5.29) and (5.30), $\mathbf{p}^+$ satisfies the following property for every sample point $\omega \in \Omega$: for any finite interval $I \subset (0, \infty)$ and for any $\rho \in (0, 1)$,

$$
\left\{ s \in I : \sup_{0 \leq t \leq \zeta(\mathbf{p}^+_s)} (1 - u_\alpha(\mathbf{p}^+_s(t))) > \rho \right\} \text{ is a finite set.} \tag{5.49}
$$

Combining this with the inequality $\alpha G^0_\alpha 1(x) \leq 1 - u_\alpha(x)$, $x \in E$, it is not hard to see that $u_\alpha(X^a_t)$, $G^0_\alpha f(X^a_t)$ and hence $v(X^a_t)$ enjoy the properties in the statement of the proposition. □

5.4 Constructing a Standard Process X on $E_0 \cup \{a\}$

Combining the given standard process X^0 on E_0 with the process X^a constructed and studied in the last two subsections, we can now construct a right process X on $E := E_0 \cup \{a\}$ whose resolvent coincides with $\{G_\alpha, \alpha > 0\}$ defined by (5.38) and (5.40). We will only do the construction of X. But obviously the analogous procedure allows us to construct out of $\widehat{X}^0$ a right process $\widehat{X}$ on E with resolvent given by (5.39) and (5.41), and these two right processes on E are in weak duality with respect to m if and only if their killing rates δ_0 and $\widehat{\delta}_0$ at a satisfy the relation (4.10).

With the preparations made in the last subsections, we can now just follow the corresponding arguments in [16, §4] without any essential change to construct the desired process X on E.

First, using the approaching time σ to a of X^0 defined in the beginning of the last subsection, we define $P_t f(x)$ for $t > 0, x \in E, f \in \mathcal{B}(E)$, as follows:

$$P_t f(a) := \mathbf{E}\left(f(X_t^a);\ t < \zeta_\omega\right), \tag{5.50}$$

$$P_t f(x) := P_t^0 f(x) + \mathbf{E}_x^0\left[P_{t-\sigma} f(a);\ \sigma \le t\right] \qquad \text{for } x \in E_0. \tag{5.51}$$

Evidently the Laplace transform of P_t equals the resolvent G_α in view of (5.37) and (5.40) and we can see exactly in the same way as the proof of [16, Lemma 4.10] that $\{P_t,\, t \ge 0\}$ is a sub-Markovian transition semigroup on E:

$$P_{t+s} = P_t P_s \quad \text{with} \quad P_t 1 \le 1 \quad \text{for } t, s > 0.$$

Proposition 5.13. (i) *$X^a = \{X_t^a, 0 \le t < \zeta_\omega, \mathbf{P}\}$ is a Markov process on E starting from a with transition semigroup $\{P_t, t > 0\}$.*

(ii) *$\mathbf{P}(\sigma_a = 0,\ \tau_a = 0) = 1$, where $\sigma_a = \inf\{t > 0 : X_t^a = a\}$ and $\tau_a = \inf\{t > 0 : X_t^a \in E_0\}$.*

Proof. The proof of [16, Proposition 4.4] still works to obtain the first assertion (i). The only places to be modified in the proof are to replace $L(m_0, \psi)$ appearing there with $L^0(\widehat{\varphi} \cdot m_0, 1 - \varphi) + \delta_0$ in the present case.

The second assertion (ii) follows from (i) and Proposition 5.12. just as the proof of [16, Lemma 4.12]. □

In §5.1, we have started with a standard process

$$X^0 = \left\{X_t^0,\ 0 \le t < \zeta^0,\ \mathbf{P}_x^0,\ x \in E_0\right\}$$

on E_0, where $\mathbf{P}_x^0$, $x \in E_0$, are probability measures on a certain sample space, say Ω^0.

In §5.2, we have constructed a cadlag process

$$X^a = \{X_t^a(\omega'),\ 0 \le t < \zeta_{\omega'},\ \mathbf{P}\}$$

on E starting from a by piecing together excursions away from a, where $\mathbf{P}$ is a probability measure on another sample space, say Ω', to define the Poisson point process with value in $(W_a \cup \{\partial\},\ \bar{\mathbf{n}})$.

For convenience, we assume that Ω^0 contains an extra path η with $\mathbf{P}_x^0(\{\eta\}) = 0$ for every $x \in E_0$, and we set $\mathbf{P}_a^0 = \delta_\eta$, η representing the constant path taking value a identically.

We now define

$$\Omega = \Omega^0 \times \Omega', \qquad \mathbf{P}_x = \mathbf{P}_x^0 \times \mathbf{P} \quad \text{for } x \in E. \tag{5.52}$$

Note that $\zeta^0(\omega^0) \le \sigma(\omega^0)$ and $\zeta^0(\omega^0) = \sigma(\omega^0)$ when $\sigma(\omega^0) < \infty$. For $\omega = (\omega^0, \omega') \in \Omega$, let us define $X_t = X_t(\omega)$ as follows:

(1) When $\omega^0 \in \Omega^0 \setminus \{\eta\}$,

$$X_t(\omega) = \begin{cases} X_t^0(\omega^0) & 0 \le t < \zeta^0(\omega^0) \le \sigma(\omega^0) \le \infty \\ X_{t-\sigma(\omega^0)}^a(\omega') & \sigma(\omega^0) \le t < \sigma(\omega^0) + \zeta_{\omega'}, \text{ if } \sigma(\omega^0) < \infty. \end{cases} \tag{5.53}$$

(2) When $\omega^0 = \eta$,

$$X_t(\omega) = X_t^a(\omega') \quad \text{for } 0 \le t < \zeta_{\omega'}. \tag{5.54}$$

The lifetime $\zeta(\omega)$ of $X_t(\omega)$ is defined by

$$\zeta(\omega) = \begin{cases} \zeta^0(\omega^0) & \text{if } \sigma(\omega^0) = \infty, \\ \sigma(\omega^0) + \zeta_{\omega'} & \text{if } \sigma(\omega^0) < \infty. \end{cases} \tag{5.55}$$

Combining Proposition 5.13.(i) with the Markov property of $\{X_t^0,\ t \ge 0,\ \mathbf{P}_x^0,\ x \in E_0\}$, we readily get as in [16, Lemma 4.13] the next lemma:

Lemma 5.14. $X = \{X_t,\ 0 \le t < \zeta,\ \mathbf{P}_x,\ x \in E\}$ *is a Markov process on* E *with transition semigroup* $\{P_t, t \ge 0\}$ *defined by* (5.50) *and* (5.51).

The resolvent $\{G_\alpha,\ \alpha > 0\}$ of the Markov process X is defined by

$$G_\alpha f(x) = \mathbf{E}_x \left[\int_0^\infty e^{-\alpha t} f(X_t) dt \right], \qquad x \in E,\ \alpha > 0,\ f \in \mathcal{B}(E). \tag{5.56}$$

The resolvent of X^0 is denoted by G_α^0.

Theorem 5.15. *The process* X *enjoys the following properties:*

(i) X *is a right process on* E. *Its sample path* $\{X_t,\ 0 \le t < \zeta\}$ *is cadlag on* $[0, \infty)$, *continuous when* $X_t = a$ *and satisfies*

$$X_{\zeta-} \in \{a, \Delta\} \quad \text{when} \quad \zeta < \infty.$$

(ii) *The point* a *is regular for itself with respect to* X *in the sense that for the hitting time* $\sigma_a = \inf\{t > 0 : X_t = a\}$

$$\mathbf{P}_a(\sigma_a = 0) = 1.$$

(iii) X^0 *is identical in law with the subprocess of* X *killed upon hitting* a.
(iv) *The resolvent* $G_\alpha f$ *admits the expression* (5.38) *and* (5.40) *for* $f \in \mathcal{B}(E)$.
(v) *If* X^0 *is a diffusion on* E_0, *then* X *is a diffusion on* E.

Proof. (iv) follows from Lemma 5.14. and a statement next to (5.51).
(i). On account of **(A.1)**, we may assume that

$$X_t^0(\omega^0) \text{ is cadlag in } t \in [0, \zeta^0(\omega^0)) \quad \text{and}$$
$$X^0_{\zeta^0(\omega^0)-}(\omega^0) \in \{a \cup \Delta\} \text{ when } \zeta^0(\omega^0) < \infty,$$

for every $\omega^0 \in \Omega^0$. We have already chosen Ω' in a way that $\{X_t^a(\omega'), 0 \le t < \zeta_{\omega'}\}$ has the property (5.36). Hence the sample path $t \mapsto X_t(\omega)$ has the stated property in (i).

Take a countable linear subspace $\mathbf{L}$ of $C_b(E)$ such that, for any open set $G \subset E$, there exist functions $f_n \in \mathbf{L}$ increasing to I_G. We then see from the expression (5.40) of $G_\alpha f$, (5.47) and Proposition 5.12. that, for any $v = G_\alpha f$ with $f \in \mathbf{L}$, $\alpha \in Q^+$,

$$v(X_t) \text{ is right continuous in } t \in [0, \zeta) \quad \mathbf{P}_x\text{-a.s. for } x \in E.$$

Therefore X is strong Markov by [2, p. 41].
(ii) follows from Proposition 5.13.(ii).
(iii) and (v) are also evident from the construction of X. □

The right process X in the above theorem becomes a standard process if either condition **(A.1)** or **(A.4)** is replaced by the following stronger counterpart, respectively:

(A.1)' X^0 and $\widehat{X}^0$ are standard processes on E_0 in weak duality with respect to m and

$$\text{every semipolar set is } m\text{-polar for } X^0. \tag{5.57}$$

(A.4)' For any $\alpha > 0$, u_α, $\widehat{u}_\alpha \in C_b(E_0)$ and

$$G^0_\alpha(C_b(E_0)) \subset C_b(E_0), \ \widehat{G}^0_\alpha(C_b(E_0)) \subset C_b(E_0).$$

We note that condition (5.57) is automatically satisfied if X^0 is m-symmetric or more generally if the Dirichlet form of X^0 on $L^2(E_0; m_0)$ is sectorial (cf. [4]). **(A.4)'** implies **(A.4)** as we noted right after the statement of the latter. Recall that a right process is called a standard process if it is quasi-left continuous up to the lifetime.

Theorem 5.16. (i) *Suppose that the standard processes* X^0 *and* $\widehat{X}^0$ *on* E_0 *satisfy* **(A.1)**, **(A.2)**, **(A.3)**, **(A.4)'** *and additionally* **(A.5)** *in non-symmetric case and* **(A.6)** *in non-diffusion case. Then the right process* X *on* E *in* Theorem 5.15. *is quasi-left continuous up to the lifetime.*

(ii) *Suppose that the standard processes* X^0 *and* $\widehat{X}^0$ *on* E_0 *satisfy* **(A.1)'**, **(A.2)**, **(A.3)**, **(A.4)** *and additionally* **(A.5)** *in non-symmetric case and* **(A.6)** *in non-diffusion case. Then the right process* X *on* E *in* Theorem 5.15 *is quasi-left continuous up to the lifetime for* X*-q.e. starting point* $x \in E$.

Proof. (i) If condition **(A.4)'** is satisfied, then along any cadlag path of X^0, we trivially have

$$\lim_{s\uparrow t} u_\alpha(X_s^0) = u_\alpha(X_{t-}^0) \quad \text{and} \quad \lim_{s\uparrow t} G_\alpha^0 f(X_s^0) = G_\alpha^0(X_{t-}) \qquad \text{for } t \in (0, \zeta^0), \tag{5.58}$$

for any $\alpha > 0$ and $f \in C_b(E_0)$. Combining this with Lemma 5.10.(i) and Lemma 5.11., we easily see as in the proofs of Proposition 5.12. and Theorem 5.15.(i) that

$$\lim_{s\uparrow t} G_\alpha f(X_s) = G_\alpha f(X_{t-}), \quad t \in (0, \zeta), \quad \mathbf{P}_x\text{-a.s.} \tag{5.59}$$

for any $x \in E$ and for any $\alpha > 0$, $f \in C_b(E)$, from which the quasi-left continuity of X follows.

(ii) Here we use the terminologies adopted in [5]. From condition **(A.1)'**, we can deduce as in [5, Lemma 2.2] that (5.58) holds $\mathbf{P}_x^0$-a.s. for X^0-q.e. $x \in E_0$ for each $\alpha > 0$ and each $f \in C_b(E_0)$. In particular, there exists a Borel set $B \subset E_0$ with $m(B) = 0$ such that $E_0 \setminus B$ is X^0-invariant and (5.58) holds $\mathbf{P}_x^0$-a.s. for any $x \in E_0 \setminus B$ and for any $\alpha \in Q^+$, $f \in \mathbf{L}$, where $\mathbf{L}$ is a countable subfamily of $C_b(E_0)$.

Let us observe that the set $E \setminus B$ is invariant for X of Theorem 5.15. Since the restriction of X^0 to the Lusin space $E_0 \setminus B$ is a standard process again, the entrance law $\{\mu_t, t > 0\}$ uniquely characterized by the equation (5.7) is carried by $E_0 \setminus B$ for every $t > 0$ and accordingly the excursion law $\mathbf{n}$ of Proposition 5.1. is carried by the path space (5.23) with E, E_0 being replaced by $E \setminus B$, $E_0 \setminus B$ respectively. Hence $E \setminus B$ is X-invariant by the construction of X.

Now we can see by the same reasoning as in the proof of (i) that (5.59) holds for any $x \in E \setminus B$ and for any $\alpha \in Q^+$, $f \in \mathbf{L}$. Taking $\mathbf{L}$ as in the proof of Theorem 5.15.(i), we conclude that X is quasi-left continuous for every starting point $x \in E \setminus B$. □

To formulate the last theorem in this section, we need the following stronger variant **(A.2)'** of the condition of **(A.2)**:

(A.2)' For every $x \in E_0$,

$$\mathbf{P}_x^0(\zeta^0 < \infty, X_{\zeta^0-}^0 = a) > 0, \qquad \mathbf{P}_x^0(X_{\zeta^0-}^0 \in \{a, \Delta\}) = 1,$$
$$\widehat{\mathbf{P}}_x^0(\widehat{\zeta}^0 < \infty, \widehat{X}_{\widehat{\zeta}^0-}^0 = a) > 0, \qquad \widehat{\mathbf{P}}_x^0(\widehat{X}_{\widehat{\zeta}^0-}^0 \in \{a, \Delta\}) = 1.$$

Theorem 5.17. *We assume that* $m_0(U \cap E_0) < \infty$ *for some neighborhood* U *of* a *in* E. *Suppose that the pair of standard processes* X^0 *and* $\widehat{X}^0$ *on* E_0 *satisfy the conditions* **(A.1)**, **(A.2)'**, **(A.4)** *and additionally* **(A.5)** *in non-symmetric case and* **(A.6)** *in non-diffusion case. Then the integrability condition* **(A.3)** *is fulfilled by* X^0 *and* $\widehat{X}^0$.

Proof. Note that the condition **(A.3)** holds if $m_0(E_0) < \infty$. When $m_0(E_0) = \infty$, let $\gamma(x)$ be a continuous function on E_0 such that $0 < \gamma(x) \leq 1$ on E_0, $\gamma(x) = 1$ on $U \cap E_0$ and $\int_{E_0} \gamma(x) m_0(dx) < \infty$. Define for $t > 0$,

$$\tau_t := \inf\left\{ s > 0 : \int_0^s \gamma(X_r^0) dr > t \right\}$$

and

$$\widehat{\tau}_t := \inf\left\{ s > 0 : \int_0^s \gamma(\widehat{X}_r^0) dr > t \right\}.$$

Then the time changed processes $Y^0 = \{Y_t^0 := X_{\tau_t}^0,\, t \geq 0\}$ and $\widehat{Y}^0 = \{\widehat{Y}_t^0 := \widehat{X}_{\widehat{\tau}_t}^0,\, t \geq 0\}$ are standard processes on E_0 satisfying **(A.1)** with respect to the finite measure $\mu_0 = \gamma(x) m_0(dx)$. Clearly condition **(A.3)** holds for Y^0 and the reference measure μ_0. Note that since $\gamma(x) \leq 1$, we have

$$\tau_t \geq t \quad \text{and} \quad \widehat{\tau}_t \geq t \qquad \text{for every } t \geq 0.$$

Let $G_\alpha^{Y^0}$ denote the 0-order resolvent of Y^0. It is easy to check that for any non-negative Borel function f on E_0, $G^{Y^0} f = G^0(\gamma f)$. Therefore Y^0 and $\widehat{Y}^0$ inherit the conditions **(A.2)'**, **(A.4)** and in non-symmetric case **(A.5)** from X^0 and $\widehat{X}^0$.

Let (N, H) be a Lévy system of X^0. Since its defining formula (5.4) remains valid with the constant time t being replaced by any stopping time, it follows from it and a time change that Y^0 has a Lévy system (N, H^{Y^0}), where

$$H_t^{Y^0} = H_{\tau_t} \qquad \text{for every } t \geq 0.$$

According to [10, Theorem 6.2], the correspondence between PCAF and its Revuz measure is invariant under a strictly increasing time change. Therefore the Revuz measure of the PCAF of H^{Y^0} with respect to the measure μ_0 is the same as that μ_H of PCAF H of X^0 with respect to the measure m. Hence Y^0 has the same jumping measure $J_0(dx, dy) := N(x, dy)\mu_H(dy)$ as that of X^0. The same applies to $\widehat{Y}^0$. Therefore Y^0 and $\widehat{Y}^0$ also inherit the condition **(A.6)** from X^0 and $\widehat{X}^0$.

Thus by Theorem 5.15., there are duality preserving standard processes Y and $\widehat{Y}$ on $E = E_0 \cup \{a\}$ extending Y^0 and $\widehat{Y}^0$. Define for $t > 0$,

$$\sigma_t := \inf\left\{ s > 0 : \int_0^s \gamma(Y_r)^{-1} dr > t \right\}$$

and

$$\widehat{\sigma}_t := \inf\left\{ s > 0 : \int_0^s \gamma(\widehat{Y}_r)^{-1} dr > t \right\}.$$

Then $X = \{X_t := Y_{\sigma_t}, t \geq 0\}$ and $\widehat{X} = \{\widehat{X}_t := \widehat{Y}_{\widehat{\sigma}_t}, t \geq 0\}$ is a pair of standard processes on E in weak duality with respect to m. Clearly X and $\widehat{X}$ extend X^0 and $\widehat{X}^0$, they spend zero Lebesgue amount of time at $\{a\}$, and for X and Y, a is a regular point for $\{a\}$. Therefore by Proposition 4.1.(v), X^0 and $\widehat{X}^0$ must have the property **(A.3)**. □

Remark 3. In this section, we have assumed that E is a locally compact separable metric space, a is a non-isolated point of E and Δ is added to E as a one-point compactification. This assumption is used only to have (5.20) and (5.22).

The local compactness assumption on E can be relaxed and be replaced by the following conditions. Let E be a Lusin space and a a non-isolated point of E and m_0 be a σ-finite measure on $E_0 := E \setminus \{a\}$. Let Δ be a cemetery point added to E. Let X^0 and $\widehat{X}^0$ be Borel standard processes on E_0 with lifetimes ζ^0 and $\widehat{\zeta}^0$, respectively.

We say $X^0_{\zeta^0-} = a$ if $\lim_{t\uparrow\zeta^0} X_t = a$ under the topology of E, and $X^0_{\zeta^0-} = \Delta$ if the limit $\lim_{t\uparrow\zeta^0} X_t$ does not exist in the topology of E. The same applies to the process $\widehat{X}^0$.

Let $\{\mathcal{F}^0_t, t \geq 0\}$ be the minimal admissible completed σ-field generated by X^0. We assume X^0 and $\widehat{X}^0$ satisfy the conditions **(A.1)**, **(A.4)'** and additionally **(A.5)** in non-symmetric case and **(A.6)** in non-diffusion case. We also assume, instead of **(A.2)**, that

(A.2)" There is an open neighborhood U_1 of a such that its closure $\overline{U_1}$ is compact in E. Further

$$\zeta^0 \text{ is } \{\mathcal{F}^0_t\}\text{-predictable}, \ \varphi(x) > 0 \text{ on } E_0, \text{ and } \liminf_{x\to a} \varphi(x) > 0, \tag{5.60}$$

$$\widehat{\zeta}^0 \text{ is } \widehat{\mathcal{F}}^0_t\text{-predictable}, \ \widehat{\varphi}(x) > 0 \text{ on } E_0, \text{ and } \liminf_{x\to a} \widehat{\varphi}(x) > 0, \tag{5.61}$$

where φ is defined by (5.3) and $\widehat{\varphi}$ is defined analogously for $\widehat{X}^0$.

We claim that under the above assumptions, all the main results in this section, including Theorem 5.15., remain true. Note that the existence of an open neighborhood U_1 of a with $\overline{U_1}$ being compact in E guarantees the validity of (5.20). So it suffices to show that (5.22) holds almost surely under measure $\mathbf{n}$ for some neighborhood U of a under condition (5.60). As $c := \liminf_{x\to a} \varphi(x) > 0$ and φ is lower semi-continuous by **(A.4)'**, $U := \{x \in E_0 : \varphi(x) > c/2\} \cup \{a\}$ is an open neighborhood of a. On the other hand, for $x \in E_0$, we have $\mathbf{P}^0_x$-a.s. on $\{t < \zeta^0\}$,

$$\varphi(X^0_t) = \mathbf{E}_x\left[1_{\{\zeta^0<\infty \text{ and } X^0_{\zeta^0-}=a\}} \middle| \mathcal{F}^0_t \right].$$

As ζ^0 is $\{\mathcal{F}_t^0\}$-predictable, it follows that

$$\lim_{t\uparrow\zeta^0}\varphi(X_t) = 1_{\{\zeta^0<\infty \text{ and } X^0_{\zeta^0-}=a\}} \qquad \mathbf{P}_x\text{-a.s. for every } x \in E_0.$$

Hence

$$\{\zeta^0 < \infty \text{ and } X^0_{\zeta^0-} = \Delta\} \subset \{\tau_U^0 < \zeta^0\} \qquad \mathbf{P}_x\text{-a.s. for every } x \in E_0.$$

Here $\tau_U^0 := \inf\{t > 0 : X_t^0 \notin U\}$. This shows that (5.22) almost surely under measure $\mathbf{n}$. Since condition **(A.2)"** is invariant under the strict time change as in the proof of the preceding theorem, condition **(A.3)** is automatically satisfied. This proves our claim.

Note that condition (5.60) is weaker than the following condition

$$\mathbf{P}_x^0\left(\zeta^0 < \infty\right) = \mathbf{P}_x^0\left(\zeta^0 < \infty,\ X^0_{\zeta-} = a\right) \qquad \text{for every } x \in E_0. \tag{5.62}$$

□

6 Examples and Application

Several basic examples of Theorem 5.15. have been exhibited in [16, §6] when X^0 are symmetric diffusions on E_0 in which cases their extensions X are symmetric diffusions on E by [16, Theorem 4.1] there or by Theorem 5.15.(v) of the present paper. In this section, we first consider a simple case where X^0 is of pure jump type and admits no killings inside E_0. A typical example of such a process is a censored stable process on an Euclidean open set studied in [3]. We then consider the case that X^0 is an absorbing barrier non-symmetric diffusion on an Euclidean domain. As an application, we finally consider an extension of X^0 by reflecting at infinitely many holes (obstacles).

6.1 Extending Censored Stable Processes in Euclidean Domains

Let D be an open *n-set* in $\mathbb{R}^n$, that is, there exists a constant $C_1 > 0$ such that

$$m(B(x,r)) \geq C_1\, r^n \qquad \text{for all } x \in D \text{ and } 0 < r \leq 1.$$

Here m is the Lebesgue measure on $\mathbb{R}^n$, $B(x,r) := \{y \in \mathbb{R}^n : |x-y| < r\}$ and $|\cdot|$ is the Euclidean metric in $\mathbb{R}^n$. Note that bounded Lipschitz domains in $\mathbb{R}^n$ are open n-set and any open n-set with a closed subset having zero Lebesgue measure removed is still an n-set. For an n-set D (which can be disconnected), consider for $0 < \alpha < 2$ the Dirichlet space defined by

$$\mathcal{F} = \left\{u \in L^2(D;dx) : \int_{D\times D} \frac{(u(x)-u(y))^2}{|x-y|^{n+\alpha}}dxdy < \infty\right\},$$

$$\mathcal{E}(u,v) = \mathcal{A}_{n,\alpha} \int_{D\times D} \frac{(u(x)-u(y))(v(x)-v(y))}{|x-y|^{n+\alpha}} dxdy, \quad u,v \in \mathcal{F},$$

with $\mathcal{A}_{n,\alpha} = \frac{\alpha 2^{\alpha-1}\Gamma(\frac{\alpha+n}{2})}{\pi^{n/2}\Gamma(1-\frac{\alpha}{2})}$. When $D = \mathbb{R}^n$, $(\mathcal{E}, \mathcal{F})$ is just the Dirichlet form on $L^2(\mathbb{R}^n, dx)$ of the symmetric α-stable process on $\mathbb{R}^n$.

We refer the reader to [3] for the following facts. The bilinear form $(\mathcal{E}, \mathcal{F})$ is a regular irreducible Dirichlet form on $L^2(\overline{D}; 1_D(x)dx)$ and the associated Hunt process X on $\overline{D}$ may be called a *reflected α-stable process*. It is shown in [6] that X has Hölder continuous transition density functions with respect to the Lebesgue measure dx on $\overline{D}$ and therefore X can be refined to start from every point in $\overline{D}$.

The process $X^0 = (X^0_t, \mathbf{P}^0_x, \zeta^0)$ obtained from X by killing upon leaving D is called the *censored α-stable process* in D, which has been studied in detail in [3]. The process X^0 is symmetric with respect to the Lebesgue measure and its Dirichlet form on $L^2(D, dx)$ is given by $(\mathcal{E}, \mathcal{F}^0)$, where $\mathcal{F}^0$ is the closure of $C^1_0(D)$ in $\mathcal{F}$ with respect to $\mathcal{E}_1 := \mathcal{E} + (\cdot,\cdot)_{L^2(D,dx)}$. The process X^0 has no killings inside D in the sense that

$$\mathbf{P}_x\left(\zeta^0 < \infty \text{ and } X^0_{\zeta^0-} \in D\right) = 0 \qquad \text{for every } x \in D.$$

Let $\tau_D := \inf\{t > 0 : \ X_t \notin D\}$. Note that for $\beta > 0$, $u_\beta(x) = \mathbf{E}_x\left[e^{-\beta\tau_D})\right]$ is a β-harmonic function of X^0 and so it is continuous on D (see [3, (3.8)]). For any bounded measurable function f on D, we extend its definition of $\overline{D}$ by defining $f(x) = 0$ on ∂D. By [6], $G_\alpha f(x) := \mathbf{E}_x\left[\int_0^\infty e^{-\beta t} f(X_t)dt\right]$ is a continuous function on $\overline{D}$. Applying strong Markov property of X at its first exit time τ_D from D, we have for $G^0_\beta f(x) := \mathbf{E}_x\left[\int_0^{\tau_D} e^{-\beta t} f(X_t)dt\right]$,

$$G^0_\beta f(x) = G_\beta f(x) - \mathbf{E}_x\left[e^{-\beta\tau_D} G_\beta f(X_{\tau_D})\right] \qquad \text{for } x \in D.$$

Since $x \mapsto \mathbf{E}_x\left[e^{-\beta\tau_D} G_\beta f(X_{\tau_D})\right]$ is a β-harmonic function of X^0 and thus it is continuous on D, we conclude that $G^0_\beta f$ is continuous on D. Hence the conditions **(A.1)** and **(A.4)'** in §5 are always satisfied for censored α-stable process in any open n-set D. In view of [15, §5.3], a Lévy system of X^0 is given by $(N(x,dy), dt)$ with

$$N(x,dy) = 2\mathcal{A}_{n,\alpha}\, |x-y|^{-(n+\alpha)} dy$$

and the condition **(A.6)** of §5 is clearly satisfied.

Note that if D_1 is an open subset of D, then X and its subprocess killed upon leaving D_1 have the same class of m-polar sets in D_1. If a closed set $\Gamma \subset \partial D$ has a locally finite and strictly positive d-dimensional Hausdorff measure when $n \geq 2$ and is non-empty when $n = 1$, then by [3, Theorem 2.5 and Remark 2.2(i)]

$$\varphi_\Gamma(x) := \mathbf{P}^0_x(\zeta^0 < \infty,\ X^0_{\zeta^0-} \in \Gamma) > 0 \qquad \text{for every } x \in D \tag{6.1}$$

if and only if $\alpha > n - d$ when $n \geq 2$ and $\alpha > 1$ when $n = 1$.

In the following $D \subset \mathbb{R}^n$ is a proper open n-set, Γ is a closed subset of ∂D that satisfies the Hausdorff dimensional condition proceeding (6.1). The topology on $D^* = D \cup \{a\}$ will be defined in the following three special cases separately.

(i) D is an open n-set, $\Gamma = \partial D$, and $\alpha \in (n-d, n)$. Let D^* be the one point compactification of D. Note that $\varphi(x) = 1$ on D with D is bounded, and $0 < \phi < 1$ on D when D is unbounded with compact boundary.

(ii) D is an n-open set having disconnected boundary ∂D. A prototype is a bounded domain D with one or several holes in its interior. Suppose that $\partial D = \Gamma \cup \Gamma_2$, where Γ and Γ_2 are non-trivial disjoint open subsets of ∂D, with Γ being compact and satisfying the Hausdorff dimensional condition proceeding (6.1) and $\alpha \in (n-d, n)$. In this case, $0 < \varphi_\Gamma(x) \leq 1$ for $x \in D$. We prescribe a topology on D^* as follows. A subset $U \subset D^*$ containing the point $\{a\}$ is a neighborhood of a if there is an open set $U_1 \subset \mathbb{R}^d$ containing Γ_1 such that $U_1 \cap D = U \setminus \{a\}$. In other words, $D^* = D \cup \{a\}$ is obtained from D by identifying Γ into one point $\{a\}$.

(iii) $\alpha > 1 = n$, $D = (0, \infty)$ and $\Gamma = \{0\}$. In this case $\varphi_\Gamma(x) = 1$. $D^* = [0, \infty)$.

In every case, condition **(A.2)'** in §5 is fulfilled. Indeed the first half of **(A.2)'** follows from (6.1). Its second half can be also verified although the proof will be spelled out elsewhere. Consequently, condition **(A.3)** is automatically satisfied by Theorem 5.17.. Therefore, in each case, we can construct the extension X on D^* of X^0 on D satisfying the properties of Theorem 5.15. by means of the Poisson point process around $\{a\}$. X is a standard process by Theorem 5.16. but admits no jump from D to a nor from a to D.

In case **(iii)**, X coincides with the process on $[0, \infty)$ considered in the beginning of this section and may be called a reflecting α-stable process. But it differs from the two closely related processes on $[0, \infty)$ that are defined by the symmetric α-stable process x_t on $\mathbb{R}$ as

$$X_t^{(1)} = \begin{cases} x_t & t < \sigma_0 \\ x_t - \inf_{\sigma_0 \leq s \leq t} x_s & t \geq \sigma_0 \end{cases}, \qquad X_t^{(2)} = |x_t|,$$

and investigated in detail by S. Watanabe [W], because both $X^{(1)}$ and $X^{(2)}$ admit jumps from $(0, \infty)$ to 0.

Note that given an open n-set with disconnected boundary, extensions in case **(i)** and **(ii)** can be different. For example for $D = \{x \in \mathbb{R}^n : 1 < |x| < 2\}$ with $\Gamma := \{x \in \mathbb{R}^n : |x| = 1\}$, the process X in case **(ii)** is transient and gets "birth" only when X^0 approaches Γ, while in case **(i)**, the extension process is conservative and gets "birth" when X^0 approaches ∂D.

6.2 Extending Non-Symmetric Diffusions in Euclidean Domains

Let D be a proper domain in $\mathbb{R}^n$ and m be the Lebesgue measure on D. Assume that ∂D is regular for Brownian motion, or, equivalently, for $\frac{1}{2}\Delta$. Let

$$\begin{aligned}\mathcal{L} &= \frac{1}{2}\nabla \cdot (a\nabla) + b \cdot \nabla \\ &= \frac{1}{2}\sum_{i,j=1}^{n} \frac{\partial}{\partial x_i}\left(a_{ij}\frac{\partial}{\partial x_j}\right) + \sum_{i=1}^{n} b_i \frac{\partial}{\partial x_i},\end{aligned}$$

where $a : \mathbb{R}^n \to \mathbb{R}^d \otimes \mathbb{R}^n$ is a measurable, symmetric $(n \times n)$-matrix-valued function which satisfies the uniform elliptic condition

$$\lambda^{-1} I_{n\times n} \leq a(\cdot) \leq \lambda I_{n\times n}$$

for some $\lambda \geq 1$ and $b = (b_1, \cdots, b_n) : \mathbb{R}^n \to \mathbb{R}^n$ are measurable functions which could be singular such that

$$1_D |b|^2 \in \mathbf{K}(\mathbb{R}^n), \quad \sum_{i=1}^{n} \frac{\partial b_i}{\partial x_i} = 0 \text{ on } D.$$

Here $\mathbf{K}(\mathbb{R}^n)$ denote the Kato class functions on $\mathbb{R}^n$. We refer the reader to [7] for its definition. We only mention here that $L^p(\mathbb{R}^n, dx) \subset \mathbf{K}(\mathbb{R}^n)$ for $p > n/2$.

Let X^0 be the diffusion in D with infinitesimal generator $\mathcal{L}$ with Dirichlet boundary condition on ∂D. It is clearly that X^0 has a weak dual diffusion $\widehat{X}^0$ in D with respect to the Lebesgue measure m on D whose generator is $\mathcal{L}^*$, the dual operator of $\mathcal{L}$ with Dirichlet boundary condition on ∂D so that X^0 satisfies condition **(A.1)**. The conditions **(A.4)'**, **(A.5)** are satisfied by [7, Lemma 5.7 and Theorem 5.11]. Condition **(A.2)'** is also satisfied. Its first half is clear and the proof of the second half will be spelled out elsewhere. So condition **(A.3)** is automatically satisfied by Theorem 5.17. and we can apply Theorem 5.15. to construct a weak duality preserving diffusion extension X of X^0 to $D^* := D \cup \{a\}$, where the topology on D^* can be prescribed as in the three special cases **(i)**-**(iii)** in §6.1.

6.3 Extending by Reflection at Infinitely Many Holes

In this paper, we restrict ourself to consider duality preserving one-point extension of standard processes X^0 and $\widehat{X}^0$. The method of this paper allows us to do finite many points $\{a_1, \cdots, a_n\}$ or countably infinite many points $\{a_1, \cdots, a_n, \cdots\}$ extensions, with an obviously modified conditions on a_j's and with no killings at nor direct jumps between $\{a_1, a_2, \cdots\}$, provided that X^0 is symmetric (that is, $X^0 = \widehat{X}^0$). One way to do it is to do one-point extension one at a time. We leave the details to the interested reader.

Thus, for example, consider a domain $D \subset \mathbb{R}^n$ whose complement $\mathbb{R}^n \setminus D$ consists of a countable number of strictly disjoint, non-accumulating compact holes $\{K_1, K_2, \cdots\}$. Let $D^* := D \cup \{a_1, a_2, \cdots\}$ be the topological space obtained by shrinking each set K_i to a point a_i and adding all of them to D. Let $D_0^* = D$ and for each $i \geq 1$, we define $D_i^* := D_{i-1}^* \cup \{a_i\}$, the space obtained by adding K_i to D_{i-1}^* as one point just as in **(ii)** of §6.1. Given an appropriate symmetric Markov process X^0 on D, for $i \geq 1$, the extension X^i to D_i^* can be constructed from X^{i-1} on D_{i-1}^* by means of Theorem 5.15 with $\delta_0 = 0$. The extension X of X^0 on D to $D^* := D \cup \{a_1, a_2, \cdots\}$ is obtained as the limit of X^i's. The process X is then symmetric on D^* and its Dirichlet form may be described in terms of the Feller measure for X^0 on D studied in detail in [4, 13, 25].

References

1. R. M. Blumenthal, *Excursions of Markov Processes*. Birkhäuser, 1992
2. R. M. Blumenthal and R. K. Getoor, *Markov Processes and Potential Theory,* Academic Press, 1968
3. K. Bogdan, K. Burdzy and Z.-Q. Chen, Censored stable processes, *Probab. Theory Relat. Fields* **127** (2003), 89–152
4. Z.-Q. Chen, M. Fukushima and J. Ying, Traces of symmetric Markov processes and their characterizations. *Ann. Probab.* **34** (2006), 1052–1102
5. Z.-Q. Chen, M. Fukushima and J. Ying, Entrance law, exit system and Lévy system of time changed processes. *Illinois J. Math.* **50** (2006), 269–312
6. Z.-Q. Chen and T. Kumagai, Heat kernel estimates for stable-like processes on d-sets. *Stochastic Process Appl.* **108** (2003), 27–62
7. Z.-Q. Chen and Z. Zhao, Diffusion processes and second order elliptic operators with singular coefficients for lower order terms. *Math. Ann.* **302** (1995), 323–357
8. C. Dellacherie, B. Maisonneuve et P. A. Meyer, *Probabilités et potentiel,* Chap. XVII–XXIV, Hermann, Paris, 1992
9. W. Feller, On boundaries and lateral conditions for the Kolmogorov differential equations. *Ann. Math.* **65** (1957), 527–570
10. P. J. Fitzsimmons and R. K. Getoor, Revuz measures and time changes. *Math. Z.* **199** (1988), 233–256
11. P. J. Fitzsimmons and R. K. Getoor, Smooth measures and continuous additive functionals of right Markov processes. In *Itô's Stochastic Calculus and Probability Theory,* Eds. N. Ikeda, S. Watanabe, M. Fukushima and H. Kunita, pp. 31–49, Springer, 1996
12. P. J. Fitzsimmons and R. G. Getoor, Excursion theory revisited. to appear in *Illinois J. Math.* **50** (2006)
13. M. Fukushima, On boundary conditions for multi-dimensional Brownian motions with symmetric resolvent densities. *J. Math. Soc. Japan* **21** (1969), 58–93
14. M. Fukushima, P. He and J. Ying, Time changes of symmetric diffusions and Feller measures, *Ann. Probab.* **32** (2004), 3138–3166
15. M. Fukushima, Y. Oshima and M. Takeda, *Dirichlet Forms and Symmetric Markov Processes.* Walter de Gruyter, Berlin, 1994

16. M. Fukushima and H. Tanaka, Poisson point processes attached to symmetric diffusions. *Ann. Inst. H. Poincaré Probab. Statist.* **41** (2005), 419–459
17. R. K. Getoor, Markov processes: Ray processes and right processes, *Lect. Notes in Math.* **440**, Springer, 1970
18. R. K. Getoor, Duality of Lévy system. *Z. Wahrsch. verw. Gebiete* **19** (1971), 257–270
19. R. K. Getoor, Excursions of a Markov process. *Ann. Probab.* **7** (1979), 244–266
20. R. K. Getoor, *Excessive Measures.* Birkhäuser, 1990
21. R. K. Getoor and M. J. Sharpe, Naturality, standardness, and weak duality for Markov processes. *Z. Wahrsch. verw. Gebiete* **67** (1984), 1–62
22. P. Hsu: On excursions of reflecting Brownian motion. *Trans. Amer. Math. Soc.* **296** (1986), 239–264
23. N. Ikeda and S. Watanabe, *Stochastic Differential Equations and Diffusion Processes*, Second Edition, North-Holland/Kodansha, 1989
24. K. Itô, Poisson point processes attached to Markov processes, in: *Proc. Sixth Berkeley Symp. Math. Stat. Probab.* **III**, 1970, 225–239
25. H. Kunita, General boundary conditions for multi-dimensional diffusion processes, *J. Math. Kyoto Univ.* **10** (1970), 273–335
26. B. Maisonneuve, Exit systems. *Ann. Probab.* **3** (1975), 399–411.
27. P. A. Meyer, Processus de Poisson ponctuels, d'aprés K. Itô, *Séminaire de Probab.* V, in: Lecture Notes in Math., Vol. 191, Springer, Berlin, 1971, pp. 177–190
28. L. C. G. Rogers, Itô excursion theory via resolvents, *Z. Wahrsch. Verw. Gebiete* **63** (1983), 237–255
29. T. S. Salisbury, On the Itô excursion process, *Probab. Theor. Relat. Fields* **73** (1986), 319–350
30. T. S. Salisbury, Construction of right processes from excursions, *Probab. Theor. Relat. Fields* **73** (1986), 351–367
31. M. J. Sharpe, *General Theory of Markov Processes.* Academic press, 1988
32. S. Watanabe, On stable processes with boundary conditions, *J. Math. Soc. Japan* **14**(1962), 170–198

Hedging with Options in Models with Jumps

Rama Cont[1], Peter Tankov[2], and Ekaterina Voltchkova[3]

[1] CMAP-Ecole Polytechnique, France and Center for Financial Engineering, Columbia University, New York, Rama.Cont@columbia.edu
[2] Paris VII University and INRIA, Université Paris VII, Laboratoire de Probabilités et Modèles Aléatoires Case courier 7012 2, Place Jussieu, 75251 Paris, France, tankov@math.jussieu.fr
[3] Université Toulouse 1 Sciences Sociales, GREMAQ, 21, allée de Brienne, 31000 Toulouse, France, ekaterina.voltchkova@univ-tlse1.fr

Summary. We consider the problem of hedging a contingent claim, in a market where prices of traded assets can undergo jumps, by trading in the underlying asset and a set of traded options. We give a general expression for the hedging strategy which minimizes the variance of the hedging error, in terms of integral representations of the options involved. This formula is then applied to compute hedge ratios for common options in various models with jumps, leading to easily computable expressions. The performance of these hedging strategies is assessed through numerical experiments.

Keywords: quadratic hedging, option pricing, integro-differential equations, barrier option, Markov processes with jumps, Lévy process

1 Introduction

The Black–Scholes model and generalizations of it where the dynamics of prices $X_t = (X_t^1, \dots, X_t^m)$ of several assets is described by a diffusion process driven by Brownian motion

$$dX_t = X_t \sigma(t, X_t) dW_t + X_t \mu_t dt \tag{1.1}$$

have strongly influenced risk management practices in derivatives markets since the 1970s. In such models, the question of hedging a given contingent claim with payoff Y paid at a future date T can be theoretically tackled via a representation theorem for Brownian martingales: by switching to a (unique) equivalent martingale measure Q, we obtain a unique self-financing strategy ϕ_t such that

$$Y = E^Q[Y|\mathcal{F}_0] + \int_0^T \phi_t dX_t \quad Q - a.s. \tag{1.2}$$

This representation then holds almost surely under any measure equivalent to Q, thus yielding a strategy ϕ_t with initial capital $c = E^Q[Y|\mathcal{F}_0]$ which "replicates" the terminal payoff Y almost-surely. On the computational side, ϕ_t can be computed by differentiating the option price $C(t, S_t) = E^Q[Y|\mathcal{F}_t]$ with respect to the underlying asset(s) X_t. These ideas are central to the use of diffusion models in option pricing and hedging.

Stochastic processes with discontinuous trajectories are being increasingly considered, both in the research literature and in practice, as realistic alternatives to the Black–Scholes model and its diffusion-based generalizations. A natural question is therefore to examine what becomes of the above assertions in presence of discontinuities in asset prices. It is known that, except in very special cases [25], martingales with respect to the filtration of a discontinuous process X cannot be represented in the form (1.2), leading to *market incompleteness*. Far from being a shortcoming of models with jumps, this property corresponds to a genuine feature of real markets: the impossibility of "replicating" an option by trading in the underlying asset.

A natural extension, due to Föllmer and Sondermann [18], has been to *approximate* the target payoff Y by optimally choosing the initial capital c and a self-financing trading strategy $(\phi_t^1, \ldots, \phi_t^m)$ in the assets $X^1, \ldots, X^m$ in order to minimize the quadratic hedging error [7, 18]:

$$\text{minimize} \quad E\left(c + \sum_{i=1}^{m} \int_0^T \phi_t^i dX_t^i - Y\right)^2 . \tag{1.3}$$

Unlike approaches based on other (non-quadratic) loss functions, quadratic hedging has the (great) advantage of yielding linear hedging rules, which correspond to observed market practices.

The expectation in (1.3) can be understood either as being computed under an "objective" measure meant as a statistical model of price fluctuations [2, 17, 19, 27] or as being computed under a martingale ("risk-adjusted") measure [6, 7, 16, 18, 24]. Whereas the first choice may seem more natural, there are practical *and* theoretical motivations for using a risk-adjusted (martingale) measure fitted to market prices of options for computing the hedging performance.

- When X is a martingale, problem (1.3) is related to the *Kunita-Watanabe decomposition* of Y, which has well-known properties guaranteeing the existence of a solution under mild conditions [22]. By contrast, quadratic hedging with discontinuous processes under an arbitrary measure may lead to negative "prices" or not have a solution in general [2].
- Ideally, the probability measure used to compute expectations in (1.3) should reflect future uncertainty over the lifetime of the option. When using the "statistical" measure as estimated from historical data, this only holds if increments are stationary. On the other hand, the risk-adjusted measure retrieved from quoted option prices using a "calibration"

procedure [4, 9, 10] is naturally interpreted as encapsulating the market anticipation of future scenarios.

- More generally, the use of "statistical" measures of risk such as variances or quantiles computed with "statistical" models has been questioned by Aït-Sahalia and Lo [1], who advocate instead the use of corresponding quantities computed using a risk-adjusted measure, estimated non-parametrically from prices of options observed in the market. These quantities, they argue, not only reflect probabilities of occurrence of the states of nature but also the risk premia attached to them by the market so are more natural as criteria for measuring risk.

The purpose of this work is to study the quadratic hedging problem (1.3) when underlying asset prices are modeled by a process with jumps. In accordance with the above remarks, we will assume that the expectation in (1.3) is computed using a martingale measure estimated from observed prices of options. With respect to the existing literature, our contribution can be seen as follows:

- Though quadratic hedging with the underlying asset in presence of jumps has been previously studied by several authors, the corresponding expressions for hedging strategies are not always explicit and involve for instance the carré-du-champ operator [7], the Malliavin derivative [6] or various Laplace transforms and path-dependent quantities [20].
- While previous work has focused on hedging with the underlying asset(s), as we will see in Section 4, switching from naive delta-hedging to the optimal quadratic hedging strategy reduces the risk only marginally. By contrast, we study hedging strategies combining underlying assets with a set of available options that lead to an important reduction of the residual risk.
- While previous work has exclusively focused on European options without path-dependence (calls and puts), hedging exotic options is often more important in practice than hedging call and puts. We provide easy to compute expressions for hedge ratios for Asian and barrier options.
- We implement numerically the proposed hedging strategies and compare their performance based on Monte Carlo simulations.

The paper is structured as follows. In Section 2 we derive a general expression for the strategy which minimizes the variance of the hedging error, as computed under a risk-adjusted measure, in the general framework of Itô processes with jumps. Section 3 explains how the problem of hedging with options fits into the framework described in Section 2: we provide sufficient conditions under which the prices of various options possess the representation needed to apply the hedging formula. Finally, in Section 4 we apply the general hedging formula of Section 2 to construct hedging strategies for some common options and give numerical examples of their performance.

2 Minimal Variance Hedging in the Jump-Diffusion Framework

2.1 Model Setup

Consider a d-dimensional Brownian motion W and a Poisson random measure J on $[0,\infty) \times \mathbb{R}$ with intensity measure $dt \times \nu(dx)$ defined on a probability space $(\Omega, \mathcal{F}, P)$, where ν is a positive measure on $\mathbb{R}$ such that $\int_{\mathbb{R}}(1 \wedge x^2)\nu(dx) < \infty$. $\tilde{J}$ denotes the compensated version of J:

$$\tilde{J}(dt \times dz) = J(dt \times dz) - dt \times \nu(dz).$$

Let $(\mathcal{F}_t)_{t\geq 0}$ stand for the natural filtration of W and J completed with null sets.

We consider a market consisting of m traded assets X^i, $i = 1, \dots, m$ that can be used for hedging a contingent claim $Y \in \mathcal{F}_T$ with $E[Y^2] < \infty$. We suppose that the prices of traded assets are expressed using the money market account, continuously compounded at the risk-free rate, as numeraire. We assume that, using market prices of options, we have identified a pricing measure under which the prices of traded assets $X^1, \dots, X^m$ are local martingales. This can be done using for instance methods described in [4,9]. The evolution of prices under this probability measure will be described by the following stochastic integrals:

$$X_t = X_0 + \int_0^t \sigma_s dW_s + \int_{[0,t]\times\mathbb{R}} \gamma_s(z)\tilde{J}(ds \times dz). \tag{2.1}$$

We denote $Y_t = E[Y|\mathcal{F}_t]$ the value of the option and assume that Y_t can be represented by a stochastic integral:

$$Y_t = Y_0 + \int_0^{t\wedge\tau} \sigma_s^0 dW_s + \int_{[0,t\wedge\tau]\times\mathbb{R}} \gamma_s^0(z)\tilde{J}(ds \times dz). \tag{2.2}$$

The initial values X_0 and Y_0 are deterministic, τ is a stopping time which denotes the (possibly random) termination time of the contract (to account for path-dependent features such as barriers). Such a representation can be formally obtained by expressing the option price $Y_t = f(t, X_t)$ applying an Itô formula to the function f. In Section 3, we will give various conditions under which such a representation can indeed be derived, the main obstacle being the smoothness of f.

We assume the coefficients satisfy the following conditions:

(i) $\sigma : \Omega \times [0,\infty) \to \mathbb{R}^m \otimes \mathbb{R}^d$ and $\sigma^0 : \Omega \times [0,\infty) \to \mathbb{R}^d$ are càglàd $\mathcal{F}_t$-adapted processes.
(ii) $\gamma : \Omega \times [0,\infty) \times \mathbb{R} \to \mathbb{R}^m$ and $\gamma^0 : \Omega \times [0,\infty) \times \mathbb{R} \to \mathbb{R}$ are càglàd $\mathcal{F}_t$-adapted processes such that, $\forall t \in [0,T]$, $\forall z \in \mathbb{R}$,

$$\|\gamma_t(z)\|^2 \leq \rho(z)A_t \qquad \text{and} \quad |\gamma_t^i(z)\gamma_t^0(z)| \leq \rho(z)A_t, \quad i=1,\dots,m,$$

hold almost surely for some finite-valued adapted process A and some deterministic function ρ satisfying $\int_{\mathbb{R}} \rho(z)\nu(dz) < \infty$.

(iii) We fix a time horizon T and assume

$$E\int_0^T (\|\sigma_s\|^2 + A_s)ds < \infty.$$

These assumptions imply in particular that the stochastic integrals (2.1)–(2.2) exist and define square-integrable martingales. Below we give several examples of stock price models satisfying (2.1) and the assumptions (i)–(iii) and in Section 3 we will show that European and many exotic options can indeed be represented in the form (2.2).

Example 1 (Exponential Lévy models). Let L be a Lévy process with characteristic triplet (σ,ν,γ). For e^L to be a martingale, the characteristic triplet must satisfy

$$\int_{|y|>1} e^y\nu(dy) < \infty, \quad \text{and} \quad \gamma + \frac{\sigma^2}{2} + \int (e^y - 1 - y1_{|y|\leq 1})\nu(dy) = 0.$$

In this case, $X_t = X_0 e^{L_t}$ satisfies the following stochastic differential equation:

$$X_t = X_0 + \int_0^t \sigma X_s dW_s + \int_{[0,t]\times\mathbb{R}} X_{s-}(e^z - 1)\tilde{J}_L(ds\times dz),$$

where $\tilde{J}_L$ is the compensated jump measure of L. From the Lévy-Khinchin formula,

$$E[X_t^2] = X_0^2 \exp\{t\sigma^2 + t\int_{\mathbb{R}} (e^x-1)^2\nu(dx)\},$$

hence, X_t is square integrable if $\int_{|x|\geq 1} e^{2x}\nu(dx) < \infty$.

Example 2 (Markov jump diffusions). Let $\sigma : [0,\infty)\times\mathbb{R}^m \to \mathbb{R}^m \otimes \mathbb{R}^d$ and $\gamma : [0,\infty)\times\mathbb{R}\times\mathbb{R}^m \to \mathbb{R}^m$ be deterministic functions satisfying the conditions of Lipschitz continuity and sublinear growth (see [23, Theorem III.2.32]). Then the following stochastic differential equation

$$X_t = X_0 + \int_0^t \sigma(s,X_{s-})dW_s + \int_{[0,t]\times\mathbb{R}} \gamma(s,z,X_{s-})\tilde{J}(ds\times dz), \tag{2.3}$$

admits a unique strong solution X satisfying the assumptions (i)–(iii) above. Processes of this type are referred to as (martingale) Markov jump diffusions.

Some authors define jump-diffusions by allowing the intensity measure ν to depend on the state [14]. However, whenever the intensity measure ν in Equation (2.3) is infinite and has no atom, a model with state-dependent intensity measure can be transformed to the form (2.3) by choosing appropriate coefficients [21, Theorem 14.80].

Example 3. (Barndorff-Nielsen and Shephard stochastic volatility model). Under the martingale probability the stochastic volatility model proposed by Barndorff-Nielsen and Shephard [3] has the following form:

$$dY_t = (-l(\rho) - \frac{1}{2}\sigma_t^2)dt + \sigma_t dW_t + \rho dZ_t \tag{2.4}$$

$$d\sigma_t^2 = -\lambda\sigma_t^2 dt + dZ_t, \quad \sigma_0^2 > 0 \tag{2.5}$$

where $l(\theta) = \log E(e^{\theta Z_1})$, $\rho \leq 0$, $\lambda > 0$ are constant parameters, W is a standard Brownian motion and Z is a subordinator without drift, independent from W. The stock price process $X_t = X_0 e^{Y_t}$ satisfies the following:

$$X_t = X_0 + \int_0^t \sigma_s X_s dW_s + \int_0^t \int_0^\infty X_{s-}(e^{\rho z} - 1)\tilde{J}(ds \times dz),$$

where $\tilde{J}$ is the compensated jump measure of Z. To check the integrability of X, we use the formula for the Laplace transform of Y_t [8, p. 489]:

$$E[X_t^2] = X_0^2 E[e^{2Y_t}] = X_0^2 \exp\left(-2l(\rho)t + \sigma_0^2\varepsilon(\lambda,t) + \int_0^t l(2\rho + \varepsilon(\lambda, t-s))ds\right)$$

where $\varepsilon(\lambda,t) = \frac{1-e^{-\lambda t}}{\lambda}$. A sufficient condition for this to be finite is

$$l(2\rho + 1/\lambda) < \infty \quad \Longleftrightarrow \quad \int_1^\infty e^{(2\rho+1/\lambda)x}\nu(dx) < \infty,$$

where ν is the Lévy measure of Z. Under this condition Barndorff-Nielsen and Shephard's stochastic volatility model satisfies the hypotheses (i)–(iii) above.

2.2 Minimal Variance Hedging

Consider an agent who has sold at $t = 0$ the contingent claim with terminal payoff Y for the price c and wants to hedge the associated risk by trading in assets $(X^1, \ldots, X^m) = X$. We call an *admissible hedging strategy* a predictable process $\phi : \Omega \times [0,T] \to \mathbb{R}^m$ such that $\int_0^\cdot \phi_t dX_t$ is a square integrable martingale. Denote by A the set of such strategies. The residual hedging error of $\phi \in A$ at time T is then given by

$$\epsilon_T(c,\phi) = c - Y + \int_0^T \phi_t dX_t. \tag{2.6}$$

Proposition 4. *Let $Y \in \mathcal{F}_T$ be a square integrable contingent claim and denote $Y_t = E[Y|\mathcal{F}_t]$. Suppose that X_t and Y_t admit representations* (2.1) *and* (2.2) *satisfying the hypotheses (i)–(iii) on page 200. Suppose in addition that the matrix*

$$M_t = \sigma_t \sigma_t^* + \int_{\mathbb{R}} \nu(dz)\gamma_t(z)\gamma_t(z)^*$$

is almost surely nonsingular for all $t \in [0,T]$*, where the star denotes the matrix transposition. Then the minimal variance hedge* $(\hat{c}, \hat{\phi})$*, solution of*

$$E[\epsilon_T(\hat{c},\hat{\phi})^2] = \inf_{(c,\phi)\in\mathbb{R}\times A} E[\epsilon_T(c,\phi)^2],$$

is given by

$$\hat{c} = E[Y] = Y_0 \tag{2.7}$$

$$if \quad t > \tau(\omega) \tag{2.8}$$

$$\hat{\phi}_t = M_t^{-1}\left(\sigma_t^0\sigma_t^* + \int_{\mathbb{R}} \nu(dz)\gamma_t^0(z)\gamma_t(z)^*\right) 1_{[0,\tau]}(t). \tag{2.9}$$

Proof. First, for every admissible strategy ϕ,

$$E[(\epsilon_T(c,\phi))^2] = (c - E[Y])^2 + E\left(E[Y] - Y + \int_0^T \phi_t dX_t\right)^2.$$

This shows that the initial capital is given by $\hat{c} = E[Y]$. Substituting $c = \hat{c}$ yields

$$\begin{aligned} E[\epsilon_T(\hat{c},\phi)^2] = &\int_0^{T\wedge\tau} E\|\phi_t\sigma_t - \sigma_t^0\|^2 dt + \int_{T\wedge\tau}^T E\|\phi_t\sigma_t\|^2 dt \\ &+ \int_0^{T\wedge\tau} dt \int_{\mathbb{R}} \nu(dz)E(\phi_t\gamma_t(z) - \gamma_t^0(z))^2 \\ &+ \int_{T\wedge\tau}^T dt \int_{\mathbb{R}} \nu(dz)E(\phi_t\gamma_t(z))^2. \end{aligned}$$

This expression is clearly minimized by the strategy $\hat{\phi}$. Moreover, under the assumptions of this proposition, almost surely, $(\hat{\phi}_t)_{0\le t\le T}$ is *càglàd* and therefore admissible.

Remark 5. The left-continuity of hedging strategies in other settings, in particular when explicit representations are not available, is discussed in [24].

Remark 6 (Tikhonov regularization). Although in the above result we suppose that the matrix M_t is nonsingular, in some cases it may be badly conditionned leading to numerically unstable results. To avoid this problem, one can regularize M by adding to it some fraction of the unit matrix: this corresponds to minimizing

$$J(\phi) = E[(\epsilon_T(\hat{c}, \phi))^2] + \alpha E \int_0^T \|\phi_t\|^2 dt$$

for some $\alpha > 0$. It is easy to check that the solution to the minimization problem is then given by

$$\hat{\phi}_t^{reg}(\omega) = \{M_t + \alpha I\}^{-1} \times \left(\sigma_t^0 \sigma_t^* + \int_{\mathbb{R}} \nu(dz) \gamma_t^0(z) \gamma_t(z)^* \right) 1_{[0,\tau]}(t).$$

This procedure is also equivalent to adding α to each eigenvalue of M. Following the literature on regularization of inverse problems [15], we choose the regularization parameter α in such way that the hedging error with the regularized strategy $E[(\epsilon_T(\hat{c}, \hat{\phi}^{reg}))^2]$ is at its highest acceptable level.

3 Martingale Representations for Option Prices

In this section we obtain martingale representations of type (2.1) for the prices of various options. This will allow us to apply the general formula for hedge ratios (2.9) in the case when the asset to be hedged and/or the traded assets used for hedging are options on other assets.

To obtain explicit formulas for martingale representations, we assume that the price process X is a Markov process of the form (2.3). When we need to mention explicitly the starting value of a Markov process, we denote by $(X_t^x)_{t \geq 0}$ the process started from the initial value $X_0 = x$ and by $(X_t^{(\tau,x)})_{t \geq \tau}$ the same process started from the value $X_\tau = x$ at time $t = \tau$.

In some cases (Asian options, stochastic volatility, ...), one has to introduce additional non-traded factors $\tilde{X} \in \mathbb{R}^{\tilde{m}}$ such that the extended state process $(X, \tilde{X})$ is Markovian:

$$X_t = X_0 + \int_0^t \sigma(s, X_{s-}, \tilde{X}_{s-}) dW_s + \int_{[0,t] \times \mathbb{R}} \gamma(s, z, X_{s-}, \tilde{X}_{s-}) \tilde{J}(ds \times dz),$$

$$\begin{aligned}
\tilde{X}_t = \tilde{X}_0 &+ \int_0^t \tilde{\mu}(s, X_s, \tilde{X}_s) ds + \int_0^t \tilde{\sigma}(s, X_{s-}, \tilde{X}_{s-}) dW_s \\
&+ \int_{[0,t] \times \{z:|z| \leq 1\}} \tilde{\gamma}(s, z, X_{s-}, \tilde{X}_{s-}) \tilde{J}(ds \times dz) \\
&+ \int_{[0,t] \times \{z:|z| > 1\}} \tilde{\gamma}(s, z, X_{s-}, \tilde{X}_{s-}) J(ds \times dz).
\end{aligned}$$

Note that the components of $\tilde{X}$ are not necessarily martingales because they do not represent prices of tradables. For simplicity, unless otherwise mentioned, in the rest of this section we assume there are no non-traded factors and that the price process is one-dimensional ($m = 1$). We treat separately the case of general Markov jump diffusions and the case of Lévy processes.

3.1 European Options

Let X be defined by (2.3) and H be a measurable function with $E[H(X_T)^2] < \infty$. The price of a European-type contingent claim is then a deterministic function of time t and state X_t:

$$C_t = E[H(X_T)|\mathcal{F}_t] = C(t, X_t), \tag{3.1}$$

where $C(t,x) = E[H(X_T^{(t,x)})]$.

Suppose that the option price $C(t,x)$ is continuously differentiable with respect to t and twice continuously differentiable with respect to x. The Itô formula can then be applied to show that the price of a European option satisfies a stochastic differential equation of type (2.3):

$$\begin{aligned} dC_t = {} & \frac{\partial C(t,X_t)}{\partial x}\sigma(t,X_t)dW_t \\ & + \int_{\mathbb{R}} (C(t, X_{t-} + \gamma(t,z,X_{t-})) - C(t,X_{t-}))\tilde{J}(dt \times dz) \\ & + \left\{ \frac{\partial C(t,X_t)}{\partial t} + \frac{1}{2}\sigma(t,X_t)^2 \frac{\partial^2 C(t,X_t)}{\partial x^2} \right. \\ & \left. + \int_{\mathbb{R}} \left(C(t, X_t + \gamma(t,z,X_t)) - C(t,X_t) - \gamma(t,z,X_t)\frac{\partial C(t,X_t)}{\partial x} \right) \nu(dz) \right\} dt \end{aligned} \tag{3.2}$$

Note that the first line of the above expression is a local martingale. On the other hand, from (3.1), C_t itself is a martingale. Therefore, the sum of the second and the third line is a finite variation continuous local martingale. This means that it is zero and we obtain the following martingale representation for C_t:

$$\begin{aligned} C(T,X_T) \equiv H(X_T) = C(0,X_0) + \int_0^T \frac{\partial C(t,X_t)}{\partial x}\sigma(t,X_t)dW_t \\ + \int_0^T \int_{\mathbb{R}} (C(t, X_{t-} + \gamma(t,z,X_{t-})) - C(t,X_{t-}))\tilde{J}(dt \times dz). \end{aligned} \tag{3.3}$$

Despite the simplicity of this heuristic argument, a rigorous proof of this formula requires some work. We start with the case of Lévy processes and discontinuous payoffs.

Proposition 7. *Let H be a measurable function with at most polynomial growth: $\exists p \geq 0$, $|H(x)| \leq K(1+|x|^p)$, and X be a Lévy process with characteristic triplet (σ, ν, γ) satisfying the following conditions:*

$$(i)\ \sigma > 0 \quad or \quad \exists \beta \in (0,2), \quad \liminf_{\varepsilon \downarrow 0} \frac{1}{\varepsilon^{2-\beta}} \int_{-\varepsilon}^{\varepsilon} |y|^2 \nu(dy) > 0; \tag{3.4}$$

$$(ii)\ \int_{|y|>1} |y|^{p+1}\nu(dy) < \infty. \tag{3.5}$$

Denote $X_t^x \equiv x + X_t$. *Then*

1. *The European option price* $C(t,x) = E[H(X_{T-t}^x)]$ *belongs to the class* $C^\infty([0,T)\times\mathbb{R})$ *with* $\left|\frac{\partial^{n+m}C}{\partial x^n \partial t^m}(x)\right| \le K(1+|x|^p)$, *for all* $n, m \ge 0$.[4]
2. *Suppose in addition that the set of discontinuities of* H *has Lebesgue measure zero and that*

$$\int_{|y|>1} |y|^{2p}\nu(dy) < \infty. \tag{3.6}$$

Then the process $(C(t,X_t^x))_{0\le t\le T}$ *is a square integrable martingale with the following representation*

$$\begin{aligned} C(t,X_t^x) &= C(0,x) + \int_0^t \frac{\partial C(s,X_s^x)}{\partial x}\sigma dW_s \\ &\quad + \int_0^t \int_{\mathbb{R}} (C(s,X_{s-}^x + z) - C(s,X_{s-}^x))\tilde{J}(ds\times dz). \end{aligned} \tag{3.7}$$

Proof. Part 1. Let $\phi_t(u) = E[e^{iuX_t}]$. Condition (3.4) implies

$$|\phi_t(u)| \le K_1 \exp(-K_2|u|^\alpha) \tag{3.8}$$

for some positive constants K_1, K_2, α and all $t > 0$, and therefore that X_t has a C^∞ density $p_t(x)$ for all $t > 0$. For $\sigma > 0$ this is straightforward and for $\sigma = 0$ see [26, Proposition 28.3].

The derivatives of C can now be estimated as follows (we denote $\tau = T - t$)

$$\begin{aligned} \left|\frac{\partial^{n+m}C(t,x)}{\partial x^n \partial t^m}\right| &\le \int |H(x+z)| \left|\frac{\partial^{n+m}p_\tau(z)}{\partial z^n \partial \tau^m}\right| dz \\ &\le K\int (1+|x+z|^p)\left|\frac{\partial^{n+m}p_\tau(z)}{\partial z^n \partial \tau^m}\right| dz \\ &\le K(1+|x|^p)\left\|(1+|z|^p)\frac{\partial^{n+m}p_\tau(z)}{\partial z^n \partial \tau^m}\right\|_{L^1} \\ &\le K(1+|x|^p)\left\|\frac{1}{1+|z|}\right\|_{L^2}\left\|(1+|z|^{p+1})\frac{\partial^{n+m}p_\tau(z)}{\partial z^n \partial \tau^m}\right\|_{L^2} \\ &\le K(1+|x|^p)\left(\left\|u^n\frac{\partial^m \phi_\tau(u)}{\partial \tau^m}\right\|_{L^2} + \left\|u^n\frac{\partial^{p+1+m}\phi_\tau(u)}{\partial u^{p+1}\partial \tau^m}\right\|_{L^2}\right). \end{aligned} \tag{3.9}$$

---

[4] Here and in all proofs, K denotes a constant which may depend on n, m and τ and vary from line to line.

From the Lévy-Khinchin formula, $\phi_t(u) = e^{t\psi(u)}$ with $|\psi(u)| \leq K(1 + |u|^2)$. Moreover,

$$\psi'(u) = -\sigma^2 u + i\gamma + \int iy(e^{iyu} - 1_{|y|\leq 1})\nu(dy),$$

$$\psi''(u) = -\sigma^2 + \int (iy)^2 e^{iyu}\nu(dy),$$

$$\psi^{(k)}(u) = \int (iy)^k e^{iyu}\nu(dy), \qquad 3 \leq k \leq p+1.$$

Due to the condition (3.5), the integrals in the above expressions are finite and we have $|\psi'(u)| \leq K_1(1 + |u|)$ and $|\psi^{(q)}(u)| \leq K_q$, $2 \leq q \leq p+1$. Therefore,

$$\left|\frac{\partial^{p+1+m}\phi_\tau(u)}{\partial u^{p+1}\partial\tau^m}\right| \leq K(1 + |u|^{p+1+2m})|\phi_\tau(u)|$$

and by (3.8), both terms in (3.9) are finite.

Part 2. By Part 1, representation (3.7) is valid for every $t < T$. By Corollary 25.8 in [26], (3.6) implies that $E[H^2(X_T^x)] < \infty$. Denote

$$M_t = \int_0^t \frac{\partial C(s, X_s^x)}{\partial x}\sigma dW_s + \int_0^t \int_{\mathbb{R}} (C(s, X_{s-}^x + z) - C(s, X_{s-}^x))\tilde{J}(ds \times dz).$$

Then, by Jensen's inequality,

$$E\langle M\rangle_t = E\left(C(t, X_t^x) - C(0, x)\right)^2 \leq 2E[H^2(X_T^x)].$$

This implies that the stochastic integrals

$$\int_0^T \frac{\partial C(s, X_s^x)}{\partial x}\sigma dW_s \quad \text{and} \quad \int_0^T \int_{\mathbb{R}} (C(s, X_{s-}^x + z) - C(s, X_{s-}^x))\tilde{J}(ds \times dz)$$

exist and since X has no jumps at fixed times, $M_t \to M_T$ a.s. when $t \to T$ and one can pass to the limit $t \to T$ in the right-hand side of (3.7).

It remains to prove that $\lim_{t\to T} C(t, X_t^x) = C(T, X_T^x) \equiv H(X_T^x)$ a.s. Let Z be a Lévy process independent from $\mathcal{F}_T$ and with the same law as X. We need to show $\lim_{t\to T} E[H(X_t + Z_{T-t})|\mathcal{F}_T] = H(X_T)$. Since X has no jumps at fixed times, $X_t + Z_{T-t} \to X_T$ a.s. Recalling from part 1 that X_T has an absolutely continuous density and since the set of discontinuities of H has Lebesgue measure zero, we see that $H(X_t + Z_{T-t}) \to H(X_T)$ a.s. Now the polynomial bound on H enables us to use the dominated convergence theorem and conclude that $\lim_{t\to T} E[H(X_t + Z_{T-t})|\mathcal{F}_T] = H(X_T)$ a.s.

The above result covers, for example, digital and put options in exponential Lévy models with either a non-zero diffusion component or stable-like behavior of small jumps (e.g. the tempered stable process [8]) and can be trivially extended to call options using the put-call parity. To treat other exponential Lévy models more regularity is needed for the payoff. The following result applies to Lévy processes with no diffusion component and finite second moment, but also to more general Markov processes with jumps:

Proposition 8. *Let X be as in (2.3) with $m = 1$, $\sigma(s,x) \equiv 0$ and $\gamma(s,z,x)$ satisfying*

$$|\gamma(s,z,x) - \gamma(s,z,x')| \leq \rho(z)|x - x'|$$
$$|\gamma(s,z,x)| \leq \rho(z)(1 + |x|) \quad \textit{with} \quad \int_{\mathbb{R}} \rho^2(z)\nu(dz) < \infty.$$

Suppose that the payoff function H is Lipschitz continuous: $|H(x) - H(y)| \leq K|x-y|$. Then the process $(C(t, X_t^x))_{0 \leq t \leq T}$ with $C(t,x) = E[H(X_T^{(t,x)})]$, is a square integrable martingale with the representation

$$C(t, X_t^x) = C(0,x) + \int_0^t \int_{\mathbb{R}} (C(s, X_{s-}^x + \gamma(s,z,X_{s-}^x)) - C(s, X_{s-}^x))\tilde{J}(ds \times dz). \tag{3.10}$$

Proof. By Theorem III.2.32 in [23], the stochastic differential equation (2.3) admits a non-explosive solution. Since

$$E[(X_t^x)^2] = x^2 + \int_0^t \int_{\mathbb{R}} E[\gamma^2(s,z,X_s^x)]\nu(dz)ds \leq x^2 + K\int_0^t E[1 + |X_s^x|^2]ds,$$

it follows from Gronwall's inequality that

$$E[(X_t^x)^2] \leq (x^2 + Kt)e^{Kt}. \tag{3.11}$$

We also note for future use that the same method can be used to obtain

$$E[(X_T^{(t,x)} - X_T^{(t,y)})^2] \leq (x-y)^2 e^{K(T-t)}. \tag{3.12}$$

The estimate (3.11) implies that $(C(t, X_t^x))_{0 \leq t \leq T}$ is a square integrable martingale and by Theorems III.4.29, III.2.33 in [23], it admits a martingale representation: there exists a measurable function $Z : \Omega \times \mathbb{R} \times [0,T] \to \mathbb{R}$ such that

$$C(t, X_t^x) = C(0,x) + \int_0^t \int_{\mathbb{R}} Z_s(z)\tilde{J}(ds \times dz). \tag{3.13}$$

If we are able to show the existence of

$$\tilde{C}_t = C(0,x) + \int_0^t \int_{\mathbb{R}} (C(s, X_{s-}^x + \gamma(s,z,X_{s-}^x)) - C(s, X_{s-}^x))\tilde{J}(ds \times dz), \tag{3.14}$$

then (3.10) will follow since the jumps of (3.13) and (3.14) are indistinguishable and hence $C = \tilde{C}$ (note that this part of the argument does not carry over to the case $\sigma > 0$). By Jensen's inequality and (3.12),

$$(C(t,x) - C(t,y))^2 \le E[(H(X_T^{(t,x)}) - H(X_T^{(t,y)}))^2]$$
$$\le KE[(X_T^{(t,x)} - X_T^{(t,y)})^2] \le K(x-y)^2.$$

The existence and square integrability of (3.14) now follows from

$$\int_0^t \int_{\mathbb{R}} E(C(s, X_s^x + \gamma(s,z,X_s^x)) - C(s, X_s^x))^2 \nu(dz) ds$$
$$\le K \int_0^t \int_{\mathbb{R}} E(\gamma^2(s,z,X_s^x))\nu(dz)ds \le K \int_0^t E(1+|X_s^x|)^2 ds < \infty.$$

3.2 Asian Options

Exotic options can be introduced via the process of non-traded factors $\tilde{X}$ (see the beginning of this section): for Asian options, one can take

$$\tilde{X}_t = \int_0^t X_s ds$$

and the option's price is then given by

$$C_t = E[H(\tilde{X}_T)|\mathcal{F}_t] = C(t, X_t, \tilde{X}_t),$$

that is, the option's price is now a function of time and (extended) state and one can use the theory developed for European options.

3.3 Barrier Options

The value of a knock-out barrier option can be represented as

$$C_t^B = E[H(X_T)1_{\tau > T}|\mathcal{F}_t],$$

where τ is the first exit time of X from an interval B. For example, in the case of an up-and-out option with barrier b we have $\tau = \inf\{t \ge 0 : X_t > b\}$. If X is a Markov jump-diffusion then, conditionally on the event that the barrier has not been crossed, the price of a knock-out barrier option only depends on time and state. Therefore,

$$C_t^B = 1_{\tau > t} C^B(t, X_t),$$
$$\text{where} \quad C^B(t,x) = \begin{cases} 0, & x \notin B \\ E[H(X_T^{(t,x)})1_{\tau_t > T}], & x \in B \end{cases} \tag{3.15}$$

and τ_t is defined by $\tau_t = \inf\{s \ge t : X_s^{(t,x)} \notin B\}$. Furthermore, the above is equivalent to $C_t^B = C^B(t \wedge \tau, X_{t\wedge\tau})$. Supposing that $C^B(t,x)$ possesses the required differentiability properties, we can apply the Itô formula *up to time* τ obtaining an SDE of type (2.2):

$$C_t^B = C_0^B + \int_0^{t\wedge\tau} \frac{\partial C^B(s,X_s)}{\partial x}\sigma(s,X_s)dW_s + \int_{[0,t\wedge\tau]\times\mathbb{R}} (C^B(s,X_{s-}+\gamma(s,z,X_{s-})) - C^B(s,X_{s-}))\tilde{J}(ds\times dz). \tag{3.16}$$

However, in the case of barrier options the proof of regularity is much more involved [12] than for European ones. The following result, based on Bensoussan and Lions [5], allows to obtain a martingale representation for barrier options under further assumptions:

Proposition 9. *Let X be as in (2.3) with $m = 1$ and σ and γ satisfying the following hypotheses:*

(i) There exist K_1, K_2 with $0 < K_1 < K_2$ such that $K_1 < \sigma(t,x) < K_2$ and $\left|\frac{\partial\sigma(t,x)}{\partial x}\right| < K_2$ for all t, x.
(ii) There exists a Radon measure m on $\mathbb{R}\setminus\{0\}$ such that

$$\int_{\mathbb{R}} |z|m(dz) < \infty$$

and $\forall A \in \mathcal{B}(\mathbb{R})$, $\forall(t,x)$

$$m(A) \geq \nu(\{z : \gamma(t,z,x) \in A\})$$

(iii) B is a bounded open interval on $\mathbb{R}$.
(iv) The payoff function H satisfies $H \in W_0^{1,p}(B)$, $4 < p < \infty$. The space $W_0^{1,p}(B)$ denotes the $W^{1,p}(B)$-closure of $C_0^\infty(B)$, the space of smooth functions with compact support in B. This implies that the payoff must tend to zero as one approaches the barrier.

Then the barrier option price $C^B(t,x)$ defined by (3.15) belongs to the space

$$W^{1,2,p} = \{z \in L^p([0,T]\times B) : \frac{\partial z}{\partial t}, \frac{\partial z}{\partial x}, \frac{\partial^2 z}{\partial x^2} \in L^p([0,T]\times B)\}$$

and the process $(C_t^B)_{0\leq t\leq T}$ is a square integrable martingale satisfying the representation (3.16).

Proof. The regularity result is a corollary of Theorems 3.4 and 8.2 in [5]. To obtain the martingale representation, we can approximate $C^B(t,x)$ in $W^{1,2,p}$ by a sequence of smooth functions (C_n^B), apply the Itô formula to each C_n and then pass to the limit using the inequality (III.7.32) in [5]:

$$\left| E\left[\int_t^T f(s,X_s)ds\right]\right| \leq C_{T,p}|f|_{L^p} \quad \forall f \in L^p(\mathbb{R}^d), \quad p > 2, \tag{3.17}$$

where X satisfies the hypotheses (i) and (ii) of the above proposition.

Corollary 10. *Let X be a Lévy process with $\sigma > 0$ and $\int |x|\nu(dx) < \infty$, let B be a bounded open interval and suppose that the payoff function H satisfies $H \in W_0^{1,p}(B)$, $4 < p < \infty$. Then the barrier option price $C^B(t,x)$ defined by* (3.15) *belongs to the space $W^{1,2,p}$ and the process $(C_t^B)_{0\le t\le T}$ is a square integrable martingale satisfying the representation* (3.16).

4 Hedging with Options: Examples and Applications

In this section we apply the general hedging formula (2.9) to options and analyze numerically the performance of the minimal variance hedging strategy in different settings.

Hedging with the Underlying in an Exponential Lévy Model

Suppose that the price of the underlying asset is given by the exponential of a Lévy process:

$$dX_t = X_t\sigma dW_t + \int_{\mathbb{R}} X_{t-}(e^z - 1)\tilde{J}(dt \times dz).$$

The European option price (3.1) can then be written as the expectation of a function of a Lévy process $Z_t = \log(X_t/X_0)$: $C(t,x) = E[H(xe^{Z_{T-t}})]$. If the model parameters and the payoff function $H(.)$ satisfy either the hypotheses of Proposition 7 or those of Proposition 8, we can compute a martingale representation for $C(t, X_t)$. Applying the general formula (2.9) we then obtain the following hedge ratio:

$$\phi_t = \frac{\sigma^2 \frac{\partial C}{\partial X}(t, X_{t-}) + \frac{1}{X_{t-}} \int \nu(dz)(e^z - 1)[C(t, X_{t-}e^z) - C(t, X_{t-})]}{\sigma^2 + \int (e^z - 1)^2 \nu(dz)}. \quad (4.1)$$

Note that the above equation makes sense in a much more general setting than for instance the delta-hedging strategy which requires that the option price be differentiable, a property which can fail in pure-jump models [12].

Case of a Single Jump Size

Assume that the stock price process X^1 follows an exponential Lévy model with a non-zero diffusion component and a single possible jump size:

$$dX_t^1 = X_t^1 \sigma dW_t + X_{t-}^1(e^{z_0} - 1)d\tilde{N}_t,$$

where $\tilde{N}$ is a compensated Poisson process with intensity λ. We want to hedge a European option $Y_t = C(t, X_t^1)$ with the stock and another European option $X_t^2 = C^*(t, X_t^1)$. In this case, Proposition 7 applies due to the presence of a non-degenerate diffusion component. Denoting $\Delta X = (e^{z_0} - 1)X$ and $\Delta C(t,X) = C(t, Xe^{z_0}) - C(t,X)$ we obtain the following hedge ratios:

$$\phi_t^1 = \frac{\Delta C^*(t,X_{t-}^1)\frac{\partial C(t,X_{t-}^1)}{\partial X} - \Delta C(t,X_{t-}^1)\frac{\partial C^*(t,X_{t-}^1)}{\partial X}}{\Delta C^*(t,X_{t-}^1) - \Delta X_{t-}^1\frac{\partial C^*(t,X_{t-}^1)}{\partial X}},$$

$$\phi_t^2 = \frac{\Delta C(t,X_{t-}^1) - \Delta X_{t-}^1\frac{\partial C(t,X_{t-}^1)}{\partial X}}{\Delta C^*(t,X_{t-}^1) - \Delta X_{t-}^1\frac{\partial C^*(t,X_{t-}^1)}{\partial X}}.$$

It is easy to see that with these hedge ratios *the residual hedging error* $\epsilon_T(\phi)$ *is equal to zero.*

When the jump size ΔX^1 is small, the optimal hedge is approximated by delta-gamma hedge ratios

$$\delta_t = \frac{\partial C(t,X_{t-}^1)}{\partial X} - \frac{\partial C^*(t,X_{t-}^1)}{\partial X}\left(\frac{\partial^2 C(t,X_{t-}^1)}{\partial X^2}\Big/\frac{\partial^2 C^*(t,X_{t-}^1)}{\partial X^2}\right)$$

$$\gamma_t = \frac{\partial^2 C(t,X_{t-}^1)}{\partial X^2}\Big/\frac{\partial^2 C^*(t,X_{t-}^1)}{\partial X^2},$$

obtained by setting to zero the first and the second derivative of the hedged portfolio with respect to the stock price. Note however that in general (jump size not small) the delta-gamma hedging strategy does not eliminate the risk completely, although the optimal quadratic hedging strategy does.

Barndorff-Nielsen and Shephard Model

Let us reconsider the BNS model introduced in Example 3. This model is not covered by results of Section 3 but we can check the differentiability of the option price $C(t,x,\sigma_0^2)$ directly along the lines of the proof of Proposition 7, using the explicit form of the Fourier transform of the log-price Y [8, p. 489]:

$$\phi_t(u) = E\{e^{iuY_t}\} = \exp\Big\{-iut\,l(\rho) - \frac{\sigma_0^2}{2}(iu+u^2)\varepsilon(\lambda,t) + \int_0^t l\left(i\rho u - \frac{1}{2}(iu+u^2)\varepsilon(\lambda,s)\right)ds\Big\}.$$

Because the diffusion coefficient is bounded from below on any finite time interval,

$$|\phi_t(u)| \le \exp\Big\{-\frac{\sigma_0^2 u^2}{2}\varepsilon(\lambda,t) + \int_0^t l\left(-\frac{1}{2}u^2\varepsilon(\lambda,s)\right)ds\Big\} \le \exp\Big\{-\frac{\sigma_0^2u^2}{2}\varepsilon(\lambda,t)\Big\}.$$

In addition if $\int_1^\infty z^{m+n}\nu(dz) < \infty$, we can show, as in the proof of Proposition 7, that

$$\left|\frac{\partial^{m+n}\phi_t(u)}{\partial u^m \partial t^n}\right| \le C(t)(1+|u|^{m+2n})\exp\Big\{-\frac{\sigma_0^2u^2}{2}\varepsilon(\lambda,t)\Big\}$$

and this in turn implies that the price of an option whose payoff does not grow at infinity faster than $|x|^{m-1}$ will be n times differentiable in t and infinitely differentiable in x (at every point with $\sigma > 0$). The derivatives with respect to σ_0^2 can be estimated similarly. Applying Proposition 4 yields the following optimal ratio for hedging with the underlying:

$$\phi_t = \frac{\sigma_{t-}^2 \frac{\partial C}{\partial X} + \frac{1}{X_{t-}} \int \nu(dz)(e^{\rho z} - 1)[C(t, X_{t-}e^{\rho z}, \sigma_{t-}^2 + z) - C(t, X_{t-}, \sigma_{t-}^2)]}{\sigma_{t-}^2 + \int (e^{\rho z} - 1)^2 \nu(dz)}.$$

When there are no jumps in the stock price ($\rho = 0$) the optimal hedging strategy is just delta-hedging: $\phi_t = \frac{\partial C}{\partial X}$; even though there are jumps in the option price, they cannot be hedged. On the other hand, when $\rho \neq 0$, the above formula has the same structure as equation (4.1) for exponential Lévy models, with the difference that we also have to take into account the effect of jumps in $\sigma(t)$ on the option price. The impact of the stochastic volatility on the optimal hedging strategy with the underlying asset is thus rather limited: for example, the mean reversion parameter λ does not appear in the hedging formula.

Numerical Example: Hedging a European put in Merton's Model

In this example we suppose that the asset X^1 follows the Merton (1976) model, which is an exponential Lévy model with $\sigma > 0$ and $\nu(x) = \frac{\lambda}{\delta\sqrt{2\pi}} e^{-\frac{(x-\theta)^2}{2\delta^2}}$. We simulate 10000 trajectories of stock in this model with two different parameter sets given below:

		μ	σ	λ	Jump mean	Jump stddev
Model 1:	Risk-neutral	–	0.1	5	−0.05	0.1
Bullish market	Historical	0.2	0.1	5	−0.05	0.1
Model 2:	Risk-neutral	–	0.1	10	−0.2	0.2
Fear of crash	Historical	0.2	0.1	5	−0.05	0.1

The option to be hedged is an out-of-the-money European put with strike $K = 1.2$ and time to maturity $T = 1$. For each price trajectory, we compute the residual error for hedging this option using three different strategies: delta hedging, optimal quadratic hedging with stock only, and optimal quadratic hedging with stock and another European put option with $K = 1$ and $T = 1$.

It is important to note that, in this and the following example, the hedge ratios were precomputed on a grid of time and stock price values with formula (2.9) before simulating the price trajectories; details of computations can be found in [11]. The option prices were evaluated using the following result [12]:

Proposition 11. *Let the payoff function H verify the Lipschitz condition and let $h(x) = H(S_0 e^x)$ have polynomial growth at infinity. Then forward value*

$$f_e(t, x) = E[h(x + X_t)]$$

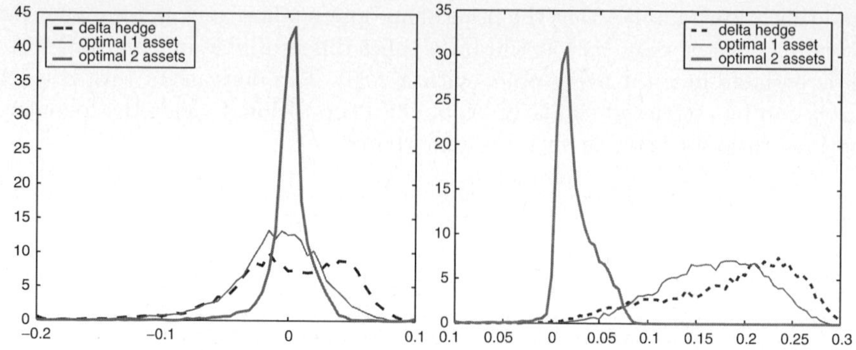

Fig. 1. Histograms of the residual hedging error for a European put with strike $K = 1.2$

of a European option is a viscosity solution of the Cauchy problem

$$\frac{\partial f}{\partial t} = \frac{\sigma^2}{2}\left[\frac{\partial^2 f}{\partial x^2} - \frac{\partial f}{\partial x}\right] + \int_{\mathbb{R}} \nu(dy)\left[f(t, x+y) - f(t,x) - (e^y - 1)\frac{\partial f}{\partial x}(t,x)\right] \tag{4.2}$$

with the initial condition $f(0, x) = h(x)$.

We have used a similar representation for the price of a barrier option [12] and the numerical scheme proposed in [13] for solving the associated PIDE (4.2).

The histograms of the hedging error are shown in Figure 1, left graph, for model 1 and in the right graph for model 2. The table below gives the variance of the residual hedging error for the two models and the three hedging strategies used.

	Bullish market (left)	Fear of crash (right)
Delta hedging:	0.0464	0.1974
Optimal 1 asset:	0.0373	0.1762
Optimal 2 assets:	0.0182	0.0319

First, one can observe that the performance of the optimal quadratic hedging strategy using the underlying only is very similar to that of delta hedging in both models. The performance of both strategies is very sensitive to the difference of Lévy measures under the historical and the risk-neutral probability: when this difference is important as in model 2, both strategies have a very poor performance. On the other hand, this numerical example shows that using options for hedging allows to reduce this sensitivity and achieve an acceptable performance even in presence of an important jump risk premium, that is, when the Lévy measure is very different under the "objective" and the risk-neutral probability.

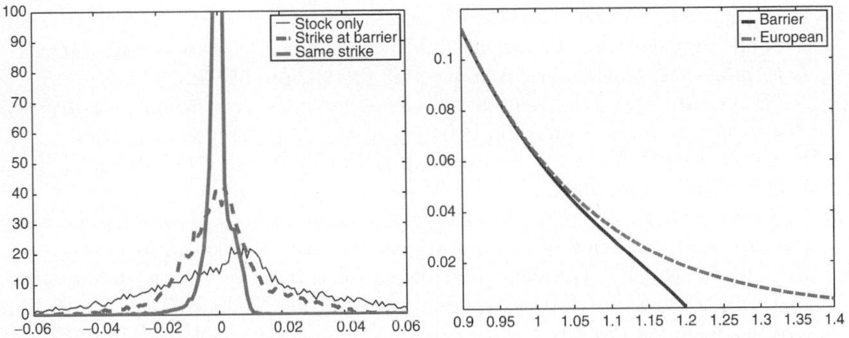

Fig. 2. Left: histograms of the residual hedging error for an up and out barrier put with strike $K = 1$ and barrier $B = 1.2$. Right: option price profiles at $T = 0.5$

Hedging a Barrier Option in Merton's Model

In this example we continue to work in Merton's model and we want to hedge a barrier put with strike $K = 1$ and barrier at $B = 1.2$ using the underlying and a European put option. Figure 2, left graph, depicts the histograms of the residual hedging error for three strategies: hedging with the underlying asset only; hedging with the underlying asset and a European put with strike at the barrier; hedging with stock and a European put with strike $K = 1$. The model parameters correspond to model 1 of previous example. This example shows that a much better hedging performance is achieved by using a European option with the same strike as that of the barrier option. The right graph shows the option price profiles of these two options at time $T = 0.5$. Using a European option for hedging allows to better reproduce the convexity of the barrier option price but it does not take into account the discontinuity of derivative at the barrier.

References

1. Y. Aït-Sahalia and A. Lo, *Nonparametric Risk Management and Implied Risk Aversion*, Journal of Econometrics, 94 (2000), pp. 9–51.
2. T. Arai, *An extension of mean-variance hedging to the discontinuous case*, Finance and Stochastics, 9 (2005), pp. 129–139.
3. O. Barndorff-Nielsen and N. Shephard, *Modelling by Lévy processes for financial econometrics*, in Lévy Processes — Theory and Applications, O. Barndorff-Nielsen, T. Mikosch, and S. Resnick, eds., Birkhäuser, Boston, 2001.
4. D. Belomestny and M. Reiss, *Spectral calibration of exponential Lévy models*, Finance and Stochastics, 10 (2006), pp. 449–474.
5. A. Bensoussan and J.-L. Lions, *Impulse Control and Quasi-Variational Inequalities*, Gauthier-Villars, 1984.

6. F. E. Benth, G. Di Nunno, A. Lôkka, B. Øksendal, and F. Proske, *Explicit representation of the minimal variance portfolio in markets driven by Lévy processes*, Mathematical finance, 13 (2003), pp. 55–72.
7. N. Bouleau and D. Lamberton, *Residual risks and hedging strategies in Markovian markets*, Stochastic Process. Appl., 33 (1989), pp. 131–150.
8. R. Cont and P. Tankov, *Financial Modelling with Jump Processes*, Chapman & Hall/CRC Press, 2004.
9. R. Cont and P. Tankov, *Nonparametric calibration of jump-diffusion option pricing models*, Journal of Computational Finance, 7 (2004), pp. 1–49.
10. R. Cont and P. Tankov, *Retrieving Lévy processes from option prices: Regularization of an ill-posed inverse problem*, SIAM Journal on Control and Optimization, 45 (2006).
11. R. Cont, P. Tankov and E. Voltchkova, *Hedging options in presence of jumps*, Working Paper.
12. R. Cont and E. Voltchkova, *Integro-differential equations for option prices in exponential Lévy models*, Finance and Stochastics, 9 (2005), pp. 299–325.
13. R. Cont and E. Voltchkova, *Finite difference methods for option pricing in jump-diffusion and exponential Lévy models*, SIAM Journal on Numerical Analysis, 43 (2005).
14. D. Duffie, J. Pan, and K. Singleton, *Transform analysis and asset pricing for affine jump-diffusions*, Econometrica, 68 (2000), pp. 1343–1376.
15. H. W. Engl, M. Hanke, and A. Neubauer, *Regularization of Inverse Problems*, vol. 375, Kluwer Academic Publishers Group, Dordrecht, 1996.
16. H. Föllmer and R. Elliott, *Orthogonal martingale representation*, in Stochastic Analysis, E. Mayer-Wolf, E. Merzbach, and A. Schwartz, eds., Academic Press: Boston, 1991, pp. 139–152.
17. H. Föllmer and M. Schweizer, *Hedging of contintgent claims under incomplete information*, in Applied Stochastic Analysis, M. H. A. Davis and R. J. Elliott, eds., Gordon and Breach, 1991, pp. 389–414.
18. H. Föllmer and D. Sondermann, *Hedging of non-redundant contingent claims*, in Contributions to Mathematical Economics, W. Hildenbrand and A. Mas-Colell, eds., North Holland, 1986, pp. 205–224.
19. D. Heath, E. Platen, and M. Schweizer, *Numerical comparison of local risk minimization and mean-variance hedging*, in Option Pricing, Interest Rates and Risk Management, J. Cvitanic, E. Jouini, and M. Musiela, eds., Cambridge University Press, 2001.
20. F. Hubalek, J. Kallsen and L. Krawczyk, *Variance-Optimal Hedging for Processes with Stationary Independent Increments*, Working Paper, 2004.
21. J. Jacod, *Calcul Stochastique et Problèmes de Martingales*, vol. 714 of Lecture Notes in Math., Springer, Berlin, 1979.
22. J. Jacod, S. Méléard, and P. Protter, *Explicit form and robustness of martingale representations*, Annals of Probability, 28 (2000), pp. 1747–1780.
23. J. Jacod and A. N. Shiryaev, *Limit Theorems for Stochastic Processes*, Springer, Berlin, 2nd ed., 2003.
24. J. Ma, P. Protter, and J. Zhang, *Explicit representations and path regularity for martingale representations*, in Lévy Processes — Theory and Applications, O. Barndorff-Nielsen, T. Mikosch, and S. Resnick, eds., Birkhäuser, Boston, 2001.
25. P. Protter and M. Dritschel, *Complete markets with discontinuous security prices*, Finance Stoch., 3 (1999), pp. 203–214.

26. K. SATO, *Lévy Processes and Infinitely Divisible Distributions*, Cambridge University Press, Cambridge, UK, 1999.
27. M. SCHWEIZER, *A guided tour through quadratic hedging approaches*, in Option Pricing, Interest Rates and Risk Management, J. Cvitanic, E. Jouini, and M. Musiela, eds., Cambridge University Press, 2001.

Power Variation Analysis of Some Integral Long-Memory Processes

José Manuel Corcuera

Facultat de Matemàtiques Universitat de Barcelona Gran Via, 585 E-08007 Barcelona, Spain

Summary. We show some results about the asymptotic behavior of the power variation and how they can be used for statistical purposes in the context of some integral long-memory processes. These processes are obtained as integrals with respect to a fractional Brownian motion with Hurst parameter $H > 1/2$.

1 Introduction

Let $\{Z_t, t \geq 0\}$ be a stochastic process. The realized power variation of order $p > 0$ is defined as

$$V_p^n(Z)_t = \sum_{i=1}^{[nt]} \left| Z_{i/n} - Z_{(i-1)/n} \right|^p .$$

For $p = 2$ we have the realized quadratic variation that has been widely used in statistics of random processes.

For any $p > 0$ the p-variation of a real valued function f on an interval $[a, b]$ is defined as

$$\mathrm{Var}_p(f; [a, b]) = \sup_\pi \left(\sum_{i=1}^n |f(t_i) - f(t_{i-1})|^p \right)^{1/p} ,$$

where the supremum runs over all partitions $\pi = \{a = t_0 < t_1 < \cdots < t_n = b\}$.

Young (1936) proved that the Riemann–Stieltjes integral $\int_a^b f dg$ exists if f and g have finite p-variation and finite q-variation, respectively, in the interval $[a, b]$ and $\frac{1}{p} + \frac{1}{q} > 1$. Moreover, the following inequality holds

$$\left| \int_a^b f dg - f(a)(g(b) - g(a)) \right| \leq c_{p,q} \mathrm{Var}_p(f; [a, b]) \, \mathrm{Var}_q(g; [a, b]) ,$$

where $c_{p,q} = \zeta(\frac{1}{q} + \frac{1}{p})$, with $\zeta(s) := \sum_{n \geq 1} n^{-s}$.

We consider processes of the form $Z_t = \int_0^t u_s dB_s^H$, where B^H is a fractional Brownian motion with Hurst parameter $H > \frac{1}{2}$, and u is a stochastic process with paths of finite q-variation, $q < \frac{1}{1-H}$. The integral is a pathwise Riemann–Stieltjes integral and we are interested in the asymptotic behavior of the realized power variation conveniently scaled

$$n^{-1+pH} \sum_{i=1}^{[nt]} \left| Z_{i/n} - Z_{(i-1)/n} \right|^p = n^{-1+pH} \sum_{i=1}^{[nt]} \left| \int_{(i-1)/n}^{i/n} u_s dB_s^H \right|^p .$$

If $B^H = \{B_t^H, t \geq 0\}$ is a fBm with Hurst parameter $H \in \left(\frac{1}{2}, 1\right)$, then it is a zero mean Gaussian process with covariance function

$$E(B_t^H B_s^H) = \frac{1}{2}\left(t^{2H} + s^{2H} - |t-s|^{2H}\right), s, t \geq 0.$$

Also B^H is self-similar with exponent H and the fractional Gaussian noise: $\{B_n^H - B_{n-1}^H, n \in \mathbf{N}\}$ is a ergodic sequence with positive correlation function:

$$\rho_H(n) = \frac{(n+1)^{2H} + (n-1)^{2H} - 2n^{2H}}{2} \sim cn^{2H-2}. \qquad (1.1)$$

2 The Results

Theorem 1. *Suppose that* $u = \{u_t, t \in [0,T]\}$ *is a stochastic process with finite q-variation, where* $q < \frac{1}{1-H}$. *Set* $Z_t = \int_0^t u_s dB_s^H$. *Then,*

$$n^{-1+pH} V_p^n(Z)_t \underset{n\to\infty}{\overset{P}{\longrightarrow}} c_p \int_0^t |u_s|^p ds.$$

where $c_p = E\left(|N(0,1)|^p\right) = \dfrac{2^{p/2}\Gamma(\frac{p+1}{2})}{\Gamma(1/2)}$.

Proof. (A simple case) Assume first that $u_s \equiv 1$. Then $Z_t = B_t^H$ and

$$\begin{aligned} n^{-1+pH} V_p^n(Z)_t &= \left(\frac{1}{n}\right)^{1-pH} \sum_{i=1}^{[nt]} \left| B_{\frac{i}{n}}^H - B_{\frac{i-1}{n}}^H \right|^p \\ &= \frac{1}{n} \sum_{i=1}^{[nt]} \left| \frac{B_{\frac{i}{n}}^H - B_{\frac{i-1}{n}}^H}{\left(\frac{1}{n}\right)^H} \right|^p \\ &\sim \frac{1}{n} \sum_{i=1}^{[nt]} \left| B_i^H - B_{i-1}^H \right|^p \quad \text{(self-similarity)} \\ &\underset{L^1}{\overset{a.s}{\to}} tE\left(|B_1^H|^p\right) = c_p t \quad \text{(ergodicity)} \end{aligned}$$

For the general case we can consider two step sizes $1/m$ and $1/n$, nt integer, then for any $m \geq n$ we have the following decomposition

$$
\begin{aligned}
& m^{-1+pH} V_p^m(Z)_t - c_p \int_0^t |u_s|^p ds \\
&= m^{-1+pH} \sum_{j=1}^{[mt]} \left(\left| \int_{\frac{j-1}{m}}^{\frac{j}{m}} u_s dB_s^H \right|^p - \left| u_{\frac{j-1}{m}} (B_{\frac{j}{m}}^H - B_{\frac{j-1}{m}}^H) \right|^p \right) \\
&\quad + m^{-1+pH} \sum_{j=1}^{[mt]} \left| u_{\frac{j-1}{m}} (B_{\frac{j}{m}}^H - B_{\frac{j-1}{m}}^H) \right|^p - \sum_{i=1}^{nt} \left| u_{\frac{i-1}{n}} \right|^p \sum_{j \in I(i)} \left| B_{\frac{j}{m}}^H - B_{\frac{j-1}{m}}^H \right|^p \\
&\quad + m^{-1+pH} \sum_{i=1}^{nt} \left| u_{\frac{i-1}{n}} \right|^p \sum_{j \in I(i)} \left| B_{\frac{j}{m}}^H - B_{\frac{j-1}{m}}^H \right|^p - c_p n^{-1} \sum_{i=1}^{nt} \left| u_{\frac{i-1}{n}} \right|^p \\
&\quad + c_p n^{-1} \sum_{i=1}^{nt} \left| u_{\frac{i-1}{n}} \right|^p - c_p \int_0^t |u_s|^p ds \\
&= A_m + B_{n,m} + C_{n,m} + D_n, \text{ where } I(i) = \left\{ j : \frac{j}{m} \in \left(\frac{i-1}{n}, \frac{i}{n} \right) \right\}, 1 \leq i \leq nt,
\end{aligned}
$$

and we have to show that each term goes to zero as n, m goes to infinity (see Corcuera *et al.* (2005) for the details).

Corollary 2. *Consider a stochastic process $Y = \{Y_t, t \geq 0\}$ such that*

$$n^{-1+pH} V_p^n(Y)_t \xrightarrow{P} 0$$

as n tends to infinity. Then

$$n^{-1+pH} V_p^n(Z+Y)_t \xrightarrow{P} c_p \int_0^t |u_s|^p ds,$$

as n tends to infinity.

Proof. (For $p \leq 1$). By the triangular inequality and the fact that

$$\left| V_p^n(Z+Y)_t - V_p^n(Z)_t \right| < V_p^n(Y)_t$$

For $H \in (\frac{1}{2}, \frac{3}{4}]$ the fluctuations of the power variation, properly normalized, have conditionally Gaussian asymptotic distributions. Set

$$v_1^2 := \lim_{n\to\infty} \operatorname{Var} \left(\frac{1}{\sqrt{n}} \sum_{i=1}^{n} \left| B_i^H - B_{i-1}^H \right|^p \right).$$

It is not difficult to see that

$$v_1^2 = \delta_p + 2 \sum_{j \geq 1} \left(\gamma_p(\rho_H(j)) - \gamma_p(0) \right),$$

with $\rho_H(n)$ given by (1.1),

$$\delta_p = 2^p \left(\frac{1}{\sqrt{\pi}} \Gamma\left(p + \frac{1}{2}\right) - \frac{1}{\pi} \Gamma\left(\frac{p+1}{2}\right)^2 \right)$$

and $\gamma_p(x) = (1 - x^2)^{p+\frac{1}{2}} \frac{2^p}{\pi} \Gamma(\frac{p+1}{2})^2 \ {}_1\mathrm{F}_1(\frac{p+1}{2}; \frac{1}{2}; x^2)$, where ${}_1\mathrm{F}_1$ the confluent hyper-geometric function, that is

$${}_1F_1(a; b; z) = 1 + \frac{az}{b} + \frac{a(a+1)}{b(b+1)} \frac{z^2}{2!} + \dots$$

Theorem 3. *Fix* $p > 0$. *Assume* $1/2 < H < 3/4$. *Then*

$$\left(B_t^H, n^{-1/2+pH} \ V_p^n(B^H)_t - c_p t n^{1/2}\right) \xrightarrow{\mathcal{L}} \left(B_t^H, v_1 W_t\right),$$

as n *tends to infinity, where* $W = \{W_t, t \in [0,T]\}$ *is a Brownian motion independent of the process* B^H, *and the convergence is in the space* $\mathcal{D}([0,T])^2$ *equipped with the Skorohod topology.*

Proof. (Sketch of the proof) First we show the convergence of the finite dimensional distributions. Let $J_k = (a_k, b_k]$, $k = 1, \dots, N$ be pairwise disjoint intervals contained in $[0,T]$. Define the random vectors $B = (B_{b_1}^H - B_{a_1}^H, \dots, B_{b_N}^H - B_{a_N}^H)$ and $X^{(n)} = (X_1^{(n)}, \dots, X_N^{(n)})$, where

$$X_k^{(n)} = n^{-1/2+pH} \sum_{[na_k] < j \le [nb_k]} \left| B_{j/n}^H - B_{(j-1)/n}^H \right|^p - n^{1/2} c_p |J_k|,$$

$k = 1, \dots, N$ and $|J_k| = b_k - a_k$. We claim that

$$(B, X^{(n)}) \xrightarrow{\mathcal{L}} (B, V),$$

where B and V are independent and V is a Gaussian random vector with zero mean and with independent components of variances $v_1^2 |J_k|$. By the self-similarity of the fBm, the convergence is equivalent to the convergence in distribution of $(B^{(n)}, Y^{(n)})$ to (B, V), where

$$B_k^{(n)} = n^{-H} \sum_{[na_k] < j \le [nb_k]} X_j, \ 1 \le k \le N$$

$$Y_k^{(n)} = \frac{1}{\sqrt{n}} \sum_{[na_k] < j \le [nb_k]} H(X_j), \ 1 \le k \le N.$$

$X_j = B_j^H - B_{j-1}^H$ and $H(x) = |x|^p - c_p$. The function $H(x)$ can be expanded in the form

$$H(x) = \sum_{m=2}^{\infty} c_m H_m(x),$$

where H_m is the mth Hermite polynomial. Let $\mathcal{H}_1$ be the closed subspace of L^2 generated by $\{X_j\}$, and for any $m \geq 2$ denote by $\mathcal{H}_m$ the closed subspace of L^2 generated by $H_m(X)$, where $X \in \mathcal{H}_1$, $E(X^2) = 1$. Let $\mathcal{H}_1^{\odot m}$ be the symmetric tensor product equipped with the norm $\sqrt{m!}\,\|\cdot\|_{\mathcal{H}_1^{\otimes m}}$. We know that the mapping

$$I_m : \mathcal{H}_1^{\odot m} \to \mathcal{H}_m$$

defined by $I_m(X^{\otimes m}) = H_m(X)$, is a linear isometry. We will denote by J_m the projection operator on $\mathcal{H}_m(X)$.

Now by the works of Nualart and Peccati (2005), Peccati and Tudor (2005) and Hu and Nualart (2005) we simply have to check that:

For any $m \geq 2$ and $k = 1, \ldots, N$, the limit $\lim_{n\to\infty} E(|J_m Y_k^{(n)}|^2) = \sigma_{m,k}^2$ exists and $\sum_{m=2}^{\infty} \sup_n E(|J_m Y_k^{(n)}|^2) < \infty$.

For any $m \geq 2$ and $k \neq h$ $\lim_{n\to\infty} E(J_m Y_k^{(n)} J_m Y_h^{(n)}) = 0$.

For any $m \geq 2$, $k = 1, \ldots, N$ and $1 \leq p \leq m-1$,

$$\lim_{n\to\infty} I_m^{-1} J_m Y_k^{(n)} \otimes_p I_m^{-1} J_m Y_k^{(n)} = 0,$$

where $\otimes_p$ denotes the contraction of p indices. (i), (ii) and (iii) are true because

$$\sum_{j=1}^{\infty} \rho_H(j)^m \sim \sum_{j=1}^{\infty} j^{(2H-2)m} < \infty$$

since $m \geq 2$ and $1/2 < H < 3/4$. For the tightness condition is sufficient to show that the sequence of processes

$$Z_t^{*(n)} = n^{-1/2+pH}\ V_p^n(B^H)_t - c_p[nt]/n^{1/2}.$$

is tight in $\mathcal{D}([0,T])$. Then we can compute for $s < t$

$$E\left(\left|Z_t^{*(n)} - Z_s^{*(n)}\right|^4\right) = n^{-2} E\left(\left|\sum_{j=[ns]+1}^{[nt]} H(X_j)\right|^4\right).$$

By Taqqu (1977) we know that, for all $N \geq 1$

$$\frac{1}{N^2} E\left(\left|\sum_{j=1}^{N} H(X_j)\right|^4\right) \leq K\left(\sum_{u=0}^{\infty} \rho_H^2(u)\right)^2.$$

As a consequence,

$$\sup_n E\left(\left|Z_t^{*(n)} - Z_s^{*(n)}\right|^4\right) \leq C|t-s|^2,$$

and by Billingsley (1968) we get the desired tightness property.

As a consequence we have the following Theorem.

Theorem 4. *Fix $p > 0$. Let B^H be a fBm with Hurst parameter $H \in (1/2, 3/4)$. Suppose that $u = \{u_t, t \in [0,T]\}$ is a stochastic process measurable with respect to $\mathcal{F}_T^H$, and with Hölder continuous trajectories of order $a > \frac{1}{2(p\wedge 1)}$. Set $Z_t = \int_0^t u_s dB_s^H$. Then*

$$n^{-1/2+pH} V_p^n(Z)_t - c_p\sqrt{n}\int_0^t |u_s|^p ds \xrightarrow{\mathcal{L}} v_1 \int_0^t |u_s|^p dW_s,$$

as n tends to infinity, where $W = \{W_t, t \in [0,T]\}$ is a Brownian motion independent of $\mathcal{F}_T^H$, and the convergence is stable *and in $\mathcal{D}([0,T])$.*

For the notion of *stable converge* see Aldous and Eagleson (1978). The following corollary gives the distributional effect of adding a process Y to the process Z, see also Corollary 2 above.

Corollary 5. *Assume the same conditions as in the previous Theorem. Consider a stochastic process $Y = \{Y_t, t \in [0,T]\}$ such that*

$$n^{-\frac{1}{2}+pH} V_p^n(Y)_T \xrightarrow{P} 0,$$

as n tends to infinity. Then,

$$n^{-1/2+pH} V_p^n(Y+Z)_t - c_p\sqrt{n}\int_0^t |u_s|^p ds \xrightarrow{\mathcal{L}} v_1 \int_0^t |u_s|^p dW_s$$

as n tends to infinity, where $W = \{W_t, t \geq 0\}$ is a Brownian motion independent of the process B^H, and the convergence is stable and in $\mathcal{D}([0,T])$.

We can also derive the following convergence in distribution for the fluctuations of the power variation of stochastic integrals, in the case $H = \frac{3}{4}$.

Theorem 6. *Suppose that $H = 3/4$ and $u = \{u_t, t \in [0,T]\}$ is a stochastic process measurable with respect to $\mathcal{F}_T^H$ with Hölder continuous trajectories of the order $a > \frac{1}{2(p\wedge 1)}$. Then,*

$$(\log n)^{-1/2}\left(n^{-1/2+pH} V_p^n(Z)_t - c_p\sqrt{n}\int_0^t |u_s|^p ds\right) \xrightarrow{\mathcal{L}} v_2 \int_0^t |u_s|^p dW_s,$$

as $n \to \infty$, where $W = \{W_t, t \in [0,T]\}$ is a Brownian motion independent of $\mathcal{F}_T^H$ and v_2 is given by

$$v_2^2 := \lim_{n\to\infty} Var\left(\frac{1}{\sqrt{n\log n}}\sum_{i=1}^n \left|B_i^H - B_{i-1}^H\right|^p\right).$$

Where the converge is stable in the Skorohod space $\mathcal{D}([0,T])$.

If $H > \frac{3}{4}$, the fluctuations of the power variation converge to a process in the second chaos which is called the Rosenblatt process.

Theorem 7. *Fix $p > 0$ and assume that $\frac{3}{4} < H < 1$. Then, in $\mathcal{D}([0,T])$,*

$$n^{2-2H}(n^{-1+pH}V_p^n(B^H)_t - c_p t) \xrightarrow{\mathcal{L}} Z_t$$

where

$$Z_t = \frac{1}{\Gamma(2-2H)\cos((1-H)\pi)} d_p \times \int_0^\infty \int_0^{x_2} \frac{e^{i(x_1+x_2)t} - 1}{i(x_1+x_2)} |x_1|^{1/2-H} |x_2|^{1/2-H} dW_{x_1} dW_{x_2},$$

is the Rosenblatt process, $\{W_t, t \in [0,T]\}$ is a Brownian motion and

$$d_p = E(|B_1^H|^{2+p}) - E(|B_1^H|^p).$$

3 Applications

Many statistical analysis of financial and temperature data have shown the presence of significant power at low frequencies in their spectral analysis, which means long-range dependence (see Willinger *et al.* (1999), Cutland *et al.* (1995), Brody *et al.* (2002) and the references in Shiryaev (1999). But these investigations have produced controversies, specially because if we use these models to describe the evolution of stock prices, then the resulting market has arbitrage opportunities and the lack of arbitrage is a paradigm in the modern financial economics (see Rogers (1997)).

We have seen that, under certain assumptions, the values of $V_p^n(Z)_t$ oscillate around cn^{pH-1}. This can be used to give a consistent estimator of H. Then, we shall study the behavior of this estimator by simulating a fractional Brownian motion and a geometric fractional Brownian motion and we will try to corroborate the theoretical results of the previous section. In the case the model is not completely specified by H because the process u is unknown, we will estimate H by a regression of $\log V_p^n(Z)_t$ against $\log n$ for different values of n. We shall consider real data of the stocks prices in the Spanish financial market and we will compare the results with the power variation analysis with the results using the well known R/S analysis.

3.1 The Method

Let $Z_{1/n}, Z_{2/n}, \ldots, Z_1$ be, a sample of n observations. Then by Theorem 1 we have

$$V_p^n(Z)_1 \sim c_p \left(\int_0^1 |u_s|^p ds \right) n^{1-pH}$$

Therefore a consistent estimator of H is given by

$$\hat{H} = \frac{1}{p} - \frac{\log V_p^n(Z)_1 - \log c_p - \log \int_0^1 |u_s|^p ds}{p \log n}.$$

Note that this estimator requires the process u to be observable to evaluate $\int_0^1 |u_s|^p ds$, this is not true in general, however if the process is, for instance, a solution of the stochastic differential equation

$$dZ_t = Z_t(bdt + \sigma dB_t^H)$$

with σ known then $u = \sigma Z$. Nevertheless Z is observed in discrete times $1/n$, $2/n, \ldots$, so $\int_0^1 |u_s|^p ds$ has to be estimated by, for instance, $\frac{1}{n}\sum_{i=1}^n |Z_{\frac{i}{n}}|^p$. It can be easily seen that if $H > \frac{1}{2(p\wedge 1)}$ this estimation does not affect to the asymptotic distribution of $\hat{H}$.

Asymptotic Behavior of $\hat{H}$

By Theorem 4 and assuming that the estimation of $\int_0^1 |u_s|^p ds$ does not affect the asymptotic behavior, we have the conditional converge

$$\frac{n^{-1/2+pH} V_p^n(Z)_1 - c_p \frac{1}{\sqrt{n}} \sum_{i=1}^n |u_{\frac{i}{n}}|^p}{\sqrt{v_1^2 \frac{1}{n} \sum_{i=1}^n |u_{\frac{i}{n}}|^{2p}}} \xrightarrow{\mathcal{L}} N(0,1),$$

from here it is easy to obtain the approximate confidence interval of coefficient γ

$$\hat{H} \pm \frac{k_\gamma v_1}{p c_p \log n} \frac{\sqrt{\sum_{i=1}^n |u_{\frac{i}{n}}|^{2p}}}{\sum_{i=1}^n |u_{\frac{i}{n}}|^p} \tag{3.1}$$

where $k_\gamma = \Phi^{-1}(\frac{1-\gamma}{2})$ and Φ denotes the c.d.f. of the standard normal distribution.

Note that v_1 depends on p as it is shown in Figure 1.

The following table gives the values of $\hat{H}$ for different values of p and H in the case we have a sample of size $n = 2000$ of equally spaced observations of fBm, in parenthesis we have the radius of a confidence interval with $\gamma = 0.95$.

	H		
p	0.6	0.65	0.7
0.75	0.602(0.005)	0.651(0.005)	0.701(0.007)
1	0.603(0.004)	0.652(0.005)	0.702(0.006)
2	0.605(0.005)	0.654(0.004)	0.703(0.004)

The following table gives the values of $\hat{H}$ for different values of p and H in the case that Z is a geometrical fBm with $\sigma = 1$ and $b = 0$, the sample size of equally spaced observations is $n = 2000$. In parenthesis we have the radius of a confidence interval with $\gamma = 0.95$.

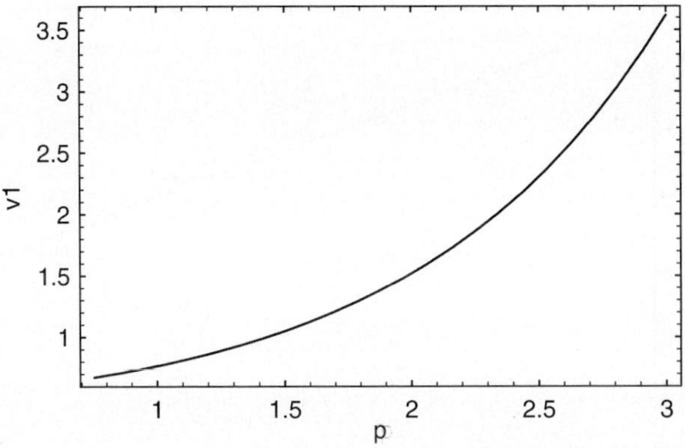

Fig. 1. Behavior of v_1 with p for $H = 0.7$

	H		
p	0.6	0.65	0.7
0.75	0.602(0.005)	0.651(0.005)	0.701(0.007)
1	0.603(0.004)	0.652(0.005)	0.702(0.006)
2	0.605(0.005)	0.654(0.004)	0.703(0.004)

3.2 Estimation of *H* When *u* is Unknown

In case of the process u is not known we can consider the statistics

$$V_p^{[n/m]}(Z)_1 = \sum_{i=1}^{[n/m]} \left| Z_{im/n} - Z_{(i-1)m/n} \right|^p$$

for different values of m, $1 \leq m \leq m_u$. The results of the previous section imply that, under the assumptions on the process $Z_t = \int_0^t u_s dB_s^H, 0 \leq t \leq 1$,

$$V_p^{[n/m]}(Z)_1 \sim c_p \left(\int_0^1 |u_s|^p ds \right) \left(\frac{m}{n} \right)^{pH-1}$$

whenever n/m is large enough. This is why we only consider values of $m \leq m_u$. Then this can be used to estimate H by a log-log plot also called *pox plot.* We shall denote $\tilde{H}$ the corresponding estimator.

Figure 2 depicts the pox plot of the power variation of order 1 corresponding to a sample of size 2.000 of a fractional Brownian motion of Hurst parameter 0.7 series and results in an estimate of H = 0.698. Figure 3 is similar but considering a geometrical fractional Brownian motion of Hurst parameter 0.7 and results also in an estimate of H = 0.697. In both cases we have taken $m_u = 40$ and $p = 1$. We have use the method of Davies and Harte (1987) for simulating a Gaussian stationary sequence.

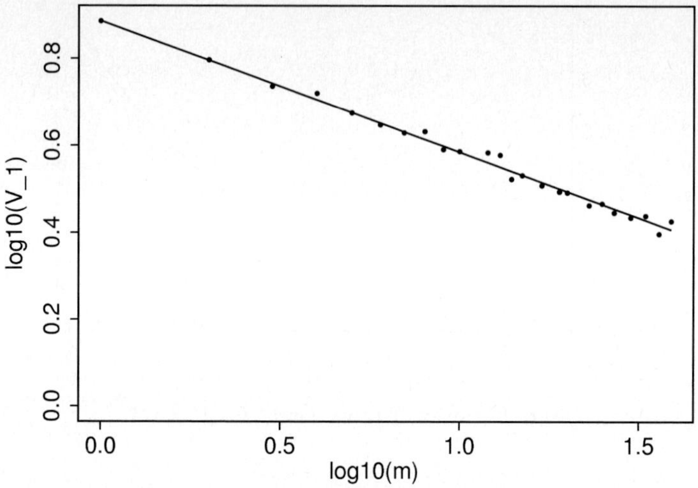

Fig. 2. Data corresponding to a fBm with $H = 0.7$, $\tilde{H} = 0.698$

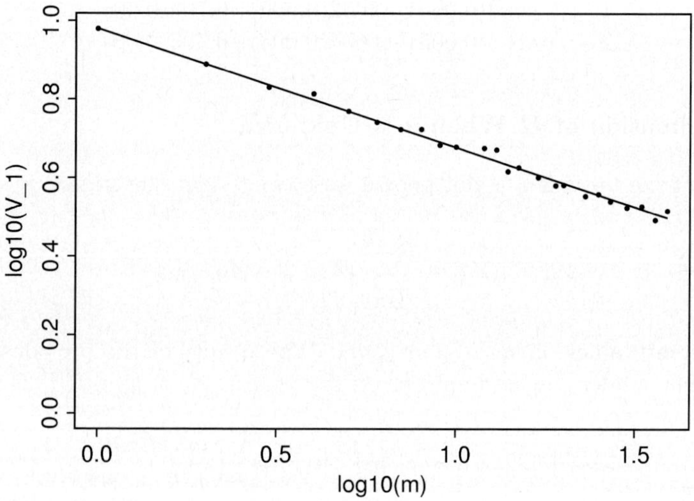

Fig. 3. Data corresponding to a geometrical fBm with $H = 0.7$, $\tilde{H} = 0.697$

In the following examples we consider the prices of certain stocks in the Spanish market as the process Z, and we estimate H by a log-log plot. Figure 4 corresponds to the index IBEX35, the estimation of H is $\tilde{H} = 0.533$, Figure 5 corresponds to the shares of the bank BBVA, resulting in $\tilde{H} = 0.517$, and Figure 6 to the shares of the Spanish Telephone company "Telefónica" and $\tilde{H} = 0.513$.

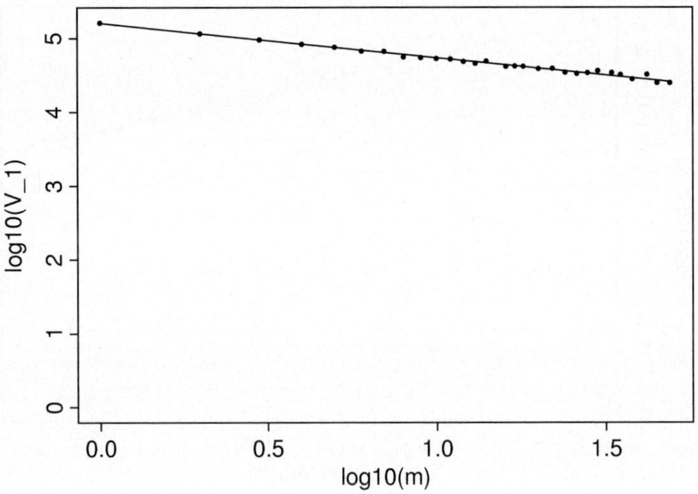

Fig. 4. Data corresponding to the Spanish index Ibex35 from 1992–2001, $n = 2465$, $\tilde{H} = 0.533$

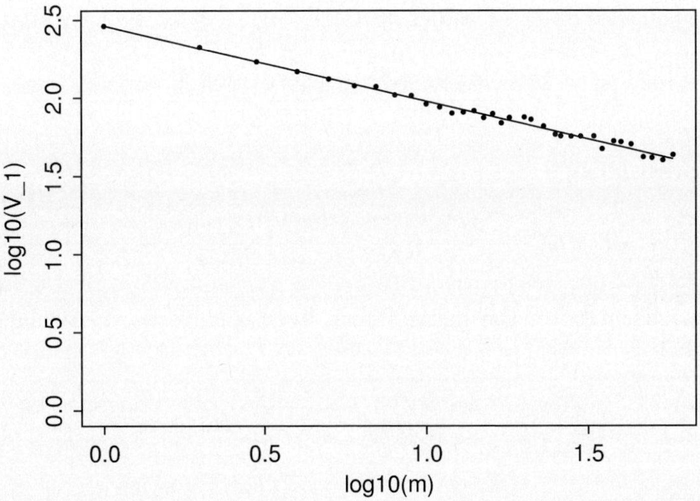

Fig. 5. Daily data for the price of the shares of the Spanish bank Bbva, from 1990–2001, $n = 2909$, $\tilde{H} = 0.517$

Asymptotic Behavior of $\tilde{H}$

The problem with the previous method is that it provides only a point estimation of H but we do not have a confidence interval. However we can try to relate $\tilde{H}$ with the estimations $\hat{H}_n$ for different values of n and from here we can get a confidence interval for $\tilde{H}$.

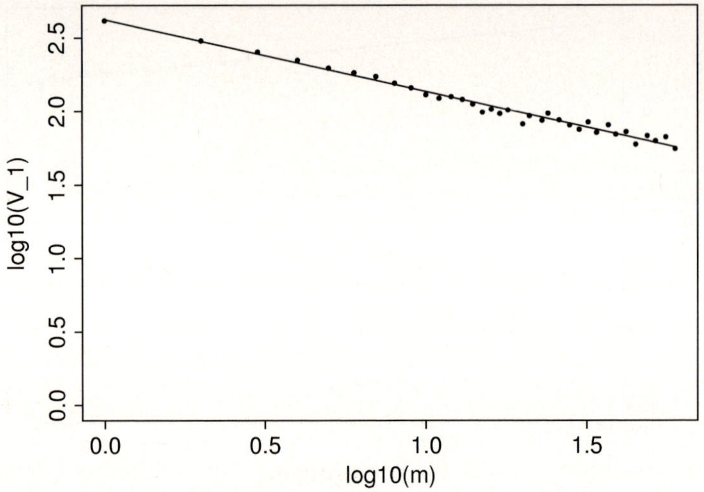

Fig. 6. Daily data for the price of shares of "Telefónica", from 1990–2001, $n = 3009$, $\tilde{H} = 0.513$

If we consider different values of n, $n_1 \leq n_2 \ldots \leq n_r$ we have that

$$\hat{H}_{n_i} = \frac{1}{p} - \frac{\log V_p^{n_i}(Z)_1 - \log c_p - \log \int_0^1 |u_s|^p ds}{p \log n_i}.$$

On the other hand

$$1 - p\tilde{H} = \frac{\sum_{i=1}^r (\log V_p^{n_i}(Z)_1 - \log \bar{V_p^n}(Z)_1)(\log n_i - \overline{\log n})}{\sum_{i=1}^r (\log n_i - \overline{\log n})^2},$$

where the bar denotes the mean. Then, by straightforward calculations, we obtain that

$$\tilde{H} = \sum_{i=1}^r \hat{H}_{n_i} \frac{(\log n_i - \overline{\log n})^2}{\sum_{i=1}^r (\log n_i - \overline{\log n})^2} + \overline{\log n} \sum_{i=1}^r \hat{H}_{n_i} \frac{(\log n_i - \overline{\log n})}{\sum_{i=1}^r (\log n_i - \overline{\log n})^2}$$

and consequently

$$\begin{aligned}\tilde{H} - H = &\sum_{i=1}^r (\hat{H}_{n_i} - H) \frac{(\log n_i - \overline{\log n})^2}{\sum_{i=1}^r (\log n_i - \overline{\log n})^2} \\ &+ \overline{\log n} \sum_{i=1}^r (\hat{H}_{n_i} - H) \frac{(\log n_i - \overline{\log n})}{\sum_{i=1}^r (\log n_i - \overline{\log n})^2}\end{aligned}$$

Then if Δ_{n_1} is the radius of a confidence interval for $\hat{H}_{n_1}$, and we take only two values of n, $n_1 = [n/2]$ and $n_2 = n$, we obtain

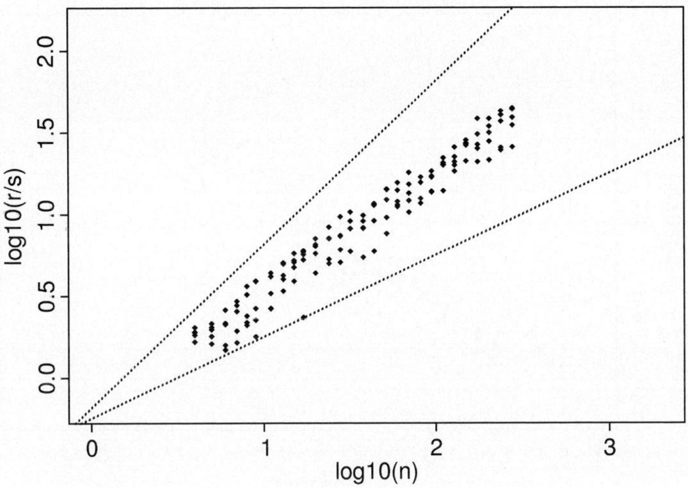

Fig. 7. Data corresponding to a fBm with $H = 0.7$, $\hat{H} = 0.716$

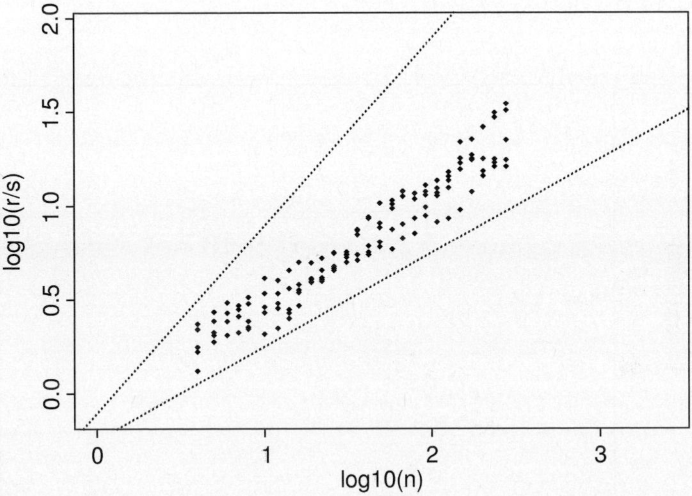

Fig. 8. Data corresponding to the Spanish Index Ibex35, $\hat{H} = 0.588$

$$\Delta = \frac{\log n}{\log 2}(\Delta_{[n/2]} + \Delta_n).$$

and Δ_{n_1} can be obtained from (3.1) if we know u except for a scale factor. This is the case of a geometric fractional Brownian motion. Note that we have to use the union-intersection principle to determine the confidence of the interval.

By considering that the data follow a geometric fractional Brownian motion and taking logarithms to estimate H we obtain the following estimations and, in parenthesis, we estimate the radius of a confidence interval

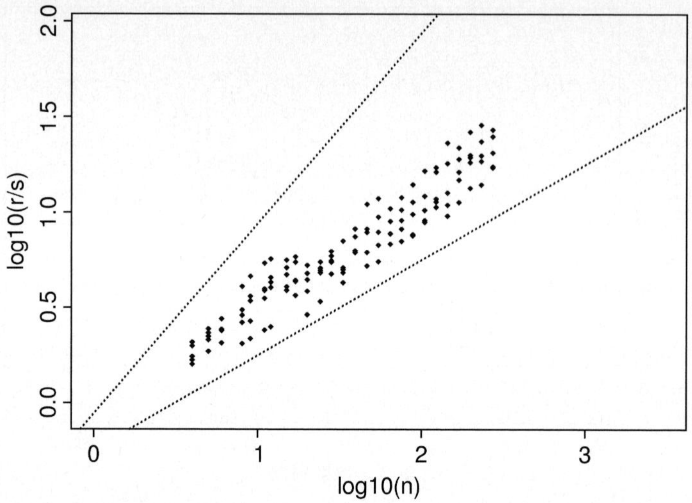

Fig. 9. Data corresponding to the shares of the Spanish Telephone company "Telefónica", $\hat{H} = 0.561$

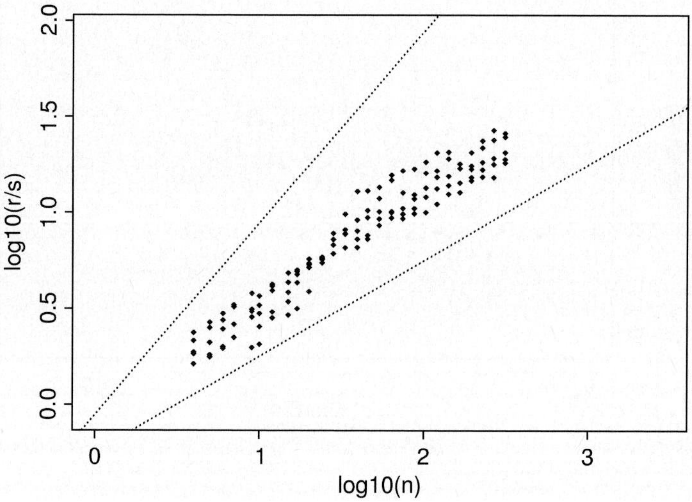

Fig. 10. Data corresponding to the shares of the Spanish Bank "Bbva", $\hat{H} = 0.582$

with $\gamma = 0.95$. The data obtained for a simulation of a fractional Brownian motion with $H = 0.7$ serve as a control.

Ibex35	Telefónica	Bbva	Fbm ($H = 0.7$)
0.538 (0.123)	0.550 (0.112)	0.574 (0.115)	0.703 (0.179).

Then, here data do not show evidence in favor of the "fractality", but the intervals we obtain are very conservative.

3.3 The R/S Method

The graphical implementation of the classical R/S-statistic given by

$$\frac{\max_{k\leq n}(\sum_{i=1}^{k} Z_i - \frac{k}{n}\sum_{i=1}^{n} Z_i) - \min_{k\leq n}(\sum_{i=1}^{k} Z_i - \frac{k}{n}\sum_{i=1}^{n} Z_i)}{\sqrt{\frac{1}{n}\sum_{i=1}^{n} Z_i^2 - \left(\frac{1}{n}\sum_{i=1}^{n} Z_i\right)^2}},$$

exploit the fact that if Z is a fractional Gaussian noise with $H > 1/2$ for large n its values oscillate around cn^H and we can also use this to estimate H by a log-log plot.

More specifically, given sample of n observations is subdivided into k blocks, each of size $[n/k]$. Then, for each *lag* $n_i, n_i \leq n$, estimates $R(k_m, n_i)/S(k_m, n_i)$ of $R(n_i)/S(n_i)$ are computed by starting at the points, $k_m = (m-1)[n/k]+1, m = 1, 2, \dots, k$, and such that $k_m + n_i \leq nN$. Thus, for any given m, all the data points before $k_m = (m-1)[n/k]+1$ are ignored. For values of n_i smaller than $[n/k]$, there are k different estimates of $R(n)/S(n)$; for values of n_i approaching n, there are fewer values, as few as 1 when $n \geq n - [n/k]$.

The graphical R/S-approach consists of calculating $R(k_m, n_i)/S(k_m, n_i)$ for logarithmically spaced values of n_i, and plotting $\log R(k_m, n_i)/S(k_m, n_i)$ versus $\log(n_i)$, for all starting points k_m. This results in the *rescaled adjusted range plot,* also known as the *pox plot of R/S.* See Willinger *et al.* (1999) for more details. The problem with the R/S method is that we do not have a distributional theory to give confidence intervals. The next figures show this method applied to the same data we used in the power variation analysis. We also denote by $\hat{H}$ the R/S estimator. Note that the results are quite similar to that using the power variation.

References

1. Aldous, D.J. and Eagleson, G.K. (1978) On Mixing and Stability of Limit Theorems. *The Annals of Probability,* **6**(2), 325–331.
2. Billingsley, P. (1968). *Convergence of probability measures.* New York: Wiley and Sons.
3. Brody, D.C., Syroka, J., and Zervos, M. (2002) Dynamical pricing of weather derivatives. Quantitative Finance, volume 2, pages 189–198. Institute of physics publishing.
4. Corcuera, J.M., Nualart, D. and Woerner, J. (2006) Power variation of some integral fractional proceses. To appear in Bernoulli.
5. Cutland, N.J., Kopp, P.E., Willinger, W. (1995) Stock price returns and the Joseph effect: a fractional version of the Black-Scholes model. In E. Bolthausen, M. Dozzi, F. Russo (eds.) Seminar on Stochastic Analysis, Random Fields and Applications. Boston: Birkhäuser, pp 327–351.
6. Davies, R.B. and Harte, D.S. (1987) Test for Hurst effect, *Biometrika,* **66**, 153–155.

7. Hu, Y. and Nualart, D. (2005) Renormalized self-intersection local time for fractional Brownian motion. *Ann. Probab.* **33**, 948–983.
8. Nualart, D. and Peccati, G. (2005) Central limit theorems for sequences of multiple stochastic integrals. *Ann. Probab.* **33**, 177–193.
9. Peccati, G. and Tudor, C.A. (2005) Gaussian limits for vector-valued multiple stochastic integrals. *Lecture Notes in Math.* Séminaire de Probabilités XXXVIII, 247–262.
10. Rogers, L.C.G. (1997) Arbitrage with fractional Brownian motion. *Mathematical Finance,* **7**(1), 95–105.
11. Shiryaev, A.N. (1999) *Essentials of Stochastic Finance.* Singapore: World Scientific.
12. Taqqu, M.S. (1977) Law of the Iterated Logarithm for Sums of Non-Linear Functions of Gaussian Variables that Exhibit a Long Range Dependence. *Z. Wahrsch. verw. Geb.*, **40**, 203–238.
13. Willinger, W., Taqqu, M.S., Teverovsky V (1999) Stock market prices and long-range dependence. *Finance and Stochastics,* **3**, 1–13.
14. Young, L.C. (1936) An inequality of Hölder type connected with Stieltjes Integration. Acta Math, 67(1936), 251–282.

Kolmogorov Equations for Stochastic PDE's with Multiplicative Noise

Giuseppe Da Prato

Scuola Normale Superiore Palzzo della Carovana Piazza dei Cavalieri, 7 56126 Pisa, Italia

1 Introduction

We are here concerned with the following stochastic differential equation in the Hilbert space $H = L^2(0,1)$,

$$\begin{cases} dX(t,\xi) = D_\xi^2 X(t,\xi)dt + g(X(t,\xi))dW(t,\xi), & t \geq 0, \ \xi \in [0,1], \\ X(t,0) = X(t,1) = 0, & t \geq 0, \\ X(0,\xi) = x(\xi), & x \in H, \ \xi \in [0,1], \end{cases} \tag{1.1}$$

where g is a real function of class C^2 bounded together with its derivatives of order less or equal to 2 and W is a cylindrical Wiener process in H (see below for a precise definition). Existence and uniqueness of a solution of (1.1) are well known, see [7, 16]. Let us denote by R_t the corresponding transition semigroup,

$$R_t\varphi(x) = \mathbb{E}[\varphi(X(t,x))], \quad t \geq 0, \ x \in H, \tag{1.2}$$

where φ is a real bounded Borel function and $X(t,x)$ is the solution of equation (1.1).

The Kolmogorov equation corresponding to (1.1) reads as follows,

$$\begin{cases} u_t(t,x) = \dfrac{1}{2} \operatorname{Tr} [\sigma^2(x) u_{xx}(t,x)] + \langle Ax, u_x(t,x)\rangle, & t \geq 0, \ x \in D(A), \\ u(0,x) = \varphi(x), & x \in H, \end{cases} \tag{1.3}$$

where A is the linear operator,

$$Ax = D_\xi^2, \quad x \in D(A) = H^2(0,1) \cap H_0^1(0,1),$$

and for any $x \in L^2(0,1)$ the symmetric Nemitskii operator $\sigma(x) \in L(L^2(0,1))$ is defined by

$$[\sigma(x)y](\xi) = g(x(\xi))y(\xi), \quad y \in L^2(0,1),\ \xi \in [0,1]. \tag{1.4}$$

There is an increasing interest on infinite dimensional Kolmogorov equations, see the monographs [5,8,13,15] (and references therein) and the papers [1–3,17].

In particular, in [13] the case when the noise is additive is mainly considered with the exception of Chapters 6 and 7. More precisely, Chapter 6 is devoted to Hölder continuous perturbations of the infinite dimensional Heat semigroup, see also some recent developements in this direction in [2,3]. In Chapter 7 of [13] the case when coefficients are of class C^3 is considered. Notice that in equation (1.1) the multiplicative noise can be written as $\sigma(x)dW(t)$ where for any $x \in L^2(0,1)$ the operator $\sigma(x) \in L(L^2(0,1))$ is the operator defined by (1.4). Thus, in spite of the fact that g is C^2, the mapping $\sigma(x)$ is only once Gateaux differentiable (except when g is constant). So, the method in [13, Chapter 7] does not work and some new technique has to be used.

The first main result of the paper is that if $\varphi \in C_b^3(H)$ there is a smooth solution of (1.3), see Theorem 13. It is well known that a candidate for the solution of (1.3) is given by

$$u(t,x) := R_t\varphi(x) = \mathbb{E}[\varphi(X(t,x))], \tag{1.5}$$

where R_t is the transition semigroup defined by (1.2). So, the proof of existence of a solution of (1.3) will consists in showing (by justifying the chain rule) that $u(t,x)$ is twice differentiable, that the trace of $\sigma^2 D^2 u(t,x)$ is finite and that equation (1.3) is fulfilled. However, these computations are not straightforward since σ is neither regular nor of trace class. This idea was applied in the case of reaction–diffusion equations with additive noise, see [6] and in the completely different situation of Navier–Stokes equations with additive noise, see [9].

We notice that the regularity we get for the solution of (1.3) is not enough to apply the Itô formula and so, to prove the uniqueness of a smoth solution of (1.3). This would require an additional job which we plan to make in a future paper.

In the second part of the paper we assume that $1/g$ is bounded so that there is a unique invariant measure μ, see [16]. We study here the Kolmogorov equation (1.2) in the space $L^2(H,\mu)$. It is well known that the transition semigroup R_t can be uniquely extended to a strongly continuous semigroup of contractions in $L^2(H,\mu)$. We shall still denote by R_t this extension and by L_μ the infinitesimal generator of R_t.

As a second main result of the paper we construct a core Γ for the generator L_μ consisting of regular functions and show that on Γ the operator L_μ is in fact a differential operator given by

$$L_\mu\varphi(x) = \frac{1}{2}\,\mathrm{Tr}\,[\sigma^2(x)D^2\varphi(x)] + \langle Ax, D\varphi(x)\rangle, \quad x \in D(A), \quad \varphi \in \Gamma.$$

As it was pointed out in several previous situations, see e.g. [8] and references therein, to have an explicit expression of L_μ on the core Γ allows to prove easily the so–called "identité du carré des champs",

$$\int_H L_\mu\varphi\,\varphi\,d\mu = -\frac{1}{2}\int_H |\sigma(x)D\varphi|^2 d\mu, \qquad \varphi\in D(L_\mu). \tag{1.6}$$

Using (1.6) it is possible to prove that the derivative operator is closable in $L^2(H,\mu)$. This allows to define the Sobolev space $W^{1,2}(H,\mu)$ and to show that the domain of L_μ is included in $W^{1,2}(H,\mu)$.

Finally, we prove the Poincaré inequality, by generalizing a result proved in [10] in the case of additive noise. As a standard consequence we obtain the spectral gap of L_μ and the exponential convergence to equilibrium of R_t.

It would be interesting to consider the more general problem,

$$\begin{cases} dX(t,\xi) = (D_\xi^2 X(t,\xi) + f(X(t,\xi)))dt + g(X(t,\xi))dW(t,\xi), \quad \xi\in[0,1],\\[2mm] X(t,0)=X(t,1)=0, \quad t\ge 0\\[2mm] X(0,\xi)=x(\xi), \quad x\in H,\ \xi\in[0,1], \end{cases} \tag{1.7}$$

where f is a suitable real function. However, problem (1.7) does not seem to be a straightforward generalization of (1.1). It will be the object of a future research.

1.1 Notations

We denote by H the Hilbert space $L^2(0,1)$ (norm $|\cdot|$, inner product $\langle\cdot,\cdot\rangle$). When there is the danger of confusion between the norm and the absolute value of a function x we shall write $|x|_{L^2(0,1)}$ instead of $|x|$. Moreover $L(H)$ (norm $\|\cdot\|$) will represent the Banach algebra of all linear bounded operators in H and $L_1(H)$ (norm $\|\cdot\|_{L_1(H)}$) the space of all trace class operators in H. We recall that

$$\|T\| = \sup\{|Tx| : \ x\in H,\ |x|=1\}, \quad T\in L(H).$$

For any Hilbert space K (norm $|\cdot|$, inner product $\langle\cdot,\cdot\rangle$), we denote by $C_b(H;K)$ the linear space of all continuous and bounded mappings $\varphi\colon H\to K$. $C_b(H;K)$ endowed with the norm

$$\|\varphi\|_0 = \sup_{x\in K}|\varphi(x)|, \quad \varphi\in C_b(H;K),$$

is a Banach space.

Moreover $C_b^1(H;K)$ will represent the subspace of $C_b(H;K)$ of all functions $\varphi\colon H\to K$ which are Fréchet differentiable on H with a continuous and bounded derivative $D\varphi$. The space $C_b^k(H;K)$ for $k\ge 2$ are defined analogously. We shall write $C_b^i(H;\mathbb{R}) = C_b^i(H)$, $i\in\mathbb{N}$.

If $\varphi \in C_b^1(H)$ and $x \in H$, we shall identify $D\varphi(x)$ with the unique element h of H such that

$$D\varphi(x)y = \langle h, y\rangle, \quad x, y \in H.$$

If $\varphi \in C_b^2(H)$ and $x \in H$, we shall identify $D^2\varphi(x)$ with the unique linear operator $T \in L(H)$ such that

$$D\varphi(x)(y,z) = \langle Ty, z\rangle, \quad x, y, z \in H.$$

1.2 An Extension of Gronwall's Lemma

The following result is a generalization of a well known result.

Lemma 1. *Assume that $f\colon [0,+\infty) \to [0,+\infty)$ fulfills the inequality,*

$$f(t) \le a(t) + b\int_0^t (t-s)^{-1/2} f(s)ds, \quad t \ge 0, \tag{1.8}$$

where a is continuous nonnegative and b is a nonnegative constant. Then we have,

$$\begin{aligned} f(t) &\le a(t) + b\int_0^t (t-s)^{-1/2} a(s)ds \\ &+ \int_0^t e^{(t-s)\pi b^2}\left[a(s) + b\int_0^s (s-\sigma)^{-1/2} a(\sigma)d\sigma\right]. \end{aligned} \tag{1.9}$$

If, in particular, $a(t) = a$ we have

$$f(t) \le ae^{\pi b^2 t} + 2ab\int_0^t s^{-1/2} e^{\pi b^2(t-s)} ds \quad t \ge 0. \tag{1.10}$$

and

$$f(t) \le 3ae^{\pi b^2 t} \quad t \ge 0. \tag{1.11}$$

Proof. We write (1.8) as

$$f \le a + b\psi_{-1/2} * f, \tag{1.12}$$

where $\psi_{-1/2}(t) = t^{-1/2}$ and $*$ denotes the convolution ([1]). Taking the convolution of both sides of (1.10) with $\psi_{-1/2}$ and taking into account that

$$(\psi_{-1/2} * \psi_{-1/2})(t) = \int_0^t (t-s)^{-1/2} s^{-1/2} ds = \pi,$$

yields

$$\psi_{-1/2} * f \le a * \psi_{-1/2} + \pi b(1 * f). \tag{1.13}$$

[1] $(f * g)(t) = \int_0^t f(t-s)g(s)ds$.

Substituting this in (1.12) yields

$$f \leq a + b(a * \psi_{-1/2}) + \pi b^2(1 * f), \tag{1.14}$$

which is equivalent to

$$f(t) \leq a(t) + b\int_0^t (t-s)^{-1/2} a(s)ds + \pi b^2 \int_0^t f(s)ds. \tag{1.15}$$

Consequently

$$\begin{aligned} f(t) \leq a(t) &+ b\int_0^t (t-s)^{-1/2} a(s)ds \\ &+ \int_0^t e^{(t-s)\pi b^2}\left[a(s) + b\int_0^s (s-\sigma)^{-1/2} a(\sigma)d\sigma\right] \end{aligned}$$

Now (1.9) follows from the classical Gronwall lemma. Finally, (1.10) is clear and by (1.10) we have

$$\begin{aligned} f(t) &\leq a e^{\pi b^2 t}\left(1 + 2b\int_0^t s^{-1/2} e^{-\pi b^2 s} ds\right) \\ &\leq a e^{\pi b^2 t}\left(1 + 2b\int_0^\infty s^{-1/2} e^{-\pi b^2 s} ds\right), \end{aligned}$$

which yields (1.11). □

2 Existence and Uniqueness of Solutions

2.1 The Abstract Setting

Let us write problem (1.1) in an abstract form introducing the linear self–adjoint operator $A\colon D(A) \subset H \to H$,

$$\begin{cases} Ax = D_\xi^2 x, \quad x \in D(A), \\ D(A) = H^2(0,1) \cap H_0^1(0,1), \end{cases}$$

where $H^i(0,1),\ i = 1, 2$ denote the usual Sobolev spaces and

$$H_0^1(0,1) = \{x \in H_0^1(0,1) : \ x(0) = x(1) = 0\}.$$

We define moreover the (generally) nonlinear operator $\sigma\colon H \to L(H)$ by setting,

$$[\sigma(x)y](\xi) = g(x(\xi))y(\xi), \quad \xi \in [0,1],\ x, y \in H.$$

We denote by (e_k) the complete orthonormal sistem in H consisting of the eigenfunctions of A,

$$e_k(\xi) = \sqrt{\frac{2}{\pi}}\ \sin k\pi\xi, \quad \xi \in [0,1],\ k \in \mathbb{N},$$

so that

$$Ae_k = -k^2\pi^2 e_k, \quad k \in \mathbb{N}.$$

Notice that

$$\|e^{tA}\| \le e^{-\pi^2 t}, \quad t \ge 0.$$

Finally, we introduce the cylindrical white noise,

$$W(t) = \sum_{k=1}^{\infty} e_k \beta_k(t), \quad t \ge 0, \tag{2.1}$$

where (β_k) is a sequence of mutually independent standard Brownian motions on a filtered probability space $(\Omega, \mathcal{F}, (\mathcal{F}_t)_{t\ge 0}, \mathbb{P})$.

Now we can write problem (1.1) as follows,

$$\begin{cases} dX &= AXdt + \sigma(X)\, dW(t) = AXdt + \sum_{k=1}^{\infty} [g(X)e_k]\, d\beta_k(t), \\ X(0) = x. \end{cases} \tag{2.2}$$

We shall solve equation (2.2) in the space $C_W([0,T],H)$ of all mean square continuous adapted (to the filtration $(\mathcal{F})_{t\ge 0}$) stochastic process $X(\cdot)$ defined in $[0,T]$ and taking values in H. It is well known that $C_W([0,T],H)$, endowed with the norm

$$\|X\|_{C_W([0,T],H)} = \left(\sup_{t\in[0,T]} \mathbb{E}(|X(t)|^2) \right)^{1/2},$$

is a Banach space.

Definition 2. *A* mild solution *of equation* (2.2) *is a process* $X \in C_W([0,T],H)$ *such that*

$$X(t) = e^{tA}x + \int_0^t e^{(t-s)A}\sigma(X(s))dW(s), \quad t \ge 0,\ x \in H. \tag{2.3}$$

In the following we shall denote by $X(\cdot, x)$ the solution of (2.3).

An important rôle will be played by the *stochastic convolution*,

$$W_X(t) = \int_0^t e^{(t-s)A}\sigma(X(s))dW(s) = \sum_{k=1}^{\infty} \int_0^t e^{(t-s)A}[g(X(s))e_k]d\beta_k(s),$$

where $X \in C_W([0,T],H)$.

As we shall see, though the the cylindrical white noise (2.1) does not leave in H (see e.g. [11, §4.3.1]), the stochastic convolution $W_X(t)$ does.

In order to study basic properties of $W_X(t)$ it is useful to introduce the function

$$F(t) = \sum_{h=1}^{\infty} e^{-th^2}, \quad t > 0. \tag{2.4}$$

Notice that

$$\begin{aligned} F(t) &\le e^{-t} + \int_1^{\infty} e^{-tx^2} dx = e^{-t}\left(1 + \int_1^{\infty} e^{-t(x^2-1)} dx\right) \\ &= e^{-t}\left(1 + \int_0^{\infty} e^{-ty^2} \frac{y}{\sqrt{y^2+1}}\, dy\right) \le e^{-t}\left(1 + \int_0^{\infty} e^{-ty^2} dy\right). \end{aligned}$$

So,

$$F(t) \le e^{-t}\left(1 + 2t^{-1/2}\right) \le 4t^{-1/2} e^{-t/2}, \quad t > 0. \tag{2.5}$$

Lemma 3. *Let $X \in C_W([0,T], H)$. Then we have*

$$\mathbb{E}(|W_X(t)|^2) = \int_0^t \sum_{h=1}^{\infty} e^{-2\pi^2 h^2 (t-s)} \mathbb{E}(|g(X(s))e_h|^2) ds. \tag{2.6}$$

Proof. Taking into account the independence of the (β_k) we have,

$$\mathbb{E}(|W_X(t)|^2) = \sum_{k=1}^{\infty} \int_0^t \mathbb{E}(|e^{(t-s)A}[g(X(s))e_k]|^2) ds.$$

Using the Parseval identity we find,

$$\begin{aligned} \sum_{k=1}^{\infty} |e^{(t-s)A}[g(X(s))e_k]|^2 &= \sum_{h,k=1}^{\infty} |\langle e^{(t-s)A}[g(X(s))e_k], e_h\rangle|^2 \\ &= \sum_{h,k=1}^{\infty} e^{-2(t-s)\pi^2 h^2} |\langle g(X(s))e_k, e_h\rangle|^2 \\ &= \sum_{h,k=1}^{\infty} e^{-2(t-s)\pi^2 h^2} |\langle e_k, g(X(s))e_h\rangle|^2 \\ &= \sum_{h=1}^{\infty} e^{-2(t-s)\pi^2 h^2} |\sigma(X(s))e_h|^2. \end{aligned}$$

So, (2.6) follows. □

Proposition 4. *Let $X \in C_W([0,T],H)$. Then we have*

$$\mathbb{E}(|W_X(t)|^2) \le 8\sqrt{\pi}\,\|g\|_0^2, \quad t \ge 0. \tag{2.7}$$

Moreover, for all $X, Y \in C_W([0,T],H)$ we have,

$$\mathbb{E}(|W_X(t) - W_Y(t)|^2) = \frac{8\|g\|_1^2}{\pi\sqrt{2\pi}} \int_0^t e^{-(t-s)\pi^2}(t-s)^{-1/2}\mathbb{E}[|X(s) - Y(s)|^2)]ds. \tag{2.8}$$

Proof. By (2.6) we have, recalling that $|e_k(\xi)|^2 \le \frac{2}{\pi}$ and taking into account (2.5),

$$\mathbb{E}|W_X(t)|^2 \le \frac{2\|g\|_0^2}{\pi} \int_0^t F(2\pi^2 s)ds \le \frac{8\|g\|_0^2}{\pi} \int_0^\infty e^{-\pi^2 s}s^{-1/2}ds$$

and (2.7) follows. The proof of (2.8) is similar. □

2.2 Existence and Uniqueness

The following result is well known, see e.g. [16], we present however the short proof for the reader's convenience.

Proposition 5. *For any $x \in H$ there exists a unique solution $X(\cdot,x)$ of equation (2.3).*

Proof. Write equation (2.3) in the form

$$X = e^{tA}x + \Lambda(X), \quad X \in C_W([0,T],H),$$

where

$$\Lambda(X)(t) = W_X(t), \quad t \in [0,T].$$

Then by (2.7) it follows that Λ maps $C_W([0,T],H)$ in itself. Let moreover $X, Y \in C_W([0,T],H)$. Then by (2.8) it follows that,

$$|\Lambda(X)(t) - \Lambda(Y)(t)| \le \frac{8\|g\|_1^2}{\pi\sqrt{\pi}} \int_0^t e^{-(t-s)\pi^2}(t-s)^{-1/2}ds\|X - Y\|_{C_W([0,T],H)}$$

$$\le \frac{16\|g\|_1^2}{\pi\sqrt{2\pi}}\, t^{1/2}\, \|X - Y\|_{C_W([0,T],H)}.$$

Now let $T_1 \in (0,T]$ be such that $\frac{16\|g\|_1^2}{\pi\sqrt{2\pi}}\, T_1^{1/2} < 1$. Then Γ is a contraction on $C_W([0,T_1],H)$. Therefore, equation (2.3) has a unique solution on $[0,T_1]$. By a similar argument, one can show existence and uniqueness on $[T_1, 2T_1]$ and so on. □

2.3 Galerkin Approximations

It is useful to consider Galerkin approximations of equation (2.3). For any $n \in \mathbb{N}$ we denote by P_n the projector

$$P_n x = \sum_{k=1}^{n} \langle x, e_k \rangle e_k, \quad x \in H$$

and set $A_n = AP_n$. Then we consider the equation

$$X^n(t,x) = e^{tA_n}x + \sum_{k=1}^{n} \int_0^t e^{(t-s)A_n}[g(X^n(s,x))e_k]d\beta_k(s). \tag{2.9}$$

The following result is standard.

Proposition 6. *For any $T > 0$, $x \in H$ and $n \in \mathbb{N}$, there exists a unique solution $X^n(\cdot, x)$ of equation (2.9). Moreover,*

$$\lim_{n\to\infty} X^n(\cdot, x) = X(\cdot, x), \quad \text{in } C_W([0,T], H), \tag{2.10}$$

where $X(\cdot, x)$ is the solution of (2.3).

3 Kolmogorov Equation

3.1 Setting of the Problem

We are here concerned with the following *Kolmogorov equation*,

$$\begin{cases} u_t(t,x) = \dfrac{1}{2} \operatorname{Tr}\,[\sigma^2(x) u_{xx}(t,x)] + \langle Ax, u_x(t,x) \rangle, & t \geq 0,\ x \in D(A), \\[2ex] u(0,x) = \varphi(x), \quad x \in H. \end{cases} \tag{3.1}$$

We are going to show that when the initial datum φ is sufficiently regular equation (3.1) has a solution in a classical sense. As it is well known, a candidate for the solution $u(t,x)$ of (3.1) is provided by the formula

$$u(t,x) = \mathbb{E}[\varphi(X(t,x))], \quad \varphi \in C_b(H),\ t \geq 0,\ x \in H, \tag{3.2}$$

where $X(t,x)$ is the solution of (2.3). We shall check that, under suitable assumptions on φ, formula (3.2) produces in fact a solution of (3.1).

We shall need to consider the approximating equation

$$\begin{cases} u_t^n(t,x) = \dfrac{1}{2} \operatorname{Tr}\,[P_n \sigma^2(x) u_{xx}^n(t,x)] + \langle A_n x, u_x^n(t,x) \rangle \\[2ex] u^n(0,x) = \varphi(P_n x), \end{cases} \tag{3.3}$$

which has a unique strict solution which we denote by $u^n(t,x)$.

3.2 Estimates for Derivatives of $X(t,x)$

This subsection is devoted to establish some estimates concerning the derivatives X_x and X_{xx}, which will be used later. We start from the directional derivative

$$\eta^z(t,x) := X_x(t,x)z = \lim_{\epsilon\to 0} \frac{1}{\epsilon}\,(X(t,x+\epsilon z) - X(t,x))$$

where $z \in H$. By using Galerkin approximations it is not difficult to show that the diretional derivative $\eta^z(t,x)$ does exist and it is the solution of the equation,

$$\eta^z(t,x) = e^{tA}z + \sum_{k=1}^{\infty} \int_0^t e^{(t-s)A}[g'(X(s,x))\eta^z(s,x)e_k]d\beta_k(s). \tag{3.4}$$

Lemma 7. *There exists two positive constants a_1 and λ_1 such that*

$$\mathbb{E}(|\eta^z(t,x)|^2) \le a_1|z|^2 e^{\lambda_1 t} \quad t\ge 0,\ x\in H. \tag{3.5}$$

Proof. We have

$$\mathbb{E}(|\eta^z(t,x)|^2) = |e^{tA}z|^2 + \sum_{k=1}^{\infty} \mathbb{E}\int_0^t |e^{(t-s)A}[g'(Y(s,x))\eta^z(s,x)e_k]|^2 ds.$$

Arguing as in the proof of Lemma 3 and taking into account (2.6), we see that,

$$\begin{aligned}\mathbb{E}(|\eta^z(t,x)|^2) &\le |e^{tA}z|^2 + \frac{2}{\pi}\,\|g'\|_0^2 \int_0^t F(2\pi^2(t-s))\mathbb{E}(|\eta^z(s,x)|^2)ds\\ &\le |e^{tA}z|^2 + \frac{8\|g'\|_0^2}{\sqrt{2}\pi^2}\int_0^t (t-s)^{-1/2}e^{(t-s)/2}\mathbb{E}(|\eta^z(s,x)|^2)ds.\end{aligned}$$

By Lemma 1 it follows that

$$\mathbb{E}(|\eta^z(t,x)|^2) \le |e^{tA}z|^2 e^{\frac{16\|g'\|_0^2}{\sqrt{2}\pi^2}\,t^{1/2}}, \quad t\ge 0$$

and the conclusion follows. □

We want now to estimate $\zeta^z(t,x) := X_{xx}(t,x)(z,z)$ where

$$X_{xx}(t,x)(z,z) = \lim_{\epsilon\to 0}\frac{1}{\epsilon}\,(\eta^z(t,x+\epsilon z) - \eta^z(t,x))$$

and $z\in H$. Formally $\zeta^z(t,x)$ is the solution of the equation,

$$\begin{aligned}\zeta^z(t,x) &= \sum_{k=1}^{\infty}\int_0^t e^{(t-s)A}[g'(Y(s,x))\zeta^z(s,x)e_k]d\beta_k(s)\\ &\quad + \sum_{k=1}^{\infty}\int_0^t e^{(t-s)A}[g''(Y(s,x))(\eta^z(s,x))^2e_k]d\beta_k(s).\end{aligned} \tag{3.6}$$

Here a problem arises since the term

$$g''(Y(s,x))(\eta^z(s,x))^2 e_k, \tag{3.7}$$

which appears in the second integral, belongs to $L^4(0,1)$ and not to $L^2(0,1)$ in general. For this reason we first need an estimate for

$$\mathbb{E}(|\eta^z(t,x)|^4_{L^4(0,1)}) = \mathbb{E}(|[\eta^z(t,x)]^2|^2_{L^2(0,1)}).$$

To get this estimate we shall proceed in two steps.

(i) We shall estimate $\mathbb{E}(|\eta^z(t,x)|^4_{L^2(0,1)})$.

(ii) We shall estimate $\mathbb{E}(|(-A)^{1/8}\eta^z(t,x)|^4_{L^2(0,1)})$.

Then we notice that, by the Sobolev embedding theorem, we have

$$D((-A)^{1/8}) \subset L^4(0,1)$$

and so we end up with the required estimate for $\mathbb{E}(|\eta^z(t,x)|^4_{L^4(0,1)})$.

Let us write (3.4) as

$$\eta^z(t,x) = e^{tA}z + \int_0^t \Phi(t-s)dW(s), \tag{3.8}$$

where

$$\Phi(t-s) = e^{(t-s)A}\sigma(s) = \sum_{k=1}^{\infty} e^{(t-s)A}[g'(X(s,x))\eta^z(s,x)e_k]. \tag{3.9}$$

We shall use the following Burkholder estimate, see [11].

$$\mathbb{E}\left[\left|\int_0^t \Phi(t-s)dW(s)\right|^4\right] \le c\mathbb{E}\left[\left(\int_0^t \|\Phi(t-s)\|^2_{HS}ds\right)^2\right], \tag{3.10}$$

where c is a given positive constant and

$$\begin{aligned}
\|\Phi(t-s)\|^2_{HS} &= \sum_{h,k=1}^{\infty} \langle e^{(t-s)A}[g'(X(s,x))\eta^z(s,x)e_k], e_h\rangle|^2 \\
&= \sum_{h,k=1}^{\infty} e^{-2\pi^2h^2(t-s)}\langle g'(X(s,x))\eta^z(s,x)e_k, e_h\rangle|^2 \\
&= \sum_{h=1}^{\infty} e^{-2\pi^2h^2(t-s)}|g'(X(s,x))\eta^z(s,x)e_h\rangle|^2 \\
&\le \frac{2}{\pi}\,\|g\|_1^2 F(2\pi^2(t-s))|\eta^z(s,x)|^2 \\
&\le \frac{32}{\pi^2}\,\|g\|_1^2(t-s)^{-1/2}|\eta^z(s,x)|^2.
\end{aligned} \tag{3.11}$$

where we have used (2.6). Now we are ready to prove

Lemma 8. *Let $z \in L^2(0,1)$. Then there exists constants $a_2 > 0$ and $\lambda_2 > 0$ such that*

$$\mathbb{E}(|\eta^z(t,x)|^4_{L^2(0,1)}) \le a_2 e^{\lambda_2 t}|z|^4_{L^2(0,1)}, \quad t \ge 0,\ x \in H. \tag{3.12}$$

Proof. Let $z \in L^2(0,1)$. By (3.8) we have

$$\mathbb{E}(|\eta^z(t,x)|^4_{L^2(0,1)}) \le 8|e^{tA}z|^4_{L^2(0,1)} + 8\mathbb{E}\left[\left|\int_0^t \Phi(t-s)dW(s)\right|^4_{L^2(0,1)}\right].$$

which, taking into account (3.10), yields

$$\mathbb{E}(|\eta^z(t,x)|^4_{L^2(0,1)}) \le 8|e^{tA}z|^4_{L^2(0,1)} + 8c\mathbb{E}\left[\left(\int_0^t \|\Phi(t-s)\|^2_{HS}ds\right)^2\right]. \tag{3.13}$$

Let us estimate the term $J := \mathbb{E}\left[\left(\int_0^t \|\Phi(t-s)\|^2_{HS}ds\right)^2\right]$. We have by (3.11)

$$\begin{aligned}
J &\le \frac{2^{10}}{\pi^4}\,\|g\|_1^4 \mathbb{E}\left[\left(\int_0^t (t-s)^{-1/2}|\eta^z(s,x)|^2_{L^2(0,1)}ds\right)^2\right]\\
&= \frac{2^{10}}{\pi^4}\,\|g\|_1^4 \int_0^t\int_0^t (t-s)^{-1/2}(t-s_1)^{-1/2}\\
&\quad \mathbb{E}\left[|\eta^z(s,x)|^2_{L^2(0,1)}|\eta^z(s_1,x)|^2_{L^2(0,1)}\right] dsds_1\\
&\le \frac{2^9}{\pi^4}\,\|g\|_1^4 \int_0^t\int_0^t (t-s)^{-1/2}(t-s_1)^{-1/2}\mathbb{E}\left[|\eta^z(s,x)|^4_{L^2(0,1)}\right] dsds_1\\
&\quad + \frac{2^9}{\pi^4}\,\|g\|_1^4 \int_0^t\int_0^t (t-s)^{-1/2}(t-s_1)^{-1/2}\mathbb{E}\left[|\eta^z(s_1,x)|^4_{L^2(0,1)}\right] dsds_1\\
&\le \frac{2^{11}}{\pi^4}\,\|g\|_1^4 t^{1/2}\int_0^t (t-s)^{-1/2}\mathbb{E}\left[|\eta^z(s,x)|^4_{L^2(0,1)}\right] ds.
\end{aligned}$$

Substituting in (3.13) yields

$$\begin{aligned}
\mathbb{E}(|\eta^z(t,x)|^4_{L^2(0,1)}) &\le 8|e^{tA}z|^4_{L^2(0,1)}\\
&\quad + c\,\frac{2^{14}}{\pi^4}\,\|g\|_1^4 t^{1/2}\int_0^t (t-s)^{-1/2}\mathbb{E}\left[|\eta^z(s,x)|^4_{L^2(0,1)}\right] ds.
\end{aligned}$$

Now by the Gronwall Lemma it follows that there exist positive constants ρ, l such that

$$\mathbb{E}(|\eta^z(t,x)|^4_{L^2(0,1)}) \le 8|e^{tA}z|^4_{L^2(0,1)} + \rho\int_0^t (t-s)^{-1/2}e^{l(t-s)}|e^{sA}z|^4_{L^2(0,1)}ds, \tag{3.14}$$

which implies the conclusion. □

Lemma 9. *Let $z \in L^2(0,1)$. Then there exists constants $a_3 > 0$ and $\lambda_3 > 0$ such that*

$$\mathbb{E}(|(-A)^{1/8}\eta^z(t,x)|^4_{L^2(0,1)}) \le a_3 e^{\lambda_3 t}|(-A)^{1/8}e^{tA}z|^4_{L^2(0,1)}, \quad t \ge 0,\ x \in H. \tag{3.15}$$

Proof. Let $z \in L^2(0,1)$. Then we have

$$\mathbb{E}(|(-A)^{1/8}\eta^z(t,x)|^4_{L^2(0,1)}) \le 8|(-A)^{1/8}e^{tA}z|^4_{L^2(0,1)} + 8\mathbb{E}\left[\left|\int_0^t \Phi_1(t-s)dW(s)\right|^4_{L^2(0,1)}\right].$$

where

$$\Phi_1(t-s) = (-A)^{1/8}e^{(t-s)A}\sigma(s) = \sum_{k=1}^\infty (-A)^{1/8}e^{(t-s)A}[g'(X(s,x))\eta^z(s,x)e_k].$$

We have

$$\begin{aligned}\|\Phi_1(t-s)\|^2_{HS} &= \sum_{h,k=1}^\infty \langle(-A)^{1/8}e^{(t-s)A}[g'(X(s,x))\eta^z(s,x)e_k], e_h\rangle|^2 \\ &= \sum_{h,k=1}^\infty (\pi h)^{1/2}e^{-2\pi^2h^2(t-s)}\langle g'(X(s,x))\eta^z(s,x)e_k, e_h\rangle|^2 \\ &= \sum_{h=1}^\infty (\pi h)^{1/2}e^{-2\pi^2h^2(t-s)}|g'(X(s,x))\eta^z(s,x)e_h\rangle|^2.\end{aligned}$$

It is not difficult to show that there is a positive constant d_1 such that

$$\|\Phi_1(t-s)\|^2_{HS} \le d_1(t-s)^{-3/4}|\eta^z(s,x)|^2. \tag{3.16}$$

Now we have

$$\mathbb{E}(|(-A)^{1/8}\eta^z(t,x)|^4_{L^2(0,1)}) \le 8|(-A)^{1/8}e^{tA}z|^4_{L^2(0,1)} + 8c\mathbb{E}\left[\left(\int_0^t \|\Phi_1(t-s)\|^2_{HS}ds\right)^2\right]. \tag{3.17}$$

By proceeding as in the proof of the previous lemma we find that there is $d_2 > 0$ such that

$$\mathbb{E}\left[\left(\int_0^t \|\Phi_1(t-s)\|^2_{HS}ds\right)^2\right] \le d_2\,\|g\|_1^4 t^{1/4}\int_0^t (t-s)^{-3/4}\mathbb{E}\left[|\eta^z(s,x)|^4_{L^2(0,1)}\right]ds$$

which yields

$$\mathbb{E}(|(-A)^{1/8}\eta^z(t,x)|^4_{L^2(0,1)}) \le 8|(-A)^{1/8}e^{tA}z|^4_{L^2(0,1)}$$
$$+d_2t^{1/2}\int_0^t (t-s)^{-3/4}\mathbb{E}\left[|\eta^z(s,x)|^4_{L^2(0,1)}\right] ds.$$

So, the conclusion follows from Lemma 8. □

Now, using the Sobolev embedding $D((-A)^{1/8}) \subset L^4(0,1)$ we can conclude that

Lemma 10. *Let $z \in L^2(0,1)$. Then there exists constants $a_4 > 0$ and $\lambda_4 > 0$ such that*

$$\mathbb{E}(|\eta^z(t,x)|^4_{L^4(0,1)}) \le a_4 e^{\lambda_4 t}|z|^4_{L^2(0,1)}, \quad t \ge 0,\ x \in H. \tag{3.18}$$

Now we are in position to estimate ζ^z for $z \in L^2(0,1)$.

Lemma 11. *There exists constants $a_5 > 0$ and $\lambda_5 > 0$ such that for all $z \in L^2(0,1)$, we have*

$$\mathbb{E}(|\zeta^z(t,x)|^2) \le a_5|z|^4_{L^2(0,1))}e^{\lambda_5 t}, \quad t \ge 0,\ x \in H. \tag{3.19}$$

Proof. Let $z \in L^2(0,1)$. Using Galerkin approximations we can show that (3.6) holds. From (3.6) we deduce that,

$$\mathbb{E}(|\zeta^z(t,x)|^2) \le 2\mathbb{E}\sum_{k=1}^\infty \int_0^t |e^{(t-s)A}[g'(X(s,x))\zeta^z(s,x)e_k]|^2 ds$$
$$+2\mathbb{E}\sum_{k=1}^\infty \int_0^t |e^{(t-s)A}[g''(X(s,x))(\eta^z(s,x))^2 e_k]|^2 ds:$$
$$= 2J_1 + 2J_2.$$

Arguing as in the proof of Lemma 3 we find

$$\begin{aligned} J_1 &= \mathbb{E}\textstyle\sum_{h,k=1}^\infty |\langle e^{(t-s)A}[g'(X(s,x))\zeta^z(s,x)e_k], e_h\rangle|^2 \\ &= \mathbb{E}\sum_{h,k=1}^\infty e^{-2(t-s)\pi^2h^2}|\langle g'(X(s,x))\zeta^z(s,x)e_k], e_h\rangle|^2 \\ &= \mathbb{E}\sum_{h=1}^\infty e^{-2(t-s)\pi^2h^2}|g'(X(s,x))\zeta^z(s,x)e_h|^2 \\ &\le \frac{2}{\pi}\ \|g\|_1^2 F(2(t-s)\pi^2)\mathbb{E}|\zeta^z(s,x)|^2 \le \frac{4}{\pi^2}(t-s)^{-1/2}\ \mathbb{E}|\zeta^z(s,x)|^2. \end{aligned} \tag{3.20}$$

Similarly for J_2 we find, taking into account Lemma 10,

$$
\begin{aligned}
J_2 &= \mathbb{E}\sum_{h,k=1}^\infty |\langle e^{(t-s)A}[g''(X(s,x))(\eta^z(s,x))^2 e_k]|, e_h\rangle|^2 \\
&= \mathbb{E}\sum_{h,k=1}^\infty e^{-2(t-s)\pi^2 h^2}|\langle g'(X(s,x))(\eta^z(s,x))^2 e_k], e_h\rangle|^2 \\
&= \mathbb{E}\sum_{h=1}^\infty e^{-2(t-s)\pi^2 h^2}|g'(X(s,x))(\eta^z(s,x))^2 e_h|^2 \qquad (3.21)\\
&\le \frac{2}{\pi}\, \|g\|_1^2 F(2(t-s)\pi^2)\mathbb{E}|(\eta^z(s,x))^2|^2 \\
&\le \frac{4}{\pi^2}(t-s)^{-1/2}\, \mathbb{E}|(\eta^z(s,x))^2|^2 \le a_4 \frac{4}{\pi^2}(t-s)^{-1/2}\, e^{\lambda_4 s}|z|^4_{L^2(0,1)}.
\end{aligned}
$$

Now the conclusion follows from (3.20), (3.21) and the Gronwall lemma. □

Remark 12. By Lemma 11 it follows that if $\varphi \in C_b^2(H)$ the function $u(t,\cdot)$ possesses bounded second order derivatives in all directions of H for any $t \ge 0$. So, it is Frécher differentiable and belongs to $C_b^1(H)$ (more precisely to $C_b^{1+\varepsilon}(H)$ for all $\varepsilon \in (0,1)$).

3.3 Strict Solutions of the Kolmogorov Equation

We are now in position to show existence of a strict solution $u(t,x)$ (in the sense that $u(t,x)$ fulfills conditions (i)-(iv) of Proposition 13 below) of equation (3.1) for all $\varphi \in C_b^2(H)$. Let us define,

$$\overline{\lambda} = \frac{1}{2}\ \max\{\lambda_i : i = 1,..,5\}$$

and

$$\kappa = \frac{1}{2}\ \max\{a_i^{1/2} : i = 1,..,5\}$$

Theorem 13. *Assume that $\varphi \in C_b^2(H)$, $D^2\varphi(x)$ is of trace class for any $x \in H$ and Tr $[D^2\varphi] \in C_b(H)$. Let*

$$u(t,x) = \mathbb{E}[\varphi(X(t,x))], \quad t \ge 0,\ x \in H,$$

where $X(t,x)$ is the mild solution of (2.2). Then the following statements hold.

(i) For all $t \ge 0$, $u(t,\cdot) \in C_b^1(H)$ and possesses second order derivatives in all directions of H.
(ii) For all $t > 0$ and any $x \in H$ we have

$$|u_x(t,x)| \le \kappa\, e^{\overline{\lambda}t}\|\varphi\|_1 \qquad (3.22)$$

and

$$|u_{xx}(t,x)| \le \kappa\, e^{\overline{\lambda}t}\|\varphi\|_2. \qquad (3.23)$$

(iii) There exists $\kappa_1 > 0$ such that for all $t \geq 0$ and any $x \in H$ we have

$$\left|\mathrm{Tr}\,[\sigma^2(x)u_{xx}(t,x)]\right| = \left|\sum_{k=1}^{\infty}\langle u_{xx}(t,x)(\sigma(x)e_k,\sigma(x)e_k)\rangle\right| \\ \leq \kappa_1 e^{\overline{\lambda}t}\|\varphi\|_2(1+\sup_{x\in H}\|D^2\varphi(x)\|_{L_1(H)}). \tag{3.24}$$

(iv) For all $x \in D(A)$, $u(\cdot,x)$ is differentiable in $(0,+\infty)$ and fulfills (3.1).

Proof. Let us prove (i). For any $x, z \in H$ and $t \geq 0$ we have

$$\langle u_x(t,x), z\rangle = \mathbb{E}[\langle D\varphi(X(t,x), \eta^z(t,x)\rangle]$$

Therefore

$$|\langle u_x(t,x), z\rangle| \leq \|\varphi\|_1|\eta^z(t,x)|.$$

By Lemma 7 it follows that

$$|\langle u_x(t,x), z\rangle| \leq \|\varphi\|_1 a_1^{1/2}e^{\lambda_1 t/2}.$$

Thus (3.22) follows from the arbitrariness of z. Moreover $u(t,\cdot) \in C_b^1(H)$ in view of Remark 12.

Let us prove (ii). For any $x, z \in H$ and $t \geq 0$ we have

$$\langle u_{xx}(t,x)z,z\rangle = \mathbb{E}[\langle D\varphi(X(t,x),\zeta^z(t,x)\rangle] + \mathbb{E}[\langle D^2\varphi(X(t,x)\eta^z(t,x),\eta^z(t,x)\rangle].$$

Therefore

$$|\langle u_{xx}(t,x)z,z\rangle| \leq \|\varphi\|_1\mathbb{E}[|\zeta^z(t,x)|] + \|\varphi\|_2\mathbb{E}[|\eta^z(t,x)|^2] \\ \leq \|\varphi\|_1\sqrt{a_5}\, e^{\lambda_5 t/2}|z|^2 + \|\varphi\|_2 a_1\, e^{\lambda_1 t/2}|z|^2$$

and (ii) follows.

Let us prove (iii). For any $x, z \in H$ and $t \geq 0$ we have

$$\begin{aligned}\mathrm{Tr}\,[\sigma^2(x)u_{xx}(t,x)] &= \sum_{k=1}^{\infty}\langle u_{xx}(t,x)(\sigma(x)e_k,\sigma(x)e_k)\rangle \\ &= \sum_{k=1}^{\infty}\mathbb{E}[\langle D\varphi(X(t,x),\zeta^{\sigma(x)e_k}(t,x)\rangle] \\ &\quad + \mathrm{Tr}\,\mathbb{E}[\sigma^2(x)X_x(t,x)D^2\varphi(X(t,x)X_x^*(t,x)] \\ &:= J_1(t,x) + J_2(t,x).\end{aligned}$$

Therefore

$$|J_2(t,x)| \leq a_1 e^{\lambda_1 t}\|D^2\varphi\|_{HS}.$$

Concerning J_1 we have

$$|J_1(t,x)| \leq \|\varphi\|_1 |T(t,x)|$$

where

$$T(t,x) = \sum_{k=1}^{\infty} \zeta^{\sigma(x)e_k}(t,x).$$

Then one can checks that $T(t,x)$ is the solution of the equation

$$\begin{aligned} T(t,x) = & \int_0^t e^{(t-s)A}[g'(X(s,x))T(s,x)]dW(s) \\ & + \int_0^t e^{(t-s)A}[g''(X(s,x))K(s,x)]dW(s), \end{aligned} \tag{3.25}$$

where

$$K(t,x) = \sum_{k=1}^{\infty} (\eta^{\sigma(x)e_k}(t,x))^2.$$

Using estimate (3.14), it is not difficult to show that equation (3.25) has a solution and estimate (3.24) holds.

Let us prove finally (iv). Assume that $x \in D(A)$ and let $u^n(t,x)$ be the solution of (3.3). Then, taking into account that estimates from Lemmas 7 and 11 can also be proved for the function u^n with constants independent of n, it is not difficult to check that,

$$\lim_{n\to\infty} u^n(t,x) = u(t,x), \quad t > 0,\ x \in H,$$

$$\lim_{n\to\infty} u^n_x(t,x) = u_x(t,x), \quad t > 0,\ x \in H,$$

and

$$\lim_{n\to\infty} \text{Tr}\ [\sigma(x)C\sigma(x)u^n_{xx}(t,x)] = \text{Tr}\ [\sigma(x)C\sigma(x)u_{xx}(t,x)], \quad t > 0,\ x \in H.$$

Consequently,

$$\lim_{n\to\infty} u^n_t(t,x) = u_t(t,x), \quad t > 0,\ x \in D(A),$$

and the conclusion follows. □

We consider finally the elliptic Kolmogorov equation

$$\lambda\varphi(x) - \frac{1}{2}\ \text{Tr}\ [\sigma^2(x)\varphi_{xx}(x)] - \langle Ax, \varphi_x(x)\rangle = f(x), \quad x \in D(A), \tag{3.26}$$

where $\lambda > 0$ and $f \in C_b(H)$ are given.

Theorem 14. *Assume that $\lambda > \overline{\lambda}$, $f \in C_b^2(H)$, $D^2 f(x)$ is of trace class for any $x \in H$ and Tr $[D^2 f] \in C_b(H)$. Define*

$$\varphi(x) = \int_0^\infty e^{-\lambda t}\mathbb{E}[f(X(t,x))]dt, \quad t \geq 0, \ x \in H,$$

Then the following statements hold.

(i) $\varphi \in C_b^1(H)$ and possesses second order derivatives in all directions of H.
(ii) For all $x \in H$ we have

$$|u_x(t,x)| \leq \frac{\kappa}{\lambda - \overline{\lambda}} \, \|\varphi\|_1 \tag{3.27}$$

and

$$|u_{xx}(t,x)| \leq \frac{\kappa}{\lambda - \overline{\lambda}} \, \|\varphi\|_2. \tag{3.28}$$

(iii) There exists $\kappa_1 > 0$ such that for all $x \in H$ we have

$$\begin{aligned} \left|\mathrm{Tr}\,[\sigma^2(x) u_{xx}(t,x)]\right| &= \left|\sum_{k=1}^\infty \langle u_{xx}(t,x)(\sigma(x)e_k, \sigma(x)e_k)\rangle\right| \\ &\leq \frac{\kappa_1}{\lambda - \overline{\lambda}} \, \|\varphi\|_2 (1 + \|D^2\varphi\|_{HS}). \end{aligned} \tag{3.29}$$

(iv) We have

$$|\varphi_x(x)| \leq \frac{\sqrt{a_1}}{\lambda - \overline{\lambda}} \, \|f\|_1, \quad x \in H, \tag{3.30}$$

and

$$\left|\mathrm{Tr}\,[\sigma^2(x)\varphi_{xx}(t,x)]\right| \leq \frac{\kappa}{(\lambda - \overline{l})} \, \|f\|_2, \quad x \in H. \tag{3.31}$$

(v) For all $x \in D(A)$ the equation (3.26) is fulfilled.

Proof. The conclusion follows from Proposition 13 and estimates (3.22), (3.23) and (3.24). □

3.4 The Kolmogorov Operator

It is well known that the semigroup R_t is not in general strongly continuous in $C_b(H)$. However, we can define its infinitesimal generator by proceeding as in [4]. Namely, for any $\lambda > 0$ and any $f \in C_b(H)$ we define

$$F_\lambda(f)(x) = \int_0^\infty e^{-\lambda t} R_t f(x) dt, \quad x \in H.$$

Proposition 15. *For any $f \in C_b(H)$ and any $\lambda > 0$ we have $F_\lambda(f) \in C_b(H)$ and the following estimate holds*

$$\|F_\lambda(f)\|_0 \le \frac{1}{\lambda}\,\|f\|_0. \tag{3.32}$$

Moreover there exists a unique closed operator $L\colon D(L) \subset C_b(H) \to C_b(H)$ such that for any $\lambda > 0$ and any $f \in C_b(H)$ we have $F_\lambda(f) = R(\lambda, L)f$.

Proof. Let first $f \in C_b^1(H)$; then it is obvious that if $F_\lambda(f) \in C_b(H)$ the inequality (3.32) holds. Moreover for all $x, y \in H$ we have

$$\begin{aligned} |F_\lambda(f)(x) - F_\lambda(f)(y)| &\le \int_0^\infty e^{-\lambda t}\mathbb{E}(|f(X(t,x)) - f(X(t,y))|)dt \\ &\le \|f\|_1 \int_0^\infty e^{-\lambda t}\mathbb{E}|X(t,x) - X(t,y)|dt. \end{aligned} \tag{3.33}$$

On the other hand we have

$$X(t,x) - X(t,y) = \int_0^1 X_x(t,(1-r)x + ry)(x-y)dr,$$

so that, recalling Lemma 7, we find

$$|X(t,x) - X(t,y)| \le \sqrt{a_1}\, e^{\frac{1}{2}\,\lambda_1 t}|x-y|. \tag{3.34}$$

Now, substituting this inequality in (3.33) yields

$$|F_\lambda(f)(x) - F_\lambda(f)(y)| \le \sqrt{a_1}\,\int_0^\infty e^{(\frac{1}{2}\,\lambda_1 - \lambda)t}dt|x-y|.$$

Thus, if $\lambda > \frac{1}{2}\,\lambda_1$ we have proved that $F_\lambda(f) \in C_b(H)$ (it is even Lipschitz). Since $C_b^1(H)$ is dense in $C_b(H)$ we can conclude that $F_\lambda(f) \in C_b(H)$ for all $f \in C_b(H)$ (and $\lambda > \frac{1}{2}\,\lambda_1$).

Now it is easy to see that F_λ fulfills the resolvent identity

$$F_\lambda - F_\mu = (\mu - \lambda)F_\lambda F_\mu, \quad \lambda, \mu > 0.$$

So, by a classical result, see e. g. [18], there exists a unique closed operator $L\colon D(L) \subset C_b(H) \to C_b(H)$ such that for any $\lambda > \frac{1}{2}\,\lambda_1$ and any $f \in C_b(H)$ we have $F_\lambda(f) = R(\lambda, L)f$.

Finally, by (3.32) we see that L is m-dissipative so that condition $\lambda > \frac{1}{2}\,\lambda_1$ can be replaced by $\lambda > 0$. □

Remark 16. Assume that $f \in C_b^2(H)$, $D^2 f(x)$ is of trace class for any $x \in H$ and $\mathrm{Tr}\,[D^2 f] \in C_b(H)$. Let moreover $\lambda > 0$ and $\varphi = R(\lambda, L)$. Then by Proposition 14 it follows that

$$L\varphi(x) = \frac{1}{2}\,\mathrm{Tr}\,[\sigma^2(x)\varphi_{xx}(x)] + \langle Ax, \varphi_x(x)\rangle, \quad x \in D(A). \tag{3.35}$$

Now we are going to prove, following an argument in [10], that if $\varphi \in R(\lambda, L)(C_b^2(H))$ we have $\varphi^2 \in D(L)$.

Proposition 17. *Assume that $f \in C_b^2(H)$, $D^2 f(x)$ is of trace class for any $x \in H$ and Tr $[D^2 f] \in C_b(H)$. Let moreover $\lambda > 0$ and $\varphi = R(\lambda, L)$. Then $\varphi^2 \in D(L)$ and*

$$L(\varphi^2) = 2\varphi\, L\varphi + |\sigma D\varphi|^2. \tag{3.36}$$

Proof. Let L_n be the approximating Kolmogorov operator

$$L_n\varphi(x) = \frac{1}{2}\,\text{Tr}\,[\sigma^2(x)P_n\varphi_{xx}(x)] + \langle A_n x, \varphi_x(x)\rangle, \quad \varphi \in C_b(H),\ x \in H \tag{3.37}$$

and let $\varphi^n = R(\lambda, L_n)f$. Then, by a straightforward computation, it follows that

$$L_n((\varphi^n)^2) = 2\varphi^n\, L_n\varphi^n + |\sigma P_n D\varphi^n|^2.$$

Now, multiplying both sides of the equation $\lambda\varphi^n - L_n\varphi^n = f$ by φ^n, yields

$$\lambda(\varphi^n)^2 - L_n\varphi^n\, \varphi^n = f\varphi^n,$$

which is equivalent to

$$2\lambda(\varphi^n)^2 - L_n((\varphi^n)^2) = 2f\varphi^n - |\sigma P_n D\varphi^n|^2.$$

Therefore,

$$(\varphi^n)^2 = R(2\lambda, L_n)(2f\varphi^n - |\sigma P_n D\varphi^n|^2).$$

Letting $n \to \infty$ yields

$$\varphi^2 = R(2\lambda, L)(2f\varphi - |\sigma D\varphi|^2).$$

Consequently

$$2\lambda\varphi^2 - L\varphi^2 = 2f\varphi - |\sigma(x)D\varphi|^2,$$

which yields (3.36). □

4 Invariant Measures

4.1 Existence and Uniqueness

We denote by $\mathcal{P}(H)$ the set of all Borel probability measures on H. We recall that a probability measure $\mu \in \mathcal{P}(H)$ is said to be *invariant* for the transition semigroup R_t defined by (1.7) if

$$\int_H R_t\varphi d\mu = \int_H \varphi d\mu \quad \text{for all } \varphi \in C_b(H). \tag{4.1}$$

Theorem 18. *There is an invariant measure* μ *for* R_t. *Moreover, for any* $\beta \in [0, 1/4)$ *we have*

$$\int_H |(-A)^\beta x|^2 \mu(dx) < +\infty. \tag{4.2}$$

Finally, if $1/g$ *is bounded the invariant measure* μ *is unique.*

Proof. Let $X(t, x)$ be the solution of (2.3). Using Lemma 3 and inequality (2.6), we find that

$$\begin{aligned}
\mathbb{E}(|X(t,x)|^2) &\le 2e^{-2\pi^2 t}|x|^2 + 2\sum_{h=1}^{\infty}\int_0^t e^{-2\pi^2 h^2(t-s)}\mathbb{E}(|\sigma(X(s,x))e_h|^2)ds \\
&\le 2e^{-2\pi^2 t}|x|^2 + \|g\|_0^2 \int_0^t F(2\pi^2(t-s))ds \\
&\le 2e^{-2\pi^2 t}|x|^2 + 4\|g\|_0^2 \int_0^t e^{-\pi^2(t-s)}(2\pi^2(t-s))^{-1/2}ds.
\end{aligned}$$

So,

$$\mathbb{E}(|X(t,x)|^2) \le 2e^{-2\pi^2 t}|x|^2 + \frac{4}{\sqrt{\pi}}\,\|g\|_0^2. \tag{4.3}$$

Now let $\beta \in (0, 1/4)$. Using the well known estimate

$$\|(-A)^\beta e^{tA}\| \le c_\beta t^{-\beta} e^{-\pi^2 t}, \quad t \ge 0, \tag{4.4}$$

where c_β is a suitable constants, we find,

$$\begin{aligned}
\mathbb{E}(|(-A)^\beta X(t,x)|^2) &\le 2c_\beta t^{-2\beta} e^{-2\pi^2 t}|x|^2 \\
&\quad + 2\int_0^t \sum_{h=1}^{\infty}(\pi h)^{4\beta} e^{-2\pi^2 h^2(t-s)}\mathbb{E}(|\sigma(X(s,x))e_h|^2)ds \\
&\le 2c_\beta t^{-2\beta} e^{-2\pi^2 t}|x|^2 + 2\|g\|_0^2 \int_0^t F_\beta(2\pi^2(t-s))ds
\end{aligned}$$

where F_β is defined by

$$F_\beta(t) = \sum_{h=1}^{\infty} h^{4\beta} e^{-h^2 t}, \quad t > 0.$$

It is not difficult to show that there is $k_\beta > 0$ such that

$$F_\beta(t) \le k_\alpha t^{-1/2-2\beta} e^{-t}, \quad t \ge 0.$$

Now we have

$$\begin{aligned}\mathbb{E}(|(-A)^{\beta}X(t,x)|^2) &\le 2c_\beta t^{-2\beta}e^{-2\pi^2 t}|x|^2\\ &\quad + 2\|g\|_0^2 k_\beta \int_0^t (2\pi^2 s)^{-1/2-2\beta}e^{-2\pi^2 s}ds \\ &\le 2c_\beta t^{-2\beta}e^{-2\pi^2 t}|x|^2 + c_\beta k_\beta \|g\|_0^2.\end{aligned} \tag{4.5}$$

Since the embedding $D(A) \subset H$ is compact, the existence of an invariant measure follows from the Krylov–Bogoliubov theorem.

Let us show (4.2). Let $\gamma > 0$ and set

$$\varphi_\gamma(x) = \frac{|x|^2}{1+\gamma|x|^2}, \quad x \in H.$$

Then $\varphi_\gamma \in C_b(H)$ and, proceeding as in the proof of (4.3) we see that there exists a constant $\kappa > 0$ (independent on λ) such that,

$$R_t(\varphi_\gamma)(x) = \mathbb{E}[\varphi_\gamma(X(t,x))] \le e^{-2\pi^2 t}\varphi_\gamma(x) + \kappa. \tag{4.6}$$

Integrating both sides of (4.6) with respect to x over H and taking into account the invariance of μ yields,

$$\int_H \varphi_\gamma(x)\mu(dx) \le e^{-2\pi^2 t}\int_H \varphi_\gamma(x)\mu(dx) + \kappa.$$

Therefore, there exists $\kappa_1 > 0$ (independent on λ) such that

$$\int_H \varphi_\gamma(x)\mu(dx) \le \kappa_1.$$

Letting γ tend to 0 yields,

$$\int_H |x|^2\mu(dx) \le \kappa_1. \tag{4.7}$$

Now, integrating both sides of (4.5) with respect to x over H and taking into account again the invariance of μ yields,

$$\int_H |(-A)^{-\beta}x|^2\mu(dx) \le 2c_\beta t^{-2\beta}e^{-2\pi^2 t}\kappa_1 + c_\beta\|g\|_0^2,$$

and so, (4.2) follows. Finally, if $1/g$ is bounded the uniqueness of μ follows from from the Doob Theorem since R_t is irreducible and strong Feller by [16]. □

Remark 19. More general results of existence of invariant measures can be found in the paper [7].

4.2 Existence of a Core of Smooth Functions for L_μ

Let us fix an invariant measure μ for R_t. It is well known that R_t can be uniquely extended to a strongly continuous semigroup of contractions on $L^2(H,\mu)$ which we shall still denote by R_t. The infinitesimal generator of R_t in $L^2(H,\mu)$ will be denoted by L_μ. Since R_t is a contraction semigroup, L_μ is m-dissipative in $L^2(H,\mu)$.

In this subsection we want to define a core of L_μ consisting of regular functions.

Proposition 20. *Set*

$$\Lambda = \{\varphi \in C_b^2(H) : \; D^2\varphi(x) \in L_1(H) \text{ for all } x \in H \text{ and } \mathrm{Tr}\,[D^2\varphi] \in C_b(H)\}$$

and

$$\Gamma := \bigcup_{\lambda>0} R(\lambda, L)(\Lambda).$$

Then Γ is a core for L_μ.

Moreover if $\varphi \in \Gamma$ we have $\varphi^2 \in D(L_\mu)$ and the following identity holds

$$L_\mu(\varphi^2) = 2\varphi\, L_\mu\varphi + |\sigma D\varphi|^2. \tag{4.8}$$

Proof. Let $\lambda > 0$. It is clear that any $\varphi \in D(L)$ belongs to $D(L_\mu)$ as well, so that we have

$$(\lambda - L_\mu)(\Gamma) = (\lambda - L)(\Gamma) \supset \Lambda.$$

Since Λ is dense in $L^2(H,\mu)$ (by a standard argument of monotone classes), we can conclude that $(\lambda - L_\mu)(\Gamma)$ is dense in $L^2(H,\mu)$. Now the Lumer-Phillips theorem implies that Γ is a core for L_μ.

Finally, the last statement follows from (3.36). □

5 The Basic Integration by Parts Formula

In this section we assume that g^{-1} is bounded. We recall that in this case μ is the unique invariant measure of the transition semigroup R_t.

Proposition 21. *The operator*

$$D : \Gamma \subset L^2(H,\mu) \to L^2(H,\mu;H), \; \varphi \mapsto D\varphi, \tag{5.1}$$

is uniquely extendible to a linear bounded operator $D\colon D(L_\mu) \to L^2(H,\mu;H)$, where $D(L_\mu)$ is endowed with the graph norm of L_μ. Moreover, the following identity holds

$$\int_H L_\mu\varphi\,\varphi\, d\mu = -\frac{1}{2}\int_H |\sigma D\varphi|^2 d\mu, \quad \varphi \in D(L_\mu). \tag{5.2}$$

Identity (5.2) is called in French "identité du carré du champs". It will play an important rôle in what follows.

Proof. Let $\varphi \in \Gamma$, then $\varphi^2 \in D(L_\mu)$ in view of Proposition 20 we have

$$L_\mu \varphi^2 = 2\varphi L_\mu \varphi + |gD\varphi|^2.$$

Integrating this identity with respect to μ over H and taking into account that

$$\int_H L_\mu \varphi^2 d\mu = 0$$

by the invariance of μ, implies (5.2) when $\varphi \in \Gamma$. Let now $\varphi \in D(L_\mu)$. Since Γ is a core for L_μ, there exists a sequence $\{\varphi_n\} \subset \Gamma$ such that

$$\varphi_n \to \varphi, \quad L_0 \varphi_n \to L_\mu \varphi \quad \text{in } L^2(H,\mu).$$

By (5.2) it follows that

$$\int_H |\sigma D(\varphi_n - \varphi_m)|^2 d\mu = -2\int_H L_\mu(\varphi_n - \varphi_m)\,(\varphi_n - \varphi_m)\,d\mu.$$

So, the sequence $\{\sigma D\varphi_n\}$ is Cauchy in $L^2(H,\mu;H)$ and the conclusion follows. □

Proposition 22. *Let $\varphi \in L^2(H,\mu)$ and $t \geq 0$. Then, for any $T > 0$, the linear operator*

$$\sigma D R_t : D(L_\mu) \subset L^2(H,\mu) \to L^2(0,T;L^2(H,\mu;H)), \ \varphi \to \sigma D R_t \varphi,$$

is uniquely extendible to a linear bounded operator, still denoted by $\sigma D R_t$, from $L^2(H,\mu)$ into $L^2(0,T;L^2(H,\mu;H))$. Moreover the following identity holds

$$\int_H (R_t\varphi)^2\, d\mu + \int_0^t ds \int_H |\sigma D R_s \varphi|^2 d\mu = \int_H \varphi^2\, d\mu. \tag{5.3}$$

Proof. We first establish (5.3) for $\varphi \in D(L_\mu)$. In this case we have

$$\frac{d}{dt} R_t \varphi = L_\mu \varphi.$$

Multiplying scalarly this identity by $R_t\varphi$, integrating with respect to μ over H and using (5.3), yields,

$$\frac{d}{dt}\int_H (R_t\varphi)^2\, d\mu + \int_H |\sigma D R_s \varphi|^2 d\mu = 0. \tag{5.4}$$

Now (5.3) follows integrating (5.4) with respect to t. The case when $\varphi \in L^2(H,\mu)$ can be handled by approximating φ by elements of $D(L_\mu)$. □

5.1 The Sobolev Space $W^{1,2}(H,\mu)$

To define the Sobolev space we first show that the mapping

$$D_\mu : \Gamma \subset L^2(H,\mu) \to L^2(H,\mu;H), \ \varphi \to D_\mu\varphi \tag{5.5}$$

is closable. Notice the difference between the map D defined by (5.1) and the map D_μ. The first one is a bounded operator in $D(L_\mu)$ (endowed with the graph norm of L_μ) whereas the second will be a closable operator in $L^2(H,\mu)$.

To prove closability of D_μ we recall the following estimate from Lemma 7.

$$\mathbb{E}(|X_x(t,x)h|^2) \le a_1 e^{-2\omega t}|h|^2 \quad t \ge 0, \ h, x \in H, \tag{5.6}$$

where $\omega = \pi^2 - 8\pi^{-2}\|g'\|_0^2$.

Lemma 23. *Let $\{\varphi_n\} \subset \Gamma$ and let $G \in L^2(H,\mu;H)$ be such that*

$$\lim_{n\to\infty} D\varphi_n = G \quad \text{in } L^2(H,\mu;H).$$

Then, for any $t \ge 0$ we have

$$\lim_{n\to\infty} DR_t\varphi_n = \mathbb{E}[X_x^*(t,x)G(X(t,x))] \quad in\ L^2(H,\mu;H).$$

In particular, if $D\varphi_n \to 0$ in $L^2(H,\mu;H)$ we have $DR_t\varphi_n \to 0$ in $L^2(H,\mu;H)$ for all $t > 0$.

Proof. Write

$$DR_t\varphi_n(x) = E[X_x^*(t,x)D\varphi_n(X(t,x)], \quad t \ge 0, \ x \in H.$$

Taking into account estimate (5.6) and the invariance of μ, yields

$$\begin{aligned}
&\int_H \left|DR_t\varphi_n(x) - \mathbb{E}[X_x^*(t,x)G(X(t,x))]\right|^2 \mu(dx)\\
&= \int_H \left|\mathbb{E}[X_x^*(t,x)(D\varphi_n(X(t,x)) - G(X(t,x))]\right|^2 \mu(dx)\\
&\le a_1 e^{-2\omega t}\int_H \mathbb{E}[|D\varphi_n(X(t,x)) - G(X(t,x))|^2]\mu(dx)\\
&= a_1 e^{-2\omega t}\int_H R_t(|D\varphi_n - G|^2)(x)\mu(dx)\\
&= a_1 e^{-2\omega t}\int_H |D\varphi_n(x) - G(x)|^2\mu(dx).
\end{aligned}$$

The conclusion of the lemma follows. □

Proposition 24. *D_μ is closable. Moreover, if φ belongs to the domain of the closure $\overline{D_\mu}$ of D_μ and $\overline{D_\mu}\varphi = 0$ we have that $\overline{D_\mu} R_t\varphi = 0$ for any $t > 0$.*

Proof. Let $\{\varphi_n\} \subset \Gamma$ and $G \in L^2(H, \mu; H)$ be such that

$$\varphi_n \to 0 \quad \text{in } L^2(H, \mu), \quad D\varphi_n \to G \quad \text{in } L^2(H, \mu; H).$$

By (5.3) we have that

$$\int_H (R_t\varphi_n)^2 \, d\mu + \int_0^t ds \int_H |\sigma DR_s\varphi_n|^2 d\mu = \int_H \varphi_n^2 \, d\mu.$$

Letting $n \to \infty$ and taking into account that g is bounded below, yields

$$\lim_{n\to\infty} \int_0^t ds \int_H |DR_s\varphi_n|^2 d\mu = 0.$$

Consequently, by Lemma 23, it follows that

$$\int_0^t ds \int_H (\mathbb{E}[X_x^*(s,x)G(X(s,x)])^2 \, \mu(dx) = 0, \quad h \in H.$$

Then for almost all $t \geq 0$ we have that

$$\mathbb{E}[X_x^*(t,x)G(X(t,x))] = 0. \tag{5.7}$$

Now fix $h \in H$. Then we have,

$$\begin{aligned} |\mathbb{E}[\langle G(X(t,x)), h\rangle]| &\leq |\mathbb{E}[\langle G(X(t,x)), X_x(t,x) \cdot h\rangle]| \\ &\quad + |\mathbb{E}[\langle G(X(t,x)), h - X_x(t,x) \cdot h\rangle]| \\ &= |\mathbb{E}[\langle G(X(t,x)), h - X_x(t,x) \cdot h\rangle]|. \end{aligned}$$

Taking into account the invariance of μ and (5.7), we find that

$$\begin{aligned} &\int_H |R_t(\langle G(x), h\rangle)| \mu(dx) \\ &= \int_H |\mathbb{E}[\langle G(X(t,x)), h\rangle]| \mu(dx) \\ &= \int_H |\mathbb{E}[\langle G(X(t,x)), h - X_x(t,x) \cdot h\rangle]| \mu(dx) \\ &\leq \left[\int_H \mathbb{E}[|G(X(t,x))|^2] \mu(dx)\right]^{1/2} \left[\int_H \mathbb{E}[|h - X_x(t,x) \cdot h|^2] \mu(dx)\right]^{1/2}. \end{aligned}$$

Therefore, as $t \to 0$ we find by the strong continuity of R_t in $L^1(H, \mu)$

$$\int_H |\langle G(x), h\rangle| \mu(dx) = 0$$

and by the arbitrariness of h it follows that $G = 0$ as required. Finally, the last statement follows from by Lemma 23. □

By Proposition 24 it follows that the mapping

$$D_\mu : \Gamma \subset L^2(H, \mu) \to L^2(H, \mu; H), \ \varphi \mapsto D\varphi,$$

is closable, let $\overline{D_\mu}$ its closure. We shall denote by $W^{1,2}(H, \mu)$ the domain of $\overline{D_\mu}$ and, if there is not possibility of confusion, we shall set $\overline{D_\mu} = D$.

Proposition 25. *We have $D(L_\mu) \subset W^{1,2}(H, \mu)$ with continuous embedding. Moreover, the following identity holds*

$$\int_H L_\mu \varphi \, \varphi \, d\mu = -\frac{1}{2} \int_H |\sigma D\varphi|^2 d\mu, \quad \varphi \in D(L_\mu). \tag{5.8}$$

Proof. Let $\varphi \in D(L_\mu)$. Since Γ is a core for L_μ, there exists a sequence $\{\varphi_n\} \subset \Gamma$ such that

$$\varphi_n \to \varphi, \quad L_0 \varphi_n \to L_\mu \varphi \quad \text{in } L^2(H, \mu).$$

By (5.2) it follows that

$$\int_H |\sigma D(\varphi_n - \varphi_m)|^2 d\mu \le 2 \int_H |L_0(\varphi_n - \varphi_m)| \, |\varphi_n - \varphi_m| \, d\mu.$$

Therefore the sequence $(D\varphi_n)$ is Cauchy in $L^2(H, \mu; H)$. Since D is closed it follows that $\varphi \in W^{1,2}(H, \mu)$ as required. □

5.2 The Poincaré Inequality

Since $1/g$ is bounded, by Theorem 18 there is a unique invariant measure μ for R_t and by the Doob theorem, see e.g. [12], we have that,

$$\lim_{n \to \infty} R_t \varphi(x) = \int_H \varphi(y) \mu(dy), \quad x \in H, \tag{5.9}$$

for all $\varphi \in C_b(H)$.

Let us prove now the Poincaré inequality.

Proposition 26. *Assume that $\|g'\|_0^2 \le \frac{1}{8}\, \pi^4$. Then, for any $\varphi \in W^{1,2}(H, \nu)$ we have*

$$\int_H |\varphi - \overline{\varphi}|^2 d\mu \le \frac{a_1}{\omega} \, \|g\|_0^2 \, \|1/g\|_0^2 \int_H |g(x) D\varphi|^2 d\mu, \tag{5.10}$$

where $\overline{\varphi} = \int_H \varphi d\mu$ and $\omega = \pi^2 - 8\pi^{-2} \|g'\|_0^2$.

Proof. Let first $\varphi \in \Gamma$. Then by (5.3) we have

$$\int_H |R_t\varphi(x) - \overline{\varphi}|^2 \mu(dx) = \int_0^t ds \int_H |\sigma D R_s\varphi|^2 d\mu. \tag{5.11}$$

Moreover by (5.6) it follows that

$$\begin{aligned}\mathbb{E}[|DR_s\varphi(x)|^2] &\le \mathbb{E}\left[|D\varphi(X(s,x))|^2\, |X_x(s,x)|^2\right] \\ &\le a_1 e^{-2\omega s}\mathbb{E}\left[|D\varphi(X(s,x))|^2\right] = a_1 e^{-2\omega s} R_s(|D\varphi|^2)(x).\end{aligned}$$

Taking into account (5.9) and the invariance of μ we obtain,

$$\begin{aligned}&\int_H |R_t\varphi(x) - \overline{\varphi}|^2 \mu(dx) \\ &\le a_1 \|g\|_0^2 \int_0^{+\infty} e^{-2\omega s} ds \int_H R_s(|D\varphi|^2)(x)\mu(dx) \\ &\le a_1 \|g\|_0^2\, \|1/g\|_0^2 \int_0^{+\infty} e^{-2\omega s} ds \int_H |g(x) D\varphi(x)|^2 \mu(dx),\end{aligned}$$

and the conclusion follows. If $\varphi \in W^{1,2}(H,\mu)$, we proceed by density. □

Remark 27. If g is constant the condition $\|g'\|_0^2 \le \frac{1}{8}\,\pi^4$ is trivially fulfilled and we recover a result in [10].

Remark 28. It is well known, see e.g. [8], that the Poincaré inequality implies that the spectrum $\sigma(L_\mu)$ of L_μ consists of 0 and a set included in the half-space

$$\{\lambda \in \mathbb{C} : \Re e\, \lambda \le -\omega_1\},$$

(*spectral gap*) where ω_1 is a positive constant.

The spectral gap in turn implies an exponential convergence of $R_t\varphi$ to the equilibrium

$$\int_{\mathbb{R}} |R_t\varphi - \overline{\varphi}|^2 d\nu \le c e^{-2\omega_1 t} \int_{\mathbb{R}} |\varphi|^2 d\nu, \quad \varphi \in L^2(\mathbb{R},\nu), \tag{5.12}$$

where c is a suitable constant.

References

1. S. Albeverio and M. Röckner, *Stochastic differential equations in infinite dimensions: solutions via Dirichlet forms,* Probab. Theory Relat. Fields, **89**, 347–386, 1991.
2. S. Athreya, R. Bass, M. Gordina and E. Perkins, *Infinite dimensional stochastic differential equations of Ornstein–Uhlenbeck tipe,* Stochastic Process. Appl. 116, 3, pp. 381–406, 2006.

3. S. Athreya, R. Bass and E. Perkins, *Hölder norm estimates for elliptic operators on finite and infinite-dimensional spaces*, Trans. Amer. Math. Soc. to appear.
4. S. Cerrai, A Hille-Yosida theorem for weakly continuous semigroups, *Semigroup Forum*, **49**, 349–367, 1994.
5. S. Cerrai, *Second order PDE's in finite and infinite dimensions. A probabilistic approach*, Lecture Notes in Mathematics, **1762**, Springer-Verlag, 2001.
6. S. Cerrai, *Classical solutions for Kolmogorov equations in Hilbert spaces*, Seminar on Stochastic Analysis, Random Fields and Applications, III (Ascona, 1999), 55–71, Progr. Probab., 52, Birkhäuser, 2002.
7. S. Cerrai, *Stochastic reaction-diffusion systems with multiplicative noise and non-Lipschitz reaction term*, Probab. Theory Relat. Fields, **125**, 271–304, 2003.
8. G. Da Prato, *Kolmogorov equations for stochastic PDEs*, Birkhäuser, 2004.
9. G. Da Prato and A. Debussche, *Ergodicity for the 3D stochastic Navier–Stokes equations*, Journal Math. Pures Appl. **82**, 877–947, 2003.
10. G. Da Prato, A. Debussche and B. Goldys, *Invariant measures of non symmetric dissipative stochastic systems,* Probab. Theory Relat. Fields, **123**, 3, 355–380.
11. G. Da Prato and J. Zabczyk, *Stochastic equations in infinite dimensions*, Cambridge University Press, 1992.
12. G. Da Prato and J. Zabczyk, *Ergodicity for infinite dimensional systems*, London Mathematical Society Lecture Notes, **229**, Cambridge University Press, 1996.
13. G. Da Prato and J. Zabczyk, *Second Order Partial Differential Equations in Hilbert Spaces*, London Mathematical Society, Lecture Notes, **293**, Cambridge University Press, 2002.
14. G. Lumer and R. S. Phillips, *Dissipative operators in a Banach space*, Pacific J. Math. **11**, 679–698, 1961.
15. Z. M. Ma and M. Röckner, *Introduction to the Theory of (Non Symmetric) Dirichlet Forms,* Springer-Verlag, 1992.
16. S. Peszat and J. Zabczyk, *Strong Feller property and irreducibility for diffusions on Hilbert spaces,* Ann. Probab., **23**, 157–172, 1995.
17. M. Röckner, *L^p-analysis of finite and infinite dimensional diffusions*, Lecture Notes in Mathematics, **1715**, G. Da Prato (editor), Springer-Verlag, 65–116, 1999.
18. K. Yosida, *Functional analysis*, Springer-Verlag, 1965.

Stochastic Integrals and Adjoint Derivatives

Giulia Di Nunno[1] and Yuri A. Rozanov[2]

[1] Centre of Mathematics for Applications, Department of Mathematics, University of Oslo, P.O. Box 1053 Blindern, N-0316 Oslo, Norway, giulian@math.uio.no
[2] IMATI-CNR, Via E. Bassini 15, I-20133 Milano, Italy, rozanov@infinito.it

Summary. In a systematic study form, the present paper treats topics of stochastic calculus with respect to stochastic measures with independent values. We focus on the integration and differentiation with respect to these measures over general space-time products.

Mathematics Subject Classification (2000): 60H05, 60H07

Keywords: Itô non-anticipating integral, non-anticipating derivative, Malliavin derivative, Skorohod integral, deFinetti-Kolmogorov law

Contents

1 Stochastic Measures and Functions

1.1 Space-time Products

In this paper we consider some elements of a stochastic calculus for random fields over general space-time products. The various space components which may be specified in the possible applications are here considered altogether and denoted by Θ. We consider Θ to be a general space equipped with some countable σ-algebra. The time-component is an interval $\mathbb{T}$. To simplify notations, we fix $\mathbb{T} = (0, T]$.

In the sequel we use the partitions of the involved spaces as basic tool. As for the interval $\mathbb{T}$, its *partition* with level of refinement n, from now on named n^{th}-*partition*, is represented by the corresponding finite n^{th}-*series* of intervals of type $(s, u]$ such that

$$\mathbb{T} = \sum (s,u] : \qquad \max_{(s,u]}(u-s) \longrightarrow 0\,, \qquad n \to \infty \tag{1.1}$$

(note that here and in the sequel we denote the disjoint union of sets by $\sum$). The partitions are such that for $n = 1, 2, \ldots$, the $(n+1)^{th}$-series is obtained by partitioning the intervals of the previous n^{th}-series. The family of all the sets of all the nth-partitions, $n = 1, 2, \ldots$, generates the Borel σ-algebra of $\mathbb{T}$.

Hereafter we introduce the nth-partitions ($n = 1, 2, \ldots$) for the standard product $\Theta \times \mathbb{T}$ of the measurable spaces Θ and $\mathbb{T}$. The *σ-algebra of* $\Theta \times \mathbb{T}$ will be treated as generated by these n^{th}-partitions. Since in the sequel we are dealing with the general σ-finite measure $M = M(\Delta)$, $\Delta \subseteq \Theta \times \mathbb{T}$, on the σ-algebra of $\Theta \times \mathbb{T}$, the *n^{th}-partitions of* $\Theta \times \mathbb{T}$ are going to be selected for the increasing sequence (which can be any) of sets

$$\Theta_n \times \mathbb{T}\,, \quad n = 1, 2, \ldots, \quad \text{such that} \quad \lim_{n\to\infty} \Theta_n \times \mathbb{T} := \bigcup_n \Theta_n \times \mathbb{T} = \Theta \times \mathbb{T}\,,$$

and with

$$M(\Theta_n \times \mathbb{T}) < \infty\,, \qquad n = 1, 2, \ldots\,. \tag{1.2}$$

The nth-*partition of* $\Theta \times \mathbb{T}$ is then actually a partition of

$$\Theta_n \times \mathbb{T} = \sum \Delta : \quad \Delta \subseteq \Theta \times (s,u] \tag{1.3}$$

given by the corresponding finite series of sets Δ, related to the nth-*series* of time-intervals $(s, u] \subseteq \mathbb{T}$ in (1.1). Any set Δ in the nth-series is the (disjoint) union of some elements of the $(n+1)^{th}$-series. Clearly $M(\Delta) < \infty$ for any element Δ of the partitions of $\Theta \times \mathbb{T}$.

We assume that the measure M satisfies

$$M(\Theta \times [t]) = \lim_{n\to\infty} M(\Theta_n \times [t]) = 0 \tag{1.4}$$

whatever the point-set $[t] \subseteq \mathbb{T}$ be. This implies, in particular, that any set $\Delta \subseteq \Theta \times \mathbb{T} : M(\Delta) < \infty$, is *infinitely-divisible* in the sense that Δ admits the nth-partitions.

We will refer to the finite (disjoint) unions of sets belonging to the same n^{th}-series of partitions (1.3) as the *simple sets* in $\Theta \times \mathbb{T}$. Note that for *any* set $\Delta \subseteq \Theta \times \mathbb{T} : M(\Delta) < \infty$, we have

$$\Delta = \sum \Delta \cap \Big(\Theta \times (s,u]\Big) \ : \quad \sum (s,u] = \mathbb{T} \,,$$

where the nth-series of time-intervals (1.1) have been used. Since $\max(u-s) \to 0$, $n \to \infty$, we have

$$\max_{(s,u]} M\Big(\Delta \cap (\Theta \times (s,u])\Big) \longrightarrow 0 \,, \qquad n \to \infty \,. \tag{1.5}$$

We write $M(d\theta\, dt)$, $(\theta, t) \in \Theta \times \mathbb{T}$, for M as integrator.

Any $\Delta \subseteq \Theta \times \mathbb{T} : M(\Delta) < \infty$, can be approximated by simple sets $\Delta^{(n)}$, $n = 1, 2, \ldots$, in the sense that

$$\Delta = \lim_{n\to\infty} \Delta^{(n)}, \text{ i.e. } \quad M\Big((\Delta \setminus \Delta^{(n)}) \cup (\Delta^{(n)} \setminus \Delta)\Big) \longrightarrow 0 \,, \qquad n \to \infty \,. \tag{1.6}$$

Note that for any finite number of *disjoint* sets $\Delta_1, \ldots, \Delta_m$: $M(\Delta_j) < \infty$, $j = 1, \ldots, m$, the approximation above can be given by the corresponding sequences of *disjoint* simple sets $\Delta_1^{(n)}, \ldots, \Delta_m^{(n)}$ $(n = 1, 2, \ldots)$.

1.2 Stochastic Measures with Independent Values

For the complete probability space $(\Omega, \mathfrak{A}, P)$, let $L_2(\Omega)$ be the standard (complex) space of random variables $\xi = \xi(\omega)$, $\omega \in \Omega$, with finite norm

$$\|\,\xi\,\| = \Big(E\,|\xi|^2\Big)^{1/2} \,. \tag{1.7}$$

We write $\mu = \mu(\Delta)$, $\Delta \subseteq \Theta \times \mathbb{T}$, for the *additive* set-function with the *real* values $\mu(\Delta) \in L_2(\Omega)$ and such that $E\mu(\Delta) = 0$, $E\mu(\Delta)^2 = M(\Delta)$. Here the variance $M = M(\Delta)$, $\Delta \in \Theta \times \mathbb{T}$, is a measure which satisfies the conditions (1.2)-(1.4). The additive set-function μ is considered on all the sets $\Delta : M(\Delta) < \infty$. The values of μ on *disjoint* sets are *independent* random variables.

Note that μ, initially considered just on the simple sets in $\Theta \times \mathbb{T}$ (related to some partitions), can be extended on all $\Delta \subseteq \Theta \times \mathbb{T} : M(\Delta) < \infty$, via the limits

$$\mu(\Delta) = \lim_{n\to\infty} \mu(\Delta^{(n)}), \text{ i.e. } \quad \big\|\mu(\Delta) - \mu(\Delta^{(n)})\big\| \longrightarrow 0 \,, \qquad n \to \infty \,, \tag{1.8}$$

where the simple sets $\Delta^{(n)}$, $n = 1, 2, \ldots$, approximate Δ, i.e.

$$M\left((\Delta \setminus \Delta^{(n)}) \cup (\Delta^{(n)} \setminus \Delta)\right) = \left\|\mu(\Delta) - \mu(\Delta^{(n)})\right\|^2 \longrightarrow 0, \quad n \to \infty.$$

Cf. (1.6). We refer to μ as *the stochastic measure with indepenent values of the type*

$$E\,\mu = 0\,, \qquad E\,\mu^2 = M\,. \tag{1.9}$$

And we write $\mu(d\theta\, dt)$, $(\theta, t) \in \Theta \times \mathbb{T}$, for μ as the integrator.

Let μ_k, $k = 1, \ldots, K$ $(K \leq \infty)$, be *independent* stochastic measures of type $E\mu_k = 0$, $E\mu_k^2 = M_k$ on the corresponding space-time products $\Theta_k \times \mathbb{T}$. Let

$$\Theta \times \mathbb{T} := \sum_k (\Theta_k \times \mathbb{T}).$$

The *mixture* of μ_k, $k = 1, \ldots, K$, is a stochastic measure μ on the space-time product $\Theta \times \mathbb{T}$ formally introduced above defined as

$$\mu(\Delta) := \sum_k \mu_k\Big(\Delta \cap (\Theta_k \times \mathbb{T})\Big)\,, \qquad \Delta \subseteq \Theta \times \mathbb{T}\,. \tag{1.10}$$

This stochastic measure is of the type $E\mu = 0$, $E\mu^2 = M$, where

$$M(\Delta) = \sum_k M_k\Big(\Delta \cap (\Theta_k \times \mathbb{T})\Big)\,, \qquad \Delta \subseteq \Theta \times \mathbb{T}\,.$$

Cf. (1.9). Naturally in the expression above the sets $\Theta_k \times \mathbb{T}\,(k = 1, \ldots, K)$ formally represent some partition sets of $\Theta \times \mathbb{T}$. To illustrate, let μ_k, $k = 1, \ldots, K$ $(K \leq \infty)$, be stochastic measures on the time interval $\mathbb{T}$. Then the space-time product $\Theta \times \mathbb{T}$ that can be applied has space component $\Theta = \{1, \ldots, K\}$.

1.3 The Events Generated

Let $\mu = \mu(\Delta)$, $\Delta \in \Theta \times \mathbb{T}$, be a general stochastic measure with independent values of the type (1.9). In particular, it can be the *mixture* of a *number* of independent components - cf. (1.10). We write

$$\mathfrak{A}_\Delta\,, \qquad \Delta \subseteq \Theta \times \mathbb{T}\,, \tag{1.11}$$

for the σ-algebras generated by μ over the subsets of Δ and augmented by all the events of zero probability. To be more precise, $\mathfrak{A}_\Delta$ is the minimal augmented σ-algebra containing all the standard events $\{\mu(\Delta') \in B\}$ for all $B \subseteq \mathbb{R}$ and the subsets $\Delta' \subseteq \Delta$. To simplify notations and terminology, we assume that the σ-algebra

$$\mathfrak{A} = \mathfrak{A}_{\Theta \times \mathbb{T}} \tag{1.12}$$

represents all the events $A \subseteq \Omega$.

We remark that, in some sense, the σ-algebras $\mathfrak{A}_\Delta$ are *continuous* with respect to the sets $\Delta \subseteq \Theta \times \mathbb{T}$. To explain, on one hand we have

$$\lim_{n\to\infty} \mathfrak{A}_{\Delta^{(n)}} := \bigvee_n \mathfrak{A}_{\Delta^{(n)}} = \mathfrak{A}_\Delta , \tag{1.13}$$

for any sequence of *increasing* sets $\Delta^{(n)}$, $n = 1, 2, \ldots$, such that $\lim_{n\to\infty} \Delta^{(n)} = \Delta$, i.e. $M\left((\Delta^{(n)} \setminus \Delta) \cup (\Delta \setminus \Delta^{(n)})\right) \to 0$, $n \to \infty$ (here, the sign $\bigvee$ defines the minimal σ-algebra containing the involved components)-cf. (1.8). On the other hand, we have the following result. See e.g. [13].

Theorem 1.1. *Let $\Delta^{(n)}$, $n = 1, 2, \ldots$, be a sequence of decreasing sets and let $\Delta = \bigcap_n \Delta^{(n)}$, then we have*

$$\lim_{n\to\infty} \mathfrak{A}_{\Delta^{(n)}} := \bigcap_n \mathfrak{A}_{\Delta^{(n)}} = \mathfrak{A}_\Delta . \tag{1.14}$$

Proof. Note that

$$\mathfrak{A}_{\Delta^{(1)}} = \mathfrak{A}_\Delta \bigvee \mathfrak{A}_{\Delta^{(1)}\setminus\Delta} \text{ where } \mathfrak{A}_{\Delta^{(1)}\setminus\Delta} = \bigvee_n \mathfrak{A}_{\Delta^{(1)}\setminus\Delta^{(n)}} .$$

Cf. (1.13). Accordingly, we have

$$H_{\Delta^{(1)}} = H_\Delta \bigvee H_{\Delta^{(1)}\setminus\Delta} \text{ where } H_{\Delta^{(1)}\setminus\Delta} = \bigvee_n H_{\Delta^{(1)}\setminus\Delta^{(n)}} .$$

for the subspaces in $L_2(\Omega)$ of random variables measurable with respect to the corresponding σ-algebras (here above, the sign $\bigvee$ defines the linear closure of the involved components). The products $\xi\cdot\xi' \,:\, \xi \in H_\Delta, \xi' \in H_{\Delta^{(1)}\setminus\Delta^{(n)}}$, $n>1$, constitute a complete system in $H_{\Delta^{(1)}}$. Hence, the orthogonal projections

$$\xi\cdot\xi' - E\left(\xi\cdot\xi'|\mathfrak{A}_\Delta\right) = \xi(\xi' - E\,\xi'),$$

on the orthogonal complement $H_{\Delta^{(1)}} \ominus H_\Delta$ to the subspace $H_\Delta \subseteq H_{\Delta^{(1)}}$, constitute a complete system in $H_{\Delta^{(1)}} \ominus H_\Delta$. For the subspace H_Δ^+ of the random variables in $H_{\Delta^{(1)}}$ measurable with respect to the σ-algebra $\mathfrak{A}_\Delta^+ := \bigcap_n \mathfrak{A}_{\Delta^{(n)}}$, any $\xi^+ \in H_\Delta^+$ is independent from all $\xi' \in H_{\Delta^{(1)}\setminus\Delta^{(n)}}$, $n > 1$, and this implies that

$$E\left(\xi^+\cdot\xi(\xi' - E\xi')\right) = E\left(\xi^+\cdot\xi\right)\cdot E\left(\xi' - E\xi'\right) = 0 .$$

Thus, ξ^+ is orthogonal to all the elements $\xi \cdot (\xi' - E\xi')$ of the complete system in $H_{\Delta^{(1)}} \ominus H_\Delta$. Accordingly, $\xi^+ \in H_\Delta$. This justifies that $H_\Delta^+ = H_\Delta$ and $\mathfrak{A}_\Delta^+ = \mathfrak{A}_\Delta$. □

The σ-algebras

$$\mathfrak{A}_t := \mathfrak{A}_{\Theta\times(0,t]} , \qquad t \in \mathbb{T} \tag{1.15}$$

- cf. (1.11), represent the flow of events in the course of time on $\mathbb{T} = (0,T]$. Thanks to the condition (1.4), *the σ-algebras* $\mathfrak{A}_t$ *are continuous with respect to* $t \in \mathbb{T}$:

$$\lim_{s\to t-0} \mathfrak{A}_s := \bigvee_{s<t} \mathfrak{A}_s = \mathfrak{A}_t \quad (0 < t \le T)\,,$$
$$\lim_{u\to t+0} \mathfrak{A}_u := \bigcap_{u>t} \mathfrak{A}_u = \mathfrak{A}_t \quad (0 \le t < T)\,. \tag{1.16}$$

Cf. (1.13)-(1.14). Note that, here above, for $t = 0$ we have the *trivial* σ-algebra $\mathfrak{A}_0$. We remark that, for any t, the values $\mu(\Delta)$, $\Delta \in \Theta \times (0,t]$, are $\mathfrak{A}_t$-measurable and the values $\mu(\Delta)$, $\Delta \in \Theta \times (t,T]$, are independent of $\mathfrak{A}_t$.

1.4 The de Finetti-Kolmogorov Infinitely-Divisible Law

Let us set

$$\mathbb{R} \setminus [0] := (-\infty, 0) \cup (0, \infty)\,.$$

Similar to the stochastic processes with independent increments (cf. e.g. [51], see also e.g. [3,49]) the stochastic measure $\mu = \mu(\Delta)$, $\Delta \subseteq \Theta \times \mathbb{T}$, of the type $E\mu = 0$, $E\mu^2 = M$ - cf. (1.9), can be characterized as follows. We can refer to [18] for the following result. We also refer for example to [30] for some results in this direction with respect to random measures.

Theorem 1.2. *The values* $\mu(\Delta)$, $\Delta \subseteq \Theta \times \mathbb{T}$, *obey the infinitely-divisible law*

$$log E\, e^{i\lambda\mu(\Delta)} = \iint_\Delta \Big[-\frac{\lambda^2}{2}\sigma^2(\theta,t) \tag{1.17}$$
$$+ \int_{\mathbb{R}\setminus[0]} (e^{i\lambda x} - 1 - i\lambda x)\, L(dx,\theta,t)\Big] \times M(d\theta\, dt), \;\; \lambda \in \mathbb{R}$$

with

$$\sigma^2(\theta,t) + \int_{\mathbb{R}\setminus[0]} x^2\, L(dx,\theta,t) \equiv 1\,.$$

Proof. The values $\mu(\Delta)$ are infinitely-divisible random variables - cf. (1.5). Hence, according to the de Finetti [8] and Kolmogorov [34] law (see [33,36]), we have

$$log E\, e^{i\lambda\mu(\Delta)} = -\frac{\lambda^2}{2}\sigma_\Delta^2 + \int_{\mathbb{R}\setminus[0]} (e^{i\lambda x} - 1 - i\lambda x)\, L_\Delta(dx)\,, \quad \lambda \in \mathbb{R} \tag{1.18}$$

with

$$\sigma_\Delta^2 + \int_{\mathbb{R}\setminus[0]} x^2\, L_\Delta(dx) \equiv M(\Delta)\,, \qquad \Delta \subseteq \Theta \times \mathbb{T}\,, \tag{1.19}$$

where the constant σ_Δ^2 and the Borel measure $L_\Delta = L_\Delta(B)$, $B \subseteq \mathbb{R} \setminus [0]$, depend on $\Delta \subseteq \Theta \times \mathbb{T}$ as additive set-functions. Taking the relationship (1.19) with the variance measure $M = M(\Delta)$, $\Delta \subseteq \Theta \times \mathbb{T}$, into account we can see that σ^2_Δ, L_Δ admit the integral representations

$$\sigma^2_\Delta = \iint_\Delta \sigma^2(\theta,t)\, M(d\theta\, dt)\,, \qquad L_\Delta(B) = \iint_\Delta L(B,\theta,t)\, M(d\theta\, dt)\,.$$

The integrands $\sigma^2(\theta,t)$ and $L(B,\theta,t)$, $(\theta,t) \in \Theta \times \mathbb{T}$, are elements of the standard L_1-space (with respect to the measure M). Moreover they are *additive* in their dependence on the Borel sets $B \subseteq \mathbb{R} \setminus [0]$. The above stochastic function $L(B,\theta,t)$, $B \subseteq \mathbb{R} \setminus [0]$, $(\theta,t) \in \Theta \times \mathbb{T}$, can be modified on a set of *zero* M-measure in a way that yields a lifting to a *new* equivalent integrand such that, whatever $(\theta,t) \in \Theta \times \mathbb{T}$ be, the set-function $L(B,\theta,t)$, $B \subseteq \mathbb{R} \setminus [0]$, is a *measure* on $\mathbb{R} \setminus [0]$ (see e.g. [24]). So, the probability law (1.18) admits a representation in the form (1.17). □

Example 1.1. The *Gaussian stochastic measure* μ, having Gaussian random variables as values $\mu(\Delta)$, $\Delta \subseteq \Theta \times \mathbb{T}$, corresponds to the probability law (1.17) with $\sigma^2 \equiv 1$ and $L \equiv 0$.

Example 1.2. The *Poisson (centred) stochastic measure* μ, having values

$$\mu(\Delta) = \nu(\Delta) - E\nu(\Delta)\,, \qquad \Delta \subseteq \Theta \times \mathbb{T}\,,$$

where $\nu(\Delta)$, $\Delta \subseteq \Theta \times \mathbb{T}$, are Poisson random variables, corresponds to the probability law (1.17) with $\sigma^2 \equiv 0$ and L concentrated the point $x = 1$ in $\mathbb{R} \setminus [0]$ with unit mass, i.e.

$$logE\, e^{i\lambda\mu(\Delta)} = (e^{i\lambda} - 1 - i\lambda)\, M(\Delta)\,, \qquad \lambda \in \mathbb{R}.$$

We recall that the non-negative additive set-function $\nu = \nu(\Delta)$, $\Delta \subseteq \Theta \times \mathbb{T}$, has values $\nu(\Delta) \in L_2(\Omega)$ which are integer random variables $\nu(\Delta) = \nu(\Delta,\omega)$, $\omega \in \Omega$. In the case Θ is a *complete separable metric space* equipped with the σ-algebra of its Borel sets, we have that $\nu = \nu(\Delta)$, $\Delta \in \Theta \times \mathbb{T}$, *admits an equivalent modification*, which is referred to as

$$\nu = \nu(\cdot,\omega),\ \omega \in \Omega\,:\, E\nu = M\,,$$

with values $\nu(\Delta) = \nu(\Delta,\omega)$, $\omega \in \Omega$, *representing the measures* $\nu(\cdot,\omega) = \nu(\Delta,\omega)$, $\Delta \subseteq \Theta \times \mathbb{T}$, *depending on* $\omega \in \Omega$ *as parameter.* For some (which can be any) sequence of increasing sets $\Theta_n \times \mathbb{T}$, such that $M(\Theta_n \times \mathbb{T}) < \infty$, $n = 1, 2, \ldots$, and $\lim_{n\to\infty} \Theta_n \times \mathbb{T} = \Theta \times \mathbb{T}$, the measures $\nu(\cdot,\omega)$ can be defined in a way that $\nu(\Theta_n \times \mathbb{T},\omega) < \infty$, $n = 1, 2, \ldots$, and all the finite values $\nu(\Delta,\omega)$ are integers. So, the measures $\nu(\cdot,\omega)$ are *purely discrete*, concentrated on the corresponding atoms $(\theta_\omega, t_\omega) \in \Theta \times \mathbb{T}$. In particular for $\Delta \subseteq \Theta \times \mathbb{T}$ with $M(\Delta) < \infty$, the possibility of having one atom $(\theta_\omega, t_\omega) \subseteq \Delta$ with $\nu(\theta_\omega, t_\omega, \omega) > 1$ or of having a couple of atoms in Δ with the *same* time components occur with zero probability. To explain, the limit

$$\lim_{n\to\infty} P\Big\{ \max_{(s,u]} \mu(\Delta \cap (\Theta \times (s,u])) > 1 \Big\} = 0$$

holds true for the nth-series of partitions of $\mathbb{T}$ - cf. (1.1), $\mathbb{T} = \sum(s, u]$: $\max(s - u) \to 0$, $n \to \infty$, and the corresponding partitions of $\Theta \times \mathbb{T}$:

$$\Delta = \sum \Delta \cap (\Theta \times (s, u]), \qquad n = 1, 2, \ldots,$$

- cf. (1.2)-(1.5). Hence, *all the atoms* $(\theta_\omega, t_\omega)$ *are in one-to-one correspondence*

$$\Theta \times \mathbb{T} \ni (\theta_\omega, t_\omega) \Longleftrightarrow t_\omega \in \mathbb{T}$$

with their time components and we have

$$\nu(\theta_\omega, t_\omega, \omega) \equiv 1.$$

1.5 Non-anticipating and Predictable Stochastic Functions

We write $L_2(\Theta \times \mathbb{T} \times \Omega)$ for the standard (complex) space of the stochastic functions $\varphi = \varphi(\theta, t)$, $(\theta, t) \in \Theta \times \mathbb{T}$, with values $\varphi(\theta, t) = \varphi(\theta, t, \omega)$, $\omega \in \Omega$, in $L_2(\Omega)$:

$$\|\varphi\|_{L_2} = \left(\iiint_{\Theta \times \mathbb{T} \times \Omega} |\varphi|^2 M(d\theta\, dt) \times P(d\omega) \right)^{1/2}$$
$$= \left(\iint_{\Theta \times \mathbb{T}} \|\varphi\|^2 M(d\theta\, dt) \right)^{1/2}. \qquad (1.20)$$

Cf. (1.7). Here, $P = P(\mathcal{A})$, $\mathcal{A} \in \mathfrak{A}$, is the probability on the σ-algebra $\mathfrak{A} = \mathfrak{A}_{\Theta \times \mathbb{T}}$ of all events $\mathcal{A} \subseteq \Omega$ and the product-measure $M \times P$ on $\Theta \times \mathbb{T} \times \Omega$ is considered on the σ-algebra generated by the product-sets

$$\Delta \times \mathcal{A} : \quad \Delta \subseteq \Theta \times (s, u], \quad \mathcal{A} \in \mathfrak{A}. \qquad (1.21)$$

The component $M = M(\Delta)$, $\Delta \subseteq \Theta \times \mathbb{T}$, satisfies (1.2)-(1.4). For the product-sets (1.21) we have

$$\iiint_{\Delta \times \mathcal{A}} M(d\theta\, dt) \times P(d\omega) = M(\Delta) \cdot P(\mathcal{A}).$$

We say that φ is a *simple function* if it admits the representation

$$\varphi = \sum \varphi \cdot 1_\Delta$$

where the sum is taken on some finite series of disjoint sets $\Delta \subseteq \Theta \times \mathbb{T}$: $M(\Delta) < \infty$, and the indicated element $\varphi \in L_2(\Omega)$ in each component $\varphi \cdot 1_\Delta$ is the value of the simple function on Δ. Note that the simple functions represented by the indicators

$$1_{\Delta \times \mathcal{A}} = 1_{\mathcal{A}} \cdot 1_\Delta : \quad \Delta \subseteq \Theta \times (s, u], \quad \mathcal{A} \subseteq \Omega,$$

with Δ belonging to the partitions of $\Theta \times \mathbb{T}$ - see (1.3), constitute a complete system in $L_2(\Theta \times \mathbb{T} \times \Omega)$. Cf. (1.3)-(1.6) and (1.21).

Let us turn to the σ-algebras

$$\mathfrak{A}_t, \quad t \in \mathbb{T},$$

characterized in (1.15) which represent the flow of events in the course of time. The *non-anticipating simple function* φ is characterized by the representation

$$\varphi = \sum \varphi \cdot 1_\Delta \tag{1.22}$$

where each component $\varphi \cdot 1_\Delta$ has $\Delta \subseteq \Theta \times (s, u]$ and the indicated value $\varphi \in L_2(\Omega)$ on Δ is an $\mathfrak{A}_s$-measurable random variable.

In general, we refer to $\varphi \in L_2(\Theta \times \mathbb{T} \times \Omega)$ as a *non-anticipating function* if its values $\varphi(\theta, t) \in L_2(\Omega)$ in the course of time are determined by the "past" events. To be more precise, for any $t \in \mathbb{T}$, the random variable $\varphi(\theta, t)$ is measurable with respect to the σ-algebra $\mathfrak{A}_t$.

Let us consider also the functions $\varphi \in L_2(\Theta \times \mathbb{T} \times \Omega)$ measurable with respect to the σ-algebra generated by the product-sets

$$\Delta \times \mathcal{A} : \quad \Delta \subseteq \Theta \times (s, u], \quad \mathcal{A} \in \mathfrak{A}_s \tag{1.23}$$

- cf. (1.21). Following the common terminology (see e.g. [11]), we refer to the above functions φ as the *predictable functions* and the σ-algebra generated by the sets (1.23) as the *predictable σ-algebra.* Note that all the *non-anticipating simple* functions are predictable. We remark that all the predictable functions are non-anticipating. The following result details the study of the converse relationship. Note that this coming result holds thanks to the left-continuity of the flow of σ-algebras $\mathfrak{A}_t$, $t \in \mathbb{T}$ - cf. (1.16).

Theorem 1.3. *Any non-anticipating function $\varphi \in L_2(\Theta \times \mathbb{T} \times \Omega)$ can be identified with the corresponding predictable function given by the limit*

$$\varphi = \lim_{n\to\infty} \varphi^{(n)}, \ i.e. \quad \left\| \varphi - \varphi^{(n)} \right\|_{L_2} \longrightarrow 0, \qquad n \to \infty,$$

of the non-anticipating simple functions $\varphi^{(n)}$, $n = 1, 2, \ldots$, defined along the nth-series of sets $\Delta \subseteq \Theta \times (s, u]$ of the partitions of $\Theta \times \mathbb{T}$ - cf. (1.3), *as*

$$\varphi^{(n)} = \sum \varphi^{(n)} \cdot 1_\Delta, \ with \ \varphi^{(n)} = \frac{1}{M(\Delta)} E\left(\iint_\Delta \varphi \, M(d\theta\, dt) \,|\mathfrak{A}_s \right). \tag{1.24}$$

Proof. At first, let us show that *any* function $\varphi \in L_2(\Theta \times \mathbb{T} \times \Omega)$ is the limit $\varphi = \lim_{n\to\infty} \varphi^{(n)}$ of simple approximations of the form

$$\varphi^{(n)} = \sum \varphi^{(n)} \cdot 1_\Delta \text{ with } \varphi^{(n)} = \frac{1}{M(\Delta)} \iint_\Delta \varphi \, M(d\theta\, dt), \tag{1.25}$$

where the sum is taken on the sets $\Delta \subseteq \Theta \times (s, u]$ of the nth-series of the partitions of $\Theta \times \mathbb{T}$. For $\varphi \in L_2(\Theta \times \mathbb{T} \times \Omega)$, there are some simple functions

$$\psi^{(n)} = \sum \psi^{(n)} 1_\Delta \text{ with } \Delta \subseteq \Theta \times (s, u],$$

such that

$$\varphi = \lim_{n\to\infty} \psi^{(n)}, \text{ i.e. } \quad \|\varphi - \psi^{(n)}\|_{L_2} \longrightarrow 0\,, \qquad n \to \infty\,.$$

For the indicated values $\varphi^{(n)}$, $\psi^{(n)}$ on the nth-series sets Δ, we have

$$\begin{aligned} \|\varphi^{(n)} - \psi^{(n)}\|^2 &= \Big\|\frac{1}{M(\Delta)} \iint_\Delta (\varphi - \psi^{(n)})\, M(d\theta\, dt)\Big\|^2 \\ &\le \frac{1}{M(\Delta)} \iint_\Delta \|\varphi - \psi^{(n)}\|^2\, M(d\theta\, dt)\,. \end{aligned} \tag{1.26}$$

So, we also have

$$\|\varphi^{(n)} - \psi^{(n)}\|_{L_2}^2 \le \sum \iint_\Delta \|\varphi - \psi^{(n)}\|^2\, M(d\theta\, dt) \le \|\varphi - \psi^{(n)}\|_{L_2}^2$$

which implies that

$$\|\varphi - \varphi^{(n)}\|_{L_2} \le 2\,\|\varphi - \psi^{(n)}\|_{L_2} \longrightarrow 0\,, \qquad n \to \infty\,.$$

Next, let us turn to the non-anticipating functions φ such that, on some (which can be any) n^{th}-series sets $\Delta \subseteq \Theta \times (s,u]$ of the considered (1.3)-partitions, the values $\varphi(\theta,t)$ for $(\theta,t) \in \Delta$ are measurable with respect to the σ-algebras $\mathfrak{A}_s$. For these functions, when n is large enough ($n \to \infty$), the approximations (1.24) are identical to the approximations (1.25). Any non-anticipating function φ admits its approximations in $L_2(\Theta \times \mathbb{T} \times \Omega)$ by the above type functions. To explain, for any $(\theta,t) \in \Theta \times \mathbb{T}$ and any set $\Delta \subseteq \Theta \times (s,u]$ of the nth-series of partitions of $\Theta \times \mathbb{T}$ such that $(\theta,t) \in \Delta$, the corresponding increasing σ-algebras $\mathfrak{A}_s$ have limit $\lim_{n\to\infty} \mathfrak{A}_s = \mathfrak{A}_t$. Thus

$$\varphi(\theta,t) = E\Big(\varphi(\theta,t)|\mathfrak{A}_t\Big) = \lim_{n\to\infty} E\Big(\varphi(\theta,t)\,|\,\mathfrak{A}_s\Big)$$

in $L_2(\Omega)$ and

$$\varphi = \lim_{n\to\infty} \sum E\Big(\varphi(\theta,t)|\mathfrak{A}_s\Big)\cdot \mathbf{1}_\Delta$$

in $L_2(\Theta \times \mathbb{T} \times \Omega)$. Cf. (1.3)-(1.5) and (1.16). So, we can see that φ is the limit $\varphi = \lim_{n\to\infty} \psi^{(n)}$ in $L_2(\Theta \times \mathbb{T} \times \Omega)$ of *some* non-anticipating simple functions $\psi^{(n)}$, $n = 1, 2, \ldots$, of the form

$$\psi^{(n)} = \sum \psi^{(n)} \cdot \mathbf{1}_\Delta$$

related to the sets $\Delta \subseteq \Theta \times (s,u]$ of the nth-series of the partitions of $\Theta \times \mathbb{T}$ - see (1.3). Hence, by the same arguments applied for the approximations (1.25), we can conclude that φ is the limit $\varphi = \lim_{n\to\infty} \varphi^{(n)}$ of the non-anticipating simple functions (1.24). □

Now, let us consider the simple functions $\varphi = \sum \varphi \cdot 1_\Delta$ where for each component $\varphi \cdot 1_\Delta$ the indicated value φ on $\Delta \subseteq \Theta \times \mathbb{T}$ is measurable with respect to the corresponding σ-algebra

$$\mathfrak{A}_{]\Delta[} : \quad]\Delta[= \Theta \times \mathbb{T} \setminus \Delta , \tag{1.27}$$

generated by the stochastic measure μ over the complement set $]\Delta[$ to Δ - cf. (1.11). We have the following result - see [18].

Theorem 1.4. *Any function $\varphi \in L_2(\Theta \times \mathbb{T} \times \Omega)$ is the limit*

$$\varphi = \lim_{n\to\infty} \varphi^{(n)}, \ i.e. \quad \left\|\varphi - \varphi^{(n)}\right\|_{L_2} \longrightarrow 0 , \quad n \to \infty,$$

of the simple functions $\varphi^{(n)}$, $n = 1, 2, \ldots$, defined along the sets of the nth-series of the partitions (1.3) *as*

$$\varphi^{(n)} = \sum \varphi^{(n)} \cdot 1_\Delta \ \textit{with} \quad \varphi^{(n)} = \frac{1}{M(\Delta)} E\left(\iint_\Delta \varphi \, M(d\theta\, dt) \,\Big|\, \mathfrak{A}_{]\Delta[}\right). \tag{1.28}$$

Proof. The proof uses the same arguments as in the proof of Theorem 1.3. Here, to explain, we just note that for any $(\theta, t) \in \Theta \times \mathbb{T}$ and any set Δ of the nth-series of the partitions of $\Theta \times \mathbb{T}$ such that $(\theta, t) \in \Delta$, we have

$$\varphi(\theta, t) = \lim_{n\to\infty} E\big(\varphi(\theta, t)|\mathfrak{A}_{]\Delta[}\big).$$

In fact the increasing σ-algebras $\mathfrak{A}_{]\Delta[}$ have limit $\lim_{n\to\infty} \mathfrak{A}_{]\Delta[} = \mathfrak{A}$, where $\mathfrak{A} = \mathfrak{A}_{\Theta\times\mathbb{T}}$ represents *all* the events in Ω. Cf. (1.3)-(1.5) and (1.13). □

2 The Itô Non-anticipating Integral

2.1 A General Definition and Related Properties

The Itô integration scheme [26] (see also e.g. [39]) can be applied to the non-anticipating integration on the general space-time product $\Theta \times \mathbb{T}$ with respect to the stochastic measure $\mu = \mu(d\theta\, dt)$, $(\theta, t) \in \Theta \times \mathbb{T}$, of type (1.9): $E\mu = 0$, $E\mu^2 = M$. In particular, it can be applied in the modeling of stochastic processes of the form

$$\xi(t) = \iint_{\Theta\times(0,t]} \varphi \, \mu(d\theta\, ds) , \qquad t \in \mathbb{T} . \tag{2.1}$$

The term *non-anticipating* is referred to the family of σ-algebras

$$\mathfrak{A}_t, \quad t \in \mathbb{T},$$

which represent the flow of events in time - cf. (1.15). The integrands φ in (2.1) are the *non-anticipating* stochastic functions treated as elements of the functional space $L_2(\Theta \times \mathbb{T} \times \Omega)$ - cf. (1.20). The non-anticipating functions $\varphi = \varphi(\theta, t)$, $(\theta, t) \in \Theta \times \mathbb{T}$, with $\varphi(\theta, t) \in L_2(\Omega)$ $\mathfrak{A}_t$-measurable, for any (θ, t), constitute the subspace

$$L_2^I(\Theta \times \mathbb{T} \times \Omega) \subseteq L_2(\Theta \times \mathbb{T} \times \Omega) \tag{2.2}$$

of all the integrands. To be more precise, this subspace is the closure of all the non-anticipating simple functions (1.24). Cf. Theorem 1.3.

Let us consider non-anticipating *simple* functions

$$\varphi = \sum \varphi \cdot 1_\Delta$$

where the sum is taken on a finite series of disjoint sets $\Delta \subseteq \Theta \times (s, u]$ $M(\Delta) < \infty$, and, for each component $\varphi \cdot 1_\Delta$, the $\mathfrak{A}_s$-measurable random variable $\varphi \in L_2(\Omega)$ is the value of μ on the indicated set Δ. The integration results in the random variable

$$I\varphi = \sum \varphi \cdot \mu(\Delta) \tag{2.3}$$

belonging to $L_2(\Omega)$. And here we have

$$\|I\varphi\| = \|\varphi\|_{L_2} .$$

Cf. (1.7) and (1.20). So, the integration formula (2.3) defines the *isometric* linear operator I:

$$L_2^I(\Theta \times \mathbb{T} \times \Omega) \ni \varphi \Longrightarrow I\varphi \in L_2(\Omega)$$

on the domain of all the non-anticipating simple functions, dense in $L_2^I(\Theta \times \mathbb{T} \times \Omega)$. The standard extension of this linear operator on $L_2^I(\Theta \times \mathbb{T} \times \Omega)$ is the *non-anticipating integral*

$$I\varphi = \iint_{\Theta \times \mathbb{T}} \varphi \, \mu(d\theta \, dt) .$$

Namely, for any $\varphi \in L_2^I(\Theta \times \mathbb{T} \times \Omega)$, i.e. the limit

$$\varphi = \lim_{n \to \infty} \varphi^{(n)} \text{ i.e. } \left\|\varphi - \varphi^{(n)}\right\|_{L_2} \longrightarrow 0 , \quad n \to \infty,$$

of the non-anticipating simple functions $\varphi^{(n)}$, $n = 1, 2, \ldots$, we have

$$I\varphi = \lim_{n \to \infty} I\varphi^{(n)}, \text{ i.e. } \|I\varphi - I\varphi^{(n)}\| \longrightarrow 0 \; n \to \infty. \tag{2.4}$$

In particular, the integration can be carried through via the standard non-anticipating simple approximations of type (1.24).

For all the integrands, the integral

$$\iint_\Delta \varphi \, \mu(d\theta \, dt) := \iint_{\Theta \times \mathbb{T}} (\varphi \cdot 1_\Delta) \, \mu(d\theta \, dt) , \quad \Delta \subseteq \Theta \times \mathbb{T}, \tag{2.5}$$

is well-defined. Cf. (2.3)-(2.4).

In this line, all the functions φ of form

$$\varphi = \sum_k \varphi_k = \sum_k \varphi 1_{\Theta_k \times \mathbb{T}} \text{ where } \sum_k \Theta_k \times \mathbb{T} = \Theta \times \mathbb{T},$$

with the components $\varphi_k = \varphi 1_{\Theta_k \times \mathbb{T}}$, $k = 1, \dots, K$ $(K \le \infty)$, represent the integrands with respect to the measure μ as integrator on $\Theta_k \times \mathbb{T}$ - cf. (1.10).

We remark that

$$E\left(\iint_\Delta \varphi\, \mu(d\theta\, dt) \Big| \mathfrak{A}_s \right) = 0 \tag{2.6}$$

and

$$E\left(\iint_\Delta \varphi\, \mu(d\theta\, dt) \cdot \iint_{\Delta'} \varphi'\, \mu(d\theta\, dt) \Big| \mathfrak{A}_s \right) = \iint_{\Delta \cap \Delta'} E\left(\varphi \cdot \varphi' | \mathfrak{A}_s\right) M(d\theta\, dt) \tag{2.7}$$

for the integrands φ, φ' and Δ, $\Delta' \subseteq \Theta \times (s, T]$, $0 \le s < T$.

The non-anticipating integration on general product spaces with the time component $\mathbb{T}$ was considered in [15]. With respect to a particular generalization of the Itô stochastic integral on the product space of the form $\mathbb{T} \times \mathbb{T}$ we can refer for example to [5].

Example 2.1. Let us consider the *optional* (*stopping*) time τ:

$$\{\tau \le t\} \in \mathfrak{A}_t, \quad t \in \mathbb{T},$$

and the *optional* σ-algebra $\mathfrak{A}_\tau$ of the events $\mathcal{A} \subseteq \Omega$ such that

$$\mathcal{A} \cap \{\tau \le t\} \in \mathfrak{A}_t, \quad t \in \mathbb{T}.$$

Thanks to the *right-continuity* of $\mathfrak{A}_t$, $t \in \mathbb{T}$, we have that the stochastic functions

$$\xi \cdot 1_{(\tau, T]} \cdot \varphi \in L_2(\Theta \times \mathbb{T} \times \Omega)$$

are integrands whatever $\mathfrak{A}_t$-measurable random variables ξ and integrands φ be applied. Moreover we have

$$\iint_{\Theta \times \mathbb{T}} \xi \cdot 1_{(\tau, T]} \varphi\, \mu(d\theta dt) = \xi \cdot \iint_{\Theta \times \mathbb{T}} 1_{(\tau, T]} \varphi\, \mu(d\theta dt).$$

2.2 The Stochastic Poisson Integral

As continuation of Example 1.2, we specify the Itô non-anticipating integral with respect to the *Poisson (centred) stochastic measure* $\mu := \nu - E\nu$ treated through its Poisson components

$$\nu = \nu(\cdot, \omega), \ \ \omega \in \Omega : \quad E\nu = M\,.$$

Here, the pure discrete measures

$$\nu(\cdot, \omega) = \nu(\Delta, \omega)\,, \qquad \Delta \subseteq \Theta \times \mathbb{T}\,, \tag{2.8}$$

which depend on $\omega \in \Omega$ as parameter, are concentrated on the atoms

$$(\theta_\omega, t_\omega): \quad \nu(\theta_\omega, t_\omega, \omega) \equiv 1.$$

All these atoms are in one-to-one correspondence

$$\Theta \times \mathbb{T} \ni (\theta_\omega, t_\omega) \Longleftrightarrow t_\omega \in \mathbb{T}$$

with their time components.

The integrands φ are *predictable* functions in $L_2(\Theta \times \mathbb{T} \times \Omega)$. We assume that they satisfy

$$\iint_{\Theta\times\mathbb{T}} (E\,|\varphi|)\, M(d\theta\, dt) < \infty. \tag{2.9}$$

Note that (2.16) holds for all the integrands in the case M is a finite measure - cf. (1.20). Now let us consider the Poisson stochastic measure

$$\nu = \nu(\Delta, \omega)\,, \qquad \Delta \subseteq \Theta \times \mathbb{T} \qquad (\omega \in \Omega),$$

and the product-measure $\nu \times P$ on $\Theta \times \mathbb{T} \times \Omega$ with values

$$(\nu \times P)(\Delta \times \mathcal{A}) = \iiint_{\Delta\times\mathcal{A}} \nu(d\theta\, dt, \omega) \times P(d\omega)$$

on the product-sets $\Delta \times \mathcal{A}$: $\Delta \subseteq \Theta \times \mathbb{T}$, $\mathcal{A} \subseteq \Omega$ - cf. (1.21). In particular we can see that

$$\nu \times P \equiv M \times P$$

on the predictable σ-algebra, i.e. the σ-algebra generated by the product-sets $\Delta \times \mathcal{A}$: $\Delta \subseteq \Theta \times (s, u]$, $\mathcal{A} \subseteq \mathfrak{A}_s$ - cf. (1.23). To explain, we have

$$\begin{aligned}
&\iiint_{\Delta\times\mathcal{A}} \nu(d\theta\, dt, \omega) \times P(d\omega) \\
&\quad = E\left(1_{\mathcal{A}} \times \nu(\Delta)\right) = E\, 1_{\mathcal{A}} \times E\, \nu(\Delta) = \iiint_{\Delta\times\mathcal{A}} M(d\theta\, dt) \times P(d\omega)
\end{aligned}$$

since the values $\nu(\Delta)$: $\Delta \subseteq \Theta \times (s, u]$, are *independent* from the events $\mathcal{A} \in \mathfrak{A}_s$. For the predictable function φ which, we recall, is a function measurable with respect to the predictable σ-algebra, the condition (2.16) says that

$$\iiint_{\Theta\times\mathbb{T}\times\Omega} |\varphi|\, \nu(d\theta\, dt, \omega) \times P(d\omega) = \iint_{\Theta\times\mathbb{T}} (E|\varphi|)\, M(d\theta\, dt) < \infty\,.$$

Accordingly, the *stochastic Poisson integral*

$$\iint_{\Delta} \varphi\, \nu(d\theta\, dt) := \iint_{\Delta} \varphi(\cdot, \omega)\, \nu(d\theta\, dt, \omega)\,, \qquad \omega \in \Omega\,,$$

is well-defined via the realizations (trajectories) $\varphi(\cdot, \omega) = \varphi(\theta, t, \omega)$, $(\theta, t) \in \Theta \times \mathbb{T}$, integrable with respect to the measures $\nu(\cdot, \omega)$ for almost all $\omega \in \Omega$.

In this scheme, we can see that the Itô non-anticipating integral with respect to the integrator $\mu = \nu - M$ is related to the stochastic Poisson integral in the following way:

$$\iint_\Delta \varphi\, \mu(d\theta\, dt) = \iint_\Delta \varphi\, \nu(d\theta\, dt) - \iint_\Delta \varphi\, M(d\theta\, dt)\,, \qquad \Delta \subseteq \Theta \times \mathbb{T}\,. \tag{2.10}$$

This is obvious for the non-anticipating simple functions φ and it is true in general via the limit

$$\varphi = \lim_{n\to\infty} \varphi^{(n)}, \text{ i.e. } \quad \left\|\varphi - \varphi^{(n)}\right\|_{L_2} \longrightarrow 0\,, \qquad n \to \infty\,,$$

of the non-anticipating simple functions $\varphi^{(n)}$, $n = 1, 2, \ldots$, thanks to the identity

$$\iiint_{\Theta\times\mathbb{T}\times\Omega} \left|\varphi - \varphi^{(n)}\right|^2 M(d\theta\, dt) \times P(d\omega)$$
$$= \iiint_{\Theta\times\mathbb{T}\times\Omega} \left|\varphi - \varphi^{(n)}\right|^2 \nu(d\theta\, dt, \omega) \times P(d\omega).$$

With respect to the representation (2.10), we remark that the stochastic Poisson integral is actually

$$\iint_\Delta \varphi\, \nu(d\theta\, dt) = \sum_{(\theta_\omega, t_\omega)\in\Delta} \varphi(\theta_\omega, t_\omega, \omega)\,, \qquad \omega \in \Omega, \tag{2.11}$$

and the above stochastic series converges absolutely with

$$\int_\Omega \left[\sum_{(\theta_\omega, t_\omega)\in\Delta} \left|\varphi(\theta_\omega, t_\omega, \omega)\right|\right] P(d\omega) = \iint_\Delta (E|\varphi|)\, M(d\theta\, dt) < \infty\,.$$

Cf. (2.8)-(2.16).

2.3 The Jumping Stochastic Processes

To continue the scheme (2.8)-(2.11), we apply it to the cadlag stochastic processes

$$\xi(t) = \iint_{\Theta\times(0,t]} \varphi \mu(d\theta\, ds)\,, \qquad t \in \mathbb{T}\,,$$

- cf. (2.1), where $\varphi = \varphi(\theta, t, \omega)$, $(\theta, t, \omega) \in \Theta \times \mathbb{T} \times \Omega$, are the real predictable integrands with respect to the Poisson (centred) stochastic measure μ of the type $E\mu = 0$, $E\mu^2 = M$, treated as $\mu = \nu - E\nu$ through its Poisson component $\nu = \nu(\cdot, \omega)$, $\omega \in \Omega$: $E\nu = M$. Recall that the involved pure discrete measures $\nu(\cdot, \omega) = \nu(\Delta, \omega)$, $\Delta \subseteq \Theta \times \mathbb{T}$, have atoms $(\theta_\omega, t_\omega) \in \Theta \times \mathbb{T}$ which are in one-to-one correspondence $(\theta_\omega, t_\omega) \Leftrightarrow t_\omega$ with the times $t_\omega \in \mathbb{T}$. Hence we can see that *all* the jumps of the realizations (trajectories) of the above process $\xi(t)$, $t \in \mathbb{T}$:

$$\xi(t,\omega) = \sum_{0<t_\omega \leq t} \varphi(\theta_\omega, t_\omega, \omega) - \iint_{\Theta\times(0,t]} \varphi(\cdot,\omega)\, M(d\theta\, dt)\,, \qquad t \in \mathbb{T} \quad (\omega \in \Omega)\,, \tag{2.12}$$

are

$$\rho_\omega := \xi(t_\omega,\omega) - \xi(t_\omega - 0,\omega) \equiv \varphi(\theta_\omega, t_\omega, \omega)\,, \qquad (\theta_\omega, t_\omega) \in \Theta \times \mathbb{T}\,.$$

Cf. (2.10)-(2.11). Accordingly, whatever real function

$$F(x,\cdot) = F(x,\theta,t,\omega)\,, \qquad (x,\theta,t,\omega) \in \mathbb{R} \times \Theta \times \mathbb{T} \times \Omega\,,$$

be considered such that $F(\varphi,\cdot)$ is a predictable integrand with

$$\iint_{\Theta\times\mathbb{T}} \Big(E\big|F(\varphi,\cdot)\big|\Big)\, M(d\theta\, dt) < \infty$$

- cf. (2.16), we obtain that, for *all* the jumps $\rho_\omega := \xi(t_\omega,\omega) - \xi(t_\omega - 0,\omega)$ of the trajectories (2.17), the corresponding trajectories

$$\eta(t,\omega) = \sum_{0<t_\omega \leq t} F(\rho_\omega, \theta_\omega, t_\omega, \omega) - \iint_{\Theta\times(0,t]} F(\varphi,\cdot)\, M(d\theta\, dt), \quad t \in \mathbb{T} \quad (\omega \in \Omega), \tag{2.13}$$

represent the stochastic process

$$\eta(t) = \iint_{\Theta\times(0,t]} F(\varphi,\cdot)\, \mu(d\theta\, ds), \quad t \in \mathbb{T}.$$

Now let φ and $F(\varphi,\cdot) = F(\varphi(\theta,t),\theta,t)$, $(\theta,t) \in \Theta \times \mathbb{T}$, be *deterministic* real functions. Then $\eta(t)$, $t \in \mathbb{T}$, here above is the process with independent increments characterized by the infinitely-divisible probability law

$$\log e^{i\lambda\eta(t)} = \iint_{\Theta\times(0,t]} (e^{i\lambda F(\varphi,\cdot)} - 1 - i\lambda F(\varphi,\cdot))\, M(d\theta\, ds), \quad \lambda \in \mathbb{R} \quad (t \in \mathbb{T}).$$

Example 2.2. In relation to the probability law (1.17), let us turn to the Poisson (centred) stochastic measure μ on the space-time product $(\mathbb{R} \setminus [0]) \times \Theta \times \mathbb{T}$:

$$E\mu = 0\,, \qquad E\mu^2 = L \times M\,,$$

with the variance represented by the standard product-measure $L \times M$ on $(\mathbb{R} \setminus [0]) \times \Theta \times \mathbb{T}$ with the component $L = L(B,\theta,t)$, $B \subseteq \mathbb{R} \setminus [0]$ such that

$$\int_{\mathbb{R}\setminus[0]} x^2 L(dx,\theta,t) \leq 1\,.$$

Assuming that Θ is a complete separable metric space, we can apply the scheme generally described in (2.8)-(2.13) and consider the stochastic measure $\mu = \nu - E\nu$, with the Poisson component

$$\nu = \nu(\cdot, \omega), \qquad \omega \in \Omega : E\nu = L \times M$$

on $(\mathbb{R} \setminus [0]) \times \Theta \times \mathbb{T}$ represented by the pure discrete measures $\nu(\cdot, \omega)$ having atoms

$$(x_\omega, \theta_\omega, \omega) \in (\mathbb{R} \setminus [0]) \times \Theta \times \mathbb{T}.$$

For the sets

$$\Delta \subseteq \Theta \times (s, u] : \quad M(\Delta) < \infty$$

in $\Theta \times \mathbb{T}$ and

$$B \subseteq \{|x| > r\} : \quad r > \epsilon > 0$$

in $\mathbb{R} \setminus [0]$, let us consider the function

$$\varphi = 1_\Delta(\theta, t) x\, 1_{\{|x| > \epsilon\}}, \qquad (x, \theta, t) \in (\mathbb{R} \setminus [0]) \times \Theta \times \mathbb{T},$$

and the integrand

$$F(\varphi, \cdot) := 1_B(\varphi) \equiv 1_{B \times \Delta}$$

on $(\mathbb{R} \setminus [0]) \times \Theta \times \mathbb{T}$. Let

$$\xi(t) = \iiint_{(\mathbb{R}\setminus[0]) \times \Theta \times (0,t]} 1_\Delta x\, 1_{\{|x| > \epsilon\}} \mu(dx\, d\theta\, ds), \qquad t \in \mathbb{T},$$

be the corresponding cadlag process. We can see that

$$\mu(B \times \Delta) = \sum_{s < t_\omega \le u} 1_B(\rho_\omega) - \iint_\Delta L(B, \theta, t) \times M(d\theta\, dt), \tag{2.14}$$

where ρ_ω are the jumps of $\xi(t, \omega)$, $t \in \mathbb{T}$. Note that here the actually involved jumps

$$\rho_\omega = x_\omega : (x_\omega, \theta_\omega, t_\omega) \in B \times \Delta$$

are the same for all $\epsilon : 0 < \epsilon < r$. Cf. (2.13). Accordingly, formula (2.14) holds true for any $r > 0$ and $\epsilon = 0$, i.e. for

$$\rho_\omega = \xi(t_\omega, \omega) - \xi(t_\omega - 0, \omega)$$

as the jumps of the cadlag process

$$\xi(t) = \iiint_{(\mathbb{R}\setminus[0]) \times \Theta \times (0,t]} 1_\Delta x\, \mu(dx\, d\theta\, ds), \qquad t \in \mathbb{T}. \tag{2.15}$$

In fact the component

$$\xi_\epsilon(t) := \iiint_{(\mathbb{R}\setminus[0]) \times \Theta \times (0,t]} 1_\Delta x\, 1_{\{|x| \le \epsilon\}} \mu(dx\, d\theta\, ds), \qquad t \in \mathbb{T},$$

is negligible, for $\epsilon \to 0$, in the above process $\xi(t)$, $t \in \mathbb{T}$, with

$$\xi(t) = \iiint_{(\mathbb{R}\setminus[0])\times\Theta\times(0,t]} 1_\Delta x\, 1_{\{|x|>\epsilon\}}\mu(dx\, d\theta\, ds) + \iiint_{(\mathbb{R}\setminus[0])\times\Theta\times(0,t]} 1_\Delta x\, 1_{\{|x|\le\epsilon\}}\mu(dx\, d\theta\, dt).$$

2.4 Gaussian-Poisson Stochastic Measures

Let us turn to the *Gaussian* stochastic measure μ^G on $\Theta \times \mathbb{T}$:

$$E\mu^G = 0\,, \qquad E(\mu^G)^2 = \sigma^2 \cdot M\,,$$

and the *Poisson* (centred) stochastic measure μ^P on $(\mathbb{R} \setminus [0]) \times \Theta \times \mathbb{T}$:

$$E\mu^P = 0\,, \qquad E(\mu^P)^2 = L \times M\,.$$

The formula

$$\mu(\Delta) := \iint_\Delta \mu^G(d\theta\, dt) + \iiint_{(\mathbb{R}\setminus[0])\times\Delta} x\,\mu^P(dx\, d\theta\, dt)\,, \qquad \Delta \subseteq \Theta \times \mathbb{T}\,, \tag{2.16}$$

defines the stochastic measure μ on $\Theta \times \mathbb{T}$ characterized by the infinitely-divisible probability law (1.17) with the above parameters σ^2, L and M such that $E\mu = 0$ and $E\mu^2 = M$. Here, we treat μ^P as in the general framework of Example 2.2.

The Poisson (centred) stochastic measure μ^P can be determined as

$$\mu^P(B \times \Delta) = \sum_{s<t_\omega\le u} 1_B(\rho^P_\omega) - \iint_\Delta L(B,\theta,t) \times M(d\theta\, dt)\,, \tag{2.17}$$

on the sets of form

$$B \times \Delta: \quad B \subseteq \{|x| > r\}\,, \ r > 0, \quad \Delta \subseteq \Theta \times (s,u] \,:\, M(\Delta) < \infty,$$

via the jumps $\rho^P_\omega := \xi^P(t_\omega,\omega) - \xi^P(t_\omega - 0,\omega)$ of the trajectories $\xi^P(t,\omega)$, $t \in \mathbb{T}$, of the processes of type

$$\xi^P(t) := \iiint_{(\mathbb{R}\setminus[0])\times\Theta\times(0,t]} 1_\Delta x\,\mu^P(dx\, d\theta\, ds)\,, \qquad t \in \mathbb{T}\,.$$

Cf. (2.14)-(2.15). For any $\Delta \subseteq \Theta \times \mathbb{T}$: $M(\Delta) < \infty$, the above process is a component in

$$\xi(t) := \iint_{\Theta\times(0,t]} 1_\Delta \mu(d\theta\, ds) = \xi^G(t) + \xi^P(t)\,, \qquad t \in \mathbb{T}\,, \tag{2.18}$$

- cf. (2.16). Here, the other component

$$\xi^G(t) = \iint_{\Theta\times(0,t]} 1_\Delta\,\mu^G(d\theta\,ds)\,, \qquad t\in\mathbb{T}\,,$$

is a *Gaussian* process with independent increments having *continuous* variance - cf. (1.3). These Gaussian processes are similar to the Wiener process. In particular, their cadlag versions have actually *continuous* trajectories $\xi^G(t,\omega)$, $t\in\mathbb{T}$, for almost all $\omega\in\Omega$. Accordingly, the trajectories $\xi(t,\omega)$, $t\in\mathbb{T}$, of the processes (2.18) have jumps as

$$\rho_\omega := \xi(t_\omega,\omega) - \xi(t_\omega - 0,\omega) \equiv \xi^P(t_\omega,\omega) - \xi^P(t_\omega - 0,\omega) = \rho^P_\omega\,.$$

Hence, the stochastic measure μ^P can be determined through the stochastic processes (2.18) as

$$\mu^P(B\times\Delta) = \sum_{s<t_\omega\le u} 1_B(\rho_\omega) - \iint_\Delta L(B,\theta,t)\times M(d\theta\,dt), \quad \omega\in\Omega. \quad (2.19)$$

Cf. (2.17). So, the stochastic measures μ^G, μ^P in the representation (2.16) are *uniquely* determined by μ.

Moreover, let us consider the *Gaussian-Poisson mixture* $\mu^{G,P}$ on the space-time product

$$\mathbb{R}\times\Theta\times\mathbb{T} = \Big([0]\times\Theta\times\mathbb{T}\Big)\cup\Big((\mathbb{R}\setminus[0])\times\Theta\times\mathbb{T}\Big)$$

of the independent Gaussian stochastic measure μ^G on $\Theta\times\mathbb{T}$, identified with $[0]\times\Theta\times\mathbb{T}$, and the Poisson (centred) stochastic measure μ^P on $(\mathbb{R}\setminus[0])\times\Theta\times\mathbb{T}$ - cf. (1.10). We can see that

$$\mathfrak{A}_{\mathbb{R}\times\Theta\times(0,t]} \equiv \mathfrak{A}_{\Theta\times(0,t]}\,, \qquad t\in\mathbb{T}\,, \quad (2.20)$$

for the σ-algebras generated in the course of time by $\mu^{G,P}$ and μ, correspondingly. Cf. (1.15)-(1.16).

Let Θ be a general complete separable metric space. Within the framework described in (2.8)-(2.20), we obtain the following result. See e.g. [25,51] in the case of stochastic processes.

Theorem 2.1. *The representation* (2.16) *holds for a general stochastic measure with independent values characterized by the probability law* (1.17).

Proof. Let μ be a general stochastic measure with independent values characterized by the probability law (1.17). For an appropriate probability space $\widetilde{\Omega}$ there exist the independent Gaussian stochastic measure $\widetilde{\mu}^G$ and Poisson (centred) stochastic measure $\widetilde{\mu}^P$ for which

$$\widetilde{\mu}(\Delta) = \iint_\Delta \widetilde{\mu}^G(d\theta\,dt) + \iiint_{(\mathbb{R}\setminus[0])\times\Delta} x\,\widetilde{\mu}^P(dx\,d\theta\,dt)\,, \qquad \Delta\subseteq\Theta\times\mathbb{T}\,,$$

is a stochastic measure with the same probability law as μ. We have

$$\widetilde{\mathfrak{A}}_{(\mathbb{R}\setminus[0])\times\Theta\times\mathbb{T}}=\widetilde{\mathfrak{A}}_{\Theta\times\mathbb{T}}$$

for the σ-algebras generated by the Gaussian-Poisson mixture $\widetilde{\mu}^{G,P}$ over $(\mathbb{R}\setminus[0])\times\Theta\times\mathbb{T}$ and the stochastic measure $\widetilde{\mu}$ over $\Theta\times\mathbb{T}$ - cf. (2.20). For the random variables $\widetilde{\xi}$ on $\widetilde{\Omega}$, measurable with respect to the σ-algebra $\widetilde{\mathfrak{A}}_{\Theta\times\mathbb{T}}$ generated by $\widetilde{\mu}$, we have the linear isometry

$$L_2(\widetilde{\Omega})\ni\widetilde{\xi}\Longrightarrow\xi\in L_2(\Omega)\,,$$

defined through the mapping

$$\widetilde{\xi}=F\Big(\widetilde{\mu}(\Delta_1),\ldots,\widetilde{\mu}(\Delta_m)\Big)\Longrightarrow F\Big(\mu(\Delta_1),\ldots,\mu(\Delta_m)\Big)=\xi$$

of all the functions of all the values of $\widetilde{\mu}$, μ. This mapping preserves the finite-dimensional probability distributions. Hence the above isometry yields

$$\mu^G(\Delta)\,:\,\widetilde{\mu}^G(\Delta)\Longrightarrow\mu^G(\Delta)\,,\qquad\Delta\subseteq\Theta\times\mathbb{T}\,,$$

and

$$\mu^P(B\times\Delta):\quad\widetilde{\mu}^P(B\times\Delta)\Longrightarrow\mu^P(B\times\Delta),\quad B\times\Delta\subseteq(\mathbb{R}\setminus[0])\times\Theta\times\mathbb{T},$$

as the independent Gaussian and Poisson (centred) stochastic measures for which we have

$$\mu(\Delta)=\iint_\Delta\mu^G(d\theta\,dt)+\iiint_{(\mathbb{R}\setminus[0])\times\Delta}x\,\mu^P(dx\,d\theta\,dt),\quad\Delta\subseteq\Theta\times\mathbb{T}.\quad\square$$

3 The Non-anticipating Integral Representation

3.1 Multilinear Polynomials and Itô Multiple Integrals

For being able to model stochastic processes via stochastic integration in the course of time - cf. (2.1), it is fundamental to characterize the random variables $\xi\in L_2(\Omega)$ which admit the non-anticipating integral representation

$$\xi=E\xi+\iint_{\Theta\times\mathbb{T}}\varphi\,\mu(d\theta\,dt).\tag{3.1}$$

Let μ be a general stochastic measure of the type $E\mu=0$, $E\mu^2=M$ - cf. (1.9). In the sequel, we focus on the random variables which are limits in $L_2(\Omega)$ of multilinear polynomials of the values of μ (hereafter μ-values). By *multilinear polynomial* we mean a linear combination of the p-power ($p=1,2,\ldots$) multilinear forms

$$\xi=\prod_{j=1}^{p}\xi_j\ \text{ with }\quad\xi_j=\mu(\Delta_j),\quad j=1,\ldots,p,\tag{3.2}$$

of the μ-values taken on the *disjoint* sets $\Delta_j \subseteq \Theta \times \mathbb{T} \ : \ M(\Delta_j) < \infty$, $j = 1, \dots, p$, plus the constants (which formally correspond to $p = 0$).

Theorem 3.1. *The multilinear polynomials admit the non-anticipating integral representation* (3.1).

Proof. The result is immediate for the multilinear forms (3.2) of the μ-values on the sets $\Delta_j \subseteq \Theta \times (s_j, u_j]$, $j = 1, \dots, p$, related to the *disjoint* time intervals $(s_j, u_j] \subseteq \mathbb{T}$, $j = 1, \dots, p$, on $\mathbb{T} = (0, T]$. Indeed, taking these intervals ordered in time $0 < s_1 < u_1 \leq \dots \leq s_p < u_p \leq T$, we can see that

$$\xi = \iint_{\Delta_p} \left(\prod_{j=1}^{p-1} \xi_j \right) \mu(d\theta \, dt) \, .$$

The range of the non-anticipating integral, as an isometric linear operator, is closed. Cf. (2.1)-(2.4). So, ξ admits the representation (3.1) if $\xi = \lim_{n \to \infty} \xi^{(n)}$ is the limit in $L_2(\Omega)$ of the linear combinations of the above type multilinear forms of the μ-values. Hereafter we refer to the limit $\xi = \lim_{n \to \infty} \xi^{(n)}$ as the *proper approximation*, when ξ is a general multilinear form (3.2) of the μ-values on the disjoint sets Δ_j, $j = 1, \dots, p$ and $\xi^{(n)}$, $n = 1, 2, \dots$, are multilinear forms which involve only the μ-values on the subsets in $\Delta = \sum_{j=1}^{p} \Delta_j$.

In general, for the limits $\xi_k = \lim_{n \to \infty} \xi_k^{(n)}$, $k = 1, \dots, m$, with the independent approximations $\{\xi_k^{(n)}, n = 1, 2, \dots\}$, $k = 1, \dots, m$, we have

$$\prod_{k=1}^{m} \xi_k = \lim_{n \to \infty} \prod_{k=1}^{m} \xi_k^{(n)},$$

$$\text{i.e.} \quad \left\| \prod \xi - \prod \xi^{(n)} \right\| \leq const \cdot \max_k \|\xi_k - \xi_k^{(n)}\| \longrightarrow 0, \ n \to \infty.$$

Keeping this in mind, let us suppose that all multilinear forms of the power $p < q$ ($q > 1$) admit a proper approximation. The claim trivially holds for $p = 1$. Considering $\xi = \prod_{k=1}^{q} \mu(\Delta_k)$ as a general q-power multilinear form (3.2) through the nth-partitions

$$\Delta_k = \sum \Big(\Delta_k \cap (\Theta \times (s, u]) \Big), \quad \sum (s, u] = \mathbb{T} \ : \ \max(u - s) \longrightarrow 0 \, , \quad n \to \infty \, ,$$

we can see that

$$\begin{aligned} \|\xi_0^{(n)}\|^2 &:= \sum \prod_{k=1}^{q} M\Big(\Delta_k \cap (\Theta \times (s, u]) \Big) \\ &\leq const \cdot \max M\Big(\Delta_k \cap (\Theta \times (s, u]) \Big) \to 0, \quad n \to \infty, \end{aligned}$$

for

$$\xi_0^{(n)} := \sum \prod_{k=1}^{q} \mu\Big(\Delta_k \cap (\Theta \times (s, u]) \Big), \quad n = 1, 2, \dots$$

- cf. (1.5). We can also see that the differences

$$\xi^{(n)} := \xi - \xi_0^{(n)} = \prod_{k=1}^{q} \Big[\sum \mu\Big(\Delta_k \cap (\Theta \times (s,u])\Big)\Big]$$

$$- \sum \prod_{k=1}^{q} \mu\Big(\Delta_k \cap (\Theta \times (s,u])\Big)$$

admit proper approximation. The same holds for ξ as the limit

$$\xi = \lim_{n\to\infty} \xi^{(n)}$$

in $L_2(\Omega)$. □

Let us now turn to $H^p \subseteq L_2(\Omega)$ as the linear closure of all the p-power multilinear forms (3.2). The subspaces H^p $(p = 1, 2, \ldots)$ are orthogonal. Let us consider

$$H := \sum_{p=0}^{\infty} \oplus H^p \tag{3.3}$$

which is the standard orthogonal sum of H^p, $p = 1, 2, \ldots$, in $L_2(\Omega)$ where H^0 represents the set of all the constants. We remark that *all random variables* $\xi \in H$ *admit the non-anticipating integral representation (3.1)*. Cf. Theorem 3.1.

The representation (3.1) of the elements in $\xi \in H^p$, $p > 1$, can be specified by means of the Itô type *multiple integrals* [27] (see also e.g. [53]). Here, we have in mind the *p-multiple integrals*

$$I^p\varphi_p = \int \cdots \int_{\{t_1<\cdots<t_p\}} \varphi_p \, \mu(d\theta_1\, dt_1) \times \cdots \times \mu(d\theta_p\, dt_p)\,, \qquad p > 1\,, \tag{3.4}$$

over the indicated domain $\{t_1 < \cdots < t_p\}$ in the p-times product $(\Theta \times \mathbb{T})^p$ which consists of

$$(\theta_1, t_1, \ldots, \theta_p, t_p) \in (\Theta \times \mathbb{T})^p : \quad t_1 < \cdots < t_p.$$

The integrator in (3.4) is the standard type *stochastic measure with orthogonal values* defined on the product-sets

$$\Delta_1 \times \cdots \times \Delta_p \subseteq \{t_1 < \cdots < t_p\}$$

as the product $\mu(\Delta_1) \times \cdots \times \mu(\Delta_p)$:

$$E\Big(\mu(\Delta_1) \times \cdots \times \mu(\Delta_p)\Big) = 0\,,$$

$$E\Big(\mu(\Delta_1) \times \cdots \times \mu(\Delta_p)\Big)^2 = M(\Delta_1) \times \cdots \times M(\Delta_p)\,.$$

The integrands φ_p are the deterministic functions

$$\varphi_p = \varphi_p(\theta_1, t_1, \dots, \theta_p, t_p), \quad (\theta_1, t_1, \dots, \theta_p, t_p) \in \{t_1 < \dots < t_p\},$$

in the standard (complex) L_2-space with the norm

$$\|\varphi_p\|_{L_2} = \left(\int \cdots \int_{\{t_1 < \cdots < t_p\}} |\varphi_p|^2 \, M(d\theta_1 \, dt_1) \times \cdots \times M(d\theta_p \, dt_p) \right)^{1/2}.$$

So, in (3.4) we have the standard stochastic integral $I^p\varphi_p$: $\|I^p\varphi_p\| = \|\varphi\|_{L_2}$. The p-power multilinear forms (3.2) of the μ-values on $\Delta_j \subseteq \Theta \times (s_j, u_j]$, $j = 1, \dots, p$, related to the *disjoint* time intervals (ordered according to $0 \le s_1 < u_1 \le \cdots \le s_p < u_p \le T$) are identical to the p-multiple integrals (3.4) with the indicators $\varphi_p = 1_{\Delta_1 \times \cdots \times \Delta_p}$ as integrands. Hence, following the proof of Theorem 3.1, we can see that *all* $\xi_p \in H^p$ are represented by *all* the p-multiple integrals $I^p\varphi_p$. For $\xi_p = I^p\varphi_p$, the non-anticipating integral representation (3.1) can be given as

$$\xi_p = \iint_{\Theta \times \mathbb{T}} I^{p-1}\varphi_p(\cdot, \theta, t) \, \mu(d\theta \, dt). \tag{3.5}$$

Here the function

$$\varphi_p(\cdot, \theta, t) = \varphi_p(\theta_1, t_1, \dots, \theta_{p-1}, t_{p-1}, \theta, t),$$
$$(\theta_1, t_1, \dots, \theta_{p-1}, t_{p-1}) \in \{t_1 < \cdots < t_{p-1}\},$$

with $(\theta, t) \in \Theta \times \mathbb{T}$ as parameter, is the integrand in the $(p-1)$-multiple integral.

Of course, in the case $p = 1$, the non-anticipating integral representation (3.1) of the elements $\xi \in H^p$ is trivial:

$$\xi_1 = \iint_{\Theta \times \mathbb{T}} \varphi_1 \, \mu(d\theta \, dt),$$

with the *deterministic* integrands φ_1: $\|\varphi_1\|_{L_2} = (\iint_{\Theta \times \mathbb{T}} |\varphi_1|^2 M(d\theta \, dt))^{1/2}$. In line with the case $p > 1$, we write $\xi_1 = I^1\varphi_1$: $\varphi_1 = I^0\varphi_1(\cdot, \theta, t)$, $(\theta, t) \in \Theta \times \mathbb{T}$, for the above stochastic integral. Clearly, the representation (3.1) of all $\xi \in H$ in the subspace $H \in L_2(\Omega)$ - cf. (3.3), is obtained via the orthogonal sum

$$\xi = \sum_{p=0}^{\infty} \oplus \xi_p : \qquad \xi_p \in H^p, \quad p = 0, 1, \dots,$$

with $\xi_0 = E\xi$ and

$$\xi_p = \iint_{\Theta \times \mathbb{T}} I^{p-1}\varphi_p(\cdot, \theta, t) \, \mu(d\theta \, dt). \quad p = 1, 2, \dots,$$

This yields

$$\xi = E\xi \oplus \iint_{\Theta\times\mathbb{T}} \Big[\sum_{p=1}^{\infty} \oplus I^{p-1}\varphi_p(\cdot,\theta,t)\Big]\,\mu(d\theta\,dt). \tag{3.6}$$

Here we refer to [16] and [28]. See also e.g. [42] for some results on Lévy processes.

3.2 Integral Representations with Gaussian-Poisson Integrators

Let $\mu = \mu(\Delta)$, $\Delta \subseteq \Theta\times\mathbb{T}$, be a general *Gaussian-Poisson mixture*, i.e. the mixture of the components μ_k, $k = 0,1,\dots$, which are either Gaussian or Poisson (centred) stochastic measures multiplied by scalars - cf. (1.10) and (2.16)-(2.19). Let us consider the subspaces

$$H_q := \sum_{p=0}^{q} \oplus H^p, \quad q = 1,2,\dots,$$

in $L_2(\Omega)$. Cf. (3.3).

Theorem 3.2. *The q-power polynomials of the values of μ belong to H_q, $q = 1,2,\dots$.*

Proof. The proof is quite similar to the one of Theorem 3.1. All the q-power multilinear polynomials belong to H_q $(q = 1,2,\dots)$. So $\xi \in H_q$ if it can be represented as limit $\xi = \lim_{n\to\infty}\xi^{(n)}$ in $L_2(\Omega)$ of the q-power multilinear polynomials $\xi^{(n)}$, $n = 1,2,\dots$. Let ξ be a q-power polynomial of μ-values, we can treat ξ as the q-power polynomial of the values $\mu(\Delta_j)$, $j = 1,\dots,m$, on appropriately choosen *disjoint* sets $\Delta_j \subseteq \Theta\times\mathbb{T}$, $j = 1,\dots,m$. Accordingly, we refer to the limit $\xi = \lim_{n\to\infty}\xi^{(n)}$ of the q-power multilinear polynomials $\xi^{(n)}$, $n = 1,2,\dots$, of the values of μ just on the subsets in $\Delta = \sum_{j=1}^{m}\Delta_j$ as the *proper approximation.* The proper approximation holds for $p = 1$. Suppose it holds for the polynomials of power $p < q$, $(q > 1)$. Than we can see that the proper approximation does hold for all the q-power polynomials, if it holds for

$$\xi = \mu(\Delta)^q, \qquad \Delta \subseteq \Theta\times\mathbb{T}.$$

Moreover, note that here it is enough to consider the sets Δ where μ is either the Gaussian or the Poisson (centred) stochastic measure. Let us take the nth-partitions

$$\Delta = \sum \Delta\cap(\Theta\times(s,u]), \qquad \sum(s,u] = \mathbb{T}:\ \max(u-s) \longrightarrow 0, \quad n\to\infty,$$

into account. We can see that the limit $\lim_{n\to\infty}\xi_0^{(n)} = \xi_0$ in $L_2(\Omega)$ with

$$\xi_0^{(n)} := \sum \mu\Big(\Delta\cap(\Theta\times(s,u])\Big)^q, \qquad n = 1,2,\dots,$$

has the following form: $\xi_0 = \mu(\Delta)$, $q = 2$, or $\xi_0 = 0$, $q > 2$, if μ is Gaussian and $\xi_0 = \mu(\Delta) + M(\Delta)$ if μ is the Poisson (centred) stochastic measure. In all cases we can say that ξ_0 admits proper approximation. Following the arguments applied in the proof of Theorem 3.1, we can also see that the differences

$$\xi-\xi_0^{(n)} = \Big[\sum \mu\Big(\Delta\cap(\Theta\times(s,u])\Big)\Big]^q - \sum \mu\Big(\Delta\cap(\Theta\times(s,u])\Big)^q, \quad n=1,2,\dots,$$

admit proper approximation as well. So, such approximation holds also for $\xi = \mu(\Delta)^q$ as the limit

$$\xi = \lim_{n\to\infty} \xi^{(n)} \text{ with } \quad \xi^{(n)} = (\xi - \xi_0^{(n)}) + \xi_0, \quad n = 1,2,\dots,$$

in $L_2(\Omega)$. □

In the sequel it is important that the σ-algebra is generated by the stochastic measure μ, i.e.

$$\mathfrak{A} := \mathfrak{A}_{\Theta\times\mathbb{T}}$$

- cf. (1.12), and that the flow of events in the course of time is represented by the σ-algebras (1.15)

$$\mathfrak{A}_t := \mathfrak{A}_{\Theta\times(0,t]}, \quad t \in \mathbb{T}.$$

Note that the polynomials of the values $\mu(\Delta)$, $\Delta \subseteq \Theta \times \mathbb{T}$, are dense in $L_2(\Omega)$ when μ is a general stochastic measure with independent values which obeys the probability law (1.17) and restricted by the condition

$$\begin{aligned} Ee^{\lambda\mu(\Delta)} = \exp \iint_\Delta \Big[\frac{\lambda^2}{2}\sigma^2(\theta,t) \\ + \int_{\mathbb{R}\setminus[0]} (e^{\lambda x} - 1 - \lambda x)\, L(dx,\theta,t)\Big] M(d\theta\, dt) < \infty, \quad \lambda \in \mathbb{R}. \quad (3.7)\end{aligned}$$

To explain, for the *complete system* of functions of the form

$$e^{i\sum_{k=1}^m \lambda_k\xi_k} \qquad (\lambda_k \in \mathbb{R},\ k = 1,\dots,m),$$

with the values $\xi_k = \mu(\Delta_k)$, $k = 1,\dots,m$, taken on all finite combinations of disjoint sets in $\Theta \times \mathbb{T}$, we have

$$\left\| e^{i\sum_{k=1}^m \lambda_k\xi_k} - \sum_{p=0}^q \frac{(i\sum_{k=1}^m \lambda_k\xi_k)^p}{p!} \right\| \longrightarrow 0, \quad q \to \infty.$$

In the following result we do consider that, for a general Gaussian-Poisson mixture μ, the polynomials of the values of μ are dense in $L_2(\Omega)$. See e.g. [10, 11, 16, 28, 42].

Theorem 3.3. *All the elements* $\xi \in L_2(\Omega)$*:*

$$L_2(\Omega) = \sum_{p=0}^{\infty} \oplus H^p \,, \tag{3.8}$$

admit the non-anticipating integral representation (3.1).

Proof. Cf. (3.4) and Theorem 3.2. □

3.3 Homogeneous Integrators

In the Theorem 3.3 we have seen that, for a given Gaussian-Poisson mixture and a flow of events generated by the values of this measure in the course of time, all the elements of the corresponding L_2-space $L_2(\Omega)$ admit integral representation (3.1). However, in general, for a given stochastic measure, though with homogeneous (see below) and independent values, and a filtration generated by the measure itself, we cannot claim that *all* the elements in the corresponding $L_2(\Omega)$ space admit the representation (3.1). This fact finds evidence and consequences in many applied situations, we can refer as an example to the incompleteness of certain well-known market models in mathematical finance. The next result addresses the issue of characterizing the stochastic measures for which it is possible that *all* the elements of the corresponding $L_2(\Omega)$ admit the representation (3.1).

Let us turn our attention to the measure $\mu = \mu(\Delta)$, $\Delta \subseteq \Theta \times \mathbb{T}$, of the type $E\mu = 0$, $E\mu^2 = M$ satisfying (3.7), which is *homogeneous*, in the sense that all the values $\mu(\Delta)$ on the sets $\Delta \subseteq \Theta \times \mathbb{T}$ of the *same* measure $M(\Delta)$ obey the *same* probability law. Accordingly, they follow the infinetely-divisible law of the form (1.17) with parameters σ^2, L that do *not* depend on $(\theta, t) \in \Theta \times \mathbb{T}$. Namely we have

$$log E e^{i\lambda\mu(\Delta)} = \left[-\frac{\lambda^2}{2}\sigma^2 + \int_{\mathbb{R}\setminus[0]} (e^{i\lambda x} - 1 - i\lambda x) L(dx) \right] \cdot M(\Delta), \lambda \in \mathbb{R}, \tag{3.9}$$

with

$$\sigma^2 + \int_{\mathbb{R}\setminus[0]} x^2 L(dx) = 1,$$

where σ^2 is constant and $L(dx)$, $x \in \mathbb{R} \setminus [0]$, is a σ^2-finite measure on $\mathbb{R} \setminus [0]$. Let Θ be a complete separable metric space. For the following result see e.g. [2, 6, 15].

Theorem 3.4. *The non-anticipating integral representation* (3.1) *holds for all* $\xi \in L_2(\Omega)$ *if and only if* μ *is either Gaussian or Poisson (centred) stochastic measure multiplied by a scalar.*

Proof. Let us treat μ as

$$\mu(\Delta) = \iint_{\Delta} \mu^G(d\theta\, dt) + \iiint_{\mathbb{R}\setminus[0]\times\Delta} x\, \mu^P(dx\, d\theta\, dt), \qquad \Delta \in \Theta \times \mathbb{T},$$

in relation to the Gaussian-Poisson mixture $\mu^{G,P}$. Cf. (2.16)-(2.20) and Theorem 2.1. For all $\xi \in L_2(\Omega)$, the non-anticipating integral representation has the form

$$\xi = E\,\xi \oplus \iint_{\Theta\times\mathbb{T}} \varphi_G\, \mu^G(d\theta\, dt) \oplus \iiint_{(\mathbb{R}\setminus[0])\times\Theta\times\mathbb{T}} \varphi_P\, \mu^P(dx\, d\theta\, dt) \qquad (3.10)$$

which is here considered with respect to $\mu^{G,P}$ as integrator Cf. Theorem 3.3. For all those ξ which admit the non-anticipating integral representation

$$\begin{aligned} \xi &= E\,\xi + \iint_{\Theta\times\mathbb{T}} \varphi\mu(d\theta\, dt) \\ &= E\,\xi \oplus \iint_{\Theta\times\mathbb{T}} \varphi\mu^G(d\theta\, dt) \oplus \iint_{\Theta\times\mathbb{T}} \varphi\cdot x\, \mu^P(dx\, d\theta\, dt) \end{aligned}$$

with respect to μ, the identity

$$\varphi(\theta,t) \equiv \varphi_G(\theta,t) \equiv x^{-1}\varphi_P(x,\theta,t)\,, \qquad (x,\theta,t) \in (\mathbb{R}\setminus[0])\times\Theta\times\mathbb{T},$$

must hold for all the integrands φ_G and φ_P. Clearly, this identity can only hold for *all* the *different* integrands φ_G, φ_P as elements of the corresponding functional L_2-spaces (related to the measures $\sigma^2\cdot M$ and $L\times M$) if either $\sigma^2 = 1$, $L = 0$ or $\sigma^2 = 0$ and L is concentrated at the *single* point $x \in \mathbb{R}\setminus[0]$, with $L(x) = x^{-2}$. In other terms it means that either $\mu = \mu^G$ or $\mu = x\mu^P$. Here μ^P is the Poisson (centred) measure concentrated on the product $[x]\times\Theta\times\mathbb{T}$, which can be identified with $\Theta\times\mathbb{T}$. □

Let us now consider the cadlag processes of type

$$\xi(t) = \iint_{\Theta\times(0,t]} 1_\Delta\mu(d\theta dt), \quad t \in \mathbb{T}$$

($\Delta \subseteq \Theta\times\mathbb{T}$), for the stochastic measure μ the values of which follow the probability law (3.9). The jumps of the trajectories of these processes are

$$\rho_\omega = \xi(t_\omega,\omega) - \xi(t_\omega - 0,\omega), \quad t_\omega \in \mathbb{T}.$$

In relation to these jumps we can define the stochastic measure $\mu^F = \mu^F(\Delta)$, $\Delta \subseteq \Theta\times\mathbb{T}$: $M(\Delta) < \infty$, as

$$\mu^F(\Delta) := \sum_{t_\omega\in\mathbb{T}} F(\rho_\omega) - \int_{\mathbb{R}\setminus[0]} F\cdot L(dx)\cdot M(\Delta) \qquad (3.11)$$

by means of the deterministic real function $F = F(x)$, $x \in \mathbb{R}$, such that $F(0) = 0$ and

$$\int_{\mathbb{R}\setminus[0]} |F(x)|^p L(dx) < \infty, \quad p = 1,2.$$

Here μ^F is a homogeneous stochastic measure with independent values of form

$$\mu^F(\Delta) = \iiint_{(\mathbb{R}\setminus[0])\times\Delta} F\mu^P(dxd\theta dt), \quad \Delta \subseteq \Theta \times \mathbb{T},$$

where μ^P is the Poisson (centred) stochastic component of μ. Accordingly we have that the random variable $\mu^F(\Delta)$ has distribution characterized by

$$logEe^{i\lambda\mu^F(\Delta)} = \int_{\mathbb{R}\setminus[0]} (e^{i\lambda F(x)} - 1 - i\lambda F(x))L(dx) \cdot M(\Delta), \quad \lambda \in \mathbb{R}.$$

To explain the statement above it is enough to observe that

$$F(1_\Delta x) = F(x)1_\Delta(\theta,t), \quad (x,\theta,t) \in (\mathbb{R}\setminus[0]) \times \Theta \times \mathbb{T},$$

and refer to the argument used in (2.16)-(2.19).

Now let us consider an *orthogonal basis* F_k, $k = 1, 2, \dots$, in the standard (complex) space $L_2(\mathbb{R}\setminus[0])$:

$$\|F\|_{L_2(\mathbb{R}\setminus[0])} = \left(\int_{\mathbb{R}\setminus[0]} |F|^2 L(dx)\right)^{1/2}.$$

For each F_k, one can apply the arguments above and define the stochastic measures

$$\mu^{F_k} = \mu^{F_k}(d\theta dt), \quad (\theta,t) \in \Theta \times \mathbb{T},\ k = 1, 2, \dots .$$

Theorem 3.5. *Let* $\xi \in L_2(\Omega)$. *In the representation* (3.10) *the integral with respect to the Poisson (centred) component* μ^P *can be written as*

$$\iiint_{(\mathbb{R}\setminus[0])\times\Theta\times\mathbb{T}} \varphi_P\mu^P(dxd\theta dt) = \sum_{k=1}^{\infty} \iint_{\Theta\times\mathbb{T}} \varphi_k\mu^{F_k}(d\theta dt). \tag{3.12}$$

Proof. The integrands φ_P in the stochastic integral with respect to the Poisson (centred) stochastic measure μ_P are elements of the subspace

$$L_2^I((\mathbb{R}\setminus[0]) \times \Theta \times \mathbb{T} \times \Omega) \subseteq L_2(\mathbb{R}\setminus[0] \times \Theta \times \mathbb{T} \times \Omega)$$

in the standard L_2-space related to the integrator

$$L(dx) \times M(d\theta dt) \times P(d\omega), \quad (x,\theta,t,\omega) \in (\mathbb{R}\setminus[0]) \times \Theta \times \mathbb{T} \times \Omega$$

- cf. (1.20) and (2.2)-(2.5). The elements of the form

$$F_k \cdot \Phi : \quad \Phi \in L_1(\Theta \times \mathbb{T} \times \Omega),\ k = 1, 2, \dots,$$

constitute a complete system in the above space. Hence, the orthogonal projections

$$F_k \cdot \varphi : \quad \varphi = E(\Phi(\theta,t)|\mathfrak{A}_t),\ (\theta,t) \in \Theta \times \mathbb{T},$$

of the elements $F_k \cdot \Phi$, $k = 1, 2, \ldots$, on the subspace $L_2^I((\mathbb{R} \setminus [0]) \times \Theta \times \mathbb{T} \times \Omega)$ constitute a complete system in this subspace. Any linear combination of the above elements is represented as the orthogonal sum

$$\sum_k \oplus \varphi_k \cdot F_k : \quad \varphi_k \in L_2(\Theta \times \mathbb{T} \times \omega).$$

Thus any integrand φ_P for μ^P can be represented as the orthogonal series

$$\varphi_P = \sum_{k=1}^{\infty} \oplus \varphi_k \cdot F_k$$

in $L_2^I((\mathbb{R} \setminus [0]) \times \Theta \times \mathbb{T} \times \Omega)$ and this yields the representation (3.7), i.e.

$$\begin{aligned} \iiint_{(\mathbb{R}\setminus[0])\times\Theta\times\mathbb{T}} \varphi_P \mu^P(dx d\theta dt) &= \sum_{k=1}^{\infty} \oplus \iiint_{(\mathbb{R}\setminus[0])\times\Theta\times\mathbb{T}} (\varphi_k \cdot F_k)\mu^P(dx d\theta dt) \\ &= \sum_{k=1}^{\infty} \oplus \iint_{\Theta\times\mathbb{T}} \varphi_k \mu^{F_k}(d\theta dt), \end{aligned}$$

as the standard orthogonal series (4.3) in $L_2(\Omega)$. □

In the sequel we will introduce the *non-anticipating derivative.* Here we would however note straightaway that the integrands in the representation (3.7) are the non-anticipating derivatives $\varphi_k = D_k \xi$ of $\xi \in L_2(\Omega)$ with respect to the stochastic measure μ^{F_k}, $k = 1, 2, \ldots$.

We also would like to note that in the case the stochastic measure μ has no Gaussian component μ^G, i.e. μ is following the probability law (3.9) with $\sigma^2 = 0$, then the representation (3.7) can be applied directly with $\mu^{F_1} = \mu$ and with the μ^{F_k}, $k = 2, 3, \ldots$, given by (4.3). Here $F_1 = x$, $x \in \mathbb{R} \setminus [0]$ and F_k, $k = 2, 3, \ldots$, constitute an orthogonal system in $L_2(\mathbb{R} \setminus [0])$. The same arguments used in the proof of Theorem 3.5 lead to the following result. See also [42].

Corollary 3.1. *Let stochastic measure μ follow the probability law* (3.9) *with $\sigma^2 = 0$. All the elements $\xi \in L_2(\Omega)$ admit the following representation via the orthogonal sum*

$$\xi = E\xi \oplus \sum_{k=1}^{\infty} \oplus \iint_{\Theta\times\mathbb{T}} \varphi_k \mu^{F_k}(d\theta dt) : \quad \mu^{F_1} = \mu. \tag{3.13}$$

4 The Non-anticipating Derivative

4.1 A General Definition and Related Properties

Let us consider the non-anticipating integral as the *isometric* linear operator I:

$$L_2^I(\Theta \times \mathbb{T} \times \Omega) \ni \varphi \Longrightarrow I\varphi \in L_2(\Omega),$$

on the subspace of the non-anticipating functions φ - cf. (2.2). In relation to I, we can define the *non-anticipating derivative as the adjoint linear operator* $D = I^*$:

$$L_2(\Omega) \ni \xi \Longrightarrow D\xi \in L_2^I(\Theta \times \mathbb{T} \times \Omega) . \tag{4.1}$$

Note that we have

$$\|D\| = \|I\| = 1$$

for the operator norm of the adjoint linear operators $D = I^*$, $I = D^*$. It is $D\xi = 0$ for ξ orthogonal to all the non-anticipating integrals

$$\iint_{\Theta\times\mathbb{T}} \varphi\, \mu(d\theta\, dt)\,, \qquad \varphi \in L_2^I(\Theta \times \mathbb{T} \times \Omega).$$

Accordingly, for any random variable $\xi \in L_2(\Omega)$, the non-anticipating derivative provides the best approximation

$$\hat{\xi} = \iint_{\Theta\times\mathbb{T}} D\xi\mu(d\theta\, dt) \tag{4.2}$$

to ξ in $L_2(\Omega)$ by non-anticipating integrals, i.e.

$$\|\xi - \hat{\xi}\| = \min_{\varphi\in L_2^I(\Theta\times\mathbb{T}\times\Omega)} \left\| \xi - \iint_{\Theta\times\mathbb{T}} \varphi\mu(d\theta\, dt) \right\| .$$

We obtain the following result - cf. [14, 15], see also [22]. We can also refer to [52] for some results in this direction in the case of the Wiener process as integrator and to [45] for the space-time Brownian sheet.

Theorem 4.1. *For all $\xi \in L_2(\Omega)$, the non-anticipating differentiation can be carried through via the limit*

$$D\xi = \lim_{n\to\infty} \sum E\left[\frac{1}{M(\Delta)} E\left(\xi \cdot \mu(\Delta) | \mathfrak{A}_s\right)\right] \cdot 1_\Delta \tag{4.3}$$

in $L_2^I(\Theta \times \mathbb{T} \times \Omega)$. Here the sum is on the nth-series sets $\Delta \subseteq \Theta \times (s, u]$ of some (which can be any) partition in $\Theta \times \mathbb{T}$ - cf. (1.3).

Proof. In the representation

$$\xi = \xi^0 \oplus \iint_{\Theta\times\mathbb{T}} \varphi\, \mu(d\theta\, dt)$$

with $\varphi = D\xi$ - cf. (4.2), the component ξ^0 is orthogonal to all the non-anticipating integrals (thus $D\xi^0 = 0$). This implies

$$E\left((1_{\mathcal{A}}\mu(\Delta)) \cdot \xi^0\right) = 0, \qquad \Delta \subseteq \Theta \times (s, u]\,, \qquad \mathcal{A} \in \mathfrak{A}_s$$

- cf. (2.3), thus it is $E(\xi^0 \cdot \mu(\Delta)|\mathfrak{A}_s) = 0$. With the use of

$$E\left(\iint_{\Theta\times\mathbb{T}} \varphi\,\mu(d\theta\,dt)\cdot\mu(\Delta)\Big|\mathfrak{A}_s\right) = E\left(\iint_{\Delta} \varphi\,M(d\theta\,dt)\Big|\mathfrak{A}_s\right),\quad \Delta\subseteq\Theta\times(s,u]$$

- cf. (2.6) and (2.7), we can see that the limit (4.3) for $\varphi = D\xi$ is identical to the limit of the approximations $\varphi^{(n)}$, $n = 1, 2\ldots$, characterized in (1.24) - cf. Theorem 1.3. □

Example 4.1. Let μ be the mixture of the stochastic measures $\mu^k = \mu^k(\Delta)$, $\Delta \subseteq \Theta_k \times \mathbb{T}$, $k = 1, 2, \ldots$ - cf. (1.10). Then, whatever $\xi \in L_2(\Omega)$ be, the non-anticipating derivative $D\xi$ is

$$D\xi = \sum_k \oplus D_k\xi\, 1_{\Theta_k\times\mathbb{T}}$$

where, for any k, $D_k\xi$ is the non-anticipating derivative with respect to the measure μ^k.

Example 4.2. For a general μ following the law (1.17), the non-anticipating derivative of the element $\xi \in L_2(\Omega)$:

$$\xi = D\xi \oplus \sum_{p=1}^{\infty} \oplus I^p\varphi_p$$

can be determined by the formula

$$D\xi = \sum_{p=1}^{\infty} \oplus I^{p-1}\varphi_p(\cdot,\theta,t), \quad (\theta,t)\in\Theta\times\mathbb{T}.$$

4.2 Differentiation Formulae

The random variables $\xi \in L_2(\Omega)$ are functions of the values $\mu(\Delta)$, $\Delta \subseteq \Theta\times\mathbb{T}$, of the stochastic measure. In fact the elements ξ are measurable with respect to the σ-algebra $\mathfrak{A} = \mathfrak{A}_{\Theta\times\mathbb{T}}$ generated by the values of μ - cf. (1.12). Let us now turn to the ξ which can be treated as functions of a *finite* number of values $\mu(\Delta)$, $\Delta \subseteq \Theta\times\mathbb{T}$. Any such random variable admits the representation

$$\xi = F(\xi_1, \ldots, \xi_m) \tag{4.4}$$

as a function of the values $\xi_k = \mu(\Delta_k)$, $k = 1, \ldots, m$, on the appropriately chosen *disjoint* sets Δ_k, $k = 1, \ldots, m$, in $\Theta\times\mathbb{T}$. Of course, the representation (4.4) *is not* unique. So, for *any* finite number of *any* particular group of *disjoint* sets

$$\Delta_k \subseteq \Theta\times\mathbb{T}: \quad M(\Delta_k) < \infty, \quad k = 1, \ldots, m,$$

we consider $\xi = F$ - cf. (4.4), for the functions

$$F = F(\xi_1, \dots, \xi_m), \quad (\xi_1, \dots, \xi_m) \in \mathbb{R}^m,$$

which are characterized as follows. First of all, we assume that $F \in C^1(\mathbb{R}^m)$, and we write

$$\partial_k^x F := \begin{cases} \frac{\partial}{\partial \xi_k} F(\dots, \xi_k, \dots), & x \neq 0, \\ \frac{1}{x}\big[F(\dots, \xi_k + x, \dots) - F(\dots, \xi_k, \dots)\big], & x = 0. \end{cases}$$

According to the characterization of the stochastic measure μ by the infinitely-divisible law - cf. (1.17), we define

$$\mathcal{D}\xi(\theta, t) := \sum_{k=1}^m \Big[\partial_k^0 F \cdot \sigma^2(\theta, t) + \int_{\mathbb{R}\setminus[0]} \partial_k^x F \cdot x^2 L(dx, \theta, t)\Big] \cdot 1_{\Delta_k}(\theta, t), \quad (\theta, t) \in \Theta \times \mathbb{T}, \tag{4.5}$$

for the elements $\xi = F$ of the type above. And we assume that

$$\mathcal{D}\xi = \mathcal{D}\xi(\theta, t), \quad (\theta, t) \in \Theta \times \mathbb{T},$$

satisfy the condition

$$|||\mathcal{D}\xi|||^2 := 7\sum_{k=1}^m \iint_{\Delta_k} \Big[\|\partial_k^0 F\|^2 \cdot \sigma^2(\theta, t) + \int_{\mathbb{R}\setminus[0]} \|\partial_k^x F\|^2 \cdot x^2 L(dx, \theta, t)\Big] M(d\theta\, dt) < \infty. \tag{4.6}$$

Hence we have in particular that $\mathcal{D}\xi \in L_2(\Omega \times \Theta \times \mathbb{T})$, since

$$\|\mathcal{D}\xi\|_{L_2} \leq |||\mathcal{D}\xi|||. \tag{4.7}$$

The following result was first published in [18].

Theorem 4.2. *The non-anticipating derivative of the random variable $\xi = F$ of type* (4.4) *defined by the limit* (4.3) *can be computed by*

$$D\,\xi(\theta, t) = E\Big(\mathcal{D}\xi(\theta, t)\big|\mathfrak{A}_t\Big), \qquad (\theta, t) \in \theta \times \mathbb{T}. \tag{4.8}$$

Proof. The proof is subdivided in several steps in which the statement is shown for more and more general random variables ξ, see steps A, B, C. Finally an appropriate approximation argument leads to the conclusion, see step D.

A. Let us take $\xi = F = F(\xi_i, \dots, \xi_m)$ with

$$F(\xi_1, \dots, \xi_m) = e^{i\sum_{k=1}^m \lambda_k \xi_k} \qquad (\lambda_k \in \mathbb{R}, \quad k = 1, \dots, m) \tag{4.9}$$

into account. In this case formula (4.5) gives

$$\mathcal{D}\xi(\theta,t) = \xi \sum_{k=1}^{m} \Big[i\lambda_k \sigma^2(\theta,t) + \int_{\mathbb{R}\setminus[0]} (e^{i\lambda_k x} - 1)x\, L(dx,\theta,t)\Big] 1_{\Delta_k}(\theta,t), \quad (\theta,t) \in \Theta \times \mathbb{T}.$$

We consider $\xi_k = \mu(\Delta_k)$, $k = 1,\dots,m$, with the *disjoint simple sets* Δ_k, $k = 1,\dots,m$. Then, for $n \to \infty$, any set $\Delta \subseteq \Theta \times (s,u]$ of the nth-series of the (1.3)-partitions either belongs to some Δ_k or it is disjoint with all Δ_k, $k = 1,\dots,m$. In this last case we have

$$E\left(\xi\, \mu(\Delta) | \mathfrak{A}_{]\Delta[}\right) = \xi\, E\, \mu(\Delta) = 0$$

- cf. (1.27). Otherwise, if $\Delta \subseteq \Delta_k$, for some k, we have

$$\begin{aligned} E\Big(\xi\, \mu(\Delta) \big| \mathfrak{A}_{]\Delta[}\Big) &= e^{-i\lambda_k \mu(\Delta)} \xi\, E\left(\mu(\Delta) e^{i\lambda_k \mu(\Delta)}\right) \\ &= e^{-i\lambda_k \mu(\Delta)} \xi\, E e^{i\lambda_k \mu(\Delta)} \iint_{\Delta} \big[i\lambda_k \sigma^2(\theta,t) \\ &\quad + \int_{\mathbb{R}\setminus[0]} (e^{i\lambda_k x} - 1)x\, L(dx,\theta,t)\big] M(d\theta dt) \\ &= E\left(\iint_{\Delta} \mathcal{D}\xi(\theta,t)\, M(d\theta\, dt) \big| \mathfrak{A}_{]\Delta[} \right). \end{aligned}$$

According to Theorem 1.4 the stochastic function $\mathcal{D}\xi$ admits the representation

$$\mathcal{D}\xi = \lim_{n\to\infty} \sum \frac{1}{M(\Delta)} E\Big(\xi \cdot \mu(\Delta) \big| \mathfrak{A}_{]\Delta[}\Big) 1_{\Delta} \tag{4.10}$$

as a limit in $L_2(\Omega \times \Theta \times \mathbb{T})$. Here the sum refers to all the elements of the same n^{th}-series of partitions of $\Theta \times \mathbb{T}$. By use of an appropriate sub-sequence we have convergence in $L_2(\Omega)$ for almost all $(\theta,t) \in \Theta \times \mathbb{T}$:

$$D\xi(\theta,t) = \lim_{n\to\infty} \frac{1}{M(\Delta)} E\left(\xi \cdot \mu(\Delta) | \mathfrak{A}_s\right), \quad (\theta,t) \in \Delta$$

and

$$\mathcal{D}\xi(\theta,t) = \lim_{n\to\infty} \frac{1}{M(\Delta)} E\Big(\xi \cdot \mu(\Delta) \big| \mathfrak{A}_{]\Delta[}\Big), \quad (\theta,t) \in \Delta$$

for $\Delta \subseteq \Theta \times (s,u]$: $\Delta \ni (\theta,t)$. Moreover taking $t^- < t$, we obtain

$$E\Big(D\xi(\theta,t) \big| \mathfrak{A}_{t^-}\Big) = \lim_{n\to\infty} \frac{1}{M(\Delta)} E\Big(\xi\mu(\Delta) \big| \mathfrak{A}_{t^-}\Big) = E\Big(\mathcal{D}\xi(\theta,t) \big| \mathfrak{A}_{t^-}\Big).$$

Let $t^- \to t$ in the above relations then we have

$$\begin{aligned} D\xi(\theta,t) = E\big(D\xi(\theta,t) | \mathfrak{A}_t\big) &= \lim_{t^-\to t} E\big(D\xi(\theta,t) | \mathfrak{A}_{t^-}\big) \\ &= \lim_{t^-\to t} E\big(\mathcal{D}\xi(\theta,t) | \mathfrak{A}_{t^-}\big) = E\big(\mathcal{D}\xi(\theta,t) | \mathfrak{A}_t\big), \end{aligned}$$

since $\lim_{t-\to t}\mathfrak{A}_{t-} = \mathfrak{A}_t$ - cf. (1.16). Thus, formula (4.8) holds for $\xi = F$ with F of form (4.9) and the $\xi_k = \mu(\Delta_k)$ with Δ_k, $k = 1,\ldots,m$, as disjoint simple sets.

B. Indeed the above result holds for any group of measurable *disjoint* sets $\Delta_1,\ldots,\Delta_m$. In fact it is enough to apply an approximation argument with $\Delta_k = \lim_{n\to\infty}\Delta_k^{(n)}$, $k = 1,\ldots,m$, by disjoint simple sets $\Delta_k^{(n)}$, $k = 1,\ldots,m$ $(n = 1,2,\ldots)$, such that $\mu(\Delta_k) = \lim_{n\to\infty}\mu(\Delta_k^{(n)})$ holds true in $L_2(\Omega)$ and for almost all $\omega\in\Omega$. Cf. (1.6) and (1.8). Accordingly for ξ and $\xi^{(n)}$ of type (4.9) with $\xi_k = \mu(\Delta_k)$, $k = 1,\ldots,m$, and $\xi_k^{(n)} = \mu(\Delta_k^{(n)})$, $k = 1,\ldots,m$, respectively we have also $\xi = \lim_{n\to\infty}\xi^{(n)}$ in $L_2(\Omega)$ and

$$D\xi = \lim_{n\to\infty} D\xi^{(n)}, \qquad \mathcal{D}\xi = \lim_{n\to\infty}\mathcal{D}\xi^{(n)} \tag{4.11}$$

in $L_2(\Theta\times\mathbb{T}\times\Omega)$. Thus

$$D\xi(\theta,t) = E\left(\mathcal{D}\xi(\theta,t)\,|\,\mathfrak{A}_t\right)$$

for almost all $(\theta,t)\in\Theta\times\mathbb{T}$, i.e. formula (4.8) holds for $D\xi$ as the element in $L_2^I(\Theta\times\mathbb{T}\times\Omega)\subseteq L_2(\Theta\times\mathbb{T}\times\Omega)$.

C. Clearly, formula (4.8) is valid for *all* $\xi = F$ which are linear combinations of functions (4.9) with $\xi_k = \mu(\Delta_k)$ on disjoint measurable sets Δ_k, $k = 1,\ldots,m$.

D. The formula (4.8) can be extended on all the functions characterized in the scheme (4.4)-(4.7). Let us define the scalar functions

$$\mathbb{D}\xi := \sum_{k=1}^{m}\partial_k^x F\cdot 1_{\Delta_k}$$

on the product space $\mathbb{R}\times\Delta\times\Omega$: $\Delta = \sum_{k=1}^m \Delta_k$, equipped with the *finite* product-type measure

$$L_0(dx,\theta,t)\times M(d\theta dt)\times P(d\omega), \quad (x,\theta,t,\omega)\in\mathbb{R}\times\Delta\times\Omega.$$

Here $L_0(dx,\theta,t)$, $(\theta,t)\in\Theta\times\mathbb{T}$, is equal to $\sigma^2(\theta,t)$ at the atom $x=0$ and to $x^2L(dx,\theta,t)$ on $\mathbb{R}\setminus[0]$. The functions

$$\mathbb{D}\xi = \mathbb{D}\xi(x,\theta,t,\omega), \quad (x,\theta,t,\omega)\in\mathbb{R}\times\Delta\times\Omega,$$

are elements of the standard space $L_2(\mathbb{R}\times\Delta\times\Omega)$ with norm

$$\|\mathbb{D}\xi\|_{L_2} := \left(\iiint_{\mathbb{R}\times\Delta\times\Omega}|\mathbb{D}\xi|^2 L_0(dx,\theta,t)\times M(d\theta dt)\times P(d\omega)\right)^{1/2}.$$

We have

$$\|\mathbb{D}\xi\|_{L_2} = |||\mathcal{D}\xi|||$$

for

$$\mathcal{D}\xi = \int_{\mathbb{R}}\mathbb{D}\xi L_0(dx,\theta,t), \quad (\theta,t,\omega)\in\Delta\times\Omega.$$

Cf. (4.5)-(4.7). The key-point of the approximation argument which will be applied is that, for $\xi = F$ and $\xi^{(n)} = F^{(n)}$, $n = 1, 2, \ldots$, the convergences

$$\|\xi - \xi^{(n)}\| \longrightarrow 0 \quad \text{and} \quad \|\mathbb{D}\xi - \mathbb{D}\xi^{(n)}\|_{L_2} \longrightarrow 0, \qquad n \to \infty, \tag{4.12}$$

imply the limits

$$D\xi = \lim_{n\to\infty} D\xi^{(n)} \quad \text{and} \quad \mathcal{D}\xi = \lim_{n\to\infty} \mathcal{D}\xi^{(n)}$$

in $L_2(\Theta \times \mathbb{T} \times \Omega)$. Note that in the coming considerations, we apply dominated point-wise convergence with the appropriate corresponding majorants in order to prove the convergences (4.12). To simplify the notation, we give the proof in the case $k = 1$ ($m = 1$), i.e. for $\xi_1 = \mu(\Delta_1)$.

For $F \in C_0^\infty(\mathbb{R})$, the convergence (4.12) holds with $\xi^{(n)} = F^{(n)}$ given by the partial sums $F^{(n)} = \Phi_n(F)$ of the Fourier series of F on $|\xi_1| \le h_n$ ($h_n \to \infty$, $n \to \infty$). In fact note that, for $n \to \infty$, we have

$$\partial_1^x F^{(n)} := \partial_1^x \Phi_n(F) = \Phi_n(\partial_1^x F)$$

whatever $x \in \mathbb{R}$ be. Next, for $\xi = F$: $F \in C_0^1(\mathbb{R})$ the convergence (4.12) holds with $\xi^{(n)} = F^{(n)}$:

$$F^{(n)} := F * \delta_n = \int_{\mathbb{R}} F(\xi_1 - x_1)\delta_n(x_1)dx_1 \in C_0^\infty(\mathbb{R})$$

with $\delta_n \in C_0^\infty(\mathbb{R})$ as the standard approximations to the delta-function. Here we have

$$\partial_1^x F^{(n)} := \partial_1^x (F * \delta_n) = (\partial_1^x F) * \delta_n, \quad n = 1, 2, \ldots$$

In general, for $\xi = F$: $F \in C^1(\mathbb{R})$, the convergence (4.12) holds with $\xi^{(n)} = F^{(n)}$ as the truncations $F^{(n)} = F \cdot w_n \in C_0^1(\mathbb{R})$. Here w_n is an appropriate approximation $w_n \in C_0^1(\mathbb{R})$ of the unit. Note that

$$\partial_1^x F^{(n)} := \partial_1^x (F \cdot w_n) = (\partial_1^x F) \cdot w_n + F \cdot (\partial_1^x w_n), \quad n = 1, 2, \ldots \quad \square$$

Example 4.3. Let μ be a mixture of the Gaussian stochastic measure on $\Theta_0 \times \mathbb{T}$ and Poisson (centred) stochastic measures (multiplied by the different scalars $x \neq 0$) on the corresponding space-time products $\Theta_x \times \mathbb{T}$. So, μ is a stochastic measure on the space-time product

$$\Theta \times \mathbb{T} = (\Theta_0 \times \mathbb{T}) \cup \sum_{x \neq 0} (\Theta_x \times \mathbb{T})$$

- cf. (1.10). For $\xi = F$ as a function in $C^1(\mathbb{R}^m)$ of the values $\xi_k = \mu(\Delta_k)$, $k = 1, \ldots, m$, on the *disjoint* sets Δ_k, $k = 1, \ldots, m$, in $\Theta \times \mathbb{T}$, the formula (4.8) can be written

$$\mathcal{D}\xi = \sum_{k=1}^m \Big[\partial_k^0 F 1_{\Theta_0 \times \mathbb{T}} + \sum_{x \neq 0} \partial_k^x F 1_{\Theta_x \times \mathbb{T}}\Big] \cdot 1_{\Delta_k}.$$

Remark 4.1. The formula (4.8) is in general *not* valid if in $\xi = F$ the function F is evaluated on values of μ which are not on *disjoint* sets.

5 The Anticipating Derivative and Integral

5.1 Definition and Related Properties

Following the discussion of the previous section we can now turn our attention to *all* the random variables $\xi = F$ where F is a linear combination of the random variables considered in the scheme (4.4)–(4.7). Also these linear combinations fit the scheme (4.4)–(4.7). Moreover we write

$$dom\mathcal{D} \subseteq L_2(\Omega) \tag{5.1}$$

for the *linear* domain of all elements in $L_2(\Omega)$ of the type $\xi = F$ characterized in (4.4)-(4.7), plus the limits $\xi = \lim_{n\to\infty} \xi^{(n)}$ in $L_2(\Omega)$ of the above type elements $\xi^{(n)}$, $n = 1, 2, \ldots$, for which the corresponding limits

$$\mathcal{D}\xi := \lim_{n\to\infty} \mathcal{D}\xi^{(n)}$$

exist $L_2(\Theta \times \mathbb{T} \times \Omega)$.

Note that whatever the representation $\xi = F$ (4.4) be, the corresponding stochastic function $\mathcal{D}\xi$ is a *unique* well-defined element of $L_2(\Theta \times \mathbb{T} \times \Omega)$. Moreover, we obtain the following result. See [19].

Theorem 5.1. *For all $\xi \in domD$, the stochastic functions $\mathcal{D}\xi$ are given by the well defined closed linear operator $\mathcal{D}$:*

$$L_2(\Omega) \supseteq dom\mathcal{D} \ni \xi \implies \mathcal{D}\xi \in L_2(\Theta \times \mathbb{T} \times \Omega), \tag{5.2}$$

with domain $dom\mathcal{D}$ dense in $L_2(\Omega)$.

Proof. For *some* (which can be *any*) partitions of $\Theta \times \mathbb{T}$, let us fix a family of the elements in $L_2(\Omega)$ which are of type (4.9) with the μ-values taken on disjoint simple sets in $\Theta \times \mathbb{T}$. Any linear combination of these elements admits the representation

$$\xi = F(\xi_1, \ldots, \xi_m)$$

with F as linear combination of the *different* elements

$$e^{i\sum_{k=1}^{m} \lambda_k \xi_k}, \quad \xi_k = \mu(\Delta_k), \quad k = 1, \ldots, m,$$

Accordingly, we can see that for *all* these linear combinations the corresponding formula (4.5) is given by the limit

$$\mathcal{D}\xi = \lim_{n\to\infty} \sum \frac{1}{M(\Delta)} E(\xi\mu(\Delta)|\mathfrak{A}_{]\Delta[})1_\Delta \tag{5.3}$$

taken in $L_2(\Theta \times \mathbb{T} \times \Omega)$. The sum is here taken on the sets Δ of the nth-series of the partitions of $\Theta \times \mathbb{T}$. Cf. (4.10). Clearly, this limit defines the *linear operator* $\mathcal{D}$:

$$dom\mathcal{D} \ni \xi \implies \mathcal{D}\xi \in L_2(\Theta \times \mathbb{T} \times \Omega) \tag{5.4}$$

on the *linear* domain $dom\mathcal{D} \subseteq L_2(\Omega)$. Let us show that *this* linear operator $\mathcal{D}$ is *closable.* Let $\Delta_0 \subseteq \Theta \times \mathbb{T}$ be a *simple set* and $\varphi_0 \cdot 1_{\Delta_0}$ be a simple function the element $\varphi_0 \in L_2(\Omega)$ as $\mathfrak{A}_{]\Delta_0[}$-measurable values on Δ_0. The limit (5.3) implies that

$$\iint_{\Theta\times\mathbb{T}} E\big[(\varphi_0 1_{\Delta_0})\mathcal{D}\xi\big]\,M(d\theta dt)$$
$$= \lim_{n\to\infty}\sum_{\Delta\subseteq\Delta_0} E\big[\varphi_0 E(\xi\mu(\Delta)|\mathfrak{A}_{]\Delta[})\big] = E\big[(\varphi_0\mu(\Delta_0))\xi\big]. \tag{5.5}$$

Hence, for the elements $\xi^{(n)} = F^{(n)}$, $n = 1, 2, \ldots$ in $dom\mathcal{D}$ such that $\lim_{n\to\infty}\xi^{(n)} = 0$ in $L_2(\Omega)$ and $\lim_{n\to\infty}\mathcal{D}\xi^{(n)} = \varphi$ in $L_2(\Theta \times \mathbb{T} \times \Omega)$, we have

$$\iint_{\Theta\times\mathbb{T}} E\Big[(\varphi_0 \cdot 1_{\Delta_0})\cdot\varphi\Big]M(d\theta dt) = 0\,.$$

The considered simple functions $\varphi_0 \cdot 1_{\Delta_0}$ constitute a complete system in $L_2(\Theta \times \mathbb{T} \times \Omega)$. Cf. Theorem 1.4. Hence, the above equation implies that $\varphi = 0$. Thus the linear operator (5.3)-(5.4) is closable. Hence it admits the standard extension on *all* the random variables $\xi = F$ as functions of the type (4.9) involving all the *disjoint* sets $\Delta_k \subseteq \Theta \times \mathbb{T}$: $M(\Delta_k) < \infty$, $k = 1, \ldots, m$. Cf. (4.11). The next standard extension of the closable linear operator (5.3)-(5.4) up to the closed linear operator (5.1)-(5.2) is done by approximation arguments with respect to the limits (4.11)-(4.12), see the proof of Theorem 4.2. □

Note that formula (4.8) holds for all the elements $\xi \in dom\mathcal{D}$ in the domain of the closed linear operator (5.1)-(5.2) and, to repeat, it is

$$D\xi(\theta,t) = E\Big[\mathcal{D}\xi(\theta,t)|\mathfrak{A}_t\Big], \quad (\theta,t) \in \Theta \times \mathbb{T}. \tag{5.6}$$

According to this relationships with the non-anticipating derivative D, we call $\mathcal{D}$ the *anticipating derivative.*

Remark 5.1. In general, the formula (4.5) for the anticipating derivative $\mathcal{D}\xi$ of the random variable (4.4): $\xi = F$ as a function of the values of μ, is *non valid* if these values are taken on sets in $\Theta \times \mathbb{T}$ which are *not disjoint.*

In addition to the scheme (4.4)-(4.7), let us consider the functions $\xi = F(\xi_1, \ldots, \xi_m)$ with $F \in C^1(\mathbb{R}^m)$ and where ξ_k, $k = 2, \ldots, m$, are the stochastic integrals

$$\xi_k = \iint_{\Theta\times\mathbb{T}} \varphi_k\mu(d\theta\, dt)$$

with the deterministic integrands φ_k having the *disjoint* supports $\Delta_k = \{(\theta,t) : \varphi_k(\theta,t) \neq 0\}$, $k = 1, \ldots, m$. We introduce the stochastic functions

$$\mathcal{D}\xi := \sum_{k=1}^{m}\Big[\partial_k^0 F\,\sigma^2(\theta,t)$$
$$+ \int_{\mathbb{R}\setminus[0]} \partial_k^x F\,x^2 L(dx,\theta,t)\Big]1_{\Delta_k}(\theta,t), \quad (\theta,t) \in \Theta \times \mathbb{T}. \tag{5.7}$$

with a new definition for the functions

$$\partial_k^x F := \begin{cases} \frac{\partial}{\partial \xi_k} F(\ldots, \xi_k, \ldots)\, \varphi_k(\theta, t), & x \neq 0, \\ \frac{1}{x}\big[F(\ldots, \xi_k + x\varphi(\theta, t), \ldots) - F(\ldots, \xi_k, \ldots)\big], & x = 0. \end{cases}$$

The stochastic functions $\mathcal{D}\xi$ introduced above satisfy the condition (4.6) with the *newly* defined components $\partial_k^x F$, $k = 1, \ldots, m$. In this setting Theorem 5.1 implies the following result.

Corollary 5.1. *The elements $\xi = F \in L_2(\Omega)$, defined here above belong to dom$\mathcal{D}$ and the formula* (5.7) *represents the anticipating derivative $\mathcal{D}\xi$.*

Proof. In the case $F \in C_b^1(\mathbb{R}^m)$ and the integrands φ_k, $k = 1, \ldots, m$, are linear combinations of indicators of the *disjoint* sets Δ_{jk}, $j = 1, \ldots, m_k$ ($k = 1, \ldots, m$), i.e.

$$\varphi_k = \sum_{j=1}^{m_k} c_{jk} \cdot 1_{\Delta_{jk}},$$

the formula (5.7) gives the anticipating derivative $\mathcal{D}\xi$ of $\xi = F(\xi_1, \ldots, \xi_m)$ as function of $\xi_{jk} = \mu(\Delta_{jk})$, $j = 1, \ldots, m_k$ ($k = 1, \ldots, m$). Cf. (4.5). By standard approximation arguments with respect to the limit (4.12) - cf. also (5.1), formula (5.7) admits extension on all elements characterized by the condition (4.6) involving the newly here above defined components $\partial_k^x F$, $k = 1, \ldots, m$. □

Example 5.1. Let $I^p \varphi_p$, $p = 1, 2, \ldots$, be the Itô p-multiple integrals with respect to a general stochastic measure μ in the scheme (3.3)-(3.1). The anticipating derivative is

$$\mathcal{D} I^p \varphi_p = I^{p-1} \hat{\varphi}_p(\cdot, \theta, t), \quad (\theta, t) \in \Theta \times \mathbb{T},$$

with the integrands

$$\hat{\varphi}_p := \sum_{j=1}^{p} \varphi_p(\ldots, \theta, t, \ldots)$$

depending on $(\theta, t) \in \Theta \times \mathbb{T}$ as parameter. The couple (θ, t) comes in at the place of the corresponding couples (θ_j, t_j), $j = 1, \ldots, p$. Here we have

$$\|\mathcal{D}\xi\|_{L_2} = p^{1/2} \|\xi\|, \quad p = 1, 2, \ldots .$$

All the elements

$$\xi = \sum_{p=0}^{\infty} \oplus \xi_p : \qquad \xi_0 = E\xi,\ \xi_p = I^p \varphi_p,\ p = 0, 1, \ldots,$$

with

$$\sum_{p=1}^{\infty} p\, \|\xi_p\|^2 < \infty$$

belong to the domain $dom\mathcal{D}$ of the anticipating derivative and

$$\mathcal{D}\xi = \sum_{p=1}^{\infty} \oplus I^{p-1}\hat{\varphi}_p. \tag{5.8}$$

See e.g. [16, 19].

Example 5.2. For μ as a general Gaussian-Poisson mixture, the elements $\xi \in dom\mathcal{D}$ characterized in Example 5.1 represent the whole domain $dom\mathcal{D}$. A key-point to show this is that, for all random variables of the form $\xi = e^{i\sum_{k=1}^{m}\lambda_k\xi_k}$ - cf. (4.9), the approximations

$$\xi = \lim_{q\to\infty} \xi^{(q)}, \qquad \mathcal{D}\xi = \lim_{q\to\infty} \mathcal{D}\xi^{(q)}$$

hold with the polynomials

$$\xi^{(q)} = \sum_{p=0}^{q} \Big(i\sum_{k=1}^{m} \lambda_k\xi_k\Big)^p \in dom\mathcal{D}, \qquad q = 1, 2, \ldots,$$

of the values of μ. Cf. Theorem 3.2 and Theorem 3.3.

The anticipating derivative (5.1)-(5.2) and its relationship (5.6) with the non-anticipating derivative can be regarded as in the same line as the *Malliavin derivative* [38] and the *Clark-Haussmann-Ocone* formula [7,23,43] within the stochastic calculus for the Wiener process. Here we would like also to refer to e.g. [1, 2, 4, 9, 12, 20, 21, 32, 37, 40, 41, 44, 46–48] and references therein, for some further developments of the Malliavin calculus with respect to the Wiener process and the Poisson process and Poisson (centred) random measure. See [35] for some results in the case of Lévy processes.

5.2 The Closed Anticipating Extension of the Itô Non-anticipating Integral

In this final section we present some results of anticipating calculus. In the framework of Wiener processes, we have the Skorohod integral [50] as the adjoint operator to the Malliavin derivative. Similar arguments can be achieved in the case of Poisson (centred) random measures. We can refer e.g. [12, 29, 31, 40, 41] and references therein.

Here as usual in this paper we consider a general stochastic measure of type (1.17) on $\Theta \times \mathbb{T}$. See [19], see also [17].

Theorem 5.2. *The closed linear operator $\mathfrak{J} = \mathcal{D}^*$, adjoint to the anticipating derivtive* (5.1)-(5.2)*:*

$$L_2(\Theta \times \mathbb{T} \times \Omega) \supseteq dom\mathfrak{J} \ni \varphi \Longrightarrow \mathfrak{J}\varphi \ni L_2(\Omega) \tag{5.9}$$

represents the extension

$$\mathfrak{J}\varphi = \iint_{\Theta\times\mathbb{T}} \varphi\, \mu(d\theta\, dt)$$

of the Itô non-anticipating integral on all the stochastic functions φ in the domain $dom\mathfrak{J}$ of $\mathfrak{J}$ dense in $L_2(\Theta \times \mathbb{T} \times \Omega)$.

Proof. The key-point of the proof is equation (5.5) which can be extended by the standard approximation arguments to hold on all the simple functions of the form $\varphi \cdot 1_\Delta$ with $\mathfrak{A}_{]\Delta[}$-measurable values φ on the indicated sets $\Delta \subseteq \Theta \times \mathbb{T}$: $M(\Delta) < \infty$, i.e.

$$\iint_{\Theta\times\mathbb{T}} E\Big[(\varphi \cdot 1_\Delta) \cdot \mathcal{D}\xi\Big] M(d\theta\, dt) = E\Big[(\varphi \cdot \mu(\Delta)) \cdot \xi\Big], \qquad \xi \in dom\mathcal{D}. \quad (5.10)$$

The linear combinations of the above type simple functions are dense in $L_2(\Theta \times \mathbb{T} \times \Omega)$. Cf. Theorem 1.4. Equation (5.10) shows that the simple functions $\varphi \cdot 1_\Delta$ belong to the domain $dom\mathfrak{J}$ of the adjoint linear operator $\mathfrak{J} = \mathcal{D}^*$ and that

$$\mathfrak{J}(\varphi \cdot 1_\Delta) = \varphi \cdot \mu(\Delta). \quad (5.11)$$

Clearly, $\mathfrak{J}$ coincides with the non-anticipating integral I on the *non-anticipating* simple functions and also on any non-anticipating function thanks to the limit

$$\varphi = \lim_{n\to\infty} \varphi^{(n)}, \text{ i.e. } \quad \|\varphi - \varphi^{(n)}\|_{L_2} \longrightarrow 0, \quad n \to \infty,$$

in $L_2(\Theta \times \mathbb{T} \times \Omega)$ of the non-anticipating simple functions $\varphi^{(n)}$, $n = 1, 2, \ldots$. The corresponding limit

$$I\varphi = \lim_{n\to\infty} I\varphi^{(n)} = \lim_{n\to\infty} \mathfrak{J}\varphi^{(n)}$$

in $L_2(\Omega)$ implies that φ belongs to the domain of the *closed* linear operator $\mathfrak{J}$ and that $\mathfrak{J}\varphi = I\varphi$. Cf. (2.3)-(2.4). □

In relation to the anticipating derivative $\mathcal{D}$, we call $\mathfrak{J} = \mathcal{D}^*$ the *anticipating integral.* Note that $\mathcal{D} = \mathfrak{J}^*$ is the adjoint linear operator to $\mathfrak{J}$. Cf. the duality $D = I^*$ for the non-anticipating derivative D and the Itô non-anticipating integral $I = D^*$.

Example 5.3. In case μ is the Gaussian-Poisson mixture, the anticipating integral $\mathfrak{J}$ can be completety characterized in terms of the multiple Itô integrals $I^p\varphi_p$, $p = 2, \ldots$, and the anticipating derivatives

$$\mathcal{D}I^p\varphi_p = I^{p-1}\hat{\varphi}_p, \quad p = 1, 2, \ldots .$$

Cf. Example 5.1 and Example 5.2. Namely, the domain $dom\mathfrak{J}$ consists of all stochastic functions admitting the representation

$$\varphi = \varphi^0 \oplus \sum_{p=1}^{\infty} \oplus \mathcal{D}\xi_p, \qquad \xi_p = I^p\varphi_p, \quad p = 1, 2, \ldots$$

$$\text{such that} \quad \sum_{p=1}^{\infty} p^2 \|\xi_p\|^2 < \infty.$$

Here the components φ^0 are orthogonal to the range of the anticipating derivative $\mathcal{D}$. Correspondingly, we have

$$\mathfrak{I}\varphi = \sum_{p=1}^{\infty} \oplus p\xi_p.$$

See [16, 19].

Remark 5.2. In general, for an integrand of the form $\varphi \cdot 1_\Delta$ where its value φ on $\Delta \subseteq \Theta \times \mathbb{T}$ is an element in $L_2(\Omega)$ which is *not* $\mathfrak{A}_{]\Delta[}$-measurable, it occurs that

$$\mathfrak{I}\left(\varphi \cdot 1_\Delta\right) \neq \varphi \cdot \mu(\Delta).$$

To illustrate we can mention that

$$\mathfrak{I}\left(\mu(\Delta) \cdot 1_\Delta\right) = \mu(\Delta)^2 - M(\Delta),$$

if μ is Gaussian, and

$$\mathfrak{I}\left(\mu(\Delta) \cdot 1_\Delta\right) = \mu(\Delta)^2 - \mu(\Delta) - M(\Delta),$$

if μ is a Poisson (centred) stochastic measure. Remind that μ is of the type $E\mu = 0$, $E\mu^2 = M$.

References

1. K. Aase, B. Øksendal, N. Privault and J. Ubøe, White noise generalizations of the Clark-Haussmann-Ocone theorem with application to mathematical finance, *Finance and Stochastics*, **4**, (2000), 465–496.
2. F. E. Benth, G. Di Nunno, A. Løkka, B. Øksendal and F. Proske, Explicit representation of the minimal variance portfolio in a market driven by Lévy processes, *Math. Finance*, f 13, (2003), 54–72.
3. J. Bertoin, *Lévy Processes*, Cambridge University Press 1996.
4. K. Bichteler, J. B. Gravereaux and J. Jacod, *Malliavin Calculus for Processes with Jumps*, Gordon and Breach Science Publisher, New York, 1987.
5. R. Cairoli and J. B. Walsh, Stochastic integrals in the plane, *Acta Math.*, **134** (1975), 111–183.
6. C. S. Chou and P. A. Meyer, Sur la représentation des martingales comme intégrales stochastiques dans les processus ponctuels, *Séminaire de Probabilités IX, Lecture Notes in Math.*, **465**, (1975), 226–236, Springer, Berlin.
7. J. M. C. Clark, The representation of functionals of Brownian motion as stochastic integrals, *Ann. Math. Statist.*, **41**, (1970), 1282–1295; Correction to the paper, *Ann. Math. Statist.*, **42**, (1971), 1778.
8. B. de Finetti, Sulle funzioni ad incremento aleatorio, *Atti Accad. Naz. Lincei*, **10**, (1929), 163–168.
9. M. de Faria, M. J. Oliveira and L. Streit, A generalized Clark-Ocone formula, *Random Oper. Stochastic Equations*, **8** (2000), 163–174.

10. C. Dellacherie, Intégrales stochastiques par rapport aux processus de Wiener ou de Poisson, *Séminaire de Probabilités VIII, Lecture Notes in Math.*, **381**, (1974), 25–26, Springer, Berlin.
11. C. Dellacherie and P. A. Meyer, *Probabilities and Potential B*, North Holland, Amsterdam 1982.
12. A. Dermoune, P. Kree and L. Wu, Calcul stochastique non adapté par rapport à la mesure aléatoire de Poisson, Séminaires de Probabilités XXII, *Lect. Notes Math.*, 1321, 477–484, Springer, Berlin, 1988.
13. G. Di Nunno, On Stochastic Differentiation with Application to Minimal Variance Hedging, Tesi di Dottorato di Ricerca, Università degli Studi di Pavia, 2002.
14. G. Di Nunno, Stochastic integral representations, stochastic derivatives and minimal variance hedging, *Stochastics Stochastics Rep.*, **73**(2002), 181–198.
15. G. Di Nunno, Random fields evolution: non-anticipating integration and differentiation, *Theory of Probability and Math. Statistics*, **66** (2002), 82–94.
16. G. Di Nunno, On orthogonal polynomials and the Malliavin derivative for Lévy stochastic measures, *Preprint Series in Pure Mathematics*, University of Oslo, **10**, 2004. To appear in *SMF, Seminaires et Congrès.*
17. G. Di Nunno, Random fields: Skorohod integral and Malliavin derivative, *Preprint Series in Pure Mathematics*, University of Oslo, **36**, 2004.
18. G. Di Nunno, Random fields evolution: non-anticipating derivative and differentiation formulae, *Preprint Series in Pure Mathematics*, University of Oslo, **1**, 2006.
19. G. Di Nunno, Random fields: the de Finetti-Kolmogorov law and related stochastic calculus, *Manuscript*, University of Oslo, 2006.
20. G. Di Nunno, B. Øksendal and F. Proske, White noise analysis for Lévy processes, *Journal of Functional Analysis*, **206** (2004), 109–148.
21. G. Di Nunno, B. Øksendal and F. Proske, Malliavin calculus for Lévy processes with Applications to Finance, Springer, *Manuscript* (version June 2006).
22. G. Di Nunno and Yu. A. Rozanov, On stochastic integration and differentiation, *Acta Appl. Math.*, **58**, (1999), 231–235.
23. U.G. Haussmann, On the integral representation of functionals of Itô processes, *Stochastics*, **3**, (1979), 17–28.
24. A. Ionescu Tulcea and C. Ionescu Tulcea, *Topics in the Theory of Lifting*, Springer-Verlag 1969.
25. K. Itô, On stochastic processes I. Infinitely divisible laws of probability, *Jap. J. Math.*, **18** (1942), 252–301.
26. K. Itô, Stochastic integrals, *Proc. Imp. Acad. Japan*, **20** (1944), 519–524.
27. K. Itô, Multiple Wiener integral, *J. Math. Soc. Japan*, **3** (1951), 157–169.
28. K. Itô, Spectral type of the shift transformation of differential processes with stationary increments, *Trans. Am. Math. Spoc.*, **81** (1956), 253–263.
29. Y. Kabanov, On extended stochastic integrals. Theory of Probability and its Applications 20 (1975), 710–722.
30. O. Kallenberg, *Random measures*, Academic Press Inc. 1986.
31. A. D. Kaminsky, Extended stochastic calculus for the Poisson random measures. Nats. Akad. Nauk Ukrain, Inst. Mat. Preprint 15, 1996.
32. I. Karatzas, D. Ocone and J. Li, An extension of Clark's formula, *Stochastics Stochastics Rep.*, **37**, (1991), 127–131.
33. A. Ya. Khinchine, A new derivation of a formula of Paul Lévy, *Bull. Moscow Gov, Univ.*, **1** (1937), 1–5.

34. A. N. Kolmogorov, Sulla forma generale di un processo stocastico omogeneo, *Atti Accad. Naz. Lincei*, **15**, (1932), 805–808, 866–869.
35. J. A. Léon, J. L. Solé, F. Utzet and J. Vives, On Lévy processes, Malliavin calculus and market models with jumps, *Finance and Stochastics*, **6** (2002), 197–225.
36. P. Lévy, Sur les intégrales dont les éléments sont des variables aléatoires indépendantes. *Ann. Scuola Norm. Sup. Pisa*, **3** (1934), 337–366, **4** (1934), 217–218.
37. A. Løkka, Martingale representation and functionals of Lévy processes. Stochastic Analysis and Applications 22 (2004), 867–892.
38. P. Malliavin, *Stochastic Analysis*, Springer-Verlag, New York 1997.
39. H. P. Mc Kean, *Stochastic Integrals*, Academic Press, New York - London 1969.
40. D. Nualart, *The Malliavin Calculus and Related Topics*, Springer-Verlag, New York 1995.
41. D. Nualart and E. Pardoux, Stochastic calculus with anticipating integrands, *Probab. Theory Related Fields*, **78**, (1988), 535–581.
42. D. Nualart and W. Schoutens, Chaotic and predictable representations for Lévy processes, *Stochastic Process. Appl.*, **90**, (2000), 109–122.
43. D. Ocone, Malliavin's calculus and stochastic integral representation of functionals of diffusion processes, *Stochastics*, **12**, (1984), 161–185.
44. D. Ocone and I. Karatzas, A generalized Clark representation formula with application to optimal portfolios, *Stochastics Stochastics Rep.*, **34**, (1991), 187–220.
45. G. Peccati, Explicit formulae for time-space Brownian chaos, *Bernoulli, 9* (2003), 25–48.
46. J. Picard, On the existence of smooth densities for jump processes, *Probab. Theory Relat. Fields*, **105**, 1996, 481–511.
47. N. Privault, J. L. Solé and J. Vives, Chaotic Kabanov formula for the Azéma martingales. Bernoulli 6 (2000), 633–651.
48. N. Privault and J. Wu, Poisson stochastic integration in Hilbert spaces. Ann. Math. Blaise Pascal 6 (1999), 41–61.
49. K. Sato, *Lévy Processes and Infinitely Divisible Distributions*, Cambridge University Studies in Advanced Mathematics, Cambridge University Press, Cambridge 1999.
50. A. V. Skorohod, On a generalization of stochastic integral, *Theor. Probability Appl.*, **20** (1975), 223–238.
51. A. V. Skorohod, *Random Processes with Independent Increments*, Kluwer, Dordrecht, 1991.
52. D. W. Stroock, Homogeneous chaos revisited, *Séminaires de Probabilités XXI*, Lecture Notes in Math. 1247, 1–8. Springer-Verlag 1987.
53. D. Surgailis, On multiple Poisson stochastic integrals and associated Markov semigroups. Probab. Math. Statist. 3 (1984), 217–239.

An Application of Probability to Nonlinear Analysis

Eugene B. Dynkin

Department of Mathematics, Cornell University, Ithaca, NY 14853, USA.
ebd1@cornell.edu. Partially supported by National Science Foundation Grant DMS-0503977

Summary. The Martin boundary theory allows to describe all positive solutions of a linear elliptic equation in an arbitrary domain E of a Euclidean space $\mathbb{R}^d$. Our goal is to describe all positive solutions of a semilinear equation $Lu = \psi(u)$. As a result of efforts of probabilists and analysts since early 1990s, now we have a solution of this problem for the equation $\Delta u = u^\alpha$ with $1 < \alpha \leq 2$ in a bounded smooth domain E. The present article contains an exposition of the theory developed to obtain this solution.[1] The central role is played by the boundary trace theory. A survey of this theory is given in Part One. In Part Two we outline the principal steps needed to construct an arbitrary positive solution starting from its trace.

Our main probabilistic tool is (L, ψ)-superdiffusions.

Part One. Trace Theory

The trace theory is applicable to a general equation

$$Lu = \psi(u) \quad \text{in } E \tag{I.1}$$

where L is a second order elliptic operator, E is an arbitrary domain in $\mathbb{R}^d$ and ψ is a continuously differentiable convex function on $[0, \infty)$ subject to the conditions

(i) $\psi(u) > 0$ for $u > 0, \psi(0) = 0$.
(ii) There is a constant a such that $\psi(2u) \leq a\psi(u)$ for all u.
(iii) $\int_N^\infty ds \, \left[\int_0^s \psi(u) \, du\right]^{-1/2} < \infty$ for some $N > 0$.

Under these conditions the class $\mathcal{U}$ of all positive solutions of (I.1) is closed under the pointwise convergence.

The trace of a solution u is a pair (Γ, ν) where Γ is a Borel subset of ∂E and ν is a σ-finite measure on $\partial E \setminus \Gamma$. [For a smooth domain E, ∂E is the geometrical boundary of E; in general, this is the Martin boundary.]

[1] Complete proofs can be found in the books [Dyn02] and [Dyn04].

A rough version of the trace used in earlier work of Le Gall, Dynkin–Kuznetsov and Marcus–Véron is adequate for small dimensions d: in this case, a solution is uniquely defined by its rough trace. However an example due to Le Gall shows that, in general, infinite many solutions can have the same rough trace. In 1998 Dynkin and Kuznetsov introduced a concept of the fine trace. The solutions in Le Gall's example have distinct fine traces. In [DK98] all values of the fine trace were described and a 1-1 correspondence was established between them and a class of solutions which we call σ-moderate.[2] Proofs of these results are presented in Chapter 11 of [Dyn02]. In the Epilogue to [Dyn02], a crucial outstanding question was formulated:

Are all the solutions σ-moderate?

In the case of the equation $\Delta u = u^2$ in a domain of class C^4, a positive answer to this question was given in the thesis of Mselati [Mse02] - a student of J.-F. Le Gall.[3] However his principal tool - the Brownian snake - is not applicable to more general equations. In a series of publications by Dynkin and Kuznetsov, Mselati's result was extended, by using a superdiffusion instead of the snake, to the equation $\Delta u = u^\alpha$ with $1 < \alpha \leq 2$. A systematic presentation of the proofs is contained in the book [Dyn04].

In Section 1 we give the definition of the fine trace and formulate its fundamental properties. In Section 2 we explain how these properties can be established by using probabilistic tools: superdiffusions and their relation to conditional diffusions.

Since we consider only the fine trace, we drop the word fine.

1 Definition and Properties of Trace

1.1 Moderate and σ-Moderate Solutions

We denote by $\mathcal{U}$ the set of all positive solutions of the equation (I.1) and by $\mathcal{H}$ the set of all positive solutions of the equation

$$Lh = 0 \quad \text{in } E. \tag{1.1}$$

We call solutions of (1.1) *harmonic functions.*

If E is smooth[4] and if $k(x, y)$ is the *Poisson kernel*[5] of L in E, then the formula

$$h_\nu(x) = \int_{\partial E} k(x, y)\nu(dy) \tag{1.2}$$

establishes a 1-1 correspondence between the set $\mathcal{M}(\partial E)$[6] and the set $\mathcal{H}$.

[2] The definition of this class is given in Section 1.

[3] The dissertation of Mselati was published in 2004 (see [Mse04]).

[4] We use the name smooth for open sets of class $C^{2,\lambda}$ unless another class is indicated explicitely.

[5] For an arbitrary domain, $k(x, y)$ should be replaced by the Martin kernel and ∂E should be replaced by a certain Borel subset E' of the Martin boundary (see Chapter 7 in [Dyn02]).

[6] We denote by $\mathcal{M}(S)$ the set of all finite measures on S.

A solution u is called *moderate* if it is dominated by a harmonic function. There exists a 1-1 correspondence between the set $\mathcal{U}_1$ of all moderate solutions and a subset $\mathcal{H}_1$ of $\mathcal{H}$: $h \in \mathcal{H}_1$ is the minimal harmonic function dominating $u \in \mathcal{U}_1$, and u is the maximal solution dominated by h. We put $\nu \in \mathcal{N}_1$ if $h_\nu \in \mathcal{H}_1$. We denote by u_ν the element of $\mathcal{U}_1$ corresponding to h_ν.

An element u of $\mathcal{U}$ is called *σ-moderate solutions* if there exist $u_n \in \mathcal{U}_1$ such that $u_n(x) \uparrow u(x)$ for all x. The labeling of moderate solutions by measures $\nu \in \mathcal{N}_1$ can be extended to σ-moderate solutions by the convention: if $\nu_n \in \mathcal{N}_1$, $\nu_n \uparrow \nu$ and if $u_{\nu_n} \uparrow u$, then put $\nu \in \mathcal{N}_0$ and $u = u_\nu$.

1.2 Lattice Structure in $\mathcal{U}$

We write $u \le v$ if $u(x) \le v(x)$ for all $x \in E$. This determines a partial order in $\mathcal{U}$. For every $\tilde{\mathcal{U}} \subset \mathcal{U}$, there exists a unique element u of $\mathcal{U}$ with the properties: (a) $u \ge v$ for every $v \in \tilde{\mathcal{U}}$; (b) if $\tilde{u} \in \mathcal{U}$ satisfies (a), then $u \le \tilde{u}$.[7] We denote this element Sup $\tilde{\mathcal{U}}$.

For every $u, v \in \mathcal{U}$, we put $u \vee v = \text{Sup}\{u, v\}$ and we put $u \oplus v = \text{Sup}\, W$ where W is the set of all $w \in \mathcal{U}$ such that $w \le u + v$. Note that $u \oplus v$ and $u \vee v$ are moderate if u and v are moderate and they are σ-moderate if so are u and v.

In general, Sup $\tilde{\mathcal{U}}$ does not coincide with the pointwise supremum (the latter does not belong to $\mathcal{U}$). However, both are equal if $u \vee v \in \tilde{\mathcal{U}}$ for every $u, v \in \tilde{\mathcal{U}}$. Moreover, in this case there exist $u_n \in \tilde{\mathcal{U}}$ such that $u_n(x) \uparrow u(x) = \text{Sup}\, \tilde{\mathcal{U}}$ for all $x \in E$. Therefore, if $\tilde{\mathcal{U}}$ is closed under $\vee$ and if it consists of moderate solutions, then Sup $\tilde{\mathcal{U}}$ is σ-moderate. Since $u \vee v$ is moderate for all moderate u and v, to every Borel subset Γ of ∂E there corresponds a σ-moderate solution

$$u_\Gamma = \text{Sup}\{u_\nu : \nu \in \mathcal{N}_1, \nu \text{ is concentrated on } \Gamma\}. \tag{1.3}$$

We also associate with Γ another solution w_Γ. First, we define w_K for closed K by the formula

$$w_K = \text{Sup}\{u \in \mathcal{U} : u = 0 \quad \text{on } \partial E \setminus K\}. \tag{1.4}$$

For every Borel subset Γ of ∂E, we put

$$w_\Gamma = \text{Sup}\{w_K : \text{closed } K \subset \Gamma\}. \tag{1.5}$$

Proving that $u_\Gamma = w_\Gamma$ was a key part of the program outlined in [Dyn02]. A sketch of the proof will be presented in Section 4.

---

[7] The existence is proved in Section 8, Chapter 5 in [Dyn02].

1.3 Singular Points of a Solution u

We consider classical solutions of (I.1) which are twice continuously differentiable in E. However they can tend to infinity as $x \to y \in \partial E$. We say that y is a *singular point of* u if it is a point of rapid growth of $\psi'(u)$. [A special role of $\psi'(u)$ is due to the fact that the tangent space to $\mathcal{U}$ at point u is described by the equation $Lv = \psi'(u)v$.] An analytic definition of rapid growth involves the Poisson kernel (or Martin kernel) $k_\ell(x,y)$ of the operator $Lu - \ell u$. Namely, $y \in \partial E$ is a point of rapid growth for a positive continuous function ℓ if $k_\ell(x,y) = 0$ for all $x \in E$.

A transparent probabilistic definition of singular points is given in Section 2.5.

We say that a Borel subset Γ of ∂E is *f-closed* if Γ contains all singular points of the solution u_Γ defined by (1.3).

1.4 Definition and Properties of Trace

The trace of $u \in \mathcal{U}$ (which we denote $\mathrm{Tr}(u)$) is defined as a pair (Γ, ν) where Γ is the set of all singular points of u and ν is a measure on $\partial E \setminus \Gamma$ given by the formula

$$\nu(B) = \sup\{\mu(B) : \mu \in \mathcal{N}_1, \mu(\Gamma) = 0, u_\mu \le u\}. \tag{1.6}$$

We have

$$u_\nu = \mathrm{Sup}\{\text{moderate } u_\mu \le u \text{ with } \mu(\Gamma) = 0\}$$

and therefore u_ν is σ-moderate.

The trace of every solution u has the following properties:[8]

1.4.A. Γ is a Borel f-closed set;[9] ν is a σ-finite measure of class $\mathcal{N}_0$ such that $\nu(\Gamma) = 0$ and all singular points of u_ν belong to Γ.

1.4.B. If $\mathrm{Tr}(u) = (\Gamma, \nu)$, then

$$u \ge u_\Gamma \oplus u_\nu. \tag{1.7}$$

Moreover, $u_\Gamma \oplus u_\nu$ is the maximal σ-moderate solution dominated by u.

1.4.C. Suppose that (Γ, ν) is an arbitrary pair subject to the condition 1.4.A. If $\mathrm{Tr}(u_\Gamma \oplus u_\nu) = (\Gamma', \nu)$, then the symmetric difference between Γ and Γ' is not charged by any measure $\mu \in \mathcal{N}_1$. Moreover, $u_\Gamma \oplus u_\nu$ is the minimal solution with this property and the only one which is σ-moderate.

[8] See Theorems 7.1–7.2 in Chapter 11 of [Dyn02].

[9] This part will be also proved in Section 2.5 below.

2 Diffusions and Superdiffusions

2.1 L-Diffusion and Its Transformations

A diffusion describes a random motion of a particle. An example is the Brownian motion in $\mathbb{R}^d$. This is a Markov process with continuous paths and with the transition density

$$p_t(x,y) = (2\pi t)^{-d/2} e^{-|x-y|^2/2t}$$

which is the fundamental solution of the heat equation

$$\frac{\partial u}{\partial t} = \frac{1}{2}\Delta u.$$

A Brownian motion in a domain E can be obtained by killing the path at the first exit time from E. By replacing $\frac{1}{2}\Delta$ by an elliptic operator L, we define a Markov process (ξ_t, Π_x) called *L-diffusion.*

Suppose that (ξ_t, Π_x) is an L-diffusion in E with the transition density $p_t(x,y)$. To every $h \in \mathcal{H}$ there corresponds a finite measure Π_x^h such that, for all $0 < t_1 < \cdots < t_n$ and every Borel subsets $B_1, \dots, B_n$ of E,

$$\begin{aligned}&\Pi_x^h\{\xi_{t_1} \in B_1, \dots, \xi_{t_n} \in B_n\}\\ &\quad = \int_{B_1} dz_1 \dots \int_{B_n} dz_n\, p_{t_1}(x,z_1) p_{t_2-t_1}(z_1,z_2) \dots p_{t_n-t_{n-1}}(z_{n-1},z_n) h(z_n).\end{aligned} \tag{2.1}$$

Note that $\Pi_x^h(\Omega) = h(x)$ and therefore $\hat{\Pi}_x^h = \Pi_x^h / h(x)$ is a probability measure. $(\xi_t, \hat{\Pi}_x^h)$ is a Markov process with continuous paths and with the transition density

$$p_t^h(x,y) = \frac{1}{h(x)} p_t(x,y) h(y).$$

For every $y \in \partial E$, we put $\Pi_x^y = \Pi_x^h$ with $h(x) = k(x,y)$. The process $(\xi_t, \hat{\Pi}_x^y)$ can be interpreted as an L-diffusion conditioned to exit from E at point y:

$$\hat{\Pi}_x^y\{C\} = \Pi_x\{C | \xi_{\tau_E} = y\}$$

where τ_E is the first exit time of ξ_t from E.

2.2 (L, ψ)-Superdiffusion

An (L, ψ)-superdiffusion is a model of random evolution of a cloud of particles. Each particle performs an L-diffusion. It dies at a random time leaving a random offspring of size controlled by the function ψ. All children move independently of each other (and of the family history) with the same transition and procreation mechanism as the parent.

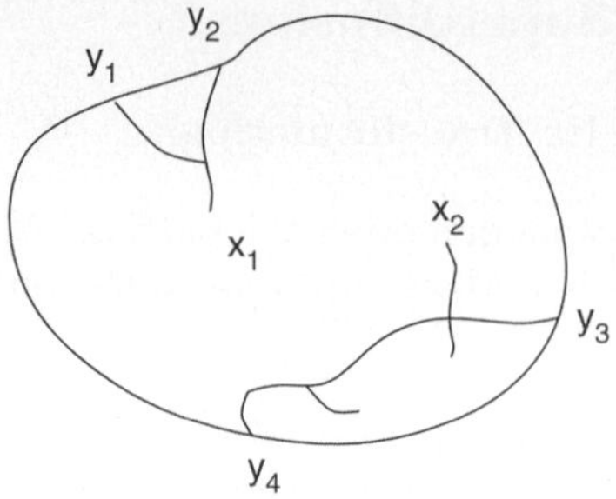

Fig. 1.

Superdiffusions appeared, first, (under the name "continuous state branching processes") in a pioneering paper of S. Watanabe [Wat68]. Important contributions to the theory of these processes were made by Dawson and Perkins.

We consider a superdiffusion as a family of the exit measures (X_D, P_μ) from open sets $D \subset E$. An intuitive picture of (X_D, P_μ) is explained on Figure 1 (borrowed from [Dyn02]).

Here we have a scheme of a process started by two particles located at points x_1, x_2 in D. The first particle produces at its death time two children that survive until they reach ∂D at points y_1, y_2. The second particle has three children. One reaches the boundary at point y_3, the second one dies childless and the third one has two children. Only one of them hits ∂D at point y_4. The initial and exit measure are described by the formulae

$$\mu = \sum \delta_{x_i}, \quad X_D = \sum \delta_{y_i}.$$

To get an (L, ψ)-superdiffusion, we pass to the limit as the mass of each particle and its expected life time tend to 0 and an initial number of particles tends to infinity. We refer for details to [Dyn02].

2.3 Superdiffusions as a Special Class of Branching Exit Markov Systems

The concept of a *branching exit Markov [BEM] system* (in a more general setting) is introduced in [Dyn02], Chapter 3. Suppose that to every $D \subset E$ and to every $\mu \in \mathcal{M}(E)$ there corresponds a random measure (X_D, P_μ). We say that this family is a BEM system if $X_D \in \mathcal{M}(E)$ for all D and if:

2.3.A. [Continuous branching property] For all positive Borel functions $f_1, \dots, f_n$, all subdomains $D_1, \dots, D_n$ of E and every $\mu \in \mathcal{M}(E)$,

$$\log P_\mu e^{-Z} = \int \log P_y e^{-Z} \mu(dy) \tag{2.2}$$

where

$$Z = \sum_{1}^{n} \langle f_i, X_{D_i} \rangle \tag{2.3}$$

and $P_y = P_{\delta_y}$.

2.3.B. [Markov property.] The σ-algebra $\mathcal{F}_{\subset D}$ generated by $X_{D'}, D' \subset D$ and the σ-algebra $\mathcal{F}_{\supset D}$ generated by $X_{D''}, D'' \supset D$ are conditionally independent given X_D.

2.3.C. For all μ and D,

$$P_\mu\{X_D(D) = 0\} = 1.$$

2.3.D. If $\mu(D) = 0$, then

$$P_\mu\{X_D = \mu\} = 1.$$

Condition 2.3.A implies that

$$P_\mu e^{-Z} = \prod P_{\mu_n} e^{-Z}$$

if $\mu = \sum \mu_n$.

A BEM system is an (L, ψ)-*superdiffusion* if

$$u(x) = -\log P_x e^{-\langle f, X_D \rangle}$$

satisfies the equation

$$u(x) + \Pi_x \int_0^{\tau_D} \psi[u(\xi_t)]dt = \Pi_x f(\xi_{\tau_D}) \tag{2.4}$$

where (ξ_t, Π_x) is an L-diffusion. If D is smooth and bounded and f is continuous and bounded, then (2.4) is equivalent to the conditions

$$\begin{aligned} Lu &= \psi(u) \quad \text{in } D, \\ u &= f \quad \text{on } \partial D. \end{aligned} \tag{2.5}$$

[The problem (2.5) has a unique solution.]

The existence of an (L, ψ)-superdiffusion is proved, in particular, for

$$\psi(u) = bu^2 + \int_0^\infty (e^{-tu} - 1 + tu)N(dt) \tag{2.6}$$

under the conditions

$$b \geq 0, \int_1^\infty tN(dt) < \infty, \int_0^1 t^2 N(dt) < \infty. \tag{2.7}$$

An important special case is the function

$$\psi(u) = u^\alpha, 1 < \alpha \le 2 \tag{2.8}$$

corresponding to $b = 0$ and

$$N(dt) = \ell t^{-1-\alpha} dt$$

where

$$\ell = \left[\int_0^\infty (e^{-\lambda} - 1 + \lambda)\lambda^{-1-\alpha} d\lambda \right]^{-1}.$$

The class $\mathcal{U}$ under investigation can be characterized probabilistically by the following *mean value property*: $u \in \mathcal{U}$ if and only if

$$P_x e^{-\langle u, X_D \rangle} = e^{-u(x)} \tag{2.9}$$

for all D such that $\bar{D} \subset E$.[10]

2.4 Stochastic Boundary Values

Denote by $\mathcal{M}_c(E)$ the set of all finite measures on E concentrated on compact subsets of E. Suppose that, for all $\mu \in \mathcal{M}_c(E)$

$$\langle u, X_{D_n} \rangle \to Z \quad P_\mu\text{-a.s.} \tag{2.10}$$

for every sequence of domains D_n such that $\bar{D}_n \subset D_{n+1}$ and E is the union of D_n. Then we say that Z is the *stochastic boundary value of* u and we write $Z = \mathrm{SBV}(u)$. The stochastic boundary values exist for all $u \in \mathcal{U}$ and for all $h \in \mathcal{H}$. Put

$$Z_u = \mathrm{SBV}(u), \quad Z_\nu = \mathrm{SBV}(u_\nu).$$

It follows from (2.10) and the mean value property (2.9) that

$$P_x e^{-Z_u} = e^{-u(x)} \quad \text{for every } u \in \mathcal{U}. \tag{2.11}$$

In particular,

$$P_x e^{-Z_\nu} = e^{-u_\nu(x)} \quad \text{for every } \nu \in \mathcal{N}_1. \tag{2.12}$$

We have

$$Z_{u \oplus v} = Z_u + Z_v, \tag{2.13}$$

$$Z_{cu} = c Z_u \quad \text{for any constant } c \ge 0, \tag{2.14}$$

$$Z_{u_n} \uparrow Z_u \quad \text{if } u_n \uparrow u. \tag{2.15}$$

[See Section 1.3, Chapter 9 in [Dyn02].]

[10] See [Dyn02], Chapter 8, 2.1.D.

2.5 Relation between Superdiffusions and Conditional Diffusions

We start with the promised probabilistic definition of singular points of a solution u. Put

$$\Phi_u = \int_0^{\tau_E} \psi'[u(\xi_t)]dt. \tag{2.16}$$

A point $y \in \partial E$ is singular for u if $\Phi_u = \infty$ Π_x^y-a.s. for every $x \in E$.

The following relation plays a fundamental role for the developing the trace theory. For every $u \in \mathcal{U}$ and every $\nu \in \mathcal{N}_1$,

$$P_x Z_\nu e^{-Z_u} = e^{-u(x)} \Pi_x^\nu e^{-\Phi_u} \tag{2.17}$$

where

$$\Pi_x^\nu = \int \nu(dy) \Pi_x^y.$$

[See Theorem 3.1 in Chapter 9 of [Dyn02].] The formula (2.17) is a key tool for proving the properties 1.4.A–1.4.C of the trace. To illustrate how it is applied, we prove that the set Γ of all singular points of u is f-closed. If ν is concentrated on Γ, then $\Phi_u = \infty$ $\mathcal{P}_x^\nu$-a.s. By (2.17), P_x-a.s., $P_x Z_\nu e^{-Z_u} = 0$. Hence, P_x-a.s., either $Z_\nu = 0$ or $Z_u = \infty$. In both cases $P_x\{Z_\nu \leq Z_u\} = 1$. By (2.11) and (2.12), this implies $u_\nu \leq u$ and, by (1.3), $u_\Gamma \leq u$. Hence, every singular point of u_Γ is a singular point of u that is it belongs to Γ.

Remark. To apply (2.17) we need to assume the existence of (L, ψ)-superdiffusion. The original version of the trace theory was developed under this assumption. Later the theory was extended to more general ψ[11] by using an inequality which follows from (2.17) but can be proved without assuming the existence of (L, ψ)-superdiffusion. The price is less transparent and more lengthy arguments.

Part Two. Representation of Solutions in Terms of their Traces

Suppose that $\mathrm{Tr}(u) = (\Gamma, \nu)$. We claim that u can be represented by the formula

$$u = u_\Gamma \oplus u_\nu \tag{II.1}$$

where u_ν is defined in Section 1 and u_Γ is defined in Section 1.2.

By 1.4.B,

$$u \geq u_\Gamma \oplus u_\nu. \tag{II.2}$$

Since u_Γ and u_ν are σ-moderate, (II.1) implies that u is σ-moderate.

Formula (II.1) will follow if we prove that

[11] See Chapter 11 in [Dyn02].

$$w_\Gamma = u_\Gamma, \tag{II.3}$$

and

$$u \le w_\Gamma \oplus u_\nu. \tag{II.4}$$

We establish (II.3) for a bounded smooth domain E and $\psi(u) = u^\alpha$ where $1 < \alpha \le 2$.[12] The bound (II.4) is proved under an additional assumption that $L = \frac{1}{2}\Delta$ (that is a superdiffusion is the super-Brownian motion).

In Section 3 we prepare tools for proving (II.3): $\mathbb{N}$-measures, range of a superdiffusion and Poisson capacities. A special role is played by an inequality (3.8) relating superdiffusions in two domains $D \subset E$. We call it (D, E)-*inequality.*

Section 4 is devoted to proof of (II.3) and Section 5 to proof of (II.4).

3 Tools

3.1 $\mathbb{N}$-Measures

An introduction of **measures** $\mathbb{N}_x$ in parallel to measures P_x is a recent enhancement of the superdiffusion theory. First, $\mathbb{N}$-measures appeared as excursion measures of the Brownian snake introduced by Le Gall. These measures were used by him and his school for investigating the equation $\Delta u = u^2$. In particular, they played a key role in Mselati's dissertation. In Le Gall's theory, measures $\mathbb{N}_x$ are defined on the space of continuous paths. We define their analog on the same space Ω as measures P_μ.

The measures $\mathbb{N}_x$ are constructed by using the integral representation of infinitely divisible random measures (X_D, P_x). They are related to P_x by the formula

$$\mathbb{N}_x(1 - e^{-Z}) = -\log P_x e^{-Z} \tag{3.1}$$

for every Z of form (2.3) and for $Z = Z_u$ where $u \in \mathcal{U}$. In particular, for every bounded smooth domain E and every continuous function f,

$$u(x) = \mathbb{N}_x(1 - e^{-\langle f, X_E\rangle})$$

is a solution of (I.1) with the boundary value f.

In contrast to probability measures P_x, measures $\mathbb{N}_x$ are infinite (but $\mathbb{N}_x Z_\nu < \infty$ for all $\nu \in \mathcal{N}_1$).

For every $\nu \in \mathcal{N}_1$,

$$\mathbb{N}_x\{Z_\nu \ne 0\} = \lim_{n\to\infty} u_{n\nu}(x). \tag{3.2}$$

Indeed, $\mathbb{N}_x\{1 - e^{-nZ_\nu}\} \to \mathbb{N}_x\{Z_\nu \ne 0\}$ as $n \to \infty$ and therefore (3.2) follows from (3.1), (2.11) and (2.14).

[12] By using purely analytic method, Marcus and Véron proved in [MV04] that in the case $L = \Delta$ the equality (II.3) holds for all $\alpha > 1$.

An increasing sequence $n\nu$ tends to a measure $\infty \cdot \nu$ equal to 0 on sets of ν-measure 0 and equal to ∞ on the rest of Borel sets. Note that $\infty \cdot \nu \in \mathcal{N}_0$ and, by (3.2),

$$\mathbb{N}_x\{Z_\nu \neq 0\} = u_{\infty\cdot\nu}(x). \tag{3.3}$$

3.2 Range

The **range of a superdiffusion** X is the area hit by X. More precisely, the range is a closed set $\mathcal{R} = \mathcal{R}(\omega)$ with the properties:

(i) For every $D \subset E$ and every $\mu \in \mathcal{M}(E)$, X_D is concentrated, P_μ-a.s. on $\mathcal{R}$.
(ii) If $\tilde{\mathcal{R}} = \tilde{\mathcal{R}}(\omega)$ is a closed set such that, for some $\mu \in \mathcal{M}(E)$ and for all $D \subset E$, X_D are concentrated, P_μ-a.s., on $\tilde{\mathcal{R}}$, then, P_μ-a.s., $\tilde{\mathcal{R}} \supset \mathcal{R}$.
(iii) For every $D \subset E$ and every $x \in D$, X_D is concentrated, $\mathbb{N}_x$-a.s. on $\mathcal{R}$.
(iv) If $\tilde{\mathcal{R}} = \tilde{\mathcal{R}}(\omega)$ is a closed set such that, for some $x \in E$ and for all $D \subset E$ which contain x, X_D are concentrated, $\mathbb{N}_x$-a.s., on $\tilde{\mathcal{R}}$, then, $\mathbb{N}_x$-a.s., $\tilde{\mathcal{R}} \supset \mathcal{R}$.

The existence of $\mathcal{R}$ is proved for all superdiffusions. It is also proved that

$$w_\Gamma(x) = -\log P_x\{\mathcal{R} \cap \Gamma = \emptyset\} = \mathbb{N}_x\{\mathcal{R} \cap \Gamma \neq \emptyset\}. \tag{3.4}$$

Moreover, for every $\nu \in \mathcal{N}_0$,

$$\begin{aligned} w_\Gamma \oplus u_\nu &= -\log P_x\{\mathcal{R} \cap \Gamma = \emptyset, e^{-Z_\nu}\} \\ &= \mathbb{N}_x\{\mathcal{R} \cap \Gamma \neq \emptyset\} + \mathbb{N}_x\{\mathcal{R} \cap \Gamma = \emptyset, 1 - e^{-Z_\nu}\}.^{13} \end{aligned} \tag{3.5}$$

In combination with (II.1) and (II.3) this formula provides a probabilistic representation of a solution with the trace (Γ, ν).

3.3 Poisson Capacities

To every constant $\alpha > 1$ there corresponds the Poisson capacity[14] defined by the formula

$$\mathrm{Cap}(\Gamma) = \sup\{\mathcal{E}(\nu)^{-1} : \nu \in \mathcal{P}(\Gamma)\}$$

[13] See Theorem 3.4 in Section 4, [Dyn04]. Writing $P\{A, X\}$ means $\int_A X dP$.

[14] Analysts work with the Bessel capacity $\mathrm{Cap}_{2/\alpha,\alpha'}$. Both settings are equivalent because, by [DK07], there exists a constant $C > 0$ such that, for all Γ,

$$C^{-1}\,\mathrm{Cap}(\Gamma) \leq \mathrm{Cap}_{2/\alpha,\alpha'} \leq C\,\mathrm{Cap}(\Gamma).$$

where $\mathcal{P}(\Gamma)$ is the set of all probability measures on Γ and

$$\mathcal{E}(\nu) = \int_E d(y, \partial E) dy [h_\nu(y)]^\alpha.$$

($d(x, K)$ stands for the distance from x to K. Function h_ν is given by (1.2).) We also use the capacities

$$\operatorname{Cap}_x(\Gamma) = \sup\{\mathcal{E}_x(\nu)^{-1} : \nu \in \mathcal{P}(\Gamma)\}$$

where

$$\mathcal{E}_x(\nu) = \int_E g(x, y) dy [h_\nu(y)]^\alpha. \tag{3.6}$$

and g is the Green function in E for L.

We establish the following relation between $\operatorname{Cap}(K)$ and $\operatorname{Cap}_x(K)$. Put

$$E_K = \{x \in E : d(x, K) \geq \frac{1}{4} \operatorname{diam}(K), \quad \varphi(x, K) = d(x, \partial E) d(x, K)^{-d}$$

where $\operatorname{diam}(K)$ means the diameter of K. There exists a constant C such that

$$\operatorname{Cap}(K) \leq C \varphi(x, K) \operatorname{Cap}_x(K) \tag{3.7}$$

for all K and all $x \in E_K$.

3.4 (D, E)-Inequality

The (D, E)-**inequality** involves $\mathbb{N}$-measures, the range, the stochastic boundary values Z_ν of u_ν and the integrals (3.6).

Suppose that $D \subset E$ are bounded smooth domains. Put

$$D^* = \{x \in \bar{D} : d(x, E \setminus D) > 0\} = D \cup L$$

where $L = \{x \in \partial E : d(x, E \setminus D) > 0\}$. For every $\nu \in \mathcal{N}_1$ and every $x \in E$,

$$\mathbb{N}_x\{\mathcal{R} \subset D^*, Z_\nu \neq 0\} \geq \text{const.} \, \mathbb{N}_x\{\mathcal{R} \subset D^*, Z_\nu\}^{\alpha/(\alpha-1)} \mathcal{E}_x(\nu)^{-1/(\alpha-1)}. \tag{3.8}$$

[This is Theorem 1.1 in Chapter 7 of [Dyn04].]

4 Proof of Equation (II.3)

4.1 Reduction to $\mathbb{N}$-Inequality

First, we prove that (II.3) can be deduced from the following proposition which we call the $\mathbb{N}$-*inequality*:

($\mathbb{N}$) *For every K, there exists a constant C with the property: for every x, there exists a measure $\nu \in \mathcal{M}(K)$ such that*

$$\mathbb{N}_x\{\mathcal{R} \cap K \neq \emptyset\} \leq C \mathbb{N}_x\{Z_\nu \neq 0\}. \tag{4.1}$$

By (3.4) and (3.3), this inequality is equivalent to

$$w_K \leq C u_{\infty \cdot \nu} \tag{4.2}$$

If $\nu \in \mathcal{M}(\partial E)$ and $\mathcal{E}_x(\nu) < \infty$, then $\nu \in \mathcal{N}_1$.[15] Denote by $\mathcal{N}_1(K)$ the class of all $\nu \in \mathcal{N}_1$ concentrated on K.

It follows easily from the definitions of u_Γ and w_Γ [(1.3)–(1.5)] that:

(i) If (II.3) is true for compact subsets of ∂E, then it is true for all Borel $\Gamma \subset \partial E$.
(ii) $w_K \geq u_K$ for all compact K, and so it is sufficient to prove that $w_K \leq u_K$.

The relation $u_\mu \vee u_\nu = u_{\mu \vee \nu}$ implies that $\mathcal{N}_1(K)$ is closed under $\vee$ and therefore, according to (1.3), for every $x \in E$, $u_K(x)$ is equal to $\sup u_\nu(x)$ over $\nu \in \mathcal{N}_1(K)$. For every $\nu \in \mathcal{N}_1(K)$, $u_{n\nu} \leq u_K$ and therefore $u_{\infty \cdot \nu}(x) \leq u_K(x)$. To prove that $w_K \leq u_K$ it is sufficient to demonstrate that, for every x, there exists $\nu \in \mathcal{N}_1(K)$ such that $w_K(x) \leq u_{\infty \cdot \nu}(x)$. Put $\eta = \infty \cdot \nu$. It follows from (2.14) that $Z_\eta = Z_{C\eta} = CZ_\eta$. Therefore the bound $w_K(x) \leq u_{\infty \cdot \nu}(x)$ will follow from (2.11) if we prove that (4.2) holds with C independent of x.

4.2 Proof of the $\mathbb{N}$-Inequality

We establish a number of estimates in terms of $\operatorname{Cap}_x(K)$.
A. An upper bound for $w_K(x) = \mathbb{N}_x\{\mathcal{R} \cap K \neq \emptyset\}$.
B. A lower bound (for sufficiently large n) for

$$\mathbb{N}_x\{\mathcal{R} \subset B_n(x, K), Z_\nu\}$$

where

$$B_n(x, K) = \{z : |x - z| < nd(x, K)\}$$

C. A lower bound (for sufficiently big n) for

$$\mathbb{N}_x\{\mathcal{R} \subset B_n(x, K), Z_\nu \neq 0\}.$$

Part A is based on an estimate

$$w_K(x) \leq C\varphi(x, K) \operatorname{Cap}(K)^{1/(\alpha-1)}, \tag{4.3}$$

where the constant C does not depend on K and x.[16] It follows from (4.3) and (3.7) that

$$w_K(x) \leq C[\varphi(x, K)^\alpha \operatorname{Cap}_x(K)]^{1/(\alpha-1)}. \tag{4.4}$$

[15] This follows from 2.1.A, Chapter 12 in [Dyn02].
[16] The bound (4.3) was proved for $\alpha = 2$ by Mselati [Mse04] and for $1 < \alpha < 2$ by Kuznetsov [Kuz04].

In part B we use the relations between superdiffusions and conditional diffusions and bounds for conditional diffusions involving first exit times from E and from a ball of radius r centered at x. As a result, we prove the existence of C and n such that, for all K, all $x \in E_K$ and all $\nu \in \mathcal{P}(K)$ such that $\mathcal{E}_x(\nu) < \infty$,

$$\mathbb{N}_x\{\mathcal{R} \subset B_n(x,K), Z_\nu\} > C\varphi(x,K). \tag{4.5}$$

Part C is deduced from the definition of $\mathrm{Cap}_x(K)$ and from (4.5) by (D,E)-inequality (3.8) applied to $D = E \cap B_n(x,K)$. We prove this way the existence of C and n with the property: for every K and every $x \in E_K$,

$$\mathbb{N}_x\{\mathcal{R} \subset B_n(x,K), Z_\nu \neq 0\} \geq C[\varphi(x,K)^\alpha \,\mathrm{Cap}_x(K)]^{1/(\alpha-1)} \tag{4.6}$$

for some $\nu \in \mathcal{P}(K)$ such that $\mathcal{E}_x(\nu) < \infty$.

It follows from (3.4), (4.4) and (4.6) that:

($\mathbb{M}$) There exist constants C and n such that, for every K and every $x \in E_K$, there is a $\nu \in \mathcal{N}_1(K)$ with the property

$$\mathbb{N}_x\{\mathcal{R} \cap K \neq \emptyset\} \leq C\mathbb{N}_x\{\mathcal{R} \subset B_n(x,K), Z_\nu \neq 0\}. \tag{4.7}$$

It remains to deduce ($\mathbb{N}$) from ($\mathbb{M}$). In both propositions we have upper estimates for $\mathbb{N}_x\{\mathcal{R} \cap K \neq \emptyset\}$. However (4.7) holds only for $x \in E_K$ and (4.1) holds for all $x \in E$. On the other hand, C in (4.1) depends on K and in (4.7) it is independent of K. Following Mselati [Mse02] and [Mse04],[17] we cover K by closed sets K_m to which we can apply $\mathbb{M}$. We get this way measures $\nu_m \in \mathcal{P}(K_m)$ with $\mathcal{E}_x(\nu_m) < \infty$. Their sum ν satisfies (4.1).

To realize this plan, we fix $x \in E$ and $K \subset \partial E$ and we put

$$K_m = \begin{cases} \{z \in K : |x - z| \leq 2\delta\} & \text{for } m = 1, \\ \{z \in K : 2^{m-1} \leq \delta \leq 2^m\delta\} & \text{for } m > 1 \end{cases}$$

where $\delta = d(x,K)$. The set M of m such that K_m is not empty is finite and $x \in E_{K_m}$ for every $m \in M$. By ($\mathbb{M}$), there exist constants C, n and measures $\nu_m \in \mathcal{N}_1(K_m)$ such that $\mathcal{E}_x(\nu_m) < \infty$ and

$$\mathbb{N}_x\{\mathcal{R} \cap K_m \neq \emptyset\} \leq C\mathbb{N}_x\{\mathcal{R} \subset B_n(x,K_m), Z_{\nu_m} \neq 0\}. \tag{4.8}$$

If $2^p > n$, then, for every positive m, $B_n(x,K_m) \subset B_{2^{p+m}}(x,K)$ and, by (4.8),

$$\mathbb{N}_x\{\mathcal{R} \cap K_m \neq \emptyset\} \leq C\mathbb{N}_x(Q_m)$$

where

$$Q_m = \{\mathcal{R} \subset B_{2^{p+m}}(x,K)\}.$$

[17] See also [Dyn04].

The sum ν of ν_m is a finite measure, $\mathcal{E}_x(\nu) < \infty$ and

$$\mathbb{N}_x\{\mathcal{R} \cap K \neq \emptyset\} \leq \sum_M \mathbb{N}_x\{\mathcal{R} \cap K_m \neq \emptyset\} \leq C \sum_1^\infty \mathbb{N}_x(Q_m).$$

Now we need to bound the right side from above. First, we prove that

$$\mathbb{N}_x\{Q_m \cap Q_{m'}\} = 0 \quad \text{if } m' \geq m + p + 1. \tag{4.9}$$

Indeed,

$$Q_m \cap Q_{m'} \subset \{\mathcal{R} \cap K_{m'} = \emptyset, Z_{\nu_{m'}} \neq 0\}.$$

Since $\nu_{m'}$ is concentrated on $K_{m'}$,

$$\mathbb{N}_x\{\mathcal{R} \cap K_{m'} = \emptyset, Z_{\nu_{m'}} \neq 0\} = 0$$

which implies (4.9).

Every integer $m \geq 1$ has a unique representation $m = n(p+1) + j$ where $j = 1, \dots, p+1$ and therefore

$$\mathbb{N}_x\{\mathcal{R} \cap K \neq \emptyset\} \leq C_\kappa \sum_{j=1}^{p+1} \sum_{n=0}^\infty \mathbb{N}_x(Q_{n(p+1)+j}). \tag{4.10}$$

It follows from (4.9) that $\mathbb{N}_x\{Q_{n(p+1)+j} \cap Q_{n'(p+1)+j}\} = 0$ for $n' > n$. Therefore, for every j,

$$\begin{aligned} \sum_{n=0}^\infty \mathbb{N}_x\{Q_{n(p+1)+j}\} &= \mathbb{N}_x\left\{\bigcup_{n=0}^\infty Q_{n(p+1)+j}\right\} \\ &\leq \mathbb{N}_x\left\{\sum_{n=0}^\infty Z_{\nu_{n(p+1)+j}} \neq 0\}\right\} \leq \mathbb{N}_x\{Z_\nu \neq 0\} \end{aligned} \tag{4.11}$$

because

$$\sum_{n=0}^\infty Z_{\nu_{n(p+1)+j}} \leq \sum_{m=1}^\infty Z_{\nu_m} = Z_\nu.$$

The bound (4.1) follows from (4.10) and (4.11).

5 Proof of Bound (II.4)

In this section we assume that u is a positive solution of the equation

$$\Delta u = u^\alpha \quad \text{in } E \tag{5.1}$$

with the trace (Γ, ν) and that $1 < \alpha \leq 2$ and we investigate the class $\mathfrak{E}$ of all domains E for which the bound (II.4) is true. The final result is: *all bounded domains of class C^4 belong to $\mathfrak{E}$.*

The main steps in the proof are:

A. There is a class $\mathfrak{E}_1 \subset \mathfrak{E}$ with the property: $E \in \mathfrak{E}_1$ if, for every $y \in \partial E$, there exists a domain $D \in \mathfrak{E}_1$ such that $D \subset E$ and $\partial D \cap \partial E$ contains a neighborhood of y in ∂E.
B. $\mathfrak{E}_1$ contains all star domains.[18]
C. If E is a C^4 domain, then, for every $y \in \partial E$, there exists a star domain $D \subset E$ such that $\partial D \cap \partial E$ contains a neighborhood of y in ∂E.

Here is the definition of class $\mathfrak{E}_1$: $E \in \mathfrak{E}_1$ if, for every $v \in \mathcal{U}(E)$ and every $\Gamma \subset \partial E$, the conditions $\mathrm{Tr}(v) = (\Lambda, \mu)$, $\Lambda \subset \Gamma$ and $\mu(\Gamma \backslash \Lambda) = 0$ imply that $v \leq w_\Gamma$.

In part A we use connections between $\mathrm{Tr}(v)$ and $\mathrm{Tr}(v')$ where $v \in \mathcal{U}(E)$ and v' is the restriction of v to $D \subset E$.

Step B is based on a self-similarity property of the equation $\Delta u = u^\alpha$: if E is a star domain relative to 0, then, for every $0 < r \leq 1$,

$$u_r(x) = r^{2/(\alpha-1)} u(rx)$$

also belongs to $\mathcal{U}(E)$. A crucial role is played by the following absolute continuity result which is also of independent interest: if $A \in \mathcal{F}_{\supset D}$, then either $P_x(A) = 0$ for all $x \in D$ or $P_x(A) > 0$ for all $x \in D$. In other words, on the σ-algebra $\mathcal{F}_{\supset D}$, P_{x_1} is absolutely continuous with respect to P_{x_2} for all $x_1, x_2 \in D$.

Step C is based on elementary arguments of differential geometry.

References

[DK98] E.B. Dynkin and S.E. Kuznetsov, *Fine topology and fine trace on the boundary associated with a class of quasilinear differential equations*, Comm. Pure Appl. Math. **51** (1998), 897–936.

[DK07] ———, *Bessel capacities on compact manifolds and their relation to Poisson capacities*, J. Functional Analysis **242** (2007), 281–294.

[Dyn02] E.B. Dynkin, *Diffusions, Superdiffusions and Partial Differential Equations*, Amer. Math. Soc., Providence, R.I., 2002.

[Dyn04] E.B. Dynkin, *Superdiffusions and Positive Solutions of Nonlinear Partial Differential Equations*, American Mathematical Society, Providence, R.I., 2004.

[Kuz04] S.E. Kuznetsov, *An upper bound for positive solutions of the equation $\Delta u = u^\alpha$*, Amer. Math. Soc., Electronic Research Announcements **10** (2004), 103–112.

[MV04] M. Marcus and L. Véron, *Capacitary estimates of positive solutions of semilinear elliptic equations with absorbtion*, J. Eur. Math. Soc. **6** (2004), 483–527.

[18] A domain E is called a star domain relative to a point c if, for every $x \in E$, the line segment $[c, x]$ connecting c and x is contained in E.

[Mse02] B. Mselati, *Classification et représentation probabiliste des solutions positives de $\Delta u = u^2$ dans un domaine*, Thése de Doctorat de l'Université Paris **6**, 2002.

[Mse04] ——, *Classification and probabilistic representation of the positive solutions of a semilinear elliptic equation*, Memoirs of the American Mathematical Society **168, Number 798** (2004).

[Wat68] S. Watanabe, *A limit theorem on branching processes and continuous state branching processes*, J. Math. Kyoto Univ. **8** (1968), 141–167.

The Space of Stochastic Differential Equations

K. David Elworthy

Mathematics Institute, University of Warwick, Coventry CV4 7AL, England
kde@maths.warwick.ac.uk

1 Stochastic Differential Equations, Hormander Representations, and Stochastic Flows

1.1 Introduction

One of the main tools arising from Itô's calculus is the theory of stochastic differential equations, now with applications to many areas of science, economics and finance. This article is a remark on some aspects of the geometry and topology of certain spaces of stochastic differential equations, making no claims to relevance to the actual theory or its applications. It is based on work with Yves LeJan & Xue-Mei Li reported in [ELL99, ELJL04] and in preparation in [ELJL]. It was stimulated by contacts with Steve Rosenberg and his article with Sylvie Paycha, [PR04]. However the topological constructions and remarks, in all except 2.4 (which is taken from [ELJL]), are essentially well known and any novelty arises from their interpretation in terms of stochastic differential equations and flows.

1.2 The Spaces

We shall consider Stratonovich equations on a compact, connected, finite dimensional manifold M. We shall write them as:

$$dx_t = X(x_t) \circ dB_t + A(x_t)dt \tag{1.1}$$

where A is a vector field on M and for each $x \in M$ we have a continuous linear map

$$X(x) : H \to T_x M$$

of a fixed, real, separable Hilbert space H into the tangent space to M at x. Our "noise" $\{B_t : t \geqslant 0\}$ is a standard Brownian motion on H, cylindrical if H is infinite dimensional. We are only interested in the case where X and H are sufficiently smooth for a there to be unique solutions for a given initial point and a solution flow of diffeomorphisms, see [Elw82, IW89, Kun90].

The solutions to equation (1.1) form a diffusion process with generator the diffusion operator $\mathcal{A}$ for

$$\mathcal{A} = 1/2 \sum_j \mathcal{L}_{X^j}\mathcal{L}_{X^j} + \mathcal{L}_A. \tag{1.2}$$

where $\mathcal{L}_{X^j}$ denotes Lie differentiation by the vector field X^j given by

$$X^j(x) = X(x)(e_j)$$

for $e_j, j = 1, 2, ...$ any orthonormal basis of H. Commonly H is finite dimensional, $H = \mathbb{R}^m$, say. However this can be included in the infinite dimensional case by taking $X(x)$ to vanish on some finite codimensional subspace for all x, and we know from [Bax84] that to obtain all stochastic flows we need to allow infinite dimensional noise.

We shall fix a diffusion generator $\mathcal{A}$ which is smooth (so has smooth coefficients in local coordinates) and *assume that the principal symbol of $\mathcal{A}$ has constant rank in TM*. This latter assumption is equivalent to the existence of a smooth subbundle E of the tangent bundle TM such that for any SDE such as equation (1.1) the map $X(x)$ maps H onto the fibre E_x of E over x. In particular it holds when $\mathcal{A}$ is elliptic, in which case $E = TM$. Note that $X(x)$ or equivalently the symbol of $\mathcal{A}$ determines an inner product, $\langle -, - \rangle_x$ on E_x for each $x \in M$ giving it a Riemannian structure. Without specifying the regularity or giving topologies at this stage let $Hor_{\mathcal{A}}$ denote the set of Hormander form representations, as equation (1.2) of $\mathcal{A}$, and $SDE_{\mathcal{A}}$ the space of SDE's whose solutions are $\mathcal{A}$-diffusions. Since the natural map from SDE's to Hormander forms depends only on a choice of basis, any such basis determines a bijection

$$Hor_{\mathcal{A}} \cong SDE_{\mathcal{A}}. \tag{1.3}$$

Moreover, since the choice of the noise coefficient X in an element of $SDE_{\mathcal{A}}$ determines the vector field A, both spaces are naturally in one-one correspondence with the space $SDE(E)$ of vector bundle maps $X : M \times H \to E$, of the trivial H-bundle onto E, which induce the given Riemannian metric on E. It is this space which we shall examine in more detail below.

Closely related to these spaces is the space $Flow_{\mathcal{A}}$ of stochastic flows of diffeomorphisms of M whose one point motions are $\mathcal{A}$-diffusions. Following Baxendale, [Bax84], these can be considered as Wiener processes on the diffeomorphism group, $DiffM$ of M, and determine and are determined by a Hilbert space $\mathcal{H}_\gamma$ of sections of E with the property that the evaluation map $ev_x : \mathcal{H}_\gamma \to E_x$ is surjective and induces the given inner product, for each $x \in M$. In turn this is determined by a suitable *reproducing kernel* $k_\gamma(x, y) : E_x^* \to E_y$, [Bax76, ELL99], defined by

$$k_\gamma(x, -) = (ev_x)^* : E_x^* \to H_\gamma. \tag{1.4}$$

Let $RKH(E)$ denote the space of such Hilbert subspaces and $RK(E)$ the, isomorphic, space of their reproducing kernels. Using the inner product on E_x to identify it with its dual space, the latter can be identified with the space of those sections $k^\sharp$ of the bundle of linear maps $\mathbb{L}(E; E)$ over $M \times M$ such that $k^\sharp(x, y) : E_x \to E_y$ satisfies

(i) $k^\sharp(x, y) = k^\sharp(y, x)^*$;
(ii) $k^\sharp(x, x) = identity : E_x \to E_x$;
(iii) for any finite set $x_1, ..., x_q$ of elements of M we have

$$\sum_{i,j=1}^{q} \langle k^\sharp(x_i, x_j)u_i, u_j\rangle_{x_j} \geqslant 0$$

for all $\{u_j\}_{j=1}^q$ with $u_j \in E_{x_j}$.

It is easy to see that these form a convex subset of the space of all sections. It is natural to identify the space of smooth flows $Flow^\infty_\mathcal{A}$ in $Flow_\mathcal{A}$ with the space of smooth elements of $RK(E)$ with topology induced from the C^∞ topology on the sections of the bundle $\mathbb{L}(E; E)$ over $M \times M$. This topology is a reasonable topology for the space of flows: for example if K is a smooth compact manifold with a map $f : K \to Flow^\infty_\mathcal{A}$ which is smooth in the sense that it is smooth when identified with a map into every Sobolev space of sections of $\mathbb{L}(E; E)$, then there is a smooth stochastic flow on $K \times M$ which restricts to $f(k)$ on each of the leaves $\{k\} \times M, k \in M$. Convexity tells us that given any two flows in $Flow^\infty_\mathcal{A}$ there is a (canonical) smooth flow on $[0, 1] \times M$ which restricts to a flow on each $\{k\} \times M$ in $Flow^\infty_\mathcal{A}$ agreeing with the given ones at $k = 0, 1$. In this sense:

- *The space of smooth stochastic flows on M whose one point motions have $\mathcal{A}$ as generator, is contractible.*

There is the natural map taking an SDE to its flow. It corresponds to the map

$$\mathcal{H} : SDE(E) \to RKH(E) \tag{1.5}$$

given by $\mathcal{H}(X) = \{X(-)(e) : e \in H\}$ with inner product induced from H. When H is infinite dimensional this is surjective. Note that given some $\mathcal{H}_\gamma$ in $RKH(E)$ we can obtain an SDE in $SDE(E)$ which maps to $\mathcal{H}_\gamma$ by choosing a linear map $U : \mathcal{H}_\gamma \to H$ which is an isometry into H and defining $X(x)e = U^*(e)(x)$, for $e \in H, x \in M$. Thus $\mathcal{H}^{-1}(\mathcal{H}_\gamma)$ is not connected in general. See also the end of Section 2.2 below.

Our main interest is in C^∞ equations and flows. To do differential calculus on the various manifolds of C^∞ mappings which will arise it would be natural to use the Froelicher-Kriegl calculus, see [KM97], as Michor in [Mic91]. However Banach manifolds are more familiar and we will generally consider manifolds of Sobolev spaces of mappings of sufficiently high differentiability class. Taking s very large compared to the dimension of M let $SDE(E)^s$, $RK(E)^s$, etc., denote the relevant subsets of Sobolev spaces of mappings of class H^s, (i.e. those whose weak derivatives of order s lie in L^2, see [Pal68]). In particular let $\mathcal{D}^s$ denote the Hilbert manifold of all diffeomorphisms of class H^s. By standard approximation techniques the homotopy class of these spaces does not depend on s given that s is large enough.

2 Induced Connections and the Action of the Gauge Group

2.1 The Gauge Group and its Universal Bundle

For $\mathcal{A}$ and E as above let q be the fibre dimension of E. Suppose that H is infinite dimensional. Consider the Grassmanian $G(q, H)$ of all q-dimensional linear subspaces of H, the space $V(q, H)$ of all q-frames in H, and the natural projection $p : V(q, H) \to G(q, H)$. Identify $\mathbb{R}^q$ with a subspace of H and let $H^{\infty-q}$ be its orthogonal complement. Let $O(H)$, $O(\infty - q)$, and $O(q)$ be the orthogonal groups of H, $H^{\infty-q}$, and $\mathbb{R}^q$, respectively. Then $V(q, H)$, which is naturally the space of all isometries of $\mathbb{R}^q$ into H, can be identified with the homogeneous space $O(H)/O(\infty - q)$ with the natural right action of $O(q)$ making p a smooth, even real analytic, principal $O(q)$-bundle, [KM97]. Here we can furnish $G(q, H)$ with the manifold structure it inherits as a homogeneous space or, equivalently, as a manifold modelled on the Hilbert space of continuous linear maps $\mathbb{L}(\mathbb{R}^q; H^{\infty-q})$. Thus both $G(q, H)$ and the total space $V(q, H)$ are modelled on Hilbert spaces.

By Kuiper's theorem $O(H)$ and $O(\infty - q)$ are contractible, and so therefore is $V(q, H)$, making p a universal $O(q)$-bundle, as is frequently used. This means that if $p' : B \to M$ is any smooth principal $O(q)$-bundle over M there is a smooth map $\chi : M \to G(q, H)$ classifying p' in the sense that the the pull back by χ of $V(q, H)$ is equivalent to B; in other words there is a diagram of smooth maps:

$$
\begin{array}{ccc}
B & \xrightarrow{\ \bar{\chi}\ } & V(q,H) \\
{\scriptstyle p'}\downarrow & & \downarrow{\scriptstyle p} \\
M & \xrightarrow[\ \chi\ ]{} & G(q,H)
\end{array}
$$

where $\bar{\chi}$ is a diffeomorphism on the fibres and is equivariant with respect to the right actions of $O(q)$. Such a lift $\bar{\chi}$ exists over any smooth map homotopic to χ, e.g see [Ste51]. It is not uniquely determined by χ; the space of all such lifts is $\{\chi \circ \alpha : \alpha \in \mathcal{G}\}$ where $\mathcal{G}$ is the *gauge group* of B, i.e. the group of all smooth $O(q)$ equivariant diffeomorphisms $\alpha : B \to B$ over the identity map of M. Following Atiyah & Bott, [AB83], let $\mathbb{H}^s_B(M;G(q,H))$ be the space of H^s maps classifying B and $\mathbb{H}^s_{O(q)}(B;V(q,H))$ the space of equivariant maps of B into $V(q,H)$ of class H^s. There is the natural projection

$$p^{\mathcal{G}}:\mathbb{H}^s_{O(q)}(B;V(q;H))\to\mathbb{H}^s_B(M;G(q,H))$$

say, which coincides with the quotient map by the right action of $\mathcal{G}^s$, the H^s version of $\mathcal{G}$. Note that $\mathbb{H}^s_B(M;G(q,H))$ is a smooth manifold with Hilbert model since it is a connected component of the space of all H^s maps of M into $G(q,H)$; that $\mathcal{G}^s$ is, and is a Lie group, is shown in [MV81]; while $\mathbb{H}^s_{O(q)}(B;V(q,H))$ is the fixed point set of the natural action of the compact group $O(q)$ on the Hilbert manifold $\mathbb{H}^s(B;V(q,H))$, and so a smooth submanifold of $\mathbb{H}^s(B;V(q,H))$ by [Pal79].

Atiyah & Bott observe that $\mathbb{H}^s_{O(q)}(B;V(q;H))$ is contractible and hence $p^{\mathcal{G}}$ is a universal $\mathcal{G}^s$-bundle, so that $\mathbb{H}^s_B(M;G(q,H))$ is a classifying space for $\mathcal{G}^s$-bundles. To see this contractibility it suffices, by a theorem of J.H.C.Whitehead, to prove that any two continuous maps $f_j, j=1,2$ of a finite dimensional complex K, say, into $\mathbb{H}^s_{O(q)}(B;V(q,H))$ are homotopic. However such maps determine a bundle map to $V(q,H)$ of the restriction to $\{0,1\}\times K\times M$ of the $O(q)$-bundle $\mathbf{I}\times\mathbf{I}\times B$ over $[0,1]\times K\times M$. By the universal property of $p:V(q,H)\to G(q,H)$ this extends over the whole bundle projecting down to give the required homotopy, c.f. the proof of Theorem 19.3 in [Ste51]. There is also a proof in [Hus94].

2.2 Stochastic Differential Equations, their Filtrations, and Gauge Equivalence

Now take B to be the orthonormal frame bundle, $O(E)$, of our subbundle E of TM. Note that an element X in $SDE(E)^s$ is equally determined by the H-valued one-form Y on E given by the its adjoint map: $Y_x = X(x)^*: E_x\to H$. From this we obtain the diagram, [ELL99],

$$\begin{array}{ccc} O(E) & \xrightarrow{\;\;\Phi\;\;} & V(q,H) \\ {\scriptstyle p'}\downarrow & & \downarrow{\scriptstyle p} \\ M & \xrightarrow[\;\;\Phi_0\;\;]{} & G(q,H) \end{array}$$

defined by: $\Phi_0(x) = Image Y_x, x\in M$, and $\Phi(u) = (Y_x u(e_1),...,Y_x u(e_q))$, for $u\in O(E)$ where $e_1,...,e_q$ is an orthonormal base for $\mathbb{R}^q$. In particular we obtain Φ belonging to $\mathbb{H}^s_{O(q)}(B;V(q,H))$ and so a smooth map $\kappa^s: SDE(E)^s\to\mathbb{H}^s_{O(q)}(B;V(q,H))$. Elements of $\mathcal{G}$ can be considered as automorphisms of the Riemannian bundle E and so act on the right on $SDE(E)^s$ by $(X,\alpha)\mapsto\alpha^{-1}\circ X(\cdot)$ so that multiplication by α maps Y to $Y\circ\alpha$. This action is free and we see that κ^s is an equivariant diffeomorphism which descends to give an isomorphism of $\mathcal{G}^s$-bundles:

$$\begin{array}{ccc} SDE(E)^s & \xrightarrow{\kappa^s} & \mathbb{H}^s_{O(q)}(B; V(q,H)) \\ \downarrow{\scriptstyle proj.} & & \downarrow{\scriptstyle p} \\ SDE(E)^s/\mathcal{G}^s & \xrightarrow[\kappa_0]{} & \mathbb{H}^s_{O(E)}(M; G(q,H)) \end{array}$$

We will say that two stochastic differential equations determined by X and X' in $SDE(E)^s$ are *gauge equivalent* if they are in the same orbit of $\mathcal{G}^s$ i.e. if there exists some $\alpha : E \to E$ in $\mathcal{G}^s$ such that $X'(x) = \alpha X(x)$ for all x in M.

This leads to one of our main observations:

- *Let $\mathcal{A}$ be a smooth diffusion generator on a compact manifold M whose symbol has constant rank. Then for all sufficiently large s the space of stochastic differential equations, $SDE^s_{\mathcal{A}}$, whose solutions are $\mathcal{A}$-diffusions is contractible. Moreover the natural right action of the group of H^s-automorphisms, $\mathcal{G}^s$, on $SDE(E)^s$ makes the latter into the total space of a universal bundle for $\mathcal{G}^s$. In particular the space of equivalence classes of elements in $SDE(E)^s$ under gauge equivalence has a natural topology which makes it a classifying space for $\mathcal{G}^s$-bundles. The corresponding results hold for smooth stochastic differential equations.*

In fact each gauge equivalence class corresponds to a map from M to $G(q,H)$, namely that given by the map Φ_0 above. Intuitively it tells us which part of the cylindrical noise is acting infinitesimally at a given point of M. It may be illuminating to consider the following, rather artificial, problem: suppose we are given a smooth map Θ of the product $K \times M$ of M with a compact connected manifold K, into the Grassmanian $G(q,H)$, and wish to construct a smooth family of stochastic differential equations in $SDE^s_{\mathcal{A}}$ parametrised by K so that at each point (k,x) the SDE is driven by the noise in the subspace $\Theta(k,x)$; what conditions on Θ are needed? From above we know that for each k in K we must have $x \mapsto \Theta(k,x)$ in the correct homotopy class of maps, $\mathbb{H}^s_{O(E)}(M; G(q,H))$, to classify E. To get a family of SDE's continuous in K we also need the resulting map $\theta : K \to \mathbb{H}^s_O(E)(M; G(q,H))$ to lift to a continuous map of K into $\mathbb{H}^s_{O(q)}(O(E); V(q,H))$. This holds if and only if θ is homotopic to a constant. The fibre over a point $k \in K$ of the pull back by θ of $\mathbb{H}^s_{O(q)}(O(E); V(q,H))$ can be identified with the space of all those stochastic differential equations which use the noise in the subspaces determined by $\Theta(k,-)$, and a section of the pull back bundle will give us the required family.

For another equivalence relation with more standard probabilistic significance it will be convenient to fix a probability space $\{\Omega, \mathcal{F}, \mathbb{P}\}$ on which our cylindrical noise B is defined. For each equation in $SDE^s_{\mathcal{A}}$ we obtain the (completed) filtration, $\mathcal{F}_t^X : 0 \leqslant t < \infty$, say, determined by its solution flow, where X is the corresponding element in $SDE(E)^s$. Clearly gauge equivalent equations give the same filtration. On the other hand the filtration is the same as the filtration of the, possibly cylindrical, Brownian motion of $\mathcal{H}(X)$, and so using the martingale representation theorem for cylindrical Brownian motions, as in [AH04], we see

- *Two stochastic differential equations X and X' in $SDE_{\mathcal{A}}^s$ give the same filtrations if and only if the kernels of their induced maps $H \to \mathcal{H}(X)$ and $H \to \mathcal{H}(X')$ are the same.*

Thus the space of all possible such filtrations can be identified with the set $\{(q, F) : q \in \mathbb{Z} \cup \{\infty\}, q \geqslant r(E) \& F$ is a q-dimensional subspace of $H\}$, where $r(E)$ is the minimal fibre dimension of a trivial bundle over M which contains a copy of E. In other words it can be identified with $\bigcup_{\infty \geqslant q \geqslant r(E)} G(q, H)$, the space of all closed linear subspaces of H of dimension at least $r(E)$. If we give this space the topology corresponding to strong convergence of the corresponding orthogonal projections, the Wijsman topology, [Tsi], it will agree with the usual topology on the finite dimensional Grassmanians. Also, any such filtration is immersed in that of our underlying cylindrical Brownian motion $\{B_t : t \geqslant 0\}$ in the sense of Tsirelson, [Tsi], and so is determined by the σ-algebra $\mathcal{F}_\infty^X$. This shows that this description fits in with the much more general discussion of filtrations in [Tsi].

From this we can also return to Equation (1.5) and observe that a stochastic differential equation in $SDE_{\mathcal{A}}$ is determined, up to a right action of $O(q) \times I_{\infty - q}$, by its flow and its filtration, where the filtration is determined by (q, F) and the group is considered as the subgroup of $O(H)$ which acts as the identity on the orthogonal complement of F.

2.3 The Connection Induced on *E*

Narasimhan & Ramanan showed in [NR61] that there is a "universal connection", ϖ, say, on any universal $O(q)$-bundle and given a metric connection on E, or equivalently any connection ϖ_E on $O(E)$, there is a classifying map which pulls ϖ back to ϖ_E. In fact they show this holds for the finite dimensional Stiefel bundles, where H is replaced by a sufficiently high dimensional Euclidean space. The universal connection in this situation is described in an Appendix in [ELL99]. In particular we can use any X in $SDE(E)^s$ to obtain a connection $(\kappa(X))^*(\varpi)$ on $O(E)$ and any metric connection on E is obtained that way. The covariant derivative operator $\breve{\nabla}$ on sections of E corresponding to $(\kappa(X))^*(\varpi)$ has the very simple expression

$$\breve{\nabla}_v(U) = X(x)d[y \mapsto Y_y(U(y))](v) \tag{2.1}$$

and in [ELL99] this connection was called the *LeJan-Watanabe connection* of the flow since a special case had been noted in the context of stochastic flows in [LW84], see also [AMV96]. A direct proof that all metric connections on E can be obtained by a suitable X with H finite dimensional is in [Qui88].

The right action of $\mathcal{G}$ on $RK(E)$ given by $(k,\alpha) \mapsto k^\alpha$ with $(k^\alpha)^\sharp(x,y) = \alpha(y)^{-1}k^\sharp(x,y)\alpha(x)$ determines a right action of $\mathcal{G}$ on the space of smooth flows $Flow^\infty_{\mathcal{A}}$, though it seems far from clear if it has any significance for the behaviour of the flows. The map from SDE to flows is equivariant with respect to this action since the reproducing kernel k^X say of $\mathcal{H}(X)$ is given by $(k^X)^\sharp(x,y) = X(y)Y_x$. We see we have a factorisation by equivariant maps:

$$SDE_{\mathcal{A}} \to Flow^\infty_{\mathcal{A}} \to \mathcal{C}_E$$

of the map $X \mapsto (\kappa(X))^*(\varpi)$ into the space of smooth metric connections $\mathcal{C}_E$ on E. (Note that by its contractibility, observed in 1.2 we can also consider the quotient of $Flow^\infty_{\mathcal{A}}$ by the action of $\mathcal{G}$ as a classifying space for $\mathcal{G}$). Each part of this factorisation is surjective. In the final section, next, we lift results from [ELJL], see also [ELJL04], which give information about the fibres of the second map.

2.4 The Induced Semi-connection on the Diffeomorphism Bundle

Fix some point x_0 of M and let $\pi : \mathcal{D}^s \to M$ be the evaluation map $\pi(\theta) = \theta(x_0)$. We shall think of this as a principal bundle with group $\mathcal{D}^s_{x_0}$, those H^s-diffeomorphisms which fix x_0, acting on the right by composition. Since the action is not smooth we need to be careful; alternatively we can consider smooth diffeomorphisms using the approach in [KM97, Mic91].

Consider a smooth stochastic flow with corresponding element $k \in RK(E)$. From it we obtain a smooth *horizontal lift map*:

$$\Xi_\theta : E_{\pi(\theta)} \to T_\theta\mathcal{D}^s$$

given by

$$\Xi_\theta(u)(y) = k^\sharp(\theta(x_0), \theta(y))(u) \in E_{\theta(y)}$$

for $u \in E_{\theta(x_0)}, y \in M, \theta \in \mathcal{D}^s$, where we identify the tangent space $T_\theta\mathcal{D}^s$ at θ to the diffeomorphism group with the space of H^s-maps of M into TM which lie over θ. This is invariant under the action of $\mathcal{D}^s_{x_0}$ on $\mathcal{D}^s$. We call such an object a *semi-connection on* $\mathcal{D}^s$ *over* E and let $SC_E(\mathcal{D}^s)$ denote the set of all of these objects. They are also called "partial connections" or "connections over E", see [Gro96]. In the elliptic case, $E = TM$, they are the usual connections. They give a procedure for obtaining horizontal lifts $\tilde{\sigma} : [0,T] \to \mathcal{D}^s$ of those smooth curves $\sigma : [0,T] \to M$ with the property that $\dot{\sigma}(t) \in E_{\sigma(t)}$ for all t. For the semi-connection determined by our kernel k this lift, starting from a given diffeomorphism θ with $\theta(x_0) = \sigma(0)$, is the composition $\widetilde{\sigma}(t) = \Psi(t) \circ \theta$ where Ψ is the flow of the time dependent dynamical system on M,

$$\dot{z}(t) = k^\sharp(\sigma(t), z(t))\dot{\sigma}.$$

Our diffeomorphism bundle can be considered as a universal natural bundle on M, and each element of $SC_E(\mathcal{D}^s)$ determines a semi-connection over E on each natural bundle over M, (see [KMS93]). In particular it gives an element of $SC_E(GL(M))$ the space of semi-connections on the full linear frame bundle of E: for this the lift of our curve σ to $GL(M)$ starting at a frame u is just $T\widetilde{\sigma} \circ u$, the composition of the derivative of our lift $\widetilde{\sigma}$ with the frame. This determines a partial covariant derivative operator ∇', say, which allows us to differentiate arbitrary smooth vector fields but only in E-directions, i.e a "semi-connection over E on TM" as defined in [ELL99]. There is a map between connections on E itself and such semi-connections: to ∇ the covariant derivative of a connection on E there corresponds the semi-connection with covariant derivative ∇' given by

$$\nabla'_u(V) = \nabla_v(U) - [V, U](x)$$

for U a smooth section of E, V a smooth vector field, $x \in M$ and $U(x) = u, V(x) = v$. Following Driver for the case $E = TM$, we say the semi-connection and connection are "adjoints", [ELL99]. From [ELJL04] we have:

- *The semi-connection on $GL(M)$ induced by a stochastic flow is the adjoint of the metric connection on E determined by the flow.*

In [ELJL] there is the following:

- *The map described from smooth flows with $\mathcal{A}$ as generator of their one point motions to smooth semi-connections over E on the diffeomorphism bundle is injective.*

From this the induced semi-connection must contain all information about the flow. We can rephrase some of these statements to:

- *The adjoint semi-connection of a metric connection ϖ_E on E has many "prolongations" to a semi-connection on the diffeomorphism bundle and so to a coherent system of semi-connections on all natural bundles over M. Some of these are induced by a stochastic flow, (necessarily unique), and from then by the choice of a classifying map for the bundle E into the infinite dimensional Grassmanian. The latter will pull back the universal connection to the given connection ϖ_E.*

We can summarise some of these observations in the following diagram:

$$\begin{array}{ccccc}
C^\infty_{O(q)}(O(E); V(q,H)) & \xrightarrow{(\kappa^\infty)^{-1}} & SDE(E) & \xrightarrow{\mathcal{H}} & Flow_{\mathcal{A}} \cong RK(E) \\
\Big\downarrow{\scriptstyle NR} & & & & \Big\downarrow{\scriptstyle \Xi} \\
\mathbf{C}_E & \xrightarrow{adjoint} & SC_E(TM) & \longleftarrow & SC_E(Diff M)
\end{array}$$

The maps on the top row are $\mathcal{G}$-equivariant and surjective, with $(\kappa^\infty)^{-1}$ bijective; the map NR refers to the pull-back of Narasimhan & Ramanan's universal connection and so is surjective and $\mathcal{G}$-equivariant.

Acknowledgments

This grew out of joint work with Yves LeJan and Xue-Mei Li, and the last section drew freely from that joint work; John Rawnsley's expertise was invaluable; thanks are also due to the Abel Memorial Fund, the organisers of this Symposium, and especially to Giulia di Nunno.

References

[AB83] M. F. Atiyah and R. Bott. The Yang-Mills equations over Riemann surfaces. *Philos. Trans. Roy. Soc. London Ser. A*, 308(1505):523–615, 1983.

[AH04] Abdulrahman Al-Hussein. Martingale representation theorem in infinite dimensions. *Arab J. Math. Sci.*, 10(1):1–18, 2004.

[AMV96] L. Accardi, A. Mohari, and Centro V. Volterra. On the structure of classical and quantum flows. *J. Funct. Anal.*, 135(2):421–455, 1996.

[Bax76] Peter Baxendale. Gaussian measures on function spaces. *Amer. J. Math.*, 98(4), 1976.

[Bax84] P. Baxendale. Brownian motions in the diffeomorphism groups I. *Compositio Math.*, 53:19–50, 1984.

[ELJL] K. D. Elworthy, Yves Le Jan, and Xue-Mei Li. A geometric approach to filtering of diffusions. In preparation.

[ELJL04] K. D. Elworthy, Yves Le Jan, and Xue-Mei Li. Equivariant diffusions on principal bundles. In *Stochastic analysis and related topics in Kyoto*, volume 41 of *Adv. Stud. Pure Math.*, pages 31–47. Math. Soc. Japan, Tokyo, 2004.

[ELL99] K. D. Elworthy, Y. LeJan, and X.-M. Li. *On the geometry of diffusion operators and stochastic flows, Lecture Notes in Mathematics 1720.* Springer, 1999.

[Elw82] K. D. Elworthy. *Stochastic Differential Equations on Manifolds.* LMS Lecture Notes Series 70, Cambridge University Press, 1982.

[Gro96] Mikhael Gromov. Carnot-Carathéodory spaces seen from within. In *Sub-Riemannian geometry*, volume 144 of *Progr. Math.*, pages 79–323. Birkhäuser, Basel, 1996.

[Hus94] Dale Husemoller. *Fibre bundles;third edition*, volume 20 of *Graduate Texts in Mathematics.* Springer-Verlag, New York, 1994.

[IW89] N. Ikeda and S. Watanabe. *Stochastic Differential Equations and Diffusion Processes, second edition.* North-Holland, 1989.

[KM97] A. Kriegl and P. W. Michor. *The convenient setting of global analysis. Mathematical Surveys and Monographs. 53.* American Mathematical Society, 1997.

[KMS93] I. Kolar, P. W. Michor, and J. Slovak. *Natural operations in differential geometry.* Springer-Verlag, Berlin, 1993.

[Kun90] H. Kunita. *Stochastic Flows and Stochastic Differential Equations*, volume 24 of *Cambridge Studies in Advanced Mathematics.* Cambridge University Press, 1990.

[LW84] Y. LeJan and S. Watanabe. Stochastic flows of diffeomorphisms. In *Stochastic analysis (Katata/Kyoto, 1982), North-Holland Math. Library, 32,*, pages 307–332. North-Holland, Amsterdam, 1984.

[Mic91] Peter W. Michor. *Gauge theory for fiber bundles*, volume 19 of *Monographs and Textbooks in Physical Science. Lecture Notes.* Bibliopolis, Naples, 1991.

[MV81] P. K. Mitter and C.-M. Viallet. On the bundle of connections and the gauge orbit manifold in Yang-Mills theory. *Comm. Math. Phys.*, 79(4):457–472, 1981.

[NR61] M. S. Narasimhan and S. Ramanan. Existence of universal connections. *American J. Math.*, 83:563–572, 1961.

[Pal68] Richard S. Palais. *Foundations of global non-linear analysis.* W. A. Benjamin, Inc., New York-Amsterdam, 1968.

[Pal79] Richard S. Palais. The principle of symmetric criticality. *Comm. Math. Phys.*, 69(1):19–30, 1979.

[PR04] Sylvie Paycha and Steven Rosenberg. Traces and characteristic classes on loop spaces. In *Infinite dimensional groups and manifolds*, volume 5 of *IRMA Lect. Math. Theor. Phys.*, pages 185–212. de Gruyter, Berlin, 2004.

[Qui88] D. Quillen. Superconnections; character forms and the Cayley transform. *Topology*, 27(2):211–238, 1988.

[Ste51] N. Steenrod. *The topology of fibre bundles, Princeton Mathematical Series No. 14.* Princeton University Press, (1951).

[Tsi] B. Tsirelson. Filtrations of random processes in the light of classification theory. (i). a topological zero-one law. arXiv:math.PR/0107121.

Extremes of supOU Processes

Vicky Fasen and Claudia Klüppelberg

Center for Mathematical Sciences, Munich University of Technology, D-85747 Garching, Germany, fasen@ma.tum.de, cklu@ma.tum.de

Summary. Barndorff-Nielsen and Shephard [3] investigate supOU processes as volatility models. Empirical volatility has tails heavier than normal, long memory in the sense that the empirical autocorrelation function decreases slower than exponential, and exhibits volatility clusters on high levels. We investigate supOU processes with respect to these stylized facts. The class of supOU processes is vast and can be distinguished by its underlying driving Lévy process. Within the class of convolution equivalent distributions we shall show that extremal clusters and long range dependence only occur for supOU processes, whose underlying driving Lévy process has regularly varying increments. The results on the extremal behavior of supOU processes correspond to the results of classical Lévy-driven OU processes.

Mathematics Subject Classification (2000): primary: 60G70, 91B70, secondary: 60G10, 91B84

Keywords: convolution equivalent distribution, extreme value theory, independently scattered random measure, Lévy process, long range dependence, point process, regular variation, shot noise process, subexponential distribution, supOU process, extremal cluster

1 Introduction

We investigate the extremal behavior of stationary supOU processes *(superposition of Ornstein-Uhlenbeck processes)* of the form

$$V_t = \int_{\mathbb{R}_+ \times \mathbb{R}} \mathrm{e}^{-r(t-s)} \mathbf{1}_{[0,\infty)}(t-s)\, d\Lambda(r, \lambda s) \quad \text{for } t \geq 0, \tag{1.1}$$

where $\lambda > 0$ and Λ is an *infinitely divisible independently scattered random measure* (i. d. i. s. r. m.). Such models coincide under weak regularity conditions with models introduced under the same acronym by Barndorff-Nielsen [1] aiming at volatility modelling. They allow for non-trivial extensions of OU (*Ornstein-Uhlenbeck*) type processes of the form

$$V_t = \int_{-\infty}^{t} \mathrm{e}^{-\lambda(t-s)} dL_{\lambda s} \quad \text{for } t \geq 0, \tag{1.2}$$

where $\lambda > 0$ and L is a Lévy process. The time-change by λ yields marginal distributions independent of λ. To guarantee that the volatility process V is positive, the Lévy process L is chosen as subordinator. The resulting price process has martingale term $dS_t = \sqrt{V_t}\, dB_t$, where B_t is a Brownian motion, independent of the volatility driving Lévy process. This model has been analyzed by Barndorff-Nielsen and Shephard [3].

An alternative continuous-time model has been suggested by Klüppelberg, Lindner and Maller [14]. In the COGARCH$(1,1)$ model, which is a continuous-time version of the GARCH$(1,1)$ process, the price process has martingale term $dS_t = \sqrt{V_t}\, dL_t$, where L is some arbitrary Lévy process and the volatility is given as solution of the SDE

$$dV_{t+} = (b - aV_t)\, dt + cV_t\, d[L, L]_t^{(d)} \tag{1.3}$$

for parameters $a, b > 0$ and $c \geq 0$, where $([L, L]_t^{(d)})_{t \geq 0}$ is the discrete part of the quadratic variation process of L.

Interestingly, although the two types of models seem at first sight to be quite different, they share many properties; see Klüppelberg, Lindner and Maller [15]. The models differ, however, in their extreme behavior. Whereas the large fluctuations in terms of the tail behavior of the volatility in the Barndorff-Nielsen and Shephard model (1.2) is inherited from the tail behavior of the increments of the Lévy process, the COGARCH model (1.3) exhibits under weak regularity conditions always Pareto-like tails. It has also been shown in Fasen, Klüppelberg and Lindner [12] that both models can only model volatility clusters, if they have Pareto-like tails; i. e. the COGARCH model always does (under weak regularity conditions), and the OU-type model does, if the Lévy process has Pareto-like increments.

Besides volatility clustering, another issue in volatility modelling is the fact that many financial time series exhibit zero autocorrelation in the data, but a long range dependence effect in the volatility. Despite the ongoing debate for the origins of this effect, the modelling issue cannot just be ignored. Unfortunately, the autocovariance functions of both volatility models, the OU-type model and the COGARCH$(1,1)$ decrease exponentially fast.

Barndorff-Nielsen [1] suggests as a remedy the generalization of V to a supOU process. In this paper we want to investigate the extremal behavior of model (1.1) with respect to volatility clustering. As empirical findings indicate and economic reasoning supports, financial data can be modelled by a normal mixture model with tails ranging from exponential to Pareto. Consequently, it is indeed interesting to identify models with such tail behavior, long range dependence effect and volatility clusters in the extremes.

Our paper is organized as follows. We start in Section 2 with an introduction into supOU processes as given in (1.1) including necessary and sufficient conditions for the existence of a stationary version of (1.1). Moreover, we compare our definition with Barndorff-Nielsen's [1] slightly different definition and show that they coincide. In the context of extreme value theory we prefer working with representation (1.1) as it allows us to apply results for mixed MA processes as derived in Fasen [10, 11]. As we shall show in Section 2.2 supOU processes can model a wide range of correlation functions from exponential to polynomial decrease. Poisson shot noise processes as introduced in Section 2.3 present the basic structure for studying the extremal behavior. In Section 2.4 we present the class of convolution equivalent distributions, which will serve as models for the Lévy increments of supOU processes.

The extremal behavior of a supOU process, whose underlying driving Lévy process is in the class of convolution equivalent distributions, is classified by the tail behavior of the random variable $L_1 = \Lambda(\mathbb{R}_+ \times [0,1])$, so that we have to distinguish between different regimes for L_1. In Section 3 we investigate the link between the tail behavior of the Lévy increments in the class of convolution equivalent distributions, represented by L_1, the stationary distribution V_0 of the supOU process, and $\sup_{0\le t\le 1} V_t$. In Section 4 we study the extremal behavior of V via marked point processes, which characterize the distributions of the locations of extremes on high levels. Moreover, we derive the distribution of cluster sizes of high level extremes and the normalizing constants of running maxima. Our findings are summarized in Section 5.

As not to disturb the flow of arguments we postpone classical definitions and concepts to an Appendix.

Throughout the paper we shall use the following notation. We abbreviate distribution function by d. f. and random variable by r. v. For any d. f. F we denote its tail $\overline{F} = 1 - F$ and $F * G$ for the convolution of F with the d. f. G. For two r. v. s X and Y with d. f. s F and G we write $X \stackrel{d}{=} Y$ if $F = G$, and by $\stackrel{n\to\infty}{\Longrightarrow}$ we denote weak convergence for $n \to \infty$. For two functions f and g we write $f(x) \sim g(x)$ as $x \to \infty$, if $\lim_{x\to\infty} f(x)/g(x) = 1$. We also denote $\mathbb{R}_+ = (0, \infty)$. For $x \in \mathbb{R}$, we define $x^+ = \max\{x, 0\}$.

2 The Model

Let $\mathcal{T}$ be a σ-ring on $\mathbb{R}_+ \times \mathbb{R}$ (i. e. countable unions of sets in $\mathcal{T}$ belong to $\mathcal{T}$ and if $A, B \in \mathcal{T}$ with $A \subset B$ then $B \backslash A \in \mathcal{T}$) and let $\Lambda = \{\Lambda(A) : A \in \mathcal{T}\}$ be an i. d. i. s. r. m., which means by definition that all finite dimensional distributions are infinitely divisible and for all disjoint sets $(A_n)_{n\in\mathbb{N}}$ in $\mathcal{T}$ we have that $(\Lambda(A_n))_{n\in\mathbb{N}}$ is an independent sequence and $\Lambda\left(\bigcup_{n=1}^{\infty} A_n\right) = \sum_{n=1}^{\infty} \Lambda(A_n)$ almost surely (a. s.). We work with i. d. i. s. r. m. s, whose characteristic function can be written in the form

$$\mathbb{E}\exp(iu\Lambda(A)) = \exp(\psi(u)\Pi(A)) \quad \text{for } u \in \mathbb{R}, \tag{2.1}$$

where Π is a measure on $\mathbb{R}_+ \times \mathbb{R}$, which is the product of a probability measure π on $\mathbb{R}_+$ and the Lebesgue measure on $\mathbb{R}$, and

$$\psi(u) = ium - \frac{1}{2}u^2\sigma^2 + \int_{\mathbb{R}} \left(\mathrm{e}^{iux} - 1 - iu\kappa(x)\right) \nu(dx) \quad \text{for } u \in \mathbb{R}$$

with $\kappa(x) = \mathbf{1}_{[-1,1]}(x)$. The function ψ is the cumulant generating function of an infinitely divisible r. v. with *generating triplet* (m, σ^2, ν), where $m \in \mathbb{R}$, $\sigma^2 \geq 0$, and ν is a measure on $\mathbb{R}$, called *Lévy measure*, satisfying $\nu(\{0\}) = 0$ and $\int_{\mathbb{R}}(1 \wedge |x|^2)\,\nu(dx) < \infty$. The *generating quadruple* (m, σ^2, ν, π) determines completely the distribution of Λ.

The *underlying driving Lévy process*

$$L_t = \Lambda(\mathbb{R}_+ \times [0,t]) \quad \text{for } t \geq 0 \tag{2.2}$$

has generating triplet (m, σ^2, ν).

2.1 Existence and Stationarity of the Model

The following result guarantees existence, infinite divisibility and stationarity of the model and ensures the equivalence of (1.1) and the supOU model as defined in Barndorff-Nielsen [1]. For the comparison we recall first that integrals of the form $\int_{\mathbb{R}_+\times\mathbb{R}} \mathrm{e}^{-r(t-s)}\mathbf{1}_{[0,\infty)}(t-s)\, d\Lambda(r, \lambda s)$ are defined for each fixed $t \geq 0$ as limit in probability of simple functions (cf. Rajput and Rosinski [16], Theorem 2.7). Hence, V_t is defined a. s. for each fixed t.

Proposition 1. *Let (m, σ^2, ν) be the generating triplet of an infinitely divisible distribution with*

$$\int_{|x|>1} \log(1 + |x|)\, \nu(dx) < \infty\,. \tag{2.3}$$

Define $T : \mathbb{R}_+ \times \mathbb{R} \to \mathbb{R}_+ \times \mathbb{R}$ by $T(r,s) = (r, r^{-1}s)$. Then the following hold:

(a) *Let $\widetilde{\pi}$ be a probability measure on $\mathbb{R}_+$ with $\lambda := \int_{\mathbb{R}_+} r\,\widetilde{\pi}(dr) < \infty$ and $\widetilde{\Lambda}$ be an i. d. i. s. r. m. with generating quadruple $(\widetilde{m}, \widetilde{\sigma}^2, \widetilde{\nu}, \widetilde{\pi})$. Then $\Lambda = \widetilde{\Lambda} \circ T^{-1}$ is an i. d. i. s. r. m. with generating quadruple $(\lambda\widetilde{m}, \lambda\widetilde{\sigma}^2, \lambda\widetilde{\nu}, \pi)$, where $\pi(dr) = \lambda^{-1} r \widetilde{\pi}(dr)$.*

(b) *Let π be a probability measure on $\mathbb{R}_+$ with $\lambda^{-1} := \int_{\mathbb{R}_+} r^{-1}\,\pi(dr) < \infty$ and Λ be an i. d. i. s. r. m. with generating quadruple (m, σ^2, ν, π). Then $\widetilde{\Lambda} = \Lambda \circ T$ is an i. d. i. s. r. m. with generating quadruple $(\lambda^{-1}m, \lambda^{-1}\sigma^2, \lambda^{-1}\nu, \widetilde{\pi})$, where $\widetilde{\pi}(dr) = \lambda r^{-1}\pi(dr)$.*

(c) *For Λ and $\widetilde{\Lambda}$ as in (a) and (b) define for $t \geq 0$,*

$$V_t = \int_{\mathbb{R}_+\times\mathbb{R}} \mathrm{e}^{-r(t-s)}\mathbf{1}_{[0,\infty)}(t-s)\, d\Lambda(r, \lambda s),$$

$$X_t = \int_{-\infty}^{\infty} \mathrm{e}^{-rt} \int_{-\infty}^{rt} \mathrm{e}^{s}\, d\widetilde{\Lambda}(r, \lambda s)\,.$$

Then, $V_t = X_t$ *a. s. for* $t \geq 0$ *and, hence,* V *is a version of* X *and vice versa. Furthermore,* V *(and hence* X*) has a stationary version. For* $d \in \mathbb{N}$ *let* $-\infty = t_0 < t_1 < \ldots < t_d < \infty$ *and* $u_1, \ldots, u_d \in \mathbb{R}$*. The finite dimensional distributions of the stationary process* V *have the cumulant generating function*

$$\log \mathbb{E} \exp(i(u_1 V_{t_1} + \ldots + u_d V_{t_d})) \\ = \sum_{m=1}^{d} \int_0^\infty \int_{t_{m-1}}^{t_m} \lambda\psi \left(\sum_{j=m}^{d} u_j \mathrm{e}^{-r(t_j - s)} \right) ds\, \pi(dr). \tag{2.4}$$

The results *(a)* and *(b)* follow by simple calculations of the characteristic functions of the finite dimensional distributions of Λ and $\widetilde{\Lambda}$. Statement *(c)* is the consequence of the change of variables in *(a)* and *(b)*, respectively, and Barndorff-Nielsen [1], Theorem 3.1 (cf. Rajput and Rosinski [16], Proposition 2.6). Condition (2.3) and $\int_{\mathbb{R}_+} r^{-1}\, \pi(dr) < \infty$ are necessary and sufficient for the existence of a stationary version of V (and hence X).

Throughout this paper we shall assume that V is a measurable, separable and stationary version of the supOU process as given in (1.1) and that $\mathbb{P}(\sup_{0 \leq t \leq 1} |V_t| < \infty) = 1$.

Remark 2. (i) By (2.4) the cumulant generating function of the stationary distribution is given by

$$\begin{aligned} \log \mathbb{E} \exp(iuV_0) &= \int_0^\infty \int_{-\infty}^0 \lambda\psi\left(u\mathrm{e}^{rs}\right) ds\, \pi(dr) \\ &= \int_{-\infty}^0 \psi\left(u\mathrm{e}^{s}\right) ds \qquad \text{for } u \in \mathbb{R}. \end{aligned} \tag{2.5}$$

This is the cumulant generating function of a stationary OU-type process (1.2) driven by the underlying driving Lévy process L as given in (2.2). Then, V_0 has absolutely continuous Lévy measure ν_V with

$$\nu_V\,(dx) = x^{-1} \nu\,[x, \infty)\, dx \qquad \text{for } x > 0, \tag{2.6}$$

and is *selfdecomposable* (Proposition A.5). Note that the stationary distribution of V_0 is independent of π.

(ii) Positivity of V, which is needed for volatility processes, can be guaranteed by choosing L as a *subordinator*; i. e. ν has only support on $\mathbb{R}_+$ with $\int_{(0,\infty)} (1 \wedge x)\, \nu(dx) < \infty$, $\sigma^2 = 0$ and $m = \int_0^\infty \kappa(x)\, \nu(dx)$.

The following examples serve as motivation.

Example 3. (a) If π has only support in some $\lambda > 0$, i. e. $\pi(\{\lambda\}) = 1$, then (2.4) reduces to the cumulant generating function of the d-dimensional distribution of an OU-type process. Thus, (1.1) defines the usual OU-type process (1.2).

(b) Let π be a discrete probability measure with $\pi(\{\lambda_k\}) = p_k$ for $k \in \mathbb{N}$ and $\lambda_k > 0$. Then the assumption $\lambda^{-1} := \int_{\mathbb{R}_+} r^{-1}\,\pi(dr) < \infty$ is equivalent to $\sum_{k=1}^{\infty} p_k \lambda_k^{-1} < \infty$. By (2.4) the cumulant generating function of the d-dimensional distribution is given by

$$\log \mathbb{E} \exp(i(u_1 V_{t_1} + \ldots + u_d V_{t_d})) = \sum_{k=1}^{\infty} \sum_{m=1}^{d} \int_{t_{m-1}}^{t_m} \lambda p_k \psi \left(\sum_{j=m}^{d} u_j \mathrm{e}^{-\lambda_k (t_j - s)} \right) ds.$$

Consequently, V has the same distribution as the superposition of independent OU processes,

$$\sum_{k=1}^{\infty} \int_{-\infty}^{t} \mathrm{e}^{-\lambda_k (t-s)}\, dL^{(k)}_{\lambda s} \quad \text{for } t \geq 0\,,$$

where $(L^{(k)})_{k\in\mathbb{N}}$ are independent Lévy processes with characteristic triplets $(p_k m, p_k \sigma^2, p_k \nu)$.

2.2 Dependence Structure

Provided the underlying driving Lévy process has finite second moment the autocorrelation function ρ of the stationary supOU process (1.1) can be calculated taking derivatives with respect to u_1 and u_2 in (2.4) and taking the limit for $u_1, u_2 \to 0$. We obtain

$$\rho(h) = \lambda \int_0^{\infty} r^{-1} \mathrm{e}^{-hr}\, \pi(dr) \quad \text{for } h \geq 0. \tag{2.7}$$

For a discrete probability measure π as given in Example 3 we obtain

$$\rho(h) = \lambda \sum_{k=1}^{\infty} p_k \lambda_k^{-1} \mathrm{e}^{-h\lambda_k} \quad \text{for } h \geq 0. \tag{2.8}$$

Remark 4. On the one hand the correlation function (2.7) of a supOU process depends only on the probability measure π and is independent of the generating triplet (m, σ^2, ν) of the underlying driving Lévy process. On the other hand the stationary distribution V_0 depends only on (m, σ^2, ν) and is independent of π, represented by the cumulant generating function given in (2.5). Thus, supOU processes can model the stationary distribution and the correlation function independently. This opens the way to a simple statistical fitting of such models. More about supOU models and applications to financial data can be found in Barndorff-Nielsen and Shephard [2, 3].

There are various notions of long range dependence, all having in common that the correlation function should decrease slower than exponential. We shall work with the following definition.

Definition 5. *A stationary process with correlation function ρ exhibits* long range dependence, *if there exists a $H \in (0, 1/2)$ and a slowly varying function l (see Definition A.2), such that*

$$\rho(h) \sim l(h)h^{-2H} \quad \textit{for } h \to \infty.$$

We observe that long range dependence implies that $\int_0^\infty \rho(h)\,dh = \infty$.

The following result explains how long range dependence can be introduced into supOU models. Essentially, the measure π needs sufficient mass near 0. We write $\pi(r)$ for $\pi((0, r])$.

Proposition 6. *Let V be a stationary supOU process as in (1.1) and L be as in (2.2) with $\mathbb{E}L_1^2 = 1$. We denote by ρ the correlation function of V. Suppose l is slowly varying and $H > 0$. Then*

$$\widetilde{\pi}(r) \sim (2H)^{-1} l(r^{-1}) r^{2H} \quad \textit{for } r \to 0, \tag{2.9}$$

if and only if

$$\rho(h) \sim \Gamma(2H) l(h) h^{-2H} \quad \textit{for } h \to \infty. \tag{2.10}$$

If

$$\pi(r) \sim \lambda^{-1}(2H+1)^{-1} l(r^{-1}) r^{2H+1} \quad \textit{for } r \to 0, \tag{2.11}$$

then (2.9) and, hence, (2.10) follow. The converse, i. e. (2.10) implies (2.11) holds, provided that π is absolutely continuous with density π', and $r^{-1}\pi'(r)$ is monotone on $(0, r_0)$ for some $r_0 > 0$.

Proof. The equivalence of (2.9) and (2.10) is a consequence of Karamata's Tauberian theorem (Theorem 1.7.1' in Bingham, Goldie and Teugels [4]) and $\rho(h) = \int_0^\infty \mathrm{e}^{-hr}\, \widetilde{\pi}(dr)$; cf. (2.7). Furthermore, if (2.11) holds, then by Proposition 1 (b) and $\widetilde{\pi}(dr) = \lambda r^{-1}\pi(dr)$, Karamata's theorem (Theorem 1.5.11 in [4]) yields

$$\begin{aligned} \widetilde{\pi}(r) &= \lambda \int_0^r s^{-1}\, \pi(ds) = \lambda r^{-1}\pi(r) + \lambda \int_{r^{-1}}^\infty \pi(s^{-1})\, ds \\ &\sim \lambda(2H+1)(2H)^{-1} r^{-1}\pi(r) \end{aligned}$$

for $r \to 0$. Hence, statements (2.9) and (2.10) follow.

If $r^{-1}\pi'(r)$ is monotone on $(0, r_0)$ for some $r_0 > 0$, and invoking the monotone density theorem (Theorem 1.7.2b in [4]), we get from (2.9)

$$r^{-1}\pi'(r) \sim \lambda^{-1} l(r^{-1}) r^{2H-1} \quad \text{for } r \to 0.$$

Hence, Theorem 1.6.1 in [4] yields $\pi(r) \sim \lambda^{-1}(2H+1)^{-1} l(r^{-1}) r^{2H+1}$ for $r \to 0$. □

Example 7. A typical example of π to generate long range dependence in a supOU process is a gamma distribution with density $\pi(dr) = \Gamma(2H+1)^{-1}r^{2H}\mathrm{e}^{-r}\,dr$ for $r>0$ and $H>0$. Then $\lambda = 2H$ and

$$\rho(h) = \Gamma(2H)^{-1}\int_0^\infty r^{2H-1}\mathrm{e}^{-r(h+1)}\,dr = (h+1)^{-2H} \quad \text{for} \quad h \geq 0.$$

Remark 8. CARMA processes as reviewed by Brockwell [5] can be interpreted as a superposition of OU-type processes. These models correspond to linear combinations of OU processes driven by one single Lévy process. This mechanism creates only processes with asymptotically exponentially decreasing correlation functions.

2.3 Positive Shot Noise Process

The structure of a supOU process can be well understood when considering the following example.

Let Λ be a *positive compound Poisson random measure* in the sense that it has generating quadruple $(\mu\mathbb{P}_F((0,1]), 0, \mu\mathbb{P}_F, \pi)$, where $\mu > 0$, $\mathbb{P}_F$ is a probability measure on $\mathbb{R}_+$ with corresponding d. f. F, and π is a probability measure on $\mathbb{R}_+$ with $\lambda^{-1} := \int_{\mathbb{R}_+} r^{-1}\,\pi(dr) < \infty$. Then Λ has the representation

$$\Lambda(A) = \sum_{k=-\infty}^{\infty} Z_k \mathbf{1}_{\{(R_k,\Gamma_k)\in A\}} \quad \text{for } A \in \mathcal{T}, \tag{2.12}$$

where $(\Gamma_k)_{k\in\mathbb{Z}}$ constitute the jump times of a Poisson process $N = (N_t)_{t\in\mathbb{R}}$ on $\mathbb{R}$ with intensity $\mu > 0$. The process N is independent of the i. i. d. sequence of positive r. v. s $(Z_k)_{k\in\mathbb{Z}}$ with d. f. F. Finally, the i. i. d. sequence $(R_k)_{k\in\mathbb{Z}}$ with distribution π is independent of all other quantities.

The resulting supOU process is then the positive shot noise process

$$\begin{aligned} V_t &= \int_{\mathbb{R}_+\times\mathbb{R}} \mathrm{e}^{-r(t-s)}\mathbf{1}_{[0,\infty)}(t-s)\,d\Lambda(r,\lambda s) \\ &= \sum_{k=-\infty}^{N_{\lambda t}} \mathrm{e}^{-R_k(t-\Gamma_k/\lambda)} Z_k \quad \text{for } t \geq 0, \end{aligned} \tag{2.13}$$

and from (2.6) we get, if $\mathbb{E}\log(1+Z_1) < \infty$ (which is the analogue of (2.3) in this model),

$$\nu_V\,[x,\infty) = \mu\int_x^\infty y^{-1}\overline{F}(y)\,dy \quad \text{for } x > 0$$

and a stationary version of V exists.

The qualitative extreme behavior of this supOU process can be seen in Figure 1 in detail. The supOU process jumps upwards, whenever $(N_{\lambda t})_{t\geq 0}$

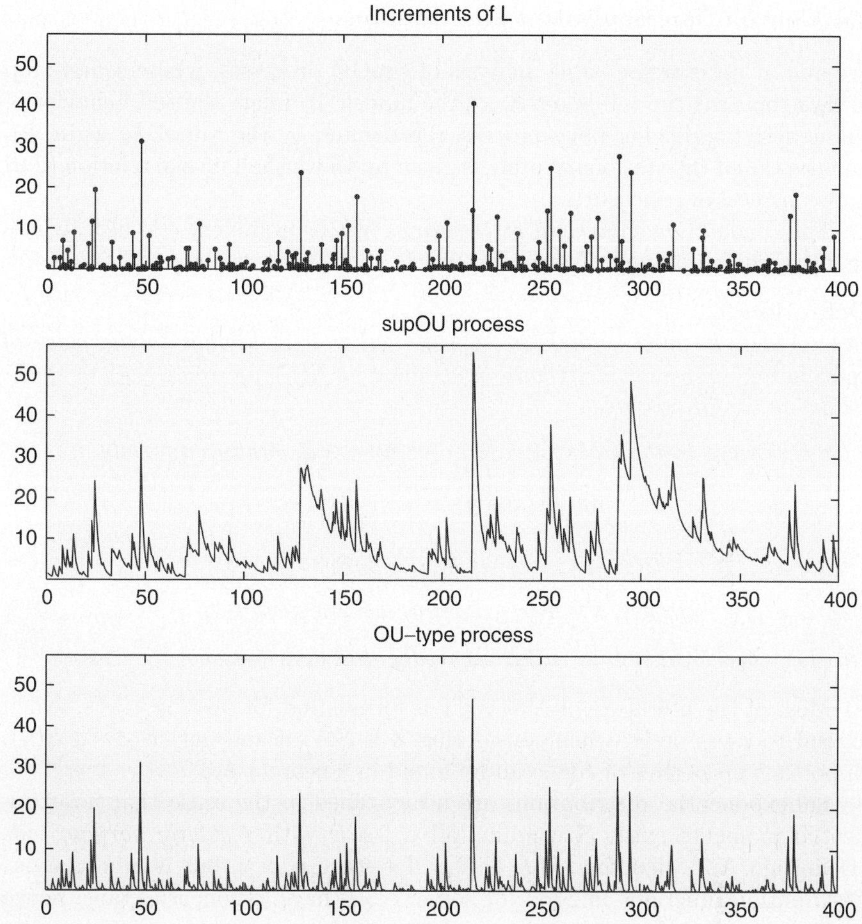

Fig. 1. Sample path of a supOU process $V_t = \sum_{k=-\infty}^{N_{\lambda t}} \mathrm{e}^{-R_k(t-\Gamma_k/\lambda)} Z_k$ as in Section 2.3 and, for comparison, the OU-type process $V_t = \sum_{k=-\infty}^{N_{\lambda t}} \mathrm{e}^{-\lambda t+\Gamma_k} Z_k$ for $0 \leq t \leq 400$, with $\lambda = 1/3$, $\mu = 1/3$, $F(x) = 1 - \exp(-x^{1/2})$ for $x > 0$ and $\pi(r) = r^{3/2}$ for $r \in (0,1)$. In the first plot we show the increments of the underlying driving Lévy process $L_{\lambda t} = \sum_{k=1}^{N_{\lambda t}} Z_k$ for $0 \leq t \leq 400$

jumps and decreases continuously between two jumps. This means in particular that V has local suprema exactly at the jump times Γ_k/λ (and $t = 0$). Consequently, it is the discrete-time skeleton of V at points Γ_k/λ that determines the extreme behavior of the shot noise process. Although the underlying driving Lévy process L of the supOU process as given in (2) and the driving Lévy process of the OU-type process are the same, we see the influence of $(R_k)_{k\in\mathbb{N}}$ on the exponential decrease of V for the simple OU-type process, which governs the memory of the supOU process.

2.4 Convolution Equivalent Distributions

We aim at an extreme value analysis of supOU processes, where a first step always concerns the tail behavior of the model. To relate the tail behavior of the underlying driving Lévy process, represented by the tail of L_1 as in (2), and the tail of the stationary process given by V_0 we shall invoke relation (2.6) between the Lévy measures.

The *convolution equivalent distributions* play a prominent role here, where we distinguish different classes.

Definition 9.
(a) A d.f. F on $\mathbb{R}$ with $F(x) < 1$ for all $x \in \mathbb{R}$ belongs to the class of convolution equivalent distributions denoted by $\mathcal{S}(\gamma)$ for some $\gamma \geq 0$, if the following conditions hold:

(i) F belongs to the class $\mathcal{L}(\gamma)$, i. e. for all $y \in \mathbb{R}$ locally uniformly

$$\lim_{x\to\infty} \overline{F}(x+y)/\overline{F}(x) = \exp(-\gamma y).$$

(ii) $\lim_{x\to\infty} \overline{F*F}(x)/\overline{F}(x)$ exists and is finite.

If Z is a r. v. with d. f. $F \in \mathcal{S}(\gamma)$, then we also write $Z \in \mathcal{S}(\gamma)$.

(b) The class $\mathcal{S}(0) = \mathcal{S}$ is called subexponential distributions.

Most of the literature on this topic is formulated for positive r. v. s, which extend to r. v. s on $\mathbb{R}$, when considering $Z \in \mathcal{S}(\gamma)$ if and only if $Z^+ \in \mathcal{S}(\gamma)$. Important properties of $\mathcal{S}(\gamma)$ can be found in Theorem A.3.

Subexponential distributions are heavy-tailed in the sense that no exponential moments exist. $\mathcal{S}$ contains all d. f. s F with *regularly varying tails* (Definition A.2), denoted by $\overline{F} \in \mathcal{R}_{-\alpha}$ for some $\alpha > 0$, but is much larger. Distribution functions in $\mathcal{S}(\gamma)$ for some $\gamma > 0$ have exponential tails, hence are lighter tailed than subexponential distributions.

Next we present two different regimes governed by extreme value theory, which classifies distributions according to their *maximum domain of attraction*. The maximum domain of attraction condition is an assumption on the tail behavior of a d. f. F. Suppose we can find sequences of real numbers $a_n > 0$ and $b_n \in \mathbb{R}$ such that

$$\lim_{n\to\infty} n\overline{F}(a_n x + b_n) = -\log G(x) \quad \text{for } x \in \mathbb{R},$$

for some non-degenerate d. f. G. Then we say F is in the maximum domain of attraction of G ($F \in \mathrm{MDA}(G)$). The Fisher-Tippett Theorem A.1 says that G is either a Fréchet (Φ_α, $\alpha > 0$), Gumbel (Λ) or Weibull (Ψ_α, $\alpha > 0$) distribution. Convolution equivalent distributions can be in two different maximum domains of attraction, since they have unbounded support to the right (thus excluding the Weibull distribution). All d. f. s such that $\overline{F} \in \mathcal{R}_{-\alpha}$ for some $\alpha > 0$ are subexponential and belong to $\mathrm{MDA}(\Phi_\alpha)$. Other convolution equivalent distributions may belong to $\mathrm{MDA}(\Lambda)$.

Example 10. Typical examples of d. f. s in $\mathcal{S} \cap \mathrm{MDA}(\Lambda)$ have density functions

$$g(x) \sim \text{const.}\, x^{\beta} \mathrm{e}^{-x^{\alpha}} \quad \text{as } x \to \infty$$

for some $\beta \in \mathbb{R}$, $\alpha \in (0,1)$, like the heavy-tailed Weibull distributions. Distribution functions, whose probability density satisfies

$$g(x) \sim \text{const.}\, x^{\beta-1} \mathrm{e}^{-\gamma x} \quad \text{as } x \to \infty \tag{2.14}$$

for $\beta < 0$ are an important subclass of $\mathcal{S}(\gamma) \cap \mathrm{MDA}(\Lambda)$. The papers of Cline [7] and Goldie and Resnick [13] investigate criteria for d. f. s to be in $\mathcal{S}(\gamma) \cap \mathrm{MDA}(\Lambda)$.

We present here some important examples satisfying (2.14), which are also used for financial modelling; we refer to Schoutens [20] for an overview of these d. f. s.

(a) $GIG(\beta, \delta, \gamma)$ (generalized inverse Gaussian distribution) with $\beta < 0$, $\delta > 0$ and $\gamma \geq 0$, is in $\mathcal{S}(\gamma^2/2)$ with probability density

$$g(x) = \text{const.}\, x^{\beta-1} \exp\left(-\left(\delta^2 x^{-1} + \gamma^2 x\right)/2\right) \quad \text{for } x > 0.$$

A special case is for $\beta = -1/2$ the inverse Gaussian distribution $IG(\delta, \gamma)$.

(b) $NIG(\alpha, \beta, \delta, \mu)$ (normal inverse Gaussian distribution) is for $\beta, \delta, \mu \in \mathbb{R}$ and $\alpha > |\beta|$ in $\mathcal{S}(\alpha - \beta)$ and

$$g(x) \sim \text{const.}\, x^{-3/2} \exp(-x(\alpha - \beta)) \quad \text{as } x \to \infty.$$

(c) $GH(\alpha, \beta, \delta, \mu, \gamma)$ (generalized hyperbolic distribution) is for $\beta, \delta, \mu \in \mathbb{R}$, $\alpha > |\beta|$, $\gamma < 0$ in $\mathcal{S}(\alpha - \beta)$ and

$$g(x) \sim \text{const.}\, x^{\gamma-1} \exp(-x(\alpha - \beta)) \quad \text{as } x \to \infty.$$

For $\gamma = -1/2$ the GH distribution is the NIG distribution, while the hyperbolic distribution occurs for $\gamma = 1$.

(d) $CGMY(C, G, M, Y)$ for $C, M, G > 0$, $Y \in (-\infty, 2]$, introduced by Carr, Geman, Madan and Yor [6]. For $0 < Y < 2$ it belongs to $\mathcal{S}(M)$ with Lévy density

$$\nu(dx) = C|x|^{-1-Y} \exp\left(\frac{G-M}{2}x - \frac{G+M}{2}|x|\right) \quad \text{for } x \in \mathbb{R}\backslash\{0\}.$$

All these distributions are selfdecomposable, which means that they are possible stationary distributions of OU-type processes and, hence, also of supOU processes. We summarize in Proposition A.5 necessary and sufficient conditions of d. f. s to be selfdecomposable.

3 Tail Behavior

We use extensively the fact that for every infinitely divisible convolution equivalent distribution the tail of the distribution function and the tail of its Lévy measure are asymptotically equivalent; see Theorem A.3 *(i)*.

Proposition 11 (Tail Behavior of V).
Let V be a stationary supOU process as in (1.1) *and L be the underlying driving Lévy process* (2.2).

(a) Then $L_1 \in \mathcal{R}_{-\alpha}$ if and only if $V_0 \in \mathcal{R}_{-\alpha}$. In this case

$$\mathbb{P}(V_0 > x) \sim \alpha^{-1}\mathbb{P}(L_1 > x) \quad \text{for } x \to \infty.$$

(b) If $L_1 \in \mathcal{S}(\gamma) \cap \mathrm{MDA}(\Lambda)$ with tail representation as given in (A.1)*, then also $V_0 \in \mathcal{S}(\gamma) \cap \mathrm{MDA}(\Lambda)$,*

$$\mathbb{P}(V_0 > x) \sim \frac{a(x)}{x} \frac{\mathbb{E}\mathrm{e}^{\gamma V_0}}{\mathbb{E}\mathrm{e}^{\gamma L_1}} \mathbb{P}(L_1 > x) \quad \text{for } x \to \infty,$$

and $\mathbb{P}(V_0 > x) = o(\mathbb{P}(L_1 > x))$ for $x \to \infty$.

Proof. Recall from Remark 2 that the stationary distribution of a supOU process driven by an i. d. i. s. r. m. with generating quadruple (m, σ^2, ν, π) coincides with the stationary distribution of an OU-type process (1.2) driven by the Lévy process L with generating triplet (m, σ^2, ν). Thus, applying Proposition 3.2 and Proposition 3.9 in Fasen et al. [12] we obtain sufficiency in (a) and (b). To prove the converse of (a) assume that $V_0 \in \mathcal{R}_{-\alpha}$. Since $\nu_V(x, \infty) = \int_x^\infty y^{-1}\nu(y, \infty)\,dy$ for $x > 0$, and $\nu_V(x, \infty) \sim \mathbb{P}(V_0 > x)$ for $x \to \infty$, we obtain by Bingham et al. [4], Theorem 1.7.2, that $\nu(x, \infty) \sim \alpha\nu_V(x, \infty)$ for $x \to \infty$. Hence, by Theorem A.3 (i) we conclude

$$\mathbb{P}(L_1 > x) \sim \alpha\mathbb{P}(V_0 > x) \quad \text{for } x \to \infty.$$ □

Lemma 12. *Let V be a stationary supOU process as in* (1.1) *with absolutely continuous Lévy density $\nu_V(dx) = u(x)\,dx$, where*

$$u(x) \sim const.\, x^{\beta-1}\mathrm{e}^{-\gamma x} \quad \text{for } x \to \infty$$

for $\gamma > 0$, and let L be the underlying driving Lévy process (2.2)*. Then $V_0 \in \mathcal{S}(\gamma) \cap \mathrm{MDA}(\Lambda)$ if and only if $\beta < 0$, and $L_1 \in \mathcal{S}(\gamma) \cap \mathrm{MDA}(\Lambda)$ if and only if $\beta < -1$.*

Proof. Using (2.6) we obtain $\nu(x, \infty) = xu(x)$ for $x > 0$. Thus,

$$\frac{\nu(dx)}{dx} = -u(x) - xu'(x) \sim \mathrm{const.}\, \gamma\, x^\beta \mathrm{e}^{-\gamma x} \quad \text{for } x \to \infty.$$

The result follows then from Rootzén [18], Lemma 7.1, and Theorem A.3 *(i)*. □

The next proposition follows from Fasen [11], Proposition 3.3, and [10], Theorem 3.3.

Proposition 13 (Tail Behavior of $M(h)$).
Let V be a supOU process and define $M(h) = \sup_{0 \le t \le h} V_t$ for $h > 0$.
(a) If $L_1 \in \mathcal{R}_{-\alpha}$, then also $M(h) \in \mathcal{R}_{-\alpha}$ and

$$\mathbb{P}(M(h) > x) \sim \left(\lambda h + \alpha^{-1}\right) \mathbb{P}(L_1 > x) \quad \text{for } x \to \infty.$$

(b) If $L_1 \in \mathcal{S}(\gamma) \cap \mathrm{MDA}(\Lambda)$, then also $M(h) \in \mathcal{S}(\gamma) \cap \mathrm{MDA}(\Lambda)$ and

$$\mathbb{P}(M(h) > x) \sim \lambda h \frac{\mathbb{E}e^{\gamma V_0}}{\mathbb{E}e^{\gamma L_1}} \mathbb{P}(L_1 > x) \quad \text{for } x \to \infty.$$

Remark 14. *(i)* From Lemma 12 follows immediately that for $\beta \in [-1, 0)$, $V_0 \in \mathcal{S}(\gamma) \cap \mathrm{MDA}(\Lambda)$ but $L_1 \notin \mathcal{S}(\gamma)$.

(ii) Proposition 13 implies that the tail of the maximum of a supOU process driven by an i. d. i. s. r. m. with generating quadruple (m, σ^2, ν, π) behaves like the tail of the maximum of an OU-type process driven by a Lévy process with generating triplet (m, σ^2, ν). From this we conclude immediately that the long memory property of supOU processes does not affect the tail behavior of $M(h)$.

4 Extremal Behavior of supOU Processes

For a general i. d. i. s. r. m. Λ we decompose

$$\Lambda = \Lambda^{(1)} + \Lambda^{(2)} \tag{4.1}$$

into two independent i. d. i. s. r. m. s.

$\Lambda^{(1)}$ has only jumps greater than 1; i.e. it has generating quadruple $(0, 0, \nu_1, \pi)$ with $\nu_1(x, \infty) = \nu(1 \vee x, \infty)$ for $x > 0$ and $\nu_1(-\infty, 1] = 0$. Consequently, $\Lambda^{(1)}$ is a positive compound Poisson random measure with representation (2.12) whose underlying driving Lévy process $L^{(1)}$ is a compound Poisson process with intensity $\nu(1, \infty)$, jump times $-\infty < \cdots < \Gamma_{-1} < \Gamma_0 < 0 < \Gamma_1 < \cdots < \infty$ and jump sizes Z_k with probability measure $\nu_1/\nu(1, \infty)$.

$\Lambda^{(2)}$ summarizes all other features of the model; i.e. it has generating quadruple $(m, \sigma^2, \nu_2, \pi)$ with $\nu_2(-\infty, -x) = \nu(-\infty, -x)$ and $\nu_2(x, \infty) = \nu(1 \wedge x, 1]$ for $x > 0$. This means that all the small positive jumps, the negative jumps, the Gaussian component and the drift are summarized in $\Lambda^{(2)}$.

For $d \in \mathbb{N}_0$ let $t_1, \ldots, t_d \ge 0$, and define

$$M_k = \sup_{t \in [\Gamma_k/\lambda, \Gamma_{k+1}/\lambda)} V_t \quad \text{and} \quad \mathbf{V}(\Gamma_k) = (V_{\Gamma_k + t_1}, \ldots, V_{\Gamma_k + t_d}) \quad \text{for } k \in \mathbb{N}.$$

For a Radon measure ϑ we write $\mathrm{PRM}(\vartheta)$ for a *Poisson random measure* with intensity measure ϑ, see Definition A.7. In our set-up ϑ will be a Radon measure on either of the spaces $S_F = [0, \infty) \times (0, \infty] \times [-\infty, \infty]^d$ or

$S_G = [0,\infty) \times (-\infty,\infty] \times [-\infty,\infty]^d$, and $M_P(S_F)$ and $M_P(S_G)$ will denote the spaces of all point measures on S_F and S_G, respectively. For details on point processes see Resnick [17].

The following proposition is a consequence of Fasen [9], Theorem 2.5.1 and [10], Theorem 4.1.

Proposition 15 (Point Process Behavior).
Let V be a stationary supOU process as in (1.1) and L be the underlying driving Lévy process (2.2). Decompose Λ as in (4.1).

(a) Let $L_1 \in \mathcal{R}_{-\alpha}$ with norming constants $a_T > 0$ such that

$$\lim_{T\to\infty} T\mathbb{P}(L_1 > a_T x) = x^{-\alpha} \qquad \text{for } x > 0.$$

Suppose $\sum_{k=1}^{\infty} \varepsilon_{(s_k,P_k)}$ is a PRM(ϑ) with

$$\vartheta(dt \times dx) = dt \times \alpha x^{-\alpha-1}\,dx$$

independent of the i. i. d. sequences $(\Gamma_{k,j})_{j\in\mathbb{N}}$ for $k \in \mathbb{N}$ with $(\Gamma_{k,j})_{j\in\mathbb{N}} \stackrel{d}{=} (\Gamma_j)_{j\in\mathbb{N}}$ and independent of the i. i. d. sequence $(R_k)_{k\in\mathbb{N}}$ with probability distribution π. Define $\Gamma_{k,0} = 0$ for $k \in \mathbb{N}$. Then, in the space $M_P(S_F)$,

$$\sum_{k=1}^{\infty} \varepsilon_{\left(\Gamma_k/(\lambda n),\, a_{\lambda n}^{-1} M_k,\, a_{\lambda n}^{-1}\mathbf{V}(\Gamma_k/\lambda)\right)}$$
$$\stackrel{n\to\infty}{\Longrightarrow} \sum_{k=1}^{\infty}\sum_{j=0}^{\infty} \varepsilon_{\left(s_k,\, P_k \mathrm{e}^{-R_k\Gamma_{k,j}/\lambda},\, P_k\left(\mathrm{e}^{-R_k(\Gamma_{k,j}/\lambda+t_1)},\ldots,\mathrm{e}^{-R_k(\Gamma_{k,j}/\lambda+t_d)}\right)\right)}.$$

(b) Let $L_1 \in \mathcal{S}(\gamma) \cap \mathrm{MDA}(\Lambda)$ with norming constants $a_T > 0$ and $b_T \in \mathbb{R}$ such that

$$\lim_{T\to\infty} T\mathbb{P}(L_1 > a_T x + b_T) = \exp(-x) \qquad \text{for } x \in \mathbb{R}.$$

Suppose $\sum_{k=1}^{\infty} \varepsilon_{(s_k,P_k)}$ is a PRM(ϑ) with

$$\vartheta(dt \times dx) = dt \times [\mathbb{E}\mathrm{e}^{\gamma L_1}]^{-1}\mathbb{E}\mathrm{e}^{\gamma V_0}\mathrm{e}^{-x}\,dx$$

independent of the i. i. d. sequence $(R_k)_{k\in\mathbb{N}}$ with probability distribution π. Then, in the space $M_P(S_G)$,

$$\sum_{k=1}^{\infty} \varepsilon_{(\Gamma_k/(\lambda n),\, a_{\lambda n}^{-1}(M_k - b_{\lambda n}),\, b_{\lambda n}^{-1}\mathbf{V}(\Gamma_k/\lambda))} \stackrel{n\to\infty}{\Longrightarrow} \sum_{k=1}^{\infty} \varepsilon_{(s_k,\, P_k,\, (\mathrm{e}^{-R_k t_1},\ldots,\mathrm{e}^{-R_k t_d}))}.$$

We give an interpretation of the point process results. In both parts of Proposition 15 the limit relations of the first two components show that the local suprema M_k of V around Γ_k/λ, normalized by the constants determined via L_1, converge weakly to the same extreme value distribution as L_1. The third vector component indicates that for instance for $d = 1$ and $t_1 = 0$ that the second and third component have the same limiting behavior; i. e. the M_k behave like $V_{\Gamma_k/\lambda}$. The results show also that local extremes of V on high levels happen at the jump times Γ/λ of the Lévy process $(L^{(1)}_{\lambda t})_{t\geq 0}$. Thus, the various features of L, which are modelled in Λ_2, have no influence on the location of local extremes on high levels. Moreover, the third vector component indicates that, if the supOU process has an exceedance over a high threshold, then it decreases after this event exponentially fast with a random rate R_k and the distribution π of R_k governs the short/long range dependence of the model.

As for OU-type processes there is an essential difference between the models (a) and (b). In the second component and the third vector component of the limit point process in *(a)* all points $\Gamma_{k,j}/\lambda$ influence the limit, whereas in *(b)* only $\Gamma_{k,0} = 0$ does. This phenomenon certainly originates in the very large jumps caused by regular variation of the underlying driving Lévy process. Even though the behavior of the supOU process between the large jumps has the tendency to decrease exponentially fast (this comes from the shot-noise process generated by $\Lambda^{(1)}$ and may be overlaid by small positive jumps, negative jumps, a drift and a Gaussian component), huge positive jumps can have a long lasting influence on excursions above high thresholds. This is in contrast to the semi-heavy tailed case in *(b)*.

Result *(b)* can be interpreted that local extremes of models in $\mathcal{S}(\gamma) \cap \mathrm{MDA}(\Lambda)$ show no extremal clusters. The constant $[\mathbb{E}e^{\gamma L_1}]^{-1}\mathbb{E}e^{\gamma V_0}$ in the intensity of the Poisson random measure, which is 1 for $\gamma = 0$, reflects that for $\gamma > 0$ the small jumps of L have a certain influence on the size of the local extremes of V, which is in contrast to subexponential models in *(a)* and *(b)* with $\gamma = 0$. Although $(V_{\Gamma_k/\lambda})_{k\in\mathbb{N}}$ is not a stationary sequence $V_{\Gamma_k/\lambda} \stackrel{k\to\infty}{\Longrightarrow} V_0 + Z_1$ (recall that Z_1 has d. f. $\nu_1/\nu(1,\infty)$). Furthermore,

$$\nu(1,\infty)\mathbb{P}(V_0 + Z_1 > x) \sim [\mathbb{E}e^{\gamma L_1}]^{-1}\mathbb{E}e^{\gamma V_0}\mathbb{P}(L_1 > x) \quad \text{for } x \to \infty.$$

Thus *(b)* implies that the exceedances of $(V_{\Gamma_k/\lambda})_{k\in\mathbb{N}}$ at times $(\Gamma_k/\lambda)_{k\in\mathbb{N}}$ behave like those of an i. i. d. sequence with distribution $V_0 + Z_1$. We have seen this constant $[\mathbb{E}e^{\gamma L_1}]^{-1}\mathbb{E}e^{\gamma V_0}$ already earlier in Proposition 13.

Corollary 16 (Point Process of Exceedances).
Let V satisfy the assumptions of Proposition 15 and decompose Λ as in (4.1).

(a) Let $L_1 \in \mathcal{R}_{-\alpha}$. Suppose $(s_k)_{k\in\mathbb{N}}$ are the jump times of a Poisson process with intensity $x^{-\alpha}$ for fixed $x > 0$. Let $(\zeta_k)_{k\in\mathbb{N}}$ be i. i. d. discrete r. v. s, independent of $(s_k)_{k\in\mathbb{N}}$, with probability distribution

$$q_k = \mathbb{P}(\zeta_1 = k) = \mathbb{E}\exp(-\alpha R_0 \Gamma_k/\lambda) - \mathbb{E}\exp(-\alpha R_0 \Gamma_{k+1}/\lambda) \quad \textit{for } k \in \mathbb{N}.$$

Then

$$\sum_{k=1}^{\infty} \varepsilon_{(\Gamma_k/(\lambda n), a_{\lambda n}^{-1} M_k)}(\cdot \times (x,\infty)) \overset{n\to\infty}{\Longrightarrow} \sum_{k=1}^{\infty} \zeta_k \varepsilon_{s_k} \quad \text{in } M_P([0,\infty)).$$

(b) Let $L_1 \in \mathcal{S}(\gamma) \cap \mathrm{MDA}(\Lambda)$. *Suppose* $(s_k)_{k\in\mathbb{N}}$ *are the jump times of a Poisson process with intensity* $[\mathbb{E}\mathrm{e}^{\gamma L_1}]^{-1}\mathbb{E}\mathrm{e}^{\gamma V_0}\,\mathrm{e}^{-x}$ *for fixed* $x \in \mathbb{R}$. *Then*

$$\sum_{k=1}^{\infty} \varepsilon_{\left(\Gamma_k/(\lambda n), a_{\lambda n}^{-1}(M_k - b_{\lambda n})\right)}(\cdot \times (x,\infty)) \overset{n\to\infty}{\Longrightarrow} \sum_{k=1}^{\infty} \varepsilon_{s_k} \quad \text{in } M_P([0,\infty)).$$

Again the qualitative difference of the two regimes is visible. For a regularly varying underlying driving Lévy process L the limiting process is a compound Poisson process, where at each Poisson point a cluster appears, whose size is random with distribution $(q_k)_{k\in\mathbb{N}}$. In contrast to this, in the MDA(Λ) case, the limit process is simply a homogeneous Poisson process; no clusters appear in the limit.

The next proposition follows immediately from Proposition 15.

Proposition 17 (Running Maxima).
Let V *be a stationary supOU process as in* (1.1) *and* L *be the underlying driving Lévy process* (2). *Define* $M(T) = \sup_{0\le t\le T} V_t$ *for* $T > 0$.

(a) Let $L_1 \in \mathcal{R}_{-\alpha}$ *with norming constants* $a_T > 0$ *such that*

$$\lim_{T\to\infty} T\mathbb{P}(L_1 > a_T x) = x^{-\alpha} \quad \text{for } x > 0.$$

Then

$$\lim_{T\to\infty} \mathbb{P}\left(a_{\lambda T}^{-1} M(T) \le x\right) = \exp(-x^{-\alpha}) \quad \text{for } x > 0.$$

(b) Let $L_1 \in \mathcal{S}(\gamma) \cap \mathrm{MDA}(\Lambda)$ *with norming constants* $a_T > 0$ *and* $b_T \in \mathbb{R}$, *such that*

$$\lim_{T\to\infty} T\mathbb{P}(L_1 > a_T x + b_T) = \exp(-x) \quad \text{for } x \in \mathbb{R}.$$

Then

$$\lim_{T\to\infty} \mathbb{P}\left(a_{\lambda T}^{-1}(M(T) - b_{\lambda T}) \le x\right) = \exp\left(-\left[\mathbb{E}\mathrm{e}^{\gamma L_1}\right]^{-1} \mathbb{E}\mathrm{e}^{\gamma V_0}\mathrm{e}^{-x}\right) \quad \text{for } x \in \mathbb{R}.$$

Definition 18 (Extremal Index Function).
Let $(V_t)_{t\ge 0}$ *be a stationary process. Define the sequence* $M_k(h) = \sup_{(k-1)h\le t\le kh} V_t$ *for* $k \in \mathbb{N}$, $h > 0$. *Let* $\theta(h)$ *be the extremal index (Definition A.8) of the sequence* $(M_k(h))_{k\in\mathbb{N}}$. *Then we call the function* $\theta : (0,\infty) \to [0,1]$ *extremal index function.*

The idea is to divide the positive real line into blocks of length h. By taking local suprema of the process over these blocks the natural dependence of the continuous-time process is weakened, in certain cases it even disappears. However, for fixed h the extremal index function is a measure for the expected cluster sizes among these blocks. For an extended discussion on the extremal index in the context of discrete- and continuous-time processes see Fasen [9], pp. 83.

Corollary 19 (Extremal Index Function).
Let V be a stationary supOU process as in (1.1) *and L be the underlying driving Lévy process* (2).

(a) If $L_1 \in \mathcal{R}_{-\alpha}$, then $\theta(h) = \lambda h\alpha/(\lambda h\alpha + 1)$ for $h > 0$.
(b) If $L_1 \in \mathcal{S}(\gamma) \cap \mathrm{MDA}(\Lambda)$, then $\theta(h) = 1$ for $h > 0$.

Regularly varying supOU processes exhibit clusters among blocks, since $\theta(h) < 1$. So they have the potential to model both features: heavy tails and high level clusters. This is in contrast to supOU processes in $\mathcal{S}(\gamma) \cap \mathrm{MDA}(\Lambda)$, where no extremal clusters occur.

5 Conclusion

In this paper we have investigated the extremal behavior of supOU processes, whose underlying driving Lévy process is in the class of convolution equivalent distributions. In contrast to OU-type and COGARCH processes (cf. [14]), regardless of the driving Lévy process they can model long memory. We have concentrated on models with tails ranging from exponential to regularly varying; i. e. tails as they are found in empirical volatility. The stochastic quantities characterizing the extreme behavior for such models, which we have derived in this paper, include

- the tail of the stationary distribution of the supOU process V_0 and $M(h) = \sup_{0 \le t \le h} V_t$, and the relation to the tail of the distribution governing the extreme behavior,
- the asymptotic distribution of the running maxima, i. e. their MDA and the norming constants,
- the cluster behavior of the model on high levels.

We want to indicate that long memory of a supOU process represented by π has no influence on the existence of extremal clusters, only on the cluster sizes. SupOU processes in $\mathcal{S}(\gamma) \cap \mathrm{MDA}(\Lambda)$ cannot model clusters on high levels. In contrast to that, regularly varying supOU processes exhibit extremal clusters, which can be described quite precisely by the distribution of the cluster sizes, which depends on π; see Corollary 16. In terms of the tail behavior of V_0, $M(h)$ and the running maxima the results for a supOU process coincide with the results of an OU-type process. Again they are not affected by the long memory property.

Appendix

A Basic Notation and Definition

We summarize some definitions and concepts used throughout the paper. For details and further references see Embrechts, Klüppelberg and Mikosch [8].

The following is the fundamental theorem in extreme value theory.

Theorem A.1 (Fisher-Tippett Theorem).
Let $(X_n)_{n\in\mathbb{N}}$ be an i. i. d. sequence with d. f. F and denote

$$M_n = \max_{k=1,\ldots,n} X_k.$$

Suppose we can find sequences of real numbers $a_n > 0$, $b_n \in \mathbb{R}$ such that

$$\lim_{n\to\infty} \mathbb{P}(a_n^{-1}(M_n - b_n) \le x) = \lim_{n\to\infty} F^n(a_n x + b_n) = G(x) \quad \text{for } x \in \mathbb{R}$$

and some non-degenerate d. f. G (we say F is in the maximum domain of attraction *of G and write $F \in \mathrm{MDA}(G)$). Then there are $a > 0$, $b \in \mathbb{R}$ such that $x \mapsto G(ax+b)$ is one of the following three extreme value d. f. s:*

- *Fréchet:* $\Phi_\alpha(x) = G(ax+b) = \begin{cases} 0, & x \le 0, \\ \exp(-x^{-\alpha}), & x > 0, \end{cases}$ *for* $\alpha > 0$.
- *Gumbel:* $\Lambda(x) = G(ax+b) = \exp(-\mathrm{e}^{-x}), \quad x \in \mathbb{R}$.
- *Weibull:* $\Psi_\alpha(x) = G(ax+b) = \begin{cases} \exp(-(-x)^{\alpha}), & x \le 0, \\ 1, & x > 0, \end{cases}$ *for* $\alpha > 0$.

Definition A.2. *A positive measurable function $u : \mathbb{R} \to \mathbb{R}_+$ is called regularly varying with index α, denoted by $u \in \mathcal{R}_\alpha$ for $\alpha \in \mathbb{R}$, if*

$$\lim_{t\to\infty} \frac{u(tx)}{u(t)} = x^\alpha \quad \text{for } x > 0.$$

The function u is said to be slowly varying if $\alpha = 0$.

Theorem A.3. *Let F be a d. f. with $F(x) < 1$ for all $x \in \mathbb{R}$ and $\widehat{f}(\gamma) = \int_{-\infty}^{\infty} \mathrm{e}^{\gamma x}\, F(dx)$.*

(i) Let F be infinitely divisible with Lévy measure ν and $\gamma \ge 0$. Then

$$F \in \mathcal{S}(\gamma) \iff \nu(1,\cdot]/\nu(1,\infty) \in \mathcal{S}(\gamma).$$

(ii) Suppose $F \in \mathcal{S}(\gamma)$, $\lim_{x\to\infty} \overline{G}(x)/\overline{F}(x) = q \ge 0$ and $\widehat{f}_G(\gamma) < \infty$. Then

$$\lim_{x\to\infty} \frac{\overline{F*G}(x)}{\overline{F}(x)} = \widehat{f}_2(\gamma) + q\widehat{f}_1(\gamma)$$

and $F * G \in \mathcal{S}(\gamma)$. If $q > 0$, then also $G \in \mathcal{S}(\gamma)$.

(iii) $F \in \mathcal{L}(\gamma)$, $\gamma \geq 0$, *has the representation*

$$\overline{F}(x) = c(x) \exp\left[-\int_0^x \frac{1}{a(y)}\,dy\right] \qquad \text{for } x > 0, \tag{A.1}$$

where $a, c : \mathbb{R}_+ \to \mathbb{R}_+$ *and* $\lim_{x\to\infty} c(x) = c > 0$ *and* a *is absolutely continuous with* $\lim_{x\to\infty} a(x) = \gamma^{-1}$ *and* $\lim_{x\to\infty} a'(x) = 0$.

The following concept has proved useful in comparing tails.

Definition A.4 (Tail-equivalence).
Two d. f. s F *and* G *(or two measures* μ *and* ν*) are called* tail-equivalent *if both have support unbounded to the right and there exists some* $c > 0$ *such that*

$$\lim_{x\to\infty} \overline{F}(x)/\overline{G}(x) = c \quad \text{or} \quad \lim_{x\to\infty} \nu(x,\infty)/\mu(x,\infty) = c\,.$$

For two tail-equivalent d. f. s in MDA(G) *for some* G *one can choose the same norming constants.*

Proposition A.5. *Let* X *be a r. v. The following conditions are equivalent:*

(a) X *is selfdecomposable.*
(b) There exists a Lévy process L *such that* $X \stackrel{d}{=} \int_0^\infty \mathrm{e}^{-s}\,dL_s$.
(c) X *is infinitely divisible with absolutely continuous Lévy measure given by*

$$\nu(dx) = \frac{k(x)}{|x|}\,dx \qquad \text{for } x \in \mathbb{R}\backslash\{0\},$$

$k(x) \geq 0$, *and* $k(x)$ *is increasing on* $(-\infty, 0)$ *and decreasing on* $(0,\infty)$.

Remark A.6. The integral in (b) exists if and only if (2.3) holds. The above proposition is presented and discussed in Barndorff-Nielsen and Shephard [2], where also further references can be found. It can also be found e.g. in Sato [19], Cor. 15.11 and Theorem 17.5.

Definition A.7 (Poisson Random Measure).
Let $(A, \mathcal{A}, \vartheta)$ *be a measurable space, where* ϑ *is* σ*-finite, and* $(\Omega, \mathcal{F}, \mathbb{P})$ *be a probability space. A* Poisson random measure N with intensity measure ϑ, *denoted by* PRM(ϑ), *is a collection of r. v. s* $(N(A))_{A\in\mathcal{A}}$, *where* $N(A)$: $(\Omega, \mathcal{F}, \mathbb{P}) \to (\mathbb{N}_0, \mathcal{B}(N_0))$, *with* $N(\emptyset) = 0$, *such that:*

(a) Given any sequence $(A_n)_{n\in\mathbb{N}}$ *of mutually disjoint sets in* $\mathcal{A}$:

$$N\left(\bigcup_{n\in\mathbb{N}} A_n\right) = \sum_{n\in\mathbb{N}} N(A_n) \qquad a.\,s.$$

(b) $N(A)$ *is Poisson distributed with intensity* $\vartheta(A)$ *for every* $A \in \mathcal{A}$.
(c) For mutually disjoint sets $A_1, \ldots, A_n \in \mathcal{A}$, $n \in \mathbb{N}$, *the r. v. s* $N(A_1), \ldots,$ $N(A_n)$ *are independent.*

Definition A.8 (Extremal Index).
Let $X = (X_n)_{n\in\mathbb{Z}}$ be a strictly stationary sequence and $\theta \geq 0$. If for every $x > 0$ there exists a sequence $u_n(x)$ with

$$\lim_{n\to\infty} n\mathbb{P}(X_1 > u_n(x)) = x \quad \text{and} \quad \lim_{n\to\infty} \mathbb{P}\left(\max_{k=1,\ldots,n} X_n \leq u_n(x)\right) = \exp(-\theta x),$$

then θ is called the extremal index *of X and has value in* $[0,1]$.

Acknowledgement

We thank Ole Barndorff-Nielsen for various remarks on a former version of the paper, which corrected some errors and improved the presentation. Financial support of V.F. from the Deutsche Forschungsgemeinschaft through the graduate program "Angewandte Algorithmische Mathematik" at the Munich University of Technology is gratefully acknowledged.

References

1. O. E. Barndorff-Nielsen. Superposition of Ornstein–Uhlenbeck type processes. *Theory Probab. Appl.*, 45(2):175–194, 2001.
2. O. E. Barndorff-Nielsen and N. Shephard. Modelling by Lévy processes for financial econometrics. In O. E. Barndorff-Nielsen, T. Mikosch, and S. I. Resnick, editors, *Lévy Processes: Theory and Applications*, pages 283–318, Boston, 2001. Birkhäuser.
3. O. E. Barndorff-Nielsen and N. Shephard. Non-Gaussian Ornstein-Uhlenbeck based models and some of their uses in financial economics (with discussion). *J. Roy. Statist. Soc. Ser. B*, 63(2):167–241, 2001.
4. N. H. Bingham, C. M. Goldie, and J. L. Teugels. *Regular Variation.* Cambridge University Press, Cambridge, 1987.
5. P. J. Brockwell. Continuous-time ARMA process. In C. R. Rao and D. N. Shanbhag, editors, *Handbook of Statistics: Stochastic Processes, Theory and Methods*, pages 249–276. Elsevier, Amsterdam, 2001.
6. P. Carr, H. Geman, D. R. Madan, and M. Yor. The fine structure of asset returns: An empirical investigation. *J. of Business*, 75:305–332, 2002.
7. D. B. H. Cline. Convolution tails, product tails and domains of attraction. *Probab. Theory Related Fields*, 72:529–557, 1986.
8. P. Embrechts, C. Klüppelberg, and T. Mikosch. *Modelling Extremal Events for Insurance and Finance.* Springer, Berlin, 1997.
9. V. Fasen. *Extremes of Lévy Driven MA Processes with Applications in Finance.* PhD thesis, Munich University of Technology, December 2004.
10. V. Fasen. Extremes of mixed MA processes in the class of convolution equivalent distributions. Preprint, available at `http://www.ma.tum.de/stat/`, 2007.
11. V. Fasen. Extremes of regularly varying mixed moving average processes. *Adv. in Appl. Prob.*, 37:993–1014, 2005.

12. V. Fasen, C. Klüppelberg, and A. Lindner. Extremal behavior of stochastic volatility models. In A. N. Shiryaev, M. D. R. Grossinho, P. E. Oliviera, and M. L. Esquivel, editors, *Stochastic Finance*, pages 107–155, New York., 2006. Springer.
13. C. M. Goldie and S. Resnick. Distributions that are both subexponential and in the domain of attraction of an extreme-value distribution. *Adv. Appl. Probab.*, 20(4):706–718, 1988.
14. C. Klüppelberg, A. Lindner, and R. Maller. A continuous time GARCH process driven by a Lévy process: stationarity and second order behaviour. *J. Appl. Probab.*, 41(3):601–622, 2004.
15. C. Klüppelberg, A. Lindner, and R. Maller. Continuous time volatility modelling: COGARCH versus Ornstein-Uhlenbeck models. In Y. Kabanov, R. Lipster, and J. Stoyanov, editors, *From Stochastic Calculus to Mathematical Finance. The Shiryaev Festschrift*, pages 393–419, Berlin, 2006. Springer.
16. B. S. Rajput and J. Rosinski. Spectral representations of infinitely divisible processes. *Probab. Theory Related Fields*, 82(3):453–487, 1989.
17. S. I. Resnick. *Extreme Values, Regular Variation, and Point Processes.* Springer, New York, 1987.
18. H. Rootzén. Extreme value theory for moving average processes. *Ann. Probab.*, 14(2):612–652, 1986.
19. K. Sato. *Lévy Processes and Infinitely Divisible Distributions.* Cambridge University Press, Cambridge, 1999.
20. W. Schoutens. *Lévy Processes in Finance.* Wiley, Chichester, 2003.

Gaussian Bridges

Dario Gasbarra[1], Tommi Sottinen[2], and Esko Valkeila[3]

[1] Department of Mathematics and Statistics, P.O. Box 68, 00014 University of Helsinki, Finland, dario.gasbarra@rni.helsinki.fi
[2] Department of Mathematics and Statistics, P.O. Box 68, 00014 University of Helsinki, Finland, tommi.sottinen@helsinki.fi
[3] Institute of Mathematics, P.O. Box 1100, 02015 Helsinki University of Technology, Finland, esko.valkeila@hut.fi

Summary. We consider Gaussian bridges; in particular their dynamic representations. We prove a Girsanov theorem between the law of Gaussian bridge and the original Gaussian process, which holds with natural assumptions. With some additional conditions we obtain dynamical representation for a Gaussian bridge. We discuss briefly the initial enlargement of filtrations in this context.

Mathematics Subject Classification (2000): 60G15, 60G18, 60G25, 60G44

Keywords and Phrases: Gaussian processes, Brownian bridge, pinned Gaussian processes, tied down Gaussian processes, enlargement of filtration, fractional Brownian motion, fractional Brownian bridge

1 Introduction

Motivation

Let X be a continuous Gaussian process such that $X_0 = 0$ and $\mathbf{E}(X_t) = 0$. Fix $T > 0$ and define the bridge of X $U^{T,0}$ by

$$U_t^{T,0} = X_t - \frac{t}{T} X_T. \tag{1.1}$$

It is clear that the process $U^{T,0}$ is a Gaussian process. Moreover, it is a bridge in the sense that $U_0^{T,0} = U_T^{T,0} = 0$. If X is a standard Brownian motion, then it is known that the law of the process $U^{T,0}$ defined by (1.1) is the same as the conditional law of the standard Brownian motion:

$$\mathbf{P} - \text{Law}\left((X_t)_{0\le t\le T} | X_T = 0\right) = \mathbf{P} - \text{Law}\left(\left(U_t^{T,0}\right)_{0\le t\le T}\right).$$

It is well-known that in the case of standard Brownian motion the bridge process $U^{T,0}$ has a representation as solution to the differential equation (1.2). We refer to the next subsection for more information on Brownian bridge.

We study the properties of the bridge process $U^{T,0}$ of X in the case of arbitrary Gaussian process X. We define the bridge process using the conditional law of X. It turns out that it is quite easy to obtain the analog of (1.1) for the arbitrary Gaussian process X; see Proposition 4 for the exact result. If the Gaussian process X is a martingale, then it is quite easy to to describe the bridge process $U^{T,0}$ as a solution to a differential equation analogous to (1.2). But if the process X is a fractional Brownian motion, then the corresponding differential equation contains Volterra operators.

Representations for the Brownian Bridge

Fix $T > 0$ and let $W = (W_t)_{t\in[0,T]}$ be a standard Brownian motion on a probability space $(\Omega, \mathcal{F}, \mathbf{P})$ starting from $W_0 = \xi$.

Let (T, θ) be a "conditioning". Then the notation $W^{T,\theta}$ means that the process W is conditioned to be θ at time T. That is $W^{T,\theta}$ is a bridge from $(0, \xi)$ to (T, θ).

For the Brownian bridge $W^{T,\theta}$ from $(0, \xi)$ to (T, θ) one finds in the literature the following three equivalent definitions

$$\mathrm{d}Y_t^{T,\theta} = \mathrm{d}W_t + \frac{\theta - Y_t^{T,\theta}}{T-t}\,\mathrm{d}t, \qquad Y_0^{T,\theta} = \xi, \tag{1.2}$$

$$Y_t^{T,\theta} = \xi + (\theta - \xi)\frac{t}{T} + (T-t)\int_0^t \frac{\mathrm{d}W_s}{T-s}, \tag{1.3}$$

$$W_t^{T,\theta} = \theta\frac{t}{T} + \left(W_t - \frac{t}{T}W_T\right). \tag{1.4}$$

The representation (1.3) is just the solution of the (stochastic or pathwise) differential equation (1.2). So, the equations (1.2) and (1.3) define the same process $Y^{T,\theta}$. The equation (1.4), however, does not define the same process as the equations (1.2) and (1.3). The equality between representations (1.2)–(1.3) and (1.4) is only an equality in law: $\mathrm{Law}(Y^{T,\theta}; \mathbf{P}) = \mathrm{Law}(W^{T,\theta}; \mathbf{P})$. That the processes $Y^{T,\theta}$ and $W^{T,\theta}$ are different is obvious from the fact that the process $Y^{T,\theta}$ is adapted to the filtration of W while the process $W^{T,\theta}$ is not. Indeed, to construct $W_t^{T,\theta}$ by using (1.4) we need information of the random variable W_T. The fact that the two processes $Y^{T,\theta}$ and $W^{T,\theta}$ have the same law is also obvious, since they have the same covariance and expectation. It is also worth noticing that if the Brownian bridge $Y^{T,\theta}$ is given by the equation (1.3) then the original Brownian motion W may be recovered from the bridge $W^{T,\theta}$ by using the equation (1.2). In particular, this means that in this case the filtration of this Brownian bridge is the same as the filtration of the Brownian motion: $\mathbf{F}^{Y^{T,\theta}} = \mathbf{F}^W$.

The non-adapted representation (1.4) comes from the orthogonal decomposition of Gaussian variables. Indeed, the conditional law of process $(W_t)_{t\in[0,T]}$ given the variable W_T is Gaussian with

$$\mathbf{E}(W_t|W_T) = \frac{t}{T}(W_T - \xi) + \xi,$$
$$\mathbf{Cov}(W_t, W_s|W_T) = t \wedge s - \frac{ts}{T}.$$

The second-order structure of the Brownian bridge is easily calculated from the representation (1.4):

$$\mathbf{E}\left(W_t^{T,\theta}\right) = \xi + (\theta - \xi)\frac{t}{T}, \tag{1.5}$$

$$\mathbf{Cov}\left(W_t^{T,\theta}, W_s^{T,\theta}\right) = t \wedge s - \frac{ts}{T}. \tag{1.6}$$

Girsanov Theorem and Brownian Bridge

We know that Brownian bridge is defined only up to distribution. Put $\mathbf{P}^{T,\theta} := \mathrm{Law}(W^{T,\theta}; \mathbf{P})$. We have that $\mathbf{P}^{T,\theta} = \mathbf{P}(\cdot|W_T = \theta)$, where $\mathbf{P}$ is the law of the Brownian motion W. Consider now the restrictions of the measures $\mathbf{P}^{T,\theta}$ and $\mathbf{P}$ on the sigma-algebra $\mathcal{F}_t^W$: denote the restriction by $\mathbf{P}_t$ and $\mathbf{P}_t^{T,\theta}$. We know that $\mathbf{P}_t^{T,\theta} \sim \mathbf{P}_t$ for all $t \in [0, T)$, but, of course, $\mathbf{P}_T^{\theta} \perp \mathbf{P}_T$. From (1.2) we get, by Girsanov theorem, that

$$\frac{d\mathbf{P}_t^{T,\theta}}{d\mathbf{P}_t} = \exp\left(\int_0^t \frac{\theta - Y_s^{T,\theta}}{T-s} dW_s - \frac{1}{2}\int_0^t \left(\frac{\theta - Y_s^{T,\theta}}{T-s}\right)^2 ds\right).$$

This is a key observation for the non-anticipative representation in the general case.

Non-anticipative and Anticipative Representations

Let now $X = (X_t)_{t\in[0,T]}$ be a Gaussian process on $(\Omega, \mathcal{F}, \mathbf{P})$ with $X_0 = \xi$. We want to understand what is the corresponding bridge $X^{T,\theta}$ from $(0, \xi)$ to (T, θ). If one merely replaces the Brownian motion W with the process X in representations (1.2)–(1.4) then the "X-bridges" obtained from the first two representations of course coincide. However, the bridge obtained from the last one does not coincide with the first two ones. The following example, communicated to us by M. Lifshits, elaborates this point.

Example 1. Let $(f_n)_{n\geq 1}$ be a sequence of smooth isomorphisms of $[0, T]$ onto itself. Take

$$X_{n,t} := W_{f_n(t)}$$

and set

$$X_{n,t}^{1,T,\theta} := \theta\frac{t}{T} + X_{n,t} - \frac{t}{T}X_{n,T},$$

$$X_{n,t}^{2,T,\theta} := \theta\frac{t}{T} + (T-t)\int_0^t \frac{\mathrm{d}X_{n,s}}{T-s}.$$

Then

$$\begin{aligned}\mathbf{Cov}_{n,1}(s,t) &:= \mathbf{Cov}\Big(X_{n,t}^{1,T,\theta}, X_{n,s}^{1,T,\theta}\Big)\\ &= f_n(s\wedge t) + st - sf_n(t) - tf_n(s),\\ \mathbf{Cov}_{n,2}(s,t) &:= \mathbf{Cov}\Big(X_{n,t}^{2,T,\theta}, X_{n,s}^{2,T,\theta}\Big)\\ &= (T-t)(T-s)\int_0^{s\wedge t}\frac{\mathrm{d}f_n(u)}{(T-u)^2}.\end{aligned}$$

The covariances $\mathbf{Cov}_{n,1}$ and $\mathbf{Cov}_{n,2}$ are not the same in general. Indeed, let $f_n \to \mathbf{1}_{\{1\}}$. Then for all $s,t<1$ we have that as $n\to\infty$, $\mathbf{Cov}_{n,1}(s,t)\to st$ while $\mathbf{Cov}_{n,2}(s,t)\to 0$.

Structure of the Paper

We will study Gaussian bridges. After the definition of Gaussian bridge we obtain the anticipative representation of the Gaussian bridge, which is a generalisation of the representation (1.4). Next we give the density between the bridge measure $\mathbf{P}^{T,\theta}$ and the original measure $\mathbf{P}$ and give an abstract version of the non-anticipative representation (1.3) in the general setup. In the section three we study bridges of Gaussian martingales, and this part is an easy generalisation of the Brownian bridge. In the next sections we study bridges of certain special Gaussian processes: Wiener predictable process, Volterra process and fractional Brownian motion. We end the paper by giving the connection to the enlargement of filtrations theory, where the enlargement is an initial enlargement with the final value of the Gaussian process X_T.

2 Gaussian Bridges in General

2.1 Definition of the X-bridge

The fact that for Brownian motion the Brownian bridge in unique up to law only suggests the following definition in the case of an arbitrary Gaussian process.

Definition 2. *Let X be a Gaussian stochastic process with $X_0 = \xi$. Then the Gaussian process $X^{T,\theta}$ is an X-*bridge *from $(0,\xi)$ to (T,θ) if*

$$\mathrm{Law}\big(X^{T,\theta};\mathbf{P}\big) \;=\; \mathrm{Law}\big(X;\mathbf{P}^{T,\theta}\big), \tag{2.1}$$

where the measure $\mathbf{P}^{T,\theta}$ on $(\Omega,\mathcal{F})$ is defined by

$$\mathbf{P}^{T,\theta} \;=\; \mathbf{P}(\,\cdot\,|X_T=\theta). \tag{2.2}$$

Remark 3. The definition above assumes that the process $X^{T,\theta}$ exists in the *original* space $(\Omega,\mathcal{F},\mathbf{P})$. Also, we have

$$1 \;=\; \mathbf{P}(X_T=\theta|X_T=\theta) \;=\; \mathbf{P}^{T,\theta}(X_T=\theta) \;=\; \mathbf{P}(X_T^{T,\theta}=\theta),$$

as we should. Note that in (2.2) we condition on a set of zero measure. However, we can define (2.2) as a regular conditional distribution in the Polish space of continuous functions on $[0,T]$ (see Shiryaev [9, pp. 227–228]).

In what follows we denote by μ and R the mean and covariance of X, respectively.

2.2 Anticipative Representation

The anticipative representation corresponding to (1.4) is easily obtained from the orthogonal decomposition of X with respect to X_T. Indeed, $\mathrm{Law}(X|X_T)$ is Gaussian with

$$\mathbf{E}(X_t|X_T) = \big(X_T-\mu(T)\big)\frac{R(T,t)}{R(T,T)} + \mu(t),$$

$$\mathbf{Cov}(X_t,X_s|X_T) = R(t,s) - \frac{R(T,t)R(T,s)}{R(T,T)}.$$

Thus, we have an anticipative representation for any Gaussian bridge.

Proposition 4. *Let X be a Gaussian process with mean μ and covariance R. Then the X-bridge $X^{T,\theta}$ from $(0,\mu(0))$ to (T,θ) admits a representation*

$$\begin{aligned} X_t^{T,\theta} &= \theta\frac{R(T,t)}{R(T,T)} + X_t^{T,0} \\ &= \theta\frac{R(T,t)}{R(T,T)} + \left(X_t - \frac{R(T,t)}{R(T,T)}X_T\right). \end{aligned} \tag{2.3}$$

Moreover,

$$\mathbf{E}\Big(X_t^{T,\theta}\Big) = (\theta-\mu(T))\frac{R(T,t)}{R(T,T)} + \mu(t), \tag{2.4}$$

$$\mathbf{Cov}\Big(X_t^{T,\theta},X_s^{T,\theta}\Big) = R(t,s) - \frac{R(T,t)R(T,s)}{R(T,T)}. \tag{2.5}$$

Example 5. Let X be a centered fractional Brownian motion. The bridge process $Z_t := X_t - \frac{t}{T}X_T$ is a H- self similar process, but it is not a 'fractional Brownian bridge' in the sense of Definition 2.

The correct fractional Brownian bridge in the sense of the Definition 2 is

$$X_t^{T,\theta} = X_t - \frac{t^{2H} + T^{2H} - |t-T|^{2H}}{2T^{2H}} X_T.$$

X-bridge and Drift

Let ${}^aW$ is a Brownian motion with drift $a \in \mathbb{R}$, i.e. $W_t := {}^aW_t - at$ is a standard Brownian motion starting from ξ. Then from (1.4) it easy to see that the Brownian bridge is invariant under this drift: ${}^aW^{T,\theta} = W^{T,\theta}$.

Consider now a general centered Gaussian process X, and let μ be a deterministic function with $\mu(0) = 0$. Define ${}^\mu X$ by ${}^\mu X_t := X_t + \mu(t)$. Transform ${}^\mu X$ to ${}^\mu X^{T,\theta}$ by (2.3). Then ${}^\mu X$ is a Gaussian process with the same covariance R as X and with mean μ. When does ${}^\mu X^{T,\theta}$ define the same bridge as $X^{T,\theta}$ in the sense of Definition 2? From (2.3) it follows that an invariant mean function μ must satisfy the equation

$$\mu(t) \quad = \quad \frac{R(T,t)}{R(T,T)}\mu(T).$$

So, such an invariant mean μ may depend on the time of the conditioning T. Indeed,

$$\mu(t) \quad = \quad \mu_T(t) \quad = \quad aR(T,t)$$

for some $a \in \mathbb{R}$. In particular, we see that μ is independent of T if and only if

$$R(t,s) \quad = \quad f(t \wedge s)$$

for some function f. But this means that X has independent increments, or in other words that $X - \mathbf{E}(X)$ is a martingale.

X-bridge and Self-similarity

The Brownian motion W starting from $W_0 = 0$ is 1/2-self-similar. i.e.

$$\mathrm{Law}\Big((W_t)_{t\in[0,T]}\ ;\mathbf{P}\Big) \quad = \quad \mathrm{Law}\Big(\Big(T^{1/2}W_\tau\Big)_{\tau\in[0,1]}\ ;\mathbf{P}\Big).$$

Consequently, we have for the Brownian bridge the scaling property

$$\mathrm{Law}\Big(\Big(W_t^{T,\theta}\Big)_{t\in[0,T]}\ ;\mathbf{P}\Big) \quad = \quad \mathrm{Law}\Big(\Big(T^{1/2}W_\tau^{1,\theta T^{-1/2}}\Big)_{\tau\in[0,1]}\ ;\mathbf{P}\Big).$$

From (2.3) it is easy to see that if the process $X = (X_t)_{t\in[0,T]}$ is H-self-similar, i.e.

$$\mathrm{Law}\Big((X_t)_{t\in[0,T]}\,;\mathbf{P}\Big) \;=\; \mathrm{Law}\Big(\big(T^H X_\tau\big)_{\tau\in[0,1]}\,;\mathbf{P}\Big)$$

then the corresponding bridge satisfies the scaling property

$$\mathrm{Law}\left(\left(X_t^{T,\theta}\right)_{t\in[0,T]}\;;\mathbf{P}\right) \;=\; \mathrm{Law}\left(\left(T^H X_\tau^{1,\theta T^{-H}}\right)_{\tau\in[0,1]}\;;\mathbf{P}\right).$$

So, we may represent the bridge $X^{T,\theta}$ as

$$\begin{aligned}X_t^{T,\theta} &= X_s^{1,\theta T^{-H}}\\ &= \theta\frac{R(1,\tau)}{R(1,1)} + T^H X_\tau - \frac{R(1,\tau)}{R(1,1)}T^H X_1,\end{aligned}$$

where $\tau = t/T \in [0,1]$.

2.3 Density Between the Bridge Measure $\mathbf{P}^{T,\theta}$ and $\mathbf{P}$

When we look for analogies for the non-anticipative, or dynamic, representation (1.3) and the corresponding differential equation (1.2), then the main idea is to work with the prediction martingale of X and to use the Girsanov's theorem.

We introduce some notation. Let $X^{T,\theta}$ and $\mathbf{P}^{T,\theta}$ be as in (2.1) and (2.2). Let $\hat{X}_{T|\cdot} = (\hat{X}_{T|t})_{t\in[0,T]}$ be the prediction martingale of X. I.e.

$$\hat{X}_{T|t} \;:=\; \mathbf{E}\big(X_T|\mathcal{F}_t^X\big).$$

For the incremental bracket of the Gaussian martingale $\hat{X}_{T|\cdot}$ we use the shorthand notation

$$\begin{aligned}\langle \hat{X}_{T|\cdot}\rangle_{T,t} &:= \langle \hat{X}_{T|\cdot}\rangle_T - \langle \hat{X}_{T|\cdot}\rangle_t\\ &:= \langle \hat{X}_{T|\cdot}, \hat{X}_{T|\cdot}\rangle_T - \langle \hat{X}_{T|\cdot}, \hat{X}_{T|\cdot}\rangle_t.\end{aligned}$$

(Note that since $\hat{X}_{t|\cdot}$ is a Gaussian martingale it has independent increments, and consequently its bracket $\langle \hat{X}_{T|\cdot}\rangle$ is deterministic.) Denote

$$\mathbf{P}_t \;:=\; \mathbf{P}|\mathcal{F}_t^X \qquad \text{and} \qquad \mathbf{P}_t^{T,\theta} \;:=\; \mathbf{P}^{T,\theta}|\mathcal{F}_t^X.$$

Let α_T^t denote the regular conditional law of X_T given the information $\mathcal{F}_t^X$ and let $\alpha_T = \alpha_T^0$ be the law of X_T. So, if p denotes the Gaussian density

$$p(\theta;\mu,\sigma^2) \;=\; \frac{1}{\sqrt{2\pi}\sigma}e^{-\frac{1}{2}\left(\frac{\theta-\mu}{\sigma}\right)^2},$$

it is easy enough to see that

$$\alpha_T^t(\mathrm{d}\theta) = p\left(\theta; \hat{X}_{T|t}, \langle \hat{X}_{T|}\cdot\rangle_{T,t}\right) \mathrm{d}\theta,$$
$$\alpha_T(\mathrm{d}\theta) = p\left(\theta; \mu(T), \langle \hat{X}_{T|}\cdot\rangle_{T}\right) \mathrm{d}\theta.$$

Now, by using Bayes' rule we have that

$$\begin{aligned}\frac{\mathrm{d}\mathbf{P}_t^{T,\theta}}{\mathrm{d}\mathbf{P}_t} &= \frac{\mathrm{d}\alpha_T^t}{\mathrm{d}\alpha_T}(\theta) \\ &= \frac{p\left(\theta; \hat{X}_{T|t}, \langle \hat{X}_{T|}\cdot\rangle_{T,t}\right)}{p\left(\theta; \mu(T), \langle \hat{X}_{T|}\cdot\rangle_{T}\right)} \\ &= \sqrt{\frac{\langle \hat{X}_{T|}\cdot\rangle_T}{\langle \hat{X}_{T|}\cdot\rangle_{T,t}}} \exp\left(-\frac{1}{2}\frac{\left(\theta - \hat{X}_{T|t}\right)^2}{\langle \hat{X}_{T|}\cdot\rangle_{T,t}} + \frac{1}{2}\frac{\left(\theta-\mu(T)\right)^2}{\langle \hat{X}_{T|}\cdot\rangle_{T}}\right). \end{aligned} \tag{2.6}$$

Since we want to use the Girsanov's theorem later we need to assume that the prediction martingale $\hat{X}_{T|}\cdot$ is continuous. Another way of stating this assumption is the following:

(A0) The history of X is continuous, i.e. $\mathcal{F}_{t-}^X = \mathcal{F}_{t+}^X$; here $\mathcal{F}_{t-} = \vee_{s<t}\mathcal{F}_s$ and $\mathcal{F}_{t+} = \cap_{u>t}\mathcal{F}_u$.

Also, in order for the calculations above to make sense we need the to assume that $\mathbf{P}_t^{T,\theta} \ll \mathbf{P}_t$ for all $t < T$. Or, since the both measures are Gaussian, we may as well assume that:

(A1) $\mathbf{P}_t \sim \mathbf{P}_t^{T,\theta}$ for all $t < T$.

From equation (2.6) we see that assumption (A1) says that $\langle \hat{X}_{T|}\cdot\rangle_t < \langle \hat{X}_{T|}\cdot\rangle_T$ for all $t < T$. So, another way of stating assumption (A1) is that the value of X_T cannot be predicted for certain by using the information $\mathcal{F}_t^X$ only. Indeed,

$$\langle \hat{X}_{T|}\cdot\rangle_{T,t} = \mathbf{Var}\left(\hat{X}_{T|t}\right)$$

is the prediction error of $\hat{X}_{T|t}$. Let us note that in general the measures $\mathbf{P}_T$ and $\mathbf{P}_T^{T,\theta}$ are of course singular, since $X^{T,\theta}$ is degenerate at T.

In what follows $\beta_{T,\theta}$ is a non-anticipative functional acting on Gaussian (prediction) martingales m:

$$\beta_{T,\theta}(m)_t := \frac{\theta - m_t}{\langle m\rangle_{T,t}}.$$

The following proposition is the key tool in finding a non-anticipative representation.

Proposition 6. *Let X be a Gaussian process on $(\Omega, \mathcal{F}, \mathbf{P})$ satisfying the assumptions (A0) and (A1). Then the bridge measure $\mathbf{P}^{T,\theta}$ on $(\Omega, \mathcal{F})$ may be represented as*

$$\mathrm{d}\mathbf{P}_t^{T,\theta} = L_t^{T,\theta}\,\mathrm{d}\mathbf{P}_t,$$

where

$$L_t^{T,\theta} = \exp\left(\int_0^t \beta_{T,\theta}\big(\hat{X}_{T|\cdot}\big)_s\,\mathrm{d}\hat{X}_{T|s} - \frac{1}{2}\int_0^t \beta_{T,\theta}\big(\hat{X}_{T|\cdot}\big)_s^2\,\mathrm{d}\langle\hat{X}_{T|\cdot}\rangle_s\right).$$

Proof. The claim follows from equation (2.6). Indeed, just use Itô's formula with the martingale $\hat{X}_{T|\cdot}$ to the function

$$g(t,x) := -\frac{1}{2}\frac{(\theta - x)^2}{\langle\hat{X}_{T|\cdot}\rangle_{T,t}},$$

and there you have it. □

2.4 Non-anticipative Representation

In order to come back from the "prediction martingale level" to the actual process we still need one assumption.

(A2) The non-anticipative linear mapping F_T sending the path of the Gaussian process X to the path of its prediction martingale $\hat{X}_{T|\cdot}$ is injective.

The assumption (A2) says simply that the process X may be recovered from $\hat{X}_{T|\cdot}$ by $X = F_T^{-1}(\hat{X}_{T|\cdot})$. Also, note that the assumption (A2) implies that the prediction filtration and the original filtration are the same: $\mathbf{F}^X = \mathbf{F}^{\hat{X}_{T|\cdot}}$.

Let m be a Gaussian martingale. We denote by $S_{T,\theta}(m)$ the unique solution of the differential equation

$$\mathrm{d}m_t^{T,\theta} = \mathrm{d}m_t + \beta_{T,\theta}\big(m^{T,\theta}\big)_t\,\mathrm{d}\langle m^{T,\theta}\rangle_t \tag{2.7}$$

with initial condition $m_0^{T,\theta} = \zeta$, i.e.

$$\begin{aligned} m_t^{T,\theta} &= S_{T,\theta}(m)_t \\ &= \zeta + (\theta - \zeta)\frac{\langle m^{T,\theta}\rangle_t}{\langle m^{T,\theta}\rangle_T} + \langle m^{T,\theta}\rangle_{T,t}\int_0^t \frac{\mathrm{d}m_s}{\langle m^{T,\theta}\rangle_{T,s}}. \end{aligned}$$

In order to see that $S_{T,\theta}(m)$ is indeed the solution to (2.7) one just uses the integration by parts. It is also worth noticing that classical theory of differentiation applies here: The differential equation (2.7) may be understood pathwise. Finally, note that by the Girsanov's theorem the brackets of the Gaussian martingales m and $m^{T,\theta}$ coincide: $\langle m\rangle = \langle m^{T,\theta}\rangle$.

Let us now abuse the notation slightly and set

$$S_T(m) \;:=\; S_{T,0}(m).$$

Then we have the decomposition

$$S_{T,\theta}(m) \;=\; \theta K_T(m) + S_T(m),$$

where

$$K_T(m) \;:=\; \frac{\langle m\rangle}{\langle m\rangle_T}$$

and S_T is independent of θ.

The following theorem is the analogy of the non-anticipative representation (1.3).

Theorem 7. *Let X be a Gaussian process with mean μ and covariance R satisfying (A0), (A1) and (A2). Then the bridge $X^{T,\theta}$ from $(0,\mu(0))$ to (T,θ) admits the non-anticipative representation*

$$X_t^{T,\theta} = \left(F_T^{-1}S_{T,\theta}F_T\right)(X)_t \tag{2.8}$$

$$= \theta\left(F_T^{-1}K_TF_T\right)(X)_t + \left(F_T^{-1}S_TF_T\right)(X)_t \tag{2.9}$$

$$= \theta\frac{R(T,t)}{R(T,T)} + X_t^{T,0}. \tag{2.10}$$

Moreover, the original process X may be recovered from the bridge $X^{T,\theta}$ by

$$X_t \;=\; \left(F_T^{-1}S_{T,\theta}^{-1}F_T\right)\left(X^{T,\theta}\right)_t. \tag{2.11}$$

Proof. Let us first prove the equations (2.8)–(2.10). By the equation (2.3) we already know the contribution coming from θ. Indeed, we must have

$$\left(F_T^{-1}K_TF_T\right)(X)_t \;=\; \theta\frac{R(T,t)}{R(T,T)}.$$

So, we may assume that $\theta = 0$ and consider the corresponding bridge $X^{(T,0)}$. Now, we map X to its prediction martingale $F_T(X)$. Then $(S_TF_T)(X)$ is the solution of the stochastic differential equation (2.7) with $m = F_T(X)$ and the initial condition $\zeta = \mu(T)$. Consequently, the Girsanov's theorem and Proposition 6 tells us that

$$\mathrm{Law}\Big(\left(S_TF_T\right)(X)\,;\,\mathbf{P}\Big) \;=\; \mathrm{Law}\big(F_T(X)\,;\,\mathbf{P}^{T,\theta}\big). \tag{2.12}$$

So, the claim (2.8) follows simply by recovering the process X by using the map F_T^{-1} on both sides of the equation (2.12).

The equation (2.11) is now obvious, since $S_{T,\theta}$ is invertible. Indeed,

$$\begin{aligned} S_{T,\theta}^{-1}\left(F_T\left(X^{T,\theta}\right)\right)_t &= S_{T,\theta}^{-1}\left(\hat{X}^{T,\theta}_{T|\cdot}\right)_t \\ &= \hat{X}^{T,\theta}_{T|t} + \int_0^t \beta\left(\hat{X}^{T,\theta}_{T|\cdot}\right)_s \mathrm{d}\langle \hat{X}^{T,\theta}_{T|\cdot}\rangle_s \end{aligned}$$

This finishes the proof. □

Remark 8. For the differential equation (1.2) have the following formal analogy

$$X_t^{T,\theta} = X_t + F_T^{-1}\left(\int_0^s \frac{\theta - F_T(X^{T,\theta})_u}{\langle F_T(X)\rangle_{T,u}} \mathrm{d}\langle F_T(X)\rangle_u \,;\, s \le t\right)_t .$$

In the following sections we consider some special Gaussian bridges and give the somewhat abstract Theorem 7, and highly abstract Remark 8, more concrete forms. In particular, we consider cases where the operators F_T and F_T^{-1} may be represented as Wiener integrals.

3 Bridges of Gaussian Martingales

The case of Gaussian martingales is extremely simple. Indeed, the analogy to the Brownian case is complete.

Proposition 9. *Let M be a continuous Gaussian martingale with strictly increasing bracket $\langle M\rangle$ and $M_0 = \xi$. Then the M-bridge $M^{T,\theta}$ admits the representations*

$$\mathrm{d}M_t^{T,\theta} = \mathrm{d}M_t + \frac{\theta - M_t^{T,\theta}}{\langle M\rangle_{T,t}} \mathrm{d}\langle M\rangle_t, \qquad M_0^{T,\theta} = \xi, \tag{3.1}$$

$$M_t^{T,\theta} = \xi + (\theta - \xi)\frac{\langle M\rangle_t}{\langle M\rangle_T} + \langle M\rangle_{T,t}\int_0^t \frac{\mathrm{d}M_s}{\langle M\rangle_{T,s}}, \tag{3.2}$$

$$M_t^{T,\theta} = \theta\frac{\langle M\rangle_t}{\langle M\rangle_T} + \left(M_t - \frac{\langle M\rangle_t}{\langle M\rangle_T} M_T\right). \tag{3.3}$$

Moreover, we have

$$\mathbf{E}M_t^{T,\theta} = \xi + (\theta - \xi)\frac{\langle M\rangle_t}{\langle M\rangle_T},$$

$$\mathbf{Cov}(M_t^{T,\theta}, M_s^{T,\theta}) = \langle M\rangle_{t\wedge s} - \frac{\langle M\rangle_t \langle M\rangle_s}{\langle M\rangle_T}.$$

Proof. Since M is continuous and $\langle M \rangle$ is strictly increasing the assumption (A0) and (A1) are satisfied. The assumption (A2) is trivial in this case. Now, the solution of (3.1) is (3.2) and this is just the equation (2.8) where F_T is the identity operator. Representation (3.3) as well as the mean and covariance functions come from the representation (2.3). Indeed, for Gaussian martingales we have $R(t,s) = \langle M \rangle_{t \wedge s}$. □

Remark 10. Actually, one can deduce the result of Proposition 9 without using the "Bayes–Itô–Girsanov machinery" introduced in Section 2. Indeed, the result follows quite easily from equations (1.2)–(1.4) and the representation of the Gaussian martingale M as the time-changed Brownian motion $W_{\langle M \rangle}$.

4 Bridges of Wiener Predictable Processes

Let us first consider abstract Wiener-integration with respect to Gaussian processes. The *linear space* $\mathcal{H}_t$ of a Gaussian process X is the closed Gaussian subspace of $L^2(\Omega, \mathcal{F}, \mathbf{P})$ generated by the random variables X_s, $s \le t$. For the prediction martingale of X it is well known that $\hat{X}_{T|t} \in \mathcal{H}_t$. Let $\mathcal{E}_t$ denote the space of elementary functions over $[0,t]$ equipped with the inner product generated by the covariance of X:

$$\left\langle\!\left\langle \mathbf{1}_{[0,s)}, \mathbf{1}_{[0,u)} \right\rangle\!\right\rangle \;:=\; R(s,u),$$

Let Λ_t be the completion of $\mathcal{E}_t$ in the inner product $\langle\!\langle \cdot, \cdot \rangle\!\rangle$. Now the mapping

$$\mathcal{I}_t : \mathbf{1}_{[0,s)} \mapsto X_s$$

extends to an isometry between Λ_t and $\mathcal{H}_t$. We call this extension the *abstract Wiener integral.*

Alas, the space Λ_t is not in general a space of functions (or more precisely a space of equivalence classes of functions). However, we can find a subspace of it whose elements may be identified as (equivalence classes) of functions. Viz. the space $\tilde{\Lambda}_t$ which consists of such function f that

$$\sup_{\pi} \sum_{s_i, s_j \in \pi} f(s_{i-1}) f(s_{j-1}) \left\langle \mathbf{1}_{[s_{i-1}, s_i)}, \mathbf{1}_{[s_{j-1}, s_j)} \right\rangle \;<\; \infty.$$

Here the supremum is taken over all partitions π of the interval $[0,t]$. The reason to take a supremum instead of letting the mesh of the partition go to zero is that the $\langle\!\langle \cdot, \cdot \rangle\!\rangle$-norm of a function may increase when multiplied by an indicator function. For details of this phenomenon in the case of fractional Brownian motion see Bender and Elliot [1].

If $f \in \tilde{\Lambda}_t$ then we write

$$\int_0^t f(s)\, \mathrm{d}X_s \;:=\; \mathcal{I}_t[f]. \tag{4.1}$$

So, the Wiener integral (4.1) of a function $f \in \tilde{\Lambda}_t$ is defined as a $\langle\!\langle\cdot,\cdot\rangle\!\rangle$-limit of simple functions. Note that if $t \le T$ then $\tilde{\Lambda}_t \subset \tilde{\Lambda}_T$ and $\mathcal{I}_t[f] = \mathcal{I}_T\left[f\mathbf{1}_{[0,t)}\right]$ for $f \in \tilde{\Lambda}_T$.

Since the operator F_T is linear and non-anticipative we have

$$\hat{X}_{T|t} \;=\; \mathcal{I}_t\left[p_{T,t}\right]$$

for some $p_{T,t} \in \Lambda_t$. We assume now that this prediction kernel $p_{T,t}$ is actually a function in $\tilde{\Lambda}_t$:

(A3) There exists a Volterra kernel p_T such that $p_T(t,\cdot) \in \tilde{\Lambda}_t$ for all t and m may be represented as the Wiener integral

$$\hat{X}_{T|t} \;=\; \int_0^t p_T(t,s)\,\mathrm{d}X_s. \tag{4.2}$$

Representation (4.2) suggests that, if we are lucky enough, the inverse operator F_T^{-1} may be represented as a Wiener integral with respect to $\hat{X}_{T|\cdot}$. This is the meaning of the next assumption we make.

(A4) There exists a Volterra kernel p_T^* such that the original Gaussian process X may be reconstructed from the prediction martingale m as a Wiener integral

$$X_t \;=\; \int_0^t p_T^*(t,s)\,\mathrm{d}\hat{X}_{T|s}. \tag{4.3}$$

Remark 11. The Wiener integral in (A4) may understood as an abstract Wiener integral or, as well, as the stochastic integral with respect to the martingale m. Indeed, in this case Λ_t is the function space $L^2([0,t],\,\mathrm{d}\langle\hat{X}_{T|\cdot}\rangle)$. Also, assumption (A4) gives us an alternative way of defining the Wiener integral (4.2). Indeed, let the operator P_T^* be the linear extension of the map $\mathbf{1}_{[0,t)} \mapsto p_T^*(t,\cdot)$. Then the assumption (A3) may be restated as:

> The operator P_T^* has the indicator functions $\mathbf{1}_{[0,t)}$, $t \in (0,T]$, in its image.

In this case we may define Wiener integrals with respect to X as

$$\int_0^t f(s)\,\mathrm{d}X_s \;:=\; \int_0^t P_T^*[f](s)\,\mathrm{d}\hat{X}_{T|s}$$

for such f that $P_T^*[f] \in L^2([0,t],\,\mathrm{d}\langle\hat{X}_{T|\cdot}\rangle)$. Moreover, in this case

$$p_T(t,\cdot) \;=\; \left(P_T^*\right)^{-1}\left[\mathbf{1}_{[0,t)}\right].$$

Indeed, this is the approach taken in the next section.

Remark 12. Obviously (A4) implies (A2). Also, we have implicitly assumed that X is centred with $X_0 = 0$. However, adding a mean function to X causes no difficulties. Indeed, let $\tilde{m}$ be the prediction martingale of the centred process $X - \mu$ and let $\tilde{p}_T$ and $\tilde{p}_T^*$ be the kernels associated to this centred process. Then

$$\hat{\tilde{X}}_{T|t} = \hat{X}_{T|t} - \mu(T),$$

$$X_t = \int_0^t \tilde{p}_T^*(t,s)\,\mathrm{d}\hat{\tilde{X}}_{T|s} + \mu(t),$$

$$\hat{X}_{T|t} = \int_0^t \tilde{p}_T(t,s)\,\mathrm{d}(X_s - \mu(s)) + \mu(T).$$

Remark 13. The relation (4.3) says that the covariance R of X may be written as

$$R(t,s) \;=\; \int_0^{t\wedge s} p_T^*(t,u)p_T^*(s,u)\;\mathrm{d}\langle\hat{X}_{T|\cdot}\rangle_u. \tag{4.4}$$

So, p_T^* is a "square root" of R. Note, however, that in general a decomposition like (4.4) is by no means unique, even if the measure is given. This means that from an equation like (4.4) we cannot deduce the kernel p_T^* even if we knew the measure $\mathrm{d}\langle\hat{X}_{T|\cdot}\rangle$ induced by the bracket $\langle\hat{X}_{T|\cdot}\rangle$.

We have the following analogue of representations (1.2) and (1.3).

Proposition 14. *Let X be a Gaussian process with covariance R satisfying (A0), (A1), (A3) and (A4). Then the bridge $X^{T,\theta}$ satisfies the integral equation*

$$X_t^{T,\theta} \;=\; X_t + \int_0^t \left\{\theta - \int_0^s p_T(s,u)\,\mathrm{d}X_u^{T,\theta}\right\} \frac{p_T^*(t,s)}{\langle\hat{X}_{T|\cdot}\rangle_{T,s}}\,\mathrm{d}\langle\hat{X}_{T|\cdot}\rangle_s. \tag{4.5}$$

Moreover $X^{T,\theta}$ admits the non-anticipative representation

$$X_t^{T,\theta} \;=\; \theta\frac{R(T,t)}{R(T,T)} + X_t - \int_0^t \phi_T(t,s)\,\mathrm{d}X_s, \tag{4.6}$$

where

$$\phi_T(t,s) \;=\; \int_s^t \left\{\int_s^u \frac{p_T(v,s)}{\langle\hat{X}_{T|\cdot}\rangle_{T,v}^2}\,\mathrm{d}\langle\hat{X}_{T|\cdot}\rangle_v - \frac{p_T(u,s)}{\langle\hat{X}_{T|\cdot}\rangle_{T,u}}\right\} p_T^*(t,u)\,\mathrm{d}\langle\hat{X}_{T|\cdot}\rangle_u.$$

Remark 15. Note that unlike the equations (1.2) and (3.1) the equation (4.5) is not of differential form. Indeed, it is clear by now that the differential connection is characteristic to the martingale case.

Proof. [Proposition 14] Consider the prediction martingale $\hat{X}_{T|\cdot}$. Using the relation (2.7) i.e.

$$\mathrm{d}\hat{X}^{T,\theta}_{T|t} = \mathrm{d}\hat{X}_{T|t} + \frac{\theta - \hat{X}^{T,\theta}_{T|t}}{\langle \hat{X}_{T|\cdot}\rangle_{T,t}} \mathrm{d}\langle \hat{X}_{T|\cdot}\rangle_t$$

with (4.3) yields

$$X^{T,\theta}_t = X_t + \int_0^t \left\{\theta - \hat{X}^{T,\theta}_{T|s}\right\} \frac{p^*_T(t,s)}{\langle \hat{X}_{T|\cdot}\rangle_{T,s}} \mathrm{d}\langle \hat{X}_{T|\cdot}\rangle_s. \tag{4.7}$$

The integral equation (4.5) follows now from (4.7) and (4.2).

Let us now derive the non-anticipative representation (4.6). Inserting the solution $\hat{X}^{T,\theta}_{T|\cdot} = S_{T,\theta}(\hat{X}_{T|\cdot})$ to the equation (4.7) we obtain

$$\begin{aligned}
X^{T,\theta}_t &= X_t + \int_0^t \frac{p^*_T(t,s)}{\langle \hat{X}_{T|\cdot}\rangle_{T,s}} \left\{\theta \frac{\langle \hat{X}_{T|\cdot}\rangle_{T,s}}{\langle \hat{X}_{T|\cdot}\rangle_T} + \langle \hat{X}_{T|\cdot}\rangle_{T,s} \int_0^s \frac{\mathrm{d}\hat{X}_{T|u}}{\langle \hat{X}_{T|\cdot}\rangle_{T,u}}\right\} \mathrm{d}\langle \hat{X}_{T|\cdot}\rangle_s \\
&= X_t + \frac{\theta}{\langle \hat{X}_{T|\cdot}\rangle_T} \int_0^t p^*_T(t,s)\, \mathrm{d}\langle \hat{X}_{T|\cdot}\rangle_s \\
&\quad + \int_0^t \int_0^s \frac{\mathrm{d}\hat{X}_{T|u}}{\langle \hat{X}_{T|\cdot}\rangle_{T,u}} p^*_T(t,s)\, \mathrm{d}\langle \hat{X}_{T|\cdot}\rangle_s \\
&=: X_t + \theta f_T(t) + \Phi_T\big(\hat{X}_{T|\cdot}\big)_t.
\end{aligned}$$

Note now that

$$X_T = \hat{X}_{T|T} = \int_0^T p^*_T(T,s)\, \mathrm{d}\hat{X}_{T|s},$$

which implies that $p^*_T(T,s) = \mathbf{1}_{[0,T)}(s)$. Consequently, by (4.4)

$$\int_0^t p^*_T(t,s)\, \mathrm{d}\langle \hat{X}_{T|\cdot}\rangle_s = \int_0^{T\wedge t} p^*_T(T,s) p^*_T(t,s)\, \mathrm{d}\langle \hat{X}_{T|\cdot}\rangle_s = R(T,t),$$

and, since $\langle \hat{X}_{T|\cdot}\rangle_T = R(T,T)$, we have

$$f_T(t) = \frac{R(T,t)}{R(T,T)}$$

(a fact that we actually knew already by (2.4)). Now we want to express $\Phi_T\big(\hat{X}_{T|\cdot}\big)$ in terms of X. We proceed by integrating by parts:

$$\int_0^s \frac{\mathrm{d}\hat{X}_{T|u}}{\langle \hat{X}_{T|\cdot}\rangle_{T,u}} = \frac{\hat{X}_{T|s}}{\langle \hat{X}_{T|\cdot}\rangle_{T,s}} - \int_0^s \frac{\hat{X}_{T|u}}{\langle \hat{X}_{T|\cdot}\rangle^2_{T,u}} \mathrm{d}\langle \hat{X}_{T|\cdot}\rangle_u. \tag{4.8}$$

Using the assumption (A4) to (4.8) and changing the order of integration we obtain

$$\begin{aligned}\int_0^s \frac{\mathrm{d}\hat{X}_{T|u}}{\langle \hat{X}_{T|\cdot}\rangle_{T,u}} &= \frac{1}{\langle \hat{X}_{T|\cdot}\rangle_{T,s}} \int_0^s p_T(s,u)\,\mathrm{d}X_u \\ &\quad - \int_0^s \frac{1}{\langle \hat{X}_{T|\cdot}\rangle_{T,u}^2} \int_0^u p_T(u,v)\,\mathrm{d}X_v\,\mathrm{d}\langle \hat{X}_{T|\cdot}\rangle_u \\ &= \int_0^s \left\{ \frac{p_T(s,u)}{\langle \hat{X}_{T|\cdot}\rangle_{T,s}} - \int_u^s \frac{p_T(v,u)}{\langle \hat{X}_{T|\cdot}\rangle_{T,v}^2}\,\mathrm{d}\langle \hat{X}_{T|\cdot}\rangle_v \right\} \mathrm{d}X_u.\end{aligned}$$

Thus,

$$\begin{aligned}&\Phi_T(\hat{X}_{T|\cdot})_t \\ &= \int_0^t \int_0^s \left\{ \frac{p_T(s,u)}{\langle \hat{X}_{T|\cdot}\rangle_{T,s}} - \int_u^s \frac{p_T(v,u)}{\langle \hat{X}_{T|\cdot}\rangle_{T,v}^2}\,\mathrm{d}\langle \hat{X}_{T|\cdot}\rangle_v \right\} \mathrm{d}X_u p_T^*(t,s)\,\mathrm{d}\langle \hat{X}_{T|\cdot}\rangle_s \\ &= \int_0^t \int_s^t \left\{ \frac{p_T(u,s)}{\langle \hat{X}_{T|\cdot}\rangle_{T,u}} - \int_s^u \frac{p_T(v,s)}{\langle \hat{X}_{T|\cdot}\rangle_{T,v}^2}\,\mathrm{d}\langle \hat{X}_{T|\cdot}\rangle_v \right\} p_T^*(t,u)\,\mathrm{d}\langle \hat{X}_{T|\cdot}\rangle_u\,\mathrm{d}X_s \\ &= -\int_0^t \phi_T(t,s)\,\mathrm{d}X_s.\end{aligned}$$

This proves the decomposition (4.6). □

5 Bridges of Volterra Processes

The result of the previous section is still rather implicit. Indeed, we have no explicit relation between the covariance R of X and the bracket $\langle \hat{X}_{T|\cdot}\rangle$ of the prediction martingale m. Moreover, in general there is no simple way of finding, or even insuring the existence, of the kernels p_T^* and p_T. In this section we consider a model where these connections are clear, although the formulas turn out to be rather complicated.

(A5) There exists a Volterra kernel k and a continuous Gaussian martingale M with strictly increasing bracket $\langle M\rangle$ such that X admits a representation

$$X_t = \int_0^t k(t,s)\,\mathrm{d}M_s. \tag{5.1}$$

Remark 16. Since M is continuous, $\langle M\rangle$ is also continuous. Also, if $\langle M\rangle$ is not strictly increasing on an interval $[a,b]$, say, then nothing happens on that interval. Consequently, we could just remove it.

Remark 17. The connection between the covariance R and the kernel k is

$$R(t,s) \;=\; \int_0^{t\wedge s} k(t,u)k(s,u)\,\mathrm{d}\langle M\rangle_u. \tag{5.2}$$

Moreover, if R admits the representation (5.2) with some measure $\mathrm{d}\langle M\rangle$, then X admits the representation (5.1).

Now we define the Wiener integral with respect to X by using the way described in Remark 11. Let K extend the relation $\mathrm{K}:\mathbf{1}_{[0,t)}\mapsto k(t,\cdot)$ linearly. So, we have

$$\int_0^T f(t)\,\mathrm{d}X_t = \int_0^T \mathrm{K}[f](t)\,\mathrm{d}M_t, \tag{5.3}$$

$$\int_0^T g(t)\,\mathrm{d}M_t = \int_0^T \mathrm{K}^{-1}[g](t)\,\mathrm{d}X_t$$

for any $g\in L^2([0,T],\mathrm{d}\langle M\rangle)$ and such functions f that are in the preimage of $L^2([0,T],\mathrm{d}\langle M\rangle)$ under K.

We need to have the inverse K^{-1} defined for a large enough class of functions. Thus, we assume

(A6) For any $t\le T$ the equation

$$\mathrm{K}f \;=\; \mathbf{1}_{[0,t)}$$

has a solution in f.

(A7) For any $t\le T$ the equation

$$\mathrm{K}g \;=\; \mathbf{1}_{[0,t)}k(T,\cdot)$$

has a solution in g.

By the assumption (A6), we have a reverse representation to (5.1). Indeed,

$$M_t \;=\; \int_0^t k^*(t,s)\,\mathrm{d}X_s, \tag{5.4}$$

where we have denoted

$$k^*(t,s) \;:=\; \mathrm{K}^{-1}\left[\mathbf{1}_{[0,t)}\right](s).$$

Since M is a martingale we have

$$\mathrm{d}m_t \;=\; k(T,t)\,\mathrm{d}M_t.$$

By assumption (A6) we have $k(t,s) \neq 0$ for $s < t$ $\mathrm{d}\langle M\rangle$-almost everywhere (and as $\langle M\rangle$ is strictly increasing also $\mathrm{d}t$-almost everywhere). So, we may write

$$\mathrm{d}M_t = \frac{\mathrm{d}\hat{X}_{T|t}}{k(T,t)}.$$

Thus,

$$X_t = \int_0^t \frac{k(t,s)}{k(T,s)}\,\mathrm{d}\hat{X}_{T|s}$$

and we have the assumption (A4) satisfied with

$$p_T^*(t,s) = \frac{k(t,s)}{k(T,s)}.$$

Consequently, the assumption (A2) is also satisfied. Also, the assumption (A6) implies the assumption (A1), since

$$\mathrm{d}\langle \hat{X}_{T|\cdot}\rangle_t = k(T,t)^2\,\mathrm{d}\langle M\rangle_t.$$

Indeed, this implies that $\langle \hat{X}_{T|\cdot}\rangle$ is strictly increasing.

For the kernel p_T we find the representation by using the assumption (A7) as follows:

$$\begin{aligned}
\hat{X}_{T|t} &= \int_0^t k(T,s)\,\mathrm{d}M_s \\
&= \int_0^t \mathrm{K}\left[\mathbf{1}_{[0,T)}\right](s)\,\mathrm{d}M_s \\
&= \int_0^t \mathrm{K}^{-1}\left[\mathbf{1}_{[0,t)}\mathrm{K}\left[\mathbf{1}_{[0,T)}\right]\right](s)\,\mathrm{d}X_s \\
&= \int_0^t \left\{\mathrm{K}^{-1}\mathrm{K}\left[\mathbf{1}_{[0,t)}\right](s) + \mathrm{K}^{-1}\left[\mathbf{1}_{[0,t)}\mathrm{K}\left[\mathbf{1}_{[t,T)}\right]\right](s)\right\}\mathrm{d}X_s \\
&= X_t + \int_0^t \Psi_T(t,s)\,\mathrm{d}X_s,
\end{aligned}$$

where we have denoted

$$\Psi_T(t,s) := \mathrm{K}^{-1}\left[\mathbf{1}_{[0,t)}\mathrm{K}\left[\mathbf{1}_{[t,T)}\right]\right](s). \tag{5.5}$$

So, we have found that

$$\begin{aligned}
\mathrm{d}\langle \hat{X}_{T|\cdot}\rangle_t &= k(T,t)^2\,\mathrm{d}\langle M\rangle_t, \\
p_T(t,s) &= \mathbf{1}_{[0,t)}(s) + \Psi_T(t,s), \\
p_T^*(t,s) &= \frac{k(t,s)}{k(T,s)}
\end{aligned}$$

and we may rewrite Proposition 14 as follows.

Proposition 18. *Let X satisfy assumptions (A5), (A6) and (A7). Then the bridge $X^{T,\theta}$ satisfies the integral equation*

$$X_t^{T,\theta} = X_t + \int_0^t \Big\{ \theta - X_s^{T,\theta} - \int_0^s \Psi_T(s,u)\,\mathrm{d}X_u^{T,\theta} \Big\} \frac{k(T,s)k(t,s)}{\int_s^T k(T,u)^2\,\mathrm{d}\langle M\rangle_u}\,\mathrm{d}\langle M\rangle_s. \tag{5.6}$$

Moreover, the bridge $X^{T,\theta}$ admits the non-anticipative representation

$$X_t^{T,\theta} = \theta\frac{R(T,t)}{R(T,T)} + X_t - \int_0^t \varphi_T(t,s)\,\mathrm{d}X_s, \tag{5.7}$$

where

$$\varphi_T(t,s) = \int_s^t \Big\{ \int_s^u \frac{\big(1+\Psi_T(v,s)\big)k(T,v)^2}{\big(\int_v^T k(T,w)^2\,\mathrm{d}\langle M\rangle_w\big)^2}\,\mathrm{d}\langle M\rangle_v - \frac{1+\Psi_T(u,s)}{\int_u^T k(T,v)^2\,\mathrm{d}\langle M\rangle_v}\Big\} k(T,u)k(t,u)\,\mathrm{d}\langle M\rangle_u.$$

6 Fractional Brownian Bridge

The fractional Brownian motion Z is a centred stationary increment Gaussian process with variance $\mathbf{E}\big(Z_t^2\big) = t^{2H}$ for some $H \in (0,1)$. Another way of charaterising the fractional Brownian motion if to say that it is the unique (up to multiplicative constant) centred H-self-similar Gaussian process with stationary increments.

In order to represent the fractional Brownian motion as a Volterra process we first recall some preliminaries of fractional calculus. For details we refer to Samko et al. [8].

Let f be a function over the interval $[0,1]$ and $\alpha > 0$. Then

$$I_\pm^\alpha[f](t) := \frac{1}{\Gamma(\alpha)}\int_0^1 \frac{f(s)}{(t-s)_\pm^{1-\alpha}}\,\mathrm{d}s$$

are the *Riemann–Liouville fractional integrals* of order α. For $\alpha \in (0,1)$,

$$D_\pm^\alpha[f](t) := \frac{\pm 1}{\Gamma(1-\alpha)}\frac{\mathrm{d}}{\mathrm{d}t}\int_0^1 \frac{f(s)}{(t-s)_\pm^{\alpha}}\,\mathrm{d}s.$$

are the *Riemann–Liouville fractional derivatives* of order α; $I_\pm^0$ and $D_\pm^0$ are identity operators.

If one ignores the troubles concerning divergent integrals and formally changes the order of differentiation and integration one obtains

$$I_{\pm}^{-\alpha} = D_{\pm}^{\alpha}.$$

We shall take the above as the definition for fractional integral of negative order and use the obvious unified notation.

Now, the fractional Brownian motion is a Volterra process satisfying assumptions (A5) and (A6). Indeed, let K be a weighted fractional integral or differential operator

$$\mathrm{K}\left[f\right](t) := c_H t^{\frac{1}{2}-H} I_{-}^{H-\frac{1}{2}} \left[s^{H-\frac{1}{2}} f(s)\right](t),$$

where

$$c_H = \sqrt{\frac{2H(H-\frac{1}{2})\Gamma(H-\frac{1}{2})^2}{\mathrm{B}(H-\frac{1}{2}, 2-2H)}}$$

and Γ and B are the gamma and beta functions. Then we have the relation (5.1) for fractional Brownian motion:

$$Z_t = \int_0^t \mathrm{K}\left[\mathbf{1}_{[0,t)}\right](s)\,\mathrm{d}W_s, \tag{6.1}$$

where W is the standard Brownian motion. Thus, the fractional Brownian motion satisfies the assumption (A5).

The operator K satisfies the assumption (A6). Indeed,

$$\mathrm{K}^{-1}[f](t) = \frac{1}{c_H} t^{\frac{1}{2}-H} I_{-}^{\frac{1}{2}-H} \left[s^{H-\frac{1}{2}} f(s)\right](t)$$

The kernel Ψ_T has been calculated e.g. in Pipiras and Taqqu [7], Theorem 7.1. Indeed, for any $H \in (0,1)$ we have

$$\Psi_T(t,s) = \frac{\sin\left(\pi(H+\frac{1}{2})\right)}{\pi} s^{\frac{1}{2}-H}(t-s)^{\frac{1}{2}-H} \int_t^T \frac{u^{H+\frac{1}{2}}(u-t)^{H+\frac{1}{2}}}{u-s}\,\mathrm{d}u.$$

As for the kernel p_T^* note that for $H \in (0,1)$ we have

$$\begin{aligned} k(t,s) &= \mathrm{K}\left[\mathbf{1}_{[0,t)}\right](s) \\ &= c_H' \left\{ \left(\frac{t}{s}\right)^{H-\frac{1}{2}} (t-s)^{H-\frac{1}{2}} - \left(H-\tfrac{1}{2}\right) s^{\frac{1}{2}-H} \int_s^t u^{H-\frac{3}{2}}(u-s)^{H-\frac{1}{2}}\,\mathrm{d}u \right\}, \end{aligned}$$

where

$$c_H' = \sqrt{\frac{2H\Gamma(\frac{3}{2}-H)}{\Gamma\left(H+\frac{1}{2}\right)\Gamma(2-2H)}}.$$

If $H > 1/2$ then we have a slightly simpler expression, viz.

$$k(t,s) \;=\; c'_H \left(H-\tfrac{1}{2}\right) s^{\frac{1}{2}-H} \int_s^t u^{H-\frac{1}{2}}(u-s)^{H-\frac{3}{2}}\,du.$$

For the derivation of these formulas see Norros et al. [6] and Jost [5].

The representations for the fractional Brownian bridge follow now by plugging in our Ψ_T and k to the formulas (5.6) and (5.7) in Proposition 18 with $M = W$ and $\mathrm{d}\langle M\rangle_t = \mathrm{d}t$. Unfortunately, it seems that there is really nothing we can do to simplify the resulting formula (except some trivial use of the H-self-similarity), even in the case $H > 1/2$. So, we do not bother to write the equations (5.6) and (5.7) again here.

7 Enlargement of Filtration Point of View

Let us denote by $\mathbf{F}^X$ be its natural (continuous) filtration of the Gaussian process X. Setting in our conditioning simply $\theta := X_T$ we may interpret the bridge measure $\mathbf{P}^{(0,X_0)\to(T,X_T)}$ as initial enlargement of the filtration $\mathbf{F}^X$ by the random variable X_T. Let $\mathbf{F}^{X,X_T}$ be this enlarged filtration. We have formally

$$(\Omega,\mathcal{F},\mathbf{F}^X,\mathbf{P}^{(0,X_0)\to(T,X_T)}) \;\simeq\; (\Omega,\mathcal{F},\mathbf{F}^{X,X_T},\mathbf{P}).$$

For the Brownian motion W we have the following: with respect to the measure $\mathbf{P}^\theta$ the Browian motion has the representation

$$W_t = W_t^\theta + \int_0^t \beta_{T,\theta}(W)_s ds = W_s^\theta + \int_0^t \frac{\theta - W_s}{T-s} ds.$$

This means that with respect to the filtration $\mathbf{F}^{W,W_T}$ and measure $\mathbf{P}$ Brownian motion W has the representation

$$W_t = W_t^{\mathbf{F}^{W,W_T}} + \int_0^t \frac{W_T - W_s}{T-s} ds,$$

where Law $(W^{\mathbf{F}^{W,W_T}}|\mathbf{P}) = \mathrm{Law}(W|\mathbf{P})$, $W^{\mathbf{F}^{W,W_T}}$ is a $(\mathbf{P},\mathbf{F}^{W,W_T})$ Brownian motion, but W is a $(\mathbf{P},\mathbf{F}^{W,W_T})$ semimartingale.

Similarly, if we have an arbitrary gaussian process such that the process has a Volterra representation (4.3)

$$X_t \;=\; \int_0^t p_T^*(t,s)\,\mathrm{d}\hat{X}_{T|s}.$$

we can use the enlargement of filtration results to give a semimartingle representation for the martingale $\hat{X}_{T|\cdot}$ with respect to $(\mathbf{P},\mathbf{F}^{X,X_T})$:

$$\hat{X}_{T|t} \;=\; \hat{X}_{T|t}^{\mathbf{F}^{X,X_T}} + \int_0^t \frac{X_T - \hat{X}_{T|s}}{\langle \hat{X}_{T|\cdot}\rangle_{T,s}} d\langle \hat{X}_{T|\cdot}\rangle_s, \tag{7.1}$$

where the $(\mathbf{P}, \mathbf{F}^{X,X_T})$-gaussian martingale $\hat{X}_{T|\cdot}^{\mathbf{F}^{X,X_T}}$ has the same law as $\hat{X}_{T|\cdot}$ (see [2, 3] for more details). We can now use (4.3), (4.6) and (7.1) to obtain the following representation for the process X

$$X_t = X_t^{\mathbf{F}^{X,X_T}} + X_T \frac{R(T,t)}{R(T,T)} - \int_0^t \phi_T(t,s) dX_s, \tag{7.2}$$

with

$$\phi_T(t,s) = \int_s^t \left\{ \int_s^u \frac{p_T(v,s)}{\langle \hat{X}_{T|\cdot} \rangle_{T,v}^2} \mathrm{d}\langle \hat{X}_{T|\cdot} \rangle_v - \frac{p_T(u,s)}{\langle \hat{X}_{T|\cdot} \rangle_{T,u}} \right\} p_T^*(t,u)\, \mathrm{d}\langle \hat{X}_{T|\cdot} \rangle_u .$$

Acknowledgements

T. Sottinen is grateful for partial support from EU-IHP network DYNSTOCH. We thank M. Lifshits for the example 1.

References

1. C. Bender, Elliott, R. On the Clark-Ocone theorem for fractional Brownian motions with Hurst parameter bigger than a half, Stoch. Stoch. Rep. 75 (2003) 391–405.
2. D. Gasbarra, E. Valkeila, Initial enlargement: a Bayesian approach, Theory Stoch. Process. 9 (2003) 26–37.
3. D. Gasbarra, E. Valkeila, L. Vostrikova, Enlargement of filtration and additional information in pricing models: a Bayesian approach. In: From Stochastic Analysis to Mathematical Finance (Y. Kabanov, R. Liptser, and J. Stoyanov, eds.) Springer (2006), 257–285.
4. T. Jeulin, Semi-martingales et Grossisement d'une Filtration., Lect. Notes Math. 1118, Springer, Berlin 1980.
5. C. Jost, Tranformation formulas for fractional Brownian motion. Stochastic Processes and their Applications, 116 (2006), 1341–1357.
6. I. Norros, E. Valkeila, J. Virtamo, An elementary approach to a Girsanov formula and other analytical results on fractional Brownian motion, Bernoulli 5 (1999) 571–587.
7. V. Pipiras, M. Taqqu, Are classes of deterministic integrands for fractional Brownian motion on an interval complete?, Bernoulli 7 (2001) 873–897.
8. S.G. Samko, A.A. Kilbas, O.I. Marichev, Fractional integrals and derivatives. Theory and applications, Gordon and Breach Science Publishers, Yverdon 1993.
9. A.N. Shiryaev, Probability. Springer, Berlin, 1984.

Some of the Recent Topics on Stochastic Analysis

Takeyuki Hida

Department of Mathematics, Meijo University, Shiogamaguchi, Tenpaku-ku, Nagoya 468-8502, Japan

Summary. First, we shall quickly explain why and how the space of generalized white noise functionals has been introduced. The space has big advantages to carry on the analysis of nonlinear functionals of white noise (or of Brownian motion) and to apply the theory to various fields. It should be noted that the introduction of generalized functionals was motivated by the Itô formula for Brownian functionals. Using this space we discuss the following two topics.

1. Path integrals. To formulate Lagrangian path integrals, we have to concretize the expressions of the Lagrangian in terms of paths. We propose that quantum mechanical paths (trajectories) are expressed as a sum of the classical paths and fluctuation which is taken to be a Brownian bridge. It is possible to give a plausible reason why a Brownian bridge is fitting in this case. With this choice of possible trajectories, there arises a difficulty that the kinetic energy becomes a generalized functional of a Brownian motion. It is now possible to overcome this difficulty to take our favorable space of generalized white noise functionals. Then follows the integration. Our method can be applied to a wider class of dynamics, for instance, to those cases with singular potentials and to some fields over non-euclidean space.

2. Infinite dimensional rotation group and unitary group. It is well known that the "infinite dimensional rotation group" has naturally been introduced in connection with white noise, and the group describes certain invariance of the white noise measure. Hence, we may say that the white noise analysis should have an aspect of an infinite dimensional harmonic analysis. It seems natural, in fact by many reasons, to complexify the rotation group to have "infinite dimensional unitary group". Thus complexified group has various interesting applications to the analysis of nonlinear functionals of complex white noise. In addition, we can find good connections with Lie group theory and theory of quantum dynamics, to which we can give new interpretations.

1 Introduction

We are interested in essentially infinite dimensional analysis and discuss functionals of the form

$$f(B(s), s \in T, t) \;\to\; \varphi(\dot{B}(s), s \in T, t),$$

where $\{\dot{B}(t)\}$ is a white noise and is a system of *idealized elemental random variables.*

Flow Diagram

1. Start with the **Itô formula**, the simplest case:
$$(dB(t))^2 = dt.$$
Magnify $(dB(t))^2 - E[(dB(t)^2]$ (multiply by $\frac{1}{(dt)^2}$ to have : $\dot{B}(t)^2$:).
Generally, renormalized monomials in $\dot{B}(t)$'s.
2. Space of **generalized white noise functionals** by using
   i) Wiener-Itô Decomposition of (L^2),
   ii) S-transform.
3. **Analysis**
Harmonic Analysis arising from rotation group.
4. **Two proposed directions**
   i) Path integrals,
   ii) Infinite dimensional unitary group.

2 Our Idea of White Noise Analysis

The idea to discuss random complex systems is based on the **"Reductionism"**. Actual implementation is to construct the *innovation* by extracting necessary and sufficient information from the given random complex system. This is the first step of our mathematical approach to the study of random complex systems. The standard innovation can be obtained as the time derivative of a Lévy process. We are thus given an elemental random system. This choice is quite reasonable for our purpose.

Then, follows the next step **"Synthesis"**. There the given random complex system should be expressed as a functional (which is non-random and, in general, nonlinear functional) of the innovation that has just been obtained in the step of the reduction. Thus, we have an analytic representation of the random complex phenomenon in question by using functions known in functional analysis.

Finally, we are ready to study the **"Analysis"** of the functionals, in fact, nonlinear functionals of the innovation. It can be proceeded having been suggested by the ordinary functional analysis. Further various applications can be discussed, and even one can see beautiful interplay between our theory and the studies of actual problems in various fields of science.

To be added, there are interesting applications in various fields of science, some of which will be presented in the present notes.

We are sure that the **innovation approach** is one of the most efficient and legitimate directions to the study of stochastic processes and random fields, or more generally to random complex systems. See [7].

We now focus our attention on the case, where the innovation is Gaussian.

3 Generalized White Noise Functionals

Having had the system of variables to be $\{\dot{B}(t)\}$, we are naturally led to introduce basic functions of the $\dot{B}(t)$'s. We had a naive observation on the square of $\dot{B}(t)$. A particular case of the Itô formula $(dB(t))^2 = dt$ derives renormalized variable $: \dot{B}(t)^2 :$. It makes sense as a generalized functional of white noise.

Remark. L. Accardi has done profound research on powers of $\dot{B}(t)$'s.

With these facts in mind we define classes of *generalized white noise functionals*, which are the most important concepts in white noise theory.

Our starting point is the Fock space for (L^2), that involves ordinary white noise functionals with finite variance,

$$(L^2) = \sum_0^\infty \bigoplus H_n,$$

where H_n is the space of multiple Wiener integrals in Itô sense of degree n.

There are two typical ways of introducing generalized white noise functionals.

3.1 Use of the Sobolev Spaces $K^m(R^n)$ over R^n of Degree m

$$\widehat{K^m(R^n)} \subset \widehat{L^2(R^n)} \subset \widehat{K^{-m}(R^n)},$$

where the both inclusions are continuous injections.

Our favorable choice of the degree m of the Sobolev space is $(n+1)/2$. For one thing, members in $K^{(n+1)/2}(R^n)$ are continuous and the restriction of them to an $(n-1)$-dimensional hyperplane belongs to the space $\widehat{K^{n/2}(R^{n-1})}$, namely the degree decreases by $1/2$ when the restriction is made to a subspace of one dimension lower.

Let $m = (n+1)/2$ in the above triple involving symmetric Sobolev spaces. We have

$$\widehat{K^{(n+1)/2}(R^n)} \subset \widehat{L^2(R^n)} \subset \widehat{K^{-(n+1)/2}(R^n)},$$

and we form

$$H_n^{(n)} \subset H_n \subset H_n^{(-n)},$$

each space of this triple is isomorphic to the corresponding symmetric Sobolev space. The norms in those spaces are denoted by $\|\cdot\|_n$, $\|\cdot\|$, and $\|\cdot\|_{-n}$. Thus, we have a Hilbert space $H_n^{(-n)}$ of *generalized white noise functionals* of degree n.

There remains a freedom on how to sum up the spaces $H_n^{(-n)}, n \geq 0$. Choose an increasing sequence $c_n > 0$, and form a Hilbert space $(L^2)^+$ by the direct sum

$$(L^2)^+ = \bigoplus c_n H_n^{(n)}.$$

The direct sum forms a Hilbert space and its dual space is expressed in the form

$$(L^2)^- = \bigoplus c_n^{-1} H_n^{(-n)}.$$

Naturally, we are given a triple

$$(L^2)^+ \subset (L^2) \subset (L^2)^-.$$

The space $(L^2)^+$ consists of test functionals, while $(L^2)^-$ is the space of *generalized white noise functionals* (For details, see [8], Chap. 2).

3.2 An Analogue of the Schwartz Space

Take the parameter space R and an operator A:

$$A = -\frac{d^2}{du^2} + u^2 + 1$$

acting on $L^2(R)$. Then, apply the second quantization technique to introduce the operator $\Gamma(A)$ acting on the space $\bigoplus \widehat{L^2(R^n)}$. By using the isomorphism π defined in I:

$$\pi : H_n \longrightarrow \widehat{L^2(R^n)},$$

we can easily define the operator $\tilde{\Gamma}(A) = \pi^{-1}\Gamma(A)\pi$. It is proved, as in I, that for $\varphi \in H_n$

$$\|\varphi\| = \sqrt{n!}\|\pi\varphi\|_{L^2(R^n)}.$$

Unless no confusion occurs, $\tilde{\Gamma}(A)$ is also denoted by $\Gamma(A)$. It acts on (L^2).

Set $(S)_n = \mathcal{D}(\Gamma(A)^n)$ and set

$$(S) = \bigcap_n (S)_n.$$

The projective limit topology is introduced to (S).

The dual space of $(S)_n$ is denoted by $(S)_{-n}$ and its inductive limit:

$$(S)^* = \lim (S)_{-n}$$

is formed. The $(S)^*$ is the space of *generalized white noise functionals.* It is often called the space of *white noise distributions.*

Remind the T-transform for $\varphi \in (L^2)$:

$$(T\varphi)(\xi) = \int_{E^*} \exp[i < x, \xi >]\varphi(x)d\mu(x),$$

and the S-transform

$$(S\varphi)(\xi) = C(\xi) \int_{E^*} \exp[< x, \xi >]\varphi(x) d\mu(x).$$

They give us visualized and convenient representations of generalized white noise functionals.

4 White Noise Approach to Path Integrals

Our method for path integrals in quantum dynamics is to take white noise measure to define average (or expectation) of functionals, and to take generalized functionals in the integrand to have visualized expression. The path integral method has been originated by R. Feynman, with some motivation due to Dirac's idea, and it is viewed as a third method of quantization different from those by W. Heisenberg and E. Schrödinger. Our method of path integrals follows mainly the Feynman's idea in methodology, however, some new techniques are introduced.

In quantum dynamics there are many possible paths (trajectories) of a particle, and each of them is viewed as a sum of the classical trajectory and fluctuation. We assert that the amount of the fluctuation is expressed as a Brownian bridge.

First, we need to give a characterization of a Brownian bridge $X(t), t \in [0, 1]$, over the unit time interval, since it plays a key role in our setup. A Brownian bridge is a Gaussian Markov process with mean 0 and covariance function $E(X(t)X(s)) = \Gamma(t, s) = (t \wedge s)(1 - t \vee s)$, $s, t \in [0, 1]$.

Heuristically speaking, it was 1981 when we proposed a white noise approach to path integrals to have quantum mechanical propagators (see [13] appeared later in 1983). We had, at that time, some idea in mind for the use of a Brownian bridge and had many good examples that have quite wider class of potentials, and we obtained various satisfactory results.

Now it is time to recall the original idea by characterizing the Brownian bridge taking some physical intuition (see e.g. [13]) into account, so that we can explain why a Brownian bridge is fitting for describing the fluctuation around the classical path. We now have a theorem:

Theorem. *The Brownian bridge $X(t)$ over the interval $[0, 1]$ is characterized (up to constant) by the conditions*

i) *$X(t)$ is a Gaussian Markov process that has the canonical representation,*
ii) *$X(0) = X(1) = 0$ (bridged), and $E(X(t)) \equiv 0$,*
iii) *the normalized process $Y(t)$ enjoys the projective invariance,*
iv) *the local continuity of $Y(t)$ as $t \to 0$ in terms of covariance function is the same as that of the normalized Brownian motion $B(t)/\sqrt{t}$.*

Proof. Assumptions i) - iii) proves that the covariance function $\Gamma(t,s)$ of the process to be determined has to be of the form

$$\Gamma(t,s) = f\left(\frac{s}{1-s}\right) / f\left(\frac{t}{1-t}\right). \tag{4.1}$$

The assumption iv) asserts that f is the square root of a variable. Thus, the theorem is proved.

The Brownian bridge determined above has a canonical representation that is expressed in the form (up to constant):

$$X(t) = (1-t)\int_0^t \frac{1}{(1-u)}\dot{B}(u)du.$$

The covariance function $\Gamma(t,s)$ of the normalized process $Y(t) = X(t)/\sqrt{E(X(t)^2)}$ is of the form

$$\Gamma(t,s) = \sqrt{(0,1;s,t)},$$

where $(0,1;s,t)$ denotes the anharmonic ratio, that is $\frac{(1-t)/t}{(1-s)/s}$.

By observing the expression of the covariance function, it is obvious that a Brownian bridge is reversible in time, and further the theorem implies, as was announced before, that the Brownian bridge is fitting for describing the amount of fluctuation of a classical trajectory when we formulate a rigorous Feynman path integral following the idea due to Dirac and Feynman. This fact is illustrated probabilistically in what follows.

The actual expression and computations of the propagator are given successively as follows:

We follow the Lagrangian dynamics. The possible trajectories are sample paths $y(s), s \in [0,1]$, expressed in the form

$$y(s) = x(s) + \sqrt{\frac{\hbar}{m}}B(s),$$

where $B(t)$ is an ordinary Brownian motion. Hence the action S is expressed in the form

$$S = \int_0^t L(y(s), \dot{y}(s))ds.$$

Note that the bridged effect is done by putting the delta-function $\delta_0(y(t)-y_2)$, where $y_2 = x(t)$.

We have

Theorem. *The quantum mechanical propagator $G(0,t;y_1,y_2)$ is given by the following average*

$$G(0,t;y_1,y_2) = \left\langle N\exp\left[\frac{i}{\hbar}\int_0^t L(y,\dot{y})ds + \frac{1}{2}\int_0^t \dot{B}(s)^2 ds\right]\delta_o(y(t)-y_2)\right\rangle, \tag{4.2}$$

where N is the normalizing constant.

Actual computations for given potentials (including those have some singularity at the boundary) give us the propagator.

It should be noted that there are generalized white noise functionals in the above expectation. Namely, they are delta functions, in fact the Donsker's delta function $\delta_o(y(t)-y_2)$ and Gauss kernels, one of which implicitly appears in the action and the other is $\exp[\frac{1}{2}\int_0^t \dot{B}(s)^2 ds]$ with a multiplicative renormalizing constant. This generalized functional serves for flattening effect of the white noise measure. One may ask why the latter functional is so. An intuitive answer to this question is as follows: If we write a Lebesgue measure (exists only virtually) on E^* by dL, the white noise measure μ may be expressed in the form $\exp[-\frac{1}{2}\int_0^t \dot{B}(s)^2 ds]dL$. Hence, the factor in question is put to make the measure μ to be the flat measure dL. In fact, this makes sense eventually.

Returning to the formula (4.2), it is important to note that the integrand (inside the angular bracket) is integrable, in other words, to see that it is a bilinear form of a generalized functional and a test functional.

In general, the integrand is expressed as a product of a test functional and a functional of the form $\varphi(x) \cdot \delta(\langle x, f\rangle - a)$, $f \in L^2(R)$, $a \in R$. To this end, we have to prepare some notes.

The first one is short. Since a sample function x is a generalized function, so the canonical bilinear form $\langle x, \xi\rangle$ is defined for $\xi \in E$ (pointwise in x and ξ). For our purpose it is necessary to extend the bilinear form to $\langle x, f\rangle$, where f is in $L^2(R)$. This can be defined as a stochastic bilinear form, although it is no more continuous in f.

The next note is important. In general, the formula involves a product of functionals of the form $\varphi(x)\cdot\delta(\langle x, f\rangle - a), f \in L^2(R), a \in \mathbf{C}$. To give a correct interpretation to the expectation (4.2), it should be checked that it can be regarded as a bilinear form of a test functional and a generalized functional. The following assertion answers this question.

Proposition (Streit et al [14]). *Let $\varphi(x)$ be a generalized white noise functional. Assume that the T-transform $(T\varphi)(\xi), \xi \in E$, of φ is extended to a functional on $L^2(R)$, in particular function of $\xi + \lambda f$, and that $(T\varphi)(\xi - \lambda f)$ is an integrable function of λ for any fixed ξ and λ. If the Fourier transform of $(T\varphi)(\xi - \lambda f)$ is a U-functional, then the pointwise product $\varphi(x)\cdot\delta(\langle x, f\rangle - a)$ is defined and is a generalized white noise functional.*

Proof. First a formula for the δ-function is provided.

$$\delta_a(t) = \delta(t-a) = \frac{1}{2\pi}\int e^{ia\lambda}e^{-i\lambda x}d\lambda$$

(in distribution sense).

Hence, for $\varphi \in (S)^*$ and $f \in L^2(R)$ we have

$$T(\varphi(x)\delta(\langle x, f\rangle - a))\xi) = \frac{1}{2\pi}\int e^{ia\lambda}e^{-i\lambda\langle x,f\rangle}e^{i\langle x,\xi\rangle}\varphi(x)d\mu(x)d\Lambda$$
$$= \frac{1}{2\pi}\int e^{ia\lambda}(T\varphi)(\xi^\lambda f)d\lambda.$$

By assumption this determines a U-functional, which means the product $\varphi(x)\cdot\delta(\langle x, f\rangle - a)$ makes sense and it is a generalized white noise functional.

Example. A Gauss kernel $\varphi_c(x) = N\exp[c\int x(t)^2dt]$.
The following cases are fitting.

i) The case c is real and $c < 0$.
ii) The case $c = \frac{1}{2} + ia, a \in R$.

The same expression as in i), and it is shown that Proposition is applied.

Recent Developments. Now we should like to mention that there are many successful computations of various propagators. It is easy to see that in the cases i) free particle, ii) simple harmonic oscillator, iii) the Albeverio-Hoegh-Krohn potential which is the Fourier transform of a measure, the results obtained by our method are in agreement with the known propagators, respectively. In addition, some more interesting cases, including those with much singular potentials and time depending potentials, we have satisfactory results in the recent developments.

Example 1. Kuna, Streit and Westerkampf [14] obtained explicit formulae in the cases:

1) A time depending Lagrangian of the form

$$L(x(t), \dot{x}(t), t) = \frac{1}{2}m(t)\dot{x}(t)^2 - k(t)^2x(t)^2 - f(t)x(t),$$

where $m(t), k(t)$ and $f(t)$ are smooth functions.
2) A singular potential $V(x)$ of the form

$$V(x) = \sum_n c^{-n^2}\delta_n(x), \quad c > 0,$$

and others.

Example 2. C. Bernidos' results [15] on polymer entanglements.

Chern-Simons Functional Integral. See Albeverio-Sengupta [2].

The paper [2] gives us an interesting problem, where white noise analysis is applied. Namely, the authors propose a 3-dimensional gauge theory based on the Chern-Simons action $CS(A)$, where A is a connection over a 3-manifold that runs through $\mathcal{A}$. Namely, there appears a functional integral of the form

$$\int_{\mathcal{A}} \exp[iCS(A)\phi(A)DA.$$

Under various assumptions, A can be expressed in the form

$$A = a_0 dx_0 + a_1 dx_1 + a_2 dx_2,$$

where a_i is a Lie algebra-valued function on R^3.

The problem, if we understand correctly, is to define the integral

$$\frac{1}{N}\int e^{i\kappa/(2\pi)\langle a_0, f_1\rangle}\phi(a_0, f_1)Da_0 Df_1,$$

where $f_1 = \partial_2 a_1$.

Now it is interesting to consider a functional $e^{c\langle a_0, f_1\rangle}$. It has some similarity to the Gauss kernel discussed before, but it poses a new problem regarding the inner product of two independent white noises. We note that this problem can be discussed within our framework.

5 Infinite Dimensional Rotation Group

This section is devoted to the harmonic analysis that comes from the infinite dimensional rotation group, which is one of our favorite tools in white noise analysis.

Definition. *A continuous linear homeomorphism g acting on E is called a rotation of E if the following equality holds for every $\xi \in E$;*

$$\|g\xi\| = \|\xi\|.$$

The collection of all rotations forms a group under the usual product, and is denoted by $O(E)$.

Definition. *Introduce compact-open topology to $O(E)$ to have infinite dimensional rotation group.*

The adjoint transformation g^* on E^* is defined, and the collection $O^*(E^*)$ $= \{g^*; g \in O(E)\}$ forms a group and isomorphic to $O(E)$. We know that the white noise measure is invariant under $g^* \in O^*(E^*)$.

We are now ready to discuss harmonic analysis arising from $O^*(E^*)$.

Since our infinite dimensional rotation group is quite big, indeed it is neither compact nor locally compact, it seems useful to take two subclasses of significant subgroups and investigate their roles. We are particularly interested in essentially infinite dimensional rotations. There is a criterion for this property.

We define *average power* $a.p.(g_\pi)$ of a rotation g_π which comes from a permutation π:

$$a.p.(g_\pi)(x) = \lim\sup_{N\to\infty} \frac{1}{N}\sum_1^N \langle x, \xi_{\pi(n)} - \xi_n\rangle^2.$$

Definition. *If $a.p.(g)(x)$ is positive μ-a.e., then we call g_π essentially infinite dimensional. Contrary to this case, if $a.p.(g_\pi)(x) = 0$ almost surely, then g_π is said to be approximated by finite dimensional rotations.*

We can see that there are many members in the Lévy group (see Example below) that are essentially infinite dimensional.

Example. Pairwise permutation of the coordinates. The average power is equal to 2.

It is interesting to note that there should be an intimate connection between the Lévy group and the Lévy Laplacian (e.g. the forthcoming paper by Si Si and the author).

The Windmill Subgroup

There is another subgroup of $O(E)$ that contains essentially infinite dimensional transformation. It is a windmill subgroup $\mathcal{W}$, which is defined in the following manner. Take E to be the Schwartz space S and take a sequence $n(k)$ of positive integers satisfying the condition

$$(n(k+1) - n(k))\frac{n(k+1)}{n(k)} \le K, \ (K > 1).$$

Let $\xi_n, n \ge 0$, be the complete orthonormal system in $L^2(R)$ such that ξ_n is the eigenfunction of A as defined before: $A\xi_n = 2(n+1)\xi_n$. Denote by E_k the $(n(k+1) - n(k))$-dimensional subspace of $E = S$ that is spanned by $\{\xi_{n(k)+1}, \xi_{n(k)+2}, \cdots, \xi_{n(k+1)}\}$. Let G_k be the rotation group acting on E_k. Then, $\mathcal{W} = \mathcal{W}(\{n(k)\})$ is defined by

$$\mathcal{W} = \otimes_k G_k.$$

Definition. *The subgroup $\mathcal{W}$ is called a windmill subgroup.*

We can show that $\mathcal{W}$ describes interesting properties of our harmonic analysis.

6 Infinite Dimensional Unitary Group

We may assume that a complexification of white noise (E_c^*, ν) is known. Let $z \in E_c^*$ be of the form:

$$z = x + iy, \quad x, y \in E.$$

The complex white noise measure ν is the product of white noise measures μ_1 and μ_2 with variance 1/2:

$$\nu = \mu_1 \times \mu_2.$$

Now the unitary group $U(E_c)$ is defined. It is a collection of all transformations g on E_c such that

1. g is a linear homeomorphism of E_c,
2. g preserves the complex $L^2(R)$-norm:

$$\|g\eta\| = \|\eta\|, \quad \zeta \in E_c.$$

Definition. *The topological group $U(E_c)$ is called the infinite dimensional unitary group.*

The adjoint g^* of g in $U(E_c)$ is defined and we see that

$$g^*\nu = \nu.$$

Hence, we are given a unitary operator U_g defined by

$$U_g\varphi(z) = \varphi(g^*z), \varphi \in (L_c^2).$$

Under the usual product the collection of U_g's forms a topological group that is isomorphic to the group $U(E_c)$.

It is noted that the infinite dimensional rotation group $O(E)$ may be identified with a subgroup of $U(E_c)$.

6.1 Subgroups of $U(E_c)$

Many results below are known, but reviews or rephrasements are useful.

1. Conformal group.
   In the R^d-parameter case, if the basic nuclear space is taken to be $D_{0,c}$, then we are given the conformal group $C(d)$ which is a subgroup of $O(E)$ as was briefly mentioned before. Hence, the complex form of $C(d)$, denote it by $C_c(D_0)$, acting on the space $D_{0,c}$, is a subgroup of $U(D_{0,c})$. We call it a complex conformal group. It is locally isomorphic to the (real) linear group $SO(d+1,1)$ and is generated by one-parameter groups including whiskers as many as $\frac{(d+1)(d+2)}{2}$. Their generators are as follows:

$$s = -\frac{d}{du_i}, \quad i = 1, 2, \ldots, d,$$

$$\tau = r\frac{d}{dr} + \frac{d}{2}, \quad r = |u|,$$

$$r_{j,k} = u_j\frac{d}{du_k} - u_k\frac{d}{du_j}, \quad 1 \leq j \neq k \leq d,$$

$$\kappa_j = u_j^2\frac{d}{du_j} + u_j, \ j = 1, 2, \ldots, d.$$

They correspond to the shifts, isotopic dilation, rotations on R^d and special conformal transformations, respectively.

2. Heisenberg group.
From now on, one can see the effective use of complex white noise. Now take E_c to be $\mathcal{S}_c = \mathcal{S} + i\mathcal{S}$, $\mathcal{S}$ being the Schwartz space.
2.1) The gauge transformation I_t is defined by

$$I_t : \zeta(u) \longrightarrow I_t\zeta(u) = e^{it}\zeta(u).$$

Obviously I_t is a member of $U(E_c)$, and $\{I_t\}$ forms a continuous one parameter subgroup, periodic with period 2π. Let I be the identity.

$$\begin{aligned} I_t I_s &= I_{t+s}, \quad t, s \in R, \\ I_{t+2\pi} &= I_t, \\ I_t &\to I \text{ as } t \to 0. \end{aligned}$$

The group $\{I_t, t \in R\}$ is called the *gauge group*. Let U_t be defined by U_{I_t}. This unitary group has only point spectrum on the subspace H_n. The eigenspace belonging to the eigenvalue $-n+2k$ is $H_{(n-k,k)}$. Hence, the space $H_n, n > 1$, is classified by I_t into its subspaces $H_{(n-k,k)}$. The generator of the gauge group is iI.

Remark. The operator I_t extends to a more general gauge transformation, where it is replaced by $itf, f \in E$, so that we have a collection of generators that span a space isomorphic to a nuclear space E, hence a nuclear Lie algebra is given.

2.2) The shifts S_t^j with generators

$$-\frac{\partial}{\partial u_j}, \quad j = 1, 2, \cdots, d.$$

2.3) Multiplication $\pi_t^j, j = 1, 2, \cdots, d$. Let them be defined to be the conjugate to the shifts via the Fourier transform $\mathcal{F}$:

$$\pi_t^j = \mathcal{F} S_t^j \mathcal{F}^{-1}.$$

Definition. *The subgroup generated by the gauge group, the shifts and the multiplication is called the Heisenberg group.*

3. The Fourier-Mehler transforms $\mathcal{F}_\theta$ $(d = 1)$.
It is possible to consider the fractional power of the ordinary Fourier transform. It is defined by the integral kernel $K_\theta(u, v)$:

$$K_\theta(u, v) = (\pi(1 - \exp[2i\theta]))^{-1/2} \exp\left[-\frac{i(u^2 + v^2)}{2tan\theta} + \frac{iuv}{sin\theta}\right].$$

It defines an operator $\mathcal{F}_\theta$ by writing

$$(\mathcal{F}_\theta \zeta)(u) = \int_{-\infty}^{\infty} K_\theta(u,v)\zeta(v)dv,$$

where $\theta \neq \frac{1}{2}k\pi,\ k \in Z$. Particular choices of θ give

$$\mathcal{F}_{\pi/2} = \mathcal{F},\ \ \mathcal{F}_{(3/2)\pi} = \mathcal{F}^{-1}.$$

Thus, we have obtained a periodic one-parameter unitary group including the Fourier transform and its inverse.

The infinitesimal generator of $\mathcal{F}_\theta$ is denoted by if and is expressed in the form

$$if = -\frac{1}{2}i\left(\frac{d^2}{du^2} - u^2 + I\right).$$

Observing the commutation relations among the generators, so as to have a finite dimensional Lie algebra, either real or complex form, we are given a generator σ' of the form

$$\sigma' = \frac{1}{2}\left(\frac{d^2}{du^2} + u^2\right).$$

We are now interested in probabilistic roles or meanings of this operator in quantum dynamics (as the *repulsive oscillator*).

It is convenient to take $\sigma = \sigma' + \frac{i}{2}I$, namely we have

$$\sigma = \frac{1}{2}\left(\frac{d^2}{du^2} + u^2 + iI\right).$$

A one-parameter group with the generator σ can be defined locally in space-time. See e.g. [16].

6.2 Lie Algebras of Infinitesimal Generators

We have so far various infinitesimal generators. For simplicity we consider the case $d = 1$, that is the case of one-dimensional parameter complex white noise.

The algebraic structure of the space spanned by the generators is helpful for applications to quantum dynamics and differential geometry.

We list the generators so far obtained (for the case $d = 1$).

$$I, s, \tau, \kappa, \pi, f, \sigma$$

Proposition. *Based on the set of operators*

$$\{iI, s, i\pi, \tau, f, \sigma\}$$

we have 6-dimensional complex Lie algebra $\mathbf{g}$.

Proposition. *The algebra generated by* $\{iI, s, i\pi\}$ *is a radical of* $\mathbf{g}$.

In the multi-dimensional, say d-dimensional, case the rotations $r_{j,k}$, $1 \le j$, $k \le d$, are involved in the algebra for conformal group:

$$r_{j,k} = u_j \frac{\partial}{\partial u_k} - u_k \frac{\partial}{\partial u_j}, \quad 1 \le j \ne k \le d.$$

The algebraic structure of the Lie algebra involving the $r_{j,k}$ does not make much difference from the case $d = 1$.

Now it seems necessary to give an interpretation to the fact that the generator κ of the special conformal transformation is a particular one, being excluded from $\mathbf{g}$.

1. The reason why the κ has been taken.
   i) It is a good candidate to be introduced among the possible expressions of generators expressed in the form $a(u)\frac{d}{du} + \frac{1}{2}a'(u)$. If the basic nuclear space E is taken to be D_0, the κ is acceptable with $a(u) = u^2$. As a result, we have proved that the algebra generated by those possible generators is isomorphic to $sl(2, R)$.
   ii) Similar to s, the κ is transversal to τ, which defines a flow of the Ornstein-Uhlenbeck process (flow).
2. On the other hand, there are crucial reasons why κ should not be involved in the algebra $\mathbf{g}$.
   i) In order to introduce the κ we need particular space like D_0, instead of a familiar space S_c.
   ii) It does not satisfy favorable commutation relations with other favorable ones.

References

1. L. Accardi et al, Selected papers of Takeyuki Hida. World Scientific Pub. Co. 2001.
2. S. Albeverio and A. Sengupta, The Chern-Simons functional integral as an infinite dimensional distribution. Nonlinear Analysis, Theory, Methods and Applications. 30 (1997), 329–335.
3. T. Hida, Analysis of Brownian functionals. Carleton Math. Notes no.13, 1975.
4. T. Hida, Brownian motion. Springer-Verlag, 1980.
5. K. Itô, Multiple Wiener integrals. J. Math. Soc. Japan. 3 (1951), 157–169.
6. T. Hida, H.-H. Kuo, J. Potthoff and L. Streit, White noise. An Infinite dimensional calculus. Kluwer Academic Pub. 1993.
7. Hida and Si Si, Innovation approach to random fields: An application of white noise theory. World Sci. Pub. Co. 2004.
8. T. Hida and Si Si, Lectures on white noise functionals. Monograph to appear in World Sci. Pub. Co.
9. P. Lévy, Processu stochastiques et mouvement brounien. Gauthier-Villars, 1848.
10. P. Lévy, Problèmes concrets d'analyse fonctionelle. Gauthier-Villars, 1951.

11. P. Lévy, Random functions: General theory with special reference to Lapalcian random functions. Univ. of Calif. Pub. in Statistics. I, 12 (1953), 331–388.
12. Si Si, Effective determination of Poisson noise. IDAQP 6 (2003), 609–617.
13. L. Streit and T. Hida, Generalized Brownian functional and Feynman integral. Stochastic Processes and Appl. 16 (1983), 55–69.
14. T. Kuna, L. Streit and W. Westerkampf, Feynman integrals for a class of exponentially growing potentials. J. Math. Phys. 39 (1998), 4476–4491.
15. C.C. Bernido and M.V. Carpio-Bernido, White noise functional approach to polymer entanglements. Proc. Stochastic Analysis: Classical and Quantum. World Sci. Pub. Co. (2005), 1–12.
16. P. Topping, Repulsion and quantization in almost harmonic maps, and asymptotics of the harmonic map flow. Ann. Math. 159 (2004), 465–534.

Differential Equations Driven by Hölder Continuous Functions of Order Greater than 1/2

Yaozhong Hu and David Nualart

Department of Mathematics, University of Kansas, 405 Snow Hall, Lawrence, Kansas 66045-2142.

Summary. We derive estimates for the solutions to differential equations driven by a Hölder continuous function of order $\beta > 1/2$. As an application we deduce the existence of moments for the solutions to stochastic differential equations driven by a fractional Brownian motion with Hurst parameter $H > \frac{1}{2}$.

1 Introduction

We are interested in the solutions of differential equations on $\mathbb{R}^d$ of the form

$$x_t = x_0 + \int_0^t f(x_r) dy_r, \tag{1.1}$$

where the driving force $y : [0, \infty) \to \mathbb{R}^m$ is a Hölder continuous function of order $\beta > 1/2$. If the function $f : \mathbb{R}^d \to \mathbb{R}^{md}$ has bounded partial derivatives which are Hölder continuous of order $\lambda > \frac{1}{\beta} - 1$, then there is a unique solution $x : \mathbb{R}^d \to \mathbb{R}$, which has bounded $\frac{1}{\beta}$-variation on any finite interval. These results have been proved by Lyons in [2] using the p-variation norm and the technique introduced by Young in [8]. The integral appearing in (1.1) is a Riemann-Stieltjes integral.

In [9] Zähle has introduced a generalized Stieltjes integral using the techniques of fractional calculus. This integral is expressed in terms of fractional derivative operators and it coincides with the Riemann-Stieltjes integral $\int_0^T f dg$, when the functions f and g are Hölder continuous of orders λ and μ, respectively and $\lambda + \mu > 1$ (see Proposition 1 below). Using this formula for the Riemann-Stieltjes integral, Nualart and Răşcanu have obtained in [3] the existence of a unique solution for a class of general differential equations that includes (1.1). Also they have proved that the solution of (1.1) is bounded on a finite interval $[0, T]$ by $C_1 \exp(C_2 \|y\|_{0,T,\beta}^{\kappa})$, where $\kappa > \frac{1}{\beta}$ if f is bounded

Y. Hu is supported in part by the National Science Foundation under Grant No. DMS0204613 and DMS0504783

and $\kappa > \frac{1}{1-2\beta}$ is f has linear growth. Here $\|y\|_{0,T,\beta}$ denotes the β-Hölder norm of y on the time interval $[0,T]$. These estimates are based on a suitable application of Gronwall's lemma. It turns out that the estimate in the linear growth case is unsatisfactory because κ tends to infinity as β tends to $1/2$.

The main purpose of this paper is to obtain sharper estimates for the solution x_t in the case where f is bounded or has linear growth using a direct approach based on formula (2.8). In the case where f is bounded we estimate $\sup_{0\le t\le T}|x_t|$ by

$$C\left(1+\|y\|_{0,T,\beta}^{\frac{1}{\beta}}\right)$$

and if f has linear growth we obtain the exponential bound

$$C_1\exp\left(C_2\|y\|_{0,T,\beta}^{\frac{1}{\beta}}\right).$$

In Theorem 2 we provide explicit dependence on f and T for the constants C, C_1 and C_2. We also establish estimates for the solution of a linear equation with rough time dependent coefficient (Theorem 3.2).

Another novelty of this paper is that we establish stability type of results for the solution x_t to (1.1) on the initial condition x_0, the driving control y and the coefficient f (Theorem 3.2).

As an application we deduce the existence of moments for the solutions to stochastic differential equations driven by a fractional Brownian motion with Hurst parameter $H > \frac{1}{2}$. We also discuss the regularity of the solution in the sense of Malliavin Calculus, improving the results of Nualart and Saussereau [4], and we apply the techniques of the Malliavin calculus to establish the smoothness of the density of the solution under suitable non-degeneracy conditions. More precisely, Theorem 3.2 allows us to show that the solution of a stochastic differential equation is unique.

2 Fractional Integrals and Derivatives

Let $a,b\in\mathbb{R}$ with $a<b$. Let $f\in L^1(a,b)$ and $\alpha>0$. The left-sided and right-sided fractional Riemann-Liouville integrals of f of order α are defined for almost all $x\in(a,b)$ by

$$I_{a+}^{\alpha}f(t)=\frac{1}{\Gamma(\alpha)}\int_a^t (t-s)^{\alpha-1}f(s)\,ds$$

and

$$I_{b-}^{\alpha}f(t)=\frac{(-1)^{-\alpha}}{\Gamma(\alpha)}\int_t^b (s-t)^{\alpha-1}f(s)\,ds,$$

respectively, where $(-1)^{-\alpha}=e^{-i\pi\alpha}$ and $\Gamma(\alpha)=\int_0^\infty r^{\alpha-1}e^{-r}dr$ is the Euler gamma function. Let $I_{a+}^{\alpha}(L^p)$ (resp. $I_{b-}^{\alpha}(L^p)$) be the image of $L^p(a,b)$ by the operator I^{α}_{a+} (resp. I^{α}_{b-}). If $f \in I^{\alpha}_{a+}(L^p)$ (resp. $f \in I^{\alpha}_{b-}(L^p)$) and $0 < \alpha < 1$ then the Weyl derivatives are defined as

$$D^{\alpha}_{a+}f(t) = \frac{1}{\Gamma(1-\alpha)}\left(\frac{f(t)}{(t-a)^{\alpha}} + \alpha\int_a^t \frac{f(t)-f(s)}{(t-s)^{\alpha+1}}ds\right) \tag{2.1}$$

and

$$D^{\alpha}_{b-}f(t) = \frac{(-1)^{\alpha}}{\Gamma(1-\alpha)}\left(\frac{f(t)}{(b-t)^{\alpha}} + \alpha\int_t^b \frac{f(t)-f(s)}{(s-t)^{\alpha+1}}ds\right) \tag{2.2}$$

where $a \le t \le b$ (the convergence of the integrals at the singularity $s = t$ holds point-wise for almost all $t \in (a,b)$ if $p = 1$ and moreover in L^p-sense if $1 < p < \infty$).

For any $\lambda \in (0,1)$, we denote by $C^{\lambda}(a,b)$ the space of λ-Hölder continuous functions on the interval $[a,b]$. We will make use of the notation

$$\|x\|_{a,b,\beta} = \sup_{a\le\theta<r\le b} \frac{|x_r - x_\theta|}{|r-\theta|^{\beta}},$$

and

$$\|x\|_{a,b,\infty} = \sup_{a\le r\le b} |x_r|,$$

where $x : \mathbb{R}^d \to \mathbb{R}$ is a given continuous function.

Recall from [6] that we have:

- If $\alpha < \frac{1}{p}$ and $q = \frac{p}{1-\alpha p}$ then
$$I^{\alpha}_{a+}(L^p) = I^{\alpha}_{b-}(L^p) \subset L^q(a,b).$$
- If $\alpha > \frac{1}{p}$ then
$$I^{\alpha}_{a+}(L^p) \cup I^{\alpha}_{b-}(L^p) \subset C^{\alpha-\frac{1}{p}}(a,b).$$

The following inversion formulas hold:

$$I^{\alpha}_{a+}\left(D^{\alpha}_{a+}f\right) = f, \qquad \forall f \in I^{\alpha}_{a+}(L^p) \tag{2.3}$$
$$I^{\alpha}_{a-}\left(D^{\alpha}_{a-}f\right) = f, \qquad \forall f \in I^{\alpha}_{a-}(L^p) \tag{2.4}$$

and

$$D^{\alpha}_{a+}\left(I^{\alpha}_{a+}f\right) = f, \quad D^{\alpha}_{a-}\left(I^{\alpha}_{a-}f\right) = f, \quad \forall f \in L^1(a,b)\,. \tag{2.5}$$

On the other hand, for any $f, g \in L^1(a,b)$ we have

$$\int_a^b I^{\alpha}_{a+}f(t)g(t)dt = (-1)^{\alpha}\int_a^b f(t)I^{\alpha}_{b-}g(t)dt\,, \tag{2.6}$$

and for $f \in I^{\alpha}_{a+}(L^p)$ and $g \in I^{\alpha}_{a-}(L^p)$ we have

$$\int_a^b D^{\alpha}_{a+}f(t)g(t)dt = (-1)^{-\alpha}\int_a^b f(t)D^{\alpha}_{b-}g(t)dt. \tag{2.7}$$

Suppose that $f \in C^\lambda(a,b)$ and $g \in C^\mu(a,b)$ with $\lambda + \mu > 1$. Then, from the classical paper by Young [8], the Riemann-Stieltjes integral $\int_a^b f dg$ exists. The following proposition can be regarded as a fractional integration by parts formula, and provides an explicit expression for the integral $\int_a^b f dg$ in terms of fractional derivatives (see [9]).

Proposition 1. *Suppose that $f \in C^\lambda(a,b)$ and $g \in C^\mu(a,b)$ with $\lambda + \mu > 1$. Let $\lambda > \alpha$ and $\mu > 1 - \alpha$. Then the Riemann Stieltjes integral $\int_a^b f dg$ exists and it can be expressed as*

$$\int_a^b f dg = (-1)^\alpha \int_a^b D_{a+}^\alpha f(t)\, D_{b-}^{1-\alpha} g_{b-}(t)\, dt, \tag{2.8}$$

where $g_{b-}(t) = g(t) - g(b)$.

3 Estimates for the Solutions of Differential Equations

Suppose that $y : [0,\infty) \to \mathbb{R}^m$ is a Hölder continuous function of order $\beta > 1/2$. Fix an initial condition $x_0 \in \mathbb{R}^d$ and consider the following differential equation

$$x_t = x_0 + \int_0^t f(x_r) dy_r, \tag{3.1}$$

where $f : \mathbb{R}^d \to \mathbb{R}^{md}$ is given function. Lyons has proved in [2] that Equation (3.1) has a unique solution if f is continuously differentiable and it has a derivative f' which is bounded and locally Hölder continuous of order $\lambda > \frac{1}{\beta} - 1$.

Our aim is to obtain estimates on x_t which are better than those given by Nualart and Răşcanu in [3].

Theorem 2. *Let f be a continuously differentiable function such that f' is bounded and locally Hölder continuous of order $\lambda > \frac{1}{\beta} - 1$.*

(i) Assume $\|f'\|_\infty > 0$. There is a constant k depending only on β, such that for all T,

$$\sup_{0\le t\le T} |x_t| \le 2^{1+kT\left[\|f'\|_\infty \vee |f(0)|\right]^{1/\beta} \|y\|_{0,T,\beta}^{1/\beta}} (|x_0| + 1). \tag{3.2}$$

(ii) Assume that f is bounded. Then, there is a constant k, which depends only on β, such that for all T,

$$\sup_{0\le t\le T} |x_t| \le |x_0| + k\|f\|_\infty \left(T^\beta \|y\|_{0,T,\beta} \vee T \|f'\|_\infty^{\frac{1-\beta}{\beta}} \|y\|_{0,T,\beta}^{\frac{1}{\beta}} \right). \tag{3.3}$$

Proof. Without loss of generality we assume that $d = m = 1$. Set $\|y\|_\beta = \|y\|_{0,T,\beta}$. We can assume that $\|y\|_\beta > 0$, otherwise the inequalities are obvious. Let $\alpha < 1/2$ such that $\alpha > 1-\beta$. Henceforth k will denote a generic constant depending only on β.

Step 1. Assume first that f is bounded. First we use the fractional integration by parts formula given in Proposition 1 to obtain for all $s, t \in [0,T]$,

$$\left|\int_s^t f(x_r)dy_r\right| \leq \int_s^t \left|D_{s+}^\alpha f(x_r)\, D_{t-}^{1-\alpha} y_{t-}(r)\right| dr.$$

From (2.1) and (2.2) it is easy to see

$$|D_{t-}^{1-\alpha} y_{t-}(r)| \leq k\|y\|_{r,t,\beta}(t-r)^{\alpha+\beta-1} \leq k\|y\|_\beta(t-r)^{\alpha+\beta-1} \tag{3.4}$$

and

$$|D_{s+}^\alpha f(x_r)| \leq k\left[\|f\|_\infty (r-s)^{-\alpha} + \|f'\|_\infty \|x\|_{s,t,\beta}(r-s)^{\beta-\alpha}\right]. \tag{3.5}$$

Therefore

$$\begin{aligned}\left|\int_s^t f(x_r)dy_r\right| &\leq k\|y\|_\beta \int_s^t \left[\|f\|_\infty (r-s)^{-\alpha}(t-r)^{\alpha+\beta-1}\right.\\ &\qquad \left.+ \|f'\|_\infty \|x\|_{s,t,\beta}(r-s)^{\beta-\alpha}(t-r)^{\alpha+\beta-1}\right] dr\\ &\leq k\|y\|_\beta \left[\|f\|_\infty (t-s)^\beta + \|f'\|_\infty \|x\|_{s,t,\beta}(t-s)^{2\beta}\right].\end{aligned}$$

Consequently, we have

$$\|x\|_{s,t,\beta} \leq k\|y\|_\beta \left[\|f\|_\infty + \|f'\|_\infty \|x\|_{s,t,\beta}(t-s)^\beta\right].$$

Choose Δ such that

$$\Delta = \left(\frac{1}{2k\,\|f'\|_\infty \|y\|_\beta}\right)^{\frac{1}{\beta}}.$$

Then, for all s and t such that $t-s \leq \Delta$ we have

$$\|x\|_{s,t,\beta} \leq 2k\|y\|_\beta \|f\|_\infty. \tag{3.6}$$

Therefore,

$$\|x\|_{s,t,\infty} \leq |x_s| + \|x\|_{s,t,\beta}(t-s)^\beta \leq |x_s| + 2k\|y\|_\beta \|f\|_\infty \Delta^\beta. \tag{3.7}$$

If $\Delta \geq T$ we obtain the estimate

$$\|x\|_{0,T,\infty} \leq |x_0| + 2k\|y\|_\beta \|f\|_\infty T^\beta. \tag{3.8}$$

Assume $\Delta < T$. Then, from (3.7) we get

$$\|x\|_{s,t,\infty} \leq |x_s| + \|f\|_\infty \|f'\|_\infty^{-1}. \tag{3.9}$$

Divide the interval $[0, T]$ into $n = [T/\Delta] + 1$ subintervals (where $[a]$ denotes the largest integer bounded by a). Applying the inequality (3.9) for $s = 0$ and $t = \Delta$ we obtain

$$\sup_{0 \le t \le \Delta} |x_t| \le |x_0| + \|f\|_\infty \|f'\|_\infty^{-1} .$$

Then, applying the inequality (3.9) on the intervals $[\Delta, 2\Delta], \ldots, [(n-1)\Delta, n\Delta]$ recursively, we obtain

$$\begin{aligned} \sup_{0 \le t \le T} |x_t| &\le |x_0| + n \|f\|_\infty \|f'\|_\infty^{-1} \le |x_0| + \Delta^{-1}(T + \Delta) \|f\|_\infty \|f'\|_\infty^{-1} \\ &\le |x_0| + Tk \|f\|_\infty \|f'\|_\infty^{\frac{1-\beta}{\beta}} \|y\|_\beta^{\frac{1}{\beta}} . \end{aligned} \tag{3.10}$$

The inequality (3.3) follows from (3.8) and (3.10).

Step 2. In the general case, assuming $\|f'\|_\infty > 0$, instead of (3.5) we have

$$|D_{s+}^\alpha f(x_r)| \le k \left[(|f(0)| + \|f'\|_\infty |x_r|) (r-s)^{-\alpha} + \|f'\|_\infty \|x\|_{s,t,\beta} (r-s)^{\beta-\alpha} \right] .$$

As a consequence,

$$\|x\|_{s,t,\beta} \le k \|y\|_\beta \left[|f(0)| + \|f'\|_\infty \|x\|_{s,t,\infty} + \|f'\|_\infty \|x\|_{s,t,\beta} (t-s)^\beta \right] .$$

Suppose that Δ satisfies

$$\Delta \le \left(\frac{1}{3k \|f'\|_\infty \|y\|_\beta} \right)^{\frac{1}{\beta}} . \tag{3.11}$$

Then, for all s and t such that $t - s \le \Delta$ we have

$$\|x\|_{s,t,\beta} \le \frac{3}{2} k \|y\|_\beta \left(|f(0)| + \|f'\|_\infty \|x\|_{s,t,\infty} \right) .$$

Therefore,

$$|x_t| \le |x_s| + \frac{3}{2} k \|y\|_\beta \left(|f(0)| + \|f'\|_\infty \|x\|_{s,t,\infty} \right) \Delta^\beta ,$$

and

$$\|x\|_{s,t,\infty} \le |x_s| + \frac{3}{2} k \|y\|_\beta \left(|f(0)| + \|f'\|_\infty \|x\|_{s,t,\infty} \right) \Delta^\beta .$$

Using again (3.11) we get

$$\|x\|_{s,t,\infty} \le 2|x_s| + 2k \|y\|_\beta |f(0)| \Delta^\beta .$$

Assume also that

$$\Delta \le \left(\frac{1}{k \, |f(0)| \, \|y\|_\beta} \right)^{\frac{1}{\beta}} . \tag{3.12}$$

Then

$$\|x\|_{s,t,\infty} \leq 2\left(|x_s|+1\right).$$

Hence,

$$\sup_{0\leq r\leq t}|x_r| \leq 2\left(\sup_{0\leq r\leq s}|x_r|+1\right). \tag{3.13}$$

As before, divide the interval $[0,T]$ into $n=[T/\Delta]+1$ subintervals, and use the estimate (3.13) in every interval to obtain

$$\sup_{0\leq t\leq T}|x_t| \leq 2^n\left(|x_0|+1\right). \tag{3.14}$$

Choose

$$\Delta = \left(k\|y\|_\beta\left(\|f'\|_\infty \vee |f(0)|\right)\right)^{-\frac{1}{\beta}},$$

in such a way that (3.11) and (3.12) hold. Then, (3.14) implies

$$\sup_{0\leq t\leq T}|x_t| \leq 2^{1+kT\left[\|f'\|_\infty\vee|f(0)|\right]^{1/\beta}\|y\|_\beta^{1/\beta}}\left(|x_0|+1\right).$$

The proof of the theorem is now complete.

Consider now the following system of equations

$$x_t = x_0+\int_0^t f(x_r)dy_r,$$

$$z_t = z_0+\int_0^t g(x_r)z_r dy_r,$$

where $y:[0,\infty)\to\mathbb{R}^m$ is a Hölder continuous function of order $\beta>1/2$. $f:\mathbb{R}^d\to\mathbb{R}^{md}$ and $g:\mathbb{R}^d\to\mathbb{R}^{M^2d}$ are given functions and $x_0\in\mathbb{R}^m$, $z_0\in\mathbb{R}^M$. We make the following assumptions:

H1) f is bounded with a bounded derivative f' which is locally Hölder continuous of order $\lambda>\frac{1}{\beta}-1$.
H2) g is bounded with bounded derivative.

Theorem 3. *Assume conditions H1) and H2). Then, there is a constant k depending only on β, such that for all T,*

$$\sup_{0\leq t\leq T}|z_t| \leq 2^{1+kT\left[\|f'\|_\infty\vee\left(\|g\|_\infty+\sqrt{\|g'\|_\infty\|f\|_\infty}\right)\right]^{1/\beta}\|y\|_{0,T,\beta}^{1/\beta}}|z_0|. \tag{3.15}$$

Proof. Without loss of generality we assume that $d=m=M=1$. Set $\|y\|_\beta=\|y\|_{0,T,\beta}$. We can assume that $\|y\|_\beta>0$, otherwise the inequality is obvious. Let $\alpha<1/2$ such that $\alpha>1-\beta$.

If we choose Δ such that

$$\Delta^\beta \le \frac{1}{2k\,\|f'\|_\infty\,\|y\|_\beta},$$

by (3.6) for all s and t such that $t-s\le\Delta$ we have

$$\|x\|_{s,t,\beta} \le 2k\|y\|_\beta\,\|f\|_\infty\,. \tag{3.16}$$

On the other hand, using the fractional integration by parts formula we obtain for all $s,t\in[0,T]$,

$$\left|\int_s^t g(x_r)z_r dy_r\right| \le \int_s^t |D_{s+}^\alpha\,\left(g(x_r)z_r\right)\,D_{t-}^{1-\alpha}y_{t-}(r)|dr. \tag{3.17}$$

From (2.1) we get

$$\left|D_{s+}^\alpha\left(g(x_r)z_r\right)\right| \le k\,\left(\|g\|_\infty\,\|z\|_{s,r,\infty}(r-s)^{-\alpha} + \int_s^r \frac{|g(x_r)z_r - g(x_\theta)z_\theta|}{|r-\theta|^{\alpha+1}}d\theta\right).$$

Now if $0\le s\le r\le t\le T$, then

$$\begin{aligned}\int_s^r \frac{|g(x_r)z_r - g(x_\theta)z_\theta|}{|r-\theta|^{\alpha+1}}d\theta &\le k\,\|g\|_\infty \int_s^r \|z\|_{s,r,\beta}|r-\theta|^{\beta-\alpha-1}d\theta \\ &\quad + k\,\|g'\|_\infty \int_s^r \|z\|_{s,r,\infty}\|x\|_{s,r,\beta}|r-\theta|^{\beta-\alpha-1}d\theta \\ &\le k\left(\|g\|_\infty\,\|z\|_{s,t,\beta} + \|g'\|_\infty\,\|z\|_{s,r,\infty}\|x\|_{s,r,\beta}\right)|r-s|^{\beta-\alpha}.\end{aligned}$$

Therefore

$$\begin{aligned}\left|D_{s+}^\alpha\left(g(x_r)z_r\right)\right| &\le k(\|g\|_\infty\,\|z\|_{s,r,\infty}(r-s)^{-\alpha} \\ &\quad + \left(\|g\|_\infty\,\|z\|_{s,t,\beta} + \|g'\|_\infty\,\|z\|_{s,r,\infty}\|x\|_{s,r,\beta}\right)|r-s|^{\beta-\alpha}).\end{aligned} \tag{3.18}$$

Substituting (3.18) and (3.4) into (3.17) yields

$$\begin{aligned}\left|\int_s^t g(x_r)z_r dy_r\right| &\le k\|y\|_\beta\Bigg(\|g\|_\infty\,\|z\|_{s,t,\infty}(t-s)^\beta \\ &\quad + \left(\|g\|_\infty\,\|z\|_{s,t,\beta} + \|g'\|_\infty\,\|z\|_{s,t,\infty}\|x\|_{s,t,\beta}\right)(t-s)^{2\beta}\Bigg).\end{aligned}$$

Consequently, is $t-s\le\Delta$, applying (3.16) yields

$$\begin{aligned}\|z\|_{s,t,\beta} &\le k\|y\|_\beta\Big\{\|g\|_\infty\,\|z\|_{s,t,\infty} \\ &\quad + \left(\|g\|_\infty\,\|z\|_{s,t,\beta} + \|g'\|_\infty\,\|z\|_{s,t,\infty}\|x\|_{s,t,\beta}\right)\Delta^\beta\Big\} \\ &\le k\|y\|_\beta\Big\{\|g\|_\infty\,\|z\|_{s,t,\infty} \\ &\quad + \left(\|g\|_\infty\,\|z\|_{s,t,\beta} + \|g'\|_\infty\,\|f\|_\infty\,\|y\|_\beta\|z\|_{s,t,\infty}\right)\Delta^\beta\Big\}.\end{aligned}$$

Supposse that Δ is sufficiently small such that

$$\Delta \le \left(\frac{1}{2k\|g\|_\infty \|y\|_\beta}\right)^{\frac{1}{\beta}}. \tag{3.19}$$

Then we have

$$\|z\|_{s,t,\beta} \le 2k\|y\|_\beta \|z\|_{s,t,\infty} \left(\|g\|_\infty + \|g'\|_\infty \|f\|_\infty \|y\|_\beta \Delta^\beta\right).$$

This implies that

$$\|z\|_{s,t,\infty} \le |z_s| + k\|y\|_\beta \Delta^\beta \|z\|_{s,t,\infty} \left(\|g\|_\infty + \|g'\|_\infty \|f\|_\infty \|y\|_\beta \Delta^\beta\right).$$

If Δ satisfies

$$\|g\|_\infty \Delta^\beta + \|g'\|_\infty \|f\|_\infty \|y\|_\beta \Delta^{2\beta} \le \frac{1}{2k\|y\|_\beta} \tag{3.20}$$

then we have

$$\|z\|_{s,t,\infty} \le 2|z_s|$$

Hence,

$$\sup_{0\le r\le t} |z_r| \le 2 \sup_{0\le r\le s} |z_r|. \tag{3.21}$$

As before, divide the interval $[0,T]$ into $n = [T/\Delta] + 1$ subintervals, and use the estimate (3.21) in every interval to obtain

$$\|z\|_{0,T,\infty} \le 2^n |z_0|. \tag{3.22}$$

Notice that for (3.20) to hold it suffices that

$$\begin{aligned} \Delta^\beta \|y\|_\beta &\le \frac{\sqrt{\|g\|_\infty^2 + \frac{2}{k}\|g'\|_\infty \|f\|_\infty} - \|g\|_\infty}{2\|g'\|_\infty \|f\|_\infty} \\ &= \frac{1}{k\left(\sqrt{\|g\|_\infty^2 + \frac{2}{k}\|g'\|_\infty \|f\|_\infty} + \|g\|_\infty\right)}. \end{aligned}$$

If we choose

$$\Delta = \left[k\|y\|_\beta \max\left(\|f'\|_\infty, \|g\|_\infty + \sqrt{\|g'\|_\infty \|f\|_\infty}\right)\right]^{-\frac{1}{\beta}},$$

then (3.22) yields

$$\|z\|_{0,T,\infty} \le 2^{1+kT\left[\|f'\|_\infty \vee \left(\|g\|_\infty + \sqrt{\|g'\|_\infty \|f\|_\infty}\right)\right]^{1/\beta} \|y\|_{0,T,\beta}^{1/\beta}} |z_0|.$$

The proof is now complete.

Suppose now that we have two differential equations of the form

$$x_t = x_0 + \int_0^t f(x_s)dy_s,$$

and

$$\tilde{x}_t = \tilde{x}_0 + \int_0^t \tilde{f}(\tilde{x}_s)\tilde{y}_s\,,$$

where y and $\widetilde{y}$ are Hölder continuous functions of order $\beta > 1/2$, and f and $\widetilde{f}$ are two functions which are continuously differentiable with locally Hölder continuous derivatives of order $\lambda > \frac{1}{\beta} - 1$. Then, we have the following estimate.

Theorem 4. *Suppose in addition that f is twice continuously differentiable and f'' is bounded. Then there is a constant k such that*

$$\begin{aligned}\sup_{0\le r\le T} |x_r - \tilde{x}_r| \le{}& k2^{kD^{1/\beta}\|y\|_{0,T,\beta}^{1/\beta}T} \\ &\times\{|x_0 - \tilde{x}_0| + \|y\|_{0,T,\beta}[\|f - \tilde{f}\|_\infty + \|x\|_{0,T,\beta}\|f' - \tilde{f}'\|_\infty] \\ &+[\|\tilde{f}\|_\infty + \|\tilde{f}'\|_\infty\|x\|_{0,T,\infty}]\|y - \tilde{y}\|_{0,T,\beta}\},\end{aligned}$$

where

$$D = \|f'\|_\infty \vee \left(\|f'\|_\infty + \|f''\|_\infty(\|x\|_{0,T,\beta} + \|\tilde{x}\|_{0,T,\beta})T^\beta\right).$$

Remark 5. The above inequality is valid only when each term appeared on the right hand side is finite.

Proof. Fix $s, t \in [0, T]$. Set

$$x_t - \tilde{x}_t - (x_s - \tilde{x}_s) = I_1 + I_2 + I_3,$$

where

$$\begin{aligned}I_1 &= \int_s^t [f(x_r) - f(\tilde{x}_r)]dy_r \\ I_2 &= \int_s^t [f(\tilde{x}_r) - \tilde{f}(\tilde{x}_r)]dy_r \\ I_3 &= \int_s^t \tilde{f}(\tilde{x}_r)d[y_r - \tilde{y}_r].\end{aligned}$$

The terms I_2 and I_3 can be estimated easily. In fact, we have

$$|I_2| \le k\|y\|_\beta[\|f - \tilde{f}\|_\infty(t-s)^\beta + \|f' - \tilde{f}'\|_\infty\|\tilde{x}\|_{s,t,\beta}(t-s)^{2\beta}]$$

and

$$|I_3| \le k\|y - \tilde{y}\|_\beta[\|\tilde{f}\|_\infty(t-s)^\beta + \|\tilde{f}'\|_\infty\|\tilde{x}\|_{s,t,\beta}(t-s)^{2\beta}],$$

where $\|y\|_\beta = \|y\|_{0,T,\beta}$ and $\|y-\tilde{y}\|_\beta = \|y-\tilde{y}\|_{0,T,\beta}$. The term I_1 is a little more complicated.

$$\begin{aligned}
|I_1| &\le \int_s^t |D_{s+}^{\alpha}\left[f(x_r)-f(\tilde{x}_r)\right]||D_{t-}^{1-\alpha}y_{t-}(r)|dr \\
&\le k\int_s^t \|y\|_{s,t,\beta}(t-r)^{\alpha+\beta-1}\left[|f(x_r)-f(\tilde{x}_r)|(r-s)^{-\alpha}\right. \\
&\quad +\|f'\|_\infty\|x-\tilde{x}\|_{s,r,\beta}(r-s)^{\beta-\alpha} \\
&\quad \left.+\|f''\|_\infty\|x-\tilde{x}\|_{s,r,\infty}\left[\|x\|_{s,r,\beta}+\|\tilde{x}\|_{s,r,\beta}\right](r-s)^{\beta-\alpha}\right]dr \\
&\le k\|y\|_\beta\left\{\|f'\|_\infty\|x-\tilde{x}\|_{s,t,\infty}(t-s)^\beta+\|f'\|_\infty\|x-\tilde{x}\|_{s,t,\beta}(t-s)^{2\beta}\right. \\
&\quad \left.+\|f''\|_\infty\|x-\tilde{x}\|_{s,t,\infty}\left[\|x\|_{s,t,\beta}+\|\tilde{x}\|_{s,t,\beta}\right](t-s)^{2\beta}\right\}.
\end{aligned}$$

Therefore

$$\begin{aligned}
\|x-\tilde{x}\|_{s,t,\beta} \le k\|y\|_\beta\Big\{&\|f'\|_\infty\|x-\tilde{x}\|_{s,t,\infty}+\|f'\|_\infty\|x-\tilde{x}\|_{s,t,\beta}(t-s)^\beta \\
&+\|f''\|_\infty\|x-\tilde{x}\|_{s,t,\infty}\left[\|x\|_{s,t,\beta}+\|\tilde{x}\|_{s,t,\beta}\right](t-s)^\beta \\
&+\|f-\tilde{f}\|_\infty+\|f'-\tilde{f}'\|_\infty\|\tilde{x}\|_{s,t,\beta}(t-s)^\beta\Big\} \\
&+k\|y-\tilde{y}\|_\beta[\|\tilde{f}\|_\infty+\|\tilde{f}'\|_\infty\|\tilde{x}\|_{s,t,\beta}(t-s)^\beta].
\end{aligned}$$

Rearrange it to obtain

$$\begin{aligned}
\|x-\tilde{x}\|_{s,t,\beta} \le k(1-k\|f'\|_\infty\|y\|_\beta(t-s)^\beta)^{-1}\{&\|y\|_\beta[\|f'\|_\infty\|x-\tilde{x}\|_{s,t,\infty} \\
&+\|f''\|_\infty\|x-\tilde{x}\|_{s,t,\infty}[\|x\|_{s,t,\beta}+\|\tilde{x}\|_{s,t,\beta}](t-s)^\beta \\
&+\|f-\tilde{f}\|_\infty+\|f'-\tilde{f}'\|_\infty\|\tilde{x}\|_{s,t,\beta}(t-s)^\beta] \\
&+k\|y-\tilde{y}\|_\beta[\|\tilde{f}\|_\infty+\|\tilde{f}'\|_\infty\|\tilde{x}\|_{s,t,\beta}(t-s)^\beta]\}.
\end{aligned}$$

Set $\Delta = t-s$, and $A = k\|f'\|_\infty\|y\|_\beta$. Then

$$\begin{aligned}
\|x-\tilde{x}\|_{s,t,\infty} &\le |x_s-\tilde{x}_s|+\|x-\tilde{x}\|_{s,t,\beta}(t-s)^\beta \\
&\le |x_s-\tilde{x}_s|+k(1-A\Delta^\beta)^{-1}\Delta^\beta\{\|y\|_\beta[\|f'\|_\infty\|x-\tilde{x}\|_{s,t,\infty} \\
&\quad +\|f''\|_\infty\|x-\tilde{x}\|_{s,t,\infty}[\|x\|_{s,t,\beta}+\|\tilde{x}\|_{s,t,\beta}]\Delta^\beta \\
&\quad +\|f-\tilde{f}\|_\infty+\|f'-\tilde{f}'\|_\infty\|\tilde{x}\|_{s,t,\beta}\Delta^\beta] \\
&\quad +k\|y-\tilde{y}\|_\beta[\|\tilde{f}\|_\infty+\|\tilde{f}'\|_\infty\|\tilde{x}\|_{s,t,\beta}\Delta^\beta]\}.
\end{aligned}$$

Denote

$$B = k\|y\|_\beta\left(\|f'\|_\infty+\|f''\|_\infty(\|x\|_{0,T,\beta}+\|\tilde{x}\|_{0,T,\beta})T^\beta\right).$$

Then

$$\begin{aligned}\|x-\tilde{x}\|_{s,t,\infty} \leq & \left(1-(1-A\Delta^\beta)^{-1}\Delta^\beta B\right)^{-1} \\ & \times\{|x_s-\tilde{x}_s|+k(1-A\Delta^\beta)^{-1}\Delta^\beta \\ & \times[\|y\|_\beta[\|f-\tilde{f}\|_\infty+\|f'-\tilde{f}'\|_\infty\|\tilde{x}\|_{s,t,\beta}\Delta^\beta] \\ & +\|y-\tilde{y}\|_\beta[\|\tilde{f}\|_\infty+\|\tilde{f}'\|_\infty\|\tilde{x}\|_{s,t,\beta}\Delta^\beta]]\}.\end{aligned}$$

Let Δ satisfy

$$A\Delta^\beta \leq 1/3\,, \quad B\Delta^\beta \leq 1/3$$

Namely, we take

$$\Delta=\left(\frac{1}{3\,(A\vee B)}\right)^{1/\beta}.$$

Then

$$\|x-\tilde{x}\|_{s,t,\infty} \leq 2\left[|x_s-\tilde{x}_s|+C\Delta^\beta\right],$$

where

$$\begin{aligned}C=&\frac{3}{2}k[\|y\|_\beta[\|f-\tilde{f}\|_\infty+\|f'-\tilde{f}'\|_\infty\|\tilde{x}\|_{s,t,\beta}\Delta^\beta] \\ & +\|y-\tilde{y}\|_\beta[\|\tilde{f}\|_\infty+\|\tilde{f}'\|_\infty\|\tilde{x}\|_{s,t,\beta}\Delta^\beta]].\end{aligned}$$

Applying the above estimate recursively we obtain

$$\sup_{0\leq r\leq T}|x_r-\tilde{x}_r| \leq 2^n\left[|x_0-\tilde{x}_0|+C\Delta^\beta\right],$$

where $n=[T/\Delta]+1$. Or we have

$$\begin{aligned}\sup_{0\leq r\leq T}|x_r-\tilde{x}_r| \leq & k2^{k(\|f'\|_\infty\vee(\|f'\|_\infty+\|f''\|_\infty(\|x\|_{0,T,\beta}+\|\tilde{x}\|_{0,T,\beta})T^\beta))^{1/\beta}\|y\|_{0,T,\beta}^{1/\beta}T} \\ & \times\{|x_0-\tilde{x}_0|+\|y\|_{0,T,\beta}[\|f-\tilde{f}\|_\infty+\|\tilde{x}\|_{0,T,\beta}\|f'-\tilde{f}'\|_\infty] \\ & +[\|\tilde{f}\|_\infty+\|\tilde{f}'\|_\infty\|\tilde{x}\|_{0,T,\infty}]\|y-\tilde{y}\|_{0,T,\beta}\}.\end{aligned}$$

4 Stochastic Differential Equations Driven by a fBm

Let $B=\{B_t, t\geq 0\}$ be an m-dimensional fractional Brownian motion (fBm) with Hurst parameter $H>1/2$. That is, B is a Gaussian centered process with the covariance function $E(B_t^iB_s^j)=R_H(t,s)\delta_{ij}$, where

$$R_H(t,s)=\frac{1}{2}\left(t^{2H}+s^{2H}-|t-s|^{2H}\right).$$

Consider the stochastic differential equation on $\mathbb{R}^d$

$$X_t=X_0+\int_0^t\sigma(X_s)dB_s, \qquad (4.1)$$

where X_0 is a fixed d-dimensional random variable and the stochastic integral is is a path-wise Riemann-Stieltjes integral. ([1]). This equation has a unique solution (see [2,3]) provided σ is continuously differentiable, and σ' is bounded and Hölder continuous of order $\lambda > \frac{1}{H} - 1$.

Then, using the estimate (3.2) in Theorem 2 we obtain the following estimate for the solution of Equation (4.1), if we choose $\beta \in \left(\frac{1}{2}, H\right)$. Notice that $\frac{1}{\beta} < 2$.

$$\sup_{0\le t\le T} |X_t| \le 2^{1+kT\left(\|\sigma'\|_\infty \vee |\sigma(0)|\right)\|B\|_{0,T,\beta}^{1/\beta}} \left(|X_0| + 1\right). \tag{4.2}$$

If σ is bounded and $\|\sigma'\| \neq 0$ we can make use of the estimate (3.3) and we obtain

$$\sup_{0\le t\le T} |X_t| \le |X_0| + k\|\sigma\|_\infty \left(T^\beta \|B\|_{0,T,\beta}^{\frac{1}{\beta}} \vee T\|\sigma'\|_\infty^{\frac{1-\beta}{\beta}} \|B\|_{0,T,\beta}^{\frac{1}{\beta}} \right). \tag{4.3}$$

These estimates improve those obtained by Nualart and Răşcanu in [3] based on a suitable version of Gronwall's lemma. The estimates (4.2) and (4.3) allow us to establish the following integrability properties for the solution of Equation (4.1).

Theorem 6. *Consider the stochastic differential equation (4.1), and assume that* $E(|X_0|^p) < \infty$ *for all* $p \ge 2$. *If* σ' *is bounded and Hölder continuous of order* $\lambda > \frac{1}{H} - 1$, *then*

$$E\left(\sup_{0\le t\le T} |X_t|^p \right) < \infty \tag{4.4}$$

for all $p \ge 2$. *If furthermore* σ *is bounded and* $E\left(\exp(\lambda |X_0|^\gamma)\right) < \infty$ *for any* $\lambda > 0$ *and* $\gamma < 2H$, *then*

$$E\left(\exp \lambda \left(\sup_{0\le t\le T} |X_t|^\gamma \right)\right) < \infty \tag{4.5}$$

for any $\lambda > 0$ *and* $\gamma < 2H$.

In [4] Nualart and Saussereau have proved that the random variable X_t belongs locally to the space $\mathbb{D}^\infty$ if the function σ is infinitely differentiable and bounded together with all its partial derivatives. As a consequence, they have derived the absolute continuity of the law of X_t for any $t > 0$ assuming that the initial condition is constant and the vector space spanned by $\{(\sigma^{ij}(x_0))_{1\le i\le d}, 1 \le j \le m\}$ is $\mathbb{R}^d$.

Applying Theorem 3.2 we can show that the derivatives of X_t possess moments of all orders, and we can then derive the C^∞ property of the density. Define the matrix

$$\alpha(x) = \left(\sum_{l=1}^m \sigma^{il}(x)\sigma^{jl}(x) \right)_{1\le i,j\le d}.$$

Theorem 7. *Consider the stochastic differential equation (4.1), with constant initial condition* x_0. *Suppose that* $\sigma(x)$ *is bounded infinitely differentiable with bounded derivatives of all orders, and* $\alpha(x)$ *is uniformly elliptic. Then, for any* $t > 0$ *the probability law of* X_t *has an* C^∞ *density.*

Proof. Let us first show that X_t belongs to the space $\mathbb{D}^\infty$. From Equation (34) of [4] we have

$$D_r^j X_t^i = \sigma^{ij}(X_r) + \sum_{k=1}^d \int_0^t \sum_{l=1}^m \partial_k \sigma^{il}(X_u) D_r^j X_u^k dB_u^l. \tag{4.6}$$

As a consequence, (3.15) applied to the system formed by the equations (4.1) and (4.6) yields

$$\left| D_r^j X_t^i \right| \le 2^{1+kT\left[\|\sigma\|_\infty \vee \left(\|\sigma'\|_\infty + \sqrt{\|\sigma''\|_\infty \|\sigma\|_\infty}\right)\right]^{\frac{1}{\beta}} \|B\|_{0,T,\beta}^{1/\beta}} \|\sigma\|_\infty.$$

This implies that for all $p \ge 2$

$$E\left(\left| \sum_{j=1}^m \int_0^t \int_0^t D_s^j X_t^i D_r^j X_t^i |r-s|^{2H-2} ds dr \right|^p \right) < \infty,$$

and the random variable X_t^i belongs to the Sobolev space $\mathbb{D}^{1,p}$ for all $p \ge 2$. In a similar way, writing down the linear equations satisfied by the iterated derivatives, one can show that X_t^i belongs to the Sobolev space $\mathbb{D}^{k,p}$ for all $p \ge 2$ and $k \ge 2$.

In order to show the nongeneracy of the density we use the notation of [4] and follow the idea of [7]. By Itô's formula we have

$$\begin{aligned} D_r^j X_t^i D_{r'}^j X_t^{i'} &= \sigma^{ij}(X_r)\sigma^{i'j}(X_{r'}) + \sum_{k=1}^d \sum_{l=1}^m \int_0^t \partial_k \sigma^{i'l}(X_u) D_r^j X_u^i D_{r'}^j X_u^k dB_u^l \\ &+ \sum_{k=1}^d \sum_{l=1}^m \int_0^t \partial_k \sigma^{il}(X_u) D_{r'}^j X_u^{i'} D_r^j X_u^k dB_u^l. \end{aligned}$$

Denote

$$\beta_l(X_u) = \left(\partial_k \sigma^{i'l}(X_u) \right)_{1 \le i', k \le d}$$

$$\Gamma_t = \left(\sum_{j=1}^m \int_0^t \int_0^t |r - r'^{2H-2} D_r^j X_t^i D_{r'}^j X_t^{i'} dr dr' \right)_{1 \le i, i' \le d}.$$

Then $H(2H-1)\Gamma_t$ is the Malliavin covariance matrix of the random vector X_t, and we need to show that Γ_t^{-1} is in L_p for any $p \ge 1$ and for all $t > 0$. We have

$$\Gamma_t = \alpha_0 + \sum_{l=0}^{m} \int_0^t \left(\beta_l(X_u)\Gamma_u + \Gamma_u \beta_l^T(X_u)\right) dB_u^l,$$

where

$$\alpha_0 = \sum_{j=1}^{m} \int_0^t \int_0^t |r - r'|^{2H-2} \sigma_{ij}(X_r)\sigma_{i'j}(X_{r'}) dr dr'.$$

By using Itô formula again we have

$$\Gamma_t^{-1} = \alpha_0^{-1} - \sum_{l=0}^{m} \int_0^t \left(\Gamma_u^{-1}\beta_l(X_u) + \beta_l^T(X_u)\Gamma_u^{-1}\right) dB_u^l. \tag{4.7}$$

By the estimate (3.15) applied to the equations (4.1) and (4.7), we see that Γ_t^{-1} is in L_p for any $p \geq 1$. This proves the theorem (see [5]).

References

1. Hu, Y. Integral transformations and anticipative calculus for fractional Brownian motions. Mem. Amer. Math. Soc. 175 (2005), no. 825.
2. Lyons, T. Differential equations driven by rough signals (I): An extension of an inequality of L. C. Young. *Mathematical Research Letters* **1** (1994) 451–464.
3. Nualart, D., Răşcanu, A. Differential equations driven by fractional Brownian motion. *Collect. Math.* **53** (2002) 55–81.
4. Nualart, D., Saussereau, B. Malliavin calculus for stochastic differential equations driven by a fractional Brownian motion. Preprint.
5. Nualart, D. The Malliavin calculus and related topics. Probability and its Applications (New York). Springer-Verlag, New York, 2006.
6. Samko S. G., Kilbas A. A. and Marichev O. I. *Fractional Integrals and Derivatives. Theory and Applications.* Gordon and Breach, 1993.
7. Stroock, D. Some applications of stochastic calculus to partial differential equations. Eleventh Saint Flour probability summer school—1981 (Saint Flour, 1981), 267–382, Lecture Notes in Math., 976, Springer, Berlin, 1983.
8. Young, L. C. An inequality of the Hölder type connected with Stieltjes integration. *Acta Math.* **67** (1936) 251–282.
9. Zähle, M. Integration with respect to fractal functions and stochastic calculus. I. *Prob. Theory Relat. Fields* **111** (1998) 333–374.

On Asymptotics of Banach Space-valued Itô Functionals of Brownian Rough Paths

Yuzuru Inahama[1] and Hiroshi Kawabi[2]

[1] Department of Mathematics, Graduate School of Science and Engineering, Tokyo Institute of Technology
2-12-1, Oh-okayama, Meguro-ku, Tokyo 152-8551, Japan
inahama@math.titech.ac.jp

[2] Department of Mathematics, Faculty of Science, Okayama University
3-1-1, Tsushima-Naka, Okayama 700-8530, Japan
kawabi@math.okayama-u.ac.jp

Dedicated to Professor Kiyosi Itô on the occasion of his 90th birthday

Abstract: In this paper, we discuss asymptotics for certain Banach space-valued Itô functionals of Brownian rough paths based on the results of Inahama-Kawabi [10] and Inahama [9]. Our main tool is the Banach space-valued rough path theory of T. Lyons. As examples, we deal with heat processes on loop spaces and solutions of the stochastic differential equations (SDEs) on M-type 2 Banach spaces.

1 Introduction

Let (X, H, μ) be an abstract Wiener space, i.e., X is a real separable Banach space, H is the Cameron-Martin space and μ is the Wiener measure on X. Let Y be another real separable Banach space and $w := (w_t)_{0 \le t \le 1}$ be the X-valued Brownian motion on a completed probability space $(\Omega, \mathcal{F}, \mathbb{P})$ associated with μ. We denote by $L(X, Y)$ the space of bounded linear operators from X to Y. In this paper, we consider a class of Y-valued Wiener functionals $X^\varepsilon := (X_t^\varepsilon)_{0 \le t \le 1}$ defined through the following formal Stratonovich type stochastic differential equation (SDE) on Y:

$$dX_t^\varepsilon = \sigma(X_t^\varepsilon) \circ \varepsilon dw_t + b(\varepsilon, X_t^\varepsilon)dt, \qquad X_0^\varepsilon = 0, \tag{1.1}$$

where the coefficients σ and b take values in $L(X, Y)$ and Y, respectively, with a suitable regularity condition. Here, we note that the equation (1.1) cannot be discussed through the usual theory of SDEs when X and Y are infinite dimensional Banach spaces, because the diffusion coefficient σ takes values in $L(X, Y)$. See Section 3 for the precise formulation of our Wiener functionals X^ε. The main objective of this paper is to discuss the Freidlin-Wentzell type large deviation principle for X^ε and the asymptotic behavior of the Laplace type functional integral $\mathbb{E}\big[\exp(-F(X^\varepsilon)/\varepsilon^2)\big]$ as $\varepsilon \searrow 0$, which is called Laplace's method. For a class of continuous loop space-valued diffusion processes called heat processes, these asymptotics were studied in earlier papers Inahama-Kawabi [10] and Inahama [9], respectively. In this paper, we interpret our Wiener functionals X^ε as Itô functionals of Brownian rough paths, and show that these asymptotics hold for wider classes of (infinite dimensional) Banach space-valued Wiener functionals by using the fact that the rough path theory of T. Lyons works on any Banach space.

To establish the large deviation principle for X^ε, due to the lack of the continuity of the Itô map $w \mapsto X^\varepsilon$, Schilder's theorem and the contraction principle may not be used directly. To overcome this difficulty, Freidlin and Wentzell developed refined techniques involving the exponential continuity (see Deuschel-Stroock [7]). On the other hand, recently, Ledoux-Qian-Zhang [16] gave a new proof for the large deviation principle by using the rough path theory. The basic idea in [16] is summarized as follows: First, they show that the laws of Brownian rough paths satisfy the large deviation principle. Next, they use the contraction principle since the Itô map is continuous in the framework of the rough path theory. Hence their approach seems straightforward and much simpler than conventional proofs. In [10], it is shown that their approach is also applicable to a class of stochastic processes on infinite dimensional spaces.

As an application of the large deviation principle, Laplace's method is investigated in many research fields of probability theory and mathematical physics. In finite dimensional settings, Schilder [19] initiated the study in the case of $X^\varepsilon = \varepsilon w$ and Azencott [2] and Ben Arous [3] continued this study for (1.1). (For results concerning with more general Wiener functionals, see Kusuoka-Stroock [13, 14] and Takanobu-Watanabe [20].) In these papers, the stochastic Taylor expansion for X^ε plays an important role. The problem of [19] is rather easier because each term of the expansion is continuous, which comes from the fact that X^ε is nothing but the scaled Brownian motion. So, there is no ambiguity in the formulation. However, in general, it is very complicated to give a precise interpretation on each term of this expansion through conventional stochastic analysis because the Itô map is not a continuous Wiener functional. On the other hand, Aida [1] proposed a new approach with the rough path theory for this problem recently. In [1], he obtained the stochastic Taylor expansion with respect to the topology of the space of geometric rough paths for finite dimensional cases. Since the Itô map is continuous in the rough path sense, each term of the expansion is continuous. Hence we do not need to face the difficulty mentioned above. Based on the idea of [1], the first author [9] showed the stochastic Taylor expansion in an infinite dimensional setting.

The organization of this paper is as follows: In Section 2, we give a simple review of the rough path theory and review the Cameron-Martin theorem and

Fernique's theorem in the framework of Brownian rough paths. In Section 3, we give a framework and state our results. In Section 4, we give an outline of the proof of our results based on [9, 10]. Finally, in Section 5, we give two examples to which our results are applicable. The first example is a class of heat processes described above and the second one comes from the SDE theory on M-type 2 Banach spaces.

2 Preliminaries from the Rough Path Theory

In this section we set notations and review some basic results of the rough path theory.

First we recall the definition of spaces of geometric rough paths. Let B be a real separable Banach space. The algebraic tensor product is denoted by $B \otimes_{\mathrm{a}} B$. We consider a norm $|\cdot|$ on $B \otimes_{\mathrm{a}} B$ such that $|x \otimes y| \leq |x|_B \cdot |y|_B$ holds for all $x, y \in B$. We denote by $B \otimes B$ the completion of $B \otimes_{\mathrm{a}} B$ by this norm. We often suppress the subscripts of Banach norms when there is no fear of confusion.

Let $2 < p < 3$ be the roughness and fix it throughout this paper. A continuous map $\overline{x} = (1, \overline{x}_1, \overline{x}_2)$ from the simplex $\Delta := \{(s,t) |\ 0 \leq s \leq t \leq 1\}$ to the truncated tensor algebra $T^{(2)}(B) := \mathbb{R} \oplus B \oplus (B \otimes B)$ is said to be a B-valued rough path of roughness p if it satisfies that, for every $s \leq u \leq t$,

$$\overline{x}_1(s,t) = \overline{x}_1(s,u) + \overline{x}_1(u,t),$$
$$\overline{x}_2(s,t) = \overline{x}_2(s,u) + \overline{x}_2(u,t) + \overline{x}_1(s,u) \otimes \overline{x}_1(u,t)$$

and

$$\|\overline{x}_j\|_{p/j} := \left(\sup_D \sum_{l=1}^{n} |\overline{x}_j(t_{l-1}, t_l)|^{p/j} \right)^{j/p} < \infty \qquad \text{for } j = 1, 2,$$

where $D = \{0 = t_0 < t_1 < \cdots < t_n = 1\}$ runs over all finite partition of $[0, 1]$. For two rough paths $\overline{x}$ and $\overline{y}$, p-variation distance is defined by

$$d_p(\overline{x}, \overline{y}) = \|\overline{x}_1 - \overline{y}_1\|_p + \|\overline{x}_2 - \overline{y}_2\|_{p/2}.$$

Let $P(B) := \{x \in C([0,1], B) \mid x_0 = 0\}$. For $x \in P(B)$, we denote by $\|x\|_{P(B)} := \sup_{0 \leq t \leq 1} |x_t|_B$ and sometimes write $x(t)$ for x_t. Moreover we often write $\overline{x}_1(\cdot)$ for $\overline{x}_1(0, \cdot) \in P(B)$ for simplicity. We denote by $\mathrm{BV}(B) := \{\gamma \in P(B) \mid \|\gamma\|_1 < \infty\}$, where $\|\gamma\|_1$ denotes the total variation norm of γ. For $\gamma \in \mathrm{BV}(B)$, we set $\overline{\gamma} = (1, \overline{\gamma}_1, \overline{\gamma}_2)$ by

$$\overline{\gamma}_1(s,t) := \gamma_t - \gamma_s, \quad \overline{\gamma}_2(s,t) := \int_s^t (\gamma_u - \gamma_s) \otimes d\gamma_u, \quad 0 \leq s \leq t \leq 1,$$

where the right-hand side of $\overline{\gamma}_2$ is the Riemann-Stieltjes integral. A rough path obtained in this way is called the smooth rough path lying above γ.

A rough path obtained as the d_p-limit of a sequence of smooth rough paths is called a geometric rough path and the set of all the geometric rough paths is denoted by $G\Omega_p(B)$. It is well-known that $G\Omega_p(B)$ is a complete separable metric space.

We set

$$\mathcal{H}(B) := \Big\{ y \in P(B) \mid y_t = \int_0^t y'_s ds \text{ with } \|y\|^2_{\mathcal{H}(B)} := \int_0^1 |y'_t|^2_B dt < \infty \Big\}.$$

Clearly, there are natural continuous injections $\mathcal{H}(B) \hookrightarrow \mathrm{BV}(B) \hookrightarrow G\Omega_p(B)$. Note that $\mathcal{H}(B)$ is dense in $G\Omega_p(B)$ and has a natural Hilbert structure in the case when B is a Hilbert space.

Next we introduce Brownian rough paths on an abstract Wiener space (X, H, μ). Let $w = (w_t)_{t \geq 0}$ be the X-valued Brownian motion introduced in the previous section. For $\varepsilon > 0$, the law of εw on $P(X)$ is denoted by $\mathbb{P}'_\varepsilon$. Then $(P(X), \mathcal{H}(H), \mathbb{P}'_1)$ is also an abstract Wiener space. We write $\mathcal{H} := \mathcal{H}(H)$ for simplicity. When $|\cdot|_{X \otimes X}$ and μ satisfy the *exactness condition* (see Definition 1 in Ledoux-Lyons-Qian [15]), the Brownian rough path exists (see Theorem 3 in [15]). Let $\overline{w} = (1, \overline{w}_1, \overline{w}_2)$ be the Brownian rough path. It is the $\mathbb{P}$-almost sure limit of the $w(m)$ as $m \to \infty$ in $G\Omega_p(X)$ with respect to d_p-topology, where $w(m)$ is the m-th dyadic polygonal approximation of w. Note that $\overline{w}_1(s,t) = w_t - w_s$ for $\mathbb{P}$-almost surely. We denote by $\mathbb{P}_\varepsilon, \varepsilon > 0$, the law of the scaled Brownian rough path $\overline{\varepsilon w} = (1, \varepsilon \overline{w}_1, \varepsilon^2 \overline{w}_2)$.

Now we present a theorem of Fernique for Brownian rough paths. We set $\xi(\overline{x}) := \|\overline{x}_1\|_p + \|\overline{x}_2\|^{1/2}_{p/2}$. The following theorem is taken from Theorem 2.2 in [9].

Theorem 2.1. *There exists a positive constant β such that*

$$\mathbb{E}\big[\exp(\beta \xi^2)\big] = \int_{G\Omega_p(X)} \exp\Big(\beta \xi(\overline{w})^2\Big) \mathbb{P}_1(d\overline{w}) < \infty.$$

Finally, we give a theorem for absolute continuity of the laws of shifted Brownian rough paths. It is similar to the well-known Cameron-Martin theorem. For $\overline{x} \in G\Omega_p(X)$ and $\gamma \in \mathrm{BV}(X)$, we define the shifted rough path $\overline{x + \gamma} \in G\Omega_p(X)$ by

$$\overline{(x+\gamma)}_1(s,t) = \overline{x}_1(s,t) + \gamma_t - \gamma_s,$$
$$\overline{(x+\gamma)}_2(s,t) = \overline{x}_2(s,t) + \int_s^t \overline{x}_1(s,u) \otimes d\gamma_u + \int_s^t (\gamma_u - \gamma_s) \otimes \overline{x}_1(s,du) + \overline{\gamma}_2(s,t).$$

Here the second and the third terms on the right-hand side are Young integrals. It is well-known that the map $(\overline{x}, \gamma) \mapsto \overline{x \pm \gamma}$ is continuous from $G\Omega_p(X) \times \mathrm{BV}(X)$ to $G\Omega_p(X)$ (see Theorem 3.3.2 in [17]). The following theorem is taken from Lemma 2.3 in [9].

Theorem 2.2. *Let $\varepsilon > 0$ and $h \in \mathcal{H}$. Then for every bounded measurable function F on $G\Omega_p(X)$, it holds that*

$$\int_{G\Omega_p(X)} F(\overline{w+h})\mathbb{P}_\varepsilon(d\overline{w})$$
$$= \int_{G\Omega_p(X)} F(\overline{w}) \exp\left(\frac{1}{\varepsilon^2}\int_0^1 h'(t)d\overline{w}_1(t) - \frac{1}{2\varepsilon^2}\|h\|_{\mathcal{H}}^2\right)\mathbb{P}_\varepsilon(d\overline{w}),$$

where $\int_0^1 h'(t)d\overline{w}_1(t)$ is the stochastic integral with respect to the scaled Brownian motion $(\overline{w}_1(0,t))_{0\le t\le 1}$ definedon the probability space $(G\Omega_p(X),\mathbb{P}_\varepsilon)$. (Hereafter we denote it by $[h](\overline{w})$ for simplicity.)

3 Framework and Results

In this section, we set notations, introduce our Wiener functionals through the Itô map in the rough path sense and state our results. From now on, we only consider the projective norm on the tensor product of any pair of Banach spaces, and we always assume the *exactness condition* for $|\cdot|_{X\otimes X}$ and μ to treat Brownian rough paths.

First, we set notations for coefficients. Let $\sigma \in C_b^4(Y, L(X,Y))$ and $b_1, b_2 \in C_b^4(Y,Y)$. We set $\tilde{X} := X \oplus \mathbb{R}^2$ and define $\tilde{\sigma} \in C_b^4(Y, L(\tilde{X}, Y))$ by

$$\tilde{\sigma}(y)\big[(x,u)\big]_{\tilde{X}} := \sigma(y)x + b_1(y)u_1 + b_2(y)u_2, \quad y \in Y, x \in X, u = (u_1,u_2) \in \mathbb{R}^2.$$

Next, we consider the following differential equation in the rough path sense:

$$dy_t = \tilde{\sigma}(y_t)d\tilde{x}_t, \qquad y_0 = 0. \tag{3.1}$$

Then for any $\overline{\tilde{x}} \in G\Omega_p(\tilde{X})$, there exists a unique solution $\overline{z} \in G\Omega_p(\tilde{X} \oplus Y)$ in the rough path sense. Note that the natural projection of $\overline{z}$ onto the first component is $\overline{\tilde{x}}$. Projection of $\overline{z}$ onto the second component is denoted by $\overline{y} \in G\Omega_p(Y)$. We write $\overline{y} = \Phi(\overline{\tilde{x}})$ and call it a (unique) solution of (3.1). The map $\Phi : G\Omega_p(\tilde{X}) \to G\Omega_p(Y)$ is called the Itô map and is locally Lipschitz continuous in the sense of Theorem 6.2.2 in [17]. If $\tilde{x}_t = (\gamma_t, \lambda_t^{(1)}, \lambda_t^{(2)})$ is a $\tilde{X}$-valued continuous path of finite variation, the map $t \mapsto \Phi(\overline{\tilde{x}})_1(0,t)$ is the solution of

$$dy_t = \sigma(y_t)d\gamma_t + b_1(y_t)d\lambda_t^{(1)} + b_2(y_t)d\lambda_t^{(2)}, \qquad y_0 = 0$$

in the usual sense and $\overline{z}$ is the smooth rough path lying above $\big(\tilde{x}, \Phi(\overline{\tilde{x}})_1(0,\cdot)\big)$.

For $\lambda = (\lambda^{(1)}, \lambda^{(2)}) \in \mathrm{BV}(\mathbb{R}^2)$ and $\overline{x} \in G\Omega_p(X)$, we set $\iota(\overline{x},\lambda) \in G\Omega_p(\tilde{X})$ by $\iota(\overline{x},\lambda)_1(s,t) = \big(\overline{x}_1(s,t), \lambda_t - \lambda_s\big)$ and

$$\iota(\overline{x},\lambda)_2(s,t) = \Bigg(\overline{x}_2(s,t), \int_s^t \overline{x}_1(s,u)\otimes d\lambda_u,$$
$$\int_s^t (\lambda_u - \lambda_s)\otimes \overline{x}_1(s,du), \int_s^t (\lambda_u-\lambda_s)\otimes d\lambda_u\Bigg).$$

Here the second and the third component are Young integrals. If $\overline{h}$ is a smooth rough path lying above $h \in \mathrm{BV}(X)$, then $\iota(\overline{h}, \lambda)$ is a smooth rough path lying above $(h, \lambda) \in \mathrm{BV}(\tilde{X})$. Note that the map $\iota : G\Omega_p(X) \times \mathrm{BV}(\mathbb{R}^2) \to G\Omega_p(\tilde{X})$ is continuous. Here we also regard the Itô map defined above as a map from $\mathcal{H}(X)$ to $\mathcal{H}(Y)$. We define $\Psi_\varepsilon : \mathcal{H}(X) \to \mathcal{H}(Y)$ by $\Psi_\varepsilon(h)_t := \Phi\big(\iota(\overline{h}, \lambda^\varepsilon)\big)_1(0,t)$ for $0 \le t \le 1$. That is, $y := \Psi_\varepsilon(h)$ is the unique solution of

$$dy_t = \sigma(y_t)dh_t + b_1(y_t)\varepsilon^2 dt + b_2(y_t)dt, \qquad y_0 = 0. \tag{3.2}$$

For the X-valued Brownian motion w, let $\overline{w}$ be the Brownian rough path over X. For $\varepsilon \ge 0$, we define a Wiener functional $X^\varepsilon \in P(Y)$ by

$$X_t^\varepsilon := \Phi\big(\iota(\varepsilon\overline{w}, \lambda^\varepsilon)\big)_1(0,t), \qquad 0 \le t \le 1.$$

We investigate the asymptotic behavior of the law of X^ε as $\varepsilon \searrow 0$. First, we state a large deviation principle which is essentially shown in Theorem 4.9 of Inahama-Kawabi [10].

Theorem 3.1. *For $\varepsilon > 0$, we denote by $\mathcal{V}_\varepsilon$ the law of the process X^ε. Then, $\{\mathcal{V}_\varepsilon\}_{\varepsilon>0}$ satisfies a large deviation principle as $\varepsilon \searrow 0$ with the good rate function I, where*

$$I(\phi) = \begin{cases} \frac{1}{2}\inf\{\|\gamma\|_{\mathcal{H}}^2 \mid \phi = \Psi_0(\gamma)\}, & \text{if } \phi = \Psi_0(\gamma) \text{ for some } \gamma \in \mathcal{H}, \\ \infty, & \text{otherwise.} \end{cases}$$

More precisely, for any measurable set $K \subset P(Y)$, it holds that

$$-\inf_{\phi \in K^\circ} I(\phi) \le \liminf_{\varepsilon \searrow 0} \varepsilon^2 \log \mathcal{V}_\varepsilon(K) \le \limsup_{\varepsilon \searrow 0} \varepsilon^2 \log \mathcal{V}_\varepsilon(K) \le -\inf_{\phi \in \overline{K}} I(\phi).$$

As a consequence of Theorem 3.1., we have the following asymptotics for every bounded continuous function F on $P(Y)$:

$$\lim_{\varepsilon \searrow 0} \varepsilon^2 \log \mathbb{E}\big[\exp\big(-F(X^\varepsilon)/\varepsilon^2\big)\big] = -\inf\big\{F(\phi) + I(\phi) \mid \phi \in P(Y)\big\}.$$

This is Varadhan's integral lemma. See [7] for example. Our next concern is to investigate the exact asymptotics of the integral on the left-hand side of above quality, i.e., to find the asymptotics behavior of $\mathbb{E}\big[\exp\big(-F(X^\varepsilon)/\varepsilon^2\big)\big]$ as $\varepsilon \searrow 0$.

In this paper, we impose the following assumptions on the function F. In what follows, we especially denote by D the Fréchet derivatives on $\mathcal{H}(X)$ and $P(Y)$.

(F1): F is a real-valued bounded continuous function defined on $P(Y)$.

(F2): The function $F \circ \Psi_0 + \|\cdot\|_{\mathcal{H}}^2/2$ defined on $\mathcal{H}$ attains its minimum 0 at a unique point $\gamma_0 \in \mathcal{H}$. For this γ_0, we write $\phi_0 := \Psi_0(\gamma_0)$.

(F3): F is three times Fréchet differentiable on a neighborhood $B(\phi_0)$ of ϕ_0, and D^iF, $i = 1, 2, 3$, are bounded on $B(\phi_0) \subset P(Y)$.

(F4): We consider the Hessian $A := D^2(F\circ\Psi_0)(\gamma_0)|_{\mathcal{H}\times\mathcal{H}}$ at the point $\gamma_0 \in \mathcal{H}$. As a bounded self-adjoint operator on $\mathcal{H}$, the operator A is strictly larger than $-\mathrm{Id}_{\mathcal{H}}$ in the form sense. (By the min-max principle, it is equivalent to assume that all eigenvalues of A are strictly larger than -1.)

Now we are in a position to state our main result which is essentially due to Inahama [9]. The explicit value of α_0 will be given later (Theorem 4.9.) since we need to introduce a few more notations which we cannot introduce briefly.

Theorem 3.2. *Let X^ε be as above and assume* **(F1)**, **(F2)**, **(F3)** *and* **(F4)**. *Then there exists a positive constant α_0 such that*

$$\lim_{\varepsilon\searrow 0}\mathbb{E}\big[\exp\big(-F(X^\varepsilon)/\varepsilon^2\big)\big] = \alpha_0.$$

Remark 3.3. As a continuation of this paper, we have already established the following asymptotic expansion formula:

$$\mathbb{E}\big[\exp\big(-F(X^\varepsilon)/\varepsilon^2\big)\big] = \alpha_0 + \alpha_1\varepsilon + \cdots + \alpha_n\varepsilon^n + O(\varepsilon^{n+1}), \quad n\in\mathbb{N}. \tag{3.3}$$

The reader is referred to Inahama-Kawabi [11] for details.

4 Proof of Results

In this section, we show Theorems 3.1. and 3.2. Since the Itô map $\Phi : G\Omega_p(\tilde{X}) \to G\Omega_p(Y)$ is continuous, Theorem 3.1. is easily obtained by combining the contraction principle with the following Schilder type large deviations for the scaled Brownian rough path $\overline{\varepsilon w}$. The following result is taken from Theorem 3.2 in [10].

Theorem 4.1. *For $\varepsilon > 0$, we denote by $\mathbb{P}_\varepsilon$ the law of the scaled Brownian rough path $\overline{\varepsilon w}$ on $G\Omega_p(X)$. Then, $\{\mathbb{P}_\varepsilon\}_{\varepsilon>0}$ satisfies a large deviation principle as $\varepsilon\searrow 0$ with the good rate function I_0, where*

$$I_0(\overline{x}) = \begin{cases} \frac{1}{2}\|h\|_{\mathcal{H}}^2, & \text{if } \overline{x} = \overline{h} \text{ for some } h\in\mathcal{H},\\ \infty, & \text{otherwise}.\end{cases}$$

More precisely, for any measurable set $K \subset G\Omega_p(X)$, it holds that

$$\begin{aligned} -\inf_{\overline{x}\in K^\circ} I_0(\overline{x}) &\le \liminf_{\varepsilon\searrow 0}\varepsilon^2\log\mathbb{P}_\varepsilon(K)\\ &\le \limsup_{\varepsilon\searrow 0}\varepsilon^2\log\mathbb{P}_\varepsilon(K) \le -\inf_{\overline{x}\in\overline{K}} I_0(\overline{x}).\end{aligned}$$

In the sequel, we give an outline of the proof of Theorem 3.2. based on the arguments in [9]. We divide into several subsections.

4.1 Stochastic Taylor Expansion in the Sense of Rough Paths

In this subsection, we introduce the stochastic Taylor expansion for the differential equation (3.2) in the sense of rough paths. We remark that it is deterministic in this case. Hence the term *"stochastic Taylor expansion"* may not be appropriate anymore. In the sequel, we denote by ∇ the Fréchet derivative on Y.

Let $\gamma \in \mathcal{H}(X)$ and $\phi := \Psi_0(\gamma)$. For each $h \in \mathcal{H}(X)$, we define $\chi_t = \chi(h)_t$ and $\psi_t = \psi(h,h)_t$ by

$$d\chi_t - (\nabla\sigma)(\phi_t)[\chi_t, d\gamma_t] - (\nabla b_2)(\phi_t)[\chi_t]dt = \sigma(\phi_t)dh_t, \qquad \chi_0 = 0, \tag{4.1}$$

and

$$\begin{aligned} d\psi_t - (\nabla\sigma)(\phi_t)[\psi_t, d\gamma_t] - (\nabla b_2)(\phi_t)[\psi_t]dt = 2(\nabla\sigma)(\phi_t)[\chi_t, dh_t] \\ +(\nabla^2\sigma)(\phi_t)[\chi_t, \chi_t, d\gamma_t] + (\nabla^2 b_2)(\phi_t)[\chi_t, \chi_t]dt, \ \psi_0 = 0, \end{aligned} \tag{4.2}$$

where $\nabla^i\sigma : Y \to L^i(Y,\ldots,Y; L(X,Y))$, $\nabla^i b_2 : Y \to L^i(Y,\ldots,Y;Y)$ for $i = 1,2$. Here $L^i(B_1,\ldots,B_i;B_{i+1})$ denotes the space of bounded multi-linear maps from the product of Banach spaces $B_1 \times \cdots \times B_i$ to another Banach space B_{i+1}. All Fréchet derivatives on (4.1) and (4.2) exist and bounded. We should note that $\chi = D\Psi_0(\gamma)[h]$ and $\psi = D^2\Psi_0(\gamma)[h,h]$ hold, where $\Psi_0 : \mathcal{H}(X) \to \mathcal{H}(Y)$ is defined in Section 3.

At the beginning, we give a simple lemma to deal with differential equations such as (4.1) and (4.2). See Lemma 3.1 in [9] for details.

Lemma 4.2. *Fix $\gamma \in \mathcal{H}(X)$ and $\phi = \Psi_0(\gamma)$. Let $M : [0,1] \to L(Y,Y)$ be the solution of the differential equation*

$$dM_t = d\Omega_t M_t, \qquad M_0 = \mathrm{Id}_Y,$$

where

$$d\Omega_t := (\nabla\sigma)(\phi_t)[\,\cdot\,, d\gamma_t] + (\nabla b_2)(\phi_t)[\,\cdot\,]dt \in L(Y,Y), \quad t \geq 0.$$

Then M_t is invertible for all $t \geq 0$.

Moreover, for each $k \in \mathcal{H}(Y)$, we define $\Gamma(k) = \Gamma_\gamma(k) \in \mathcal{H}(Y)$ by

$$\Gamma(k)_t := M_t \int_0^t M_s^{-1} dk_s, \quad t \geq 0.$$

Then $\Gamma(k)$ is the unique solution of the differential equation

$$d\Gamma(k)_t - d\Omega_t \Gamma(k)_t = dk_t, \quad \Gamma(k)_0 = 0,$$

and the operator $\Gamma : \mathcal{H}(Y) \to \mathcal{H}(Y)$ can be extended to a bounded linear operator from $P(Y)$ to $P(Y)$.

By using Lemma 4.2., we have the following expressions for the solutions of the equations (4.1) and (4.2):

$$\chi(h)_t = \Gamma\left(\int_0^{\cdot} \sigma(\phi_s)dh_s\right)_t, \quad \psi(h,h)_t = \Gamma\left(\int_0^{\cdot} dC_{h,h}(s)\right)_t, \tag{4.3}$$

where

$$\begin{aligned} dC_{h,\hat{h}}(s) &= (\nabla\sigma)(\phi_s)[\chi(h)_s, d\hat{h}_s] + (\nabla\sigma)(\phi_s)[\chi(\hat{h})_s, dh_s] \\ &\quad + (\nabla^2\sigma)(\phi_s)[\chi(h)_s, \chi(\hat{h})_s, d\gamma_s] \\ &\quad + (\nabla^2 b_2)(\phi_s)[\chi(h)_s, \chi(\hat{h})_s]ds \qquad \text{for } h, \hat{h} \in \mathcal{H}(X). \end{aligned} \tag{4.4}$$

Next we give estimates of $\sup_{0\le t\le 1} |\chi(h)|_Y$ and $\sup_{0\le t\le 1} |\psi(h,h)_t|_Y$ in terms of the function ξ. See Lemmas 5.1 and 5.3 in [9] for the proof.

Lemma 4.3. *Let $\chi_t = \chi(h)_t$ and $\psi_t = \psi(h,h)_t$. Let r_0, r_1 be any positive constants. Then, there exists a positive constant $c = c(r_0, r_1)$ such that*

$$\sup_{0\le t\le 1} |\chi(h)_t|_Y \le c\xi(\overline{h}),$$
$$\sup_{0\le t\le 1} |\psi(h,h)_t|_Y \le c\xi(\overline{h})^2$$

hold for all $h \in \mathcal{H}(X)$ with $\xi(\overline{h}) \le r_0$ and for all $\gamma \in \mathcal{H}(X)$ with $\|\gamma\|_{\mathcal{H}(X)} \le r_1$. Moreover the maps $h \in \mathcal{H}(X) \mapsto \chi(h)$ and $h \in \mathcal{H}(X) \mapsto \psi(h,h)$ can be extended to continuous maps from $G\Omega_p(X)$ to $P(Y)$.

By the above lemma, we can define $\chi(\overline{w})$ and $\psi(\overline{w},\overline{w})$ by the continuous extensions of $\chi(h)$ and $\psi(h,h)$, respectively. From now, we aim to give an explicit representation of $\psi(\overline{w},\overline{w})$. We set some notations. For $K \in L^2(X,X;Y)$, we define the trace of K by

$$\mathrm{Tr}(K) := \int_X K[x,x]\mu(dx).$$

By virtue of Fernique's theorem (Theorem 3.1 in Kuo [12]), it holds that

$$|\mathrm{Tr}(K)|_Y \le \|K\|_{L^2(X,X;Y)} \cdot \int_X |x|_X^2 \mu(dw) < \infty.$$

This means that $\mathrm{Tr} : L^2(X,X;Y) \to Y$ is a bounded linear map. By recalling Itô-Nisio's theorem, we have

$$\lim_{n\to\infty} \left| \mathrm{Tr}(K) - \sum_{i=1}^{n} K[e_i, e_i] \right|_Y = 0,$$

where $\{e_i\}_{i=1}^{\infty} \subset X^*$ is a C.O.N.S. of H. We denote $Q_2(\phi_s) \in L(X,X;Y)$, $0 \le s \le 1$, by

$$Q_2(\phi_s)\big[x_1, x_2\big] := \frac{1}{2}(\nabla\sigma)(\phi_s)\big[\sigma(\phi_s)x_1, x_2\big]_{Y\times X}, \quad x_1, x_2 \in X.$$

For $\alpha \in Y^*$, $0 \le t \le 1$ and $(u,s) \in \Delta_t := \{(u,s)|\ 0 \le u \le s \le t\}$, we define a continuous map $K(\alpha)_t(u,s) : \Delta_t \to L^2(X, X; \mathbb{R})$ by

$$\begin{aligned}&K(\alpha)_t(u,s)[x_1, x_2]_{X\times X}\\&:= \alpha\Big(M_t M_s^{-1}(\nabla\sigma)(\phi_s)\big[M_s M_u^{-1}\sigma(\phi_u)x_1, x_2\big]_{Y\times X}\Big), \quad x_1, x_2 \in X.\end{aligned}$$

Then we have

Lemma 4.4. *Let $\psi = \psi(\overline{w}, \overline{w})$ be the continuous extension of $\psi = \psi(h,h)$ as in Lemma 4.3.. Then for any $\alpha \in Y^*$ and $t \in [0,1]$,*

$$\begin{aligned}\alpha(\psi(\overline{w}, \overline{w})_t) = {} & 2\int_0^t \int_0^s K(\alpha)_t(u,s)[d\overline{w}_1(u), d\overline{w}_1(s)]\\&+2\alpha\left(\Gamma\left(\int_0^{\cdot} \mathrm{Tr}(Q_2)(\phi_s)ds\right)_t\right)\\&+\alpha\Bigg(\Gamma\Bigg(\int_0^{\cdot} (\nabla^2\sigma)(\phi_s)[\chi(\overline{w})_s, \chi(\overline{w})_s, d\gamma_s]\\&+(\nabla^2 b_2)(\phi_s)[\chi(\overline{w})_s, \chi(\overline{w})_s]ds\Bigg)_t\Bigg)\end{aligned}$$

holds $\mathbb{P}_1$-almost surely $\overline{w}$, where the first term on the right-hand side is a usual stochastic iterated integral with respect to the Brownian motion $(\overline{w}_1(0,t))_{0\le t\le 1}$ on the probability space $(G\Omega_p(X), \mathbb{P}_1)$.

Proof. By (4.3) and (4.4), we have the following expression for every $h \in \mathcal{H}(X)$:

$$\begin{aligned}\psi(h,h)_t &= 2\Gamma\Big(\int_0^{\cdot}(\nabla\sigma)(\phi_s)[\chi(h)_s, dh_s]\Big)_t\\&\quad+\Gamma\Big(\int_0^{\cdot}(\nabla^2\sigma)(\phi_s)[\chi(h)_s, \chi(h)_s, d\gamma_s] + (\nabla^2 b_2)(\phi_s)[\chi(h)_s, \chi(h)_s]ds\Big)_t\\&= 2\int_0^t\int_0^s M_t M_s^{-1}(\nabla\sigma)(\phi_s)\big[M_s M_u^{-1}\sigma(\phi_u)dh_u, dh_s\big]_{Y\times X}\\&\quad+\Gamma\Big(\int_0^{\cdot}(\nabla^2\sigma)(\phi_s)[\chi(h)_s, \chi(h)_s, d\gamma_s] + (\nabla^2 b_2)(\phi_s)[\chi(h)_s, \chi(h)_s]ds\Big)_t.\end{aligned} \tag{4.5}$$

Then for $\alpha \in Y^*$, (4.5) leads us that

$$\begin{aligned}&\alpha\left(\psi(\overline{w(m)}, \overline{w(m)})_t\right)\\&= 2\int_0^t\int_0^s K(\alpha)_t(u,s)\big[dw(m)(u), dw(m)(s)\big]\\&\quad+\alpha(\Gamma(\int_0^{\cdot}(\nabla^2\sigma)(\phi_s)\big[\chi(\overline{w(m)})_s, \chi(\overline{w(m)})_s, d\gamma_s\big]\\&\quad+(\nabla^2 b_2)(\phi_s)\big[\chi(\overline{w(m)})_s, \chi(\overline{w(m)})_s\big]ds)_t).\end{aligned} \tag{4.6}$$

Here we have the convergence

$$\lim_{m\to\infty} \mathbb{E}\Bigg[\Bigg| \int_0^t \int_0^s K(\alpha)_t(u,s) \,\big[dw(m)(u), dw(m)(s)\big] \\ - \left\{ \int_0^t \int_0^s K(\alpha)_t(u,s)[dw_u, dw_s] + \frac{1}{2}\int_0^t \mathrm{Tr}\big(K(\alpha)_t(s,s)\big)ds \right\} \Bigg| \Bigg] = 0, \tag{4.7}$$

and the equality

$$\int_0^t \mathrm{Tr}\big(K(\alpha)_t(s,s)\big)ds \\ = \alpha\left(M_t \int_0^t M_s^{-1}\mathrm{Tr}(Q_2)(\phi_s)ds\right) = \alpha\left(\Gamma\left(\int_0^\cdot \mathrm{Tr}(Q_2)(\phi_s)ds\right)_t\right). \tag{4.8}$$

Hence by letting $m \to \infty$ on both sides of (4.6) and by recalling Lemma 4.3., (4.7) and (4.8), we obtain the desired assertion. □

Now we are in a position to give the stochastic Taylor expansion up to the order 2. For fixed $\gamma \in \mathcal{H}(Y)$, we recall that $\phi \in \mathcal{H}(X)$ is defined by $\phi = \Psi_0(\gamma)$. For $0 < \varepsilon \le 1$ and $h \in \mathcal{H}(X)$, $R^i_\varepsilon = R^i_\varepsilon(h)$, $i = 1, 2, 3$, are defined as follows:

$$\begin{aligned} R^1_\varepsilon(t) &:= \Psi_\varepsilon(h+\gamma)_t - \phi_t, \\ R^2_\varepsilon(t) &:= \Psi_\varepsilon(h+\gamma)_t - \phi_t - \chi(h)_t, \\ R^3_\varepsilon(t) &:= \Psi_\varepsilon(h+\gamma)_t - \phi_t - \chi(h)_t - \frac{1}{2}\psi(h,h)_t - \varepsilon^2\Gamma\left(\int_0^\cdot b_1(\phi_s)ds\right)_t . \end{aligned}$$

Then we have the following estimates for the remainder terms of the stochastic Taylor expansion. See Lemma 6.1 in [9] for details.

Lemma 4.5. *For $0 < \varepsilon \le 1$, let $R^1_\varepsilon(t), R^2_\varepsilon(t)$ and $R^3_\varepsilon(t)$ be as above. Let r_0 and r_1 be any positive constants. Then, there exists a positive constant $c = c(r_0, r_1)$ such that*

$$\sup_{0\le t\le 1} |R^i_\varepsilon(t)|_Y \le c\,\big(\xi(\overline{h}) + \varepsilon\big)^i, \qquad i = 1, 2, 3,$$

hold for all $h \in \mathcal{H}(X)$ with $\xi(\overline{h}) \le r_0$ and $\gamma \in \mathcal{H}(X)$ with $\|\gamma\|_{\mathcal{H}(X)} \le r_1$. Moreover, for each fixed ε and γ the map $h \in \mathcal{H}(X) \mapsto R^i_\varepsilon = R^i_\varepsilon(h) \in \mathcal{H}(Y), i = 1, 2, 3,$ can be extended to continuous maps from $G\Omega_p(X)$ to $P(Y)$.

4.2 Computation of the Hessian

In this subsection, we present some fundamental properties on the Hessian A defined in Section 2. First, we give an explicit representation of A. We recall the equations (4.1), (4.2) and use Lemma 4.2. Then, for $h, \hat{h} \in \mathcal{H}$, we obtain

$$
\begin{aligned}
&(Ah,\hat{h})_{\mathcal{H}} \\
&= DF(\phi_0)\left[D^2\Psi_0(\gamma_0)[h,\hat{h}]\right] + D^2F(\phi_0)\left[D\Psi_0(\gamma_0)[h], D\Psi_0(\gamma_0)[\hat{h}]\right] \\
&= DF(\phi_0)\left[M_\cdot \int_0^\cdot M_s^{-1} dC_{h,\hat{h}}(s)\right] + D^2F(\phi_0)\left[\chi(h),\chi(\hat{h})\right]. \qquad (4.9)
\end{aligned}
$$

Here we set

$$
\begin{aligned}
V(h,\hat{h})_t := M_t \int_0^t M_s^{-1} \Big\{ &(\nabla\sigma)(\phi_0(s))\left[\chi(h)_s, d\hat{h}_s\right] \\
&+(\nabla\sigma)(\phi_0(s))\left[\chi(\hat{h})_s, dh_s\right]\Big\}, \quad t \geq 0, \qquad (4.10)
\end{aligned}
$$

and define a bounded self-adjoint operator $\tilde{A}$ on $\mathcal{H}$ by

$$
(\tilde{A}h,\hat{h})_{\mathcal{H}} := DF(\phi_0)[V(h,\hat{h})] \qquad \text{for } h,\hat{h} \in \mathcal{H}. \qquad (4.11)
$$

Then by (4.9), (4.10) and (4.11), we obtain

$$
\begin{aligned}
&((A-\tilde{A})h,\hat{h})_{\mathcal{H}} \\
&= DF(\phi_0)\Big[\Gamma\Big(\int_0^\cdot (\nabla^2\sigma)(\phi_0(s))[\chi(h)_s, \chi(\hat{h})_s, d\gamma_0(s)] \\
&\quad +(\nabla^2 b_2)(\phi_0(s))[\chi(h)_s, \chi(\hat{h})_s] ds\Big)\Big] + D^2F(\phi_0)\left[\chi(h),\chi(\hat{h})\right]
\end{aligned}
$$

and it implies that

$$
\left|((A-\tilde{A})h,\hat{h})_{\mathcal{H}}\right| \leq c\|\chi(h)\|_{P(Y)} \cdot \|\chi(\hat{h})\|_{P(Y)} \leq c\|h\|_{P(X)} \cdot \|\hat{h}\|_{P(X)}
$$

holds for some constant $c > 0$. Then by applying Theorem 4.6 in [12] to an abstract Wiener space $(P(X),\mathcal{H},\mathbb{P}_1')$, we can see that $A-\tilde{A}$ is a trace class operator on $\mathcal{H}$.

Moreover, we have the following properties on the operators A and $\tilde{A}$:

Lemma 4.6. (1) *A and $\tilde{A}$ are self-adjoint Hilbert-Schmidt operators on $\mathcal{H}$.* (2) *The continuous extension of the quadratic form defined by $A-\tilde{A}$ is represented as*

$$
\begin{aligned}
\langle (A-\tilde{A})\overline{w},\overline{w}\rangle = D^2F(\phi_0)&\left[\chi(\overline{w}),\chi(\overline{w})\right] \\
&+DF(\phi_0)\Big[\Gamma\Big(\int_0^\cdot (\nabla^2\sigma)(\phi_0(s))[\chi(\overline{w})_s, \chi(\overline{w})_s, d\gamma_0(s)] \\
&+(\nabla^2 b_2)(\phi_0(s))[\chi(\overline{w})_s, \chi(\overline{w})_s] ds\Big)\Big].
\end{aligned}
$$

Proof. Firstly, we note that the unitary isometry $\mathcal{H} \cong L^2([0,1],\mathbb{R}) \otimes H$, where $\otimes$ denotes the Hilbert-Schmidt tensor product. Let $\{v_j'\}_{j=0}^\infty$ be a C.O.N.S. of $L^2([0,1],\mathbb{R})$ defined by $v_0'(t) = 1$, $v_{2j-1}'(t) = \sqrt{2}\sin(2\pi jt)$ and $v_{2j}'(t) =$

$\sqrt{2}\cos(2\pi jt)$ for $j \in \mathbb{N}$. Secondly, let $\{e_l\}_{l=1}^{\infty}$ be a C.O.N.S. of H such that $\sum_{l=1}^{\infty} |e_l|_X^2 < \infty$. (See Theorem 3.5.10 in Bogachev [4].) Then $\{v_j \otimes e_l\}_{j,l=1}^{\infty}$ is a C.O.N.S. of $\mathcal{H}$. Hence by noting $\sum_{l=1}^{\infty} |e_l|_X^2 < \infty$ and following the proof of Lemma 7.2 and Corollary 7.5 in [9], we can show the items (1) and (2). □

Next, we consider the stochastic integration of the kernel associated with $\tilde{A}$. Recall any self-adjoint Hilbert-Schmidt operator S on $\mathcal{H}$ corresponds to a kernel function $K_S \in L^2([0,1] \times [0,1], H \otimes H)$ with $K_S(u,s) = K_S(s,u)^*$ for almost all (u,s) since $\mathcal{H} \cong L^2([0,1], H) \cong L^2([0,1], \mathbb{R}) \otimes H$. The correspondence $S \mapsto K_S$ is isometric. Then for the X-valued Brownian motion $w = (w_t)_{0 \le t \le 1}$, an iterated stochastic integral $\hat{K}_S(w) := 2\int_0^1 \int_0^s K_S(u,s)[dw_u, dw_s]$ is well-defined. Clearly, this random variable is in $L^2(\mathbb{P}'_1)$ with expectation 0. The correspondence $S \mapsto \hat{K}_S \in L^2(\mathbb{P}'_1)$ is isometric. The following lemma is essentially shown in Corollary 7.3 and Lemma 7.4 in [9].

Lemma 4.7. (1) *For each $\alpha \in P(Y)^*$, $\alpha \circ V$ is a Hilbert-Schmidt symmetric bilinear form on $\mathcal{H}$. (We also denote by $\alpha \circ V$ the self-adjoint Hilbert-Schmidt operator on $\mathcal{H}$ associated with this bilinear form.)*
(2) *For any $\alpha \in P(Y)^*$, it holds that*

$$\alpha(\Theta(\overline{w})) = \hat{K}_{\alpha \circ V}(\overline{w}_1), \qquad \mathbb{P}_1\text{-almost surely,}$$

where

$$\begin{aligned}\Theta(\overline{w}) := \psi(\overline{w}, \overline{w}) &- \Gamma\left(\int_0^{\cdot} \mathrm{Tr}(Q_2)(\phi_0(s))ds\right)\\ &-\Gamma\Big(\int_0^{\cdot} (\nabla^2\sigma)(\phi_0(s))[\chi(\overline{w})_s, \chi(\overline{w})_s, d\gamma_0(s)]\\ &\quad+(\nabla^2 b_2)(\phi_0(s))[\chi(\overline{w})_s, \chi(\overline{w})_s]ds\Big).\end{aligned}$$

In particular, $DF(\phi)[\Theta(\overline{w})] = \hat{K}_{\tilde{A}}(\overline{w}_1)$ holds $\mathbb{P}_1$-almost surely.

Before closing this subsection, we present an integral formula to compute the quantity α_0. See the proof of Lemma 8.3 in [9] for details.

Lemma 4.8. *It holds that*

$$\begin{aligned}\int_{G\Omega_p(X)} \exp\left\{-\frac{1}{2}\left(\hat{K}_{\tilde{A}}(\overline{w}_1) - \langle (A-\tilde{A})\overline{w}, \overline{w}\rangle\right)\right\}\mathbb{P}_1(d\overline{w})\\ = e^{-\frac{1}{2}\mathrm{Tr}(A-\tilde{A})} \cdot \det{}_2(\mathrm{Id}_{\mathcal{H}} + A)^{-1/2},\end{aligned}$$

where $\det_2$ *denotes the Carleman-Fredholm determinant.*

4.3 Outline of the Proof of Theorem 3.2

In this subsection, we explain about the outline of the proof of Theorem 3.2. briefly. Besides we give the explicit value of α_0. For details, the reader is referred to Section 8 in [9].

We proceed with the following steps. First, we denote a neighborhood of $\overline{\gamma_0}$ and its exterior in $G\Omega_p(X)$ by $U(\overline{\gamma_0})$ and $U(\overline{\gamma_0})^c$, respectively. We divide our functional integral into

$$\begin{aligned}&\mathbb{E}\big[\exp(-F(X^\varepsilon)/\varepsilon^2)\big]\\&=\int_{U(\overline{\gamma_0})}+\int_{U(\overline{\gamma_0})^c}\exp\Big\{-\frac{1}{\varepsilon^2}F(\Phi(\iota(\overline{w},\lambda^\varepsilon))_1)\Big\}\mathbb{P}_\varepsilon(d\overline{w})=:I_1(\varepsilon)+I_2(\varepsilon).\end{aligned}$$

For the integral $I_2(\varepsilon)$, we can neglect as $\varepsilon\searrow 0$ by the large deviation principle for Brownian rough paths (Theorem 4.1.). For the integral $I_1(\varepsilon)$, we regard $\Phi(\iota(\overline{w},\lambda^\varepsilon))_1$ as the perturbation of a quadratic function on $U(\overline{\gamma_0})$.

Next, we put $\phi=\phi_0$, $\gamma=\gamma_0$ and consider

$$g_\varepsilon^1(h):=\chi(h), \tag{4.12}$$

$$g_\varepsilon^2(h):=\psi(h,h)+2\varepsilon^2\Gamma\left(\int_0^\cdot b_1\big(\phi_0(s)\big)ds\right), \tag{4.13}$$

$$R_\varepsilon^3(h-\gamma_0):=\Psi_\varepsilon(h)-\phi_0-g_\varepsilon^1(h-\gamma_0)-\frac{1}{2}g_\varepsilon^2(h-\gamma_0) \tag{4.14}$$

for $\varepsilon>0$ and $h\in\mathcal{H}(X)$. Note that $\overline{w}\in G\Omega_p(X)\mapsto\Phi\big(\iota(\overline{w},\lambda^\varepsilon)\big)_1\in P(Y)$ is the continuous extension of $h\in\mathcal{H}(X)\mapsto\Psi_\varepsilon(h)\in\mathcal{H}(Y)$. By recalling Lemmas 4.3. and 4.5., all functions on (4.14) can be extended to continuous functions on $G\Omega_p(X)$, which will be denoted by the same symbols. (For example, we write $g_\varepsilon^1(\overline{w-\gamma_0})$, etc.)

Then by combining the Taylor expansion for F and (4.14), we obtain that

$$\begin{aligned}F(\Phi(\,\iota(\overline{w},\lambda^\varepsilon))_1)-F(\phi_0)&=DF(\phi_0)\big[g_\varepsilon^1(\overline{w-\gamma_0})\big]\\&+\frac{1}{2}DF(\phi_0)\big[g_\varepsilon^2(\overline{w-\gamma_0})\big]+\frac{1}{2}D^2F(\phi_0)\big[g_\varepsilon^1(\overline{w-\gamma_0}),g_\varepsilon^1(\overline{w-\gamma_0})\big]\\&+R_\varepsilon^3(F)(\overline{w-\gamma_0}),\end{aligned} \tag{4.15}$$

where $R_\varepsilon^3(F)(\overline{w-\gamma_0})$ is the remainder term and all the functions above are continuous on $G\Omega_p(X)$.

On the other hand, by Assumption **(F2)**, the function $h\in\mathcal{H}\mapsto F(\Psi_0(h))+\|h\|_{\mathcal{H}}^2/2$ attains minimum 0 at $\gamma_0\in\mathcal{H}$. Hence for any $h\in\mathcal{H}$,

$$0=\big(\gamma_0,h\big)_{\mathcal{H}}+DF(\phi_0)[\chi(h)]$$

holds. As the continuous extension of the above equality, it holds that

$$DF(\phi_0)[g_\varepsilon^1(\overline{w-\gamma_0})]=\|\gamma_0\|_{\mathcal{H}}^2-[\gamma_0](\overline{w}),\qquad \mathbb{P}_\varepsilon\text{-almost surely.} \tag{4.16}$$

Then by combining (4.15) with (4.16), and by using Lemmas 2.2., 4.6., 4.7., 4.8., we obtain

$$\begin{aligned}
&\lim_{\varepsilon \searrow 0} I_1(\varepsilon) \\
&= \int_{G\Omega_p(X)} \exp\left\{-\frac{1}{2}\left(DF(\phi_0)[g_1^2(\overline{w})] + D^2F(\phi_0)[g_1^1(\overline{w}), g_1^1(\overline{w})]\right)\right\} \mathbb{P}_1(d\overline{w}) \\
&= \exp\left\{-\frac{1}{2}\mathrm{Tr}(A-\tilde{A}) - DF(\phi_0)\left[\Gamma\left(\int_0^{\cdot} b_1(\phi_0(s)) + \mathrm{Tr}(Q_2)(\phi_0(s))ds\right)\right]\right\} \\
&\quad \times \det{}_2(\mathrm{Id}_{\mathcal{H}} + A)^{-1/2}. \qquad (4.17)
\end{aligned}$$

Finally, by summarizing the above arguments, we can present the following theorem:

Theorem 4.9. *Let α_0 be denoted in Theorem 3.2. Then we have*

$$\begin{aligned}
\alpha_0 = \exp&\left\{-\frac{1}{2}\mathrm{Tr}(A-\tilde{A}) - DF(\phi_0)\left[\Gamma\left(\int_0^{\cdot} b_1(\phi_0(s)) + \mathrm{Tr}(Q_2)(\phi_0(s))ds\right)\right]\right\} \\
&\times \det{}_2(\mathrm{Id}_{\mathcal{H}} + A)^{-1/2}.
\end{aligned}$$

5 Examples

5.1 Heat Processes on Loop Spaces

In this subsection, we consider a class of stochastic processes on continuous loop spaces and show that the theory of rough paths is applicable to them. The processes are usually called heat processes on loop spaces and are defined by a collection of finite-dimensional SDEs. Processes of this kind were first introduced by Malliavin [18] in the case of loop groups and then were generalized by many authors.

Let $s > 1/2$ and $\mathcal{L}_0(\mathbb{R}^d) := \{x \in C([0,1], \mathbb{R}^d) |\ x(0) = x(1) = 0\}$. For $h \in \mathcal{L}_0(\mathbb{R}^d)$ of the form

$$h(\tau) = \sum_{n \neq 0} \hat{h}(n)\left(e^{2\pi\sqrt{-1}n\tau} - 1\right),$$

we set

$$\|h\|^2_{H_0^s(\mathbb{R}^d)} := \sum_{n \neq 0} |2\pi n|^{2s} |\hat{h}(n)|^2$$

and $H_0^s(\mathbb{R}^d) := \{h \in \mathcal{L}_0(\mathbb{R}^d) \,|\, \|h\|_{H_0^s(\mathbb{R}^d)} < \infty\}$. It is well-known that $H_0^s(\mathbb{R}^d)$ is a Hilbert space embedded in $\mathcal{L}_0(\mathbb{R}^d)$ and that there exists a Gaussian measure μ^s such that the triplet $(\mathcal{L}_0(\mathbb{R}^d), H_0^s(\mathbb{R}^d), \mu^s)$ becomes an abstract Wiener space. When $s = 1$, μ^1 is the usual d-dimensional pinned Wiener measure and is of particular importance. For $\tau \in [0,1]$ and $j = 1, 2, \ldots, d$, we denote by δ_τ^j the element in $\mathcal{L}_0(\mathbb{R}^d)^*$ defined by $\langle \delta_\tau^j, x\rangle = x^j(\tau)$ and set $x(\tau) := (x^1(\tau), \ldots, x^d(\tau))$. Let $(w_t)_{t \geq 0}$ be a $\mathcal{L}_0(\mathbb{R}^d)$-valued Brownian motion associated with μ^s. We set $w_t^j(\tau) := \langle \delta_\tau^j, w_t\rangle$ and $w_t(\tau) := (w_t^1(\tau), \ldots, w_t^d(\tau))$.

Now we give heat processes in a slightly generalized form. Let

$$A_j(x) = \sum_{i=1}^{r} a_{ij}(x)\frac{\partial}{\partial x_i}, \quad A_0(x) = \sum_{i=1}^{r} b_i(x)\frac{\partial}{\partial x_i}, \quad V_0(x) = \sum_{i=1}^{r} \beta_i(x)\frac{\partial}{\partial x_i}$$

be vector fields on $\mathbb{R}^r$, $j = 1, \ldots, d$. We assume the following regularities on the coefficients:

$$a_{ij}, b_i, \beta_i \in C_b^4(\mathbb{R}^r, \mathbb{R}) \qquad \text{for } 1 \le i \le r, 1 \le j \le d. \tag{5.1}$$

We write a for the $r \times d$-matrix $\{a_{ij}\}_{1\le i\le r, 1\le j\le d}$ and write b and β for the column vectors $(b_1, \ldots, b_r)^{\mathrm{T}}$ and $(\beta_1, \ldots, \beta_r)^{\mathrm{T}}$, respectively.

For each fixed space parameter $\tau \in [0,1]$ and $\varepsilon > 0$, we consider the following (finite dimensional) SDE:

$$\begin{aligned} d_t X_t^\varepsilon(\tau) &= \sum_{j=1}^{r} A_j(X_t^\varepsilon(\tau)) \circ \varepsilon d_t w_t^j(\tau) + A_0(X_t^\varepsilon(\tau))\varepsilon^2 dt + V_0(X_t^\varepsilon(\tau))dt \\ &= a(X_t^\varepsilon(\tau)) \circ \varepsilon d_t w_t(\tau) + b(X_t^\varepsilon(\tau))\varepsilon^2 dt + \beta(X_t^\varepsilon(\tau))dt. \end{aligned} \tag{5.2}$$

with the initial data $X_0^\varepsilon(\tau) = 0$. We will often write $X^\varepsilon(t,\tau) := X_t^\varepsilon(\tau)$. In Proposition 5.1. below, we will prove that $X^\varepsilon(t,\tau)$ has a bi-continuous modification. We call $X^\varepsilon = (X^\varepsilon(t,\cdot))_{0\le t\le 1}$ the heat process. X^ε can be regarded as a random variable in $P(\mathcal{L}_0(\mathbb{R}^d))$.

Next we recall that $(\mathcal{L}_0(\mathbb{R}^d), \mu^s)$ satisfies the exactness condition for all tensor norms (including the projective tensor norm) on $\mathcal{L}_0(\mathbb{R}^d) \otimes \mathcal{L}_0(\mathbb{R}^d)$. (See Lemma 4.1 in [10] for the proof.) Therefore the Brownian rough path $\overline{w} \in G\Omega_p(\mathcal{L}_0(\mathbb{R}^d))$ defined by $(w_t)_{t\ge 0}$ exists and we can deal with our heat process X^ε defined by (5.2) from the viewpoint of rough paths.

We define a Nemytski map $\tilde{\sigma} : \mathcal{L}_0(\mathbb{R}^r) \to L(\mathcal{L}_0(\mathbb{R}^d) \oplus \mathbb{R}^2, \mathcal{L}_0(\mathbb{R}^r))$ by

$$\tilde{\sigma}(y)[(x,u_1,u_2)](\tau) := a\big(y(\tau)\big)x(\tau) + b\big(y(\tau)\big)u_1 + \beta\big(y(\tau)\big)u_2, \qquad \tau \in [0,1]. \tag{5.3}$$

for $(x,u_1,u_2) \in \mathcal{L}_0(\mathbb{R}^d) \oplus \mathbb{R}^2$ and $y \in \mathcal{L}_0(\mathbb{R}^r)$. Note that the assumption (5.1) implies $\tilde{\sigma} \in C_b^4\big(\mathcal{L}_0(\mathbb{R}^r), L\big(\mathcal{L}_0(\mathbb{R}^d) \oplus \mathbb{R}^2, \mathcal{L}_0(\mathbb{R}^r)\big)\big)$. Then we can consider a random element $\Phi(\iota(\overline{\varepsilon w}, \lambda^\varepsilon))$ in $G\Omega_p(\mathcal{L}_0(\mathbb{R}^r))$ through the differential equation in the rough path sense (3.1). The following proposition is taken from Lemma 4.8 in [10]. By this proposition, we can obtain a dynamics on $\mathcal{L}_0(\mathbb{R}^r)$. In the proof, the Wong-Zakai approximation theorem plays a crucial role.

Proposition 5.1. *For each $\varepsilon > 0$, $(t,\tau) \mapsto \Phi(\iota(\overline{\varepsilon w}, \lambda^\varepsilon))_1(0,t)(\tau)$ is a bi-continuous modification of the two-parameter process $(X^\varepsilon(t,\tau))_{0\le t\le 1, 0\le\tau\le 1}$ defined in* (5.2).

Remark 5.2. In [18], Kolmogorov's criterion is used for the proof of the existence of continuous modifications. Hence Proposition 5.1. is regarded as a revisit via the rough path theory. In this paper we assume that the starting loop is the constant loop at 0 for simplicity. However it is easy to modify the proof for general starting elements in $\mathcal{L}_0(\mathbb{R}^r)$ since the initial conditions may be arbitrary in the rough path theory. In other words, we do not need a Hölder-like condition on the starting loops.

We define the heat kernel measure ν_t to be the law of $X^1(t,\cdot)$ in the case of $V_0 = 0$. This measure ν_t is supported in $\mathcal{L}_0(\mathbb{R}^r)$. As a consequence of Theorem 3.1., we easily have a large deviation principle for ν_t by noting that $(X^\varepsilon(t,\cdot))_{t\geq 0}$ and $(X^1(\varepsilon^2 t,\cdot))_{t\geq 0}$ have the same law under $V_0 = 0$. See the second item of Theorem 4.9 in [10] for details.

Theorem 5.3. *Let $V_0 = 0$. Then the heat kernel measure ν_t satisfies a large deviation principle as $t \searrow 0$ with the good rate function $\tilde{I}$, where*

$$\tilde{I}(y) = \begin{cases} \frac{1}{2}\inf\left\{ \|\gamma\|^2_{\mathcal{H}(H_0^s(\mathbb{R}^d))} \mid y = \Psi_0(\gamma)_1 \right\}, & \\ & \text{if } y = \Psi_0(\gamma)_1 \text{ for some } \gamma \in \mathcal{H}(H_0^s(\mathbb{R}^d)), \\ \infty, & \text{otherwise.} \end{cases}$$

More precisely, for any measurable set $K \subset \mathcal{L}_0(\mathbb{R}^r)$, it holds that

$$-\inf_{y\in K^\circ} \tilde{I}(y) \leq \liminf_{t\searrow 0} t\log \nu_t(K) \leq \limsup_{t\searrow 0} t \log \nu_t(K) \leq -\inf_{y\in \overline{K}} \tilde{I}(y).$$

Remark 5.4. Fang-Zhang [8] showed the large deviation principle for for heat processes and heat kernel measures on loop groups. Our Theorems 3.1. and 5.3. are regarded as generalizations of their results.

5.2 SDEs on M-type 2 Banach Spaces

The theory of SDEs in infinite dimensional Hilbert spaces has been developed and is well understood. However, for general separable Banach spaces, there exist difficulties in defining a meaningful Itô's integral. Recently, Brzeźniak and Elworthy developed a theory of SDEs for a certain class of Banach spaces called M-type 2 Banach spaces. In this subsection, we consider SDEs on M-type 2 Banach spaces. For detailed explanations and further references, we refer the reader to Brzeźniak-Carroll [5] and Brzeźniak-Elworthy [6].

Let (X, H, μ) be an abstract Wiener space and $w = (w_t)_{t\geq 0}$ be the X-valued Brownian motion. We assume that (X,μ) satisfies the *exactness condition* for the projective tensor norm on $X \otimes X$. Let Y be a M-type 2 Banach space (see Definition 2.1 in [6] for the definition). For a progressively measurable process $\xi = (\xi_t)_{0\leq t\leq 1}$ which takes values in $L(X,Y)$ and satisfies $\mathbb{E}\big[\int_0^1 |\xi_s|^p_{L(X,Y)} ds\big] < \infty$, $p > 1$, we can define the stochastic integral $I(t) := \int_0^t \xi_s dw_s$, $0 \leq t \leq 1$, as a continuous Y-valued martingale. Moreover there exists a constant c_p, independent of ξ, such that

$$\mathbb{E}\left[\sup_{0\le t\le 1}\left|\int_0^t \xi_s dw_s\right|_Y^p\right] \le c_p\left(\int_0^1 \mathbb{E}\left[|\xi_s|^2_{L(X,Y)}\right] ds\right)^{p/2}.$$

See Theorem 2.9 and Remark 2.11 in [6] for details.

Let $\sigma \in C_b^4(X, L(X,Y))$ and $b_1, b_2 \in C_b^4(Y,Y)$. We consider the following Stratonovich type SDE on Y:

$$dX_t^\varepsilon = \sigma(X_t^\varepsilon) \circ \varepsilon dw_t + b_1(X_t^\varepsilon)\varepsilon^2 dt + b_2(X_t^\varepsilon)dt, \qquad X_t^\varepsilon = 0. \tag{5.4}$$

Here we call $(X_t^\varepsilon)_{0\le t\le 1}$ is a solution to (5.4) if and only if it satisfies for each $0 \le t \le 1$,

$$\begin{aligned} X_t^\varepsilon = &\int_0^t \sigma(X_s^\varepsilon)\varepsilon dw_s + \int_0^t \mathrm{Tr}(Q_2)(X_s^\varepsilon)\varepsilon^2 ds \\ &+ \int_0^t b_1(X_s^\varepsilon)\varepsilon^2 ds + \int_0^t b_2(X_s^\varepsilon)ds, \quad \text{a.s..} \end{aligned}$$

Under our conditions for the coefficients, there exists a unique solution $(X_t^\varepsilon)_{0\le t\le 1}$ to the SDE (5.4). See Theorem 2.26 in [6] and Theorem 2 in [5] for the detail. Moreover they have already established the Wong-Zakai approximation theorem (Theorem 3 in [5]) for the SDE (5.4). Then by the same argument as in the previous subsection, we can obtain

Proposition 5.5. *For each* $\varepsilon > 0$, $t \mapsto \Phi(\iota(\overline{\varepsilon w}, \lambda^\varepsilon))_1(0,t)$ *is almost surely equal to* $(X_t^\varepsilon)_{0\le t\le 1}$ *defined in* (5.4).

Remark 5.6. Here we explain why we do not consider heat processes in the previous subsection on the Sobolev-Slobodetski space $W^{\theta,p}(S^1)$. Note that $W^{\theta,p}(S^1)$ is an M-type 2 Banach space and is continuously embedded in the space of continuous loops if $p > 1$ and $1/p < \theta < 1$. (See Section 5 of [6] for details.) In Section 6 of [6] and Section 4 of [5], it is proved that $(X_t^\varepsilon)_{t\ge 0}$ defined by the SDE (5.2) can be considered as the solution of a $W^{\theta,p}(S^1)$-valued SDE (5.4) if $p > 2$ and $1/p < \theta < 1/2$ (at least for the case $s = 1$).

For general $s > 1/2$, under a suitable condition on p and θ, we see by straightforward computation that the Gaussian measure μ_s is supported on $W^{\theta,p}(S^1)$ and see from inequality (6.8) in page 575 of [15] that $W^{\theta,p}(S^1) \otimes W^{\theta,p}(S^1)$ is exact with respect to μ_s.

Hence, one may wonder why we do not work on the Sobolev-Slobodetski space $W^{\theta,p}(S^1)$. The main reason why we avoided $W^{\theta,p}(S^1)$ is that the map σ in (5.3) is not bounded with respect to the topology of $W^{\theta,p}(S^1)$. (See Section 5 of [6].) Since we would not like to treat the Itô maps or ODEs with unbounded coefficients, we choose to work on the space of continuous loops.

Acknowledgment

The authors would like to thank Professors Shigeki Aida, Sergio Albeverio, David Elworthy, Paul Malliavin and Shinzo Watanabe for their helpful comments and encouragements. The second author is very grateful to the organizers of the Abel Symposium 2005 for giving him opportunities to talk in the conference and to contribute in this volume. The authors were supported by JSPS Research Fellowships for Young Scientists and the second author was supported by 21st century COE program "Development of Dynamic Mathematics with High Functionality" at Faculty of Mathematics, Kyushu University.

References

1. S. Aida, *Semi-classical limit of the bottom of spectrum of a Schrödinger operator on a path space over a compact Riemannian manifold*, preprint, 2006.
2. R. Azencott, *Formule de Talyor stochastique et développement asymptotique d'intégrales de Feynman*, Seminar on Probability, XVI, Supplement, pp. 237–285, Lecture Notes in Math., **921**, Springer, Berlin-New York, 1982.
3. G. Ben Arous, *Methods de Laplace et de la phase stationnaire sur l'espace de Wiener*, Stochastics **25**, (1988), no. 3, pp. 125–153.
4. V.I. Bogachev, Gaussian Measures, American Mathematical Society, Providence, RI, 1998.
5. Z. Brzeźniak and A. Carroll, *Approximations of the Wong-Zakai type for stochastic differential equations in M-type 2 Banach spaces with applications to loop spaces*, Séminaire de Probabilités XXXVII, pp. 251–289, Lecture Notes in Math., **1832**, Springer, Berlin, 2003.
6. Z. Brzeźniak and K. D. Elworthy, *Stochastic differential equations on Banach manifolds*, Methods Funct. Anal. Topology **6** (2000), no. 1, pp. 43–84.
7. J.D. Deuschel and D.W. Stroock, Large Deviations, Academic Press, Boston, 1989.
8. S. Fang and T.S. Zhang, *Large deviations for the Brownian motion on loop groups*, J. Theoret. Probab. **14** (2001), no. 2, pp. 463–483.
9. Y. Inahama, *Laplace's method for the laws of heat processes on loop spaces*, J. Funct. Anal. **232** (2006), no. 1, pp. 148–194.
10. Y. Inahama and H. Kawabi, *Large deviations for heat kernel measures on loop spaces via rough paths*, J. London Math. Society **73** (2006), no. 3, pp. 797–816.
11. Y. Inahama and H. Kawabi, *Asymptotic expansions for the Laplace approximations for Itô functionals of Brownian rough paths*, J. Funct. Anal. **243** (2007), no. 1, pp. 270–322.
12. H.H. Kuo, Gaussian Measures on Banach Spaces, Lecture Notes in Math., **463**, Springer, Berlin, 1985.
13. S. Kusuoka and D.W. Stroock, *Precise asymptotics of certain Wiener functionals*, J. Funct. Anal. **99** (1991), no. 1, pp. 1–74.
14. S. Kusuoka and D.W. Stroock, *Asymptotics of certain Wiener functionals with degenerate extrema*, Comm. Pure Appl. Math. **47** (1994), no. 4, pp. 477–501.
15. M. Ledoux, T. Lyons and Z. Qian, *Lévy area of Wiener processes in Banach spaces*, Ann. Probab. **30** (2002), no. 2, pp. 546–578.
16. M. Ledoux, Z. Qian and T.S. Zhang, *Large deviations and support theorem for diffusion processes via rough paths*, Stochastic Process. Appl. **102** (2002), no. 2, pp. 265–283.
17. T. Lyons and Z. Qian, System control and rough paths, Oxford University Press, Oxford, 2002.

18. P. Malliavin, *Hypoellipticity in infinite dimensions*, Diffusion processes and related problems in analysis, Vol. I (Evanston, IL, 1989), pp. 17–31, Progr. Probab., **22**, Birkhäuser Boston, Boston, MA, 1990.
19. M. Schilder, *Some asymptotic formulas for Wiener integrals*, Trans. Amer. Math. Soc. **125** (1966), pp. 63–85.
20. S. Takanobu and S. Watanabe: *Asymptotic expansion formulas of the Schilder type for a class of conditional Wiener functional integrations*, in "Asymptotic problems in probability theory: Wiener functionals and asymptotics" (Sanda/Kyoto, 1990), pp. 194–241, Pitman Res. Notes Math. Ser., **284**, Longman Sci. Tech., Harlow, 1993.

Continuous-Time Markowitz's Problems in an Incomplete Market, with No-Shorting Portfolios

Hanqing Jin[1] and Xun Yu Zhou[2]

[1] Department of Mathematics, National University of Singapore, Singapore, matjinh@nus.edu.sg

[2] Department of Systems Engineering and Engineering Management, The Chinese University of Hong Kong, Shatin, Hong Kong, Tel.: 852-2609-8320, fax: 852-2603-5505, xyzhou@se.cuhk.edu.hk. Supported by the RGC Earmarked Grants CUHK 4175/03E, CUHK418605, and Croucher Senior Research Fellowship. We also thank an anonymous referee for helpful comments that have led to an improved version

Dedicated to Professor Kiyosi Itô for his 90th birthday

Summary. Continuous-time Markowitz's mean–variance portfolio selection problems with finite-time horizons are investigated in an arbitrage-free yet incomplete market. Models with unconstrained and no-shorting portfolios are tackled respectively. The sets of the terminal wealths that can be replicated by admissible portfolios are characterized in explicit terms. This enables one to transfer the original dynamic portfolio selection problems into ones of static, albeit constrained, optimization problems in terms of the terminal wealth. Solutions to the latter are obtained via certain dual (static) optimization problems. When all the market coefficients are deterministic processes, mean–variance efficient portfolios and frontiers are derived explicitly.

Mathematics Subject Classification (1991): 91B28, 60H10

Keywords and Phrases: Mean–variance portfolio selection, continuous time, arbitrage-free, incomplete market, Lagrange multiplier, backward stochastic differential equation, attainable wealth set, replication

1 Introduction

Markowitz's Nobel-prize-winning work on single-period mean–variance portfolio selection [23] has laid down the foundation for modern financial portfolio theory. Nevertheless, as pointed out in a recent survey paper [28]

the mean–variance approach has received little attention in the context of dynamic investment planning, especially in the continuous time setting. Most continuous-time portfolio selection models in literature assume that the investor seeks to maximize expected utility, which is a departure from the mean–variance model. While the utility approach was theoretically justified by von Neumann and Morgenstern [25], in practice "few if any investors know their utility functions; nor do the functions which financial engineers and financial economists find analytically convenient necessarily represent a particular investor's attitude towards risk and return" [24].

Research on faithfully extending the Markowitz model to the dynamic setting has emerged in very recent years, starting with Li and Ng [17] where a discrete-time, multiperiod, mean–variance problem is solved explicitly using an embedding technique to cope with the non-applicability of dynamic programming caused by the variance term. Subsequently, in a series of papers [2,18,20,30] various continuous-time Markowitz models have been investigated thoroughly with closed-form solutions obtained for most cases. Two main approaches are exploited to solve the problems. In the first approach, which is adopted in [18, 20, 30], one transfers the underlying mean–variance problem into a family of indefinite stochastic linear–quadratic (LQ) optimal control problems, and then uses an elaborative completion-of-square technique, via one or more stochastic Riccati equations, to derive the solutions. This approach is inspired by the recent development in indefinite LQ control [4, 29], and is particularly effective when there is no constraint on the state variable (the wealth, that is). Since in this approach a mean–variance efficient portfolio is obtained dynamically and forwardly in the process of optimization, we call it a "forward approach" or "primal approach". In contrast, in the second approach which is taken in [2], an optimal *terminal* wealth is first identified by solving a *static* optimization problem, and then an efficient portfolio is obtained by *replicating* the optimal terminal wealth. This approach, which we call a "backward approach" or "dual approach", also widely known as that of equivalent (risk neutral) martingale measures, goes back to Harrison and Kreps [12] and Pliska [26]. It is particularly powerful in solving the Markowitz problem with additional wealth constraints, as demonstrated in [2].

In all the papers [2, 18, 20, 30], it is assumed that the dimension of the underlying Brownian motion, used to model the stock prices, is the same as the number of the stocks, and the covariance matrix has uniformly positive eigenvalues. This induces a complete market where the risk associated with any reasonable contingent claim can be completely hedged. Although in [2, 18] there are no-shorting or no-bankruptcy constraints, which essentially render the market incomplete, the aforementioned assumption is critical for the forward or backward approaches to work.

In this paper, we consider the continuous-time Markowitz problem in a market where the dimension of the Brownian motion is different from the number of the stocks, and all the market coefficients are random (i.e., the investment opportunity set is stochastic). In addition, we will attack the problem for two cases respectively: 1) portfolios are unconstrained; and 2) shorting is prohibited. (Strictly speaking, the unconstrained case can be regarded as a special case of the other. We treat it separately because it is simpler, and we intend to use it to first showcase the essential idea without having to involve too much technicality.) As discussed earlier none of the approaches taken in the previous related articles would work. To overcome, for each of the two cases, we will first characterize, in explicit terms, the so-called attainable terminal wealth set, namely the set of terminal wealths that can be replicated by admissible portfolios satisfying the respective constraint. Then we solve a static optimization problem on random variables of terminal wealth with the attainable terminal wealth set representing an *additional* constraint. The mean–variance efficient portfolios are then derived by replicating those optimal, attainable terminal wealths for both cases. In the general situation of random market coefficients, we will prove all the necessary existence and uniqueness results while suggesting a scheme of approaching the final solutions. This is then exemplified by the scenario of deterministic coefficients where we are able to solve the original problem in *explicit and analytical* forms.

While the present work attempts to tackle the continuous-time mean–variance portfolio selection in incomplete markets, there have been many works in literature devoted to continuous-time portfolio selection with incomplete markets, albeit in the realm of expected utility; see [5,10,11,14,15,27] among others. It should be emphasized again that the existing results in the utility framework do not at all cover the mean–variance models for the main reason that the assumptions typically imposed on a utility function are not satisfied by a mean–variance model. To be specific, a typical utility function should satisfy several conditions, especially the one that its derivative must vanish at infinity; see, e.g., Karatzas and Shreve [15, p. 94, Definition 4.1]. (If proportional portfolios – portfolios being defined as proportions of wealth allocated to different stocks – are being considered, then the utility function must be further that its derivative is infinite at 0; see, e.g., [5].) These properties are crucial in deriving all the results with the expected utility and hence have been all along *the standing assumptions* in relevant literature. Unfortunately, the quadratic utility associated with the mean-variance model (if you must "embed" the mean-variance into the utility framework!) does *not* satisfy most of these assumptions. Consequently, one cannot apply *a priori* the results from the utility model. To the best of our knowledge the only preceding paper that deals with the Markowitz problem in an incomplete market is Lim [19] where a forward approach is applied together with a completion-of-market trick of [14] and nonlinear backward stochastic differential equation theory. However, only unconstrained portfolios are considered in [19]. In comparison, in the present paper we go with the backward approach, and solve constrained problems in incomplete markets.

The remainder of the paper is organized as follows. In Section 2 we introduce the market under consideration along with some of its important properties. Section 3 is devoted to some technical results on the pricing kernels that are vital for the subsequent analysis. Section 4 sets up the continuous-time Markowitz models. In Sections 5–6 we respectively solve the two cases. Finally, Section 7 concludes the paper.

2 Market

In this paper T is a fixed terminal time and $(\Omega, \mathcal{F}, P, \{\mathcal{F}_t\}_{t\geq 0})$ is a fixed filtered complete probability space on which is defined a standard n-dimensional Brownian motion $W(t) \equiv (W^1(t), \cdots, W^n(t))'$ with $W(0) = 0$, and $\mathcal{F}_t = \sigma\{W(s) : 0 \leq s \leq t\}$ augmented by all P-null sets. We denote by $L^2_{\mathcal{F}}(0,T;\mathbf{R}^d)$ the set of all $\mathbf{R}^d$-valued, $\mathcal{F}_t$-progressively measurable stochastic processes $f(\cdot) = \{f(t) : 0 \leq t \leq T\}$ with $\| f(\cdot) \|_{L^2_{\mathcal{F}}(0,T;\mathbf{R}^d)} := (E\int_0^T |f(t)|^2 dt)^{\frac{1}{2}} < +\infty$, by $L^\infty_{\mathcal{F}}(0,T;\mathbf{R}^d)$ the set of all $\mathbf{R}^d$-valued, essentially bounded, $\mathcal{F}_t$-progressively measurable stochastic processes $f(\cdot)$ with $\| f(\cdot) \|_{L^\infty_{\mathcal{F}}(0,T;\mathbf{R}^d)} := \text{esssup}_{(t,\omega)\in[0,T]\times\Omega}|f(t,\omega)| < +\infty$, and by $L^2_{\mathcal{F}_T}(\Omega;\mathbf{R}^d)$ the set of all $\mathbf{R}^d$-valued, $\mathcal{F}_T$-measurable random variables η such that $\| \eta \|_{L^2_{\mathcal{F}_T}(\Omega;\mathbf{R}^d)} := (E|\eta|^2)^{\frac{1}{2}} < +\infty$. Throughout this paper, a (t,ω)-null set is a null-set with respect to the product of the Lebesgue measure on $[0,T]$ and P on Ω, and a.s. signifies that the corresponding statement holds true with probability 1 (with respect to P).

Notation. We use the following additional notation:

$\mathbf{Q}^d$: the set of d-dimensional vectors with rational components;
$\mathbf{R}^d_+$: the set of d-dimensional vectors with nonnegative components;
$\mathbf{R}_+ := \mathbf{R}^1_+$;
M' : the transpose of any vector or matrix M;
$|M| := \sqrt{\sum_{i,j} m_{ij}^2}$ for any matrix or vector $M = (m_{ij})$;
$\alpha^+ := \max\{\alpha, 0\}$ for any real number α;
$\alpha^- := \max\{-\alpha, 0\}$ for any real number α.

In the market under consideration in this paper, there are $m+1$ assets (or securities) being traded continuously. One of the assets is a bank account whose price process $S_0(t)$ is subject to the following differential equation:

$$\begin{cases} dS_0(t) = r(t)S_0(t)dt, \ \ t \in [0,T], \\ S_0(0) = s_0 > 0, \end{cases} \tag{2.1}$$

where the interest rate process $r(\cdot) \in L^\infty_{\mathcal{F}}(0,T;\mathbf{R})$. Note that normally one would assume that $r(t) \geq 0$; yet this assumption is not necessary in our subsequent analysis. The other m assets are stocks whose price processes $S_i(t)$, $i = 1, \cdots, m$, satisfy the following stochastic differential equation (SDE):

$$\begin{cases} dS_i(t) = S_i(t)\left[\mu_i(t)dt + \sum_{j=1}^n \sigma_{ij}(t)dW^j(t)\right], \ \ t \in [0,T], \\ S_i(0) = s_i > 0, \end{cases} \tag{2.2}$$

where $\mu_i(\cdot) \in L^\infty_\mathcal{F}(0,T;\mathbf{R})$ and $\sigma_{ij}(\cdot) \in L^\infty_\mathcal{F}(0,T;\mathbf{R})$ are the processes of appreciation and dispersion (or volatility) rates, respectively.

Denote

$$\begin{aligned} &\sigma(t) := (\sigma_{ij}(t))_{m\times n}, \\ &B(t) \equiv (b_1(t), \cdots, b_m(t))' := (\mu_1(t) - r(t), \cdots, \mu_m(t) - r(t))'. \end{aligned}$$

Consider an agent whose total wealth at time $t \geq 0$ is denoted by $x(t)$, and the dollar amount invested in stock i, $i = 1, \cdots, m$, is $\pi_i(t)$. Assume that the trading of shares takes place continuously in a self-financing fashion (i.e., there is no consumption or income) and there are no transaction costs. Then the wealth process $x(\cdot)$ satisfies

$$dx(t) = [r(t)x(t) + B(t)'\pi(t)]dt + \pi(t)'\sigma(t)dW(t), \quad x(0) = x, \tag{2.3}$$

where $\pi(t) = (\pi_1(t), \cdots, \pi_m(t))'$ is the portfolio of the agent at time t, and x is the initial wealth of the agent.

The following assumption will be imposed throughout this paper.

Basic Assumption (A): There exists $\theta \in L^\infty_\mathcal{F}(0,T,\mathbf{R}^n)$ such that $\sigma(t)\theta(t) = B(t)$, a.s., a.e.$t \in [0,T]$.

The above assumption is satisfied if $\sigma(t)'\sigma(t)$ is uniformly positive definite (i.e., there is $\delta > 0$ such that $\sigma(t)'\sigma(t) \geq \delta I_n$ a.s., a.e.$t \in [0,T]$), in which case, necessarily, $n \leq m$, and there is a unique such θ. In general, however, the process θ, if it exists, may not be unique.

Definition 1. *A portfolio (process) $\pi(\cdot)$ is said to be admissible if $\sigma(\cdot)'\pi(\cdot) \in L^2_\mathcal{F}(0,T;\mathbf{R}^n)$. The set of all admissible portfolio is denoted by Π. A pair $(x(\cdot),\pi(\cdot))$ is called an (admissible) wealth–portfolio pair if $(x(\cdot),\pi(\cdot))$ satisfies (2.3).*

Observe that under Assumption (A), for any $\pi(\cdot) \in \Pi$, $B(\cdot)'\pi(\cdot) = \theta(\cdot)'[\sigma(\cdot)'\pi(\cdot)] \in L^2_\mathcal{F}(0,T;\mathbf{R})$. Hence by standard SDE theory a unique strong solution $x(\cdot) \equiv x^\pi(\cdot)$ exists for the wealth equation (2.3).

For any $\theta \in L^\infty_\mathcal{F}(0,T,\mathbf{R}^n)$, define

$$H_\theta(t) := \exp\left\{-\int_0^t [r(s) + \frac{1}{2}|\theta(s)|^2]ds - \int_0^t \theta(s)'dW(s)\right\}. \tag{2.4}$$

Equivalently, $H_\theta(\cdot)$ can be defined as the unique solution to the following SDE

$$\begin{cases} dH_\theta(t) = -r(t)H_\theta(t)dt - H_\theta(t)\theta(t)'dW(t), \\ H_\theta(0) = 1. \end{cases} \tag{2.5}$$

It is clear that for any $\theta \in L^{\infty}_{\mathcal{F}}(0,T,\mathbf{R}^n)$ there is a constant $c = c(\| \theta \|_{L^{\infty}_{\mathcal{F}}(0,T,\mathbf{R}^n)})$ such that $E[\sup_{0\le t\le T} H_\theta(t)^2] \le c$.

Define

$$\Theta := \{\theta \in L^{\infty}_{\mathcal{F}}(0,T,\mathbf{R}^n) : \ \sigma(t)\theta(t) = B(t), \text{ a.s., a.e.} t \in [0,T]\}, \qquad (2.6)$$

and

$$\hat{\Theta} := \{\theta \in L^{\infty}_{\mathcal{F}}(0,T,\mathbf{R}^n) : \sigma(t)\theta(t) \ge B(t), \text{ a.s., a.e.} t \in [0,T]\}, \qquad (2.7)$$

where the greater or equal relation between two vectors is in the componentwise sense. It follows from Assumption (A) that $\emptyset \ne \Theta \subseteq \hat{\Theta}$.

Let $\theta \in \Theta$, and $(x(\cdot), u(\cdot))$ be an admissible wealth–portfolio pair. Then it is known ([7, p. 22, Proposition 2.2]) that

$$x(t) = H_\theta(t)^{-1} E(x(T)H_\theta(T)|\mathcal{F}_t), \text{ a.s., } \forall t \in [0,T]. \qquad (2.8)$$

Definition 2. *The market is said to be arbitrage-free if whenever a wealth process $x(\cdot)$ under an admissible portfolio satisfies $x(T) \ge 0$ a.s. and $P\{x(T) > 0\} > 0$, it must hold that $x(0) > 0$.*

Arbitrage-free is a very weak market condition, for many optimization problems would become ill-posed in a non arbitrage-free market. It is easy to show that the market is arbitrage-free under Assumption (A). Conversely, if the market is arbitrage-free, then it can be proved, as in [15, p. 12, Theorem 4.2], that there must be an $\mathcal{F}_t$-progressively measurable process θ satisfying $\sigma(t)\theta(t) = B(t)$, a.s., a.e.$t \in [0,T]$. Hence Assumption (A) is very close to the arbitrage-free assumption, a minimum condition for a "viable" market.

Let us now turn to the completeness of the market.

Definition 3. *A contingent claim $\xi \in L^2_{\mathcal{F}_T}(\Omega;\mathbf{R})$ is said to be replicable if there exists an initial wealth x and an admissible wealth–portfolio pair $(x(\cdot),\pi(\cdot))$ satisfying (2.3) with $x(T) = \xi$. The market is called complete if any contingent claim $\xi \in L^2_{\mathcal{F}_T}(\Omega;\mathbf{R})$ is replicable.*

Proposition 4. *Under Assumption (A), the market is complete if and only if* $\text{rank}(\sigma(t)) = n$, a.s., a.e.$t \in [0,T]$.

Proof. Consider the backward stochastic differential equation (BSDE) with a given $\xi \in L^2_{\mathcal{F}_T}(\Omega;\mathbf{R})$:

$$dx(t) = [r(t)x(t) + \theta(t)'z(t)]dt + z(t)'dW(t), \ \ x(T) = \xi, \qquad (2.9)$$

which admits a unique solution pair $(x(\cdot), z(\cdot)) \in L^2_{\mathcal{F}}(0,T,\mathbf{R}) \times L^2_{\mathcal{F}}(0,T,\mathbf{R}^n)$. If $\text{rank}(\sigma(t)) = n$, a.s., a.e.$t \in [0,T]$, then there exists $\pi(\cdot)$ such that $\sigma(t)'\pi(t) = z(t)$, a.s., a.e.$t \in [0,T]$. By Lemma 28, we may assume that the process $\pi(\cdot)$ is $\mathcal{F}_t$-progressively measurable. Substituting $z(t)$ by $\sigma(t)'\pi(t)$ in (2.9) we conclude that $(x(\cdot), \pi(\cdot))$ is an admissible wealth–portfolio pair with $x(T) = \xi$; hence ξ is replicable.

Conversely, assume that the market is complete. For any $z \in \mathbf{R}^n$, let $y(\cdot)$ solves the following SDE

$$dy(t) = [r(t)y(t) + \theta(t)'z]dt + z'dW(t), \quad y(0) = 0.$$

Since $y(T)$ is replicable, there exists $(x(\cdot), \pi(\cdot)) \in L^2_{\mathcal{F}}(0, T, \mathbf{R}) \times L^2_{\mathcal{F}}(0, T, \mathbf{R}^m)$ with $\sigma(\cdot)'\pi(\cdot) \in L^2_{\mathcal{F}}(0, T, \mathbf{R}^n)$ so that

$$dx(t) = [r(t)x(t) + \theta(t)'\sigma(t)'\pi(t)]dt + \pi(t)'\sigma(t)dW(t), \quad x(T) = y(T).$$

Comparing the two preceding equations and by the uniqueness of the BSDE solution we conclude that $\sigma(t)'\pi(t) = z$. This yields $\operatorname{rank}(\sigma(t)) = n$ as $z \in \mathbf{R}^n$ is arbitrary. □

Remark 5. There is a very similar result in [15, p. 24, Theorem 6.6]. However, notice that in [15] an admissible portfolio is defined to be square integrable in t almost surely in ω and tame (i.e., the corresponding wealth process is bounded below), whereas in our mean–variance setting an admissible portfolio is required to be square integrable in (t, ω) (otherwise the variance of the terminal wealth may not even be well defined). Moreover, the definition of completeness is also different there in terms of the set of contingent claims to be replicated (see [15, p. 21, Definition 6.1]). In other words, we have a *different* class of admissible portfolios and a *different* notion of market completeness which are dictated by the nature of our problem.

It should be noted that the number of stocks, m, is generally different from the dimension of the underlying Brownian motion, n, and $\operatorname{rank}(\sigma(t)) = n$ may not hold. Hence, the market is in general *incomplete* in our setup.

The following technical lemma is useful in the sequel.

Lemma 6. *Given a set $A \subseteq L^\infty_{\mathcal{F}}(0, T, \mathbf{R}^n)$. If $k\theta_1 + (1 - k)\theta_2 \in A$ whenever $\theta_1 \in A, \theta_2 \in A$ and $k \in L^1_{\mathcal{F}}(0, T; [0, 1])$, then the set $\{H_\theta(\cdot) : \theta \in A\}$ is convex.*

Proof. For any $\theta_1, \theta_2 \in A$ and $\lambda \in [0, 1]$, denote $H(\cdot) := \lambda H_{\theta_1}(\cdot) + (1 - \lambda)$ $H_{\theta_2}(\cdot)$. Then $H(0) = 1$, and

$$\begin{aligned} dH(t) &= \lambda[-r(t)H_{\theta_1}(t)dt - H_{\theta_1}(t)\theta_1(t)'dW(t)] \\ &\quad + (1 - \lambda)[-r(t)H_{\theta_2}(t)dt - H_{\theta_2}(t)\theta_2(t)'dW(t)] \\ &= -r(t)H(t)dt - H(t)[k(t)\theta_1(t) + (1 - k(t))\theta_2(t)]'dW(t), \end{aligned}$$

where $k(t) := \frac{\lambda H_{\theta_1}(t)}{\lambda H_{\theta_1}(t) + (1-\lambda)H_{\theta_2}(t)}$. Define $\theta(t) := k(t)\theta_1(t) + [1 - k(t)]\theta_2(t)$. Then $\theta \in A$. It then follows from the definition of H_θ, see (2.5), that $H(t) \equiv H_\theta(t)$. This completes the proof. □

To end this section we introduce two stochastic processes that are vital for the subsequent analysis.

Define the following processes

$$\theta^*(t) := \operatorname{argmin}_{\theta \in \{\theta \in \mathbf{R}^n : \sigma(t)\theta = B(t)\}} |\theta|^2, \tag{2.10}$$

and

$$\hat{\theta}(t) := \operatorname{argmin}_{\theta \in \{\theta \in \mathbf{R}^n : \sigma(t)\theta \geq B(t)\}} |\theta|^2. \tag{2.11}$$

Lemma 7. *We have the following conclusions:*

(i) $\theta^* \in L^\infty_\mathcal{F}(0, T, \mathbf{R}^n)$ and $\hat{\theta} \in L^\infty_\mathcal{F}(0, T, \mathbf{R}^n)$.
(ii) There exists an $\mathbf{R}^m$-valued, $\mathcal{F}_t$-progressively measurable process $u(\cdot)$ such that $\sigma(t)'u(t) = \theta^*(t)$, a.s., a.e.$t \in [0, T]$.
(iii) There exists an $\mathbf{R}^m_+$-valued, $\mathcal{F}_t$-progressively measurable process $v(\cdot)$ such that $\sigma(t)'v(t) = \hat{\theta}(t)$, a.s., a.e.$t \in [0, T]$.
(iv) For any $\theta \in \Theta$, $\theta^*(t)'\theta(t) = |\theta^*(t)|^2$, a.s., a.e.$t \in [0, T]$.
(v) For any $\theta \in \hat{\Theta}$, $\hat{\theta}(t)'\theta(t) \geq |\hat{\theta}(t)|^2 = \hat{\theta}(t)'\theta^*(t)$, a.s., a.e.$t \in [0, T]$.

The proof is relegated to the appendix.

Remark 8. When $\sigma(t)'\sigma(t)$ is uniformly positive definite (in which case the market is complete) the process θ^* is the only θ that satisfies $\sigma(t)\theta(t) = B(t)$, and θ^* is the so-called *pricing kernel.* In the present case of incomplete market, as will be demonstrated in what follows, θ^* and $\hat{\theta}$ play the same important roles of pricing kernels associated with different constraints on portfolios.

3 Markowitz's Portfolio Selection Models

Fix an initial wealth x_0. A general continuous-time Markowitz's mean–variance portfolio selection problem (with constrained portfolios) is formulated as

$$\begin{array}{ll} \text{minimize} & \operatorname{Var} x(T) \equiv Ex(T)^2 - z^2, \\ \text{subject to} & \begin{cases} Ex(T) = z, \quad \pi(\cdot) \in \Pi, \\ (x(\cdot), \pi(\cdot)) \text{ satisfies equation (2.3) with } x(0) = x_0, \\ (x(\cdot), \pi(\cdot)) \in C, \end{cases} \end{array} \tag{3.1}$$

where C is a given convex set in $L^2_\mathcal{F}(0, T, \mathbf{R}) \times \Pi$, and $z \in \mathbf{R}$ is a parameter. The optimal portfolio for this problem (corresponding to a fixed z) is called an *efficient portfolio*, and the set of all points $(\operatorname{Var} x^*(T), z)$, where $\operatorname{Var} x^*(T)$ denotes the optimal value of (3.1) corresponding to z and z runs over certain range of $\mathbf{R}$, is called the *efficient frontier*.

Remark 9. For each z, an optimal solution to (3.1) in fact gives rise to a *variance minimizing portfolio*, and the set of all points $(\text{Var}\, x^*(T), z)$ where z runs over the whole real axis is called a *variance minimizing frontier*. The financial interpretation of a variance minimizing portfolio is clear: it tries to minimize the variance, representing the risk, while specifying a targeted expected return depicted by z. A main difference between the Markowitz model and the utility one, inter alia, is the presence of this constraint on the terminal payoff. On the other hand, in the original Markowitz's definition, an efficient portfolio is both variance minimizing and *return maximizing* (that is, it maximizes the expected terminal payoff subject to a same variance level). In other words, the efficient frontier is only a certain portion of the variance minimizing frontier. This is why in the above definition the efficient frontier only corresponds to z being in certain range. There is a detailed study on this range in [2]. Here we only remark that in the case when all the market coefficients are deterministic, then the efficient range of z is $z \geq x_0 e^{\int_0^T r(t)dt}$.

In this paper, the following two cases of the constraint set C will be studied respectively:

Case 1. $C = L^2_{\mathcal{F}}(0,T,\mathbf{R}) \times \Pi$, corresponding to the case where portfolios are not constrained.

Case 2. $C = \{(x(\cdot),\pi(\cdot)) \in L^2_{\mathcal{F}}(0,T,\mathbf{R}) \times \Pi : \pi(t) \geq 0, \text{ a.s., a.e.} t \in [0,T]\}$, corresponding to the case where short-selling is prohibited.

Given a constraint set C associated with one of the two cases above, define the following *attainable terminal wealth set*:

$$\begin{aligned} A_C := \{X \in L^2_{\mathcal{F}_T}(\Omega;\mathbf{R}) :\ & \text{there exist } x \in \mathbf{R} \text{ and } \pi(\cdot) \in \Pi \text{ such that} \\ (x(\cdot),\pi(\cdot)) \text{ satisfies (2.3) with } & x(0) = x,\ x(T) = X, \text{ and } (x(\cdot),\pi(\cdot)) \in C\}. \end{aligned} \tag{3.2}$$

To solve problem (3.1), the following *static* optimization problem plays a critical role:

$$\begin{aligned} &\text{minimize} \quad EX^2 - z^2, \\ &\text{subject to} \begin{cases} EX = z, \\ E[XH_{\theta^*}(T)] = x_0, \\ X \in A_C, \end{cases} \end{aligned} \tag{3.3}$$

where θ^* is defined by (2.10). This problem is to locate the optimal attainable terminal wealth X^* in A_C. Once this is solved, an optimal portfolio for (3.1) can be obtained by replicating X^* (which is possible by the very definition of A_C along with the second constraint in (3.3)). Notice that, compared with the case of a complete market [2], the main difficulty in the present incomplete market situation is to characterize the attainable set A_C for each of the two cases before solving (3.3).

The following result verifies that in order to solve the original problem (3.1) it suffices to solve (3.3).

Theorem 10. *If $(x^*(\cdot), \pi^*(\cdot))$ is optimal for (3.1), then $x^*(T)$ is optimal for (3.3). Conversely, if $X^* \in \mathcal{A}_C$ is optimal for (3.3), then any wealth–portfolio pair $(x^*(\cdot), \pi^*(\cdot))$ satisfying (2.3) with $(x^*(\cdot), \pi^*(\cdot)) \in C$ and $x^*(T) = X^*$ is optimal for (3.1).*

Proof. This is straightforward by the definition of $\mathcal{A}_C$. □

To solve (3.3), we first transform it to an equivalent problem as stipulated in the following theorem.

Theorem 11. *If problem (3.3) admits a solution X^*, then there exists a pair of scalars (λ, μ) such that X^* is also the optimal solution for the following problem:*

$$\begin{array}{ll} \text{minimize} & E[X - (\lambda - \mu H_{\theta^*}(T))]^2, \\ \text{subject to} & X \in \mathcal{A}_C. \end{array} \tag{3.4}$$

Conversely, if there is a pair of scalars (λ, μ) such that the optimal solution X^* of (3.4) satisfies

$$\begin{cases} EX^* = z, \\ E[X^* H_{\theta^*}(T)] = x_0. \end{cases} \tag{3.5}$$

then X^* must be an optimal solution of (3.3).

Proof. It is easy to see that $\mathcal{A}_C$ is a convex set due to the convexity of C (for both cases). Hence the theorem can be proved in exactly the same fashion as [2, Theorem 4.1] (by applying a Lagrange multiplier approach [2, Proposition 4.1]). □

The preceding theorem suggests that in order to solve (3.3) one can first solve problem (3.4) for general (λ, μ), which is a problem with $\mathcal{A}_C$ being the only constraint set, and then determine the values of (λ, μ) via the equations (3.5).

To re-capture, solving the mean–variance problem (3.1) consists of the following steps:

Step 1 Solve (3.4) with parameters (λ, μ) and get solution $X^* = X^*(\lambda, \mu)$.
Step 2 Determine the values of (λ, μ) via (3.5).
Step 3 Any admissible portfolio (that satisfies the constraint specified by C) replicating $X^*(\lambda, \mu)$ is an efficient portfolio.

In the next two sections, we will study the two cases respectively. We will mainly devote ourselves to characterizing the attainable set $\mathcal{A}_C$ and solving (3.4) for each case. For the general situation when the market parameters $r(\cdot)$, $\mu_i(\cdot)$ and $\sigma_{ij}(\cdot)$ are stochastic processes, it is impossible to solve (3.4) *explicitly* in terms of (λ, μ). However, for the market when all the parameters are deterministic, we will obtain analytical solution to (3.4) and thereby get explicit solution to the original problem (3.1) for both cases.

4 Case 1: Portfolios Unconstrained

In this case the constraint set $C = L^2_{\mathcal{F}}(0, T, \mathbf{R}) \times \Pi$. Our first result characterizes the attainable terminal wealth set A_C for this constraint set.

Theorem 12. *Given $X \in L^2_{\mathcal{F}_T}(\Omega; \mathbf{R})$. The following assertions are equivalent:*

(i) $X \in A_C$.
(ii) $E[XH_\theta(T)]$ is independent of $\theta \in \Theta$.
(iii) $E[XH_\theta(T)]$ is independent of $\theta \in \Theta_1$ where

$$\Theta_1 := \{\theta \in \Theta : \parallel \theta - \theta^* \parallel_{L^\infty_{\mathcal{F}}(0,T,\mathbf{R}^n)} \leq 1\}. \tag{4.1}$$

Proof. If $X \in A_C$, then there is $x \in \mathbf{R}$ and a portfolio $\pi(\cdot) \in \Pi$ such that

$$\begin{cases} dx(t) = [r(t)x(t) + B(t)'\pi(t)]dt + \pi(t)'\sigma(t)dW(t), \\ x(0) = x, \quad x(T) = X. \end{cases}$$

Now, for any $\theta \in \Theta$,

$$\begin{aligned} dx(t) &= [r(t)x(t) + B(t)'\pi(t)]dt + \pi(t)'\sigma(t)dW(t) \\ &= [r(t)x(t) + \theta(t)'\sigma(t)'\pi(t)]dt + \pi(t)'\sigma(t)dW(t). \end{aligned}$$

Applying Itô's formula, we obtain

$$x \equiv x(0) = E[x(T)H_\theta(T)] = E[XH_\theta(T)],$$

implying that $E[XH_\theta(T)]$ is independent of the choice of $\theta \in \Theta$. This proves that (i) implies (ii).

The implication from (ii) to (iii) is trivial. To close the loop of equivalence we prove that (iii) yields (i). Assume that $E[XH_\theta(T)]$ does not depend on $\theta \in \Theta_1$. By the BSDE theory, for any $\theta \in \Theta_1$, the following equation

$$\begin{cases} dX(t) = [r(t)X(t) + \theta(t)'Z(t)]dt + Z(t)'dW(t), \\ X(T) = X \end{cases} \tag{4.2}$$

admits a unique solution pair $(X_\theta(\cdot), Z_\theta(\cdot))$, with $X_\theta(0) = E[XH_\theta(T)]$. So by the assumption $X_\theta(0)$, $\theta \in \Theta_1$, are all the same, which is denoted by x_0.

Next, let $(X_{\theta^*}(\cdot), Z_{\theta^*}(\cdot))$ solves (4.2) with $\theta = \theta^*$. We are to prove that there exists a portfolio $\pi_0(\cdot) \in \Pi$ such that

$$Z_{\theta^*}(t) = \sigma(t)'\pi_0(t), \text{ a.s., a.e.} t \in [0, T]. \tag{4.3}$$

Indeed, define

$$\pi_0(t) := \text{argmin}_{\pi \in \text{argmin}_{\pi \in \mathbf{R}^m} |\sigma(t)'\pi - Z_{\theta^*}(t)|^2} |\pi|^2.$$

Notice that the set $\text{argmin}_{\pi \in \mathbf{R}^m} |\sigma(t)'\pi - Z_{\theta^*}(t)|^2$ is nonempty due to the Frank–Wolfe theorem (Lemma 26). Moreover, $\pi \in \text{argmin}_{\pi \in \mathbf{R}^m} |\sigma(t)'\pi - Z_{\theta^*}(t)|^2$ if and only if $\sigma(t)\sigma(t)'\pi - \sigma(t)Z_{\theta^*}(t) = 0$. Thus $\pi_0(t)$ is well-defined (again by the Frank–Wolfe theorem). Furthermore, we can apply Lemma 28 to conclude that $\pi_0(\cdot)$ is an $\mathcal{F}_t$-progressively measurable stochastic process.

Set $\bar{\rho}(t) := \sigma(t)'\pi_0(t) - Z_{\theta^*}(t)$ and

$$\rho(t) := \begin{cases} 0, & \text{if } \bar{\rho}(t) = 0, \\ \bar{\rho}(t)/|\bar{\rho}(t)|, & \text{if } \bar{\rho}(t) \neq 0. \end{cases}$$

Then $\rho(\cdot) \in L^\infty_{\mathcal{F}}(0, T, \mathbf{R}^n)$. Moreover, $\sigma(t)\bar{\rho}(t) = \sigma(t)\sigma(t)'\pi_0(t) - \sigma(t) Z_{\theta^*}(t) = 0$, owing to the fact that $\pi_0(t)$ minimizes $|\sigma(t)'\pi - Z_{\theta^*}(t)|^2$. This implies that $\sigma(t)\rho(t) = 0$ and hence

$$\theta^* + \rho \in \Theta_1. \tag{4.4}$$

On the other hand, $Z_{\theta^*}(t)'\bar{\rho}(t) = [\pi_0(t)'\sigma(t)\bar{\rho}(t) - \bar{\rho}(t)'\bar{\rho}(t)] = -|\bar{\rho}(t)|^2$; thus $Z_{\theta^*}(t)'\rho(t) = -|\bar{\rho}(t)|^2$.

Define $\hat{X}(\cdot)$ to be the solution of the following (forward) SDE:

$$\begin{cases} d\hat{X}(t) = [r(t)\hat{X}(t) + (\theta^*(t) + \rho(t))'Z_{\theta^*}(t)]dt + Z_{\theta^*}(t)'dW(t), \\ \hat{X}(0) = x_0. \end{cases}$$

Itô's formula implies

$$E[\hat{X}(T)H_{\theta^*+\rho}(T)] = \hat{X}(0) = x_0 = E[X_{\theta^*}(T)H_{\theta^*+\rho}(T)], \tag{4.5}$$

where the last equality is due to (4.4) and the assumption. However,

$$d[\hat{X}(t) - X_{\theta^*}(t)] = r(t)[\hat{X}(t) - X_{\theta^*}(t)]dt + Z_{\theta^*}(t)'\rho(t)dt, \; \hat{X}(0) - X_{\theta^*}(0) = 0;$$

hence $\hat{X}(T) - X_{\theta^*}(T) = \int_0^T e^{\int_t^T r(s)ds} Z_{\theta^*}(t)'\rho(t)dt = -\int_0^T e^{\int_t^T r(s)ds}|\bar{\rho}(t)|^2 dt$. Comparing this with (4.5) we conclude that $\bar{\rho}(t) = 0$, a.s., a.e.$t \in [0, T]$, which leads to (4.3). Since $\sigma(\cdot)'\pi_0(\cdot) = Z_{\theta^*}(\cdot) \in L^2_{\mathcal{F}}(0, T, \mathbf{R}^n)$, it follows that $\pi_0(\cdot) \in \Pi$. Now, the BSDE (4.2) that $(X_{\theta^*}(\cdot), Z_{\theta^*}(\cdot))$ satisfies can be rewritten as

$$\begin{cases} dX_{\theta^*}(t) = [r(t)X_{\theta^*}(t) + B(t)'\pi_0(t)]dt + \pi_0(t)'\sigma(t)dW(t), \\ X_{\theta^*}(T) = X, \end{cases}$$

which means that X is attained by the portfolio $\pi_0(\cdot)$. □

Remark 13. Similar results have been obtained before in, e.g., [6, 8, 16], albeit in different contexts. Again, in these works, an admissible portfolio is defined to be square integrable in t almost surely in ω and/or tame, which is different from ours. There are also other technical subtleties comparing our result to the existing ones. For example, in [8] a contingent claim is assumed to be bounded above by the terminal value of a portfolio (see [8, p. 35]). This condition seems to be critical in deriving the result there. Finally, we believe our proof, based on the BSDE theory, is quite clean and simple compared with those in [6,8,16]. The same can be said of Theorem 20 for the no-shorting case.

Corollary 14. *A_C is a (nonempty) linear subspace of $L^2_{\mathcal{F}_T}(\Omega;\mathbf{R})$.*

By Theorem 12, we can rewrite problem (3.4) as follows:

$$\begin{array}{ll} \text{minimize} & E[X-(\lambda-\mu H_{\theta^*}(T))]^2, \\ \text{subject to} & \begin{cases} X \in L^2_{\mathcal{F}_T}(\Omega;\mathbf{R}), \\ E[X(H_\theta(T)-H_{\theta^*}(T))] = 0 \ \ \forall \theta \in \Theta. \end{cases} \end{array} \tag{4.6}$$

First notice that (4.6) is a convex optimization problem with a coercive, strictly convex cost function and a nonempty, closed convex constraint set; hence it must admit a unique optimal solution. However, it is generally hard to construct the optimal solution since (4.6) actually involves *infinitely many* constraints. Denote $L := \text{span}\{H_\theta(T) - H_{\theta^*}(T) : \theta \in \Theta\}$, where span$(A)$ means the minimal linear space that contains A, and consider $\bar{L}$, the closure of L in the $L^2_{\mathcal{F}_T}(\Omega;\mathbf{R})$-norm. Since each $H_\theta(T) \in L^2_{\mathcal{F}_T}(\Omega;\mathbf{R})$, it follows that $\bar{L} \subset L^2_{\mathcal{F}_T}(\Omega;\mathbf{R})$. The following theorem provides a way to finding a solution to (4.6).

Theorem 15. *For any given (λ,μ), consider the following problem*

$$\begin{array}{l} \text{minimize } \ E(\lambda-\mu H_{\theta^*}(T)-Y)^2, \\ \text{subject to } Y \in \bar{L}. \end{array} \tag{4.7}$$

We have the following conclusions:

(i) *Problem (4.7) admits a unique optimal solution. Moreover, $Y^* \in \bar{L}$ is the optimal solution to (4.7) if and only if $\lambda - \mu H_{\theta^*}(T) - Y^* \in A_C$.*
(ii) *The unique optimal solution to (4.6) can be expressed as $X^* = \lambda - \mu H_{\theta^*}(T) - Y^*$ where Y^* is the unique optimal solution to (4.7).*

Proof. (i) First of all, by the projection theorem in Hilbert spaces (refer to, e.g., [22, p. 51, Theorem 2]), (4.7) has a unique optimal solution Y^*. Moreover, Y^* is optimal for (4.7) if and only if $E[(\lambda-\mu H_{\theta^*}(T)-Y^*)Y] = 0\ \forall Y \in \bar{L}$. The latter is equivalent to that $X^* := \lambda-\mu H_{\theta^*}(T)-Y^*$ is feasible for (4.6) which, in view of Theorem 12, is further equivalent to that $\lambda-\mu H_{\theta^*}(T)-Y^* \in A_C$.

(ii) We have proved in (i) that $X^* = \lambda-\mu H_{\theta^*}(T)-Y^*$ is feasible for (4.6) if Y^* is optimal for (4.7). Now, for any feasible solution X of (4.6):

$$\begin{aligned} E[X-&(\lambda-\mu H_{\theta^*}(T))]^2 \\ &= E[X-(\lambda-\mu H_{\theta^*}(T))+Y^*]^2 \\ &\quad + 2E[Y^*(\lambda-\mu H_{\theta^*}(T))] - 2E[XY^*] - E[Y^*]^2 \\ &= E[X-(\lambda-\mu H_{\theta^*}(T)-Y^*)]^2 \\ &\quad + 2E[Y^*(\lambda-\mu H_{\theta^*}(T))] - 2E[X^*Y^*] - E[Y^*]^2 \\ &\geq E[X^*-(\lambda-\mu H_{\theta^*}(T)-Y^*)]^2 \\ &\quad + 2E[Y^*(\lambda-\mu H_{\theta^*}(T))] - 2E[X^*Y^*] - E[Y^*]^2 \\ &= E[X^*-(\lambda-\mu H_{\theta^*}(T))]^2, \end{aligned}$$

where we have used the fact that $E[XY^*] = E[X^*Y^*] = 0$ due to the constraint of problem (4.6). Hence X^* is the unique optimal solution to (4.6). □

Remark 16. The above theorem suggests that one can obtain the optimal solution to (4.6), hence that to (3.4), via the (unique) optimal solution to the projection problem (4.7). In fact, (4.7) is a dual problem of (4.6), in the sense that the cost function of the former is the conjugate of that of the latter, while the feasible regions of the two problems are orthogonal to each other. In many cases the primal–dual relation between (4.6) and (4.7) helps us in finding solutions to the both, as solving one problem may be easier than directly solving the other.

When the market parameters, $r(\cdot), \mu(\cdot)$ and $\sigma(\cdot)$, are all deterministic processes, both (4.6) and (4.7) can be solved explicitly which in turn leads to the closed-form solution to the underlying mean–variance portfolio selection problem.

Lemma 17. *If $r(\cdot), \mu(\cdot)$ and $\sigma(\cdot)$ are deterministic, then $\lambda - \mu H_{\theta^*}(T) \in A_C$ for any (λ, μ).*

Proof. Fix $\theta \in \Theta$. We have

$$
\begin{aligned}
&H_{\theta^*}(T)H_\theta(T)\\
&= \exp\{-\int_0^T [2r(t) + \frac{1}{2}(|\theta(t)|^2 + |\theta^*(t)|^2)]dt - \int_0^T [\theta(t) + \theta^*(t)]'dW(t)\}\\
&= \exp\{-\int_0^T [2r(t) - |\theta^*(t)|^2]dt\}\\
&\quad \times \exp\{-\int_0^T \frac{1}{2}|\theta(t) + \theta^*(t)|^2 dt - \int_0^T [\theta(t) + \theta^*(t)]'dW(t)\},
\end{aligned}
$$

where we have used $\theta^*(t)'\theta(t) = |\theta^*(t)|^2$; see Lemma 7-(iv). Thus,

$$
E[H_{\theta^*}(T)H_\theta(T)] = \exp\{-\int_0^T [2r(t) - |\theta^*(t)|^2]dt\}
$$

which is independent of $\theta \in \Theta$. We have then $H_{\theta^*}(T) \in A_C$ thanks to Theorem 12. The conclusion follows as $\lambda \in A_C$. □

Theorem 18. *If $r(\cdot), \mu(\cdot)$ and $\sigma(\cdot)$ are deterministic, and $\int_0^T |B(t)|dt > 0$, then $\pi(t) := [\lambda e^{-\int_t^T r(s)ds} - x(t)]u(t)$ is an efficient portfolio, in a feedback form, for the mean–variance problem (3.1) corresponding to $z \geq x_0 e^{\int_0^T r(t)dt}$, where*

$$\lambda = \frac{ze^{\int_0^T |\theta^*(t)|^2 dt} - x_0 e^{\int_0^T r(t)dt}}{e^{\int_0^T |\theta^*(t)|^2 dt} - 1}, \qquad \mu = \frac{ze^{\int_0^T r(t)dt} - x_0 e^{\int_0^T 2r(t)dt}}{e^{\int_0^T |\theta^*(t)|^2 dt} - 1}, \tag{4.8}$$

and $u(\cdot)$ is a measurable function satisfying $\sigma(t)'u(t) = \theta^*(t)$. Moreover, the efficient frontier is

$$\mathrm{Var}(x(T)) = \frac{1}{e^{\int_0^T |\theta^*(t)|^2 dt} - 1}[z - x_0 e^{\int_0^T r(t)dt}]^2, \quad z \geq x_0 e^{\int_0^T r(t)dt}. \tag{4.9}$$

Proof. Fix $z \geq x_0 e^{\int_0^T r(t)dt}$. By virtue of Lemma 17 and Theorem 15, $Y^* = 0$ is the unique optimal solution to (4.7), or $X^* = \lambda - \mu H_{\theta^*}(T)$ is the unique optimal solution to (4.6). To determine (λ, μ) so as to obtain the solution to (3.3), we apply Theorem 11 to derive the following system of equations

$$\begin{cases} \lambda - \mu E H_{\theta^*}(T) = z \\ \lambda E H_{\theta^*}(T) - \mu E[H_{\theta^*}(T)^2] = x_0. \end{cases}$$

It follows from $\int_0^T |B(t)|dt > 0$ that $e^{\int_0^T |\theta^*(t)|^2 dt} - 1 \neq 0$. Solving the preceding equations we get the expressions (4.8), noting that $EH_{\theta^*}(T) = e^{-\int_0^T r(t)dt}$, $E[H_{\theta^*}(T)^2] = e^{-\int_0^T [2r(t) - |\theta^*(t)|^2]dt}$.

An admissible portfolio is efficient corresponding to z if it replicates the terminal wealth $X^* = \lambda - \mu H_{\theta^*}(T)$. Appealing to (2.8), we can get the corresponding wealth process to be

$$\begin{aligned} x(t) &= H_{\theta^*}(t)^{-1} E\Big([\lambda - \mu H_{\theta^*}(T)] H_{\theta^*}(T) | \mathcal{F}_t\Big) \\ &= \lambda e^{-\int_t^T r(s)ds} - \mu e^{-\int_t^T (2r(s) - |\theta^*(s)|^2)ds} H_{\theta^*}(t). \end{aligned}$$

A direct computation on the above, using (2.5), yields

$$\begin{aligned} dx(t) &= [rx(t) + \mu |\theta^*(t)|^2 e^{-\int_t^T (2r(s) - |\theta^*(s)|^2)ds} H_{\theta^*}(t)]dt \\ &\quad + \mu e^{-\int_t^T (2r(s) - |\theta^*(s)|^2)ds} H_{\theta^*}(t)\theta^*(t)'dW(t) \\ &= [rx(t) + \mu e^{-\int_t^T (2r(s) - |\theta^*(s)|^2)ds} H_{\theta^*}(t)\theta^*(t)'\theta^*(t)]dt \\ &\quad + \mu e^{-\int_t^T (2r(s) - |\theta^*(s)|^2)ds} H_{\theta^*}(t)\theta^*(t)'dW(t). \end{aligned}$$

Comparing the above with the wealth equation (2.3), we conclude that a portfolio $\pi(\cdot)$ realizes the wealth process $x(\cdot)$ if and only if

$$\sigma(t)'\pi(t) = \mu e^{-\int_t^T (2r(s) - |\theta^*(s)|^2)ds} H_{\theta^*}(t)\theta^*(t). \tag{4.10}$$

By Lemma 7-(ii), there exists an $\mathcal{F}_t$-progressively measurable process $u(\cdot)$ satisfying $\sigma(t)'u(t) = \theta^*(t)$. Hence, the following portfolio

$$\pi(t) := \mu e^{-\int_t^T (2r(s)-|\theta^*(s)|^2)ds} H_{\theta^*}(t)u(t) \equiv [\lambda e^{-\int_t^T r(s)ds} - x(t)]u(t)$$

indeed satisfies (4.10), and hence is efficient.

Finally, the variance of the optimal terminal wealth is

$$\begin{aligned}\mathrm{Var}(x(T)) = \mathrm{Var}(X^*) = \mu^2 \mathrm{Var}(H_{\theta^*}(T)) &= \frac{[zEH_{\theta^*}(T) - x_0]^2}{\mathrm{Var}(H_{\theta^*}(T))} \\ &= \frac{1}{e^{\int_0^T |\theta^*(t)|^2 dt} - 1}[z - x_0 e^{\int_0^T r(t)dt}]^2.\end{aligned}$$

□

Remark 19. While the wealth process that replicates X^* is unique, there may be more than one replicating portfolios, i.e., there may be many portfolios $\pi(\cdot)$ satisfying (4.10). Hence, efficient portfolios corresponding to a same z are not unique.

5 Case 2: Shorting Prohibited

Again, we need to first characterize the attainable set A_C in this case.

Theorem 20. *For any $X \in L^2_{\mathcal{F}_T}(\Omega; \mathbf{R})$, $X \in A_C$ if and only if there exists $\bar{\theta} \in \hat{\Theta}$ such that $\sup_{\theta\in\hat{\Theta}} E[XH_\theta(T)] = E[XH_{\bar{\theta}}(T)]$. Furthermore, $\sup_{\theta\in\hat{\Theta}} E[XH_\theta(T)] = E[XH_{\theta^*}(T)]$ if $X \in A_C$.*

Proof. If $X \in A_C$, then there is $(x(\cdot), \pi(\cdot)) \in C$ satisfying (2.3) with $x_0 = E[XH_{\theta^*}(T)]$. Take any $\theta \in \hat{\Theta}$ and consider $H_\theta(\cdot)$ that satisfies (2.5). Applying Itô's formula we get easily

$$d[x(t)H_\theta(t)] = [B(t)-\sigma(t)\theta(t)]'\pi(t)H_\theta(t)dt+[\pi(t)'\sigma(t)-x(t)\theta(t)']H_\theta(t)dW(t);$$

thus $E[XH_\theta(T)] = x_0+E\int_0^T [B(t)-\sigma(t)\theta(t)]'\pi(t)H_\theta(t)dt \le x_0 = E[XH_{\theta^*}(T)]$ (here that the expectation of the stochastic integral vanishes can be proved in the same way as in proving (2.8); see [7, p. 22, Proposition 2.2]). This yields $\sup_{\theta\in\hat{\Theta}} E[XH_\theta(T)] = E[XH_{\theta^*}(T)]$.

Conversely, suppose there is $\bar{\theta} \in \hat{\Theta}$ such that $x_0 := E[XH_{\bar{\theta}}(T)] \ge E[XH_\theta(T)]\ \forall\theta \in \hat{\Theta}$. Let $(X^*(\cdot), Z^*(\cdot))$ be the unique solution to the following BSDE

$$\begin{cases} dX^*(t) = [r(t)X^*(t) + \bar{\theta}(t)'Z^*(t)]dt + Z^*(t)'dW(t), \\ X^*(T) = X. \end{cases} \tag{5.1}$$

We are to show that there exists an admissible portfolio $\pi_0(\cdot)$ satisfying the no-shorting constraint such that

$$Z^*(t) = \sigma(t)'\pi_0(t), \text{ a.s., a.e.} t \in [0, T]. \tag{5.2}$$

Indeed, define

$$\pi_0(t) := \text{argmin}_{\pi \in \text{argmin}_{\pi \in \mathbf{R}_+^m} |\sigma(t)'\pi - Z^*(t)|^2} |\pi|^2.$$

Note that $\text{argmin}_{\pi \in \mathbf{R}_+^m} |\sigma(t)'\pi - Z^*(t)|^2 \neq \emptyset$ due to the Frank–Wolfe theorem (Lemma 26). On the other hand, $\pi \in \text{argmin}_{\pi \in \mathbf{R}_+^m} |\sigma(t)'\pi - Z^*(t)|^2$ can be rewritten as $|\sigma(t)'\pi - Z^*(t)|^2 - g(t) \leq 0$, where $g(t) := \min_{\pi \in \mathbf{R}_+^m} |\sigma(t)'\pi - Z^*(t)|^2$ which is clearly $\mathcal{F}_t$-progressively measurable. Hence we can apply Lemma 28 to conclude that $\pi_0(\cdot)$ is an $\mathcal{F}_t$-progressively measurable stochastic process. Note $Z^*(t) \notin \{\sigma(t)'\pi : \pi \in \mathbf{R}_+^m\}$ whenever $\sigma(t)'\pi_0(t) - Z^*(t) \neq 0$. Thus by Lemmas 27 and 28, there is $\bar{\rho}(\cdot)$ which is $\mathcal{F}_t$-progressively measurable satisfying $\bar{\rho}(t) \neq 0, Z^*(t)'\bar{\rho}(t) < 0, \sigma(t)\bar{\rho}(t) \geq 0$, a.s., a.e.$t$ on the set where $\sigma(t)'\pi_0(t) - Z^*(t) \neq 0$. Set

$$\rho(t) := \begin{cases} 0, & \text{if } \sigma(t)'\pi_0(t) - Z^*(t) = 0, \\ \bar{\rho}(t)/|\bar{\rho}(t)|, & \text{if } \sigma(t)'\pi_0(t) - Z^*(t) \neq 0. \end{cases} \tag{5.3}$$

Then $\rho(\cdot) \in L^\infty_{\mathcal{F}}(0, T, \mathbf{R}^n)$, $\sigma(t)\rho(t) \geq 0$, and

$$Z^*(t)'\rho(t) < 0 \quad \text{whenever} \quad \sigma(t)'\pi_0(t) - Z^*(t) \neq 0. \tag{5.4}$$

Since $\sigma(t)[\bar{\theta}(t) + \rho(t)] \geq \sigma(t)\bar{\theta}(t) \geq B(t)$, we conclude $\bar{\theta} + \rho \in \hat{\Theta}$.

Define $\bar{X}(\cdot)$ to be the solution of the following SDE:

$$\begin{cases} d\bar{X}(t) = [r(t)\bar{X}(t) + (\bar{\theta}(t) + \rho(t))'Z^*(t)]dt + Z^*(t)'dW(t), \\ \bar{X}(0) = x_0. \end{cases}$$

Then

$$E[\bar{X}(T)H_{\bar{\theta}+\rho}(T)] = \bar{X}(0) = x_0 = E[X^*(T)H_{\bar{\theta}}(T)] \geq E[X^*(T)H_{\bar{\theta}+\rho}(T)]. \tag{5.5}$$

On the other hand,

$$d[\bar{X}(t) - X^*(t)] = r(t)[\bar{X}(t) - X^*(t)]dt + Z^*(t)'\rho(t)dt, \ \bar{X}(0) - X^*(0) = 0;$$

hence $\bar{X}(T) - X^*(T) = \int_0^T e^{\int_t^T r(s)ds} Z^*(t)'\rho(t)dt$. It then follows from (5.5) and (5.4) that $\sigma(t)'\pi_0(t) - Z^*(t) = 0$, a.s., a.e.$t \in [0, T]$. This proves (5.2).

Next, let $\hat{X}(\cdot)$ be the solution to the following SDE:

$$\begin{cases} d\hat{X}(t) = [r(t)\hat{X}(t) + \theta^*(t)'Z^*(t)]dt + Z^*(t)'dW(t), \\ \hat{X}(0) = x_0. \end{cases} \tag{5.6}$$

Then $E[\hat{X}(T)H_{\theta^*}(T)] = x_0 \geq E[X^*(T)H_{\theta^*}(T)]$. On the other hand,

$$d[\hat{X}(t) - X^*(t)] = r(t)[\hat{X}(t) - X^*(t)]dt + [\theta^*(t) - \bar{\theta}(t)]'Z^*(t)dt$$
$$= r(t)[\hat{X}(t) - X^*(t)]dt + [B(t) - \sigma(t)\bar{\theta}(t)]'\pi_0(t)dt,$$

where we have used the fact that $Z^*(t) = \sigma(t)'\pi_0(t)$. Hence, $\hat{X}(T) - X^*(T) = \int_0^T e^{\int_t^T r(s)ds}[B(t) - \sigma(t)\bar{\theta}(t)]'\pi_0(t)dt \leq 0$. By $E[(\hat{X}(T) - X^*(T))H_{\theta^*}(T)] \geq 0$ we have

$$B(t)'\pi_0(t) = \bar{\theta}(t)'\sigma(t)'\pi_0(t) \equiv \bar{\theta}(t)'Z^*(t), \text{ a.s., a.e.} t \in [0,T], \tag{5.7}$$

and

$$E[XH_{\bar{\theta}}(T)] \equiv x_0 \equiv E[\hat{X}(T)H_{\theta^*}(T)] = E[XH_{\theta^*}(T)]. \tag{5.8}$$

It follows from (5.1) and (5.7) that $(X^*(\cdot), \pi_0(\cdot))$ satisfies

$$\begin{cases} dX^*(t) = [r(t)X^*(t) + B(t)'\pi_0(t)]dt + \pi_0(t)'\sigma(t)dW(t), \\ X^*(T) = X, \end{cases}$$

meaning that $X \in A_C$. Finally, the second assertion of the theorem follows from (5.8). □

Corollary 21. *A_C is a (nonempty) convex subset of $L^2_{\mathcal{F}_T}(\Omega; \mathbf{R})$.*

Remark 22. The preceding theorem along with Theorem 12 imply that $X \in A_C$ if and only if the maximum of $E[XH_\theta(T)]$ over $\theta \in \hat{\Theta}$ is achieved at *any* point on Θ, the "boundary" of $\hat{\Theta}$.

By virtue of Theorem 20, problem (3.4) for Case 2 can be written as

$$\begin{array}{ll} \text{minimize} & E[X - (\lambda - \mu H_{\theta^*}(T))]^2, \\ \text{subject to} & \begin{cases} X \in L^2_{\mathcal{F}_T}(\Omega; \mathbf{R}), \\ \max_{\theta \in \hat{\Theta}} E[XH_\theta(T)] = E[XH_{\theta^*}(T)]. \end{cases} \end{array} \tag{5.9}$$

Denote $M := \left\{k\left(H_\theta(T) - H_{\theta^*}(T)\right) \in L^2_{\mathcal{F}_T}(\Omega; \mathbf{R}) : k \geq 0,\ \theta \in \hat{\Theta}\right\}$, which can be easily verified, via Lemma 6, to be a convex cone. Consider $\bar{M}$, the closure of M in the $L^2_{\mathcal{F}_T}(\Omega; \mathbf{R})$-norm, which is a closed convex cone.

The following theorem is the no-shorting counterpart of Theorem 15.

Theorem 23. *For any given (λ, μ), consider the following problem*

$$\begin{array}{ll} \text{minimize} & E(\lambda - \mu H_{\theta^*}(T) - Y)^2, \\ \text{subject to} & Y \in \bar{M}. \end{array} \tag{5.10}$$

We have the following conclusions:

(i) *Problem (5.10) admits a unique optimal solution. Moreover, $Y^* \in \bar{M}$ is the optimal solution to (5.10) if and only if*

$$E[(\lambda - \mu H_{\theta^*}(T) - Y^*)Y^*] = 0, \quad \lambda - \mu H_{\theta^*}(T) - Y^* \in A_C. \tag{5.11}$$

(ii) *The unique optimal solution to (5.9) can be expressed as* $X^* = \lambda - \mu H_{\theta^*}(T) - Y^*$ *where* Y^* *is the unique optimal solution to (5.10).*

Proof. (i) First of all, (5.10) is an optimization problem with a coercive, strictly convex cost function and a nonempty, closed convex constraint set, which therefore must admit a unique optimal solution. Moreover, $Y^* \in \bar{M}$ is optimal to (5.10) if and only if for any $Y \in \bar{M}$, $0 \in \text{argmin}_{0 \le \alpha \le 1} E[f(\alpha)]$ where $f(\alpha) := [\lambda - \mu H_{\theta^*}(T) - Y^* + \alpha(Y^* - Y)]^2$. Now, for any sufficiently small $h > 0$, we have

$$\begin{aligned}\left|\tfrac{1}{h}[f(h) - f(0)]\right| &\le \tfrac{1}{h}\left|[2(\lambda - \mu H_{\theta^*}(T) - Y^*) + h(Y^* - Y)]h(Y^* - Y)\right| \\ &\le \left|2(\lambda - \mu H_{\theta^*}(T) - Y^*) + h(Y^* - Y)\right| \cdot |Y^* - Y|.\end{aligned}$$

Thus by the dominated convergence theorem we have

$$\begin{aligned}\lim_{h\to 0+} \frac{Ef(h) - Ef(0)}{h} &= E\frac{\partial}{\partial \alpha}[\lambda - \mu H_{\theta^*}(T) - Y^* + \alpha(Y^* - Y)]^2\Big|_{\alpha=0} \\ &= 2E[(\lambda - \mu H_{\theta^*}(T) - Y^*)(Y^* - Y)].\end{aligned}$$

Consequently, $Y^* \in \bar{M}$ is optimal if and only if

$$E[(\lambda - \mu H_{\theta^*}(T) - Y^*)(Y^* - Y)] \ge 0 \ \ \forall Y \in \bar{M}. \tag{5.12}$$

To prove that (5.11) and (5.12) are equivalent, first note that (5.11) easily yields (5.12) thanks to Theorem 20. Now, suppose (5.12) holds. Taking $Y = 0 \in \bar{M}$ we get from (5.12) that $E[(\lambda - \mu H_{\theta^*}(T) - Y^*)Y^*] \ge 0$, and taking $Y = 2Y^* \in \bar{M}$ (recall that $\bar{M}$ is a cone) we get $E[(\lambda - \mu H_{\theta^*}(T) - Y^*)Y^*] \le 0$. Consequently $E[(\lambda - \mu H_{\theta^*}(T) - Y^*)Y^*] = 0$ and, together with (5.12), results in (5.11).

(ii) We have proved in (i) that $X^* = \lambda - \mu H_{\theta^*}(T) - Y^*$ is feasible for (5.9) if Y^* is optimal for (5.10). On the other hand, for any feasible solution X of (5.9):

$$\begin{aligned}&E[X - (\lambda - \mu H_{\theta^*}(T))]^2 \\ &\quad = E[X - (\lambda - \mu H_{\theta^*}(T)) + Y^*)]^2 \\ &\qquad + 2E[Y^*(\lambda - \mu H_{\theta^*}(T))] - 2E[XY^*] - E[Y^*]^2 \\ &\quad \ge E[X - (\lambda - \mu H_{\theta^*}(T) - Y^*)]^2 \\ &\qquad + 2E[Y^*(\lambda - \mu H_{\theta^*}(T))] - 2E[X^*Y^*] - E[Y^*]^2 \\ &\quad \ge E[X^* - (\lambda - \mu H_{\theta^*}(T) - Y^*)]^2 \\ &\qquad + 2E[Y^*(\lambda - \mu H_{\theta^*}(T))] - 2E[X^*Y^*] - E[Y^*]^2 \\ &\quad = E[X^* - (\lambda - \mu H_{\theta^*}(T))]^2,\end{aligned}$$

where we have used the facts that $E[XY^*] \le 0$ and $E[X^*Y^*] = 0$. This means that X^* is an optimal solution for (5.9). □

As with Case 1, we now discuss the case when all the market coefficients are deterministic and show how to apply Theorem 23 to solve the mean–variance problem. First recall the definition of the pricing kernel $\hat{\theta}$; see (2.11).

Lemma 24. *If $r(\cdot), \mu(\cdot)$ and $\sigma(\cdot)$ are deterministic, then for any given (λ, μ) with $\mu \geq 0$, $\lambda - \mu H_{\hat{\theta}}(T) \in A_C$ and (5.10) has the optimal solution $Y^* := \mu(H_{\hat{\theta}}(T) - H_{\theta^*}(T))$.*

Proof. According to Theorems 20 and 23, to prove the desired results it suffices to show that

$$\begin{aligned} &E[(\lambda - \mu H_{\hat{\theta}}(T))(H_{\hat{\theta}}(T) - H_{\theta^*}(T))] = 0, \\ &E[(\lambda - \mu H_{\hat{\theta}}(T))(H_{\theta}(T) - H_{\theta^*}(T))] \leq 0 \ \ \forall \theta \in \hat{\Theta}. \end{aligned}$$

Since in the case of deterministic coefficients the value of $E[H_\theta(T)]$ is independent of $\theta \in \hat{\Theta}$, the above is equivalent to (noting that $\mu \geq 0$)

$$\begin{aligned} &E[H_{\hat{\theta}}(T)(H_{\hat{\theta}}(T) - H_{\theta^*}(T))] = 0, \text{ and} \\ &E[H_{\hat{\theta}}(T)(H_{\theta}(T) - H_{\theta^*}(T))] \geq 0 \ \ \forall \theta \in \hat{\Theta}. \end{aligned} \tag{5.13}$$

Now,

$$\begin{aligned} &H_{\hat{\theta}}(T)H_{\theta^*}(T) \\ &\quad = \exp\{-\int_0^T [2r(t) + \frac{1}{2}(|\hat{\theta}(t)|^2 + |\theta^*(t)|^2)]dt - \int_0^T [\hat{\theta}(t) + \theta^*(t)]'dW(t)\} \\ &\quad = \exp\{-\int_0^T [2r(t) - \hat{\theta}(t)'\theta^*(t)]dt\} \\ &\qquad \times \exp\{-\int_0^T \frac{1}{2}|\theta(t) + \theta^*(t)|^2 dt - \int_0^T [\theta(t) + \theta^*(t)]'dW(t)\} \\ &\quad = \exp\{-\int_0^T [2r(t) - |\hat{\theta}(t)|^2]dt\} \\ &\qquad \times \exp\{-\int_0^T \frac{1}{2}|\theta(t) + \theta^*(t)|^2 dt - \int_0^T [\theta(t) + \theta^*(t)]'dW(t)\}, \end{aligned}$$

where we have used the identity $\hat{\theta}(t)'\theta^*(t) = |\hat{\theta}(t)|^2$; see Lemma 7-(v). Thus, $E[H_{\hat{\theta}}(T)H_{\theta^*}(T)] = \exp\{-\int_0^T [2r(t) - |\hat{\theta}(t)|^2]dt\} = E[H_{\hat{\theta}}(T)^2]$, which proves the first equality of (5.13). Next, thanks to the inequality in Lemma 7-(v), a similar calculation as above shows that, for any $\theta \in \hat{\Theta}$, $E[H_{\hat{\theta}}(T)H_\theta(T)] = \exp\{-\int_0^T [2r(t) - \hat{\theta}(t)'\theta(t)]dt\} \geq \exp\{-\int_0^T [2r(t) - |\hat{\theta}(t)|^2]dt\} = E[H_{\hat{\theta}}(T)^2]$. This, together with the proved first equality of (5.13), leads to the second inequality of (5.13). □

Theorem 25. *If $r(\cdot)$, $\mu(\cdot)$ and $\sigma(\cdot)$ are deterministic, and $\sum_{j=1}^m \int_0^T B(t)_j^+ dt > 0$ where $B(t)_j$ denotes the j-th component of $B(t)$, then $\pi(t) := [\lambda e^{-\int_t^T r(s)ds} - x(t)]v(t)$ is an efficient portfolio for the mean–variance problem (3.1) corresponding to $z \geq x_0 e^{\int_0^T r(t)dt}$, where*

$$\lambda = \frac{ze^{\int_0^T |\hat{\theta}(t)|^2 dt} - x_0 e^{\int_0^T r(t)dt}}{e^{\int_0^T |\hat{\theta}(t)|^2 dt} - 1}, \qquad \mu = \frac{ze^{\int_0^T r(t)dt} - x_0 e^{\int_0^T 2r(t)dt}}{e^{\int_0^T |\hat{\theta}(t)|^2 dt} - 1}, \tag{5.14}$$

and $v(\cdot)$ is an $\mathbf{R}_+^m$-valued measurable function satisfying $\sigma(t)'v(t) = \hat{\theta}(t)$. Moreover, the efficient frontier is

$$\mathrm{Var}(x(T)) = \frac{1}{e^{\int_0^T |\hat{\theta}(t)|^2 dt} - 1}[z - x_0 e^{\int_0^T r(t)dt}]^2, \quad z \geq x_0 e^{\int_0^T r(t)dt}. \tag{5.15}$$

Proof. By virtue of Lemma 24 and Theorem 23, $X^* := \lambda - \mu H_{\hat{\theta}}(T)$ is the unique optimal solution to (5.9), provided that $\mu \geq 0$. The system of equations (3.5) reduces to

$$\begin{cases} \lambda - \mu E H_{\hat{\theta}}(T) = z \\ \lambda E H_{\hat{\theta}}(T) - \mu E[H_{\hat{\theta}}(T)^2] = x_0, \end{cases}$$

where we have used the fact that $E[H_{\hat{\theta}}(T) H_{\theta^*}(T)] = E[H_{\hat{\theta}}(T)^2]$ which was proved in the proof of Lemma 24. When $\sum_{j=1}^m \int_0^T B(t)_j^+ dt > 0$, it must hold that $\int_0^T |\hat{\theta}(t)|^2 dt > 0$; therefore (5.14) is well-defined which gives the (only) solution pair to the above system, with $\mu \geq 0$ under the assumption that $z \geq x_0 e^{\int_0^T r(t)dt}$.

Going through exactly the same argument that leads to (4.10), we have that an admissible portfolio $\pi(\cdot)$ is efficient if and only if it satisfies the no-shorting constraint and

$$\sigma(t)'\pi(t) = \mu e^{-\int_t^T (2r(s) - |\hat{\theta}(s)|^2)ds} H_{\hat{\theta}}(t)\hat{\theta}(t). \tag{5.16}$$

By Lemma 7-(iii), there exists an $\mathbf{R}_+^m$-valued, $\mathcal{F}_t$-progressively measurable stochastic process $v(\cdot)$ satisfying $\sigma(t)'v(t) = \hat{\theta}(t)$. Hence, the following portfolio

$$\pi(t) := \mu e^{-\int_t^T (2r(s) - |\hat{\theta}(s)|^2)ds} H_{\hat{\theta}}(t) v(t) \equiv [\lambda e^{-\int_t^T r(s)ds} - x(t)]v(t)$$

indeed satisfies the no-shorting constraint as well as (5.16), and hence is efficient. The rest of the proof, in proving the form of the efficient frontier, is exactly the same as that of Theorem 18. □

6 Concluding Remarks

In this paper we have studied the mean–variance portfolio selection in a continuous-time incomplete market, with a no-shorting constraint on portfolios. One of the main results is that we have completely characterized, via some equivalent conditions, those contingent claims that are replicable by portfolios satisfying the constraint. This result per se is independent of the portfolio selection problem, and is more in the realm of risk hedging or option pricing. Nonetheless, this result has played a central role in handling the incompleteness of the market. Using a backward approach, we transferred the original mean–variance problems into static optimization problems on the terminal wealth, where all the original constraints including the budget constraint and incompleteness of market are translated into some terminal constraints. Solving these static constrained optimization problems using primal–dual convex optimization in Hilbert spaces has led to solution schemes for the underlying Markowitz problems and, in the case of deterministic opportunity set, to complete and closed-form solutions.

While the continuous-time portfolio selection models with the security price processes governed by geometric Brownian motions are considered in this paper, we believe that our results extend to semimartingale models, including the discrete-time case, with necessary technical modifications, some of which may be straightforward and some may be involved. However, we chose to use the current setup, as what we have hitherto done in our related works, for the main reason that we do not want to let unduly technicality blur the financial essence of the results and distract the reader's attention.

Appendix

A Some Lemmas

We present several technical lemmas that are useful in the main context. We start with the following result which is originally due to Frank and Wolfe [9]. A complete proof (for a more general case) can be found in [21].

Lemma 26. *If a quadratic function f: $\mathbf{R}^d \to \mathbf{R}$ is bounded below on a nonempty polyhedron S, then f attains its infimum on S.*

Lemma 27. *Given $a \in \mathbf{R}^n$ and $A \in \mathbf{R}^{m\times n}$. If $a \notin \{A'u : u \in \mathbf{R}^m_+\}$, then there exists $v \in \mathbf{R}^n \setminus \{0\}$ such that $a'v = -1$ and $Av \geq 0$.*

Proof. By the assumption $a \neq 0$. Denote $M := \{w \in \mathbf{R}^n : a'w < 0\}, N := \{w \in \mathbf{R}^n : Aw \geq 0\}$, which are both nonempty convex cones. If $M \cap N = \emptyset$, then by the convex separation theorem, there exists $y \in \mathbf{R}^n \setminus \{0\}$ with $\sup_{w\in A} y'w \leq \inf_{w\in B} y'w$. This implies

$$y'w \leq 0 \ \forall w \text{ with } a'w < 0, \tag{A.1}$$

and

$$y'w \geq 0 \ \forall w \text{ with } Aw \geq 0. \tag{A.2}$$

It follows from (A.1) that there exists $k > 0$ such that $a = ky$. On the other hand, (A.2) together with Farkas' lemma (see, e.g., [1, p. 58, Theorem 2.9.1]) yields there is $\pi \in \mathbf{R}_+^m$ such that $y = A'\pi$. So $a = ky = A'(k\pi) \in \{A'u : u \in \mathbf{R}_+^m\}$, leading to a contradiction. Hence $M \cap N \neq \emptyset$. The desired conclusion then follows immediately. □

Before we state the next lemma, we note that a set $A \subset [0,T] \times \Omega$ is said to be $\mathcal{F}_t$-progressive if the corresponding indicator function 1_A is $\mathcal{F}_t$-progressively measurable. The $\mathcal{F}_t$-progressive sets form a σ-field (see, e.g., [13, p. 99]).

Lemma 28. *Let $X \equiv \{X(t) : 0 \leq t \leq T\}$ be a given n-dimensional, $\mathcal{F}_t$-progressively measurable stochastic process. Assume that $S(t,\omega) := \{y \in \mathbf{R}^m : f(X(t,\omega), y) \leq 0\} \neq \emptyset$ for any $(t,\omega) \in [0,T] \times \Omega$, where $f : \mathbf{R}^n \times \mathbf{R}^m \to \mathbf{R}^k$ is jointly measurable in both variables and continuous in the second variable. Then the process $\alpha \equiv \{\alpha(t) : 0 \leq t \leq T\}$ defined as $\alpha(t,\omega) := \operatorname{argmin}_{y \in S(t,\omega)} |y|^2$ is also $\mathcal{F}_t$-progressively measurable.*

Proof. First of all, for each $(t,\omega) \in [0,T] \times \Omega$, $S(t,w)$ is a closed set, and the square function is strictly convex and coercive. Hence $\alpha(t,\omega)$ is well defined. Set $g(t,\omega) := |\alpha(t,\omega)|^2$. Then for any $x \in \mathbf{R}$,

$$\{(t,\omega) : g(t,\omega) < x\} = \cup_{v \in \mathbf{Q}^m, |v|^2 < x} \{(t,\omega) : f(X(t,\omega), v) \leq 0\}.$$

This shows that g is $\mathcal{F}_t$-progressively measurable.

Denote $S_n(t,\omega) := S(t,\omega) \cap \{y \in \mathbf{R}^m : |y|^2 \leq g(t,\omega) + 1/n\}$, for $(t,\omega) \in [0,T] \times \Omega$, and $n = 1, 2, \cdots$. Fix n. For any open set $O \subset \mathbf{R}^m$, we have

$$\begin{aligned}\{(t,\omega) : S_n(t,\omega) \cap O \neq \emptyset\} = \cup_{v \in O \cap \mathbf{Q}^m} \{(t,\omega) &: f(X(t,\omega), v) \leq 0, \\ &|v|^2 \leq g(t,\omega) + 1/n\},\end{aligned}$$

which is therefore an $\mathcal{F}_t$-progressive set. This shows that $S_n(t,\omega)$ satisfies the condition required in the measurable selection theorem [3, p. 281, Theorem 8.3.ii]. Hence, there exists an $\mathcal{F}_t$-progressively measurable process α_n with $\alpha_n(t,\omega) \in S_n(t,\omega)$ almost surely on $[0,T] \times \Omega$. It is clear that $\alpha_n(t,\omega) \to \alpha(t,\omega)$, almost surely, as $n \to \infty$. Thus α is $\mathcal{F}_t$-progressively measurable. □

B Proof of Lemma 7

(i) First of all, θ^* is clearly well-defined by (2.10). By Lemma 28, θ^* is an $\mathcal{F}_t$-progressively measurable process. Moreover, due to Assumption (A) we must also have $\theta^* \in L^\infty_\mathcal{F}(0,T,\mathbf{R}^n)$. Similarly, one can prove $\hat{\theta} \in L^\infty_\mathcal{F}(0,T,\mathbf{R}^n)$ (indeed $|\hat{\theta}(t)| \leq |\theta^*(t)|$ by their definitions).

(ii) Pointwisely in (t,ω) (other than those points in a (t,ω)-null set), $\theta^*(t)$ minimizes $|\theta|^2$ subject to $\sigma(t)\theta = B(t)$. Hence by the Lagrange approach there is $u \in \mathbf{R}^m$ so that $\theta^*(t)$ minimizes $|\theta|^2 - 2[\sigma(t)\theta - B(t)]'u$ over $\theta \in \mathbf{R}^n$. The zero-derivative condition then gives $\theta^*(t) = \sigma(t)'u$. This implies that $\{u \in \mathbf{R}^m : \sigma(t)'u = \theta^*(t)\} \neq \emptyset$. Define $u(t) := \text{argmin}_{u\in\{u\in\mathbf{R}^m:\sigma(t)'u=\theta^*(t)\}}|u|^2$. Then by virtue of Lemma 28 $u(\cdot)$ is the desired process.

(iii) For each fixed (t,ω) not in a (t,ω)-null set, $\hat{\theta}(t)$ minimizes $|\theta|^2$ subject to $\sigma(t)\theta \geq B(t)$. By the Kuhn–Tucker theorem there exists $v \in \mathbf{R}^m_+$ such that $\hat{\theta}(t)$ minimizes $|\theta|^2 - 2[\sigma(t)\theta - B(t)]'v$ over $\theta \in \mathbf{R}^n$. This leads to $\hat{\theta}(t) = \sigma(t)'v$. The rest of the proof is the same as in (ii) above.

(iv) For any $\theta \in \Theta$, we have, by (ii), that $\theta^*(t)'\theta(t) = u(t)'\sigma(t)\theta(t) = u(t)'B(t) = u(t)'\sigma(t)\theta^*(t) = |\theta^*(t)|^2$.

(v) We continue with the argument in proving (iii) above. It follows from the Kuhn–Tucker theorem that there exists $v \in \mathbf{R}^m_+$ such that $\hat{\theta}(t) = \sigma(t)'v$ and $v'[\sigma(t)\hat{\theta}(t) - B(t)] = 0$. So for any $\theta(\cdot) \in \hat{\Theta}$,

$$\hat{\theta}(t)'\theta(t) = v'\sigma(t)\theta(t) \geq v'B(t) = v'\sigma(t)\hat{\theta}(t) = |\hat{\theta}(t)|^2.$$

Moreover, the only inequality in the above becomes equality when $\theta(\cdot) = \theta^*(\cdot)$. This proves the desired results. □

References

1. A.V. Balakrishnan, *Applied Functional Analysis*, Springer–Verlag, New York, 1976.
2. T.R. Bielecki, H. Jin, S.R. Pliska and X.Y. Zhou, *Continuous-time mean–variance portfolio selection with bankruptcy prohibition*, Math. Finance, 15 (2005), pp. 213–244.
3. L. Cesari, *Optimization–Theory and Applications*, Springer–Verlag, New York, 1985.
4. S. Chen, X. Li and X. Zhou, *Stochastic linear quadratic regulators with indefinite control weight costs*, SIAM J. Contr. Optim., 36 (1998), pp. 1685–1702.
5. J. Cvitanic and I. Karatzas, *Convex duality in constrained portfolio optimization*, Ann. Appl. Prob., 2 (1992), pp. 767–818.
6. J. Cvitanic and I. Karatzas, *Hedging contingent claims with constrained portfolios*, Ann. Appl. Probab., 3 (1993), pp. 652–681.
7. N. El Karoui, S. Peng and M.C. Quenez, *Backward stochastic differential equations in finance*, Math. Finance, 7 (1997), pp. 1–71.
8. N. El Karoui and M.C. Quenez, *Dynamic programming and pricing of contingent claims in an incomplete market.* SIAM J. Contr. Optim. 33 (1995), pp. 29–66.
9. M. Frank and P. Wolfe, *An algorithm for quadratic programming*, Naval Res. Logistics Quart., 3 (1956), pp. 95–110.

10. H. He and N. Pearson, *Consumption and portfolio policies with incomplete markets and short-sale constraints: The finite-dimensional case*, Math. Finance, 1 (1991), pp. 1–10.
11. H. He and N. Pearson, *Consumption and portfolio policies with incomplete markets: The infinite-dimensional case*, J. Econom. Theory, 54 (1991), 259–305.
12. J.M. Harrison and D. Kreps, *Martingales and multiperiod securities market*, J. Econom. Theory, 20 (1979), pp. 381–408.
13. O. Kallenberg, *Foundations of Modern Probability*, Springer, Berlin, 1997.
14. I. Karatzas, J.P. Lehoczky, S.E. Shreve and G. Xu, *Martingale and duality methods for utility maximization in an incomplete market*, SIAM J. Control Optim., 29 (1991), pp. 702–730.
15. I. Karatzas and S.E. Shreve, *Methods of Mathematical Finance*, Springer–Verlag, New York, 1998.
16. D. Kramkov, *Optional decomposition of supermartingales and hedging contingent claims in incomplete security markets*, Probab. Theory Rel. Fields 105 (1996), pp. 459–479.
17. D. Li and W.L. Ng, *Optimal dynamic portfolio selection: Multiperiod mean–variance formulation*, Math. Finance, 10 (2000), pp. 387–406.
18. X. Li, X.Y. Zhou and A.E.B. Lim, *Dynamic mean–variance portfolio selection with no-shorting constraints*, SIAM J. Control Optim., 40 (2001), pp. 1540–1555.
19. A.E.B. Lim, *Quadratic hedging and mean–variance portfolio selection with random parameters in an incomplete market*, Math. Oper. Res., 29 (2004), pp. 132–161.
20. A.E.B. Lim and X.Y. Zhou, *Mean–variance portfolio selection with random parameters*, Math. Oper. Res., 27 (2002), pp. 101–120.
21. Z.Q. Lou and S. Zhang, *On extensions of the Frank-Wolfe theorems*, Comput. Optim. Appl., 13 (1999), pp. 87–110.
22. D.G. Luenberger, *Optimization by Vector Space Methods*, Wiley, New York, 1969.
23. H. Markowitz, *Portfolio selection*, J. of Finance, 7 (1952), pp. 77–91.
24. H. Markowitz, Private communication, 2004.
25. J. von Neumann and O. Morgenstern, *Theory of Games and Economic Behavior*, 2nd Edition, Princeton University Press, Princeton, New Jersey, 1947.
26. S.R. Pliska, *A discrete time stochastic decision model*, Advances in Filtering and Optimal Stochastic Control, edited by W.H. Fleming and L.G. Gorostiza, Lecture Notes in Control and Information Sciences, 42, Springer–Verlag, New York, 290–304, 1982.
27. W. Schachermayer, *Optimal investment in incomplete markets when wealth may become negative*, Ann. Appl. Probab., 11 (2001), pp. 694–734.
28. M.C. Steinbach, *Markowitz revisited: Mean–variance models in financial portfolio analysis*, SIAM Rev., 43 (2001), pp. 31–85.
29. J. Yong and X.Y. Zhou, *Stochastic Controls: Hamiltonian Systems and HJB Equations*, Springer, New York, 1999.
30. X.Y. Zhou and D. Li, *Continuous time mean–variance portfolio selection: A stochastic LQ framework*, Appl. Math. Optim., 42 (2000), pp. 19–33.

Quantum and Classical Conserved Quantities: Martingales, Conservation Laws and Constants of Motion

Torbjørn Kolsrud

Department of Mathematics, Royal Institute of Technology, SE-100 44 Stockholm, Sweden, kolsrud@math.kth.se

To K. Itô on the occasion of his 90th birthday

Summary. We study a class of diffusions, conjugate Brownian motion, related to Brownian motion in Riemannian manifolds. Mappings that, up to a change of time scale, carry these processes into each other, are characterised. The characterisation involves conformality and a space-time version of harmonicity. Infinitesimal descriptions are given and used to produce martingales and conservation laws. The relation to classical constants of motion is presented, as well as the relation to Noether's theorem in classical mechanics and field theory.

Introduction

The theme of the present article is *invariance properties* of a wide class of diffusions, termed *conjugate Brownian motion.* Much of the inspiration comes from the interplay between classical and quantum mechanics as expressed in the ideas of R.P. Feynman [FH] and early approaches to quantum mechanics. (See also Nelson [N].) Feynman's ideas are based on using the classical variational principle (in the free case)

$$\delta \int \frac{1}{2}|\dot{q}|^2\, dt = 0 \tag{1}$$

to explain, e.g., its quantum counterpart

$$\delta \int \frac{1}{2}|du(q)|^2\, dq = 0. \tag{2}$$

The latter case is related to *harmonic morphisms*, which, under pull back, preserve harmonic functions, and therefore Brownian motion. The basic result, due to Fuglede and Ishihara, is a characterisation of harmonic morphisms in terms of harmonicity (harmonic mappings Eells-Lemaire [EL]) and conformality. See [F1,F2,Ish]. These concepts correspond to preservation of martingales, and conformality (Darling [D] and Emery [Em]). A non-geometric treatment of related problems can be found in Øksendal [Ok].

The heat equation can obtained from the variational principle ([BK2,Gol, IK,MF])

$$\delta \iint \frac{1}{2}\Big(\theta\dot{\theta}^* - \dot{\theta}\theta^* + \langle d\theta, d\theta^*\rangle\Big) = 0, \tag{3}$$

where integration is carried out over space-time. The result is a pair of equations, viz., $\dot{\theta} + \frac{1}{2}\Delta\theta = 0$ and $\dot{\theta}^* - \frac{1}{2}\Delta\theta^* = 0$. Time-symmetry is built in: we have one backward and one forward heat equation.

The intimate relations between the classical Newton equation $\ddot{q} = 0$ and the (backward) heat equation $\dot{\theta} + \frac{1}{2}\theta'' = 0$, in one space dimension for simplicity, are fundamental. They are apparent when looking at the classical constants of motion $1, p, pt - q$, the Heisenberg algebra, and $p^2, p(pt-q), (pt-q)^2$, the Lie algebra sl_2, and comparing with the heat Lie algebra (Lie [Lie], Anderson-Ibragimov [AI], Ibragimov [Ibr2,Ibr3], Olver [Or1]). The latter consists of linear differential operators of order at most one, viz., the Heisenberg algebra $\langle 1, \partial_q, t\partial_q - q\rangle$, and

$$\mathfrak{a} = \langle \partial_t, t\partial_t + \tfrac{1}{2}q\partial_q, \tfrac{1}{2}t^2\partial_t + \tfrac{1}{2}tq\partial_q - \tfrac{1}{4}(q^2 - t)\rangle. \tag{4}$$

This is another representation of sl_2. Whereas the first five elements in the classical and the heat Lie algebras correspond via the symbol map—a kind of Laplace transform, see Sect. 6—the sixth elements differ. In the classical case, we have the function $-\frac{1}{4}t^2p^2 + \frac{t}{2}pq - \frac{1}{4}q^2$, the symbol of the PDO $\frac{1}{2}t^2\partial_t + \frac{1}{2}tq\partial_q - \frac{1}{4}q^2$. In the heat Lie algebra, there is an additional term: instead of q^2 we have $q^2 - t$. The former function satisfies the equation $\dot{u} = 0$, the second satisfies $\dot{u} + \frac{1}{2}u'' = 0$. What we observe is *Itô's formula:* $q^2 - t \equiv :q^2:$ is the *renormalised second power*, corrected to fit the heat equation. In essence, this is the difference between the equations.

The main results presented below are

- A characterisation of the mappings that preserve ordinary and *conjugate* Brownian motion; (Sect. 4)
- The corresponding infinitesimal description in terms of Lie algebras, and the identification, in the free case, of the Lie algebra in terms of classical Lie algebras; (Sect. 5)
- Analysis of conservation laws and stochastic constants of motion (martingales), and their relation to the heat Lie algebra. Relations to the classical Lie algebra. Comparison via two Noether theorems. (Sect. 6)

In Sect. 2, we provide background on diffusions in manifolds, as well as the dynamical aspects of conjugate Brownian motion. In Sect. 3 we present the background for the mappings and Lie algebras needed. Cf. [Ibr2] or [Or1].

Among other related treatments we mention Djehiche–Kolsrud [DK], Ibragimov [Ibr2], Kolsrud [K2], Kolsrud–Loubeau [KL], Loubeau [Lob] and Thieullen–Zambrini [TZ1, TZ2].

General references are [AI, Arn, AG, AKN, DKN, DNF, Ga, Go, Gol, Gorb, Ibr2, Ibr3, IW, Ko, LL, M, MF, Or1].

1 Preliminaries

We shall consider a connected manifold N of dimension n. When N is given a Riemannian structure, the metric $g = (g_{ij})$ will also be written $\langle \cdot, \cdot \rangle$. ∇ shall denote covariant (Levi-Civita) differentiation with Christoffel symbols Γ^k_{ij}, and d the outer derivative. Recall that ∇ is completely determined by being *metric*: $\nabla g = 0$, and *symmetric*: $\Gamma^k_{(ij)} := \frac{1}{2}(\Gamma^k_{ij} + \Gamma^k_{ji}) = \Gamma^k_{ij}$. μ_g, or just vol, is the volume form determined by g. We will need two Laplacians: The *Laplace-Beltrami operator*

$$\Delta = \Delta_g = g^{ij}(\partial_i \partial_j - \Gamma^k_{ij}\partial_k) = \nabla^\dagger \nabla = g^{ij}\nabla_i \nabla_j = \mathrm{Tr}_g \nabla^2, \tag{1.1}$$

and the *de Rham-Hodge Laplacian*

$$\Box = -(dd^\dagger + d^\dagger d), \tag{1.2}$$

where $d^\dagger$ is the formal $L^2(\mu_g)$-adjoint of d. Here and below we use Einstein's summation convention with respect to repeated indices, one up, one down. When acting on functions, Δ and $\Box$ coincide. The Ricci tensor will be denoted by Ric. We shall often make use of Weitzenböck's identity

$$\Delta \alpha = \Box \alpha + \mathrm{Ric}\, \alpha \tag{1.3}$$

for one-forms α, and the fact that d and $\Box$ commute:

$$[d, \Box] = 0. \tag{1.4}$$

2 Basics on Diffusions in Manifolds

2.1 Connections, Geodesics and Scalar 2nd Order Elliptic PDOs

Consider a differential operator on N of the form $Q := \frac{1}{2} g^{ij}\partial_i \partial_j + \bar{b}_k \partial_k$. We assume that for all points x in N, $g^{ij}(x)\xi_i \xi_j > 0$ whenever some $\xi_i \neq 0$. Then (g^{ij}) is the inverse of a metric $g = (g_{ij})$ on N. Let $\Delta = \Delta_g$ be the corresponding Laplace-Beltrami operator. Q may be written $Q = \frac{1}{2}\Delta + b$, where b is a vector field. Up to a sign, this is the general expression for a scalar linear second order elliptic differential operator in N, satisfying $Q1 = 0$. It is formally self-adjoint (w.r.t. to $L^2(e^{2F}\mu_g)$) precisely when b is a gradient: $b = \mathrm{grad}\, F$. Let

$$\overline{\Gamma}^k_{ij} := \Gamma^k_{ij} + \frac{2}{n-1}(\delta^k_i b_j - g_{ij} b^k), \qquad n \neq 1, \tag{2.1}$$

where $b_k = g_{ik} b^i$. Then $(\overline{\Gamma}^k_{ij})$ defines a metric connection $\overline{\nabla}$ such that $Q = \frac{1}{2}\mathrm{Tr}_g(\overline{\nabla}^2)$. $\overline{\nabla}$ is unique for $n = 2$, but not otherwise (Ikeda-Watanabe [IW], Prop. V.4.3). In general, the difference of two connections is a (1,2)-tensor. We may write

$$(\overline{\Gamma} - \Gamma)^k_{ij} = \overline{A}^k_{ij} + \overline{S}^k_{ij}, \tag{2.2}$$

where, for each k, $\overline{A}^k_{ij}$ is antisymmetric and $\overline{S}^k_{ij}$ symmetric in the lower indices. The $\overline{\nabla}$-geodesics are given by

$$\frac{\overline{D}^2 x^k}{dt^2} = \frac{D^2 x^k}{dt^2} + \overline{S}^k_{ij}\dot{x}^i\dot{x}^j = \ddot{x}^k + (\Gamma^k_{ij} + \overline{S}^k_{ij})\dot{x}^i\dot{x}^j = 0, \quad 1 \leq k \leq n, \tag{2.3}$$

independently of $\overline{A}$. $\overline{A} \neq 0$ if and only if $\overline{\nabla} \neq \nabla$, the connections being metric. Q however, only depends on $\overline{\nabla}$ through the trace of its symmetric part.

Proposition 1. *Let ∇' and ∇'' be two g-metric connections with symmetric parts S' and S'', respectively. They have the same geodesics if and only if $S' = S''$. They have the same Laplacian if and only if $Tr_g(S')^k = Tr_g(S'')^k$ for each k. Moreover, this happens if and only if their corresponding torsion tensors satisfy $T'^k_{ik} = T''^k_{ik}$.*

The last characterisation can be found in [IW], Prop. V.4.3.

The important conclusion is

Observation 1. *Two 'quantum equivalent' connections need not be classically equivalent: Laplacians and geodesics do not correspond.*

Ground State Transform

Consider now the case $Q1 \neq 0$. Write $Q = Q_0 - V$, where V is a smooth function (potential) on N, and $Q_0 = \frac{1}{2}\Delta + b$, as above. In general, multiplication by a function $\Omega > 0$ is an isometry (unitary equivalence) between $L^2(\Omega^{-2}\mu_g)$ and $L^2(\mu_g)$. Suppose also that Ω solves the equation $Q\Omega = 0$. (The case where Ω corresponds to another eigenvalue can be handled by letting $V \to V + \text{const.}$)

This is an implicit condition on V. Then (ground state transform or Doob's h-transform)

$$\overline{Q} := \Omega^{-1} Q \Omega = \Omega^{-1}(Q_0 - V)\Omega = Q_0 + \text{grad} \log \Omega, \tag{2.4}$$

independently of b. Conjugation by Ω transforms Q to an operator without constant term. Let $b = 0$ (the symmetric case when b is a gradient can be handled similarly), so that Q_0 is the Laplace-Beltrami operator. Then, for $n \neq 2$, $\overline{Q}$ is a factor times the Laplace-Beltrami operator for a new, conformally equivalent, metric:

$$\Omega^{-1}\left(\frac{1}{2}\Delta_g - V\right)\Omega = \frac{1}{2}\Delta_g + \operatorname{grad}\log\Omega = \frac{1}{2}\rho^{-1}\Delta_{\rho g}, \quad \rho = \Omega^{\frac{4}{n-2}},\ n \neq 2. \tag{2.5}$$

There seems to be no general relations between the geodesics of ρg and the solutions curves of the classical Euclidean Newton equations

$$\frac{D^2 x}{dt^2} = \operatorname{grad} V(x). \tag{2.6}$$

In other words, there is no classical counterpart of the ground state transformation. The Maupertuis principle ([Arn, Ga, LL]) shows how, given a constant energy submanifold, one can remove a potential and instead introduce another conformally equivalent metric. This new metric is, however, not the one in (2.5).

2.2 $(g, \overline{\nabla})$-Brownian Motion and Conformal Martingales

Let (N, g) be Riemannian with a g-metric connection $\overline{\nabla}$ (not necessarily the Levi-Civita). The Christoffel symbols are denoted $\overline{\Gamma}^k_{ij}$. Let Y be a continuous N-valued semi-martingale, so that in local coordinates

$$dY^k = dA^k + dM^k, \quad 1 \leq k \leq n, \tag{2.7}$$

where the A^k are processes of finite variation, and the M^k are (ordinary) martingales. Define the covariant Itô differential $d^c Y$ of Y by

$$d^c Y^k_t := dY^k_t + \frac{1}{2}\overline{\Gamma}^k_{ij}(Y_t)d[Y^i, Y^j]_t, \quad 1 \leq k \leq n, \tag{2.8}$$

where the brackets indicate compensator. Given a connection, $d^c Y$ is well defined, as observed originally by Bismut.

Definition 2. *A semimartingale Y is a martingale w.r.t. $\overline{\nabla}$ if*

$$dA^k_t = -\frac{1}{2}\overline{\Gamma}^k_{ij}(Y_t)d[Y^i, Y^j]_t, \quad 1 \leq k \leq n. \tag{2.9}$$

Definition 3. *A semimartingale Y is conformal w.r.t. g if*

$$d[Y^i, Y^j]_t = g^{ij}(Y_t)dC_t, \quad 1 \leq i, j \leq n, \tag{2.10}$$

for some strictly positive and increasing continuous process C.

Itô's formula shows that if Y satisfies (2.9) and (2.10), the time-shifted process $Y(\gamma_t)$, where γ is the inverse of C, has generator $\frac{1}{2}\overline{\Delta} := \frac{1}{2}\operatorname{Tr}\overline{\nabla}^2$. $\overline{\Delta}$ is the *Laplacian* given by the connection.

Proposition 4. *With respect to $(g, \overline{\nabla})$, any conformal martingale is a time-shift of Brownian motion, and conversely.*

2.3 Conjugate BM

Let $N = (N, g)$ and $\overline{\nabla}$ be as in Sect. 2.2 with corresponding Laplacian $\overline{\Delta}$. Let $I = [-1, 1]$, and let $\theta : I \times N \to (0, \infty)$ be smooth. Write $V = (\dot{\theta} + \frac{1}{2}\overline{\Delta}\theta)/\theta$, and $H = -\frac{1}{2}\overline{\Delta} + V$, so that $\dot{\theta} = H\theta$. Hence θ is a solution of the (backward) heat equation with potential V. By *conjugating* the generator of $\mathrm{BM}(g, \nabla)$ with θ we obtain a new, in general non-stationary, diffusion $Z = (Z_t,\, t \in I)$, called *conjugate BM*. Its (forward) *generator* (regularised forward derivative of u along Z) is

$$Du := \frac{1}{\theta}\left(\frac{\partial}{\partial t} - H\right)(\theta \cdot u) = \left(\frac{\partial}{\partial t} + \frac{1}{2}\overline{\Delta}\right)u + \frac{1}{\theta}\langle d\theta, du\rangle. \tag{2.11}$$

When θ is independent of time, this is the ground state transform as in Sect. 2.1. The (forward) *Itô equation* is

$$d^c Z_t = \operatorname{grad} \log \theta(t, Z_t)\, dt + dM_t. \tag{2.12}$$

The compensator satisfies

$$d[Z^\alpha, Z^\beta]_t = g^{\alpha\beta}(Z_t)\, dt. \tag{2.13}$$

Note that (2.11) makes sense for tensor fields:

$$D\sigma = \frac{\partial \sigma}{\partial t} + \frac{1}{2}\overline{\nabla}^i \overline{\nabla}_i \sigma + \overline{\nabla}_\xi \sigma, \tag{2.14}$$

where ξ is the vector field dual to $d\theta/\theta$. The following variant of Itô's formula is useful:

$$D(\Phi(u)) = \Phi'(u)Du + \frac{1}{2}\Phi''(u)|du|^2, \qquad \Phi \in C^2(\mathbb{R}). \tag{2.15}$$

2.4 Schrödinger Diffusions

Schrödinger (or Bernstein) diffusions are *time-symmetric* in the following sense. The forward description is a conjugate BM w.r.t. a positive solution of a backward heat equation, whereas the backward description is a conjugate backward BM w.r.t. a positive solution of the corresponding forward heat equation. (Cf. [KZ1] for details on the construction of Bernstein processes. See also [CZ].)

Consider again the situation in the preceding section, but with V given (and sufficiently smooth). For simplicity we assume that V does not depend on time. We assume that θ^* solves the usual, forward, heat equation with potential V: $\dot{\theta}^* = -H\theta^*$. The forward generator of the Bernstein diffusion Z is as in (2.11). The backward generator is

$$D^*u := \frac{1}{\theta^*}\left(\frac{\partial}{\partial t} + H\right)(\theta^* \cdot u) = \left(\frac{\partial}{\partial t} - \frac{1}{2}\Delta\right)u - \frac{1}{\theta^*}\langle d\theta^*, du\rangle. \tag{2.16}$$

There is also a backward Itô equation similar to the one in Sect. 2.3.

Generally speaking, the backward heat evolution creates irregularities. We can produce smooth positive solutions by letting $\theta(t,\cdot) = \exp\{-(1-t)H\}\chi$, where $0 \not\equiv \chi \geq 0$ is given. Similarly, we obtain solutions to the forward heat equation with potential V by $\theta^*(t,\cdot) = \exp\{-(1+t)H\}\chi^*$. In particular we may choose $\theta^*(t,\cdot) = \theta(-t,\cdot)$.

The *probability density* of Z is the product of these two functions: $\mathbb{P}(Z_t \in C) = \int_C \theta(t,\cdot)\theta^*(t,\cdot)\, d\mu_g$, $C \in \text{Borel}\,(N)$. This requires the normalisation

$$\int_N \chi^* \exp(-2H)\chi \,\text{dvol} = 1. \tag{2.17}$$

2.5 Forward Dynamics of Conjugate BMs

Drift and Momenta

If X is a continuous semi-martingale with values in a manifold with connection $\{\overline{\Gamma}^{\alpha}_{\beta\gamma}\}$, we define its *drift*, DX, by

$$d^c X_t = DX_t\, dt + dM_t, \tag{2.18}$$

where (M_t) is a martingale. DX_t is a tangent vector above X_t for each t. The drift measures the martingale deviation for X.

For conjugate BM, $DZ_t = \text{grad}\, A(t, Z_t)$, where $A = \log\theta$. We let p be the corresponding one-form, i.e. $p = dA = d\log\theta$. This is the vector of *momenta* corresponding to our process.

$p(t, Z_t)$ is a regularised forward derivative along Z. For the Levi-Civita connection, one can calculate the drift directly, but along *harmonic coordinates.* Then $Dq^\alpha = \langle p, dq^\alpha\rangle = g^{\alpha\beta}p_\beta = p^\alpha$.

The Lagrangian

Using $D\theta = |d\theta/\theta|^2 + V = |p|^2 + V$, (2.15) yields

$$DA = D\log\theta = \frac{1}{2}|p|^2 + V = L(p,\cdot), \quad L(\omega, q) := \frac{1}{2}|\omega|^2 + V(q). \tag{2.19}$$

L is the *classical Euclidean Lagrangian.* By Dynkin's formula we get the *path integral formula*

$$\theta(t, Z_t) = \exp\Big\{ -\mathbb{E}\big[\int_t^1 L(DZ_{t'}, Z_{t'})\, dt' - \log\theta(1, Z_1)|\mathcal{F}_t\big]\Big\}. \tag{2.20}$$

We therefore look at A as the *forward action density.*

We can now deduce the *(forward) regularised Newton's equations*:

Proposition 5. *For the Levi-Civita connection,*

$$Dp - \frac{1}{2}\mathrm{Ric}\, p = dV. \tag{2.21}$$

Proof. On forms, the term in the generator involving the logarithmic derivatives of θ has to be replaced by the corresponding covariant differentiation. By Weitzenböck's formula $Dp - \frac{1}{2}\mathrm{Ric}\, p = \partial_t p + \frac{1}{2}\Box p + \nabla_\xi p :=$ $D_0 p$, where ξ is the vector field dual to p. The right-hand side equals $D_0 dA = dD_0 A + [D_0, d]A = dDA + [D_0, d]A =: I + II$. Using $\nabla g = 0$ and $\xi^\alpha = g^{\alpha\gamma}\nabla_\gamma A$, we find $\nabla_\xi dA = p^\alpha \nabla_\alpha \nabla_\beta A dq^\beta = g^{\alpha\gamma}\nabla_\gamma A \nabla_\alpha \nabla_\beta A dq^\beta =$ $\frac{1}{2} d|dA|^2 = \frac{1}{2} d|p|^2$. Since ∂_t and $\Box$ commute with d, $II = [\nabla_\xi, d]A$. Obviously $d\nabla_\xi A = d(dA(\xi)) = d|dA|^2 = d|p|^2$, so $II = -\frac{1}{2} d|p|^2$. Finally, $dDA = d(\frac{1}{2}|p|^2 + V)$ according to (2.19) above.

Energy

The space-time differential

$$\bar{d}A := \dot{A}\, dt + \partial_\alpha A\, dq^\alpha = E\, dt + p\, dq, \tag{2.22}$$

of the forward action density A plays a similar role as the Poincaré invariant in classical mechanics. The second term is (forward) momentum. The first function E is the (forward) *energy* $\dot{A} = H\theta/\theta$. We have

$$E = -\frac{1}{2}|p|^2 - \frac{1}{2}\nabla^\dagger p + V. \tag{2.23}$$

The energy is a *stochastic constant of motion* in that $DE = 0$ whenever $\dot{V} = 0$. See Sect. 6.

Time Reflection

For Bernstein diffusions, all that has been said has a backward counterpart. We do not go into details here but refer to to [KZ1] and references therein.

3 Groups of Mappings and their Lie Algebras

3.1 Extensions of Diff and Vect

Let M_0 and M_1 be differentiable manifolds, without any additional structure. Let $f : M_1 \to M_0$ and $a : M_0 \to \mathbb{R}$ be C^∞. The pair $\gamma := (f, a)$ induces, by pullback and multiplication, a mapping $C^\infty(M_0) \to C^\infty(M_1)$ by

$$u \to u \cdot \gamma := a \cdot u \circ f, \tag{3.1}$$

cf. [DK].

If we have several manifolds and mappings: $f_2 : M_2 \to M_1$ and $f_1 : M_1 \to M_0$ and a_j are functions on M_j, we get

$$\gamma_1 \cdot \gamma_2 = (f_1 \circ f_2, a_2 \cdot a_1 \circ f_2). \tag{3.2}$$

Now suppose that all manifolds are one and the same, M, and denote by $\mathcal{D} := \mathrm{Diff}\,(M)$ all diffeomorphisms $M \to M$. If $f_i \in \mathcal{D}$, and the a_i are never zero, the previous identity is the composition in the group

$$\widetilde{\mathcal{D}} := \mathcal{D} \ltimes C^\infty(M)^\times, \tag{3.3}$$

where $\ltimes$ indicates semi-direct product. $\widetilde{\mathcal{D}}$ is an extension of $\mathcal{D}$, and $u \to u \cdot \gamma$ is a right-action of $\widetilde{\mathcal{D}}$ on $C^\infty(M)$.

The infinitesimal version of this is as follows. Let $\mathcal{V} := \mathrm{Vect}\,(M)$ denote all vector fields on M, and consider all first-order differential operators

$$u \to \Lambda u := X(u) + U \cdot u, \tag{3.4}$$

where $X \in \mathcal{V}$ and U is a function on M. Equipped with the natural commutator, this is the Lie algebra

$$\widetilde{\mathcal{V}} := \mathcal{V} \oplus C^\infty(M), \tag{3.5}$$

where the sum is semi-direct. In analogy with the preceding case, $\widetilde{\mathcal{V}}$ is a central extension of $\mathcal{V}$.

We shall write the elements of $\widetilde{\mathcal{V}}$ as (X, U) or simply $X + U$. By definition $X + U = 0$ if and only if both components X and U vanish. Explicitly, the bracket in $\widetilde{\mathcal{V}}$ is

$$[\Lambda_1, \Lambda_2] = [(X_1, U_1), (X_2, U_2)] := ([X_1, X_2], X_1(U_2) - X_2(U_1)), \tag{3.6}$$

where the first term on the right is the usual commutator of vector fields.

The relation between $\gamma = (a, f)$ and $\Lambda = (X, U)$ can be described thus: Let $f_\varepsilon := \exp \varepsilon X$ denote the (local) flow of X, and assume $f = f_1$. Then, by Lie's formula,

$$a = a_1 = \exp \int_0^1 U \circ f_\varepsilon \, d\varepsilon, \tag{3.7}$$

and $\gamma = \exp \Lambda$ (cf. the Feynman-Kac formula).

Conversely, differentiating a local one-parameter group $\gamma_\varepsilon = (f_\varepsilon, a_\varepsilon)$ at $\varepsilon = 0$, one obtains a first order differential operator $\Lambda = X + U$.

3.2 Orbits for PDOs

Consider again the situation in (3.1). Let K_i be linear partial differential operators on $C^\infty(M_i)$, $i = 0, 1$. Assume $\gamma = (f, a)$ satisfies

$$K_1(u \cdot \gamma) = \phi \cdot (K_0 u) \cdot \gamma, \tag{3.8}$$

where in general $\phi = \phi_\gamma \in C^\infty(M_1)^\times$ will depend on γ.

This identity implies that the pullback under γ of solutions to $K_0 u = 0$ yield solutions to $K_1 v = 0$. Except for the 'conformal' factor ϕ, this is an intertwining relation.

Composing, as in Sect. 3.1, we obtain the *cocycle identities*

$$\phi_{\gamma_1\gamma_2} = \phi_{\gamma_2}(\phi_{\gamma_1} \cdot \gamma_2) = \phi_{\gamma_2} \cdot \phi_{\gamma_1} \circ f_2, \tag{3.9}$$

$$K_2(u \cdot \gamma_1\gamma_2) = \phi_{\gamma_1\gamma_2} \cdot (K_1 u) \cdot \gamma_1\gamma_2. \tag{3.10}$$

In the group case $\gamma = (f, a) \in \widetilde{\mathcal{D}}$, with $M_i = M$ and $K_i = K$, (3.8) may be written

$$(\gamma^{-1} K \gamma) u = \phi_\gamma \circ f^{-1} \cdot Ku. \tag{3.11}$$

Thus, except for the cocycle, K and its conjugation under the inner automorphism given by γ are equal. Clearly, this identity defines a group. It may be seen as a deformed (by the cocycle ϕ) subgroup of $\widetilde{\mathcal{D}}$.

Suppose now that in (3.11) we have a (local) one-parameter group (γ_ε) with generator $\Lambda = (X, U)$ as above, and with corresponding cocycle $\phi^\varepsilon = 1 + \varepsilon\Phi + o(\varepsilon)$. Then, upon differentiating w.r.t. ε at 0 we get

$$[K, \Lambda] = \Phi \cdot K, \tag{3.12}$$

where the function Φ depends on Λ. The relation between ϕ and Φ is as in (3.7) for a and U. In particular,

$$Ke^\Lambda = \exp\{\int_0^1 \Phi \circ f_\varepsilon \, d\varepsilon\} \cdot e^\Lambda K. \tag{3.13}$$

For fixed K, the relation $[K, \Lambda] = \Phi \cdot K$ defines a Lie algebra:

Proposition 6. *Let K be a linear differential operator on M. Then all the first order linear differential operators $\Lambda = (X, U)$ such that $[K, \Lambda] = \Phi \cdot K$ for some function $\Phi = \Phi_\Lambda$, form a Lie algebra with the commutator in* (3.7). *We have*

$$\Phi_{[\Lambda_1, \Lambda_2]} = X_1(\Phi_2) - X_2(\Phi_1). \tag{3.14}$$

Remark 7. Henceforth this Lie algebra will be denoted Lie (K).

Proof (Proposition 6). Assuming (3.14) holds for Λ_1 and Λ_2, we must show that it also holds for their commutator. By the Jacobi identity, and with obvious notation,

$$[K, [\Lambda_1, \Lambda_2]] = [\Lambda_1, [K, \Lambda_2]] - [\Lambda_2, [K, \Lambda_1,]] = [\Lambda_1, \Phi_2 \cdot K] - [\Lambda_2, \Phi_1 \cdot K]. \tag{3.15}$$

We always have

$$[\Lambda, \Phi \cdot K] = X(\Phi) \cdot K + \Phi \cdot [\Lambda, K], \tag{3.16}$$

from which the conclusion is immediate.

Remark 8. It is a very general fact that the relation $K, \Lambda] = \Phi \cdot K$ defines a Lie algebra. We have only used that the bracket is the natural one, and that we have a derivation plus a multiplication.

Intertwining of Lie(K_i)

It is obvious that (3.8) relates the Lie algebras of K_0 and K_1 to one another. To makes this more clear, suppose $\gamma = (f, a)$ satisfies $K_1\gamma = \phi \cdot \gamma K_0$, and suppose Λ_i are related by

$$\Lambda_1\gamma = \gamma\Lambda_0, \tag{3.17}$$

i.e.,

$$X_1(a)u \circ f + aX_1(u \circ f) + a(U_1 u) \circ f = aX_0(u) \circ f + a(U_0 u) \circ f. \tag{3.18}$$

One finds

$$K_1\Lambda_1\gamma = K_1\gamma\Lambda_0 = \phi\gamma K_0\Lambda_0, \tag{3.19}$$

and

$$\Lambda_1 K_1\gamma = \Lambda_1(\phi \cdot \gamma K_0) = X_1(\phi)\gamma K_0 + \phi\gamma\Lambda_0 K_0. \tag{3.20}$$

Hence

$$[K_1, \Lambda_1]\gamma = \phi\gamma[K_0, \Lambda_0] + X_1(\phi)\gamma K_0. \tag{3.21}$$

Since ϕ is never zero, we see that, on the appropriate domains, Λ_i preserve the kernels of K_i simultaneously. If $\Lambda_0 \in \operatorname{Lie} K_0$, then

$$[K_1, \Lambda_1]\gamma = \phi\gamma\Phi_0 \cdot K_0 + X_1(\phi)\gamma K_0 = (\Phi_0 \circ f + X_1(\log\phi))K_1\gamma. \tag{3.22}$$

Similarly, $\Lambda_1 \in \operatorname{Lie} K_1$ implies

$$\gamma[K_0, \Lambda_0] = (\Phi_1 - X_1(\log\phi))\gamma K_0. \tag{3.23}$$

4 Heat and Harmonic Morphisms

4.1 Basic Characterisations

Let $\widetilde{M}$ and M be two manifolds, and f a map of $\widetilde{M}$ into M. Let P and K be two (scalar) linear differential operators on $C^\infty(\widetilde{M})$ and $C^\infty(M)$, respectively.

Definition 9. *f is a morphism for P and K if for each open set $\Omega \subset M$ and each $u \in C^\infty(\Omega)$,*

$$Ku = 0 \quad on \quad \Omega \qquad \Longrightarrow \qquad P(u \circ f) = 0 \quad on \quad f^{-1}\Omega. \tag{4.1}$$

Suppose now that $\widetilde{K}$ and K are differential operators on $\widetilde{M}$ and M, and $a : \widetilde{M} \to \mathbb{R} \setminus 0$. Suppose also that $\gamma = (f, a)$ satisfies (cf. (3.8))

$$\widetilde{K}\gamma = \phi \cdot \gamma K, \tag{4.2}$$

i.e.,

$$\widetilde{K}(a \cdot u \circ f) = \phi a \cdot Ku \circ f, \tag{4.3}$$

where $\phi \neq 0$. Writing

$$Pv := a^{-1}\widetilde{K}(av), \tag{4.4}$$

we obtain

$$P(u \circ f) = \phi \cdot Ku \circ f. \tag{4.5}$$

Let now $\widetilde{N}$ and N be Riemannian manifolds with metric connections $\widetilde{\nabla}$ and ∇, and corresponding Laplacians $\widetilde{\Delta}$ (= Tr $(\widetilde{\nabla}^2)$) and Δ, respectively. Let $\widetilde{I}$ and I be time intervals with variables s and t. We shall consider the following two situations:

(i) Harmonic morphisms: $\widetilde{M} = \widetilde{N}$, $M = N$, $\widetilde{K} = \widetilde{\Delta}$, and $K = \Delta$.
(ii) Heat morphisms: $\widetilde{M} = \widetilde{I} \times \widetilde{N}$, $M = I \times N$, $\widetilde{K} = \partial_s + \frac{1}{2}\widetilde{\Delta}$, and $K = \partial_t + \frac{1}{2}\Delta$.

In order not to get a zero-order term we must require

$$Ka = 0. \tag{G2}$$

When (G2) holds, we have (assuming without loss of generality $a > 0$)

$$P = \widetilde{K} + \text{grad} \log a \tag{4.6}$$

in both cases.

The associated diffusion $\widetilde{X}$ with generator P is a conjugate of BM$(\widetilde{N}, \widetilde{g}, \widetilde{\nabla})$ satisfying

$$\begin{aligned} d^c\widetilde{X}_s &= \text{grad} \log a(s, \widetilde{X}_s)\, ds + d\widetilde{\xi}_s, \\ d[\widetilde{X}^i, \widetilde{X}^j]_s &= \widetilde{g}^{ij}(\widetilde{X}_s)\, ds, \end{aligned} \tag{4.7}$$

where only in case (ii) a will depend explicitly on s. The process X corresponding to Q is in both cases BM(N, g, ∇). Let us write

$$f = (f^0, f^\alpha, 1 \leq \alpha \leq n), \quad t = f^0,$$

where f^0 does not appear in case (i).

Proposition 10. *Let P be as in* (4.4). *Then f is a morphism for P and K if and only if $P(u \circ f) = \phi \cdot Ku \circ f$.*

Proof. Clearly the condition implies that f is a morphism. The converse follows from Fuglede's beatiful argument in [F1], Remark 1, p. 129. The only thing needed is a function w satisfying $Qw > 0$ on some neighbourhood, arbitrarily small, of a given point in M, and then we may take

$$\phi := \frac{P(w \circ f)}{Kw \circ f}. \tag{4.8}$$

This proves our assertion.

We now normalise so that f is *time-preserving*:

$$\frac{dt}{ds} = \dot{f}^0 > 0. \tag{4.9}$$

As in [DK] we shall use the following further conditions on $\gamma = (f, a)$:

$$df^0 = 0; \tag{G1}$$

$$\frac{1}{2}\widetilde{\Delta} f^\alpha + a^{-1}\langle da, df^\alpha\rangle + \frac{1}{2}\Gamma^\alpha_{\beta\gamma} \circ f\langle df^\beta, df^\gamma\rangle = 0, \quad 1 \le \alpha \le n; \tag{G3i}$$

$$\dot{f}^\alpha + \frac{1}{2}\widetilde{\Delta} f^\alpha + a^{-1}\langle da, df^\alpha\rangle + \frac{1}{2}\Gamma^\alpha_{\beta\gamma} \circ f\langle df^\beta, df^\gamma\rangle = 0, \quad 1 \le \alpha \le n; \tag{G3ii}$$

$$\langle df^\beta, df^\gamma\rangle = \lambda^2 g^{\beta\gamma} \circ f; \tag{G4i}$$

$$\langle df^\beta, df^\gamma\rangle = 2\frac{dt}{ds} g^{\beta\gamma} \circ f. \tag{G4ii}$$

The roman numerals of course refer to the cases defined above.

In [DK] we showed for the heat case that (G1)–(G4) are sufficient for f to be a morphism for P and K. Following [F1, F2, Ish], in the case of harmonic morphisms, we now show that these conditions are also necessary in the heat case. We assume the normalisation (4.9). To this end, let $\widetilde{X}$ be the process in (4.7), and put $Y^\alpha_s := f^\alpha(s, \widetilde{X}_s)$, $1 \le \alpha \le n$. By Itô's formula,

$$dY^\alpha_s = (\dot{f}^\alpha + \frac{1}{2}\widetilde{\Delta} f^\alpha + a^{-1}\langle da, df^\alpha\rangle)(s, \widetilde{X}_s)\, ds + \partial_i f^\alpha(s, \widetilde{X}_s) d\widetilde{\xi}^i_s. \tag{4.10}$$

Thus

$$d[Y^\beta, Y^\gamma]_s = \langle df^\beta, df^\gamma\rangle(s, \widetilde{X}_s)\, ds. \tag{4.11}$$

We may write this as

$$\begin{aligned} d^c Y^\alpha_s = &\left(\dot{f}^\alpha + \frac{1}{2}\widetilde{\Delta} f^\alpha + a^{-1}\langle da, df^\alpha\rangle + \frac{1}{2}\Gamma^\alpha_{\beta\gamma} \circ f\langle df^\beta, df^\gamma\rangle\right)(s, \widetilde{X}_s)\, ds \\ &+ \partial_i f^\alpha(s, \widetilde{X}_s) d\widetilde{\xi}^i_s. \end{aligned} \tag{4.12}$$

To start with, Y must be a time shift of X (Sect. 2.2). Since X is a martingale, we see that condition (G3) is necessary, and then

$$d^c Y^\alpha_s = \partial_i f^\alpha(s, \widetilde{X}_s) d\widetilde{\xi}^i_s. \tag{4.13}$$

The conformality of X requires condition (G4i).

Up to now we have only considered the 'harmonic' case under a time-dependent transformation. To reach the heat equation, and condition (G4ii), we must study the time-space process $f(s, \widetilde{X}_s) = (f^0(s), Y_s)$. If u is a (local) function on M, then

$$d(u \circ f(s, \widetilde{X}_s)) = \left(\frac{dt}{ds}\dot{u} \circ f(s, \widetilde{X}_s) + \frac{1}{2}\lambda^2 \Delta u \circ f(s, \widetilde{X}_s)\right) ds + d\xi, \tag{4.14}$$

where ξ is a martingale. The coefficient of the first (ds-)term on the right-hand side must be proportional to $(\dot{u} + (1/2)\Delta u) \circ f(s, \widetilde{X}_s)$. Clearly this requires (G4ii).

We collect our findings in

Theorem 11. *Given* $\gamma = (f, a)$*, suppose* (4.9) *holds and* $a > 0$*. The following are equivalent*
– *in case (i):*

a) $\widetilde{\Delta}(a \cdot u \circ f) = \phi a \Delta u \circ f$*, with* $\phi = \lambda^2 = n^{-1} \mathrm{Tr}\, df \otimes df$*;*
b) the process $f(\widetilde{X}_s)$ *is a time shift of* BM(N, g, ∇)*;*
c) Eqs. (G2)–(G4) hold.

– *in case (ii):*

a) $(\partial_s + (1/2)\widetilde{\Delta})(a \cdot u \circ f) = \phi a(\partial_t + (1/2)a\Delta u) \circ f$*, with* $\phi = dt/ds$*;*
b) the process $f(s, \widetilde{X}_s)$ *is a time shift of* (t, X_t)*, where* X *is* BM(N, g, ∇)*;*
c) Eqs. (G1)–(G4) hold.

Remark 12. Condition (G2) means that martingales are preserved ([D, Em]). In case (i), f is a *harmonic mapping* ([F1, F2, Ish]) when $a \equiv 1$. See also the discussion on affine maps in [Em]. Condition (G4) means that the map is *horizontally conformal*, see [F1]. Together with (G2) we have (in case (ii)) a characterisation of maps preserving conformal martingales, cf. Proposition 4.

Condition (G4ii) links the time and space scales. It implies that the dilation w.r.t. the space variable is independent of the space variable, as opposed to the case of harmonic morphisms.

Remark 13. If ϕ is a harmonic morphism with constant dilation, and $dt/ds = \frac{1}{2}\lambda^2$, then (t, ϕ) is a heat harmonic morphism. This case was treated independently by Loubeau [Lob].

4.2 Morphisms for Conjugate BM

We consider case (ii) of the preceding section, and assume that (f, a) satisfy (G1)–(G4). Let $\theta > 0$ and V be as in Sect. 2.3 and put

$$\widetilde{V} := (dt/ds)V \circ f. \tag{4.15}$$

We write $\widetilde{D}_0 := \partial_s + \frac{1}{2}\widetilde{\Delta} - \widetilde{V}$ and $D_0 := \partial_t + \frac{1}{2}\Delta - V$, so that $D_0\theta = 0$. By Theorem 11 and the definition of $\widetilde{V}$ $\widetilde{D}_0(au \circ f) = a(dt/ds)D_0 u \circ f$. Defining $\widetilde{\theta} := a \cdot \theta \circ f$, also $\widetilde{D}_0\widetilde{\theta} = 0$.

We introduce the forward generators $\widetilde{D} := \widetilde{D}_{\widetilde{\theta}}$ and $D := D_\theta$ as in (2.11):

$$\widetilde{D}\widetilde{u} := \widetilde{D}_0(\widetilde{\theta}\widetilde{u})/\widetilde{\theta} \quad \text{and} \quad Du = D_0(\theta u)/\theta. \tag{4.16}$$

The following result from [DK] shows that heat harmonic morphisms also preserve conjugate BM.

Theorem 14. *Suppose* (f, a) *satisfy (G1)–(G4) and the normalisation* (4.9) *in Sect. 4.1. Then* f *is a morphism for* $\widetilde{D}$ *and* D*:*

$$\widetilde{D}(u \circ f) = \frac{dt}{ds} Du \circ f \tag{4.17}$$

for any smooth function u *on* M*.*

Proof. By what we have just seen,

$$\begin{aligned} \widetilde{D}(u \circ f) = \frac{1}{\widetilde{\theta}} \widetilde{D}_0(\widetilde{\theta} \cdot u \circ f) &= \frac{1}{a\theta \circ f} \widetilde{D}_0(a \cdot (\theta u) \circ f) \\ &= \frac{1}{\theta \circ f} \frac{dt}{ds} D_0(\theta u) \circ f = \frac{dt}{ds} Du \circ f. \end{aligned} \tag{4.18}$$

Let $\widetilde{Z}$ denote the process associated with $\widetilde{\theta}$. Then $(f(s, \widetilde{Z}_s))_{s \in \widetilde{I}}$ and $(t, Z_t)_{t \in I}$, after a time-change, have the same distribution (cf. [Ok] and Sects. 2.2–2.3).

4.3 Dynamical Invariance for Conjugate BM

We now give a result already stated in [DK] describing the transformation properties of the (forward) energy, momenta, Lagrangian and equations of motion. We only consider the situation when ∇ is the Levi-Civita connection. We assume that (f, a) satisfy (G1)–(G4) and the normalisation $\dot{f}^0 > 0$.

In the sequal $T^* f$ denotes pullback via f w.r.t. the space variables.

Theorem 15. *Let* $\Psi := \log a$ *and* $\widetilde{E}_0 := \dot{a}/a$*. Then*

$$\widetilde{E} = \widetilde{E}_0 + \frac{dt}{ds} E \circ f + \dot{f}^\alpha p_\alpha \circ f, \tag{4.19}$$

$$\widetilde{p} = T^* f p + d\Psi, \tag{4.20}$$

$$\widetilde{L} = \frac{dt}{ds} L \circ f + \widetilde{D}\Psi, \tag{4.21}$$

$$\left(\widetilde{D} - \frac{1}{2} \widetilde{\mathrm{Ric}} \right) \widetilde{p} = \frac{dt}{ds} T^* f \left(Dp - \frac{1}{2} \mathrm{Ric}\, p \right). \tag{4.22}$$

Proof. (4.20) is obvious since $\widetilde{p}_i = \partial_i \log \widetilde{\theta} = \partial_i \log a + \partial_\alpha \theta \circ f \partial_i f^\alpha = \partial_i \log a + (T^* f p)_i$. (4.19) is obtained similarly.

Squaring and summing (4.20), (G4) yields

$$\begin{aligned} \frac{1}{2} \frac{dt}{ds} |p|^2 \circ f &= \frac{1}{2} \lambda^2 (g^{\alpha\beta} p_\alpha p_\beta) \circ f = \frac{1}{2} \widetilde{g}^{ij} p_\alpha \circ f \partial_i f^\alpha p_\beta \circ f \partial_i f^\beta \\ &= \frac{1}{2} \widetilde{g}^{ij} \left(\widetilde{p}_i - \frac{\partial_i a}{a} \right) \left(\widetilde{p}_j - \frac{\partial_j a}{a} \right) \\ &= \frac{1}{2} |\widetilde{p}|^2 + \frac{1}{2} \left| \frac{da}{a} \right|^2 - \frac{1}{a} \langle \widetilde{p}, da \rangle. \end{aligned} \tag{4.23}$$

Now $\widetilde{D}a = \partial_s a + \frac{1}{2}\Delta a + \langle \widetilde{p}, da/a \rangle = \langle \widetilde{p}, da/a \rangle$, so by (2.15) $\widetilde{D} \log a = \langle \widetilde{p}, da/a \rangle - \frac{1}{2}|da/a|^2$. Hence, using the definition of $\widetilde{V}$ in Sect. 4.2 we obtain (4.21):

$$\frac{dt}{ds} L \circ f = \frac{dt}{ds}\left(\frac{1}{2}|p|^2 \circ f + V \circ f\right) = \frac{1}{2}|\widetilde{p}|^2 - \widetilde{D}\log a + \widetilde{V} = \widetilde{L} - \widetilde{D}\Psi. \quad (4.24)$$

(4.22) follows from the Newton equations (2.21) for the two processes, together with the definition of $\widetilde{V}$.

Remark 16. Concerning the extra terms on the right, e.g., in (4.21), one should recall that already in the classical case, the equations of motion are not altered when a total time differential is added to the Lagrangian. (See, e.g., [DNF], Sect. 31, p. 305.) (4.21) states the invariance of the Lagrangian time-differential up to a regularised time derivative: $\widetilde{L}\, ds - L \circ f\, dt = \widetilde{D}\Psi\, ds$.

The transformation of the energy in (4.19) may seem strange. In the example $f = (t, \phi)$ of Remark 13, however, it transforms in a less exotic way: $\widetilde{E} = \widetilde{E}_0 + \frac{dt}{ds} E \circ f$.

5 The Heat and Laplace Lie Algebras

Throughout this chapter, the underlying manifold N is Riemannian with metric g. We shall only consider the Levi-Civita connection.

For a (possibly time-dependent) vector field Q on N, we shall denote by ω its dual one-form: $\omega_k = g_{ik} Q^i$, i.e. $Q = \omega^\sharp$.

5.1 Conformal Groups and Lie Algebras

For any vector field Q we have

$$(\mathcal{L}_Q g)_{ij} = \nabla_j \omega_i + \nabla_j \omega_i = 2(\nabla \omega)_{(ij)}, \quad (5.1)$$

where $\mathcal{L}_Q$ denotes the Lie derivative along Q, and the parentheses indicate symmetrisation.

Definition 17. *The conformal Lie algebra* $\mathfrak{conf}$ *consists of all vector fields Q on N such that*

$$\mathcal{L}_Q g = \mu \cdot g, \quad (5.2)$$

for some function μ depending on Q.

The corresponding situation for maps, i.e. local flows, is

$$g \circ f(df(\xi_1), df(\xi_2)) = \kappa \cdot g(\xi_1, \xi_2), \quad (5.3)$$

for some function κ. That $\mathfrak{conf}$ is a Lie algebra follows from

$$[\mathcal{L}_{Q_1}, \mathcal{L}_{Q_2}]g = (Q_1(\mu_2) - Q_2(\mu_1)) \cdot g. \tag{5.4}$$

Equation (5.1) implies

$$\mu = -\frac{2}{n} d^\dagger \omega = \frac{2}{n} \nabla^\dagger \omega, \quad Q \in \mathfrak{conf}. \tag{5.5}$$

By, e.g., [Go], Eqs. (3.7.4) and (3.8.4) or [Ko],

$$\frac{1}{2}\Box\omega + \mathrm{Ric}\,\omega - \left(\frac{1}{2} - \frac{1}{n}\right) dd^\dagger \omega = \frac{1}{2}\Box\omega + \mathrm{Ric}\,\omega + \frac{n-2}{4} d\mu = 0. \tag{5.6}$$

Define three Lie algebras:

$$\mathfrak{k} := \{Q \in \mathfrak{conf} : \mu = 0\}, \tag{5.7}$$

$$\mathfrak{h} := \{Q \in \mathfrak{conf} : \mu = \text{const.}\}, \tag{5.8}$$

$$\mathfrak{conf}_h := \{Q \in \mathfrak{conf} : \Delta\mu = 0\}. \tag{5.9}$$

Clearly,

$$\mathfrak{k} \subset \mathfrak{h} \subset \mathfrak{conf}_h \subset \mathfrak{conf}. \tag{5.10}$$

Here $\mathfrak{k}$ stands for *Killing vector fields*, i.e., infinitesimal isometries, and $\mathfrak{h}$ indicates that the corresponding flows are *homothetic transformations.* The inclusions are obvious, and 5.1.4 shows that $\mathfrak{k}$ and $\mathfrak{h}$ are Lie algebras. Clearly $\mathfrak{h} = \mathfrak{k} \oplus \mathbb{R}$ globally.

We shall need

Lemma 18. *If $Q_1 \in \mathfrak{conf}$ and $Q_2 \in \mathfrak{conf}_h$, then*

$$\Delta(Q_1(\mu_2)) = -\frac{n-2}{2} \langle d\mu_1, d\mu_2 \rangle. \tag{5.11}$$

We assume the lemma momentarily. For $Q \in \mathfrak{conf}$, let

$$U := \frac{n-2}{4} \mu = \frac{n-2}{2n} \nabla^\dagger \omega, \tag{5.12}$$

and

$$\overline{Q} := (Q, U) := Q + U = Q + \frac{n-2}{2n} \nabla^\dagger \omega \in \widetilde{\mathcal{V}}, \tag{5.13}$$

where $\widetilde{\mathcal{V}}$ was defined in Sect. 3.1. The bracket is (by (3.7))

$$[\overline{Q}_1, \overline{Q}_2] = [Q_1, Q_2] + Q_1(U_2) - Q_2(U_1) = [Q_1, Q_2] + \frac{n-2}{4n} \mathrm{Tr}\,(\mathcal{L}_{[Q_1, Q_2]} g). \tag{5.14}$$

This way we get a map

$$\mathcal{V} \supset \mathfrak{conf} \ni Q \to \overline{Q} \in \widetilde{\mathcal{V}}. \tag{5.15}$$

We note that in dimension $n = 2$ on $\mathcal{V}$, or on $\mathfrak{k}$ in any dimension, (5.15) is the identity map: $\overline{Q} = (Q, 0)$.

Theorem 19. $\mathfrak{conf}_h$ *is a Lie algebra. The restriction of (10) to* $\mathfrak{conf}_h$ *is a Lie algebra isomorphism onto its image:*

$$[\overline{Q}_1, \overline{Q}_2] = [Q_1, Q_2] + \frac{n-2}{2n}\nabla^\dagger[\omega_1, \omega_2], \qquad X \in \mathfrak{conf}_h, \tag{5.16}$$

where $[\omega_1, \omega_2]$ *is the one-form dual to* $[Q_1, Q_2]$.

Proof. The first statement follows from Lemma 18 which clearly implies that $\Delta(Q_1(\mu_2) - Q_2(\mu_1)) = 0$ for $Q_i \in \mathfrak{conf}_h$.

We now show the displayed identity. Up to a multiplicative constant $Q_1(\mu_2)$ is equal to $\langle \omega_1, (\Box + \mathrm{Ric}\,)\omega_2\rangle$ by (5.6). Hence, by Weitzenböck, the difference $Q_1(\mu_2) - Q_2(\mu_1)$ is a constant times

$$\langle \omega_1, \Delta\omega_2\rangle - \langle \omega_2, \Delta\omega_1\rangle, \tag{5.17}$$

since the Ricci tensor is symmetric. This equals

$$g^{ij}\nabla_j(\langle \omega_1, \nabla_i\omega_2\rangle - \langle \omega_2, \nabla_i\omega_1\rangle). \tag{5.18}$$

Using that ∇ commutes with the duality $T^*N \cong TN$ given by g, the expression in the parentheses is dual to

$$\nabla_{Q_1}Q_2 - \nabla_{Q_2}Q_1 = [Q_1, Q_2], \tag{5.19}$$

the connection being torsion free. Checking the constant, one finds (5.16).

Proof (Lemma 18). Using $\nabla g = 0$ we get

$$\nabla^2 Q_1(\mu_2) = \langle \nabla^2\omega_1, d\mu_2\rangle + 2\langle \nabla\omega_1, \nabla d\mu_2\rangle + \langle \omega_1, \nabla^2 d\mu_2\rangle.$$

To obtain the Laplacian, we must take the trace. Having done this, the second term on the right is proportional to $\mu_1\Delta\mu_2$, by (5.1), (5.2). By assumption it vanishes. We get, using the Weitzenböck formula, the identity $[d, \Box] = 0$ and (5.6),

$$\begin{aligned}\Delta Q_1(\mu_2) &= \langle \Delta\omega_1, d\mu_2\rangle + \langle \omega_1, \Delta d\mu_2\rangle \\ &= \langle(\Box + \mathrm{Ric}\,)\omega_1, d\mu_2\rangle + \langle \omega_1, \mathrm{Ric}\, d\mu_2\rangle \\ &= \langle(\Box + 2\mathrm{Ric}\,)\omega_1, d\mu_2\rangle = -\frac{n-2}{2}\langle d\mu_1, d\mu_2\rangle.\end{aligned}$$

The claim follows.

5.2 Characterisation of the Heat Lie Algebra

We consider the situation in (4.7), case (ii), with the further requirement that Δ be the Laplace-Beltrami operator. For $\Lambda = (X, U)$ we write

$$X = T\frac{\partial}{\partial t} + Q^i\frac{\partial}{\partial x^i} = T\frac{\partial}{\partial t} + Q, \tag{5.20}$$

where, at this point, T and the Q^i are functions of $t \in I$ and $x \in N$.

We now characterise the heat Lie algebra in terms of PDEs. In general, the system obtained is overdetermined. It should be no surprise that the equations obtained are completely analogous to Eqs. (G1-G4) in Sect. 4.1.

Theorem 20. $\Lambda = (T, Q, U) \in \mathrm{Lie}\,(\partial_t + \frac{1}{2}\Delta - V)$ *if and only if the following equations are satisfied:*

$$dT = 0; \tag{A1}$$

$$\dot{U} + \frac{1}{2}\Delta U + \frac{\partial}{\partial t}(TV) + Q(V) = 0; \tag{A2}$$

$$\dot{\omega} + \frac{1}{2}\Box\omega + Ric\,\omega + dU = 0; \tag{A3}$$

$$(\nabla\omega)^{(ij)} = \frac{1}{2}\dot{T}g^{ij}. \tag{A4}$$

The associated cocycle is $\Phi = \Phi_\Lambda = \dot{T}$.

Proof. We start with the free case $V \equiv 0$. With $K := \partial_t + \frac{1}{2}\Delta$ we easily find

$$[K, U]u = (\dot{U} + \frac{1}{2}\Delta U)u + \langle dU, du\rangle, \tag{5.21}$$

and

$$[\partial_t, X]u = \dot{X}u = \dot{T}\dot{u} + \langle\dot{\omega}, du\rangle. \tag{5.22}$$

Furthermore,

$$\begin{aligned}[\Delta, X]u = {} & \Delta T \cdot \dot{u} + 2\langle dT, d\dot{u}\rangle + T\Delta\dot{u} \\ & + \langle\Box\omega, du\rangle + 2\mathrm{Ric}\,(\omega, du) + 2\langle\nabla\omega, \nabla du\rangle.\end{aligned} \tag{5.23}$$

To see this, note that the left-hand side is

$$\begin{aligned}\Delta(T\dot{u} + \langle\omega, du\rangle) - T\Delta\dot{u} - \langle\omega, d\Delta u\rangle = {} & \Delta T \cdot \dot{u} + 2\langle dT, d\dot{u}\rangle + T\Delta\dot{u} \\ + \langle\Delta\omega, du\rangle) + \langle\omega, \Delta du\rangle + 2\langle\nabla\omega, \nabla du\rangle & - T\Delta\dot{u} - \langle\omega, d\Delta u\rangle,\end{aligned} \tag{5.24}$$

and use the identities for Laplacians stated in Sect. 1.

¿From the above equations we now get

$$\begin{aligned}[Q, \Lambda]u = {} & (\dot{U} + \frac{1}{2}\Delta U)u + (\dot{T} + \frac{1}{2}\Delta T)\dot{u} + \langle dT, d\dot{u}\rangle \\ & + \langle\dot{\omega} + \frac{1}{2}\Box\omega + \mathrm{Ric}\,\omega + dU, du\rangle + \langle\nabla\omega, \nabla^2 u\rangle.\end{aligned} \tag{5.25}$$

We want the left-hand side to be a function times Ku. First, no constant term is allowed, so that U has to satisfy (A2). To avoid terms with mixed time and space derivatives we must also require that T only depends on time, i.e., that (A1) holds. The first-order space derivatives disappear if and only if (A3) holds.

We now have

$$[K, \Lambda]u = \dot{T}\dot{u} + \langle \nabla\omega, \nabla^2 u\rangle. \tag{5.26}$$

To get a Laplacian out of the second term, we must require that the symmetric part of $\nabla\omega$ is proportional to the inverse metric (g^{ij}), $\nabla^2 u$ being symmetric. (A4), finally, is needed to adjust the scales between time and space derivatives.

The general case, including a potential V, follows from $[V, \Lambda]u = -X(V)\cdot u$ and $[K - V, \Lambda] = [K, \Lambda] - [V, \Lambda] = \Phi(K - V) = \dot{T}(K - V)$.

Remark 21. Combining Eqs. (A2), (A4) and (A5), we see that for each t, we must have $\omega(t, \cdot)$ in $\mathfrak{h}$ (Sect. 5.1). Hence the heat Lie algebra is trivial (i.e., consists only of constants) whenever $\mathfrak{k} = 0$. This is different from the Laplace case where elements in $\mathfrak{conf}\setminus h$ may occur, e.g. in one space-dimension. See also Sect. 5.3 below.

A part of Lie (Δ) is always contained in Lie $(\partial_t + \frac{1}{2}\Delta)$, viz., when non-void, the one corresponding to constant (in time) elements in $\mathfrak{h}$. Comparing with (5.5) we see that $2c = \mu = -(2/n)d^\dagger\omega$, so we may take U constant and $T(t) = \mu t + \mu_0$. Note that when $\omega \in \mathfrak{k}$, the Killing algebra, time does not enter explicitly, i.e., T is constant.

Our next result clarifies the relation between Lie $(\partial_t + \frac{1}{2}\Delta)$ and $\mathfrak{h}$.

Theorem 22. Lie $(\partial_t + \frac{1}{2}\Delta)$ *consists of all* $\Lambda = (T, Q, U)$ *with* $Q \in C^\infty(I \to \mathfrak{h})$ *such that its dual one-form* ω *satisfies*

$$\frac{\partial^2}{\partial t^2} d^\dagger\omega = 0, \tag{5.27}$$

$$\frac{\partial}{\partial t} d\omega = 0. \tag{5.28}$$

T is a polynomial in t of degree at most 2 with $\dot{T} = -(2/n)d^\dagger\omega$. $U = \frac{1}{2}\alpha - \dot{\alpha}U_0$, where $\alpha = d^\dagger\omega = -(n/2)\dot{T}$, and U_0 satisfies $\Delta U_0 = 1$.

Proof. Note first that if ω does not depend on time, (5.27) and (5.28) are trivially fulfilled, and we are back in the case discussed in Remark 21.

Let us start from $Q \in C^\infty(I \to \mathfrak{h})$ satisfying (5.27) and (5.28). It is clear how to obtain T from ω. We shall show how to find U.

We know from (5.6) that $\frac{1}{2}\Box\omega + \mathrm{Ric}\,\omega = 0$ for each fixed time. Hence, (A3) becomes

$$\dot{\omega} + dU = 0. \tag{5.29}$$

By (5.28), this is fulfilled for some U. We must show that we can arrange so that U satisfies the anti-heat equation (A1). From $\dot{\omega} + dU = 0$ follows, invoking (5.27),

$$\Delta U = d^\dagger \dot{\omega} = \text{const.} = \varkappa. \tag{5.30}$$

If this constant vanishes, we are again back to Remark 21, and may take U constant.

If not, write $U_0 = U/\varkappa$, so that $\Delta U_0 = 1$. (Note that, locally, there are always such functions.) Without altering this, we may put $U = \frac{1}{2}\alpha - \dot{\alpha}U_0$, and then U satisfies (A2).

This proves that (5.27) and (5.28) are sufficient. Clearly condition (5.28), as well as the expression for T is necessary. It remains to deduce (5.27), and the explicit form for U. We may assume that U is non-constant. As above, we get $U(t,\cdot) = \psi(t) + \varphi(t)U_0$, for some functions φ and ψ, where U_0 is independent of time and $\Delta U_0 = 1$. Then (A2) holds if and only if $\dot{\varphi} = 0$ and $-\frac{1}{2}\dot{\psi} = \varphi$. Since also (A4) must hold we deduce $\varphi = d^\dagger \dot{\omega}$, which implies (5.27).

5.3 Characterisation of the Laplace Lie Algebra

We now consider the Lie algebra of the Laplace-Beltrami operator on (N, g), in which case $X = Q$ (cf. (5.20)).

As in Sect. 5.2 we get

Theorem 23. $\Lambda = (Q, U) \in \mathrm{Lie}\,(\Delta)$ *if and only if the following equations are satisfied:*

$$\Delta U = 0; \tag{a2}$$

$$\frac{1}{2}\Box\omega + \mathrm{Ric}\,\omega + dU = 0; \tag{a3}$$

$$(\nabla\omega)^{(ij)} = c \cdot g^{ij}. \tag{a4}$$

The cocycle is $\Phi = \Phi_\Lambda = c$.

¿From Theorem 19 we immediately get

Theorem 24. *As Lie algebras* $\mathrm{Lie}\,(\Delta) = \mathfrak{conf}_h$*, through the map* (5.15).

6 Constants of Motion, Conservation Laws and Martingales

6.1 Quantum Picture

We start from a self-adjoint Hamiltonian $H = -\frac{1}{2}\Delta + V$, where $V = V(t, q)$, and q is the coordinate in N. Let $\Lambda = T\partial_t + Q^i\partial_i + U = X + U$, where X is a (smooth) vector field on space-time M and U is a (smooth) function on M. This is the general form for a linear PDO of order ≤ 1 on M. We write $\mathfrak{D}^1(M)$ for all such operators.

Denote by

$$K := \frac{\partial}{\partial t} - H \tag{6.1}$$

the (backward) *heat operator.* If $\Lambda \in \text{Lie}\,(K)$, i.e., $[K, \Lambda] = \Phi \cdot K$ for some function Φ, then

$$K\Lambda u = ([K, \Lambda] + \Lambda K)u = (\Phi + \Lambda)Ku. \tag{6.2}$$

Hence,

$$K\Lambda u = 0 \quad \text{if} \quad Ku = 0 \quad \text{and} \quad \Lambda \in \text{Lie}\,(K), \tag{6.3}$$

i.e., the Lie algebra preserves the kernel of K.

The operator $1 : u \to u$ always belongs to Lie (K). It expresses the *conservation law*

$$\frac{d}{dt} \int_N \theta\theta^* \,\text{dvol} = 0. \tag{6.4}$$

This is a direct consequence of H being self-adjoint: the left-hand side is

$$\langle \dot{\theta}, \theta^* \rangle + \langle \theta, \dot{\theta}^* \rangle = \langle H\theta, \theta^* \rangle + \langle \theta, -H\theta^* \rangle = \langle H\theta, \theta^* \rangle - \langle H\theta, \theta^* \rangle = 0. \tag{6.5}$$

Suppose f is a smooth function on M with appropriate growth conditions. Then

$$\frac{d}{dt} \int_N f \cdot \theta\theta^* \,\text{dvol} = \int_N Df \cdot \theta\theta^* \,\text{dvol} = \int_N D^* f \cdot \theta\theta^* \,\text{dvol} \tag{6.6}$$

Suppose now that Λ is any element the heat Lie algebra. By definition $D = \theta^{-1} K \theta$, so that

$$D(\theta^{-1}\Lambda\theta) = \theta^{-1} K \Lambda \theta = \theta^{-1}(\Phi + \Lambda)K\theta = 0, \tag{6.7}$$

i.e., along the process, $\theta^{-1}\Lambda\theta$ is a martingale. This can also be expressed as the conservation law

$$\frac{d}{dt} \int_N \Lambda\theta \cdot \theta^* \,\text{dvol} = 0, \tag{6.8}$$

because the LHS is $(d/dt) \int_N \theta^{-1}\Lambda\theta \cdot \theta\theta^* \,\text{dvol} = \int_N D(\theta^{-1}\Lambda\theta) \cdot \theta\theta^* \,\text{dvol} = 0$, as we just saw.

By repetition of these arguments,

Theorem 25. *If $\Lambda_j \in \text{Lie}\,(K)$, and if $s_j \geq 0$ are integers, $1 \leq j \leq k$, then, along the process, $\theta^{-1}\Lambda_1^{s_1} \cdots \Lambda_k^{s_k}\theta$ is a martingale, and we have the conservation law*

$$\frac{d}{dt} \int_N \Lambda_1^{s_1} \cdots \Lambda_k^{s_k} \theta \cdot \theta^* \,\text{dvol} = 0. \tag{6.9}$$

We remark that by time-reflection and duality we get corresponding statements for θ^* and $\Lambda_1^{\dagger s_1} \cdots \Lambda_k^{\dagger s_k}$.

Let us now return to the function $\theta^{-1}\Lambda\theta$. With the notation $\widehat{p}_i = \partial_i\theta/\theta$ and $\widehat{E} = \dot{\theta}/\theta$, it becomes

$$\frac{\Lambda\theta}{\theta} = \widehat{E}T + \widehat{p}_i Q^i + U. \tag{6.10}$$

$\widehat{p}_i$ are the *momentum densities* and $\widehat{E}$ the *energy density* from Sect. 2.5. (We have added hats in this chapter to distinguish between classical and quantum objects.)

Hence, by the theorem,

Corollary 26.

$$D\left(\widehat{E}T + \langle\widehat{p}, Q\rangle + U\right) = 0 \tag{6.11}$$

whenever $\Lambda = (T, Q, U) \in \mathrm{Lie}\,(K)$.

We recall that with $A = \log\theta$ the coefficients $\widehat{p}_i = \partial_i\theta/\theta$ and $\widehat{E} = \dot{\theta}/\theta$ are given by

$$\overline{d}A = \widehat{E}\,dt + \widehat{p}_i\,dq^i = \widehat{E}\,dt + \widehat{p}dq, \tag{6.12}$$

where $\overline{d}A$ signifies space-time differential. The right-hand side is the restriction of the first fundamental form $\omega = Edt + pdq$ to a Lagrangian manifold: the second fundamental form $\Omega = \overline{d}\omega = dE \wedge dt + dp \wedge dq$ vanishes there, because $\overline{d}^2 = 0$.

6.2 Classical Picture

M is the *configuration space*, and the cotangent bundle T^*M is the (extended) *phase space*. The fibre coordinates are (E, p).

Definition 27. *The symbol map takes* $\Lambda \in \mathfrak{D}^1(M)$ *to the function* $F_\Lambda \in C^\infty(T^*M)$ *defined by*

$$F_\Lambda(t, q, E, p) := ET(t, q) + \langle p, Q(t, q)\rangle + U(t, q). \tag{6.13}$$

Using the first fundamental form $\omega = E\,dt + p\,dq$, $F = \omega(\pi_* X) + U$, where $X = (T, Q)$ and π is the projection $TM \to M$. The second fundamental form $\Omega = d\omega = dE \wedge dt + dp \wedge dq$ on T^*M gives rise to the *Poisson bracket*

$$\{\phi, \psi\} = \frac{\partial\phi}{\partial E}\frac{\partial\psi}{\partial t} - \frac{\partial\phi}{\partial t}\frac{\partial\psi}{\partial E} + \frac{\partial\phi}{\partial p_i}\frac{\partial\psi}{\partial q^i} - \frac{\partial\phi}{\partial q^i}\frac{\partial\psi}{\partial p_i}, \quad \phi, \psi \in C^\infty(T^*M). \tag{6.14}$$

Theorem 28. *The symbol map is 1–1 onto the space of functions in* C^∞ (T^*M) *which are of order at most one in* E *and* p. *It is a Lie algebra morphism:*

$$\{F_{\Lambda_1}, F_{\Lambda_2}\} = F_{[\Lambda_1, \Lambda_2]}. \tag{6.15}$$

Given an 'ordinary' Hamiltonian $H = H(t, q, p)$, we get an *extended Hamiltonian* K defined by

$$K = K(t, q, E, p) := E - H(t, q, p). \tag{6.16}$$

For any differentiable function F on T^*M we define

$$\frac{dF}{dt} := \{K, F\}. \tag{6.17}$$

This is equal to

$$\frac{\partial F}{\partial t} - \frac{\partial H}{\partial p_i}\frac{\partial F}{\partial q^i} + \frac{\partial H}{\partial q^i}\frac{\partial F}{\partial p_i} + \frac{\partial H}{\partial t}\frac{\partial F}{\partial E} \tag{6.18}$$

and leads to

$$\frac{dF}{dt} = \frac{\partial F}{\partial t} + p_i\frac{\partial F}{\partial q^i} + \frac{\partial V}{\partial q^i}\frac{\partial F}{\partial p_i} + \frac{\partial V}{\partial t}\frac{\partial F}{\partial E} \tag{6.19}$$

if we choose the Euclidean Hamiltonian

$$H = -\frac{1}{2}|p|^2 + V, \quad \text{where} \quad V = V(t, q). \tag{6.20}$$

¿From now on, this is our choice for H. The *equations of motion* become

$$\dot{q} = p^\sharp, \quad \frac{Dp}{dt} = dV, \quad \dot{E} = \dot{V}, \quad \dot{t} = 1. \tag{6.21}$$

Here, D/dt denotes the covariant derivative.

Definition 29. *The classical Lie algebra,* $\mathrm{Lie}_c(K)$, *consists of all* $F \in C^\infty(T^*M)$ *of order at most one in* (E, p) *which satisfy*

$$\{K, F\} = a \cdot K \tag{6.22}$$

for some (local) function $a = a_F$.

By the implicit function theorem, this is equivalent to requiring that $dF/dt = 0$ on the set where $E = H(t, q, p)$.

The next point is to determine the classical Lie algebra. We calculate dF_Λ/dt and substitute $E = H(t, q, p)$. The result is a polynomial of order three in p, with coefficients depending on (t, q). All these coefficients must vanish.

$$\begin{aligned}
&\frac{d}{dt}(ET + \langle p, Q\rangle + U) \\
=&\dot{V}T + (-\tfrac{1}{2}|p|^2 + V)(\dot{T} + \langle p, dT\rangle) + \langle dV, Q\rangle \\
&+ \langle p, \dot{Q}\rangle + \nabla Q(p, p) + \dot{U} + \langle p, dU\rangle \\
=& -\tfrac{1}{2}|p|^2\langle p, dT\rangle + \left(\nabla Q^{(ij)} - \tfrac{1}{2}\dot{T}g^{ij}\right)p_i p_j \\
&+ \langle p, V dT + \dot{\omega} + dU\rangle + \dot{U} + \dot{T}V + T\dot{V} + Q(V).
\end{aligned} \tag{6.23}$$

Here, ω is the 1-form dual to Q.

Theorem 30. *$\Lambda = (T, Q, U)$ belongs to the classical Lie algebra for $K = E - H$ precisely when the following equations hold:*

$$dT = 0; \tag{C1}$$

$$\dot{U} + \frac{\partial}{\partial t}(TV) + Q(V) = 0; \tag{C2}$$

$$\dot{\omega} + dU = 0; \tag{C3}$$

$$\nabla\omega^{(ij)} = \tfrac{1}{2}\dot{T}g^{ij}. \tag{C4}$$

We shall now connect our findings with the Noether theorem. We refer to the presentation in Ibragimov [Ibr2], pp. 236–239.

6.3 The Classical Case

In general, a classical Lagrangian $L = L(t, q, \dot{q})$ and a Hamiltonian $H = H(t, q, p)$ are related by (Euclidean conventions)

$$L = \dot{q}^i p_i + H. \tag{6.24}$$

For $H = -\frac{1}{2}|p|^2 + V$, the Lagrangian becomes $L = \frac{1}{2}|\dot{q}|^2 + V$. The Euler-Lagrange equations are

$$\frac{\delta L}{\delta q^i} = 0, \quad i = 1, \dots, n, \tag{6.25}$$

where

$$\frac{\delta L}{\delta q^i} = \frac{\partial L}{\partial q^i} - \frac{d}{dt}\frac{\partial L}{\partial \dot{q}^i}. \tag{6.26}$$

Given a vector field $X = T\partial/\partial t + Q^i\partial/\partial q^i$, we make an infinitesimal variation of the action $\int L\,dt$:

$$\delta_X \int L\,dt = \int \left(X^{(1)}(L) + L\frac{dT}{dt}\right) dt. \tag{6.27}$$

Here, the vector field X is *prolonged* ([Ibr2, Or1]) to $X^{(1)}$, so that it can act also on the variable $\dot{q}$:

$$X^{(1)} = X + \left(\frac{DQ^i}{dt} - \dot{q}^i\frac{dT}{dt}\right)\frac{\partial}{\partial \dot{q}^i} = X + W^i\frac{\partial}{\partial \dot{q}^i}. \tag{6.28}$$

After some manipulation we get

$$X^{(1)}(L) + L\frac{dT}{dt} = \frac{d}{dt}\left(TL + W^i\frac{\partial L}{\partial \dot{q}^i}\right) + W^i\frac{\delta L}{\delta q^i}. \tag{6.29}$$

Thus, *Noether's invariance condition* (invariance of the Lagrangian differential $L\,dt$ modulo exact differentials)

$$X^{(1)}(L) + L\frac{dT}{dt} = -\frac{dU}{dt}, \tag{6.30}$$

for some function $U(t, q)$, implies that

$$\frac{d}{dt}\left(TL + W^i \frac{\partial L}{\partial \dot{q}^i} + U\right) = 0, \tag{6.31}$$

i.e.,

$$\frac{d}{dt}(TE + \langle p, Q\rangle + U) = 0. \tag{6.32}$$

This is exactly the case in Sect. 6.2 and should be compared with Corollary 6.1.7. In both cases we have what perhaps should be termed a Poisson-Noether algebra. It is a Lie algebra of conserved quantities; in general, only a part of the Lie algebra of a differential equation gives rise to conserved quantities.

We remark that this case of Noether's theorem was proved by M. Lévy already in 1878.

6.4 Quantum Case

The Hamiltonian is $H = -\frac{1}{2}\Delta + V$, and the *Hamiltonian field density* is $\mathsf{H} = \frac{1}{2}(H\theta \cdot \theta^* + \theta H\theta^*)$. This is an equivalent form of $\frac{1}{2}\langle d\theta, d\theta^*\rangle + V\theta\theta^*$.

The *Lagrangian field density* is

$$\mathsf{L} = \frac{1}{2}(\theta\dot{\theta}^* - \dot{\theta}\theta^*) + \mathsf{H}. \tag{6.33}$$

Write $(\theta, \theta^*) = (\theta^0, \theta^1)$, and $\theta^a_\mu = \partial_\mu \theta^a$. In general, the Euler-Lagrange equations, obtained from the space-time variational principle

$$\delta \iint \mathsf{L}\, dt\, \mathrm{dvol} = 0, \tag{6.34}$$

are

$$\frac{\delta \mathsf{L}}{\delta \theta^a} = 0, \quad a = 0, 1, \tag{6.35}$$

where

$$\frac{\delta \mathsf{L}}{\delta \theta^a} := \frac{\partial \mathsf{L}}{\partial \theta^a} - D_\mu \frac{\partial \mathsf{L}}{\partial \theta^a_\mu}. \tag{6.36}$$

In the present case we get

$$\dot{\theta} - H\theta = 0, \qquad \dot{\theta}^* + H\theta^* = 0. \tag{6.37}$$

Write $x = (x^0, x^1, ..., x^n) = (t, q^1, .., q^n)$. Consider a vector field Λ on the first order jet bundle over M with variables x^μ and θ^a, $0 \le \mu \le n$, $a = 0, 1$:

$$\Lambda = X^\mu \frac{\partial}{\partial x^\mu} + \eta^a \frac{\partial}{\partial \theta^a} = X^\mu \partial_\mu + \eta^a \frac{\partial}{\partial \theta^a}. \tag{6.38}$$

Write $dx := dt\, d\text{vol}$. We get

$$\delta_\Lambda \int_M \mathsf{L}\, dx = \int_M (\Lambda^{(2)}(\mathsf{L}) + \mathsf{L} D_\mu X^\mu)\, dx, \tag{6.39}$$

where $D_\mu := d/dx^\mu$ denotes the total derivative w.r.t. x_μ. Λ is prolonged to the second-order jet bundle. One finds

$$\Lambda^{(2)}(\mathsf{L}) + \mathsf{L} D_\mu X^\mu = D_\mu \left(\mathsf{L} X^\mu + W^a \frac{\partial \mathsf{L}}{\partial \theta^a_\mu} \right) + W^a \frac{\delta \mathsf{L}}{\delta \theta^a}, \tag{6.40}$$

where

$$W^a = \eta^a - \theta^a_\mu X^\mu. \tag{6.41}$$

Choose a vector field of the form

$$\Lambda = T \frac{\partial}{\partial t} + Q^i \frac{\partial}{\partial q^i} + U \left(\theta^* \frac{\partial}{\partial \theta^*} - \theta \frac{\partial}{\partial \theta} \right). \tag{6.42}$$

These vector fields are just another representation of the first order PDOs $T\frac{\partial}{\partial t} + Q^i \frac{\partial}{\partial q^i} + U$ employed above. The Lie algebras are isomorphic. – As usual, T, Q^i and U only depend on t and q. In this general case, Noether's theorem states that $\Lambda^{(2)}(\mathsf{L}) + \mathsf{L} D_\mu X^\mu = D_\mu B^\mu$ for some vector field B if and only if $\mathsf{L} X^0 + W^a \partial \mathsf{L}/\partial \dot{\theta}^a$ is the density of a conservation law. The latter quantity is

$$I = I_\Lambda = T \cdot \frac{1}{2}(\dot{\theta}\theta^* - \theta\dot{\theta}^*) + Q^i \cdot \frac{1}{2}(\theta_i\theta^* - \theta\theta^*_i) + U\theta\theta^*. \tag{6.43}$$

$I/\theta\theta^*$ is just a symmetric version of the function $\widehat{E}T + \widehat{p}_i Q^i + U$ in (6.10) above. The Noether theorem gives the same densities for conservation laws as we encountered in Sect. 6.1.

6.5 Connecting the Classical and the Quantum Algebras. Examples

The total (i.e., space-time) differential $\sigma := \overline{d}A$ of the action density $A = \log\theta$ defines a section of T^*M. If $F = F_\Lambda$, then

$$\widehat{E}T + \widehat{p}_i Q^i + U = \sigma^* F = F \circ \sigma =: \widehat{F}. \tag{6.44}$$

Write $F_j = F_{\Lambda_j}$. Theorem 28 implies, with obvious notation,

$$\{F_1, F_2\widehat{\}} = \widehat{F}_{[1,2]}. \tag{6.45}$$

One readily shows the following commutator formula:

$$\widehat{F}_{[1,2]} = X_1(\widehat{F}_2) - X_2(\widehat{F}_1). \tag{6.46}$$

We now characterise the classical Lie algebra for a class of quadratic potentials. The proof is left out.

Theorem 31. *Suppose $N = \mathbb{R}^n$. All potentials on the form*

$$V(t,q) = \frac{1}{2}a(t)|q|^2 + b(t)\cdot q + c(t), \tag{6.47}$$

where a, b and c are smooth functions of t, with $a > 0$, yield isomorphic classical Lie algebras for the Hamiltonian $H = -\frac{1}{2}|p|^2 + V$. They can be represented as

$$\mathfrak{h} + \mathfrak{b}_1 + \mathfrak{b}_2, \tag{6.48}$$

where $\mathfrak{h} = \langle 1, \xi_i, \eta_j\rangle_{i,j=0}^n$ is a representation of the Heisenberg algebra: $\{\xi_i, \eta_j\} = \delta_{ij}$ and all other brackets vanish. The centre is generated by 1, $\mathfrak{h}$ is an ideal and $\mathfrak{b}_1$ and $\mathfrak{b}_2$ commute. $\mathfrak{b}_1 = \langle |\xi|^2, \xi\cdot\eta, |\eta|^2\rangle$ is a representation of sl_2, and $\mathfrak{b}_2$ is a representation of so_n. The dimension is $2n + 4 + n(n-1)/2$.

Finally, we shall compare the two Lie algebras in this case, so $V(t,q) = \frac{1}{2}a(t)|q|^2 + b(t)\cdot q + c(t)$. If we integrate (C3)–(C4), omitting the case when $\nabla\omega$ is antisymmetric, i.e., the case of infinitesimal rotations, we first find $Q = \frac{1}{2}\dot{T}q + \alpha$, where α is a function of t, and then

$$U = -\frac{1}{4}\ddot{T}|q|^2 - \dot{\alpha}\cdot q + \beta, \tag{6.49}$$

where $\beta = \beta(t)$. Equation (C2) leads to

$$-\frac{1}{4}\dddot{T} + 2a\dot{T} + \dot{a}T = 0, \quad \ddot{\alpha} - a\alpha = \frac{3}{2}\dot{T}b + T\dot{b}, \quad \dot{\beta} = -(Tc)^{\cdot} + \alpha b. \tag{6.50}$$

Now $\Delta U = -\frac{n}{2}\ddot{T}$, so the only change caused by the heat equation is that β must fulfill

$$\dot{\beta} = -(Tc)^{\cdot} + \alpha b - \frac{n}{4}\ddot{T}. \tag{6.51}$$

This is the Itô correction. It only concerns one of the elements of the sl_2-part. Changing the representation for this part, in case of the heat equation, one finds that the classical Lie algebra and the heat Lie algebra are isomorphic.

This is related to Gaussian diffusions, oscillator-like systems and the semi-classical limit. See Brandão [B1], Brandão-Kolsrud [BK1, BK2], M. Kolsrud [K0] and Kolsrud-Zambrini [KZ2].

Comments and Acknowledgements

The major part of this article is included in the unpublished article [K1], most of which was written during a stay in Lisbon 1995. I thank my friends there once more, in particular the late Augusto Brandão Correia, 1961–2004.

It was a pleasure to participate in the Abel Symposium 2005. Many thanks to my Norwegian friends for kindly inviting me to the university where, almost twenty years ago, I first learned from Raphael Høegh-Krohn the beauty of the relations between the classical and the quantum. Thanks also to my uncle Marius K., with whom I have discussed this and other areas of mathematical physics many late evenings.

References

[AI] Anderson, R.L., Ibragimov, N.H.: Lie-Bäcklund Transformations in Applications. SIAM, Philadelphia (1979)

[Arn] Arnold, V.I.: Mathematical methods in classical mechanics. Springer, Berlin Heidelberg New York (1978)

[AG] Arnold, V.I., Givental, A.B.: Symplectic Geometry. In: Arnold, V.I. and Novikov, S.P. (eds) Dynamical Systems IV, Encyclopedia of Mathematical Sciences, Vol 4. Springer, Berlin Heidelberg New York (1990)

[AKN] Arnold, V.I., Kozlov, V.V., Neishtadt, A.I.: Mathematical aspects of classical and celestial mechanics. In: Arnold, V.I. (ed) Dynamical Systems III, Encyclopedia of Mathematical Sciences, Vol 3. Springer, Berlin Heidelberg New York (1988)

[B1] Brandão, A.: Symplectic structure for Gaussian diffusions. J. Math. Phys., **39**, 4257–4283 (1998)

[BK1] Brandão, A., Kolsrud T.: Phase space transformations of Gaussian diffusions. Potential Anal., **10**, 119–132 (1999)

[BK2] Brandão, A., Kolsrud T.: Time-dependent conservation laws and symmetries for classical mechanics and heat equations. In: Harmonic morphisms, harmonic maps, and related topics (Brest, 1997). Chapman & Hall/CRC Res. Notes Math., 413, Chapman & Hall/CRC, Boca Raton FL (2000)

[CZ] Cruzeiro, A.B., Zambrini, J.-C.: Malliavin calculus and Euclidean quantum mechanics. I. Functional calculus. J. Funct. Anal., **96**, 62–95 (1991)

[D] Darling, R.W.R.: Martingales in manifolds—Definition, examples and behaviour under maps. In: Azéma, J., Yor, M. (eds) Séminaire de Probabilités XVI, Lect. Notes Math. 921, Springer-Verlag, Berlin Heidelberg New York (1982)

[DK] Djehiche, B., Kolsrud, T.: Canonical transformations for diffusions. C. R. Acad. Sci. Paris, **3**21, I, 339–44 (1995)

[DKN] Dubrovin, B.A., Krichever, I.M., Novikov, S.P.: Integrable Systems I. In: Arnold, V.I., Novikov, S.P. (eds) Dynamical Systems IV, Encyclopedia of Mathematical Sciences, Vol 4., Springer, Berlin Heidelberg New York (1990)

[DNF] Doubrovine, B., Novikov, S., Fomenko, A.: Géométrie contemporaine. Méthodes et applications. 1re partie. Traduction française. Éditions Mir, Moscou (1982)

[EL] Eells, J., Lemaire, L.: Selected Topics in Harmonic Maps. CBMS Regional Conf. Ser. in Math 50. Amer. Math. Soc., Providence RI (1983)

[Em] Emery, M.: Stochastic Calculus in Manifolds. Springer, Berlin Heidelberg New York (1989)

[FH] Feynman, R.P., Hibbs, A.R.: Quantum mechanics and path integrals. McGraw-Hill, New York (1965)

[F1] Fuglede, B.: Harmonic morphisms between Riemannian manifolds. Ann. Inst. Fourier, **28**, 107–44 (1978)

[F2] Fuglede, B., Harmonic morphisms. In: Springer Lect. Notes in Math. 747. Springer, Berlin Heidelberg New York (1979)

[Ga] Gallavotti, G.: The Elements of Mechanics (transl. of *Meccanica Elementare*. Ed. Boringhieri, Torino (1980)). Springer, Berlin Heidelberg New York (1983)

[Go] Goldberg, S.I.: Curvature and homology. Academic Press, New York (1962)

[Gol] Goldstein, H.: Classical mechanics, 2nd ed. Addison-Wesley, New York (1980)

[Gorb] Gorbatsevich, V.V., Onishchik, A.L., Vinberg, E.B.: Foundations of Lie Theory and Lie Transformation Groups. Springer, Berlin Heidelberg New York (1993)

[Ibr2] Ibragimov, N.H.: Transformation Groups Applied to Mathematical Physics. Nauka, Moscow (1983) (English translation by D. Reidel, Dordrecht (1985))

[Ibr3] Ibragimov, N.H. (ed) CRC handbook of Lie group analysis of differential equations, Vol 1, Symmetries, exact solutions and conservation laws. CRC Press, Boca Raton FL (1993)

[IK] Ibragimov, N.H., Kolsrud T.: Lagrangian approach to evolution equations: symmetries and conservation laws. Nonlinear Dynam., **36**, 29–40 (2004)

[IW] Ikeda, N., Watanabe, S.: Stochastic differential equations and diffusion processes, 2nd ed. North-Holland, Kodansha Amsterdam Tokyo (1989)

[Ish] Ishihara, T.: A mapping of Riemannian manifolds which preserves harmonic functions. J. Math. Kyoto Univ., **19**, 215–229 (1979)

[I] Itô, K.: The Brownian motion and tensor fields on a Riemannian manifold. In: Proc. Int. Congr. Math., Stockholm 1962. Inst. Mittag-Leffler, Djursholm (1963)

[Ko] Kobayashi, S.: Transformation groups in differential geometry. Springer. Berlin Heidelberg New York (1972)

[K0] Kolsrud, M.: Exact quantum dynamical solutions for oscillator-like systems. Phys. Mat. Univ. Osloensis, **28** (1965)

[K1] Kolsrud, T.: Quantum constants of motion and the heat Lie algebra in a Riemannian manifold. Preprint TRITA-MAT, Royal Institute of Technology, Stockholm (1996)

[K2] Kolsrud, T.: Symmetries for the Euclidean Non-Linear Schrödinger Equation and Related Free Equations. In: Proc. of MOGRAN X (Cyprus 2004). Also preprint TRITA-MAT, Royal Institute of Technology, Stockholm (2005)

[KL] Kolsrud, T., Loubeau, E.: Foliated manifolds and conformal heat morphisms. Ann. Global Anal. Geom., **21**, 241–67 (2002)

[KZ1] Kolsrud, T., Zambrini, J. C.: The general mathematical framework of Euclidean quantum mechanics. In: Stochastic analysis and applications (Lisbon 1989), Birkhäuser, Basel (1991)

[KZ2] Kolsrud, T., Zambrini, J.C.: An introduction to the semiclassical limit of Euclidean quantum mechanics. J. Math. Phys., **33**, 1301–1334 (1992)

[LL] Landau, L.D., Lifshitz, E.M., Course of Theoretical Physics, Vol 1, Mechanics, Vol 3, Quantum mechanics, 3rd ed. Pergamon Press, Oxford (1977)

[Lie] Lie, S.: Über die Integration durch bestimmte Integrale von einer Klasse linearer partieller Differentialgleichungen. Arch. Math., **6**, 328–368 (1881)

[Lob] Loubeau, E.: Morphisms of the heat equation. Ann. Global Anal. Geom., **15**, 487–496 (1997)

[M] Malliavin, P.: Géométrie differéntielle stochastique. Séminaire de Mathèmatiques Supérieures 64. Presses de l'Université de Montréal, Montreal (1978)

[MF] Morse, P.M., Feshbach, H.: Methods of theoretical physics, Vol. I-II. McGraw-Hill, New York (1953)

[N] Nelson, E.: Quantum Fluctuations. Princeton University Press (1985)

[Or1] Olver, P. J.: Applications of Lie groups to differential equations, 2nd ed. Springer, Berlin Heidelberg New York (1993)

[Ok] Øksendal, B.: When is a stochastic integral a time change of a diffusion? J. Th. Prob., **3**, 207–226 (1990)

[TZ1] Thieullen, M., Zambrini, J.C.: Probability and quantum symmetries I. The theorem of Noether in Schrödinger's euclidean quantum mechanics. Ann. Inst. Henri Poincaré, Phys. Théorique, **67**, 297–338 (1997)

[TZ2] Thieullen, M., Zambrini, J.C.: Symmetries in the stochastic calculus of variations. Probab. Theory Relat. Fields, **107**, 401–427 (1997)

Different Lattice Approximations for Høegh-Krohn's Quantum Field Model

Song Liang

1) Graduate School of Information Sciences, Tohoku University (Japan)
2) Institute of Applied Mathematics, University of Bonn (Germany)
3) Financially supported by Alexander von Humboldt Foundation (Germany) and Grant-in-Aid for the Encouragement of Young Scientists (No. 15740057), Japan Society for the Promotion of Science

Dedicated to Prof. Kiyosi Itô on the occasion of his 90th birthday

1 Introduction

It is a fundamental question whether one can give a rigorous meaning to quantum fields described heuristically by a "probability measure" on the space of Schwartz tempered distributions. Methods used include, for example, lattice approximation and wavelet approximation. We focus on the lattice approximation in this article.

Our main target of this article is Høegh-Krohn's quantum field model, which was first introduced in [8]. Simply speaking, Høegh-Krohn's quantum field model is the model with interaction of the form $: \exp \alpha\phi :$, here $: \cdot :$ means the Wick power.

Albeverio–Høegh-Krohn [6] showed that, for this model, when using the same cutoffs for the free and the interacting parts (see below for the precise meaning of these terms), the approximating probability measure converges as the lattice cutoff is removed. In this sense, they gave a rigorous meaning to Høegh-Krohn's quantum field model by using lattice approximation.

In this article, we consider lattice approximations with different lattice cutoffs in the free and the interacting parts.

2 Lattice Approximation of the Free Field

First, let us recall the meaning of lattice approximation of the free field. Let $m_0 > 0$ be a fixed number and let μ_0 be the (Nelson or Euclidean) free field measure on $\mathbf{R}^2$ of mass m_0, *i.e.*, Gaussian measure on $\mathcal{S}'(\mathbf{R}^2)$ with the covariance $(-\Delta+m_0^2)^{-1}$. Here $\mathcal{S}'(\mathbf{R}^2)$ denotes the space of Schwartz tempered distributions, which is given as the dual of $\mathcal{S}(\mathbf{R}^2)$, the Schwartz space of rapidly decreasing smooth test functions, in $L^2(\mathbf{R}^2)$.

For any $a > 0$, let G_a be the free lattice measure of m_0 and lattice spacing a on $a\mathbf{Z}^2$, and let $C^{(a)}(x-y) = \langle \phi_x \phi_y \rangle_{G_a}$ for $x, y \in a\mathbf{Z}^2$, where $\langle \cdot \rangle_*$ denotes the expectation with respect to $*$. G_a is thus the lattice Gaussian measure with covariance $C^{(a)}$. One has by definition (see [9])

$$C^{(a)}(x-y) = (2\pi)^{-2} \int_{[-\frac{\pi}{a}, \frac{\pi}{a}]^2} e^{ik\cdot(x-y)} \mu_a(k)^{-2} dk,$$

where $\mu_a(k) := \left(m_0^2 + 2a^{-2} \sum_{j=1}^2 (1 - \cos(ak_j))\right)^{1/2}$ for $k = (k_1, k_2)$.

Notice that G_a is a probability measure on $a\mathbf{Z}^2$. So $\{G_a\}_{a>0}$ is a family of probability measures on different spaces. In order to consider their convergence, it is useful to convert them to probability measures on a common space. This is done in the following way. Let $\mu(k) := (m_0^2 + |k|^2)^{1/2}$, where $k = (k_1, k_2)$, $|k|^2 = k_1^2 + k_2^2$, and let $f_{a,x}(\cdot)$ be the function whose Fourier transform is

$$\mathcal{F}(f_{a,x})(k) = (2\pi)^{-1} e^{-ik\cdot x} \mu_a(k)^{-1} \mu(k) 1_{[-\frac{\pi}{a}, \frac{\pi}{a}]}(k_1) 1_{[-\frac{\pi}{a}, \frac{\pi}{a}]}(k_2).$$

Denote by ϕ the coordinate process associated with μ_0 (called Nelson's or Euclidean free field): ϕ is first defined as an element of $\mathcal{S}'(\mathbf{R}^2)$, so that $\phi(g)$ is the dualization of $g \in \mathcal{S}(\mathbf{R}^2)$ with $\phi \in \mathcal{S}'(\mathbf{R}^2)$. ϕ is then extended by continuity in $L^2(d\mu_0)$ to a linear process $\phi(g)$, with g belonging to a larger space than $\mathcal{S}(\mathbf{R}^2)$. In fact, this space contains functions of the form $f_{a,x}$, and it is easy to check that

$$\langle \phi(f_{a,x}) \phi(f_{a,y}) \rangle_{\mu_0} = \langle \phi_x \phi_y \rangle_{G_a}.$$

In this sense, we can realize the above Gaussian field ϕ_x on $a\mathbf{Z}^2$ by $\phi(f_{a,x})$ defined on $\mathcal{S}'(\mathbf{R}^2)$. (See, *e.g.*, [6, 9] for details).

In this way, we say that μ_0 can be approximated by G_a, as $a \to 0$.

3 Høegh-Krohn's Model and its Lattice Approximations

Let us first give the heuristic "definition" of Høegh-Krohn's quantum field model. Let ν be any even positive measure with finite total mass and with support $\mathrm{supp}(\nu) \subset [-\alpha_0, \alpha_0]$ for some $\alpha_0 < \frac{4}{\sqrt{\pi}}$, and let Λ be any compact subset of $\mathbf{R}^2$. Høegh-Krohn's quantum field model is heuristically given by

$$Z^{-1} e^{-\lambda \int_\Lambda dx \left(\int :e^{\alpha\phi_x}: \nu(d\alpha) \right)} \mu_0(d\phi). \tag{3.1}$$

Here λ is a positive constant, Z is the normalizing constant (depending on λ, Λ and α), and $: e^{\alpha\phi_x} :$ means the Wick exponential of $\alpha\phi_x$, *i.e.*, $: e^{\alpha\phi_x} := \sum_{k=0}^{\infty}$

$\frac{\alpha^k}{k!} : \phi_x^k :$, where $: \phi_x^k :$ is the k-th Wick power of ϕ_x with respect to μ_0 (see [9] for the precise definition of the latter).

As declared, we are interested in the lattice approximation of it. We have already given the lattice approximation G_a for the free part μ_0, and we still need to consider the lattice approximation of the interacting part. More precisely, we need to approximate the integral $\int_\Lambda$ on a lattice. We can do so by using either of the following approximation:

1. $a^2 \sum_{\ell \in \mathbf{Z}^2 : a\ell \in \Lambda}$,
2. $a'^2 \sum_{\ell \in \mathbf{Z}^2 : a'\ell \in \Lambda}$ (which will be written as $\int_{a'\mathbf{Z}^2 \cap \Lambda} dx$ for the sake of simplicity), with $a' = a'(a) \geq a$ satisfying $\lim_{a \to 0} a'(a) = 0$ and $a'\mathbf{Z}^2 \subset a\mathbf{Z}^2$.

The first one means that we use same "lattice cutoff" for both the free part and the interacting part. This approximation has been discussed by [6]. However, one can also use the latter approximation given above, which corresponds to the case of different lattice cutoffs for the free and the interacting part.

Correspondingly, we can consider the probability measure $\mu_{\lambda,a,a'}$ on $\mathcal{S}'(\mathbf{R}^2)$ given by

$$\mu_{\lambda,a,a'}(d\phi) \equiv Z_{\lambda,a,a'}^{-1} e^{-\lambda \int_{a'\mathbf{Z}^2 \cap \Lambda} dx \left(\int :e^{\alpha\phi(f_{a,x})}: \nu(d\alpha) \right)} \mu_0(d\phi),$$

where $Z_{\lambda,a,a'}$ is the normalizing constant. This is the object of our present article, and we want to know whether $\mu_{\lambda,a,a'}$ converges as $a \to 0$.

The corresponding problem for the ϕ_2^4-quantum field model has been discussed in [1].

4 Motivations

The study of this "different lattice cutoffs" problem was first motivated by the attempt to understand better the 3 space-time dimensional case, which is believed to have totally different properties from the 2 dimensional case. For example, although the ϕ_2^4-quantum field model with interaction in a bounded region Λ is equivalent to the 2 dimensional free field, it is believed that the ϕ_3^4-quantum field model with a corresponding interaction is singular to the 3 dimensional free field.

It is well-known that for the free field in d space-time dimensions, the Schwinger function $S(x,y) \equiv \langle \phi_x \phi_y \rangle_{\mu_0}$ has order

$$S_2(x,y) \sim \begin{cases} |\log|x-y||, & \text{if } d = 2, \\ \frac{1}{|x-y|}, & \text{if } d = 3, \end{cases}$$

as $|x-y| \to 0$. Here $a \sim b$ stands for that $\frac{a}{b}$ converges to a constant in $(0, \infty)$.

In other words, when using the same cutoff for both the free part and the interacting part, the situations in 2 space-time dimensional case and in 3 space-time dimensional case are actually different: we are using the lattice cutoff for the interacting part of the same order as for the Schwinger functions of the free field in 3 space-time dimensional case, and are using a lattice cutoff for the interacting part of a different order from the one of the Schwinger functions of the free field in 2 space-time dimensional case.

Therefore, it will be interesting to discuss what will happen when we use the lattice cutoff for the interacting part of the order of the Schwinger functions of the free field in 2 space-time dimensional case also. This is the first motivation of our present research.

To see another motivation more clearly, let us use the ϕ_2^4-model to explain. The ϕ_2^4-quantum field model is the probability measure on $\mathcal{S}'(\mathbf{R}^2)$ heuristically described by

$$Z^{-1}e^{-\lambda\int_\Lambda :\phi_x^4:dx}\mu_0(d\phi),$$

with the notations same as before. The lattice approximation of it with different cutoffs a and $a'(a)$ in the free and the interacting parts, respectively, is given by

$$\mu_{a,a'}=Z_{a,a'}^{-1}e^{-\lambda\int_{a'\mathbf{Z}^2\cap\Lambda}dx:\phi(f_{a,x})^4:}\mu_0(d\phi),$$

where $Z_{a,a'}$ is the normalizing constant.

Since $\int_\Lambda\int_\Lambda(\log|x-y|)^4dxdy<\infty$, by a simple calculation, we have that

$$E\Big[\Big(\int_{a'\mathbf{Z}^2\cap\Lambda}dx:\phi(f_{a,x})^4:\Big)^2\Big]\sim\begin{cases}a'^2|\log a|^4, & \text{if } \lim_{a\to 0}a'|\log a|^2=+\infty,\\ 1, & \text{if } \lim_{a\to 0}a'|\log a|^2<+\infty,\end{cases}$$

as $a\to 0$. Therefore, the order of the interaction changes dramatically according to whether $a'|\log a|^2$ converges or diverges as $a\to 0$. More precisely, [5] showed that if $\lim_{a\to 0}a'|\log a|^2=+\infty$, then $\frac{1}{a'|\log a|^2}\int_{a'\mathbf{Z}^2\cap\Lambda}dx:\phi(f_{a,x})^4:$ under μ_0 converges to a Gaussian measure. (The corresponding central limit theory problem for Høegh-Krohn's quantum field model has been discussed by [4]). Therefore, when $\lim_{a\to 0}a'|\log a|^2=+\infty$, the density of the probability measure $\mu_{a,a'}$ with respect to μ_0, which is given by

$$Z_{a,a'}^{-1}\exp\Big(-\lambda a'|\log a|^2\Big\{\frac{1}{a'|\log a|^2}\int_{a'\mathbf{Z}^2\cap\Lambda}dx:\phi(f_{a,x})^4:\Big\}\Big),$$

is getting more and more singular as $a\to 0$. This gives us hope to find the limit of $\mu_{a,a'}$ as a new probability measure, different from the "classical" ϕ_2^4-field in Λ (given as the limit of $\mu_{a,a'}$ with $a'=a$), which is well-known to be equivalent with respect to μ_0 (see [3] for some related results).

In other words, by using this different cutoff approximation procedure, it seems to be hopeful to find two (singular with respect to each other) probability measures, which are given by the same heuristic one.

5 Results and the Sketch of Proof

Let us come back to our Høegh-Krohn's quantum field model in $\mathbf{R}^2$ described in Section 3. For any $m\in\mathbf{N}$, let $S_{2m}^{\lambda,a,a'}$ be the $2m$-point-function given by

$$S_{2m}^{\lambda,a,a'}(x_1,\cdots,x_{2m}) \equiv \langle \phi(f_{a,x_1})\cdots\phi(f_{a,x_{2m}})\rangle_{\mu_{\lambda,a,a'}}.$$

Then we have the following result about the existence of the limit probability measure (see [2]).

Theorem 1 (Albeverio–Liang [2]). *Assume that* $\lim_{a\to 0} a'(a)|\log a| < \infty$. *Then there exists a* $\lambda_0 > 0$ *such that for any* $\lambda \in [0, \lambda_0]$, *there exists a sequence* $\{a_n\}_{n\in\mathbf{N}}$ *with* $\lim_{n\to\infty} a_n = 0$ *(and writing* $a'(a_n) = a'_n$*) such that for any given* $m \in \mathbf{N}$ *and* $f_1, \cdots, f_{2m} \in \mathcal{S}(\mathbf{R}^2)$, *the following limit exists*

$$S_{2m}^{\lambda}(f_1,\cdots,f_{2m}) := \lim_{n\to\infty} \sum_{x_1,\cdots,x_{2m}} S_{2m}^{\lambda,a_n,a'_n}(x_1,\cdots,x_{2m}) \prod_{i=1}^{2m} a_n^2 f_i(x_i).$$

Moreover, there exists a probability measure μ_λ *on* $\mathcal{S}'(\mathbf{R}^2)$ *(which may depend on* Λ*) such that*

$$S_{2m}^{\lambda}(f_1,\cdots,f_{2m}) = \int_{\mathcal{S}'(\mathbf{R}^2)} \phi(f_1)\cdots\phi(f_{2m})\mu_\lambda(d\phi)$$

for $m \in \mathbf{N}$ *and* $f_1,\cdots,f_{2m} \in \mathcal{S}(\mathbf{R}^2)$.

The idea of the proof (see [2]) is learned from Brydges–Fröhlich–Sokal [7]. We give the sketch of the proof in the following (for details see [2]). First, we have the following skeleton inequality:

Lemma 2.

$$S_{2m}^{\lambda,a,a'}(x_1,\cdots,x_{2m}) \geq \frac{1}{(2m-1)!} \sum_{\pi\in Q_{2m}} \prod_{i=1}^{m} S_2^{\lambda,a,a'}(x_{\pi(2i-1)}, x_{\pi(2i)}),$$

$$S_{2m}^{\lambda,a,a'}(x_1,\cdots,x_{2m}) \leq \sum_{\pi\in Q_{2m}} \prod_{i=1}^{m} S_2^{\lambda,a,a'}(x_{\pi(2i-1)}, x_{\pi(2i)}),$$

with Q_{2m} *denoting the set of all pair-partitions of* $\{1, 2, \cdots, 2m\}$.

By Lemma 2, we have that the behavior of $S_{2m}^{\lambda,a,a'}$ is dominated by $S_2^{\lambda,a,a'}$. Therefore, we only need to estimate $S_2^{\lambda,a,a'}$.

For $S_2^{\lambda,a,a'}$, we first show the following relation between $S_2^{\lambda,a,a'}$ and $C^{(a)}$, the covariance of the lattice free field G_a, by using the integration by parts formula for the lattice free field.

Lemma 3.

$$\begin{aligned} &S_2^{\lambda,a,a'}(x,y) - C^{(a)}(x-y) \\ &= \lambda \int_{a'\mathbf{Z}^2\cap\Lambda} dz \langle \Big(\int \nu(d\alpha)\alpha : e^{\alpha\phi(f_{a,z})} : \Big) \phi(f_{a,y}) \rangle_{\mu_{\lambda,a,a'}} C^{(a)}(x-z). \end{aligned}$$

It is well-known that

1. $C^{(a)}(z) \le C(1 + |\log|z||), z \in a\mathbf{Z}^2 \setminus \{0\}$,
2. $C^{(a)}(0) \le C|\log a|$ as $a \to 0$.

Therefore, it is sufficient to show that the difference between $S_2^{\lambda,a,a'}(x, y)$ and $C^{(a)}(x - y)$ is small enough. Let

$$X_{\lambda,a} = \sup_{x,y \in a'\mathbf{Z}^2} |S_2^{\lambda,a,a'}(x, y) - C^{(a)}(x - y)|.$$

Then by using Lemma 2 (ii), Lemma 3 and the Taylor expansion for $e^{\alpha\phi(f_{a,z})}$, we can show the following.

Lemma 4. *Suppose that* $\lim_{a\to 0} a'|\log a| < \infty$. *Then there exists a constant* $c_1 > 0$ *independent of* a *such that*

$$X_{\lambda,a} \le \lambda c_1 (X_{\lambda,a} + 1) e^{\frac{1}{2}\alpha_0^2 X_{\lambda,a}}, \tag{5.1}$$

where α_0 *is the constant given in Section 3.*

Let $\lambda_0 = (3C_1 e^{\alpha_0^2})^{-1}$. Then for any $\lambda \in [0, \lambda_0]$, we have by (5.1) that if $X_{\lambda,a} \le 2$ then $X_{\lambda,a} \le 1$. In other words, $X_{\lambda,a}$ can never take values in $(1, 2]$. This combined with the fact that $X_{0,a} = 0$ and that $X_{\lambda,a}$ is continuous with respec to λ gives us that $X_{\lambda,a} \le 1$. Substituting this into (5.1) again, we get the following: There exist constants $\lambda_0, C_2 > 0$ independent of a such that

$$X_{\lambda,a} \le C_2\lambda, \qquad \text{for any } \lambda \in [0, \lambda_0], a > 0.$$

Therefore, $S_2^{\lambda,a,a'}(x, y)$ behaves like $|\log|x - y||$ as $|x - y| \to 0$. This completes the proof of our theorem.

6 Remarks and Open Problems

This article is just a starting of the study of "different cutoffs problem", and there are still a lot of open problems left. For example, we do not know whether the condition $\lim_{a\to 0} a'|\log a| < \infty$ is optimal for the convergence of the sequence $\mu_{\lambda,a,a'}$ as $a \to 0$; also, the uniqueness of the limit is not proved. Moreover, the following conjecture, which is one of the main motivation of the present study, is not solved yet, either.

Conjecture. The probability measure μ_λ is singular with respect to μ_0 if a' is big enough compared with a.

Acknowledgements

The author would like to express her heartful thanks to Prof. S. Albeverio, for his kindly helping all the time. The financial support by the Alexander von Humboldt Stiftung is also gratefully acknowledged.

References

1. S. Albeverio, M. S. Bernabei, and X. Y. Zhou, *Lattice approximations and continuum limits of ϕ_2^4-quantum fields,* J. Math. Phys. 45 (2004), 149–178
2. S. Albeverio and S. Liang, *A new lattice approximation for the Hoegh-Krohn quantum field model,* Rep. Math. Phys. 54, no. 2, 149–157 (2004)
3. S. Albeverio and S. Liang, *A remark on different lattice approximations and continuum limits for ϕ_2^4-fields,* Random Oper. Stochastic Equations 12, no. 4, 313–318 (2004)
4. S. Albeverio and S. Liang, *A limit theorem for the Wick exponential of the free lattice fields,* Markov Process. Related Fields 11, no. 1, 157–164 (2005)
5. S. Albeverio and X. Y. Zhou, *A Central Limit Theorem for the Fourth Wick Power of the Free Lattice Field,* Comm. Math. Phys. 181, no. 1, 1–10 (1996)
6. S. Albeverio and R. Høegh-Krohn, *The Wightman axioms and the mass gap for strong interactions of exponential type in two-dimensional space-time,* J. Functional Analysis 16, 39–82 (1974)
7. D. C. Brydges, J. Fröhlich and A. D. Sokal, *The random-walk representation of classical spin systems and correlation inequalities II. The skeleton inequalities,* Commun. Math. Phys. 91 (1983), 117–139, *A new proof of the existence and nontriviality of the continuum φ_2^4 and ϕ_3^4 quantum field theories,* Commun. Math. Phys. 91 (1983), 141–186
8. R. Høegh-Krohn, *A general class of quantum fields without cut-offs in two space-time dimensions,* Comm. Math. Phys. 21, 244–255 (1971)
9. B. Simon, *The $P(\phi)_2$ Euclidean (Quantum) Field Theory,* Princeton University Press, 1974

Itô Atlas, its Application to Mathematical Finance and to Exponentiation of Infinite Dimensional Lie Algebras

Paul Malliavin

10 rue Saint Louis en l'Isle, 75004 Paris, France, sli@ccr.jussieu.fr

Introduction

The names "charts" or "atlas" have their origin in geography; it is of importance to represent "faithfully" on a sheet of paper a town, a province, on a region of the Earth; the atlas is a collection of charts such that every point of the Earth appears at least one time in the range of a chart of this collection.

In a more mathematical way the representation by longitude and latitude is described as follows: on the two dimensional sphere S^2, choose a North Pole N and a half meridian of reference (Greenwich); denote S^2_0 the open subset of S^2 constituted by the complement of the reference half meridian; then the longitude, latitude $(\phi, \theta) \in]0, 2\pi[\times]0, \pi[$ defines a *local chart* a bijective map u_0 of the *domain of the chart,* that is the open subset $]0, 2\pi[\times]0, \pi[\subset R^2$ onto the *range of the local chart,* that is the open subset $S^2_0 \subset S^2$; by choosing another North Pole and another reference half meridian we can construct another open subset S^2_1 of S^2 such that $S^2 = S^2_0 \cup S^2_1$; denote u_1 the corresponding local chart; then the two local charts u_0, u_1 constitute an *atlas* of S^2. This representation is *faithful* in the sense that

A function f defined on S^2 is differentiable if and only if $f_i := f \circ u_i$, $i = 0, 1$ are differentiable; furthermore all differential geometric computations on S^2 can be effectively realized through the atlas.

Another possible atlas is the Mercator atlas: we denote v_N the stereographic projection from the North pole which sends a sheet of paper into S^2 in one to one correspondence with the tangent plane to S^2 at the South pole; the range of v_N is the open set of S^2 constituted by the complement of the North pole. Denote v_S the stereographic projection from the South pole; then the Mercator atlas is the collection of the two charts v_N, v_S. For the geography it is possible to study the Earth equally well in each of these atlas; the choice between them is a question of convenience: the Mercator atlas preserves the angles and the longitude-latitude atlas refer to coordinates which can be immediately obtained from observations.

Turn now to the same concepts for a d dimensional manifold $\mathcal{M}$; a chart ϕ_α will be a continuous injective map of an open subset of $O_\alpha \subset R^d$ such that its range $\phi_\alpha(O_\alpha)$ is an open set of $\mathcal{M}$; every point of $\mathcal{M}$ is in the range of a chart: $\cup_\alpha \phi_\alpha(O_\alpha) = \mathcal{M}$. Finally the differentiable structure on $\mathcal{M}$ will be defined as the "image" of the classical differential structure on O_α; more precisely a function $f : \mathcal{M} \mapsto R$ is differentiable if and only if $f \circ \phi_\alpha \in C^1(O_\alpha)\ \forall \alpha$; the implementation of this definition can be done assuming the *coherence hypothesis*, that is: $\phi_\beta^{-1} \circ \phi_\alpha \in C^1(O_\alpha)$.

A classical theory of *Banach modeled manifold* exits where the O_α are now open subsets of a fixed Banach space B.

Lie algebra appearing in mathematical physics, as the Virasoro algebra $\mathcal{V}$, are infinite dimensional; the physical prerequisite of symplecticity determines on $\mathcal{V}$ a unique Hilbertian metric which has low regularity; this low regularity implies that the exponentiation from $\mathcal{V}$ to a propective "infinite dimensional group" cannot be realized into the context of Banach modeled manifold. A new concept of differential geometry has to be built in order to fit this infinite dimensional challenge: this will be sketched in the second part of this paper.

Let us justify now the first part. It can be thought that the "state space" of a financial market is an abstract manifold $\mathcal{M}$. This manifold $\mathcal{M}$ can be given through the choice of some model. We take a phenomenological point of view: a broker, from the simple observation of the market, wants to obtain, *model free*, some information on the structure of $\mathcal{M}$. For instance he will inquire: *is it possible to compute econometrically the pathwise sensitivities?* The computation of the Greeks require the differentiation of the coefficients the driving SDE. How could be possible to differentiate when we know this SDE only on the path on $\mathcal{M}$ describing the market evolution!

Itô calculus answers these questions coming from two different worlds.

Itô's Atlas

Given a "smooth manifold" S (finite or infinite dimensional) and an "elliptic differential operator" on S, consider the sample path of the associated diffusion $s_\omega(\tau), \tau \in [0,1],\ \omega \in \Omega$ being the probability space.

An *Itô local chart*is the map $\tau \mapsto s_\omega(\tau),\ \ \tau \in [0,1]$: the domain of charts are the 1-dimensional segment $[0,1]$, the range are sample paths of the diffusion.

The "dimensionality" of the range is 1, even if S is infinitely dimensional; but the cardinality of the atlas is large: there are as many local charts as there are sample paths of the diffusion.

Theorem 1 (Localization of Derivative). *Given a smooth function f on S, its derivative can be computed in an Itô local chart from the knowledge of f on the range of this local chart*

Proof. Consider a local chart $u : O \mapsto S$, in the classical sense: O is an open subset of R^N, the range of u is an open subset of S; let $f^u := f \circ u$.

Given a semi-martingale $\varphi(\omega,\tau)$ defined on Ω, denote $<\varphi>_\tau$ the associated increasing process and define $\mathrm{Vol}_\tau(\varphi,\varphi)=\frac{d}{d\tau}(<\varphi>_\tau)$; then $\mathrm{Vol}_\tau(\varphi,\varphi)$ is *pathwise computable;* define by polarization $\mathrm{Vol}(\phi,\psi)=\frac{1}{4}(\mathrm{Vol}(\phi+\psi,\phi+\psi)-\mathrm{Vol}(\phi-\psi,\phi-\psi))$; then by Itô Calculus

$$\frac{\partial f^u}{\partial \xi^k}=\frac{\mathrm{Vol}(\phi,\psi)}{\mathrm{Vol}(\phi,\phi)},\ \phi=[u^{-1}(s_\omega)]^k,\ \psi=f(s_\omega) \tag{1}$$

We split this paper in two parts which can be read independently one of the other.

1 Computation of Pathwise Sensitivities in Mathematical Finance

Let p be the asset price, we will assume that $p(t)$ is a continuous semi-martingale satisfying the SDE

$$dp(t)=\sigma(t)\,dW(t)+b(t)\,dt \tag{1.1}$$

where W is a Brownian motion on a filtered probability space $(\Omega,(\mathcal{F}_t)_{t\in[0,T]},P)$, σ and b are *stochastic processes* such that $E[\int_0^T\sigma^4(t)dt+\int_0^T b^2(t)dt]<\infty$. The *random* function $\sigma^2(t)$ is the *spot volatility.* We want to propose a pathwise econometrical computation of the spot volatility [4, 13].

1.1 Fourier Econometrical Computation of Spot Volatility

By change of the origin of time and rescaling the unit of time we reduce ourselves to the case where the time window is $[0,2\pi]$. Given a function ϕ on the circle S^1, we consider its *Fourier transform,* defined by

$$\mathcal{F}(\phi)(k)=:\frac{1}{2\pi}\int_0^{2\pi}\phi(\vartheta)\exp(-ik\vartheta)\,d\vartheta,\quad \text{for } k\in\mathbf{Z}.$$

Define

$$\mathcal{F}(d\phi)(k)=:\frac{1}{2\pi}\int_{]0,2\pi[}\exp(-ik\vartheta)\,d\phi(\vartheta)\,,$$

then by integration by part

$$\mathcal{F}(\phi)(k)=\frac{i}{k}\times[\frac{1}{2\pi}\phi(2\pi)-\phi(0)-\mathcal{F}(d\phi)(k)].$$

Given two functions Φ,Ψ on the integers their *Bohr convolution product* is defined by

$$(\Phi*_B\Psi)(k)=:\lim_{N\to\infty}\frac{1}{2N+1}\sum_{s=-N}^{N}\Phi(s)\Psi(k-s).$$

Theorem 2. *Consider a process p satisfying assumption* (1.1)*; then we have*

$$\frac{1}{2\pi}\mathcal{F}(\sigma^2) = \Phi *_B \Phi, \text{ where } \Phi(k) := \mathcal{F}(dp)(k), \tag{1.2}$$

the limit corresponding to the definition of $\Phi *_B \Phi$ *is attained in probability.*

1.2 A Reduced Sensitivity: The Feedback Price-Volatility Rate

We shall firstly discussed a reduced sensitivity which has the advantage to be computable through an ordinary ODE, when classical sensitivities require the integration of SDE. Assume that the price of the considered asset is given by a geometric martingale:

$$dS_W(t) = \sigma(S_W(t))\, dW(t) \tag{1.3}$$

where W is a Brownian motion, σ is a fixed but unknown smooth function depending only upon the price.

The classical pathwise Greek Delta $\zeta_W(t)$ is defined as the solution of the linearized SDE

$$d\zeta_W(t) = (\sigma'(S_W(t))\, dW((t))\, \zeta_W(t)$$

We associate to $\zeta(t)$ the *rescaled variation* defined as

$$z(t) = \frac{\zeta(t)}{\sigma(S_W(t))} \tag{1.4}$$

Theorem 3 (Price-volatility Feedback Rate [5]). *The rescaled variation is a derivable function of t; its logarithmic derivative* $\lambda(t)$ *will be called the price-volatility feedback rate function. Then*

$$\lambda(t) = -\frac{1}{2}(\sigma\sigma'')(S_W(t)); \quad z(t) = \exp\left(\int_s^t \lambda(\tau)\, d\tau\right) z(s). \tag{1.4}$$

Remark: $\lambda(t) < 0$ corresponds to a liquid market.

We suppose that we do not know the explicit expression of the function σ; we want to obtain from the pathwise observation of the market evolution an econometrical computation in real time of $\lambda(t)$. Making the change of variables

$$x_W(t) = \log(S_W(t)); \quad a(x) = \exp(-x)\sigma((\exp(x)),$$

then $x_W(t)$ satisfies the following SDE:

$$dx_W(t) = a(x_W(t))\, dW(t) - \frac{1}{2}a^2(x_W(t))\, dt$$

The price-volatility feedback rate has in logarithmic coordinate the following expression:

$$\lambda = -\frac{1}{2}(a'a + aa'') \tag{1.5}$$

Denote $\star$ the Itô contraction, which can be compute pathwise by (1.2).

Theorem 4 (Econometrical Computation of the Feed-back Rate). *Define* $dx \star dx := A(t)$, $dA \star dx := B(t)$, $dB \star dx := C(t)$, *then*

$$\lambda(t) = \frac{3}{8}\frac{B^2}{A^3} - \frac{1}{4}\frac{B}{A} - \frac{1}{2}\frac{C}{A^2} \tag{1.6}$$

Remark. The computations are made in the following order; firstly compute the Fourier coefficients of x; by applying (1.2) we obtain the Fourier coefficients of A; denote α a parameter, then the Fourier coefficients of $x + \alpha A$ are known; by applying again (1.2) we get the Fourier coefficients of $\mathrm{Vol}(x + \alpha A)$; as $4B = \mathrm{Vol}(x + A) - \mathrm{Vol}(x - A)$ we get the Fourier coefficients of B; then the Fourier coefficients of $x + \alpha B$ are known and by using again (1.2) we get the Fourier coefficients of $\mathrm{Vol}(x + \alpha B)$; then the identity $4C = \mathrm{Vol}(x + B) - \mathrm{Vol}(x - B)$ gives the Fourier coefficients of C.

1.3 Outline of Computations of General Pathwise Sensitivities

We sketch some facts which are developed in [14]. Start firstly from the observation that for the infinitesimal generator $\mathcal{L}$ of the *risk free measure* the drift can be computed in term of the volatility: in the case of Black–Scholes model $\mathcal{L}$ has a vanishing drift; in the case of HJM model of the interest curve the drift is explicitly expressed in term of the volatility matrix. As the volatility matrix can be econometrically pathwise computed, it results that $\mathcal{L}$ can be pathwised computed.

Using (1) the derivatives of the coefficients of $\mathcal{L}$ can be computed; then the pathwise Greek Delta can be computed, at the price of solving numerically an SDE.

Compared to the computation of the feedback price-volatility rate, these operations have the advantage to involve one step of computation of volatility less but from the other hand there are leading to an SDE which is numerically more instable than the ODE driving the reduced variation.

2 Differential Geometry on Hölderian Jordan Curves

When in 1870–1890 Sophus Lie made his foundational work on what is known now as the *theory of finite dimensional Lie algebras*, he splitted his progression along three main theorems; he called the Third Theorem the fact that given a Lie algebra $\mathcal{G}$, then it is possible to construct a group G having for Lie algebra $\mathcal{G}$. Let us call this statement the *exponentiation problem for the Lie algebra* $\mathcal{G}$. It took about twenty years to Sophus Lie to solve in full generality the exponentiation problem for finite dimensional Lie algebras.

For infinite dimensional Lie algebras $\mathcal{G}$ coming from mathematical physics, infinitesimal representations of $\mathcal{G}$ are known from around twenty years back; the exponentiation problem can be thought as follows: given an infinitesimal representation λ of $\mathcal{G}$, find a probability measure μ_λ, such that λ integrates into a representation on $L^2_{\mu_\lambda}$; this situation is furthermore complicated by a the necessity to prove a *Cameron–Martin theorem.*

It is clear that in this situation the Banach-model manifold theory is hopeless. We shall emphasize the advantage of the Itô atlas in a CFT (conformal field theory) context.

2.1 Jordan Curves and their Parameterizations

A *Jordan curve* Γ is the range of a continuous injective map ϕ of the circle S^1 into the plane. Let $h \in H(S^1)$ be the group of homeomorphisms of S^1; then ϕ and $\phi \circ h$ define the same Jordan curve.

2.2 Holomorphic Parameterization

The Jordan curve Γ splits the complex plane in two simply connected domains Γ^+, Γ^-; by the Riemann mapping theorem there exists an holomorphic map f^+ realizing a bijective map of the unit disk D onto Γ^+; by Caratheodory f^+ has a continuous extension $\bar{f}^+$ to $\bar{D}$; the restriction $\bar{f}^+$ to ∂D, the circle S^1, defines the *holomorphic parameterization* which is unique up to $h \in \mathcal{H}(S^1)$ the Poincaré group of holomorphic automorphisms of D.

2.3 Conformal Welding Parameterization

The composition $g:\ \theta \mapsto ([\bar{f}^+]^{-1} \circ \bar{f}^-)(e^{i\theta})$ defines $g \in H(S^1)$, where $H(S^1)$ denotes the group of homeomorphisms of the circle. Denote G the group of C^∞ diffeomorphisms of S^1, then by Beurling–Ahlfors [6] the set of C^∞ Jordan curves is isomorphic to $\mathcal{H}\backslash G/\mathcal{H}$.

2.4 Canonical Hilbert Norm on $\mathcal{G}$

The Lie algebra $\mathcal{G}$ of G the group of smooth diffeomorphisms of the circle is constituted of smooth vector fields on S^1; granted the parallelism $\frac{d}{d\theta}$ we have a linear isomorphism $\mathcal{G} \simeq C^\infty(S^1)$, the vector space of smooth functions on S^1.

$\mathcal{H}$ denotes the restriction to the circle of the Poincaré group of homographic transformations

$$z \mapsto \frac{az+b}{\bar{b}z+\bar{a}}, \quad |a|^2 - |b|^2 = 1;$$

the Lie algebra of $\mathcal{H}$ is su$(1,1)$ which is generated by the three vector fields $\cos\theta,\ \sin\theta,\ 1$.

There exists [2] a *unique semi-Hilbertian metric on $\mathcal{G}$ invariant by the adjoint action of su*$(1,1)$. Therefore the set of C^∞ Jordan curves $\simeq \mathcal{H}\backslash G/\mathcal{H}$ has a unique Hilbert structure.

The associated canonical Brownian motion on $\mathcal{G}$ has the following expression:

$$x(t) = \sum_{k>1} \frac{1}{\sqrt{k^3 - k}} (x_{2k}(t) \cos k\theta + x_{2k+1}(t) \sin k\theta), \tag{2.1}$$

where $x_*(t)$ is an infinite sequence of scalar valued independent Brownian motions.

2.5 Exponentiating Brownian Motion on $\mathcal{G}$

Fix $\rho \in]0, 1[$ and use the Abel regularization:

$$^{\rho}x_t(\theta) := \sum_{k>1} \frac{\rho^k}{\sqrt{k^3 - k}} (x_{2k}(t) \cos k\theta + x_{2k+1}(t) \sin k\theta),$$

then $^{\rho}x_t(*) \in C^\infty(S^1)$; by the theory of Stochastic flow of diffeomorphisms there exits a solution of the SDE

$$d_t\ (^{\rho}g_{x_t}(\theta)) = (d_t {}^{\rho}x_t)(^{\rho}g_{x_t}(\theta)), \quad ^{\rho}g_{x_0}(\theta) = \theta \tag{2.2}$$

which is a C^∞ diffeomorphism of S^1. Furthermore by Beurling–Ahlfors is associated to the process $t \mapsto {}^{\rho}g_{x_t}$ a non markovian process $t \mapsto {}^{\rho}\Gamma_{x_t}$ on the space of C^∞ Jordan curves.

Theorem 5 ([3, 11]). *The limit when $\rho \to 1$ of ${}^{\rho}g_{x_t}$ exists $=: g_{x_t}$; and g_{x_t} is an homeomorphism which is Hölderian of exponent* $\exp(-2t)$.

Theorem 6 ([2]). *The process $t \mapsto {}^{\rho}\Gamma_{x_t}$ converges to a diffusion $t \mapsto \Gamma_{x_t}$ on the space of Hölderian Jordan curves, starting at time $t = 0$ from the unit circle.*

2.6 C^∞ Differential Geometry "on" Hölderian Jordan Curves

Denote μ_t the law of Γ_{x_t}; to study the Cameron–Martin Theorem for μ_t it is known [9, 10] that infinite dimensional Riemannian geometry has to be introduced.

It would be a non sense to try to enforce a Banach-model differentiable structure on the space of Jordan curves.

We could proceed by using the notion of *dressed up trajectory* $\tilde{\Gamma}_{x_*}$ of $\Gamma_{x,*}$ which is the lift to the C^∞ jet bundle of $\Gamma_{x,*}$ provided by iterative applications of the formula (1). As effective computations on jet bundles are difficult, even in finite dimension, we shall not follow this approach.

We shall use the lift to the orthonormal frame bundle; this lift will provide the same information that the lift to the C^∞ jet bundle and will enjoy computational flexibility.

2.7 Orthonormal Frame Bundle of Over G_ρ

Denote $\mathcal{G}_\rho$ the Hilbert space associated to $(*|*)_\rho$ the scalar product associated to the regularized metric which has been made positive definite by introducing on the 3-dimensional subspace V generated by $1, \sin\theta, \cos\theta$ the hilbertian metric generated by the $L^2(S^1)$ norm, V being furthermore orthogonal to all others trigonometrical functionals.

Denote $\mathcal{U}_\rho$ the group of unitary operators of $\mathcal{G}_\rho$. Denote $O(G_\rho)$ the *bundle of orthonormal frame of G_ρ: a frame is an hilbertian isomorphism of the* tangent plane $T_g(G_\rho)$ into the Hilbert space $\mathcal{G}_\rho$.

We have a canonical section of $O(G_\rho)$ given by

$$g \mapsto r_g^0, \quad r_g^0(\zeta) \text{ where } \zeta = \exp(\epsilon r_g^0(\zeta))\, g, \quad \epsilon \to 0; \tag{2.2}$$

in another words r_g^0 is the inverse map of the infinitesimal exponential map.

The natural action of $\mathcal{U}_\rho$ acts on $\mathcal{G}_\rho$ prolongate at the level of frame as $U \times r^0 = U \circ r^0$. In this way we have the following canonical isomorphism

$$O(G_\rho) \simeq \mathcal{U}_\rho \times G_\rho. \tag{2.3}$$

2.8 Christofell Symbols

Two parallel transports in the direction z coexist: $\mathcal{T}_z^1 :=$ the algebraic transport obtained by composition of homeomorphisms, $\mathcal{T}_z^2 :=$ the Riemannian Levi–Civita parallel transport; then $\mathcal{T}_z^2 - \mathcal{T}_z^1$ defines $\Gamma(z) \in$ Endomorphism $(\mathcal{G}_\rho)$); this key endomorphism is determined by the next Theorem.

Theorem 7 (see [8]). *The Riemannian Levi–Civita connection on G_ρ is expressed in the canonical moving frame r_*^0 by the following* $(1,2)$ *constant tensor*

$${}^\rho\Gamma_{ij}^k = \frac{1}{2}\big(([{}^\rho e_i, {}^\rho e_j] \mid {}^\rho e_k)_\rho - ([{}^\rho e_j, {}^\rho e_k] \mid {}^\rho e_i)_\rho + ([{}^\rho e_k, {}^\rho e_i] \mid {}^\rho e_j)_\rho\big); \tag{2.4}$$

denote ${}^\rho\mathbf{\Gamma}_i$ the endomorphism of $\mathcal{G}$ corresponding to the last two indices, then ${}^\rho\mathbf{\Gamma}_i \in \mathrm{so}(\mathcal{G}_\rho)$.

Proof. As the antisymmetry property implies that connection preserves the Riemannian metric we have only to prove that it has no torsion which means that

$${}^\rho\Gamma_{ij}^k - {}^\rho\Gamma_{ji}^k = ([{}^\rho e_i, {}^\rho e_j] \mid {}^\rho e_k)_\rho,$$

identity which results by direct inspection from (2.4).

Corollary 8. *The process $t \mapsto {}^\rho g_x(t)$ is the Brownian motion on G_ρ.*

As we have

$$
\begin{aligned}
2[\cos k\theta\,,\,\cos p\theta] &= (k-p)\sin(k+p)\theta + (k+p)\sin(k-p)\theta \\
2[\sin k\theta\,,\,\sin p\theta] &= (p-k)\sin(p+k)\theta + (p+k)\sin(k-p)\theta \\
2[\cos k\theta\,,\,\sin p\theta] &= (p-k)\cos(p+k)\theta + (p+k)\cos(p-k)\theta
\end{aligned}
$$

we deduce the following expression for the Christofell symbol in the complex exponential basis;

For $p,\, k \geq 2$,

$$
\begin{aligned}
{}^{\rho}\Gamma(e^{ip\theta})e^{ik\theta} &= i\,\frac{(2p+k)\alpha(k)}{\alpha(p+k)}\,e^{i(p+k)\theta} \\
{}^{\rho}\Gamma(e^{ip\theta})e^{-\,ik\theta} &= -\,i\rho^{2p}\,(p+k)\,1_{k\geq p+2}\times e^{-\,i(k-p)\theta}
\end{aligned}
\tag{2.5}
$$

where $\alpha(k) = k^3 - k$.

2.9 Riemannian Parallelism on the Frame Bundle $O(G_\rho)$

We have an algebraic parallelism on $\mathcal{U}_\rho \times G$ induced by the right invariant Lie algebra of the product group $\mathrm{so}(\mathcal{G}_\rho) \times \mathcal{G}$; a *constant vector field* in the algebraic parallelism is of the form

$$
(\zeta)_{u,g} = (\exp\epsilon\ddot{\zeta}) \times U,\;\; \exp(\epsilon\dot{\zeta}) \times g),\;\; \epsilon \to 0,
$$
$$
\ddot{\zeta} \in \mathrm{so}(\mathcal{G}_\rho),\;\; \dot{\zeta} \in \mathcal{G}_\rho
$$

The 1-differential forms describing the passage from the algebraic to the Riemannian parallelism are:

$$
\begin{aligned}
&<\zeta\,,\,\dot{\sigma}>_{U,g} = U(\dot{\zeta}), \\
&<\zeta\,,\,\ddot{\sigma}>_{U,g} = U\circ\left(\ddot{\zeta} + {}^{\rho}\mathbf{\Gamma}(U^*(\dot{\zeta}))\right)\circ U^*
\end{aligned}
\tag{2.6}
$$

where U^* is the adjoint (and the inverse) of U.

The parallelism differential forms do not depend upon g, fact which corresponds to the invariance of the Riemannian metric under the right G action.

2.10 Structural Equations [8]

The coboundary of the parallelism differential forms are

$$
\begin{aligned}
&<\zeta_1\wedge\zeta_2,\, d\dot{\sigma}> = \ddot{\sigma}(\zeta_1)(\dot{\sigma}(\zeta_2)) - \ddot{\sigma}(\zeta_2)(\dot{\sigma}(\zeta_1)) \\
&<\zeta_1\wedge\zeta_2,\, d\ddot{\sigma}> = \ddot{\sigma}(\zeta_1)\ddot{\sigma}(\zeta_2) - \ddot{\sigma}(\zeta_2)\ddot{\sigma}(\zeta_1) + {}^{\rho}\mathbf{R}(\dot{\sigma}(\zeta)\wedge\dot{\sigma}(\zeta))
\end{aligned}
\tag{2.7}
$$

where ${}^{\rho}\mathbf{R}$ is the Riemannian curvature propagated by tensorial variance from its definition in $U =$ Identity given as

$$
{}^{\rho}\mathbf{R}(\dot{\zeta}_1\wedge\dot{\zeta}_2) = {}^{\rho}\mathbf{\Gamma}(\dot{\zeta}_1){}^{\rho}\mathbf{\Gamma}(\dot{\zeta}_2) - {}^{\rho}\mathbf{\Gamma}(\dot{\zeta}_2){}^{\rho}\mathbf{\Gamma}(\dot{\zeta}_1) - {}^{\rho}\mathbf{\Gamma}([\dot{\zeta}_1,\dot{\zeta}_2]).
$$

2.11 Change of Parameterization through Horizontal Parameterization

The *canonical horizontal lifting* of ${}^{\rho}g_x(t)$ to $O(G_\rho)$ is the process $(\Omega_t, g_x(t))$ where Ω_* is obtained by solving the stochastic differential system

$$\begin{aligned} dz(t) &= \Omega_t(dx(t)) \\ d\Omega_t &= \Big(- \sum_{k\geq -1} {}^{\rho}\mathbf{\Gamma}({}^{\rho}e_k)\, dz_k(t) + \frac{1}{2} \sum_{k\geq -1} ({}^{\rho}\mathbf{\Gamma}({}^{\rho}e_k))^2\, dt \Big) \Omega_t \end{aligned} \tag{2.8}$$

with $\Omega_0 = \mathrm{Id}$.

We could look to the system (2.8) as x given and z must be determined by an implicit SDE, which is difficult to solve.

As the mapping $x \mapsto z$ is an isomorphism of probability space, it is better to take as initial data z then solve the second SDE (2.8), then obtain x by $dx(t) = \Omega_t^{-1}(dz(t))$ and finally g_x by solving $dg_x(t) = *dx(t) \circ g_x(t)$.

2.12 Stochastic Calculus of Variations on the Horizontal Flow

Let $a \mapsto {}_z\Theta_{t\leftarrow 0}(a)$ the stochastic flow on $O(G_\rho)$ defined by solving the Stratanovitch SDE

$$< *dr_z,\, \dot{\sigma} >= dz(t), \quad < *dr_z,\, \ddot{\sigma} >= 0, \quad r_z(0) = a;$$

Given $a(\epsilon)$ a differentiable curve on $O(G_\rho)$, the Jacobian flow ${}_z\Theta'_{t\leftarrow 0}$ is defined by

$$\begin{aligned} \frac{d}{d\epsilon_{\epsilon=0}} {}_z\Theta_{t\leftarrow 0}(a(\epsilon)) &=: {}_z\Theta'_{t\leftarrow 0}(a'(0)) \\ < {}_z\Theta'_{t\leftarrow 0}(a'(0)),\, \dot{\sigma} > &=: \dot{h}(t), \\ < {}_z\Theta'_{t\leftarrow 0}(a'(0)), \ddot{\sigma} > &=: \ddot{h}(t). \end{aligned}$$

Theorem 9.

$$\begin{aligned} d_t\dot{h} + \frac{1}{2} {}^{\rho}\mathrm{Ricci}_{\Omega_z(t)}(\dot{h}(t))\, dt &= \ddot{h}(t)\, dz_t, \\ d_t\ddot{h} &= {}^{\rho}\mathbf{R}_{\Omega_z(t)}(dz_t \wedge \dot{h}(t)) + {}^{\rho}\mathbf{f}\tilde{\mathbf{R}}_{\Omega_z(t)}(\ddot{h}(t))\, dt, \end{aligned} \tag{2.9}$$

where $\tilde{\mathbf{R}}$ *is the curvature tensor considered as defining an endomorphism of* $\mathrm{so}(\mathcal{G}_\rho)$*, the space of antihermitian bounded operators over* $\mathcal{G}_\rho$.

2.13 Integration by Part for the Regularized Metric

Theorem 10 (Theorem of Integration by Part ([7,9,11])). *Let* $v \in \mathcal{G}_\rho$*; assume that there exists* $\delta > 0$ *such that, in the sense of the order of hermitian operator on* $\mathcal{G}_\rho$*, the following inequality holds true* ${}^{\rho}\mathrm{Ricci} \geq -\delta \times \mathrm{Identity}$*, then*

$$\frac{d}{d\epsilon_{\epsilon=0}}\Big(E(f({}^\rho g_x(t_0)\exp(\epsilon v_0)))\Big) = E(f({}^\rho g_x(t_0))\ k(x)),\ \textit{where}$$
$$E(|k(x)|^2) \leq \frac{\exp(\delta t_0)-1}{\delta t_0^2}\|v_0\|_\rho^2. \tag{2.10}$$

Proof. We follow [12]; an infinitesimal euclidean motion of the $\mathcal{G}_\rho$-cylindrical Brownian motion is defined

$$z \mapsto y,\ y(t) = \int_0^t \exp(\epsilon q_s)\ dz(s) + \epsilon\int_0^t w(s)\ ds$$

where $s \mapsto q_s$ is an adapted functional with values in $\mathrm{so}(\mathcal{G}_\rho)$ and where the map $* \mapsto w_*$ is an adapted functional with values in $L^2([0,t_0];\mathcal{G}_\rho)$; define the derivative

$$D_{q,w}\left({}_*\Theta_{t_0\leftarrow 0}\right) := \zeta(t_0),\ \ \zeta(t) := \int_0^t {}_*\Theta'_{t_0\leftarrow s}(q_s, w_s)\ ds. \tag{a}$$

We choose (q,w) by solving the system

$$d_t v + \frac{1}{2}\mathrm{Ricci}(v) = w,\ v(0) = v_0,\ \ dq = \mathbf{R}(*dz \wedge v),\ q(0) = 0; \tag{b}$$

then taking ζ from (a) and (b) we get,

$$d_t\dot\zeta + \frac{1}{2}{}^\rho\mathrm{Ricci}_{\Omega_z(t)}(\dot\zeta(t))\ dt = (q - \ddot\zeta(t))dz_t + w\ dt,$$
$$d_t\ddot\zeta = {}^\rho\mathbf{R}_{\Omega_z(t)}(*dz_t \wedge \dot\zeta(t))$$

which leads to the following remarkable expression

$$\dot\zeta = v,\ \ddot\zeta = q. \tag{c}$$

Denote $A_{t\leftarrow 0}$ the resolvent of the first equation (b), take

$$w(t) = -\frac{1}{t_0}A_{t\leftarrow 0}(v_0); \tag{d}$$

$$w(t_0) = A_{t_0\leftarrow 0}(v_0) + \int_0^{t_0} A_{t_0\leftarrow t}h(t)\ dt = 0 \tag{e}$$

this last equality meaning that the "jump" resulting from the derivative at $t=0$ have been completely sweep out at time t_0. Let us emphasize that the equation (d) gives rise to an adapted h. As the infinitesimal rotations preserves the gaussian measure we have $D_{q,h}$ as the same formula of integration by part as $D_{0,h}$ formula which is provided by Girsanov:

$$k(z) = \frac{1}{t_0}\int_0^{t_0} (A_{t\leftarrow 0}(v_0) \mid dz_t);$$

by Itô energy identity of stochastic integral we have

$$E(|k(z)|^2) \leq \|v_0\|^2_{\mathcal{G}_\rho} \frac{1}{t_0^2} \int_0^{t_0} \exp(\delta t)\, dt,$$

expression equal to the r.h.s. of (2.10).

2.14 Cameron–Martin for the Brownian on G

Theorem 11. *Assume that there exists δ such that for all $\rho < 1$,*

$$({}^\rho\mathrm{Ricci}\zeta \mid \zeta)_\rho \geq -\delta \|\zeta\|^2_\rho \quad \forall \zeta \tag{2.11}$$

and that for all $\rho < 1$ the following SDE is resoluble and that its solution is a unitary operator

$$d\, {}^\rho\Omega_t = \left(- \sum_{k \geq -1} {}^\rho\mathbf{\Gamma}({}^\rho e_k)\, dz_k(t) + \frac{1}{2} \sum_{k \geq -1} ({}^\rho\mathbf{\Gamma}({}^\rho e_k))^2\, dt \right) {}^\rho\Omega_t. \tag{2.12}$$

Then under these hypothesis (2.13) holds true:

$$\begin{aligned} \frac{d}{d\epsilon_{\epsilon=0}} E(f(g_x(t)) \exp(\epsilon v_0))) &= E(f(g_x(t)) k(x)), \\ E(|k(x)|^2) &\leq \frac{\exp(\delta t_0) - 1}{\delta t_0^2} \|v_0\|^2. \end{aligned} \tag{2.13}$$

Remark. The unique hypothesis where uniformity relatively to ρ appears is (2.11).

2.15 Existence of the Horizontal Stochastic Transport

We shall prove, for $\rho < 1$ the existence of a solution of (2.12); remark that the SDE (2.12) stays invariant in law by a rotation of the circle.

Lemma 12.

$$\mathcal{C} := \sum_{p \geq 2} ({}^r\mathbf{\Gamma}(\frac{\cos p\theta}{\sqrt{\alpha(p)}}))^2 + ({}^r\mathbf{\Gamma}(\frac{\sin p\theta}{\sqrt{\alpha(p)}}))^2$$

is a diagonal bounded operator and the coefficients on the diagonal are given by

$$\lambda_k = \frac{1}{2} \sum_{p \geq 2} [- (p+k)^2 \frac{\alpha(k-p)}{\alpha(p)\alpha(k)} 1_{k \geq p+2} - \frac{(2p+k)^2 \alpha(k)}{\alpha(p)\alpha(p+k)}] \rho^p$$

Proof. The invariance in law of (2.12) under the action of S^1 implies that $\mathcal{C}$ commutes with the action of S^1 which is equivalent to say that $\mathcal{C}$ diagonalizes in the trigonometric basis. The expression of this diagonal matrix is obtained by direct computation.

Lemma 13. *The solution of* (2.12) *is given by the Picard series:*

$$\Omega_t = \sum_{n \geq 0} \Delta_n(t),$$

$$\Delta_0(t) = \text{Identity}, \quad \Delta_n(t) = \int_0^t {}^{\rho}\mathbf{\Gamma}(dx(s))\, \Delta_{n-1}(s)$$

Proof. Define $M_n(t) := E(\Delta_n(t)(\Delta_n(t))^*)$, then as before $M_n(t)$ is a diagonal matrix; by orthogonality of iterated Itô integrals we have: $M_n(t) = \sum_k \frac{\rho^{2k}}{\alpha(k)} A_{n,k}(t)$ where $A_{n,k}$ are the diagonal matrices defined by

$$A_{n,k}(t) : = {}^{\rho}\mathbf{\Gamma}(\cos k\theta) \int_0^t M_{n-1}(s)\, ds\; {}^{\rho}\mathbf{\Gamma}(\cos k\theta)$$
$$+ {}^{\rho}\mathbf{\Gamma}(\sin k\theta) \int_0^t M_{n-1}(s)\, ds\; {}^{\rho}\mathbf{\Gamma}(\sin k\theta)$$

2.16 Curvatures of S^1-Homeomorphism Group (see [1])

For $u, v \in \text{diff}(S^1)$, the Riemann curvature has been defined as

$$R(u,v) := \Gamma(u)\Gamma(v) - \Gamma(v)\Gamma(u) - \Gamma([u,v])$$

Theorem 14. $2\,R(\cos m\theta, \sin m\theta) \sin p\theta = -\lambda_{m,p} \cos p\theta$,

$$\lambda_{m,p} = 1_{p \geq m+2}\, \frac{(m+p)^2 \alpha(p-m)}{\alpha(p)} - \frac{(2m+p)^2 \alpha(p)}{\alpha(p+m)} + 2mp$$

$$\sum_{p \geq 2} \lambda_{mp} = -\frac{13}{6}(m^3 - m)$$

Theorem 15. *The Ricci tensor is equal to* $-\frac{13}{6} \times$ Identity.

See for related results Bowick–Rajeev.

References

1. H. Airault. Riemannian connections and curvatures on the universal Teichmuller space. *Comptes Rendus Mathématique*, 341:253–258, 2005.
2. H. Airault, P. Malliavin, and A. Thalmaier. Canonical Brownian motion on the space of univalent functions and resolution of Beltrami equations by a continuity method along stochastic flows. *J. Math. Pures et App.*, 83:955–1018, 2004.
3. H. Airault and J. Ren. Modulus of continuity of the canonic brownian motion "on" the group of diffemorphisms of the circle. *J. Funct. Analysis*, 196:395–406, 2002.

4. E. Barucci, P. Malliavin, and M. E. Mancino. Harmonic Analysis methods for non parametric estimation of volatility: theory and applications. In *Proc. Ritskumeisan Conference on Mathematical Finance*, 2005. to appear in 2006.
5. E. Barucci, P. Malliavin, M. E. Mancino, and A. Thalmaier. The price-volatility feed-back rate, an implementable indicator of market stability. *Math. Finance*, 13:17–35, 2003.
6. A. Beurling and L. Ahlfors. The boundary correspondence under quasi-conformal mappings. *Acta Mathematica*, 96:125–142, 1956.
7. J. M. Bismut. Large deviations and the Malliavin Calculus. *Progress in Mathematics*, 45, 1984. Birkhäuser, Boston.
8. A. B. Cruzeiro and P. Malliavin. Renormalized differential geometry on the path space: Structural equations, Curvatures. *J. Funct. Analysis*, 139:119–181, 1996.
9. B. Driver. Integration by part and quasi-invariance for heat measure over loop groups. *J. Funct. Analysis*, 149:470–547, 1997.
10. S. Fang. Integration by part for heat measures over loop groups. *J. Math. Pures App.*, pages 877–894, 1999.
11. S. Fang. Canonical Brownian motion on the diffeomorphism group of the circle. *J. Funct. Analysis*, 196:162–179, 2002.
12. S. Fang and P. Malliavin. Stochastic Analysis on the path space of a riemannian manifold. *J. Funct. Analysis*, 118:249–274, 1993.
13. P. Malliavin and M. E. Mancino. Fourier series method for mesurement of multivariate volatilities. *Finance Stoch.*, 6:49–61, 2002.
14. P. Malliavin and A. Thalmaier. *Stochastic Calculus of Variations in Mathematical Finance.* Springer, 2005. 142 pages.
15. P. Malliavin, M.E. Mancino, M.C. Recchioni. A non parametric calibration of the HJM geometry: an application of Itô calculus to financial statistics. Japan J. Math. 2: 23, 2007.

The Invariant Distribution of a Diffusion: Some New Aspects

Henry P. McKean

CIMS, 251 Mercer Street, New York, NY 10012, USA

1 Introduction

The subject is an old one, but the conventional discussion seems in one respect incomplete: *If you have an invariant distribution, what is it the distribution of*? M. Baldini and I have found an amusing answer to this question.

Fix 1) a standard d-dimensional Brownian motion with paths $b(t) : t \geq 0$, 2) a smooth, positive-definite diffusion coefficient σ, 3) a smooth drift coefficient m, and let $x^{\uparrow}(t,x) : t \geq 0,\ x \in \mathbb{R}^d$ be the flow determined by

1) $$dx^{\uparrow} = \sigma(x)db + m(x)dt \qquad \text{with } x^{\uparrow}(0,x) = x.$$

Here, "flow" means that $x^{\uparrow}(t,x) : t \geqslant 0, x \in \mathbb{R}^d$ is (implicity) a function of the single Brownian motion $b(t) : t \geqslant 0$. You solve 1) for $t \geqslant 0$, *simultaneously* for every *e.g.* terminating binary x with the *same* Brownian motion b. Then Kolmogorov-Centsov is used to show that this, so to say "skeleton" is continuous in the pair (t,x) and so may be extended to the whole $[0,\infty] \times \mathbb{R}^d$ so as to solve 1) identically in t & x up to a possible explosion time, with probability 1. This can be found in Kunita [1990] together with the fact that if no explosion takes place, $x(t,\bullet)$ is a diffeomorphism of $\mathbb{R}^d$ (with probability 1 of course. It is assumed that $x^{\uparrow}$ returns to every neighborhood of $\mathbb{R}^d$, over and over, and more: that it has a smooth invariant density $1/\psi^2$ of total mass $\int(1/\psi^2) = 1$. Then for nice functions f,

2′) $$\lim_{T\uparrow\infty} \frac{1}{T}\int_0^T f(x^{\uparrow})dt = \int \frac{f}{\psi^2}$$

with probability 1 for each $x^{\uparrow}(0) = x$, separately, and also

2″) $$\lim_{T\uparrow\infty} e^{T\mathfrak{g}} f = \int \frac{f}{\psi^2}$$

pointwise, $\mathfrak{g}$ being the infinitesimal operator of the diffusion

3) $$\mathfrak{g} = \frac{1}{2}\sigma^2\partial^2/\partial x^2 + m\partial/\partial x.$$

Actually, it will be best to interpret 1) in Stratonovich's way, *i.e.* in Itô's language

1′) $$dx^\uparrow = \sigma db + mdt + \frac{1}{2}\sigma'\sigma dt \quad \text{with } (\sigma'\sigma)_i = \sum_{1\le j,k\le d} \frac{\partial \sigma_{ij}}{\partial x_k}\sigma_{kj},$$

and to take

3′) $$\mathfrak{g} = \frac{1}{2}\sigma^2\partial^2/\partial x^2 + (m + \frac{1}{2}\sigma'\sigma)\partial/\partial x$$

in accord with that. Itô's language is used everywhere below.

Two allied diffusions or flows are now introduced.

Fix a time $T > 0$, recompute $x^\uparrow(T, x)$ by solving 1′) not with the original Brownian motion $b(t) : t \le T$, but with the reversed $b^\downarrow(t) = b(T-t) - b(T) : t \le T$, and record only the final position $\equiv x^\downarrow(T, x)$. The motion $x^\downarrow(t, x) : t \ge 0$ is not quite a diffusion: as a diffeomorphism $x^\downarrow(t, \bullet) : t \ge 0$ it is Markovian, but for fixed x, you have only

4) $$dx^\downarrow(t, x) = \mathfrak{g}x^\downarrow(t, x) - \frac{\partial x^\downarrow(t, x)}{\partial x}\sigma db$$

in which σ is paired with the lower variable x and with db. Obviously, $x^\downarrow(t, \bullet)$ is identical in law to $x^\uparrow(t, \bullet)$ for each fixed $t \ge 0$, separately, but their motion in time is very different, as will be seen: $x^\uparrow(T)$ is driven by the innovation $db(T)$, but for $x^\downarrow(T)$ the latter is buried in the past and its influence washes out. This is the first allied flow.

The second is the bonafide diffusion $x^\sharp(t, x) : t \ge 0, x \in R^d$ determined by

5) $$dx^\sharp = \sigma db + (-m + \frac{1}{2}\sigma'\sigma)dt$$

with reversed drift $-m$ in place of m. It is intimately related to $x^\downarrow$: for each $t \ge 0$, $x^\sharp(t, \bullet)$ is the diffeomorphism inverse to $x^\downarrow(t, \bullet)$, assuming that $x^\sharp$ does not run out to ∞ in finite time.

It is the inter-relation of these three processes; $x^\uparrow$, $x^\downarrow$, and $x^\sharp$, that I will talk about. For the matter sketched above, I refer you to Kunita [1990]; it is the best presentation.

2 Mostly Dimension 1

I will explain what happens here: With

$$\psi^2 = Z\sigma \exp\left[-\int_0^x 2m/\sigma^2\right],$$

you have

$$\mathfrak{g} = \frac{\psi^2}{2} D \frac{\sigma^2}{\psi^2} D \text{ with scale } \int_0^x \frac{\psi^2}{\sigma^2} \text{ and speed measure } \frac{2dx}{\psi^2},$$

$$\mathfrak{g}^\sharp = \frac{1}{2} \frac{\sigma^2}{\psi^2} D \psi^2 D \text{ with scale and speed measure reversed,}$$

and

$$\int \frac{1}{\psi^2} = 1 \qquad \text{by choice of } Z.$$

Here, $1/\psi^2$ is the invariant density for $x^\uparrow$. I take

$$s(-\infty, 0] = \int_{-\infty}^0 \frac{\psi^2}{\sigma^2} \quad \text{and} \quad s[0, +\infty) = \int_0^\infty \frac{\psi^2}{\sigma^2} \quad \text{both } = +\infty,$$

as is automatic if $\sigma = 1$. Then for fixed x, $x^\sharp(t, x)$ tends almost surely to $\pm\infty$ as $t \uparrow \infty$ (but not before). $\mathfrak{g}$ can also be written $\frac{1}{2}\bar{D}^2 + (m/\sigma)\bar{D}$ in the scale $\bar{x} = \int_0^x (1/\sigma)$; this will be useful in section 5. Note that m cannot vanish: otherwise, ψ^2 is effectively σ and you cannot have both $\int 1/\psi^2 < \infty$ and $\int \psi^2/\sigma^2 = \infty$.

Now the chief facts in dimension 1 are these:

1) $\lim_{t\uparrow\infty} x^\downarrow(t, x) = x^\downarrow(\infty)$ exists, independently of x; it is distributed with density $1/\psi^2$. (No surprise.)

2′) $x^\sharp(t, x) \uparrow +\infty$ if $x > x^\downarrow(\infty)$.

2″) $x^\sharp(t, x) \downarrow -\infty$ if $x < x^\downarrow(\infty)$.

3) $x^\sharp(T, x^\downarrow(\infty)) = x^\downarrow(\infty)$ recomputed for the shifted Brownian motion $b^+(t) = b(t + T) - b(T) : t \geq 0$; as such it can be made stationary for $-\infty < T < +\infty$, and if its time is then reversed, you will see the stationary version of $x^\uparrow$ with initial distribution dx/ψ^2. For this reason, $x^\downarrow(\infty)$ is called the stagnation point.

4)

$$\int \frac{dx}{\psi^2} E|x^\downarrow(\infty) - x^\downarrow(t, x)|^2, = \lim_{T\uparrow\infty} \int \frac{dx}{\psi^2} E|x^\downarrow(T, x) - x^\downarrow(t, x)|^2$$
$$= \int \frac{da}{\psi^2} \int \frac{db}{\psi^2} E|x^\uparrow(t, b) - x^\uparrow(t, a)|^2,$$

and this quantity decreases to 0 provided $\int x^2/\psi^2 < \infty$, *i.e.* $x^\uparrow$ "focuses", as found by Hasminskii-Nevelson [1971] in a different form noted later on. The decay may exponentially fast or no, as you would think if $\mathfrak{g}$ has spectrum near the origin.

5) What I would *not* have thought is that focusing always takes place pathwise exponentially fast:

$$\lim_{t\uparrow\infty} \frac{1}{t} \ell n[sox^{\uparrow}(t,b) - so(x^{\uparrow}(t,a)] = -\gamma,$$

simultaneously for every $a < b$, in which you see the natural scale $s(x) = \int_0^x \psi^2/\sigma^2$ of $x^{\uparrow}$, and γ is the (to me) mysterious number

$$0 < \gamma = 2 \int \frac{m^2}{\sigma^2} \frac{1}{\psi^2} \le \infty.$$

This γ is always bigger than, and in special cases equal to, the spectral gap g of $\mathfrak{g}$, but this gap can vanish, so that's not γ; it is also bigger than or equal to the ground state of $\mathfrak{g}^{\sharp}$, but that can vanish, too. I think that γ should have *some* spectral meaning, but don't know what it is.

The proofs of 1)-5) occupy the rest of this report.

Dimension $d \ge 2$ is *much* harder. You can put on unattractive conditions to make 1)-4) and a crude version of 5) come out; see Baldini [2006] for this. I believe you need next to nothing but am now just as far from the proof as I was 2 years ago.

3 Ornstein–Uhlenbeck

The process with $\mathfrak{g} = \frac{1}{2}D^2 - xD$, $\psi^2 = \sqrt{\pi}e^{+x^2}$, and scale $\int_0^x \sqrt{\pi}e^{y^2}dy$ will illustrate all this. Here, $dx^{\uparrow} = db - x^{\uparrow}dt$, *i.e.* $x^{\uparrow}(t,x) = e^{-t}x + e^{-t}\int_0^t e^s db$, and you have

2.1) $x^{\downarrow}(t,1x) = e^{-t}x - \int_0^t e^{-s}db$ tending to $x^{\downarrow}(\infty) = -\int_0^{\infty} e^{-s}db$,

2.2) $x^{\sharp}(t,x) = e^t x + e^t \int_0^t e^{-s}db$ tending to $\pm\infty$ according as $x > x^{\downarrow}(\infty)$ or $x < x^{\downarrow}(\infty)$,

2.3) $x^{\sharp}(t,x^{\uparrow}(\infty)) = e^t \int_t^{\infty} e^{-s}db$ is identical in law to $e^t B\left(\frac{1}{2}(e^{-2t})\right)$ with a new Brownian motion B, *i.e.* it is $x^{\uparrow}$ = Ornstein-Uhlenbeck made stationary.

2.4) $\partial x^{\uparrow}/\partial x = e^{-t}$, so $\int \frac{dx}{\psi^2} E|x^{\downarrow}(\infty) - x^{\downarrow}(t,x)|^2 = 2\int \frac{x^2}{\psi^2}e^{-2t} = e^{-2t}$,

2.5) $\lim\limits_{t\uparrow\infty} \frac{1}{t}\ell n \int_{x^{\uparrow}(t,a)}^{x^{\uparrow}(t,b)} \sqrt{\pi}e^{x^2}dx = -\gamma (= 2\int \frac{x^2}{\psi^2} = 1)$, and this number is the actual spectral gap of $\mathfrak{g}$.

4 Proofs in Dimension 1

Recall the general function

$$\psi^2 = Z\sigma \exp[-\int_0^x 2m/\sigma^2]$$

and the form of the infinitesimal operators

$$\mathfrak{g} = \frac{\sigma^2}{2^2}D^2 + (m + \frac{1}{2}\sigma'\sigma)D = \frac{\psi^2}{2}D\frac{\sigma^2}{\psi^2}D.$$

and

$$\mathfrak{g}^\sharp = \frac{\sigma^2}{2}D^2 + (-m + \frac{1}{2}\sigma'\sigma)D = \frac{1}{2}\frac{\sigma^2}{\psi^2}D\psi^2 D.$$

2.2) is obvious: $x^\sharp(t,x)$ is transient, tending to $\pm\infty$ with probability 1 for each x separately, and since $x^\sharp(t,\bullet)$ is a diffeomorphism, $x^\sharp(t,b)$ tends to $+\infty$ as soon as $x^\sharp(t,a)$ does so for any $a < b$, and so forth, the self-evident conclusion being that there is a single (random) point $x^\downarrow(\infty)$ as in 2.2): $x^\sharp(t,x)$ tends to $+\infty$ if $x > x^\downarrow(\infty)$ and to $-\infty$ if $x < x^\downarrow(\infty)$. But then, for any $\varepsilon > 0$, large L, and sufficiently large T,

$$x^\sharp(T, x^\downarrow(\infty) - \varepsilon) < -L < +L < x^\sharp(T, x^\downarrow(\infty) + \varepsilon),$$

with the implication 2.1):

$$x^\downarrow(\infty) - \varepsilon < x^\downarrow(T,-L) < x^\downarrow(T,+L) < x^\downarrow(\infty) + \varepsilon.$$

Besides, for nice f,

$$Efox^\downarrow(\infty) = \lim_{T\uparrow\infty} Efox^\downarrow(T,x) = \lim_{T\uparrow\infty} Efox^\uparrow(T,x) = \int \frac{f}{\psi^2},$$

by 1.2″), so $x^\downarrow(\infty)$ is distributed by the invariant density $1/\psi^2$.

2.3) is next: With a self-evident notation, $x^\downarrow(T,x|\mathbb{B}_0^T) = x^\downarrow(t,\bullet|\mathbb{B}_0^t)o$ $x^\downarrow(T-t,x|\mathbb{B}_t^\infty)$ for $T > t$, so

$$x^\sharp(t, x^\downarrow(T,x)) = x^\downarrow(T-t, x|\mathbb{B}_t^\infty)$$

which produces

$$x^\sharp(t, x^\downarrow(\infty)) = x^\downarrow(\infty|\mathbb{B}_t^\infty)$$

at $T = \infty$, showing that $x^\sharp(t,x^\downarrow(\infty))$ is (or rather can be made) stationary. But, $dx^\sharp = \sigma db + (-m + \frac{1}{2}\sigma'\sigma)dt$, and reversing the time as in $x^\sharp(t) \to x^\flat(t) = x^\sharp(-t)$, produces $dx^\flat(t) = \sigma db + (m + \frac{1}{2}\sigma'\sigma)dt$, which is to say that the stationary $x^\sharp(\bullet, x^\downarrow(\infty))$ reversed is a copy of the stationary $x^\uparrow(t,x)$ with x distributed by $1/\psi^2$.

2.4): For fixed $T > t$, $x^\downarrow(T,x) = x^\downarrow(t,\bullet|\mathbb{B}_0^t)ox^\downarrow(T-t,x|\mathbb{B}_t^\infty)$ is identical in law to $x^\downarrow(t,\bullet|\mathbb{B}_0^t)ox^\uparrow(T-t,x|\mathbb{B}_t^\infty)$, by the independence of the fields $\mathbb{B}_0^t$ and $\mathbb{B}_t^\infty$, so you have

$$E[x^\downarrow(T,x)|\mathbb{B}_0^t] = e^{(T-t)\mathfrak{g}}x^\downarrow(t,x)$$

with $e^{(T-t)\mathfrak{g}}$ applied to the variable x, provided $\int x^2/\psi^2 < \infty$. Now the same rule applies if f is any nice (*e.g.* smooth, compact) function:

$$E[fox^{\downarrow}(T,x)|\mathbb{B}_0^t] = e^{(T-t)\mathfrak{g}} fox^{\downarrow}(t,x),$$

so

$$\int \frac{dx}{\psi^2} E|fox^{\downarrow}(T,x) - fox^{\downarrow}(t,x)|^2$$

$$= \int \frac{dx}{\psi^2} E[f^2ox^{\downarrow}(T,x) - 2e^{T-t)\mathfrak{g}} fox^{\downarrow}(t,x) \times fox^{\downarrow}(t,x) + f^2ox^{\downarrow}(t,x)]$$

in which all the arrows can be turned up, producing

$$2\int \frac{f^2}{\psi^2} - 2\int \frac{dx}{\psi^2} E[e^{(T-t)\mathfrak{g}} fox^{\uparrow}(t,x) \times fox^{\uparrow}(t,x)]$$

$$\simeq 2\int \frac{f^2}{\psi^2} - 2\int \frac{dx}{\psi^2} E[\int fox^{\uparrow}(t,x')\frac{dx'}{\psi^2} \times fox^{\uparrow}(t,x)]$$

for $T \uparrow \infty$, by 1.2″), *i.e*

$$\int \frac{dx}{\psi^2} E|fox^{\downarrow}(\infty) - fox^{\downarrow}(t,x)|^2$$

$$= \int \frac{dx}{\psi^2} \int \frac{db}{\psi^2} E|fox^{\uparrow}(t,b) - fox^{\uparrow}(t,x)|^2$$

as in 2.4). Besides,

$$\int \frac{dx}{\psi^2} E|fox^{\downarrow}(\infty) - fox^{\downarrow}(t,x)|^2$$

$$= 2\int \frac{f^2}{\psi^2} - 2E[fox^{\downarrow}(\infty) \int \frac{fox^{\downarrow}(t,x)}{\psi^2} dx],$$

and here

$$\int fox^{\downarrow}(t,x)\frac{dx}{\psi^2} = E[fox^{\downarrow}(\infty)|\mathbb{B}_0^t]$$

is a martingale and also a projection, which is to say

$$\int \frac{dx}{\psi^2} E|fox^{\downarrow}(\infty) - fox^{\downarrow}(t,x)|^2 \downarrow 0 \quad \text{as } t \uparrow \infty.$$

The rest, which is to carry all this over to $f(x) = x$ is easy: if $f(x^{\downarrow}) = x^{\downarrow}\times$ the indicator of $|x^{\downarrow}| \leq R$, then

$$\int \frac{dx}{\psi^2} E|fox^{\downarrow}(t,x) - x^{\downarrow}(t,x)|^2 = \int\limits_{|x|>R} x^2/\psi^2$$

is small for large R, independently of $t \geq 0$.

2.5) is surprising, but its pretty easy, too. It states that, in the natural scale $s(x) = \int_0^x \psi^2/\sigma^2$, $x^{\uparrow}$ focuses pathwise, exponentially fast, at rate $\gamma = 2\int m^2/\sigma^2\psi^2$. Fix $a < b$ and write A for $sox^{\uparrow}(t,a)$ and B for $sox^{\uparrow}(t,b)$.

Step 1

The role of the scale is to make $B - A$ a (positive) super-martingale ($\mathfrak{g}s = 0$). As such, it has a limit $0 \le C < \infty$. Now

$$\lim_{T\uparrow\infty} \frac{1}{T}\int_0^T tan^{-1}(B)dt = \int tan^{-1} os(x)\frac{dx}{\psi^2} \quad \text{by 1.2}'),$$

and

$$tan^{-1}(B) \simeq tan^{-1}(A + C) \text{ for } t \uparrow \infty,$$

so also

$$\lim_{T\uparrow\infty} \frac{1}{T}\int_0^T tan^{-1}(B)dt = \int tan^{-1} o[s(x) + C]\frac{dx}{\psi^2}.$$

This is not possible unless $C = 0$, *i.e.* $B - A = o(1)$. Thus far Hasminski-Nevelson [1971: Lemma 2, Part I].

Step 2

is to compute the differential of $B - A$: with

$$F = \left[\frac{\psi^2}{\sigma} ox^{\uparrow}(t, b) - \frac{\psi^2}{\sigma} ox^{\uparrow}(t, a)\right] \times (B - A)^{-1}$$

you find

$$d(B - A) = (B - A)F \times \text{the differential db of the Brownian motion}$$

and so you may write

$$B - A = [B(0) - A(0)] \times e^{\int_0^t F db - \frac{1}{2}\int_0^t F^2 dt}.$$

Step 3

is an over-estimate. The mean-value theorem is applied to F as follows:

$$\left(\frac{\psi^2}{\sigma} os^{-1}\right)' = -\frac{2m}{\sigma^2}\frac{\psi^2}{\sigma} os^{-1} \times \left(\frac{\psi^2}{\sigma^2} o(s^{-1})\right)^{-1} = -\frac{2m}{\sigma} o(s^{-1}),$$

so

$$F = \left(\frac{\psi^2}{\sigma} os^{-1}\right)(B) - \left(\frac{\psi^2}{\sigma} os^{-1}\right)(A) \qquad \text{over } B - A$$

$$= \left(-\frac{2m}{\sigma} os^{-1}\right)(C) \qquad \text{with } C \text{ between } A \text{ and } B.$$

The peculiar instance on s^{-1} pays off as follows. Take $G < 2(m^2/\sigma^2) os^{-1}$ with bounded slope. Then $G(C) \simeq G(B)$ for $t \uparrow \infty$, by step 1, and

$$\begin{aligned}\lim_{T\uparrow\infty}\frac{1}{T}\frac{1}{2}\int_0^T F^2dt &\geq \lim_{T\uparrow\infty}\frac{1}{T}\int_0^T G(B)dt\\ &= \int Gos(x)\frac{dx}{\psi^2}\\ &> \gamma'\end{aligned}$$

for any number $\gamma' < \gamma$, by choice of G, *i.e.* by step 2,

$$\varlimsup_{t\uparrow\infty}\frac{1}{t}\ell n(B-A) \leq -\gamma$$

in view of

$$\left|\int_0^t Fdb\right| \leq \sqrt{(2+)\int_0^t F^2\ \ell n\ell n\int_0^t F^2}.$$

Step 4

is the final under-estimate: Now write

$$B-A = \int_a^b \frac{\partial X}{\partial x}dx \qquad \text{with } X = sox^{\uparrow}(t,x).$$

You have $dX = (\psi^2/\sigma)(x^{\uparrow})db$, so

$$d\frac{\partial X}{\partial x} = -\frac{2m}{\sigma}\psi^2(x^{\uparrow})\frac{\partial X}{\partial x}db$$

and

$$\frac{\partial X}{\partial x} = e^{-2\int_0^t \frac{m}{\sigma}(x^{\uparrow})db - 2\int_0^t \frac{m^2}{\sigma^2}(x^{\uparrow})dt'}.$$

Now if $\gamma = \infty$, there is nothing to do, while if $\gamma < \infty$ then, for any $\gamma' > \gamma$, Fatou's lemma implies

$$\begin{aligned}&\varliminf_{t\uparrow\infty} e^{\gamma' t}(B-A)\\ &\geq \int_a^b \varliminf_{t\uparrow\infty} e^{\gamma' t}e^{-2\int_0^t \frac{m}{\sigma}(x^{\uparrow})db - 2\int_0^t \frac{m^2}{\sigma^2}(x^{\uparrow})dt'}dx\\ &= +\infty\end{aligned}$$

in view of

$$\lim_{T\uparrow\infty}\frac{1}{T}\int_0^T 2\frac{m^2}{\sigma^2}(x^{\uparrow})dt \quad = 2\int\frac{m^2}{\sigma^2}\frac{1}{\psi^2} = \gamma,$$

i.e.

$$\varliminf_{t\uparrow\infty}\frac{1}{t}\ell n(B-A) \geq -\gamma.$$

5 More About γ

The proof of 2.1)–2.5) is finished, but what is γ? Surely, it has some spectral meaning, but I don't know what. It has a little to do with the spectral gap of $\mathfrak{g}$, which is the distance g from its ground state ($= 0$ since $\mathfrak{g}1 = 0$) to the rest of its spectrum. This is the infimum of the quadratic form

$$Q \equiv -\int f\mathfrak{g}f\frac{1}{\psi^2} = \frac{1}{2}\int\frac{f'^2\sigma^2}{\psi^2} \text{ for nice } f \text{ with } \int\frac{f^2}{\psi^2} < \infty \text{ and } \int\frac{f}{\psi^2} = 0.$$

Item 1:

$g \le \gamma$. Take $f = A(\int_0^x \frac{1}{\sigma} - B)$ on a big interval $I = [-a, b]$ and extend it to the right/left by the constant values $f(b)/f(-a)$, with B taken to make $\int f/\psi^2 = 0$ and $A > 0$ to make $\int f^2/\psi^2 = 1$. Then, with $N = \int_I 1/\psi^2$,

$$Q = \frac{1}{2}A^2N,$$

$$A/N = \int\limits_I \frac{f'\sigma}{\psi^2},$$

and so

$$Q = \frac{1}{2N}\left(\int\limits_I \frac{f'\sigma}{\psi^2}\right)^2.$$

I want to integrate by parts for which I need $\underline{\lim}_{x\uparrow\infty}\, \sigma/\psi^2 = 0$ and likewise at $-\infty$. But if, for example, $\underline{\lim}_{x\uparrow\infty}\, \sigma/\psi^2 = 2$, then $1/\psi^2 \ge 1/\sigma \ge \psi^2/\sigma^2$ far out, contradicting $\int \psi^2/\sigma^2 = \infty$. Now you can write

$$\begin{aligned} g \le Q &= \frac{1}{2N}\left(\int_{-\infty}^{+\infty} f\left(\frac{\sigma}{\psi^2}\right)'\right)^2 \\ &= \frac{1}{2N}\left(\int 2f\frac{m}{\sigma}\cdot\frac{1}{\psi^2}\right)^2 \\ &\le \frac{2}{N}\int\frac{f^2}{\psi^2}\int\frac{m^2}{\sigma^2}\frac{1}{\psi^2} \\ &= \frac{\gamma}{N}, \end{aligned}$$

and making I increase to the whole line makes $N \uparrow 1$, confirming $g \le \gamma$.

Item 2:

g can vanish so that is not the meaning of γ. Take $\sigma = 1$ and $\psi^2 = \pi(1+x^2)$. Then $m = -\psi'/\psi = -x \times (1+x^2)^{-1}$, and $\gamma = 1$, while if f is the odd function x/h for $0 \le x \le h$ and 1 beyond, then

$$\int f/\psi^2 = 0 \quad \text{and} \quad \frac{1}{2}\frac{\int f'^2/\psi^2}{\int f^2/\psi^2} \le \frac{h^{-2}\int_0^h \frac{1}{\pi(1+x^2)}}{2\int_h^\infty \frac{1}{\pi(1+x^2)}} \simeq \frac{h^{-2}}{2/\pi h} = o(1)$$

for $h \uparrow \infty$.

Item 3:

$g = \gamma$ only if $\int_{-\infty}^{0} 1/\sigma = \int_0^\infty 1/\sigma = +\infty$ and

$$\bar{x}(t,x) = \int_0^{x^\dagger(t,x)} \frac{1}{\sigma(y)}dy$$

is the standard Ornstein-Uhlenbeck process, up to scalings $x \to ax + b$ and $t \to ct$. The proof uses the second display of item 1 in the form

$$\frac{\gamma}{2} = \int \frac{m^2}{\sigma^2}\frac{1}{\psi^2} \le \frac{1}{N}\left(\int \frac{fm}{\sigma}\frac{1}{\psi^2}\right)^2 = \text{ with } N \text{ and } f \text{ as before,}$$

which is to say

$$\int f\frac{m}{\sigma}\frac{1}{\psi^2} < -\sqrt{\frac{N\gamma}{2}}$$

in view of $A/N = \int f'\sigma/\psi^2 > 0$. This permits you to estimate

$$\begin{aligned}\int (f + C\frac{m}{\sigma})^2\frac{1}{\psi^2} &\le 1 + 2C\int f\frac{m}{\sigma}\frac{1}{\psi^2} + C^2\int \frac{m^2}{\sigma^2}\frac{1}{\psi^2} \\ &< 1 - 2C\sqrt{\frac{N\gamma}{2}} + \frac{C^2}{2}\gamma \\ &= o(1)\end{aligned}$$

for I increasing to the whole line, by choice of $C = \sqrt{2/\gamma}$, so that

$$f = A\left(\int_0^x \frac{1}{\sigma} - B\right) = -\sqrt{\frac{2}{\gamma}}\frac{m}{\sigma}.$$

with an error which is small in mean-square. It follows easily that, in the limit $I = \mathbb{R}$,

$$\frac{m}{\sigma} = -A\int_0^x \frac{1}{\sigma} + B$$

with new constants $A \geq 0$ and $-\infty < B < \infty$. Now

$$\mathfrak{g} = \frac{1}{2}\sigma D \sigma D + \frac{m}{\sigma}\sigma D = \frac{1}{2}\bar{D}^2 + (-A\bar{x} + B)\bar{D}$$

in the new scale $\bar{x} = \int_0^x 1/\sigma$, which is to say that $\bar{x}(t,x) = \bar{x} o x^{\uparrow}(t,x)$ solves $d\bar{x} = db + (-A\bar{x} + B)dt$, *i.e.* it is a sort of Ornstein-Uhlenbeck process if $A > 0$, or a Brownian motion with drift *if* $A = 0$, up to the first time it comes to $\bar{x}(-\infty) = -\int_{-\infty}^0 1/\sigma$ or $\bar{x}(+\infty) = \int_0^\infty 1/\sigma$, which must be finite if either of $\bar{x}(\pm\infty)$ is finite. But this never happens since $x^{\uparrow}$ never comes to $\pm\infty$, so $\int_{-\infty}^0 1/\sigma = \int\limits_0^\infty 1/\sigma = \infty$, and $\bar{x}$ reduces to standard Ornstein-Uhlenbeck by scaling; in particular, if $\sigma = 1$, $x^{\uparrow}$ itself may be so reduced.

Item 4:

The best interpretation of γ I have found is in terms of Fisher's information $\int (f')^2/f$. Write $\mathfrak{g} = \frac{1}{2}\sigma D \sigma D + mD = \frac{1}{2}\bar{D}^2 + \frac{m}{\sigma}\bar{D}$ in the scale $\bar{x} = \int_0^x 1/\sigma$. The invariant density relative to the new scale is $f(\bar{x}) = (\sigma/\psi^2)(x)$, and from $2m/\sigma = (\sigma/\psi^2)'\psi^2 = f'/f$, you find

$$\int \frac{f'^2}{f}\, d\bar{x} = \int \sigma \left(\frac{\sigma}{\psi^2}\right)'^2 \frac{\psi^2}{\sigma}\frac{dx}{\sigma} = \int \frac{m^2}{\sigma^2}\frac{1}{\psi^2} = 2\gamma$$

Note that

$$\int \frac{f'^2}{f}\, d\bar{x} \int \bar{x}^2 f d\bar{x} \geq \left(\int f'\bar{x} d\bar{x}\right)^2 = \left(\int f d\bar{x}\right)^2 = 1$$

provided $\int \bar{x}^2 f d\bar{x} < \infty$, so

$$\gamma \geq \frac{1}{2}\left(\int \frac{\bar{x}^2}{\psi^2}\right)^{-1} \quad \text{which is} \geq \frac{1}{2}\left(\int \frac{x^2}{\psi^2}\right)^{-1} \quad \text{if } \sigma \geq 1.$$

Item 5:

Fisher's information *does* have a spectral meaning of sorts, as Varadhan suggested to me. Express $\mathfrak{g}$ in the scale $\bar{x}$ as $\frac{1}{2}D^2 + mD$ where, for simplicity, $\bar{x}$ has been replaced by x, plain, and m/σ by m. The invariant density is $f = \exp(\int 2m)/\mathbb{Z}$, and $-2\sqrt{f}\mathfrak{g}/\sqrt{f}$, which is similar to $-2\mathfrak{g}$, turns out to be $-D^2 + v$ with $v = m' + m^2$. The latter has ground state $e = \sqrt{f}$ with $\int e^2 = \int f = 1$, as is obvious from $\mathfrak{g}1 = 0$. Now, for general v, if e is the ground state of $-D^2 + v$ with eigenvalue $\lambda(v)$ and $\int e^2 = 1$, then the (convex) dual $\lambda^*(u)$ of the (convex) function $\lambda(v)$ is the minimum in respect to v of the form $-\int uv + \lambda(v)$. Take $u = f = e^2$. Then from $\text{grad}[-\int uv + \lambda(v)] = -u + e^2$, you see that

$$\lambda^*(f) = -\int e^2 v + \lambda(v) \quad \text{with} \quad v = m' + m^2 \quad \text{as above}$$

$$= -\int e'' e = \int e'^2 = \frac{1}{4}\int f'^2/f.$$

In this way, $\gamma = 2\times$ Fisher's information is related to the ground state eigenvalue of $\mathfrak{g}$, which is amusing, but, for me, γ lies still in some obscurity.

References

1. Baldini, M.: A new perspective on the invariant measure of a positive recurrent diffusion on $\mathbb{R}^d$, to appear 2006.
2. Hasminskii, R. & Nevelson, M.B.: On the stabilization of solutions of one-dimensional stochastic equations. *Sov. Math. Dokl.* **12** (1971) 1492–1496.
3. Kunita, H.: Stochastic Flows and Stochastic Differential Equations. *Camb. Studies Adv. Math.* no. 24, Cambridge U. Press, Cambridge, 1990.

Formation of Singularities in Madelung Fluid: A Nonconventional Application of Itô Calculus to Foundations of Quantum Mechanics

Laura M. Morato

Facoltà di Scienze, Università di Verona, Strada le Grazie, 37134 Verona, Italy,
morato@sci.univr.it

Summary. Stochastic Quantization is a procedure which provides the equation of motion of a Quantum System starting from its classical description and incorporating quantum effects into a stochastic kinematics. After the pioneering work by E.Nelson in 1966 the method has been developed in the eighties in various different ways. In this communication I summarize and systematize the results obtained within an approach based on a Lagrangian variational principle where 3/2 order contributions in Itô calculus are required, leading to a generalization of Madelung fluid equations where velocity fields with vorticity are allowed.

Such a vorticity induces dissipation of the energy so that the irrotational solutions, corresponding to the usual conservative solutions of Schrödinger equation, act as an attracting set. Recent numerical experiments show generation of zeroes of the density with concentration of vorticity and formation of isolated central vortex lines.

1 Introduction

This communication is concerned with an application of Itô calculus to the problem of describing the dynamical evolution of a quantum system once its classical description (which can be given in terms of forces, lagrangian or hamiltonian) is given. We know that, if the classical hamiltonian is given, the canonical quantization rules lead to Schrödinger equation, which beautifully describes the behavior of microscopical systems. But we also know that this procedure seems to fail when applied to microscopical systems interacting with a (macroscopic) measuring apparatus. This fact has been a motivation for investigating other quantization procedures.

In his pioneering work in 1966 E. Nelson proposed a Stochastic Quantization (often called Stochastic Mechanics) where, given the forces acting on the system, quantum effects are incorporated into a stochastic kinematics [18]. This approach was widely developed during the eighties, with the introduction of stochastic variational principles (see for example [2, 15, 19] and references quoted therein). I present here a synthesis of the results obtained within an approach which leads to a dissipative generalization of Schrödinger equation, the usual conservative solutions being in fact dynamical equilibrium states which form an attracting set [10, 13, 14]. The basic tool is Itô calculus where stochastic increments must be estimated to the order $\frac{3}{2}$.

For a quantum particle of mass m, subjected to a force which is the gradient of a scalar potential Φ, Schrödinger equation reads

$$\imath\hbar\partial_t\Psi = \left(-\frac{1}{2m}\hbar^2\nabla^2 + \Phi\right)\Psi \tag{1.1}$$

ψ denoting the quantum mechanical wave function.

By a change of variables Schrödinger equation can be formally written in a fluidodynamical version, the so called Madelung fluid equations.

$$\begin{cases} \partial_t\rho & = -\nabla\cdot(\rho v) \\ \partial_t v & + (v\cdot\nabla)\,v - \frac{\hbar^2}{2m^2}\nabla\left(\frac{\nabla^2\sqrt{\rho}}{\sqrt{\rho}}\right) = -\frac{1}{m}\nabla\Phi \end{cases} \tag{1.2}$$

where

$$\rho = |\psi|^2$$

$$v = \nabla S$$

S being the phase of the wave function ψ.

The equivalence is only formal if the density ρ is not strictly positive at all times.

The velocity field of Madelung fluid is irrotational in all points where the density is different from zero. In many examples solutions of Schrödinger equation which exhibit nodes correspond to solutions of Madelung fluid equations with singular velocity and isolated vortex lines.

In our setting we are led to a dissipative generalization of such equations, which allow velocity fields with a distributed vorticity. It was conjectured that such a vorticity asymptotically can concentrate in the zeroes of the density, describing the formation of the singularities and in particular of isolated vortex lines. The problem is very difficult from the analytical point of view but recent numerical results seem to confirm this conjecture [3]. It is worth stressing that arrays of isolated vortex lines are observed in quantum fluids, as liquid Helium and Bose Einstein condensates (see [1, 9, 11, 12]), but the mechanism underlying their formation is still not well understood. Describing the formation of isolated vortex lines in Madelung fluid, from smooth initial data, could represent a contribution to the solution of this problem.

2 A Stochastic Quantization Procedure

For a quantum particle of mass m in a scalar potential Φ we denote its configuration at time t by $q(t)$. We model the evolution in time of the configuration by a "smooth diffusion", in the following sense:

Definition 1. *A diffusion q is a "smooth diffusion" if*

1) Its drift v_+ is a smooth (i.e. infinitely differentiable) time dependent vector field and its diffusion coefficient is constant (in this setting equal to $\frac{\hbar}{m}$, $\hbar$ denoting Planck's constant divided by 2π)

2) There exists a probability space $(\Omega, \mathcal{F}, P)$ and a standard Brownian Motion W s.t., for $t \in [0, T]$, $T > 0$,

$$q(t) = q(0) + \int_0^t v_+ (q(s), s)\, ds + \left(\frac{\hbar}{m}\right)^{\frac{1}{2}} W(t) \tag{2.1}$$

3) There exists a reversed standard Brownian Motion W^* on $(\Omega, \mathcal{F}, P)$ and v_- s.t., for any $t \in [0, T]$,

$$q(t) = q(0) + \int_0^t v_- (q(s), s)\, ds + \left(\frac{\hbar}{m}\right)^{\frac{1}{2}} (W^*(t) - W^*(0)) \tag{2.2}$$

I recall that a reversed standard Brownian Motion W^* on the finite time interval $[0.T]$ is defined by the equality

$$W^*(t) = \hat{W}(T - t), \quad t \in [0, T] \tag{2.3}$$

$\hat{W}$ still denoting a standard Brownian Motion.

The finite energy condition is sufficient for property 3) (See [5]. An extension to the infinite dimensional case is given in [6]). We also recall that if ρ is the (time dependent) density of a smooth diffusion one has, in particular

$$\frac{v_+ - v_-}{2} = \frac{\hbar}{2m} \nabla \ln \rho \tag{2.4}$$

$$\partial_t \rho = -\nabla \cdot (\rho v) \tag{2.5}$$

were v is the "current velocity", defined as

$$v := \frac{v_+ + v_-}{2} \tag{2.6}$$

For any finite time interval $[t_a, t_b]$ and positive integer N we fix the notations

$$\Delta := \tfrac{t_b - t_a}{N}$$

$$\Delta^+ q(t_i) := q(t_{i+1}) - q(t_i) \text{ future increment}$$

$$\Delta^- q(t_i) := q(t_i) - q(t_{i-1}) \text{ past increment}$$

We now consider the following mean discretized version of the classical action functional

$$A^N_{[t_a,t_b]}[q] := \mathcal{E} \sum_{i=1}^{N} \left[\frac{1}{2} m \frac{\Delta^+ q(t_i) \cdot \Delta^+ q(t_i)}{\Delta^2} - \Phi(q(t_i)) \right] \Delta \tag{2.7}$$

were q is uniquely determined by the triple $[W, v_+, q_o]$ and $\mathcal{E}$ denotes the expectation.

By exploiting the backward representation and estimating $\Delta^+ q(t_i)$ to the order $\Delta^{\frac{3}{2}}$, which gives

$$\begin{aligned} \Delta^+ q(t) = & \left(\frac{\hbar}{m}\right)^{\frac{1}{2}} \Delta^+ W(t) + v_+ \left(q(t), t\right) \Delta \\ & + \left(\frac{\hbar}{m}\right)^{\frac{1}{2}} \sum_{k=1}^{3} \left[\partial_k v_+ \left(q(t), t\right) \int_t^{t+\Delta} \left(W_k(s) - W_k(t)\right) ds \right] \\ & + o(\Delta^{\frac{3}{2}}) \end{aligned} \tag{2.8}$$

we find

$$\begin{aligned} A^N_{[t_a,t_b]}[q] = \mathcal{E} \sum_{i=1}^{N} & \left[\frac{1}{2} m \frac{\Delta^+ q(t_i) \cdot \Delta^- q(t_i)}{\Delta^2} \right. \\ & \left. + \frac{3}{2} \frac{\hbar}{\Delta} + o(\Delta) - \Phi(q(t_i)) \right] \Delta \end{aligned} \tag{2.9}$$

In order to generalize the classical action principle, starting from the above defined functional, two methods have been considered. The former, that will be called Eulerian or Stochastic Control approach, consists in eliminating the divergent term in the discretized action and then take the limit for N going to infinity. After simple manipulations one can see that such a limit can be expressed as a simple functional of the drift field v_+. This allows to exploit stochastic control like techniques [8]. The latter, that will be called Lagrangian or path-wise approach, consists in taking pathwise variations of q for fixed W in $A^N_{[t_a,t_b]}[q]$. This eliminates the divergent term .The limit for N going to infinity is taken only at the end of the calculus of variations (see [10, 13, 14]). This is the approach considered in the following.

Definition 2. *The set of admissible test diffusions for a given W is constituted by the set of all smooth diffusions associated to W according to the previous definition.*

For the test diffusion $q(t)$ at time t let $q'(t) := q(t) + \delta q(t)$ denote the varied diffusion. We require that this is still a smooth diffusion with the same W. Therefore there must exist a smooth drift field v'_+ such that

$$q(t) = q(0) + \int_0^t v_+ (q(s), s)\, ds + \left(\frac{\hbar}{m}\right)^{\frac{1}{2}} W(t) \tag{2.10}$$

$$q'(t) = q(0) + \int_0^t v'_+ (q'(s), s)\, ds + \left(\frac{\hbar}{m}\right)^{\frac{1}{2}} W(t) \tag{2.11}$$

We introduce the variation process h and the variation of the drift f by putting, for $\epsilon > 0$,

$$\begin{cases} \varepsilon h(t) := \delta q(t) \quad \varepsilon > 0 \\ \varepsilon f := v'_+ - v_+ \end{cases} \tag{2.12}$$

Then one finds

$$\dot{h}(t) = \sum_{j=1}^{3} \partial_j v_+ (q(t), t)\, h_j(t) + f\,(q(t), t) \tag{2.13}$$

so that $h(t)$ is a differentiable stochastic process. It satisfies a first order ODE for every realization of q. As a consequence h cannot be fixed both in t_a and t_b. This fact, which has no counterpart in the classical case, comes to be a typical quantum peculiarity.

Definition 3. *A process h will be said "admissible variation" for the test diffusion q if it is solution of* (2.13) *for a smooth f.*

We want now to characterize the motions which are represented by "critical diffusions":

Definition 4. *A smooth diffusion q^* is critical with fixed initial position if, $\forall\, h$ admissible,*

$$\lim_{N\uparrow\infty} \left\{ A^N_{[t_a,t_b]} [q^* + \varepsilon h] - A^N_{[t_a,t_b]} [q^*] - \varepsilon p_{t_b} h_{t_b} \right\} = o(\varepsilon) \tag{2.14}$$

$h(t_a) = 0$ and a smooth diffusion q^* is critical with fixed final position if, $\forall\, h$ admissible,

$$\lim_{N\uparrow\infty} \left\{ A^N_{[t_a,t_b]} [q^* + \varepsilon h] - A^N_{[t_a,t_b]} [q^*] + \varepsilon p_{t_a} h_{t_a} \right\} = o(\varepsilon) \tag{2.15}$$

$h(t_b) = 0$ p_{t_a} and p_{t_b} are fixed random variables playing the role of the classical initial and final "momentum".

We can prove the following.

Theorem 5. *A sufficient condition in order a smooth diffusion q^* to be critical with fixed initial condition is*

$$q^*(t) = q^*(0) + \int_0^t v_+ (q^*(s), s)\, ds + \left(\frac{\hbar}{m}\right)^{\frac{1}{2}} W(t)$$

where:

$$v_+ = v + \frac{\hbar}{2m} \nabla \ln \rho$$

and, if the initial position is fixed,

$$\begin{aligned} &\partial_t \rho = -\nabla \cdot (\rho v) \\ &\partial_t v + (v \cdot \nabla) v - \frac{\hbar^2}{2m^2} \nabla \left(\frac{\nabla^2 \sqrt{\rho}}{\sqrt{\rho}} \right) - \frac{\hbar}{m} (\nabla \ln \rho + \nabla) \wedge (\nabla \wedge v) = -\frac{1}{m} \nabla \Phi \end{aligned} \tag{2.16}$$

with the boundary constraint

$$mv(q_{t_b}, t_b) = p_{t_b}$$

or, if the final position is fixed,

$$\begin{aligned} &\partial_t \rho = -\nabla \cdot (\rho v) \\ &\partial_t v + (v \cdot \nabla) v - \frac{\hbar^2}{2m^2} \nabla \left(\frac{\nabla^2 \sqrt{\rho}}{\sqrt{\rho}} \right) + \frac{\hbar}{m} (\nabla \ln \rho + \nabla) \wedge (\nabla \wedge v) = -\frac{1}{m} \nabla \Phi \end{aligned} \tag{2.17}$$

with the boundary constraint

$$mv(q_{t_a}, t_a) = p_{t_a}$$

Proof. (Outline)

Considering, without loss of generality, the first case, we have

$$\begin{aligned} \delta A^N_{[t_a,t_b]}[q] =& \epsilon \sum_{i=1}^{N} \frac{m}{2} \mathcal{E} \left(\frac{\Delta^+ q(t_i) \cdot \Delta^- h(t_i)}{\Delta^2} + \frac{\Delta^- q(t_i) \cdot \Delta^+ h(t_i)}{\Delta^2} + o(\Delta) \right) \Delta \\ & - \epsilon \sum_{i=1}^{N} \mathcal{E}(\nabla \Phi(q(t_i, t_i) \cdot h(t_i) \Delta - \epsilon \mathcal{E} \left(p_{t_b} \cdot h(t_b) \right) + o(\epsilon) \end{aligned} \tag{2.18}$$

The analysis to the order $\Delta^{\frac{3}{2}}$ of the finite forward and backward increments in the kinetic terms gives

$$\mathcal{E} \left(\frac{\Delta^+ q(t_i) \cdot \Delta^- h(t_i)}{\Delta^2} \right) = \frac{1}{\Delta} \mathcal{E} \left(v^+(q(t), t) \cdot \Delta^- h(t) + o(\Delta) \right) \tag{2.19}$$

and

$$\mathcal{E}\left(\frac{\Delta^- q(t_i)\cdot\Delta^+ h(t_i)}{\Delta^2}\right)=\frac{1}{\Delta}\mathcal{E}\Bigg(v^-(q(t),t)\cdot\Delta^+ h(t)+\left(\frac{\hbar}{m}\right)^{\frac{1}{2}}\Delta^- W_*(t)\cdot\dot{h}+o(\Delta)\Bigg) \tag{2.20}$$

The difference between the two kinetic terms comes from the fact that the variation process h is measurable with respect to the σ algebra generated by the past of q and not by the future, if the initial position is fixed. The proof then exploits a discrete "integration by parts" and the equality

$$\Delta^- W^*(t)=2(\frac{m}{\hbar})^{\frac{1}{2}}\nabla\ln\rho\Delta+\Delta^+ W(t-\Delta)+o(\Delta) \tag{2.21}$$

Going to the limit at the end we get, exploiting (2.4) and (2.6) (see [10] for the details)

$$\lim_{N\to\infty}\delta A^N_{[t_a,t_b]}[q]=\epsilon\,\mathcal{E}\int_{t_a}^{t_b}\Bigg[-\partial_t v-(v\cdot\nabla)v+\frac{\hbar^2}{2m^2}\nabla\left(\frac{\nabla^2\sqrt{\rho}}{\sqrt{\rho}}\right)+\frac{\hbar}{m}(\nabla\ln\rho+\nabla)\wedge(\nabla\wedge v)-\frac{1}{m}\nabla\Phi\Bigg](q(t),t)\cdot h(t)dt+\epsilon\,\mathcal{E}\left[mv(q_{t_b},t_b)-p_{t_b}\right] \tag{2.22}$$

The assertion immediately follows recalling that the continuity equation (2.5) always holds if (ρ, v) are the density and the current velocity field, respectively, of a smooth diffusion.

In the case with final fixed condition the two kinetic terms read

$$\mathcal{E}\left(\frac{\Delta^+ q(t_i)\cdot\Delta^- h(t_i)}{\Delta^2}\right)=\frac{1}{\Delta}\mathcal{E}\left(v^+(q(t),t)\cdot\Delta^- h(t)+\left(\frac{\hbar}{m}\right)^{\frac{1}{2}}\Delta^+ W(t)\cdot\dot{h}+o(\Delta)\right) \tag{2.23}$$

and

$$\mathcal{E}\left(\frac{\Delta^- q(t_i)\cdot\Delta^+ h(t_i)}{\Delta^2}\right)=\frac{1}{\Delta}\,\mathcal{E}\left(v^-(q(t),t)\cdot\Delta^+ h(t)+o(\Delta)\right) \tag{2.24}$$

Then (2.21), with t replaced by $t+\Delta$, is exploited to estimate $\Delta^+ W(t)$. This turns to change the sign in front of the term of first order in $\frac{\hbar}{m}$.

The sufficient conditions as proved in the theorem are also necessary in the following sense:

Corollary 6. *Let q be critical with fixed initial position and let ρ and v be its density and current velocity respectively. Let also $p_t = mv(q(t),t)$ for all $t \in [t_a, t_b]$. Then the equality*

$$\left[\partial_t v + (v \cdot \nabla) v - \frac{\hbar^2}{2m^2} \nabla \left(\frac{\nabla^2 \sqrt{\rho}}{\sqrt{\rho}} \right) \right.$$
$$\left. - \frac{\hbar}{m} (\nabla \ln \rho + \nabla) \wedge (\nabla \wedge v) + \frac{1}{m} \nabla \Phi \right] (q(t),t) = 0 \tag{2.25}$$

holds a.s. for all t in $[t_a, t_b]$. The analogous necessary condition holds in order q to be critical with final fixed position.

Proof. (see [14] p. 1986).

Let q be critical with initial fixed position and let (ρ, v) be its density and current velocity, respectively. Let us also denote by $F(q(t),t) = 0$ the equality (2.25) and put $\delta A_{[t^*,t]}[q] := \lim_{N \to \infty} A^N_{[t_a,t_b]}$. Then if h is the admissible variation of q which solves (2.13) for $f = F$ with $h(t^*) = 0$ we get by (2.13)

$$\delta A_{[t^*,t]}|_{t=t^*} = 0,$$
$$\frac{d}{dt} \delta A_{[t^*,t]}|_{t=t^*} = 0,$$
$$\frac{d^2}{dt^2} \delta A_{[t^*,t]}|_{t=t^*} = \mathcal{E} \left[F^2(q(t^*), t^*) \right].$$

Thus if at time t^* (2.25) does not hold with probability one then q is not critical.

So we find a generalization of Madelung fluid equations, where in particular the velocity field v is not necessarily the gradient of some scalar field.

The two systems of PDE.s (2.16) and (2.17) represent two dynamical evolutions which are one the time reversal of the other.

The second one, with the $+$ sign in front of the term of the first order in $\frac{\hbar}{m}$, turns out to be dissipative.

In fact if (ρ, v) is a smooth solution of (2.17) and ρ has a good behavior at infinity, we have, with $u := \frac{\hbar}{2m} \nabla \ln \rho$ (osmotic velocity) and introducing the energy functional

$$E[\rho, v] = \int_{\mathbb{R}^3} \left(\frac{1}{2} m v^2 + \frac{1}{2} m u^2 + \Phi \right) \rho d^3 x \tag{2.26}$$

the following equality

$$\frac{dE}{dt} = -\frac{\hbar}{2} \int_{\mathbb{R}^3} (\nabla \wedge v)^2 \rho d^3 x \tag{2.27}$$

This **Energy Theorem** was proved in [10] by a purely analytical method, exploiting the equivalence of the new system of dynamical equations with a nonlinear Schrödinger equation of electromagnetic type.

Thus we consider (2.17) as physical equations. They are related to the variational principle with final fixed position, while (2.16) are interpreted as their time reversed picture.

Concluding, (2.17) is a dissipative generalization of Madelung fluid equations and such a dissipation is caused by the vorticity of the velocity field.

Notice that the domain of definition of the two systems of PDEs (2.16) and (2.17) is by construction C^∞ since the admissible test diffusions are smooth diffusions according to definition 1. The global existence for the linear Gaussian solutions of the bidimensional harmonic oscillator was proved in [16]. The general existence and uniqueness problem is still open. If a solution (ρ, v) of (2.16) or (2.17) is irrotational dx-a.s. then the energy is conserved and it solves Madelung equation.

To be more precise, if (ρ, v) satisfy equations of motion and there exists an open set $Q \in \mathbb{R}^3$ s.t.

$$\begin{aligned} (\nabla \wedge v)(x,t) = 0 \quad \forall x \in Q, \quad \forall t \geq 0 \\ \rho(x,t) > 0 \quad \forall x \in Q, \quad \forall t \geq 0 \end{aligned}$$

then $\exists\, S$ s.t.

$$v(x,t) = \frac{1}{m}\nabla S(x,t) \quad \forall x \in Q, \quad \forall t \geq 0$$

Then putting

$$\Psi = \rho^{\frac{1}{2}} e^{\frac{\imath}{\hbar} S}, \quad (\Psi\colon Q \times [0,\infty) \to \mathbb{C})$$

we have

$$\imath\hbar\partial_t \Psi = \left(-\frac{1}{2m}\hbar^2 \Delta + \Phi\right)\Psi$$

These solutions conserve the energy (which turns to be the usual quantum mechanical expectation of the observable energy) and work as an attracting set. The case of Gaussian and linear solution for the bidimensional harmonic oscillator was studied in [16, 17]. In particular it was proved that Schrödinger solutions constitute a center manifold and that the convergence is in the sense of the relative entropy.

We also quote that a version of the Lagrangian variational principle leading to (2.16) and (2.17) with a free parameter multiplying the term of the first order in $\frac{\hbar}{m}$ is proposed in [7].

3 Concentration of Vorticity

As an example we consider a bidimensional symmetric harmonic oscillator. To be more precise we put, denoting by (r,θ) polar coordinates in the (x,y) plane and by z the third spatial coordinate,

$$\Phi := \Phi(r) = \frac{1}{2}r^2\,, \quad r = \sqrt{x^2+y^2}$$

We consider the simultaneous eigenfunctions of the Hamiltonian $\mathcal{H}$ and of the angular momentum L_z with respect to the z axis

$$\chi_{n_d,n_g} = |\chi_{n_d,n_g}| \exp[\imath(n_d - n_g)]\theta, \qquad n_d, n_g = 0, 1, 2, \dots$$

These can be easily computed recursively (see for example [4]). The eigenvalues of the Hamiltonian turn to be

$$E_{n_d,n_g} = 2(n_d + n_g + 1)$$

while those of the angular momentum read

$$\ell_{n_d,n_g} = n_d - n_g$$

The eigenfunctions χ_{n_d,n_g} correspond to the following time invariant solutions of Madelung equations, and of course of our new equations (2.17), on the open set $\mathbb{R}^2 \setminus \{0\}$

$$\rho_{n_d,n_g}(r) = \left|\chi_{n_d,n_g}(r)\right|^2$$
$$\boldsymbol{v}_{n_d,n_g}(r) = \frac{\hbar}{m}\nabla((n_d - n_g)\,\theta) = \frac{\hbar}{m}\frac{n_d - n_g}{r}\hat{\theta}$$

The vorticity of the velocity field $\boldsymbol{v}_{n_d,n_g}$, $n_d, n_g = 0, 1, 2, \dots$ is at every time equal to zero in $\mathbb{R}^2 \setminus \{0\}$ but, if $n_d - n_g$ is different from zero, the circulation around $\{0\}$ is equal to $\frac{h}{m}(n_d - n_g)$.

Indeed in this case a vortex line is present in $\{0\}$ (roughly we have an "infinite vorticity" in $\{0\}$). Notice that for all n_d and n_g, except the case $n_d + n_g = 0$, corresponding to the ground state with bivariate symmetric gaussian density centered in $\{0\}$, $\left|\chi_{n_d,n_g}(r)\right|^2$ exhibits systems of rings of zeroes (see Figure 1).

Let now (ρ_o, v_o) be initial data for (2.17), smooth on the whole plane and with distributed vorticity. We choose, for a_o, A_o and Ω_o positive constants

$$v_o(r) := a_o r\hat{r} - \Omega_o r\hat{\theta}, \qquad \rho_o(r) := \frac{A_o}{2\pi}\exp\left[-\frac{A_o}{2}r^2\right]$$

The results of numerical computations with finite elements method in finite circular domains (for adimensional variables) show formation of rings

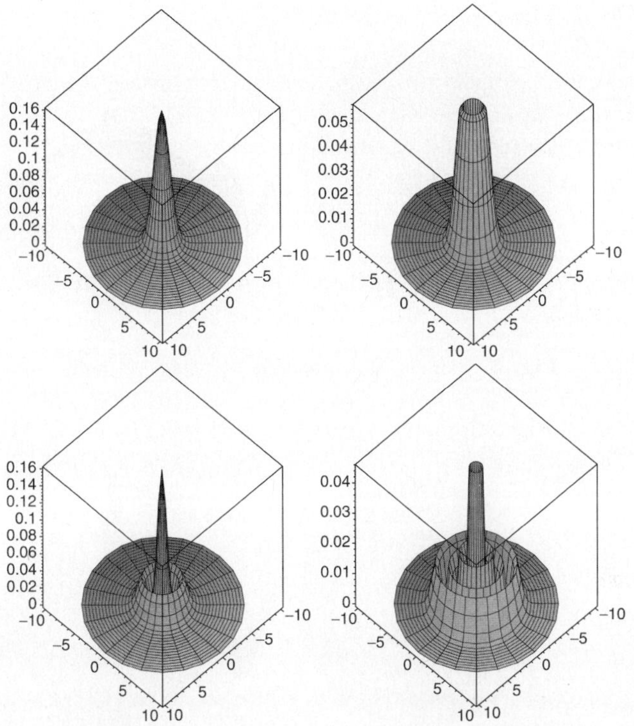

Fig. 1. Squared absolute value of some simultaneous eigenfunctions of energy and momentum operators

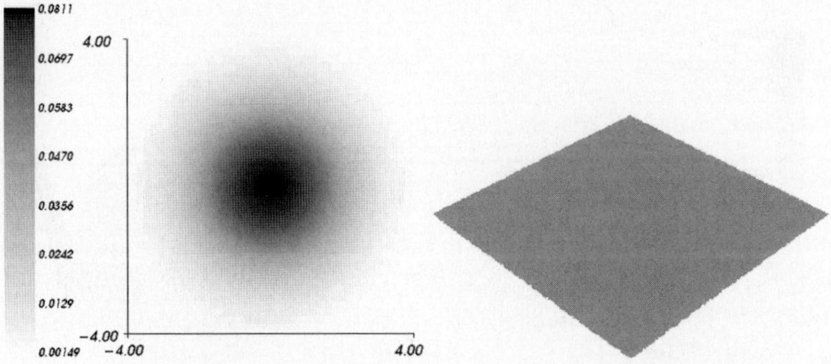

Fig. 2. ρ and $-\nabla \wedge v = 2\Omega_o$ at time $t = 0$, $E = 195$

of zeroes for the density and concentration of vorticity near such zeroes, with the approximation of an isolated vortex line in $\{0\}$ [3].

An example is given in the Figures 2–7.

We can see that vorticity tends to take oscillating relative maxima and minima in correspondence of the zeroes and maxima of the density, respectively.

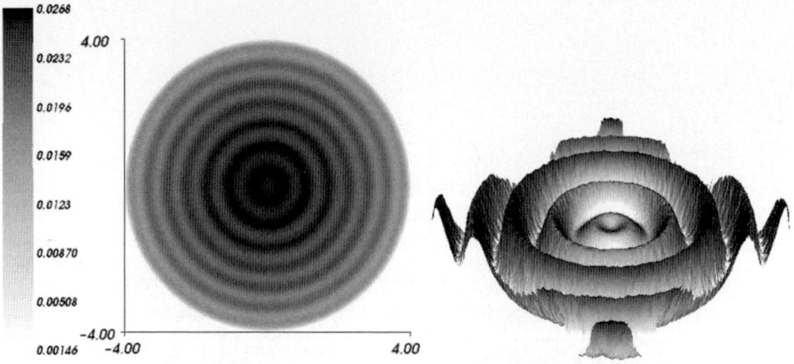

Fig. 3. ρ and $-\nabla \wedge v$ at time $t = 0.08$, $E = 43$

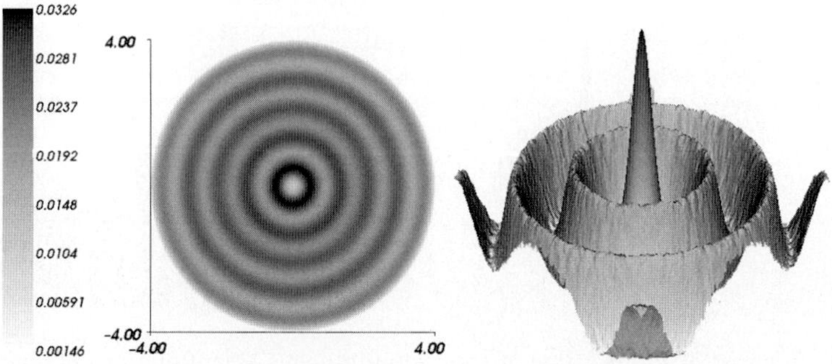

Fig. 4. ρ and $-\nabla \wedge v$ at time $t = 0.14$, $E = 17$

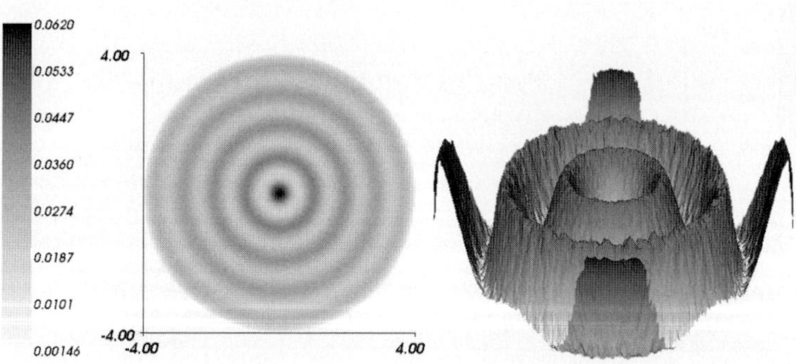

Fig. 5. ρ and $-\nabla \wedge v$ at time $t = 0.16$, $E = 15$

Fig. 6. ρ and $-\nabla \wedge v$ at time $t = 0.19$, $E = 11$

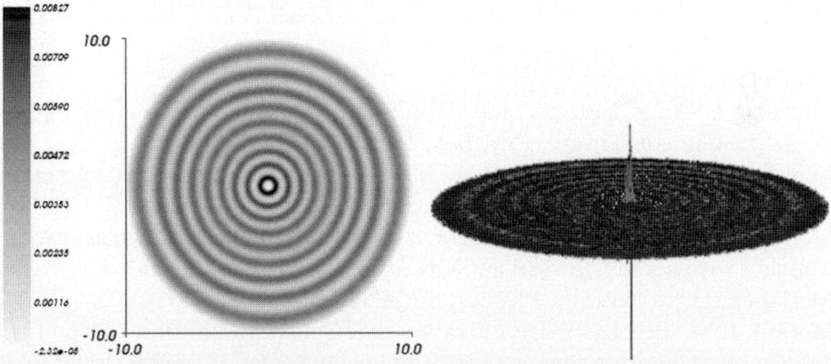

Fig. 7. ρ and $-\nabla \wedge v$ at time $t = 0.23$, $E = 9$

In particular the periodic maximum in the origin increases in time, approaching a central vortex line.

Acknowledgement

The accurate reading of the manuscript by M. Loffredo is gratefully acknowledged.

References

1. Abo-Shaeer J.R., Raman C., Vogels J.M. and Ketterle W.: Observation of vortex lattices in Bose-Einstein Condensates, Science **292**, 476, (2001)
2. Blanchard Ph., Combe Ph. and Zheng W.: Mathematical and physical aspects of Stochastic Mechanics **281**, Lecture notes in Phys. (Berlin) (1987)

3. Caliari M., Inverso G. and Morato L.M.: Dissipation caused by a vorticity field and formation of singularities in Madelung fluid, New Journal of Physics, **6**, no 69, (2005)
4. Cohen-Tannoudji C., Diu B. and Laloe F., Quantum Mechanics, John Wiley & Sons, (1978)
5. Föllmer H.: Time reversal on Wiener space, Stochastic Process in Mathematics and Physics, Lecture Notes in Math. **1158**, Springer New York, 119 (1986)
6. Föllmer H. and Wakolbinger A.: Time reversal of infinite dimensional diffusions, Stoch. Processes and Appl., **22**, 56 (1986)
7. Guerra F.: Stochastic variational principles in quantum mechanics, Ann. Inst. Henry Poincare **49**, 315 (1988)
8. Guerra F. and Morato L.: Quantization of Dynamical Systems and Stochastic Control Theory, Phys. Rev. D, **27**, 1774 (1983)
9. Yarmchuk E.J., Gordon M.J.V. and Packard R.E.: Observation of stationary vortex arrays in rotating superfluid Helium, Phy. Rev. Lett. **43**, 214 (1979)
10. Loffredo M.I. and Morato L.M.: Lagrangian variational principle in Stochastic Mechanics: gauge structure and stability, J. Math. Phys., **30**, 354 (1989)
11. Matthews M.R. et al.: Vortices in Bose-Einstein Condensate, Phy. Rev. Lett. **83**, 2498 (1999)
12. Madison K.W., Chevy F. and Wohllben W.: Vortex Formation in Stirred Bose-Einstein condensate, Phy. Rev. Lett. **84**, 806 (2000)
13. Morato, L.M.: Path–wise calculus of variations with the classical action and quantum systems, Phys. Rev. D **31**, 1982 (1985)
14. Morato, L.M.: Path-wise calculus of variations in Stochastic Mechanics, in "Stochastic Processes in Classical and Quantum Systems", ed. Albeverio, Casati and Merlini, Lecture Notes in Physics, Springer, (1985)
15. Morato, L.M.: Stochastic Quantization and Coherence, in Proc. of the Int. Conference on "Quantum-like models and Coherent effects", Erice, 1994, R. Fedele and P.K. Shuckla eds., World Scientific, 97 (1995)
16. Morato, L.M., Ugolini, S.: Gaussian solutions to the Lagrangian variational problem in Stochastic Mechanics, Annales de l'IHP **60**, 322 (1994)
17. Morato, L.M., Ugolini S.: A connection between quantum dynamics and approximation of Markov diffusions, J. Math. Phys. **9**, 4505 (1994)
18. Nelson, E., Dynamical Theories of Brownian Motion, Princeton University Press, (1966)
19. Nelson E., Quantum Fluctuation, Princeton University Press, (1985)

G-Expectation, G-Brownian Motion and Related Stochastic Calculus of Itô Type

Shige Peng

Institute of Mathematics, Institute of Finance, Shandong University, 250100, Jinan, China, peng@sdu.edu.cn*

Dedicated to Professor Kiyosi Itô for his 90th birthday

Summary. We introduce a notion of nonlinear expectation — G-expectation — generated by a nonlinear heat equation with a given infinitesimal generator G. We first discuss the notion of G-standard normal distribution. With this nonlinear distribution we can introduce our G-expectation under which the canonical process is a G-Brownian motion. We then establish the related stochastic calculus, especially stochastic integrals of Itô's type with respect to our G-Brownian motion and derive the related Itô's formula. We have also given the existence and uniqueness of stochastic differential equation under our G-expectation. As compared with our previous framework of g-expectations, the theory of G-expectation is intrinsic in the sense that it is not based on a given (linear) probability space.

Mathematics Subject Classification (2000): 60H10, 60H05, 60H30, 60J60, 60J65, 60A05, 60E05, 60G05, 60G51, 35K55, 35K15, 49L25

Keywords: g-expectation, G-expectation, G-normal distribution, BSDE, SDE, nonlinear probability theory, nonlinear expectation, Brownian motion, Itô's stochastic calculus, Itô's integral, Itô's formula, Gaussian process, quadratic variation process

1 Introduction

In 1933 Andrei Kolmogorov published his Foundation of Probability Theory (Grundbegriffe der Wahrscheinlichkeitsrechnung) which set out the axiomatic basis for modern probability theory. The whole theory is built on the Measure

* The author thanks the partial support from the Natural Science Foundation of China, grant No. 10131040. He thanks to the anonymous referee's constructive suggestions as well as typoscorrections of Juan Li. Special thanks are to the organizers of the memorable Abel Symposium 2005 for their warm hospitality and excellent work.

Theory created by Émile Borel and Henry Lebesgue and profoundly developed by Radon and Fréchet. The triple $(\Omega, \mathcal{F}, \mathbf{P})$, i.e., a measurable space $(\Omega, \mathcal{F})$ equipped with a probability measure $\mathbf{P}$ becomes a standard notion which appears in most papers of probability and mathematical finance. The second important notion, which is in fact at an equivalent place as the probability measure itself, is the notion of expectation. The expectation $\mathbf{E}[X]$ of a $\mathcal{F}$-measurable random variable X is defined as the integral $\int_\Omega XdP$. A very original idea of Kolmogorov's Grundbegriffe is to use Radon–Nikodym theorem to introduce the conditional probability and the related conditional expectation under a given σ-algebra $\mathcal{G} \subset \mathcal{F}$. It is hard to imagine the present state of arts of probability theory, especially of stochastic processes, e.g., martingale theory, without such notion of conditional expectations. A given time information $(\mathcal{F}_t)_{t\geq 0}$ is so ingeniously and consistently combined with the related conditional expectations $\mathbf{E}[X|\mathcal{F}_t]_{t\geq 0}$. Itô's calculus—Itô's integration, Itô's formula and Itô's equation since 1942 [22], is, I think, the most beautiful discovery on this ground.

A very interesting problem is to develop a nonlinear expectation $\mathbb{E}[\cdot]$ under which we still have such notion of conditional expectation. A notion of g-expectation was introduced by Peng, 1997 (see [33, 34]) in which the conditional expectation $\mathbb{E}^g[X|\mathcal{F}_t]_{t\geq 0}$ is the solution of the backward stochastic differential equation (BSDE), within the classical framework of Itô's calculus, with X as its given terminal condition and with a given real function g as the generator of the BSDE. driven by a Brownian motion defined on a given probability space $(\Omega, \mathcal{F}, \mathbf{P})$. It is completely and perfectly characterized by the function g. The above conditional expectation is characterized by the following well-known condition

$$\mathbb{E}^g[\mathbb{E}^g[X|\mathcal{F}_t]\mathbf{I}_A] = \mathbb{E}^g[X\mathbf{I}_A], \quad \forall A \in \mathcal{F}_t.$$

Since then many results have been obtained in this subject (see, among others, [4–9, 11, 12, 23–25, 35, 39, 40, 42, 44]).

In [38] (see also [37]), we have constructed a kind of filtration-consistent nonlinear expectations through the so-called nonlinear Markov chain. As compared with the framework of g-expectation, the theory of G-expectation is intrinsic, a meaning similar to the "intrinsic geometry". In the sense that it is not based on a classical probability space given a priori.

In this paper, we concentrate ourselves to a concrete case of the above situation and introduce a notion of G-expectation which is generated by a very simple one dimensional fully nonlinear heat equation, called G-heat equation, whose coefficient has only one parameter more than the classical heat equation considered since Bachelier 1900, Einstein 1905 to describe the Brownian motion. But this slight generalization changes the whole things. Firstly, a random variable X with "G-normal distribution" is defined via the heat equation. With this single nonlinear distribution we manage to introduce our G-expectation under which the canonical process is a G-Brownian motion.

We then establish the related stochastic calculus, especially stochastic integrals of Itô's type with respect to our G-Brownian motion. A new type of Itô's formula is obtained. We have also established the existence and uniqueness of stochastic differential equation under our G-stochastic calculus.

In this paper we concentrate ourselves to 1-dimensional G-Brownian motion. But our method of [38] can be applied to multi-dimensional G-normal distribution, G-Brownian motion and the related stochastic calculus. This will be given in [41].

Recently a new type of second order BSDE was proposed to give a probabilistic approach for fully nonlinear 2nd order PDE, see [10]. In finance a type of uncertain volatility model in which the PDE of Black-Scholes type was modified to a fully nonlinear model, see [3, 27, 41].

As indicated in Remark 3, the nonlinear expectations discussed in this paper are equivalent to the notion of coherent risk measures. This with the related conditional expectations $\mathbb{E}[\cdot|\mathcal{F}_t]_{t\geq 0}$ makes a dynamic risk measure: G-risk measure.

This paper is organized as follows: in Section 2, we recall the framework established in [38] and adapt it to our objective. In Section 3 we introduce 1-dimensional standard G-normal distribution and discuss its main properties. In Section 4 we introduce 1-dimensional G-Brownian motion, the corresponding G-expectation and their main properties. We then can establish stochastic integral with respect to our G-Brownian motion of Itô type and the corresponding Itô's formula in Section 5 and the existence and uniqueness theorem of SDE driven by G-Brownian motion in Section 6.

2 Nonlinear Expectation: A General Framework

We briefly recall the notion of nonlinear expectations introduced in [38]. Following Daniell (see Daniell 1918 [14]) in his famous Daniell's integration, we begin with a vector lattice. Let Ω be a given set and let $\mathcal{H}$ be a vector lattice of real functions defined on Ω containing 1, namely, $\mathcal{H}$ is a linear space such that $1 \in \mathcal{H}$ and that $X \in \mathcal{H}$ implies $|X| \in \mathcal{H}$. $\mathcal{H}$ is a space of random variables. We assume the functions on $\mathcal{H}$ are all bounded. Notice that

$$a \wedge b = \min\{a, b\} = \frac{1}{2}(a + b - |a - b|), \quad a \vee b = -[(-a) \wedge (-b)].$$

Thus $X, Y \in \mathcal{H}$ implies that $X \wedge Y$, $X \vee Y$, $X^+ = X \vee 0$ and $X^- = (-X)^+$ are all in $\mathcal{H}$.

Definition 1. ***A nonlinear expectation*** $\mathbb{E}$ *is a functional* $\mathcal{H} \mapsto \mathbb{R}$ *satisfying the following properties*

(a) Monotonicity: *If* $X, Y \in \mathcal{H}$ *and* $X \geq Y$ *then* $\mathbb{E}[X] \geq \mathbb{E}[Y]$.
(b) Preserving of constants: $\mathbb{E}[c] = c$.

In this paper we are interested in a special sublinear expectation:

(c) Sub-additivity (or self-dominated property):

$$\mathbb{E}[X] - \mathbb{E}[Y] \le \mathbb{E}[X - Y], \quad \forall X, Y \in \mathcal{H}.$$

(d) Positive homogeneity: $\mathbb{E}[\lambda X] = \lambda \mathbb{E}[X], \ \forall \lambda \ge 0, \ X \in \mathcal{H}.$

by (b) and (c) we have the well-known

(e) Constant translatability: $\mathbb{E}[X + c] = \mathbb{E}[X] + c.$

Remark 2. The above condition (d) has an equivalent form: $\mathbb{E}[\lambda X] = \lambda^+ \mathbb{E}[X] + \lambda^- \mathbb{E}[-X]$. This form will be very convenient for the conditional expectations studied in this paper (see (vi) of Proposition 16).

Remark 3. We recall the notion of the above expectations satisfying (c)–(e) was systematically introduced by Artzner, Delbaen, Eber and Heath [1, 2], in the case where Ω is a finite set, and by Delbaen [15] in general situation with the notation of risk measure: $\rho(X) = \mathbb{E}[-X]$. See also in Huber [21] for even early study of this notion $\mathbb{E}$ (called upper expectation $\mathbf{E}^*$ in Ch. 10 of [21]) in a finite set Ω. See Rosazza Gianin [44] or Peng [36], El Karoui & Barrieu [16,17] for dynamic risk measures using g-expectations. Super-hedging and super pricing (see [18, 19]) are also closely related to this formulation.

Remark 4. We observe that $\mathcal{H}_0 = \{X \in \mathcal{H}, \mathbb{E}[|X|] = 0\}$ is a linear subspace of $\mathcal{H}$. To take $\mathcal{H}_0$ as our null space, we introduce the quotient space $\mathcal{H}/\mathcal{H}_0$. Observe that, for every $\{X\} \in \mathcal{H}/\mathcal{H}_0$ with a representation $X \in \mathcal{H}$, we can define an expectation $\mathbb{E}[\{X\}] := \mathbb{E}[X]$ which still satisfies (a)–(e) of Definition 1. Following [38], we set $\|X\| := \mathbb{E}[|X|]$, $X \in \mathcal{H}/\mathcal{H}_0$. It is easy to check that $\mathcal{H}/\mathcal{H}_0$ is a normed space under $\|\cdot\|$. We then extend $\mathcal{H}/\mathcal{H}_0$ to its completion $[\mathcal{H}]$ under this norm. $([\mathcal{H}], \|\cdot\|)$ is a Banach space. The nonlinear expectation $\mathbb{E}[\cdot]$ can also be continuously extended from $\mathcal{H}/\mathcal{H}_0$ to $[\mathcal{H}]$, which satisfies (a)–(e).

For any $X \in \mathcal{H}$, the mappings

$$X^+(\omega) : \mathcal{H} \longmapsto \mathcal{H} \quad \text{and} \quad X^-(\omega) : \mathcal{H} \longmapsto \mathcal{H}$$

satisfy

$$|X^+ - Y^+| \le |X - Y| \quad \text{and} \quad |X^- - Y^-| = |(-X)^+ - (-Y)^+| \le |X - Y|.$$

Thus they are both contraction mappings under $\|\cdot\|$ and can be continuously extended to the Banach space $([\mathcal{H}], \|\cdot\|)$.

We define the partial order "$\ge$" in this Banach space.

Definition 5. *An element X in $([\mathcal{H}], \|\cdot\|)$ is said to be nonnegative, or $X \ge 0$, $0 \le X$, if $X = X^+$. We also denote by $X \ge Y$, or $Y \le X$, if $X - Y \ge 0$.*

It is easy to check that $X \ge Y$ and $Y \ge X$ implies $X = Y$ in $([\mathcal{H}], \|\cdot\|)$.

The nonlinear expectation $\mathbb{E}[\cdot]$ can be continuously extended to $([\mathcal{H}], \|\cdot\|)$ on which **(a)–(e)** still hold.

3 G-Normal Distributions

For a given positive integer n, we denote by $lip(\mathbb{R}^n)$ the space of all bounded and Lipschitz real functions on $\mathbb{R}^n$. In this section $\mathbb{R}$ is considered as Ω and $lip(\mathbb{R})$ as $\mathcal{H}$.

In the classical linear situation, a random variable $X(x) = x$ with standard normal distribution, i.e., $X \sim N(0,1)$, can be characterized by

$$E[\phi(X)] = \frac{1}{\sqrt{2\pi}} \int_{-\infty}^{\infty} e^{-\frac{x^2}{2}} \phi(x)dx, \quad \forall \phi \in lip(\mathbb{R}).$$

It is known since Bachelier 1900 and Einstein 1950 that $E[\phi(X)] = u(1,0)$ where $u = u(t,x)$ is the solution of the heat equation

$$\partial_t u = \frac{1}{2} \partial_{xx}^2 u \tag{3.1}$$

with Cauchy condition $u(0,x) = \phi(x)$.

In this paper we set $G(a) = \frac{1}{2}(a^+ - \sigma_0^2 a^-)$, $a \in \mathbb{R}$, where $\sigma_0 \in [0,1]$ is fixed.

Definition 6. *A real valued random variable X with the standard **G-normal distribution** is characterized by its G-expectation defined by*

$$\mathbb{E}[\phi(X)] = P_1^G(\phi) := u(1,0), \quad \phi \in lip(\mathbb{R}) \mapsto \mathbb{R}$$

where $u = u(t,x)$ is a bounded continuous function on $[0,\infty) \times \mathbb{R}$ which is the (unique) viscosity solution of the following nonlinear parabolic partial differential equation (PDE)

$$\partial_t u - G(\partial_{xx}^2 u) = 0, \; u(0,x) = \phi(x). \tag{3.2}$$

In case no confusion is caused, we often call the functional $P_1^G(\cdot)$ the standard G-normal distribution. When $\sigma_0 = 1$, the above PDE becomes the standard heat equation (3.1) and thus this G-distribution is just the classical normal distribution $N(0,1)$:

$$P_1^G(\phi) = P_1(\phi) := \frac{1}{\sqrt{2\pi}} \int_{-\infty}^{\infty} e^{-\frac{x^2}{2}} \phi(x)dx.$$

Remark 7. The function G can be written as $G(a) = \frac{1}{2}\sup_{\sigma_0 \le \sigma \le 1} \sigma^2 a$, thus the nonlinear heat equation (3.2) is a special kind of Hamilton–Jacobi–Bellman equation. The existence and uniqueness of (3.2) in the sense of viscosity solution can be found in, for example, [13,20,26,32,45] for $C^{1,2}$-solution if $\sigma_0 > 0$ (see also in [30] for elliptic cases). Readers who are unfamililar with the notion of viscosity solution of PDE can just consider, in the whole paper, the case $\sigma_0 > 0$, under which the solution u becomes a classical smooth function.

Remark 8. It is known that $u(t,\cdot) \in lip(\mathbb{R})$ (see e.g. [45] Ch. 4, Prop. 3.1. or [32] Lemma 3.1 for the Lipschitz continuity of $u(t,\cdot)$, or Lemma 5.5 and Proposition 5.6 in [37] for a more general conclusion). The boundedness is simply from the comparison theorem (or maximum principle) of this PDE. It is also easy to check that, for a given $\psi \in lip(\mathbb{R}^2)$, $P_1^G(\psi(x,\cdot))$ is still a bounded and Lipschitz function in x.

In general situations we have, from the comparison theorem of PDE,

$$P_1^G(\phi) \geq P_1(\phi), \ \forall \phi \in lip(\mathbb{R}). \tag{3.3}$$

The corresponding normal distribution with mean at $x \in \mathbb{R}$ and variance $t > 0$ is $P_1^G(\phi(x + \sqrt{t} \times \cdot))$. Just like the classical situation, we have

Lemma 9. *For each $\phi \in lip(\mathbb{R})$, the function*

$$u(t,x) = P_1^G(\phi(x + \sqrt{t} \times \cdot)), \quad (t,x) \in [0,\infty) \times \mathbb{R} \tag{3.4}$$

is the solution of the nonlinear heat equation (3.2) with the initial condition $u(0,\cdot) = \phi(\cdot)$.

Proof. Let $u \in C([0,\infty) \times \mathbb{R})$ be the viscosity solution of (3.2) with $u(0,\cdot) = \phi(\cdot) \in lip(\mathbb{R})$. For a fixed $(\bar{t},\bar{x}) \in (0,\infty) \times \mathbb{R}$, we denote $\bar{u}(t,x) = u(t \times \bar{t}, x\sqrt{\bar{t}} + \bar{x})$. Then $\bar{u}$ is the viscosity solution of (3.2) with the initial condition $\bar{u}(0,x) = \phi(x\sqrt{\bar{t}} + \bar{x})$. Indeed, let ψ be a $C^{1,2}$ function on $(0,\infty) \times \mathbb{R}$ such that $\psi \geq \bar{u}$ (resp. $\psi \leq \bar{u}$) and $\psi(\tau,\xi) = \bar{u}(\tau,\xi)$ for a fixed $(\tau,\xi) \in (0,\infty) \times \mathbb{R}$. We have $\psi\left(\frac{t}{\bar{t}}, \frac{x-\bar{x}}{\sqrt{\bar{t}}}\right) \geq u(t,x)$, for all (t,x) and

$$\psi\left(\frac{t}{\bar{t}}, \frac{x-\bar{x}}{\sqrt{\bar{t}}}\right) = u(t,x), \text{ at } (t,x) = \left(\tau\bar{t}, \xi\sqrt{\bar{t}} + \bar{x}\right).$$

Since u is the viscosity solution of (3.2), at the point $(t,x) = (\tau\bar{t}, \xi\sqrt{\bar{t}} + \bar{x})$, we have

$$\frac{\partial\psi\left(\frac{t}{\bar{t}}, \frac{x-\bar{x}}{\sqrt{\bar{t}}}\right)}{\partial t} - G\left(\frac{\partial^2\psi\left(\frac{t}{\bar{t}}, \frac{x-\bar{x}}{\sqrt{\bar{t}}}\right)}{\partial x^2}\right) \leq 0 \text{ (resp. } \geq 0).$$

But since G is positive homogenous, i.e., $G(\lambda a) = \lambda G(a)$, we thus derive

$$\left(\frac{\partial\psi(t,x)}{\partial t} - G\left(\frac{\partial^2\psi(t,x)}{\partial x^2}\right)\right)\Big|_{(t,x)=(\tau,\xi)} \leq 0 \text{ (resp. } \geq 0).$$

This implies that $\bar{u}$ is the viscosity subsolution (resp. supersolution) of (3.2). According to the definition of $P^G(\cdot)$ we obtain (3.4).

Definition 10. *We denote*

$$P_t^G(\phi)(x) = P_1^G(\phi(x + \sqrt{t} \times \cdot)) = u(t,x), \quad (t,x) \in [0,\infty) \times \mathbb{R}. \tag{3.5}$$

From the above lemma, for each $\phi \in lip(\mathbb{R})$, we have the following Kolmogorov–Chapman chain rule:

$$P_t^G(P_s^G(\phi))(x) = P_{t+s}^G(\phi)(x), \quad s, t \in [0, \infty), \ x \in \mathbb{R}. \tag{3.6}$$

Such type of nonlinear semigroup was studied in Nisio 1976 [28, 29].

Proposition 11. *For each $t > 0$, the G-normal distribution P_t^G is a nonlinear expectation on $\mathcal{H} = lip(\mathbb{R})$, with $\Omega = \mathbb{R}$, satisfying (a)–(e) of Definition 1. The corresponding completion space $[\mathcal{H}] = [lip(\mathbb{R})]_t$ under the norm $\|\phi\|_t := P_t^G(|\phi|)(0)$ contains $\phi(x) = x^n$, $n = 1, 2, \cdots$, as well as $x^n\psi$, $\psi \in lip(\mathbb{R})$ as its special elements. Relation (3.5) still holds. We also have the following properties:*

(1) Central symmetric: $P_t^G(\phi(\cdot)) = P_t^G(\phi(-\cdot))$;
(2) For each convex $\phi \in [lip(\mathbb{R})]$ we have

$$P_t^G(\phi)(0) = \frac{1}{\sqrt{2\pi t}} \int_{-\infty}^{\infty} \phi(x) \exp\left(-\frac{x^2}{2t}\right) dx;$$

For each concave ϕ, we have, for $\sigma_0 > 0$,

$$P_t^G(\phi)(0) = \frac{1}{\sqrt{2\pi t}\sigma_0} \int_{-\infty}^{\infty} \phi(x) \exp\left(-\frac{x^2}{2t\sigma_0^2}\right) dx,$$

and $P_t^G(\phi)(0) = \phi(0)$ for $\sigma_0 = 0$. In particular, we have

$$P_t^G((x)_{x\in\mathbb{R}}) = 0, \quad P_t^G\left((x^{2n+1})_{x\in\mathbb{R}}\right) = P_t^G\left((-x^{2n+1})_{x\in\mathbb{R}}\right), \ n = 1, 2, \cdots,$$
$$P_t^G\left((x^2)_{x\in\mathbb{R}}\right) = t, \quad P_t^G\left((-x^2)_{x\in\mathbb{R}}\right) = -\sigma_0^2 t.$$

Remark 12. Corresponding to the above four expressions, a random variable X with the G-normal distribution P_t^G satisfies

$$\mathbb{E}[X] = 0, \quad \mathbb{E}[X^{2n+1}] = \mathbb{E}[-X^{2n+1}],$$
$$\mathbb{E}[X^2] = t, \quad \mathbb{E}[-X^2] = -\sigma_0^2 t.$$

See the next section for a detail study.

4 1-Dimensional G-Brownian Motion under G-Expectation

In the rest of this paper, we denote by $\Omega = C_0(\mathbb{R}^+)$ the space of all $\mathbb{R}$-valued continuous paths $(\omega_t)_{t\in\mathbb{R}^+}$ with $\omega_0 = 0$, equipped with the distance

$$\rho(\omega^1, \omega^2) := \sum_{i=1}^{\infty} 2^{-i} \left[\left(\max_{t\in[0,i]} |\omega_t^1 - \omega_t^2|\right) \wedge 1\right].$$

We set, for each $t \in [0, \infty)$,

$$\begin{aligned}
\mathbf{W}_t &:= \{\omega_{\cdot\wedge t} : \omega \in \mathbf{\Omega}\}, \\
\mathcal{F}_t &:= \mathcal{B}_t(\mathbf{W}) = \mathcal{B}(\mathbf{W}_t), \\
\mathcal{F}_{t+} &:= \mathcal{B}_{t+}(\mathbf{W}) = \bigcap_{s>t} \mathcal{B}_s(\mathbf{W}), \\
\mathcal{F} &:= \bigvee_{s>t} \mathcal{F}_s.
\end{aligned}$$

$(\mathbf{\Omega}, \mathcal{F})$ is the canonical space equipped with the natural filtration and $\omega = (\omega_t)_{t\geq 0}$ is the corresponding canonical process.

For each fixed $T \geq 0$, we consider the following space of random variables:

$$L^0_{ip}(\mathcal{F}_T) := \{X(\omega) = \phi(\omega_{t_1}, \cdots, \omega_{t_m}), \forall m \geq 1, \\ t_1, \cdots, t_m \in [0, T], \forall \phi \in lip(\mathbb{R}^m)\}.$$

It is clear that $L^0_{ip}(\mathcal{F}_t) \subseteq L^0_{ip}(\mathcal{F}_T)$, for $t \leq T$. We also denote

$$L^0_{ip}(\mathcal{F}) := \bigcup_{n=1}^{\infty} L^0_{ip}(\mathcal{F}_n).$$

Remark 13. It is clear that $lip(\mathbb{R}^m)$ and then $L^0_{ip}(\mathcal{F}_T)$ and $L^0_{ip}(\mathcal{F})$ are vector lattices. Moreover, since $\phi, \psi \in lip(\mathbb{R}^m)$ implies $\phi \cdot \psi \in lip(\mathbb{R}^m)$ thus X, $Y \in L^0_{ip}(\mathcal{F}_T)$ implies $X \cdot Y \in L^0_{ip}(\mathcal{F}_T)$.

We will consider the canonical space and set $B_t(\omega) = \omega_t$, $t \in [0, \infty)$, for $\omega \in \Omega$.

Definition 14. *The canonical process B is called a G-**Brownian** motion under a nonlinear expectation $\mathbb{E}$ defined on $L^0_{ip}(\mathcal{F})$ if for each $T > 0$, $m = 1, 2, \cdots$, and for each $\phi \in lip(\mathbb{R}^m)$, $0 \leq t_1 < \cdots < t_m \leq T$, we have*

$$\mathbb{E}[\phi(B_{t_1}, B_{t_2} - B_{t_1}, \cdots, B_{t_m} - B_{t_{m-1}})] = \phi_m,$$

where $\phi_m \in \mathbb{R}$ is obtained via the following procedure:

$$\begin{aligned}
\phi_1(x_1, \cdots, x_{m-1}) &= P^G_{t_m - t_{m-1}}(\phi(x_1, \cdots, x_{m-1}, \cdot)); \\
\phi_2(x_1, \cdots, x_{m-2}) &= P^G_{t_{m-1} - t_{m-2}}(\phi_1(x_1, \cdots, x_{m-2}, \cdot)); \\
&\vdots \\
\phi_{m-1}(x_1) &= P^G_{t_2 - t_1}(\phi_{m-2}(x_1, \cdot)); \\
\phi_m &= P^G_{t_1}(\phi_{m-1}(\cdot)).
\end{aligned}$$

The related conditional expectation of $X = \phi(B_{t_1}, B_{t_2} - B_{t_1}, \cdots, B_{t_m} - B_{t_{m-1}})$ under $\mathcal{F}_{t_j}$ is defined by

$$\begin{aligned}
\mathbb{E}[X|\mathcal{F}_{t_j}] &= \mathbb{E}[\phi(B_{t_1}, B_{t_2} - B_{t_1}, \cdots, B_{t_m} - B_{t_{m-1}})|\mathcal{F}_{t_j}] \qquad (4.1) \\
&= \phi_{m-j}(B_{t_1}, \cdots, B_{t_j} - B_{t_{j-1}}).
\end{aligned}$$

It is proved in [38] that $\mathbb{E}[\cdot]$ consistently defines a nonlinear expectation on the vector lattice $L^0_{ip}(\mathcal{F}_T)$ as well as on $L^0_{ip}(\mathcal{F})$ satisfying (a)–(e) in Definition 1. It follows that $\mathbb{E}[|X|]$, $X \in L^0_{ip}(\mathcal{F}_T)$ (resp. $L^0_{ip}(\mathcal{F})$) forms a norm and that $L^0_{ip}(\mathcal{F}_T)$ (resp. $L^0_{ip}(\mathcal{F})$) can be continuously extended to a Banach space, denoted by $L^1_G(\mathcal{F}_T)$ (resp. $L^1_G(\mathcal{F})$). For each $0 \le t \le T < \infty$, we have $L^1_G(\mathcal{F}_t) \subseteq L^1_G(\mathcal{F}_T) \subset L^1_G(\mathcal{F})$. It is easy to check that, in $L^1_G(\mathcal{F}_T)$ (resp. $L^1_G(\mathcal{F})$), $\mathbb{E}[\cdot]$ still satisfies (a)–(e) in Definition 1.

Definition 15. *The expectation $\mathbb{E}[\cdot] : L^1_G(\mathcal{F}) \mapsto \mathbb{R}$ introduced through above procedure is called **G-expectation**. The corresponding canonical process B is called a G-Brownian motion under $\mathbb{E}[\cdot]$.*

For a given $p > 1$, we also denote $L^p_G(\mathcal{F}) = \{X \in L^1_G(\mathcal{F}) | X|^p \in L^1_G(\mathcal{F})\}$. $L^p_G(\mathcal{F})$ is also a Banach space under the norm $\|X\|_p := (\mathbb{E}[|X|^p])^{1/p}$. We have (see Appendix)

$$\|X+Y\|_p \le \|X\|_p + \|Y\|_p$$

and, for each $X \in L^p_G$, $Y \in L^q_G(Q)$ with $\frac{1}{p} + \frac{1}{q} = 1$,

$$\|XY\| = \mathbb{E}[|XY|] \le \|X\|_p \|X\|_q \, .$$

With this we have $\|X\|_p \le \|X\|_{p'}$ if $p \le p'$.

We now consider the conditional expectation introduced in (4.1). For each fixed $t = t_j \le T$, the conditional expectation $\mathbb{E}[\cdot|\mathcal{F}_t] : L^0_{ip}(\mathcal{F}_T) \mapsto L^0_{ip}(\mathcal{F}_t)$ is a continuous mapping under $\|\cdot\|$ since $\mathbb{E}[\mathbb{E}[X|\mathcal{F}_t]] = \mathbb{E}[X]$, $X \in L^0_{ip}(\mathcal{F}_T)$ and

$$\mathbb{E}[\mathbb{E}[X|\mathcal{F}_t] - \mathbb{E}[Y|\mathcal{F}_t]] \le \mathbb{E}[X - Y],$$
$$\|\mathbb{E}[X|\mathcal{F}_t] - \mathbb{E}[Y|\mathcal{F}_t]\| \le \|X - Y\| \, .$$

It follows that $\mathbb{E}[\cdot|\mathcal{F}_t]$ can be also extended as a continuous mapping $L^1_G(\mathcal{F}_T) \mapsto L^1_G(\mathcal{F}_t)$. If the above T is not fixed, then we can obtain $\mathbb{E}[\cdot|\mathcal{F}_t] : L^1_G(\mathcal{F}) \mapsto L^1_G(\mathcal{F}_t)$.

Proposition 16. *We list the properties of $\mathbb{E}[\cdot|\mathcal{F}_t]$ that hold in $L^0_{ip}(\mathcal{F}_T)$ and still hold for X, $Y \in L^1_G(\mathcal{F})$:*

(i) $\mathbb{E}[X|\mathcal{F}_t] = X$, *for* $X \in L^1_G(\mathcal{F}_t)$, $t \le T$.
(ii) *If* $X \ge Y$, *then* $\mathbb{E}[X|\mathcal{F}_t] \ge \mathbb{E}[Y|\mathcal{F}_t]$.
(iii) $\mathbb{E}[X|\mathcal{F}_t] - \mathbb{E}[Y|\mathcal{F}_t] \le \mathbb{E}[X - Y|\mathcal{F}_t]$.
(iv) $\mathbb{E}[\mathbb{E}[X|\mathcal{F}_t]|\mathcal{F}_s] = \mathbb{E}[X|\mathcal{F}_{t\wedge s}]$, $\mathbb{E}[\mathbb{E}[X|\mathcal{F}_t]] = \mathbb{E}[X]$.
(v) $\mathbb{E}[X + \eta|\mathcal{F}_t] = \mathbb{E}[X|\mathcal{F}_t] + \eta$, $\eta \in L^1_G(\mathcal{F}_t)$.
(vi) $\mathbb{E}[\eta X|\mathcal{F}_t] = \eta^+\mathbb{E}[X|\mathcal{F}_t] + \eta^-\mathbb{E}[-X|\mathcal{F}_t]$, *for each bounded* $\eta \in L^1_G(\mathcal{F}_t)$.
(vii) *For each* $X \in L^1_G(\mathcal{F}^t_T)$, $\mathbb{E}[X|\mathcal{F}_t] = \mathbb{E}[X]$,

where $L^1_G(\mathcal{F}^t_T)$ is the extension, under $\|\cdot\|$, of $L^0_{ip}(\mathcal{F}^t_T)$ which consists of random variables of the form $\phi(B_{t_1} - B_{t_1}, B_{t_2} - B_{t_1}, \cdots, B_{t_m} - B_{t_{m-1}})$, $m = 1, 2, \cdots$, $\phi \in lip(\mathbb{R}^m)$, $t_1, \cdots, t_m \in [t, T]$. Condition (vi) is the positive homogeneity, see Remark 2.

Definition 17. *An $X \in L_G^1(\mathcal{F})$ is said to be independent of $\mathcal{F}_t$ under the G-expectation $\mathbb{E}$ for some given $t \in [0,\infty)$, if for each real function Φ suitably defined on $\mathbb{R}$ such that $\Phi(X) \in L_G^1(\mathcal{F})$ we have*

$$\mathbb{E}[\Phi(X)|\mathcal{F}_t] = \mathbb{E}[\Phi(X)].$$

Remark 18. It is clear that all elements in $L_G^1(\mathcal{F})$ are independent of $\mathcal{F}_0$. Just like the classical situation, the increments of G-Brownian motion $(B_{t+s} - B_s)_{t\geq 0}$ is independent of $\mathcal{F}_s$. In fact it is a new G-Brownian motion since, just like the classical situation, the increments of B are identically distributed.

Example 19. For each $n = 0,1,2,\cdots$, $0 \leq s - t$, we have $\mathbb{E}[B_t - B_s|\mathcal{F}_s] = 0$ and, for $n = 1,2,\cdots$,

$$\mathbb{E}[|B_t - B_s|^n|\mathcal{F}_s] = \mathbb{E}[|B_{t-s}|^{2n}] = \frac{1}{\sqrt{2\pi(t-s)}}\int_{-\infty}^{\infty}|x|^n \exp\left(-\frac{x^2}{2(t-s)}\right)dx.$$

But we have

$$\mathbb{E}[-|B_t - B_s|^n|\mathcal{F}_s] = \mathbb{E}[-|B_{t-s}|^n] = -\sigma_0^n\mathbb{E}[|B_{t-s}|^n].$$

Exactly as in classical cases, we have

$$\begin{aligned}
&\mathbb{E}[(B_t - B_s)^2|\mathcal{F}_s] = t - s, \quad \mathbb{E}[(B_t - B_s)^4|\mathcal{F}_s] = 3(t-s)^2,\\
&\mathbb{E}[(B_t - B_s)^6|\mathcal{F}_s] = 15(t-s)^3, \quad \mathbb{E}[(B_t - B_s)^8|\mathcal{F}_s] = 105(t-s)^4,\\
&\mathbb{E}[|B_t - B_s||\mathcal{F}_s] = \frac{\sqrt{2(t-s)}}{\sqrt{\pi}}, \quad \mathbb{E}[|B_t - B_s|^3|\mathcal{F}_s] = \frac{2\sqrt{2}(t-s)^{3/2}}{\sqrt{\pi}},\\
&\mathbb{E}[|B_t - B_s|^5|\mathcal{F}_s] = 8\frac{\sqrt{2}(t-s)^{5/2}}{\sqrt{\pi}}.
\end{aligned}$$

Example 20. For each $n = 1,2,\cdots$, $0 \leq s \leq t < T$ and $X \in L_G^1(\mathcal{F}_s)$, since $\mathbb{E}\left[B_{T-t}^{2n-1}\right] = \mathbb{E}\left[-B_{T-t}^{2n-1}\right]$, we have, by (vi) of Proposition 16,

$$\begin{aligned}
\mathbb{E}[X(B_T - B_t)^{2n-1}] &= \mathbb{E}[X^+\mathbb{E}[(B_T - B_t)^{2n-1}|\mathcal{F}_t]\\
&\quad + X^-\mathbb{E}[-(B_T - B_t)^{2n-1}|\mathcal{F}_t]]\\
&= \mathbb{E}[|X|]\cdot\mathbb{E}\left[B_{T-t}^{2n-1}\right],\\
\mathbb{E}[X(B_T - B_t)|\mathcal{F}_s] &= \mathbb{E}[-X(B_T - B_t)|\mathcal{F}_s] = 0.
\end{aligned}$$

We also have

$$\mathbb{E}[X(B_T - B_t)^2|\mathcal{F}_t] = X^+(T-t) - \sigma_0^2X^-(T-t).$$

Remark 21. It is clear that we can define an expectation $E[\cdot]$ on $L_{ip}^0(\mathcal{F})$ in the same way as in Definition 14 with the standard normal distribution $P_1(\cdot)$ in the place of $P_1^G(\cdot)$. Since $P_1(\cdot)$ is dominated by $P_1^G(\cdot)$ in the sense

$P_1(\phi)-P_1(\psi) \leq P_1^G(\phi-\psi)$, then $E[\cdot]$ can be continuously extended to $L_G^1(\mathcal{F})$. $E[\cdot]$ is a linear expectation under which $(B_t)_{t\geq 0}$ behaves as a Brownian motion. We have

$$E[X] \leq \mathbb{E}[X], \quad \forall X \in L_G^1(\mathcal{F}). \tag{4.2}$$

In particular, $\mathbb{E}\left[B_{T-t}^{2n-1}\right] = \mathbb{E}\left[-B_{T-t}^{2n-1}\right] \geq E\left[-B_{T-t}^{2n-1}\right] = 0$. Such kind of extension under a domination relation was discussed in details in [38].

The following property is very useful

Proposition 22. *Let $X, Y \in L_G^1(\mathcal{F})$ be such that $\mathbb{E}[Y] = -\mathbb{E}[-Y]$ (thus $\mathbb{E}[Y] = E[Y]$), then we have*

$$\mathbb{E}[X+Y] = \mathbb{E}[X] + \mathbb{E}[Y].$$

In particular, if $\mathbb{E}[Y] = \mathbb{E}[-Y] = 0$, then $\mathbb{E}[X+Y] = \mathbb{E}[X]$.

Proof. It is simply because we have $\mathbb{E}[X+Y] \leq \mathbb{E}[X] + \mathbb{E}[Y]$ and

$$\mathbb{E}[X+Y] \geq \mathbb{E}[X] - \mathbb{E}[-Y] = \mathbb{E}[X] + \mathbb{E}[Y].$$

Example 23. We have

$$\begin{aligned}
\mathbb{E}[B_t^2 - B_s^2|\mathcal{F}_s] &= \mathbb{E}[(B_t - B_s + B_s)^2 - B_s^2|\mathcal{F}_s] \\
&= E[(B_t - B_s)^2 + 2(B_t - B_s)B_s|\mathcal{F}_s] \\
&= t - s,
\end{aligned}$$

since $2(B_t - B_s)B_s$ satisfies the condition for Y in Proposition 22, and

$$\begin{aligned}
\mathbb{E}[(B_t^2 - B_s^2)^2|\mathcal{F}_s] &= \mathbb{E}[\{(B_t - B_s + B_s)^2 - B_s^2\}^2|\mathcal{F}_s] \\
&= \mathbb{E}[\{(B_t - B_s)^2 + 2(B_t - B_s)B_s\}^2|\mathcal{F}_s] \\
&= \mathbb{E}[(B_t - B_s)^4 + 4(B_t - B_s)^3 B_s + 4(B_t - B_s)^2 B_s^2|\mathcal{F}_s] \\
&\leq \mathbb{E}[(B_t - B_s)^4] + 4\mathbb{E}[|B_t - B_s|^3]|B_s| + 4(t-s)B_s^2 \\
&= 3(t-s)^2 + 8(t-s)^{3/2}|B_s| + 4(t-s)B_s^2.
\end{aligned}$$

5 Itô's Integral of G-Brownian Motion

5.1 Bochner's Integral

Definition 24. *For $T \in \mathbb{R}_+$, a partition π_T of $[0, T]$ is a finite ordered subset $\pi = \{t_1, \cdots, t_N\}$ such that $0 = t_0 < t_1 < \cdots < t_N = T$. We denote*

$$\mu(\pi_T) = \max\{|t_{i+1} - t_i|, i = 0, 1, \cdots, N-1\}.$$

We use $\pi_T^N = \{t_0^N < t_1^N < \cdots < t_N^N\}$ to denote a sequence of partitions of $[0, T]$ such that $\lim_{N\to\infty} \mu(\pi_T^N) = 0$.

Let $p \geq 1$ be fixed. We consider the following type of simple processes: for a given partition $\{t_0, \cdots, t_N\} = \pi_T$ of $[0, T]$, we set

$$\eta_t(\omega) = \sum_{j=0}^{N-1} \xi_j(\omega)\mathbf{I}_{[t_j, t_{j+1})}(t),$$

where $\xi_i \in L_G^p(\mathcal{F}_{t_i})$, $i = 0, 1, 2, \cdots, N-1$, are given. The collection of these type of processes is denoted by $M_G^{p,0}(0, T)$.

Definition 25. *For an $\eta \in M_G^{1,0}(0, T)$ with $\eta_t = \sum_{j=0}^{N-1} \xi_j(\omega)\mathbf{I}_{[t_j, t_{j+1})}(t)$, the related Bochner integral is*

$$\int_0^T \eta_t(\omega)dt = \sum_{j=0}^{N-1} \xi_j(\omega)(t_{j+1} - t_j).$$

Remark 26. We set, for each $\eta \in M_G^{1,0}(0, T)$,

$$\tilde{\mathbb{E}}_T[\eta] := \frac{1}{T}\int_0^T \mathbb{E}[\eta_t]dt = \frac{1}{T}\sum_{j=0}^{N-1} \mathbb{E}[\xi_j(\omega)](t_{j+1} - t_j).$$

It is easy to check that $\tilde{\mathbb{E}}_T : M_G^{1,0}(0, T) \longmapsto \mathbb{R}$ forms a nonlinear expectation satisfying (a)–(e) of Definition 1. By Remark 4, we can introduce a natural norm $\|\eta\|_T^1 = \tilde{\mathbb{E}}_T[|\eta|] = \frac{1}{T}\int_0^T \mathbb{E}[|\eta_t|]dt$. Under this norm $M_G^{1,0}(0, T)$ can be continuously extended to $M_G^1(0, T)$ which is a Banach space.

Definition 27. *For each $p \geq 1$, we will denote by $M_G^p(0, T)$ the completion of $M_G^{p,0}(0, T)$ under the norm*

$$\left(\frac{1}{T}\int_0^T \|\eta_t^p\| \, dt\right)^{1/p} = \left(\frac{1}{T}\sum_{j=0}^{N-1} \mathbb{E}[|\xi_j(\omega)|^p](t_{j+1} - t_j)\right)^{1/p}.$$

We observe that,

$$\mathbb{E}\left[\left|\int_0^T \eta_t(\omega)dt\right|\right] \leq \sum_{j=0}^{N-1} \|\xi_j(\omega)\| \, (t_{j+1} - t_j) = \int_0^T \mathbb{E}[|\eta_t|]dt.$$

We then have

Proposition 28. *The linear mapping $\int_0^T \eta_t(\omega)dt : M_G^{1,0}(0, T) \mapsto L_G^1(\mathcal{F}_T)$ is continuous and thus can be continuously extended to $M_G^1(0, T) \mapsto L_G^1(\mathcal{F}_T)$. We still denote this extended mapping by $\int_0^T \eta_t(\omega)dt$, $\eta \in M_G^1(0, T)$. We have*

$$\mathbb{E}\left[\left|\int_0^T \eta_t(\omega)dt\right|\right] \leq \int_0^T \mathbb{E}[|\eta_t|]dt, \quad \forall \eta \in M_G^1(0, T). \tag{5.1}$$

Since $M_G^1(0, T) \supset M_G^p(0, T)$, for $p \geq 1$, this definition holds for $\eta \in M_G^p(0, T)$.

5.2 Itô's Integral of G-Brownian Motion

Definition 29. *For each $\eta \in M_G^{2,0}(0,T)$ with the form*

$$\eta_t(\omega) = \sum_{j=0}^{N-1} \xi_j(\omega)\mathbf{I}_{[t_j,t_{j+1})}(t),$$

we define

$$I(\eta) = \int_0^T \eta(s)dB_s := \sum_{j=0}^{N-1} \xi_j(B_{t_{j+1}} - B_{t_j}).$$

Lemma 30. *The mapping $I : M_G^{2,0}(0,T) \longmapsto L_G^2(\mathcal{F}_T)$ is a linear continuous mapping and thus can be continuously extended to $I : M_G^2(0,T) \longmapsto L_G^2(\mathcal{F}_T)$. In fact we have*

$$\mathbb{E}\left[\int_0^T \eta(s)dB_s\right] = 0, \tag{5.2}$$

$$\mathbb{E}\left[\left(\int_0^T \eta(s)dB_s\right)^2\right] \le \int_0^T \mathbb{E}[(\eta(t))^2]dt. \tag{5.3}$$

Definition 31. *We define, for a fixed $\eta \in M_G^2(0,T)$, the stochastic integral*

$$\int_0^T \eta(s)dB_s := I(\eta).$$

It is clear that (5.2), (5.3) still hold for $\eta \in M_G^2(0,T)$.

Proof of Lemma 30. From Example 20, for each j,

$$\mathbb{E}[\xi_j(B_{t_{j+1}} - B_{t_j})|\mathcal{F}_{t_j}] = 0.$$

We have

$$\begin{aligned}
\mathbb{E}\left[\int_0^T \eta(s)dB_s\right] &= \mathbb{E}\left[\int_0^{t_{N-1}} \eta(s)dB_s + \xi_{N-1}(B_{t_N} - B_{t_{N-1}})\right] \\
&= \mathbb{E}\left[\int_0^{t_{N-1}} \eta(s)dB_s + \mathbb{E}[\xi_{N-1}(B_{t_N} - B_{t_{N-1}})|\mathcal{F}_{t_{N-1}}]\right] \\
&= \mathbb{E}\left[\int_0^{t_{N-1}} \eta(s)dB_s\right].
\end{aligned}$$

We then can repeat this procedure to obtain (5.2). We now prove (5.3):

$$\begin{aligned}
\mathbb{E}\left[\left(\int_0^T \eta(s)dB_s\right)^2\right] &= \mathbb{E}\left[\left(\int_0^{t_{N-1}} \eta(s)dB_s + \xi_{N-1}(B_{t_N} - B_{t_{N-1}})\right)^2\right] \\
&= \mathbb{E}\left[\left(\int_0^{t_{N-1}} \eta(s)dB_s\right)^2\right. \\
&\quad + \mathbb{E}\left[2\left(\int_0^{t_{N-1}} \eta(s)dB_s\right)\xi_{N-1}(B_{t_N} - B_{t_{N-1}})\right. \\
&\quad \left.\left. + \xi_{N-1}^2(B_{t_N} - B_{t_{N-1}})^2 | \mathcal{F}_{t_{N-1}}\right]\right] \\
&= \mathbb{E}\left[\left(\int_0^{t_{N-1}} \eta(s)dB_s\right)^2 + \xi_{N-1}^2(t_N - t_{N-1})\right].
\end{aligned}$$

Thus $\mathbb{E}[(\int_0^{t_N} \eta(s)dB_s)^2] \leq \mathbb{E}\left[\left(\int_0^{t_{N-1}} \eta(s)dB_s\right)^2\right] + \mathbb{E}[\xi_{N-1}^2](t_N - t_{N-1})]$. We then repeat this procedure to deduce

$$\mathbb{E}\left[\left(\int_0^T \eta(s)dB_s\right)^2\right] \leq \sum_{j=0}^{N-1} \mathbb{E}[(\xi_j)^2](t_{j+1} - t_j) = \int_0^T \mathbb{E}[(\eta(t))^2]dt.$$

We list some main properties of the Itô's integral of G-Brownian motion. We denote for some $0 \leq s \leq t \leq T$,

$$\int_s^t \eta_u dB_u := \int_0^T \mathbf{I}_{[s,t]}(u)\eta_u dB_u.$$

We have

Proposition 32. *Let $\eta, \theta \in M_G^2(0,T)$ and let $0 \leq s \leq r \leq t \leq T$. Then in $L_G^1(\mathcal{F}_T)$ we have*

(i) $\int_s^t \eta_u dB_u = \int_s^r \eta_u dB_u + \int_r^t \eta_u dB_u$,
(ii) $\int_s^t (\alpha\eta_u + \theta_u)dB_u = \alpha \int_s^t \eta_u dB_u + \int_s^t \theta_u dB_u$, *if α is bounded and in $L_G^1(\mathcal{F}_s)$,*
(iii) $\mathbb{E}[X + \int_r^T \eta_u dB_u | \mathcal{F}_s] = \mathbb{E}[X]$, $\forall X \in L_G^1(\mathcal{F})$.

5.3 Quadratic Variation Process of G-Brownian Motion

We now study a very interesting process of the G-Brownian motion. Let π_t^N, $N = 1, 2, \cdots$, be a sequence of partitions of $[0, t]$. We consider

$$\begin{aligned}
B_t^2 &= \sum_{j=0}^{N-1}\left[B_{t_{j+1}^N}^2 - B_{t_j^N}^2\right] \\
&= \sum_{j=0}^{N-1} 2B_{t_j^N}\left(B_{t_{j+1}^N} - B_{t_j^N}\right) + \sum_{j=0}^{N-1}\left(B_{t_{j+1}^N} - B_{t_j^N}\right)^2.
\end{aligned}$$

As $\mu(\pi_t^N) \to 0$, the first term of the right side tends to $\int_0^t B_s dB_s$. The second term must converge. We denote its limit by $\langle B \rangle_t$, i.e.,

$$\langle B \rangle_t = \lim_{\mu(\pi_t^N)\to 0} \sum_{j=0}^{N-1} \left(B_{t_{j+1}^N} - B_{t_j^N}\right)^2 = B_t^2 - 2\int_0^t B_s dB_s. \tag{5.4}$$

By the above construction, $\langle B \rangle_t$, $t \geq 0$, is an increasing process with $\langle B \rangle_0 = 0$. We call it the **quadratic variation process** of the G-Brownian motion B. Clearly $\langle B \rangle$ is an increasing process. It perfectly characterizes the part of uncertainty, or ambiguity, of G-Brownian motion. It is important to keep in mind that $\langle B \rangle_t$ is not a deterministic process unless the case $\sigma = 1$, i.e., when B is a classical Brownian motion. In fact we have

Lemma 33. *We have, for each* $0 \leq s \leq t < \infty$

$$\mathbb{E}[\langle B \rangle_t - \langle B \rangle_s | \mathcal{F}_s] = t - s, \tag{5.5}$$

$$\mathbb{E}[-(\langle B \rangle_t - \langle B \rangle_s) | \mathcal{F}_s] = -\sigma_0^2(t - s). \tag{5.6}$$

Proof. By the definition of $\langle B \rangle$ and Proposition 32-(iii),

$$\begin{aligned}\mathbb{E}[\langle B \rangle_t - \langle B \rangle_s | \mathcal{F}_s] &= \mathbb{E}[B_t^2 - B_s^2 - 2\int_s^t B_u dB_u | \mathcal{F}_s] \\ &= \mathbb{E}[B_t^2 - B_s^2 | \mathcal{F}_s] = t - s.\end{aligned}$$

The last step can be check as in Example 23. We then have (5.5). (5.6) can be proved analogously with the consideration of $\mathbb{E}[-(B_t^2 - B_s^2)|\mathcal{F}_s] = -\sigma^2(t-s)$.

To define the integration of a process $\eta \in M_G^1(0,T)$ with respect to $d\langle B \rangle$, we first define a mapping:

$$Q_{0,T}(\eta) = \int_0^T \eta(s) d\langle B \rangle_s := \sum_{j=0}^{N-1} \xi_j \left(\langle B \rangle_{t_{j+1}} - \langle B \rangle_{t_j}\right) : M_G^{1,0}(0,T) \mapsto L^1(\mathcal{F}_T).$$

Lemma 34. *For each* $\eta \in M_G^{1,0}(0,T)$,

$$\mathbb{E}[|Q_{0,T}(\eta)|] \leq \int_0^T \mathbb{E}[|\eta_s|] ds. \tag{5.7}$$

Thus $Q_{0,T} : M_G^{1,0}(0,T) \mapsto L^1(\mathcal{F}_T)$ *is a continuous linear mapping. Consequently,* $Q_{0,T}$ *can be uniquely extended to* $L_{\mathcal{F}}^1(0,T)$. *We still denote this mapping by*

$$\int_0^T \eta(s) d\langle B \rangle_s = Q_{0,T}(\eta), \quad \eta \in M_G^1(0,T).$$

We still have

$$\mathbb{E}\left[\left|\int_0^T \eta(s) d\langle B \rangle_s\right|\right] \leq \int_0^T \mathbb{E}[|\eta_s|] ds, \quad \forall \eta \in M_G^1(0,T). \tag{5.8}$$

Proof. By applying Lemma 33, (5.7) can be checked as follows:

$$\begin{aligned}\mathbb{E}\left[\left|\sum_{j=0}^{N-1}\xi_j\left(\langle B\rangle_{t_{j+1}}-\langle B\rangle_{t_j}\right)\right|\right] &\leq \sum_{j=0}^{N-1}\mathbb{E}\left[|\xi_j|\cdot\mathbb{E}\left[\langle B\rangle_{t_{j+1}}-\langle B\rangle_{t_j}|\mathcal{F}_{t_j}\right]\right]\\ &=\sum_{j=0}^{N-1}\mathbb{E}[|\xi_j|](t_{j+1}-t_j)\\ &=\int_0^T \mathbb{E}[|\eta_s|]ds.\end{aligned}$$

A very interesting point of the quadratic variation process $\langle B\rangle$ is, just like the G-Brownian motion B it's self, the increment $\langle B\rangle_{t+s}-\langle B\rangle_s$ is independent of $\mathcal{F}_s$ and identically distributed like $\langle B\rangle_t$. In fact we have

Lemma 35. *For each fixed $s\geq 0$, $(\langle B\rangle_{s+t}-\langle B\rangle_s)_{t\geq 0}$ is independent of $\mathcal{F}_s$. It is the quadratic variation process of the Brownian motion $B_t^s=B_{s+t}-B_s$, $t\geq 0$, i.e., $\langle B\rangle_{s+t}-\langle B\rangle_s=\langle B^s\rangle_t$. We have*

$$\mathbb{E}[\langle B^s\rangle_t^2|\mathcal{F}_s]=\mathbb{E}[\langle B\rangle_t^2]=t^2 \tag{5.9}$$

as well as

$$\mathbb{E}[\langle B^s\rangle_t^3|\mathcal{F}_s]=\mathbb{E}[\langle B\rangle_t^2]=t^3,\quad \mathbb{E}[\langle B^s\rangle_t^4|\mathcal{F}_s]=\mathbb{E}[\langle B\rangle_t^4]=t^4.$$

Proof. The independence is simply from

$$\begin{aligned}\langle B\rangle_{s+t}-\langle B\rangle_s &= B_{t+s}^2-2\int_0^{s+t}B_rdB_r-\left[B_s^2-2\int_0^s B_rdB_r\right]\\ &=(B_{t+s}-B_s)^2-2\int_s^{s+t}(B_r-B_s)d(B_r-B_s)\\ &=\langle B^s\rangle_t.\end{aligned}$$

We set $\phi(t):=\mathbb{E}[\langle B\rangle_t^2]$.

$$\begin{aligned}\phi(t) &= \mathbb{E}\left[\left\{(B_t)^2-2\int_0^t B_udB_u\right\}^2\right]\\ &\leq 2\mathbb{E}[(B_t)^4]+8\mathbb{E}\left[\left(\int_0^t B_udB_u\right)^2\right]\\ &\leq 6t^2+8\int_0^t\mathbb{E}[(B_u)^2]du\\ &=10t^2.\end{aligned}$$

This also implies $\mathbb{E}[(\langle B\rangle_{t+s} - \langle B\rangle_s)^2] = \phi(t) \leq 14t$. Thus

$$\begin{aligned}\phi(t) &= \mathbb{E}[\{\langle B\rangle_s + \langle B\rangle_{s+t} - \langle B\rangle_s\}^2] \\ &\leq \mathbb{E}[(\langle B\rangle_s)^2] + \mathbb{E}[(\langle B^s\rangle_t)^2] + 2\mathbb{E}[\langle B\rangle_s \langle B^s\rangle_t] \\ &= \phi(s) + \phi(t) + 2\mathbb{E}[\langle B\rangle_s \mathbb{E}[\langle B^s\rangle_t]] \\ &= \phi(s) + \phi(t) + 2st.\end{aligned}$$

We set $\delta_N = t/N$, $t_k^N = kt/N = k\delta_N$ for a positive integer N. By the above inequalities

$$\begin{aligned}\phi\left(t_N^N\right) &\leq \phi\left(t_{N-1}^N\right) + \phi(\delta_N) + 2t_{N-1}^N\delta_N \\ &\leq \phi\left(t_{N-2}^N\right) + 2\phi(\delta_N) + 2\left(t_{N-1}^N + t_{N-2}^N\right)\delta_N \\ &\vdots\end{aligned}$$

We then have

$$\phi(t) \leq N\phi(\delta_N) + 2\sum_{k=0}^{N-1} t_k^N\delta_N \leq 10\frac{t^2}{N} + 2\sum_{k=0}^{N-1} t_k^N\delta_N.$$

Let $N \to \infty$ we have $\phi(t) \leq 2\int_0^t sds = t^2$. Thus $\mathbb{E}[\langle B_t\rangle^2] \leq t^2$. This with $\mathbb{E}[\langle B_t\rangle^2] \geq E[\langle B_t\rangle^2] = t^2$ implies (5.9).

Proposition 36. *Let $0 \leq s \leq t$, $\xi \in L_G^1(\mathcal{F}_s)$. Then*

$$\begin{aligned}\mathbb{E}\left[X + \xi\left(B_t^2 - B_s^2\right)\right] &= \mathbb{E}[X + \xi(B_t - B_s)^2] \\ &= \mathbb{E}[X + \xi(\langle B\rangle_t - \langle B\rangle_s)].\end{aligned}$$

Proof. By (3.4) and Proposition 22, we have

$$\begin{aligned}\mathbb{E}\left[X + \xi\left(B_t^2 - B_s^2\right)\right] &= \mathbb{E}\left[X + \xi\left(\langle B\rangle_t - \langle B\rangle_s + 2\int_s^t B_u dB_u\right)\right] \\ &= \mathbb{E}[X + \xi(\langle B\rangle_t - \langle B\rangle_s)].\end{aligned}$$

We also have

$$\begin{aligned}\mathbb{E}[X + \xi(B_t^2 - B_s^2)] &= \mathbb{E}[X + \xi\{(B_t - B_s)^2 + 2(B_t - B_s)B_s\}] \\ &= \mathbb{E}[X + \xi(B_t - B_s)^2].\end{aligned}$$

We have the following isometry:

Proposition 37. *Let $\eta \in M_G^2(0,T)$. We have*

$$\mathbb{E}\left[\left(\int_0^T \eta(s)dB_s\right)^2\right] = \mathbb{E}\left[\int_0^T \eta^2(s)d\langle B\rangle_s\right]. \tag{5.10}$$

Proof. We first consider $\eta \in M_G^{2,0}(0,T)$ with the form

$$\eta_t(\omega) = \sum_{j=0}^{N-1} \xi_j(\omega)\mathbf{I}_{[t_j,t_{j+1})}(t)$$

and thus $\int_0^T \eta(s)dB_s := \sum_{j=0}^{N-1} \xi_j(B_{t_{j+1}} - B_{t_j})$. By Proposition 22 we have

$$\mathbb{E}[X + 2\xi_j(B_{t_{j+1}} - B_{t_j})\xi_i(B_{t_{i+1}} - B_{t_i})] = \mathbb{E}[X], \text{ for } X \in L_G^1(\mathcal{F}),\ i \neq j.$$

Thus

$$\mathbb{E}\left[\left(\int_0^T \eta(s)dB_s\right)^2\right] = \mathbb{E}\left[\left(\sum_{j=0}^{N-1} \xi_j(B_{t_{j+1}} - B_{t_j})\right)^2\right]$$
$$= \mathbb{E}\left[\sum_{j=0}^{N-1} \xi_j^2(B_{t_{j+1}} - B_{t_j})^2\right].$$

This with Proposition 36, it follows that

$$\mathbb{E}\left[\left(\int_0^T \eta(s)dB_s\right)^2\right] = \mathbb{E}\left[\sum_{j=0}^{N-1} \xi_j^2\left(\langle B\rangle_{t_{j+1}} - \langle B\rangle_{t_j}\right)\right] = \mathbb{E}\left[\int_0^T \eta^2(s)d\langle B\rangle_s\right].$$

Thus (5.10) holds for $\eta \in M_G^{2,0}(0,T)$. We thus can continuously extend the above equality to the case $\eta \in M_G^2(0,T)$ and prove (5.10).

5.4 Itô's Formula for G-Brownian Motion

We have the corresponding Itô's formula of $\Phi(X_t)$ for a "G-Itô process" X. For simplification, we only treat the case where the function Φ is sufficiently regular. We first consider a simple situation.

Lemma 38. *Let $\Phi \in C^2(\mathbb{R}^n)$ be bounded with bounded derivatives and $\{\partial^2_{x^\mu x^\nu}\Phi\}_{\mu,\nu=1}^n$ are uniformly Lipschitz. Let $s \in [0,T]$ be fixed and let $X = (X^1, \cdots, X^n)^T$ be an n-dimensional process on $[s,T]$ of the form*

$$X_t^\nu = X_s^\nu + \alpha^\nu(t-s) + \eta^\nu(\langle B\rangle_t - \langle B\rangle_s) + \beta^\nu(B_t - B_s),$$

where, for $\nu = 1, \cdots, n$, α^ν, η^ν and β^ν, are bounded elements of $L_G^2(\mathcal{F}_s)$ and $X_s = (X_s^1, \cdots, X_s^n)^T$ is a given $\mathbb{R}^n$-vector in $L_G^2(\mathcal{F}_s)$. Then we have

$$\Phi(X_t) - \Phi(X_s) = \int_s^t \partial_{x^\nu}\Phi(X_u)\beta^\nu dB_u + \int_s^t \partial_{x_\nu}\Phi(X_u)\alpha^\nu du \tag{5.11}$$
$$+ \int_s^t \left[D_{x^\nu}\Phi(X_u)\eta^\nu + \frac{1}{2}\partial^2_{x^\mu x^\nu}\Phi(X_u)\beta^\mu\beta^\nu\right] d\langle B\rangle_u.$$

Here we use the Einstein convention, i.e., each single term with repeated indices μ and/or ν implies the summation.

Proof. For each positive integer N we set $\delta = (t-s)/N$ and take the partition

$$\pi_{[s,t]}^N = \{t_0^N, t_1^N, \cdots, t_N^N\} = \{s, s+\delta, \cdots, s+N\delta = t\}.$$

We have

$$\begin{aligned}\Phi(X_t) &= \Phi(X_s) + \sum_{k=0}^{N-1} \left[\Phi\left(X_{t_{k+1}^N}\right) - \Phi\left(X_{t_k^N}\right)\right] \\ &= \Phi(X_s) + \sum_{k=0}^{N-1} \left[\partial_{x^\mu}\Phi\left(X_{t_k^N}\right)\left(X^\mu_{t_{k+1}^N} - X^\mu_{t_k^N}\right)\right. \\ &\quad \left. + \frac{1}{2}\left[\partial^2_{x^\mu x^\nu}\Phi\left(X_{t_k^N}\right)\left(X^\mu_{t_{k+1}^N} - X^\mu_{t_k^N}\right)\left(X^\nu_{t_{k+1}^N} - X^\nu_{t_k^N}\right) + \eta_k^N\right]\right] \end{aligned} \quad (5.12)$$

where

$$\eta_k^N = \left[\partial^2_{x^\mu x^\nu}\Phi\left(X_{t_k^N} + \theta_k\left(X_{t_{k+1}^N} - X_{t_k^N}\right)\right) - \partial^2_{x^\mu x^\nu}\Phi\left(X_{t_k^N}\right)\right] \left(X^\mu_{t_{k+1}^N} - X^\mu_{t_k^N}\right)\left(X^\nu_{t_{k+1}^N} - X^\nu_{t_k^N}\right)$$

with $\theta_k \in [0,1]$. We have

$$\begin{aligned}\mathbb{E}\left[|\eta_k^N|\right] &= \mathbb{E}\left[\left|\left[\partial^2_{x^\mu x^\nu}\Phi\left(X_{t_k^N} + \theta_k\left(X_{t_{k+1}^N} - X_{t_k^N}\right)\right)\right.\right.\right. \\ &\quad \left.\left.\left. -\partial^2_{x^\mu x^\nu}\Phi\left(X_{t_k^N}\right)\right]\left(X^\mu_{t_{k+1}^N} - X^\mu_{t_k^N}\right)\left(X^\nu_{t_{k+1}^N} - X^\nu_{t_k^N}\right)\right|\right] \\ &\le c\mathbb{E}\left[\left|X_{t_{k+1}^N} - X_{t_k^N}\right|^3\right] \le C\left[\delta^3 + \delta^{3/2}\right],\end{aligned}$$

where c is the Lipschitz constant of $\{\partial^2_{x^\mu x^\nu}\Phi\}_{\mu,\nu=1}^n$. Thus $\sum_k \mathbb{E}\left[|\eta_k^N|\right] \to 0$. The rest terms in the summation of the right side of (5.12) are $\xi_t^N + \zeta_t^N$, with

$$\begin{aligned}\xi_t^N = \sum_{k=0}^{N-1} &\left\{\partial_{x^\mu}\Phi\left(X_{t_k^N}\right)\left[\alpha^\mu\left(t_{k+1}^N - t_k^N\right)\right.\right. \\ &\left. + \eta^\mu\left(\langle B\rangle_{t_{k+1}^N} - \langle B\rangle_{t_k^N}\right) + \beta^\mu\left(B_{t_{k+1}^N} - B_{t_k^N}\right)\right] \\ &\left. + \frac{1}{2}\partial^2_{x^\mu x^\nu}\Phi\left(X_{t_k^N}\right)\beta^\mu\beta^\nu\left(B_{t_{k+1}^N} - B_{t_k^N}\right)\left(B_{t_{k+1}^N} - B_{t_k^N}\right)\right\}\end{aligned}$$

and

$$\zeta_t^N = \frac{1}{2}\sum_{k=0}^{N-1} \partial^2_{x^\mu x^\nu}\Phi\left(X_{t_k^N}\right)\left[\alpha^\mu\left(t_{k+1}^N - t_k^N\right) + \eta^\mu\left(\langle B\rangle_{t_{k+1}^N} - \langle B\rangle_{t_k^N}\right)\right]$$

$$\times\left[\alpha^\nu\left(t_{k+1}^N - t_k^N\right) + \eta^\nu\left(\langle B\rangle_{t_{k+1}^N} - \langle B\rangle_{t_k^N}\right)\right]$$

$$+\beta^\nu\left[\alpha^\mu\left(t_{k+1}^N - t_k^N\right) + \eta^\mu\left(\langle B\rangle_{t_{k+1}^N} - \langle B\rangle_{t_k^N}\right)\right]\left(B_{t_{k+1}^N} - B_{t_k^N}\right).$$

We observe that, for each $u \in [t_k^N, t_{k+1}^N)$,

$$\mathbb{E}\left[\left|\partial_{x^\mu}\Phi(X_u) - \sum_{k=0}^{N-1}\partial_{x^\mu}\Phi\left(X_{t_k^N}\right)\mathbf{I}_{\left[t_k^N, t_{k+1}^N\right)}(u)\right|^2\right]$$

$$= \mathbb{E}\left[\left|\partial_{x^\mu}\Phi(X_u) - \partial_{x^\mu}\Phi\left(X_{t_k^N}\right)\right|^2\right]$$

$$\leq c^2\mathbb{E}\left[\left|X_u - X_{t_k^N}\right|^2\right] \leq C[\delta + \delta^2].$$

Thus $\sum_{k=0}^{N-1}\partial_{x^\mu}\Phi\left(X_{t_k^N}\right)\mathbf{I}_{\left[t_k^N, t_{k+1}^N\right)}(\cdot)$ tends to $\partial_{x^\mu}\Phi(X_\cdot)$ in $M_G^2(0,T)$. Similarly,

$$\sum_{k=0}^{N-1}\partial^2_{x^\mu x^\nu}\Phi\left(X_{t_k^N}\right)\mathbf{I}_{\left[t_k^N, t_{k+1}^N\right)}(\cdot) \to \partial^2_{x^\mu x^\nu}\Phi(X_\cdot), \text{ in } M_G^2(0,T).$$

Let $N \to \infty$, by the definitions of the integrations with respect to dt, dB_t and $d\langle B\rangle_t$ the limit of ξ_t^N in $L_G^2(\mathcal{F}_t)$ is just the right hand of (5.11). By the estimates of the next remark, we also have $\zeta_t^N \to 0$ in $L_G^1(\mathcal{F}_t)$. We then have proved (5.11).

Remark 39. We have the following estimates: for $\psi^N \in M_G^{1,0}(0,T)$ such that $\psi_t^N = \sum_{k=0}^{N-1}\xi_{t_k}^N\mathbf{I}_{\left[t_k^N, t_{k+1}^N\right)}(t)$, and $\pi_T^N = \{0 \leq t_0, \cdots, t_N = T\}$ with $\lim_{N\to\infty}\mu(\pi_T^N) = 0$ and $\sum_{k=0}^{N-1}\mathbb{E}\left[|\xi_{t_k}^N|\right]\left(t_{k+1}^N - t_k^N\right) \leq C$, for all $N = 1, 2, \ldots$, we have

$$\mathbb{E}\left[\left|\sum_{k=0}^{N-1}\xi_k^N\left(t_{k+1}^N - t_k^N\right)^2\right|\right] \to 0,$$

and, thanks to Lemma 35,

$$\mathbb{E}\left[\left|\sum_{k=0}^{N-1}\xi_k^N\left(\langle B\rangle_{t_{k+1}^N} - \langle B\rangle_{t_k^N}\right)^2\right|\right] \leq \sum_{k=0}^{N-1}\mathbb{E}\left[|\xi_k^N|\cdot\mathbb{E}\left[\left(\langle B\rangle_{t_{k+1}^N} - \langle B\rangle_{t_k^N}\right)^2|\mathcal{F}_{t_k^N}\right]\right]$$

$$= \sum_{k=0}^{N-1}\mathbb{E}\left[|\xi_k^N|\right]\left(t_{k+1}^N - t_k^N\right)^2 \to 0,$$

as well as

$$\begin{aligned}
&\mathbb{E}\left[\left|\sum_{k=0}^{N-1} \xi_k^N \left(\langle B\rangle_{t_{k+1}^N} - \langle B\rangle_{t_k^N}\right)\left(B_{t_{k+1}^N} - B_{t_k^N}\right)\right|\right] \\
&\leq \sum_{k=0}^{N-1} \mathbb{E}\left[\left|\xi_k^N\right|\right] \mathbb{E}\left[\left(\langle B\rangle_{t_{k+1}^N} - \langle B\rangle_{t_k^N}\right)\left|B_{t_{k+1}^N} - B_{t_k^N}\right|\right] \\
&\leq \sum_{k=0}^{N-1} \mathbb{E}\left[\left|\xi_k^N\right|\right] \mathbb{E}\left[\left(\langle B\rangle_{t_{k+1}^N} - \langle B\rangle_{t_k^N}\right)^2\right]^{1/2} \mathbb{E}\left[\left|B_{t_{k+1}^N} - B_{t_k^N}\right|^2\right]^{1/2} \\
&= \sum_{k=0}^{N-1} \mathbb{E}\left[\left|\xi_k^N\right|\right] \left(t_{k+1}^N - t_k^N\right)^{3/2} \to 0.
\end{aligned}$$

We also have

$$\begin{aligned}
&\mathbb{E}\left[\left|\sum_{k=0}^{N-1} \xi_k^N \left(\langle B\rangle_{t_{k+1}^N} - \langle B\rangle_{t_k^N}\right)\left(t_{k+1}^N - t_k^N\right)\right|\right] \\
&\leq \sum_{k=0}^{N-1} \mathbb{E}\left[\left|\xi_k^N\right| \left(t_{k+1}^N - t_k^N\right) \cdot \mathbb{E}\left[\left(\langle B\rangle_{t_{k+1}^N} - \langle B\rangle_{t_k^N}\right)\middle|\mathcal{F}_{t_k^N}\right]\right] \\
&= \sum_{k=0}^{N-1} \mathbb{E}\left[\left|\xi_k^N\right|\right] \left(t_{k+1}^N - t_k^N\right)^2 \to 0
\end{aligned}$$

and

$$\begin{aligned}
&\mathbb{E}\left[\left|\sum_{k=0}^{N-1} \xi_k^N \left(t_{k+1}^N - t_k^N\right)\left(B_{t_{k+1}^N} - B_{t_k^N}\right)\right|\right] \\
&\leq \sum_{k=0}^{N-1} \mathbb{E}\left[\left|\xi_k^N\right|\right] \left(t_{k+1}^N - t_k^N\right) \mathbb{E}\left[\left|B_{t_{k+1}^N} - B_{t_k^N}\right|\right] \\
&= \sqrt{\frac{2}{\pi}} \sum_{k=0}^{N-1} \mathbb{E}\left[\left|\xi_k^N\right|\right] \left(t_{k+1}^N - t_k^N\right)^{3/2} \to 0.
\end{aligned}$$

We now consider a more general form of Itô's formula. Consider

$$X_t^\nu = X_0^\nu + \int_0^t \alpha_s^\nu ds + \int_0^t \eta_s^\nu d\langle B\rangle_s + \int_0^t \beta_s^\nu dB_s.$$

Proposition 40. *Let α^ν, β^ν and η^ν, $\nu = 1, \cdots, n$, are bounded processes of $M_G^2(0,T)$. Then for each $t \geq 0$ and in $L_G^2(\mathcal{F}_t)$ we have*

$$\Phi(X_t) - \Phi(X_s) = \int_s^t \partial_{x^\nu}\Phi(X_u)\beta_u^\nu dB_u + \int_s^t \partial_{x_\nu}\Phi(X_u)\alpha_u^\nu du \tag{5.13}$$
$$+ \int_s^t \left[\partial_{x^\nu}\Phi(X_u)\eta_u^\nu + \frac{1}{2}\partial^2_{x^\mu x^\nu}\Phi(X_u)\beta_u^\mu\beta_u^\nu\right] d\langle B\rangle_u$$

Proof. We first consider the case where α, η and β are step processes of the form

$$\eta_t(\omega) = \sum_{k=0}^{N-1} \xi_k(\omega)\mathbf{I}_{[t_k,t_{k+1})}(t).$$

From the above Lemma, it is clear that (5.13) holds true. Now let

$$X_t^{\nu,N} = X_0^\nu + \int_0^t \alpha_s^{\nu,N} ds + \int_0^t \eta_s^{\nu,N} d\langle B\rangle_s + \int_0^t \beta_s^{\nu,N} dB_s$$

where α^N, η^N and β^N are uniformly bounded step processes that converge to α, η and β in $M_G^2(0,T)$ as $N \to \infty$. From Lemma 38

$$\Phi\left(X_t^{\nu,N}\right) - \Phi(X_0) = \int_s^t \partial_{x^\nu}\Phi\left(X_u^N\right)\beta_u^{\nu,N} dB_u + \int_s^t \partial_{x_\nu}\Phi\left(X_u^N\right)\alpha_u^{\nu,N} du \tag{5.14}$$
$$+ \int_s^t \left[\partial_{x^\nu}\Phi\left(X_u^N\right)\eta_u^{\nu,N} + \frac{1}{2}\partial^2_{x^\mu x^\nu}\Phi\left(X_u^N\right)\beta_u^{\mu,N}\beta_u^{\nu,N}\right] d\langle B\rangle_u$$

Since

$$\mathbb{E}\left[\left|X_t^{\nu,N} - X_t^\nu\right|^2\right] \le 3\mathbb{E}\left[\left|\int_0^t \left(\alpha_s^N - \alpha_s\right) ds\right|^2\right]$$
$$+3\mathbb{E}\left[\left|\int_0^t \left(\eta_s^{\nu,N} - \eta_s^\nu\right) d\langle B\rangle_s\right|^2\right] + 3\mathbb{E}\left[\left|\int_0^t \left(\beta_s^{\nu,N} - \beta_s^\nu\right) dB_s\right|^2\right]$$
$$\le 3\int_0^T \mathbb{E}\left[\left(\alpha_s^{\nu,N} - \alpha_s^\nu\right)^2\right] ds + 3\int_0^T \mathbb{E}\left[\left|\eta_s^{\nu,N} - \eta_s^\nu\right|^2\right] ds$$
$$+3\int_0^T \mathbb{E}\left[\left(\beta_s^{\nu,N} - \beta_s^\nu\right)^2\right] ds,$$

we then can prove that, in $M_G^2(0,T)$, we have (5.13). Furthermore

$$\partial_{x^\nu}\Phi\left(X_\cdot^N\right)\eta_\cdot^{\nu,N} + \partial^2_{x^\mu x^\nu}\Phi\left(X_\cdot^N\right)\beta_\cdot^{\mu,N}\beta_\cdot^{\nu,N} \to \partial_{x^\nu}\Phi(X_\cdot)\eta_\cdot^\nu + \partial^2_{x^\mu x^\nu}\Phi(X_\cdot)\beta_\cdot^\mu\beta_\cdot^\nu$$
$$\partial_{x_\nu}\Phi\left(X_\cdot^N\right)\alpha_\cdot^{\nu,N} \to \partial_{x_\nu}\Phi(X_\cdot)\alpha_\cdot^\nu$$
$$\partial_{x^\nu}\Phi\left(X_\cdot^N\right)\beta_\cdot^{\nu,N} \to \partial_{x^\nu}\Phi(X_\cdot)\beta_\cdot^\nu$$

We then can pass limit in both sides of (5.14) and get (5.13).

6 Stochastic Differential Equations

We consider the following SDE defined on $M_G^2(0,T;\mathbb{R}^n)$:

$$X_t = X_0 + \int_0^t b(X_s)ds + \int_0^t h(X_s)d\langle B\rangle_s + \int_0^t \sigma(X_s)dB_s,\ t\in[0,T]. \quad (6.1)$$

where the initial condition $X_0\in\mathbb{R}^n$ is given and $b,h,\sigma:\mathbb{R}^n\mapsto\mathbb{R}^n$ are given Lipschitz functions, i.e., $|\phi(x)-\phi(x')|\leq K|x-x'|$, for each x, $x'\in\mathbb{R}^n$, $\phi=b$, h and σ. Here the horizon $[0,T]$ can be arbitrarily large. The solution is a process $X\in M_G^2(0,T;\mathbb{R}^n)$ satisfying the above SDE. We first introduce the following mapping on a fixed interval $[0,T]$:

$$\Lambda_\cdot(Y) := Y \in M_G^2(0,T;\mathbb{R}^n) \longmapsto M_G^2(0,T;\mathbb{R}^n)$$

by setting Λ_t with

$$\Lambda_t(Y) = X_0 + \int_0^t b(Y_s)ds + \int_0^t h(Y_s)d\langle B\rangle_s + \int_0^t \sigma(Y_s)dB_s,\ t\in[0,T].$$

We immediately have

Lemma 41. *For each $Y,Y'\in M_G^2(0,T;\mathbb{R}^n)$, we have the following estimate:*

$$\mathbb{E}\left[|\Lambda_t(Y)-\Lambda_t(Y')|^2\right] \leq C\int_0^t \mathbb{E}\left[|Y_s-Y_s'|^2\right]ds,\ t\in[0,T],$$

where $C=3K^2$.

Proof. This is a direct consequence of the inequalities (5.1), (5.3) and (5.8).

We now prove that SDE (6.1) has a unique solution. By multiplying e^{-2Ct} on both sides of the above inequality and then integrate them on $[0,T]$. It follows that

$$\begin{aligned}\int_0^T &\mathbb{E}\left[|\Lambda_t(Y)-\Lambda_t(Y')|^2\right]e^{-2Ct}dt\\ &\leq C\int_0^T e^{-2Ct}\int_0^t \mathbb{E}[|Y_s-Y_s'|^2]dsdt\\ &= C\int_0^T\int_s^T e^{-2Ct}dt\mathbb{E}\left[|Y_s-Y_s'|^2\right]ds\\ &= (2C)^{-1}C\int_0^T (e^{-2Cs}-e^{-2CT})\mathbb{E}\left[|Y_s-Y_s'|^2\right]ds.\end{aligned}$$

We then have

$$\int_0^T \mathbb{E}\left[|\Lambda_t(Y)-\Lambda_t(Y')|^2\right]e^{-2Ct}dt \leq \frac{1}{2}\int_0^T \mathbb{E}\left[|Y_t-Y_t'|^2\right]e^{-2Ct}dt.$$

We observe that the following two norms are equivalent in $M_G^2(0,T;\mathbb{R}^n)$:

$$\int_0^T \mathbb{E}\left[|Y_t|^2\right] dt \sim \int_0^T \mathbb{E}\left[|Y_t|^2\right] e^{-2Ct} dt.$$

From this estimate we can obtain that $\Lambda(Y)$ is a contract mapping. Consequently, we have

Theorem 42. *There exists a unique solution $X \in M_G^2(0,T;\mathbb{R}^n)$ of the stochastic differential equation (6.1).*

7 Appendix

For $r > 0$, $1 < p, q < \infty$ with $\frac{1}{p} + \frac{1}{q} = 1$, we have

$$|a+b|^r \leq \max\{1, 2^{r-1}\}(|a|^r + |b|^r), \quad \forall a, b \in \mathbb{R} \tag{7.1}$$

$$|ab| \leq \frac{|a|^p}{p} + \frac{|b|^q}{q}. \tag{7.2}$$

Proposition 43.

$$\mathbb{E}[|X+Y|^r] \leq C_r(\mathbb{E}[|X|^r] + \mathbb{E}[|Y|^r]), \tag{7.3}$$

$$\mathbb{E}[|XY|] \leq \mathbb{E}[|X|^p]^{1/p} \cdot \mathbb{E}[|Y|^q]^{1/q}, \tag{7.4}$$

$$\mathbb{E}[|X+Y|^p]^{1/p} \leq \mathbb{E}[|X|^p]^{1/p} + \mathbb{E}[|Y|^p]^{1/p}. \tag{7.5}$$

In particular, for $1 \leq p < p'$, we have $\mathbb{E}[|X|^p]^{1/p} \leq \mathbb{E}[|X|^{p'}]^{1/p'}$.

Proof. (7.3) follows from (7.1). We set

$$\xi = \frac{X}{\mathbb{E}[|X|^p]^{1/p}}, \quad \eta = \frac{Y}{\mathbb{E}[|Y|^q]^{1/q}}.$$

By (7.2) we have

$$\begin{aligned}\mathbb{E}[|\xi\eta|] &\leq \mathbb{E}[\frac{|\xi|^p}{p} + \frac{|\eta|^q}{q}] \leq \mathbb{E}[\frac{|\xi|^p}{p}] + \mathbb{E}[\frac{|\eta|^q}{q}] \\ &= \frac{1}{p} + \frac{1}{q} = 1.\end{aligned}$$

Thus (7.4) follows. We now prove (7.5):

$$\begin{aligned}\mathbb{E}[|X+Y|^p] &= \mathbb{E}[|X+Y| \cdot |X+Y|^{p-1}] \\ &\leq \mathbb{E}[|X| \cdot |X+Y|^{p-1}] + \mathbb{E}[|Y| \cdot |X+Y|^{p-1}] \\ &\leq \mathbb{E}[|X|^p]^{1/p} \cdot \mathbb{E}[|X+Y|^{(p-1)q}]^{1/q} \\ &\quad + \mathbb{E}[|Y|^p]^{1/p} \cdot \mathbb{E}[|X+Y|^{(p-1)q}]^{1/q}\end{aligned}$$

This with $(p-1)q = p$ implies (7.5).

References

1. Artzner, Ph., F. Delbaen, J.-M. Eber, and D. Heath (1997), Thinking Coherently, *RISK* 10, November, 68–71.
2. Artzner, Ph., F. Delbaen, J.-M. Eber, and D. Heath (1999), Coherent Measures of Risk, *Mathematical Finance* 9, 203–228.
3. Avellaneda, M., Levy, A. and Paras, A. (1995). Pricing and hedging derivative securities in markets with uncertain volatilities. Appl. Math. Finance 2, 73–88.
4. Briand, Ph., Coquet, F., Hu, Y., Mémin J. and Peng, S. (2000) A converse comparison theorem for BSDEs and related properties of g-expectations, *Electron. Comm. Probab,* **5**.
5. Chen, Z. (1998) A property of backward stochastic differential equations, *C. R. Acad. Sci. Paris* **Sér. I Math. 326(**4), 483–488.
6. Chen, Z. and Epstein, L. (2002), Ambiguity, Risk and Asset Returns in Continuous Time, *Econometrica,* **70**(4), 1403–1443.
7. Chen, Z., Kulperger, R. and Jiang L. (2003) Jensen's inequality for g-expectation: part 1, *C. R. Acad. Sci. Paris,* **Ser. I 337**, 725–730.
8. Chen, Z. and Peng, S. (1998) A Nonlinear Doob-Meyer type Decomposition and its Application. *SUT Journal of Mathematics* (Japan), **34**(2), 197–208.
9. Chen, Z. and Peng, S. (2000), A general downcrossing inequality for g-martingales, *Statist. Probab. Lett.* **46**(2), 169–175.
10. Cheridito, P., Soner, H.M., Touzi, N. and Victoir, N., Second order backward stochastic differential equations and fully non-linear parabolic PDEs, Preprint (pdf-file available in arXiv:math.PR/0509295 v1 14 Sep 2005).
11. Coquet, F., Hu, Y., Mémin, J. and Peng, S. (2001) A general converse comparison theorem for Backward stochastic differential equations, *C. R. Acad. Sci. Paris,* **t.333,** Serie I, 577–581.
12. Coquet, F., Hu, Y., Memin J. and Peng, S. (2002), Filtration-consistent nonlinear expectations and related g-expectations, *Probab. Theory Relat. Fields,* **123**, 1–27.
13. Crandall, M., Ishii, H., and Lions, P.-L. (1992) User's Guide To Viscosity Solutions Of Second Order Partial Differential Equations, *Bulletin Of The American Mathematical Society,* **27**(1), 1-67.
14. Daniell, P.J. (1918) A general form of integral. *Annals of Mathematics,* **19**, 279–294.
15. Delbaen, F. (2002), Coherent Risk Measures (Lectures given at the Cattedra Galileiana at the Scuola Normale di Pisa, March 2000), Published by the Scuola Normale di Pisa.
16. Barrieu, P. and El Karoui, N. (2004) Pricing, Hedging and Optimally Designing Derivatives via Minimization of Risk Measures, Preprint, to appear in Contemporary Mathematics.
17. Barrieu, P. and El Karoui, N. (2005) Pricing, Hedging and Optimally Designing Derivatives via Minimization of Risk Measures, Preprint.
18. El Karoui, N., Quenez, M.C. (1995) Dynamic Programming and Pricing of Contingent Claims in Incomplete Market. *SIAM J. of Control and Optimization,* **33**(1).
19. El Karoui, N., Peng, S., Quenez, M.C. (1997) Backward stochastic differential equation in finance, *Mathematical Finance* **7**(1): 1–71.
20. Fleming, W.H., Soner, H.M. (1992) *Controlled Markov Processes and Viscosity Solutions.* Springer–Verlag, New York.

21. Huber, P.J., (1981) *Robustic Statistics*, John Wiley & Sons.
22. Itô, Kiyosi, (1942) Differential Equations Determining a Markoff Process, in Kiyosi Itô: *Selected Papers,* Edit. D.W. Strook and S.R.S. Varadhan, Springer, 1987, Translated from the original Japanese first published in Japan, Pan-Japan Math. Coll. No. 1077.
23. Jiang, L. (2004) Some results on the uniqueness of generators of backward stochastic differential equations, *C. R. Acad. Sci. Paris,* **Ser. I 338** 575–580.
24. Jiang L. and Chen, Z. (2004) A result on the probability measures dominated by g-expectation, *Acta Mathematicae Applicatae Sinica*, English Series **20**(3) 507–512.
25. Klöppel, S., Schweizer, M.: Dynamic Utility Indifference Valuation via Convex Risk Measures, Working Paper (2005) (http://www.nccr-nrisk.unizh.ch/media/pdf/wp/WP209-1.pdf).
26. Krylov, N.V. (1980) *Controlled Diffusion Processes.* Springer–Verlag, New York.
27. Lyons, T. (1995). Uncertain volatility and the risk free synthesis of derivatives. Applied Mathematical Finance 2, 117–133.
28. Nisio, M. (1976) On a nonlinear semigroup attached to optimal stochastic control. *Publ. RIMS, Kyoto Univ.*, 13: 513–537.
29. Nisio, M. (1976) On stochastic optimal controls and envelope of Markovian semi–groups. *Proc. of int. Symp. Kyoto,* 297–325.
30. Øksendal B. (1998) Stochastic Differential Equations, Fifth Edition, Springer.
31. Pardoux, E., Peng, S. (1990) Adapted solution of a backward stochastic differential equation. Systems and Control Letters, **14**(1): 55–61.
32. Peng, S. (1992) A generalized dynamic programming principle and Hamilton-Jacobi-Bellman equation. *Stochastics and Stochastic Reports*, **38**(2): 119–134.
33. Peng, S. (1997) Backward SDE and related g-expectation, in *Backward Stochastic Differential Equations*, Pitman Research Notes in Math. Series, No.364, El Karoui Mazliak edit. 141–159.
34. Peng, S. (1997) BSDE and Stochastic Optimizations, *Topics in Stochastic Analysis*, Yan, J., Peng, S., Fang, S., Wu, L.M. Ch.2, (Chinese vers.), Science Publication, Beijing.
35. Peng, P. (1999) Monotonic limit theorem of BSDE and nonlinear decomposition theorem of Doob-Meyer's type, *Prob. Theory Rel. Fields* **113**(4) 473–499.
36. Peng, S. (2004) Nonlinear expectation, nonlinear evaluations and risk measures, in K. Back, T.R. Bielecki, C. Hipp, S. Peng, W. Schachermayer, Stochastic Methods in Finance Lectures, C.I.M.E.-E.M.S. Summer School held in Bressanone/Brixen, Italy 2003, (Edit. M. Frittelli and W. Runggaldier) 143–217, LNM 1856, Springer-Verlag.
37. Peng, S. (2004) Filtration Consistent Nonlinear Expectations and Evaluations of Contingent Claims, *Acta Mathematicae Applicatae Sinica,* English Series **20**(2), 1–24.
38. Peng, S. (2005) Nonlinear expectations and nonlinear Markov chains, Chin. Ann. Math. **26B**(2), 159–184.
39. Peng, S. (2004) Dynamical evaluations, *C. R. Acad. Sci. Paris*, **Ser. I 339** 585–589.
40. Peng, S. (2005), Dynamically consistent nonlinear evaluations and expectations, in arXiv:math.PR/0501415 v1 24 Jan 2005.
41. Peng, S. (2006) Multi-dimensional G–Brownian motion and related stochastic calculus under G-expectation, Preprint, (pdf-file available in arXiv:math.PR/0601699 v1 28 Jan 2006).

42. Peng, S. and Xu, M. (2003) Numerical calculations to solve BSDE, preprint.
43. Peng, S. and Xu, M. (2005) g_Γ-expectations and the Related Nonlinear Doob-Meyer Decomposition Theorem.
44. Rosazza Giannin, E., (2002) Some examples of risk measures via g-expectations, preprint, to appear in Insurance: Mathematics and Economics.
45. Yong, J., Zhou, X. (1999) *Stochastic Controls: Hamiltonian Systems and HJB Equations.* Springer–Verlag.

Perpetual Integral Functionals of Diffusions and their Numerical Computations

Paavo Salminen[1] and Olli Wallin[2]

[1] Åbo Akademi, Mathematical Department, Fänriksgatan 3 B, FIN-20500 Åbo, Finland, phsalmin@abo.fi
[2] University of Oslo, Centre for Mathematics and Applications, P.O. Box 1053, Blindern, NO-0316 Oslo, Norway, olli.wallin@cma.uio.no

Summary. In this paper we study perpetual integral functionals of diffusions. Our interest is focused on cases where such functionals can be expressed as first hitting times for some other diffusions. In particular, we generalize the result in [24] in which one-sided functionals of Brownian motion with drift are connected with first hitting times of reflecting diffusions.

Interpreting perpetual integral functionals as hitting times allows us to compute numerically their distributions by applying numerical algorithms for hitting times. Hereby, we discuss two approaches:

- numerical inversion of the Laplace transform of the first hitting time,
- numerical solution of the PDE associated with the distribution function of the first hitting time.

For numerical inversion of Laplace tranforms we have implemented the Euler algorithm developed by Abate and Whitt. However, perpetuities lead often to diffusions for which the explicit forms of the Laplace transforms of first hitting times are not available. In such cases, and also otherwise, algorithms for numerical solutions of PDE's can be evoked. In particular, we analyze the Kolmogorov PDE of some diffusions appearing in our work via the Crank–Nicolson scheme.

Mathematics Subject Classification: 60J65, 60J60, 62E25

1 Introduction

Let $\{Y_t : t \geq 0\}$ be a regular linear diffusion taking values on an interval I. The left and right endpoints of the interval are denoted by l and r, respectively. For a locally integrable function $f : I \mapsto \mathbf{R}_+$ define the perpetual integral functional associated with f and Y via

$$\int_0^\infty f(Y_t)\, dt. \tag{1.1}$$

An important example of perpetual integral functionals is

$$\int_0^\infty \exp\left(-2a\, B_t^{(\mu)}\right) dt, \quad a > 0,$$

where $B^{(\mu)}$ is a BM with positive drift μ, studied by Dufresne in [10] in connection with risk theory and pension funding. In particular, from [10], this functional is distributed as $1/(2\,a^2\,Z_\nu)$ where Z_ν is a gamma-distributed random variable with the density function

$$f_{Z_\nu}(z) = \frac{1}{\Gamma(\nu)}\, z^{\nu-1} \mathrm{e}^{-z}, \quad \nu := \mu/a.$$

In Yor [31] (see [32] for an English translation) it is shown that

$$\int_0^\infty \exp(-2aB_s^{(\mu)})\, ds \;\stackrel{(\mathrm{d})}{=}\; H_0(R^{(\delta)}), \tag{1.2}$$

where $R^{(\delta)}$ is a Bessel process of dimension $\delta = 2(1-(\mu/a))$ started at $1/a$,

$$H_0(R^{(\delta)}) := \inf\{t : \; R_t^{(\delta)} = 0\},$$

and $\stackrel{(\mathrm{d})}{=}$ reads "is identical in law with" (in fact, $R^{(\delta)}$ can be constructed in the same probability space as $B^{(\mu)}$ and then (1.2) holds a.s.). In [24] the methodology used in [31] is developed for more general perpetual functionals for BM with positive drift and, in particular, results for one sided functionals are presented. An example of these is

$$\int_0^\infty \exp(-2aB_s^{(\mu)})\, \mathbf{1}_{\{B_s^{(\mu)}>0\}} ds \;\stackrel{(\mathrm{d})}{=}\; H_{1/a}(R^{(2\mu/a)}),$$

where the Bessel diffusion $R^{(2\mu/a)}$ is started at 0 and, in the case $0 < \mu < a$, reflected at 0. For further results and references for Dufresne's functionals, see [9, 21, 23, 24, 26].

In this paper, Section 2, we recall (from [4]) the connection between perpetual integral functionals and first hitting times. After this, the result in [24], Proposition 2.3, concerning one-sided perpetual functionals of $B^{(\mu)}$ is generalized for Y (defined via a SDE) and functionals of the type in (1.1). In Section 3, to make the paper more self contained and also as an introduction to Section 4, some basic facts about the distributions of the first hitting times are presented. Section 4 contains brief descriptions of the Euler algorithm for numerical inversion of Laplace transforms and the Crank-Nicolson scheme for solving PDE's, which we implemented in Matlab. The paper is concluded with Section 5 where the distributions of some perpetual functionals are computed numerically. In particular, we compare the one-sided functionals

$$\int_0^\infty \exp(-2B_s^{(\mu,\sigma)}) \, \mathbf{1}_{\{B_s^{(\mu,\sigma)}>0\}} \, ds$$

$$\text{and} \quad \int_0^\infty (1+\exp(B_s^{(\mu,\sigma)}))^{-2} \, \mathbf{1}_{\{B_s^{(\mu,\sigma)}>0\}} \, ds,$$

where $B_t^{(\mu,\sigma)} := \sigma B_t + \mu t$ with B the standard Brownian motion. It is also seen that some of the diffusions studied have bad singularities making the PDE's numerically troublesome to solve. In some cases this problem can, at least partly, be solved by transforming the diffusion to a new one with better behaviour. It seems to us that for a general numerical approach for calculating distributions of perpetualities, more sophisticated PDE or other methods such as Monte Carlo simulation are needed for the cases where the Laplace transform is not available for numerical inversion.

2 Perpetual Integral Functionals as First Hitting Times

Consider a diffusion Y on an open interval $I = (l, r)$ determined by the SDE

$$dY_t = \sigma(Y_t) \, dB_t + b(Y_t) \, dt, \tag{2.1}$$

where B is a standard Brownian motion defined in a complete probability space $(\Omega, \mathcal{F}, \{\mathcal{F}_t\}, \mathbf{P})$. It is assumed that σ and b are continuous and $\sigma(x) > 0$ for all $x \in I$. The diffusion Y is considered up to

$$\zeta := \inf\{t \, : \, Y_t \notin I\},$$

but it is possible that $\zeta = \infty$ a.s.

Let f be a (strictly) positive and continuous function defined on I, and consider for $t \geq 0$ the integral functional

$$A_t := \int_0^t f(Y_s) \, ds.$$

We remark that $\{A_t \, : \, t \geq 0\}$ is an additive functional of Y in the usual sense (see e.g. [2] p. 148). Taking $t = \zeta$ gives us the perpetual integral functional

$$A_\zeta := \int_0^\zeta f(Y_s) \, ds.$$

Assuming that $A_\zeta < \infty$ a.s. we are interested in the distribution of A_ζ.

A sufficient condition for finiteness is clearly that the mean of A_ζ is finite:

$$\begin{aligned} \mathbf{E}_x (A_\zeta) &= \int_0^\infty \mathbf{E}_x (f(Y_s)) \, ds \\ &= \int_l^r G_0(x,y) \, f(y) \, m(dy) < \infty, \end{aligned}$$

where G_0 denotes the Green kernel of Y and m is the speed measure (for these see, e.g., [3]). A neccessary and sufficient condition in the case of a Brownian motion with drift $\mu > 0$ is that the function f is integrable at $+\infty$ (see Engelbert and Senf [11] and Salminen and Yor [25]). We refer also to a recent paper [18] for such a condition valid for measurable and locally bounded f and a general diffusion Y.

Next proposition connects the perpetual integral functionals to the first hitting times. The result is extracted from Propositions 2.1 and 2.3 in [4] where the proof can be found. We remark also that the result generalizes Proposition 2.1 in [24].

Proposition 1. *Let Y, A, and f be as above and assume that there exists a two times continuously differentiable function g such that*

$$f(x) = \left(g'(x)\sigma(x)\right)^2, \quad x \in I. \tag{2.2}$$

Let $\{a_t : 0 \le t < A_\zeta\}$ denote the inverse of A, that is,

$$a_t := \min\{s : A_s > t\}, \qquad t \in [0, A_\zeta).$$

1. *Then the process Z given by*

$$Z_t := g\left(Y_{a_t}\right), \qquad t \in [0, A_\zeta), \tag{2.3}$$

is a diffusion satisfying the SDE

$$dZ_t = d\widetilde{B}_t + G(g^{-1}(Z_t))\, dt, \qquad t \in [0, A_\zeta).$$

where $\widetilde{B}_t$ is a Brownian motion and

$$G(x) = \frac{1}{f(x)} \left(\frac{1}{2} \sigma(x)^2 g''(x) + b(x)\, g'(x) \right). \tag{2.4}$$

2. *Let $x \in I$ and $y \in I$ be such that $\mathbf{P}_x$-a.s.*

$$H_y(Y) := \inf\{t : Y_t = y\} < \infty.$$

Then

$$A_{H_y(Y)} = \inf\{t : Z_t = g(y)\} =: H_{g(y)}(Z) \quad a.s.$$

with $Y_0 = x$ and $Z_0 = g(x)$.

3. *Suppose $g(r) := \lim_{z \to r} g(z)$ exists. Suppose also that the following statements hold a.s.*

$$(i)\ \lim_{t\to\zeta} Y_t = r, \quad (ii)\ A_\zeta := \lim_{t\to\zeta} A_t < \infty.$$

Then

$$A_\zeta = H_{g(r)}(Z) \quad a.s.$$

In [24] Proposition 2.3 one sided functionals for Brownian motion with positive drift are studied. This result is generalized here, under some assumptions, to the present case. Suppose $0 \in (l, r)$ and recall that $f(x) > 0$ for all $x \in (l, r)$. Consider the functional

$$A_\zeta^0 := \int_0^\zeta f(Y_s)\, \mathbf{1}_{\{Y_s > 0\}}\, ds.$$

Let

$$C_t := \int_0^t \mathbf{1}_{\{Y_s > 0\}}\, ds, \quad t \le \zeta,$$

and $\{c_t : 0 \le t < B_\zeta\}$ denote the inverse of C. We assume also that

$$\lim_{t \to \zeta} Y_t = r \quad \text{a.s.} \tag{2.5}$$

It is well known (see [16]) that the process

$$Y^+ := \{Y_{c_t} : 0 \le t < C_\zeta\}$$

is identical in law with Y living on $[0, r)$ and having 0 as a reflecting boundary point. Applying the random time change means that on every sample path the excursions below 0 are omitted after which the gaps created are closed by joining the excursions together. Therefore,

$$A_\zeta^0 = \int_0^{\zeta^+} f(Y_s^+)\, ds =: A_\zeta^+$$

where ζ^+ is the life time of Y^+.

Next introduce the local time of Y^+ at 0 via

$$L_t(Y^+) := \sigma^2(0) \lim_{\varepsilon \downarrow 0} (2\varepsilon)^{-1} \text{Leb}\{0 \le s \le t : Y_s^+ < \varepsilon\}.$$

Under some additional smoothness assumptions on σ and b (see McKean [20]) the pair $(Y^+, L(Y^+))$ with $Y_0^+ = x > 0$ can be viewed as the unique solution of the reflected SDE

$$dX_t = \sigma(X_t)\, dB_t + b(X_t)\, dt + dL_t(X), \quad X_0 = x,$$

such that

(a) $\lim_{t \to \zeta(X)} X(t) = r$,
(b) $0 \le X(t) < r$ for all $t < \zeta(X)$,
(c) $t \mapsto L_t(X)$ is continuous, increasing with $L_0(X) = 0$, and

$$\int_0^t \mathbf{1}_{\{0\}}(X_s)\, dL_s(X) = L_t(X).$$

We now give the promised generalization.

Proposition 2. *Let Y^+ be as given above and define for $t < \zeta^+$*

$$A_t^+ := \int_0^t f(Y_s^+)\,ds.$$

The inverse of A^+ is denoted by $\{a_t^+ : 0 \le t < A_\zeta\}$. Recall the definition of the function g in (2.2) and define the process Z^+ via

$$Z_t^+ := g\left(Y_{a_t^+}^+\right), \qquad t \in [0, A_\zeta^+). \tag{2.6}$$

Then

$$A_\zeta^+ = \inf\{t : Z_t^+ = g(r)\} \quad \text{a.s.} \tag{2.7}$$

with $Z_0 = g(x)$. Moreover, Z^+ satisfies the reflected SDE

$$dZ_t^+ = d\widetilde{B}_t + G(g^{-1}(Z_t^+))\,dt + dL_t(Z^+), \qquad t \in [0, A_\zeta^+). \tag{2.8}$$

where $\widetilde{B}_t$ is a Brownian motion,

$$L_t(Z^+) = \lim_{\varepsilon \downarrow 0} (2\varepsilon)^{-1} \text{Leb}\{0 \le s \le t : g(0) \le Z_s^+ < g(0) + \varepsilon\}, \tag{2.9}$$

and G is as in (2.4). The local time $L(Z^+)$ is related to the local time $L(Y^+)$ by

$$L_t(Z^+) = g'(0) L_{a_t^+}(Y^+). \tag{2.10}$$

Proof. Recall that since $f > 0$ it follows from (2.2) that g is monotone. Hence, to fix ideas, we assume that g is monotonically increasing. By Itô's formula for $u < \zeta$

$$\begin{aligned} g(Y_u^+) - g(Y_0^+) &= \int_0^u g'(Y_s^+)\left(\sigma(Y_s^+)\,dB_s + b(Y_s^+)\,ds + dL_s(Y^+)\right) \\ &\quad + \frac{1}{2}\int_0^u g''(Y_s^+)\sigma^2(Y_s^+)\,ds. \end{aligned}$$

Replacing u by a_t^+ yields

$$\begin{aligned} Z_t^+ - Z_0^+ &= \int_0^{a_t^+} g'(Y_s^+)\sigma(Y_s^+)\,dB_s + g'(0)\,L_{a_t^+}(Y^+) \\ &\quad + \int_0^{a_t^+} \left(g'(Y_s^+)\sigma(Y_s^+)\right)^2 G(Y_s^+)\,ds. \end{aligned}$$

Since a_t^+ is the inverse of A_t^+ and $(A_s^+)' = \left(g'(Y_s^+)\sigma(Y_s^+)\right)^2$ we have

$$(a_t^+)' = \frac{1}{\left(A_{a_t^+}^+\right)'} = \left(g'\left(Y_{a_t^+}^+\right)\sigma\left(Y_{a_t^+}^+\right)\right)^{-2}. \tag{2.11}$$

From Lévy's theorem it follows that

$$\widetilde{B}_t := \int_0^{a_t^+} g'(Y_s^+)\sigma(Y_s^+)\, dB_s, \qquad t \in [0, A_\zeta^+),$$

is a (stopped) Brownian motion. Consequently, for $t < A_\zeta^+$

$$\begin{aligned} Z_t^+ - Z_0^+ &= \widetilde{B}_t + g'(0)\, L_{a_t^+}(Y^+) + \int_0^t \left(g'\left(Y_{a_s^+}^+\right)\sigma\left(Y_{a_s^+}^+\right)\right)^2 G\left(Y_{a_s^+}^+\right) da_s^+ \\ &= \widetilde{B}_t + \int_0^t G(g^{-1}(Z_s^+))\, ds + g'(0)\, L_{a_t^+}(Y^+). \end{aligned}$$

Clearly, viewing $t \mapsto g'(0)\, L_{a_t^+}(Y^+)$ as a functional of Z^+ then this functional increases only on the set $\{t.: Z_t^+ = g(0)\}$. Moreover, since $Y_t^+ \geq 0$ for $t \geq 0$ we have $Z_t^+ \geq g(0)$ for $t \geq 0$ by monotonicity of g. Hence, $(Z^+, L(Z^+))$ can be seen as the unique solution of the reflected SDE (2.8) with $L(Z^+)$ as in (2.9) satisfying (2.10), as claimed. Finally, again by the monotonicity of g, the identity (2.7) follows from the definition (2.6) of Z^+ and the assumption (2.5).

Remark 3. Notice that the above approach yields a stronger result than in [24], i.e., the identity (2.7) holds a.s.

3 Reminder on First Hitting Times

3.1 Distribution Functions and PDEs

Let Y be a linear diffusion determined via the SDE (2.1) up to ζ. It is here assumed that Y hits r a.s. and is killed when this happens. Therefore, the boundary point r is either exit-not-entrance or regular with killing, and l is either natural or entrance-not-exit or regular with reflection (hence, in this last mentioned case Y is taking values in $[l, r)$. Letting $H_r(Y)$ denote the hitting time of r we have

$$\mathbf{P}_x(H_r(Y) > t) = \int_l^r p(t; x, y)\, m(dy), \tag{3.1}$$

where p denotes the symmetric transition density of Y with respect to its speed measure m. It is well known (see [16] p. 149 and [19]) that $(t, x) \mapsto p(t; x, y)$ satisfies for all $y \in (l, r)$ the PDE

$$\begin{aligned} \frac{\partial}{\partial t} p(t; x, y) &= \frac{1}{2}\sigma^2(x)\, \frac{\partial^2}{\partial x^2} p(t; x, y) + b(x)\, \frac{\partial}{\partial x} p(t; x, y) \\ &=: (\mathcal{G}\, p)(t; x, y) \end{aligned}$$

and the condition $\lim_{x \to r} p(t; x, y) = 0$. Moreover, in the case l is regular with reflection or entrance-not-exit we impose at l the condition

$$\lim_{x\to l}\frac{\partial}{\partial x}p(t;x,y)=0,$$

and in the case l is natural the condition

$$\lim_{x\to l}p(t;x,y)=\lim_{x\to l}\frac{\partial}{\partial x}p(t;x,y)=0.$$

See [16] or [3] for the boundary classification of linear diffusions.

Letting $\{T_t\}$ denote the semigroup associated with Y we may write from (3.1)

$$\mathbf{P}_x(H_r(Y)>t)=T_t1(x).$$

Recall from [19] (where the case with natural scale is treated) that $(t,x)\mapsto T_tg(x)$ with g bounded and continuous satisfies

$$\frac{\partial}{\partial t}(T_tg)(x)=(\mathcal{G}\,T_t)g(x).$$

Consequently, the distribution function

$$(t,x)\mapsto u(t,x):=\mathbf{P}_x(H_z(Y)<t)$$

is the unique solution of the PDE problem

$$\frac{\partial}{\partial t}u(t,x)=(\mathcal{G}\,u)(t,x) \tag{3.2}$$

with the initial condition $\lim_{t\to 0}u(t,x)=0$ for all $x\in(l,r)$ and the boundary condition $\lim_{x\to r}u(t,x)=1$ for all $t>0$. Further, if l is regular with reflection or entrance-not-exit

$$\lim_{x\to l}\frac{\partial}{\partial x}u(t,x)=0,$$

and in the case l is natural

$$\lim_{x\to l}u(t,x)=\lim_{x\to l}\frac{\partial}{\partial x}u(t,x)=0.$$

Remark 4. Using the fact (see [19]) that

$$(t,x,y)\mapsto\frac{\partial}{\partial t}p(t;x,y)$$

is continuous and satisfies the same boundary conditions as the density $p(t;x,y)$ it is easy (at least when $m(l,r)<\infty$) to deduce that

$$\frac{\partial}{\partial t}\mathbf{P}_x(H_z(Y)<t)=-\lim_{y\to z}\frac{1}{S'(y)}\frac{\partial}{\partial y}p(t;x,y),$$

where

$$S'(y)=\exp\left(-\int^y 2\,\sigma^{-2}(v)\,b(v)\,dv\right)$$

is the derivative of the scale function S (cf. [16] p. 154).

3.2 Laplace Transforms and ODEs

For the approach with the Laplace transform of $H_r(Y)$ consider the second order ODE

$$\mathcal{G}u(x) = \lambda u(x), \tag{3.3}$$

where $\lambda \geq 0$. It is known (see [12] p. 488, and [16] p. 128) that the equation (3.3) has a positive increasing solution ψ_λ and a positive decreasing solution φ_λ. In case l is natural or entrance and r is exit these solutions are unique up to multiplicative constants. When l is regular with reflection the condition $\psi'_\lambda(l) = 0$ must be posed, and when r is regular with killing the condition is $\varphi_\lambda(r-) = 0$. The Green kernel G_λ of Y can be expressed via these solutions as

$$\begin{aligned} G_\lambda(x,y) &:= \int_0^\infty \mathrm{e}^{-\lambda t}\, p(t;x,y)\, dt \\ &= \begin{cases} \frac{1}{w_\lambda}\, \psi_\lambda(x)\, \varphi_\lambda(y), & x \leq y, \\ \frac{1}{w_\lambda}\, \psi_\lambda(y)\, \varphi_\lambda(x), & y \leq x, \end{cases} \end{aligned}$$

where w_λ is the Wronskian (see e.g. [3]). Using the Green kernel the Laplace transform for the first hitting time $H_y(Y)$ is given by

$$\mathbf{E}_x\left(\mathrm{e}^{-\lambda\, H_y(Y)}\right) = \frac{G_\lambda(x,y)}{G_\lambda(y,y)}$$

and, in particular,

$$\mathbf{E}_x\left(\mathrm{e}^{-\lambda\, H_r(Y)}\right) = \frac{\psi_\lambda(x)}{\psi_\lambda(r)}. \tag{3.4}$$

4 Numerical Methods

4.1 Numerical Inversion of Laplace Transforms

There are several efficient methods for numerical inversion of the Laplace transforms of probability density functions or probability distribution functions. We have implemented a method developed by Abate and Whitt in [1]. This so called Euler-algorithm has proved to be very effective in many applications see e.g. [6–8, 13]. The main features of this method are presented below. For more details and also for further references, see [1].

Consider a non-negative random variable with density f and its Laplace transform

$$\hat{f}(\lambda) := \int_0^\infty \mathrm{e}^{-\lambda t}\, f(t)\, dt.$$

The well known inversion integral formula (called the Bromwich or also the Fourier-Mellin integral) states that

$$f(t) = \frac{1}{2\pi \mathrm{i}} \int_{a-\mathrm{i}\infty}^{a+\mathrm{i}\infty} \mathrm{e}^{\lambda t} \hat{f}(\lambda)\, d\lambda, \tag{4.1}$$

where it is assumed that $\hat{f}$ does not have singularities on or to the right of the vertical line $\lambda = a$.

Remark 5. For first hitting times of diffusions considered in Section 3 we have (cf. (3.4))

$$\hat{f}(\lambda) = \mathbf{E}_x \left(\mathrm{e}^{-\lambda H_r(Y)} \right) = \frac{\psi_\lambda(x)}{\psi_\lambda(r)}.$$

If the left boundary point l is not natural it follows from the classical theory of second order differential operators (see e.g. [17]) that $\psi_\lambda(x)$ is for every $x \in (l,r)$ an entire function of λ and the zeroes of $\lambda \mapsto \psi_\lambda(x)$ are for every $x \in (l,r)$ simple and negative. Consequently, the inversion formula (4.1) holds in this case for any $a > 0$. If l is natural we can approximate the first hitting time $H_r(Y)$ via a sequence of first hitting times $\{H_r(Y^{(n)})\}$ associated with the diffusions $Y^{(n)}$, $n = 1, 2, \ldots$, constructed from Y by reflection at $l + \frac{1}{n}$, respectively. Then $H_r(Y^{(n)}) \to H_r(Y)$ in distribution and by dominated convergence it is seen that (4.1) is valid also in this case.

Since $\mathrm{Re}(\hat{f}(a + \mathrm{i}\,u)) = \mathrm{Re}(\hat{f}(a - \mathrm{i}\,u))$, $\mathrm{Im}(\hat{f}(a + \mathrm{i}\,u)) = -\mathrm{Im}(\hat{f}(a - \mathrm{i}\,u))$, and $f(t) = 0$ for $t < 0$ the inversion integral (4.1) takes the form

$$f(t) = \frac{2\mathrm{e}^{at}}{\pi} \int_0^\infty \mathrm{Re}(\hat{f}(a + \mathrm{i}\,u)) \cos(ut)\, du. \tag{4.2}$$

Next we approximate $f(t)$ by using the trapezoidal rule for the integral (4.2) (see [1] for some comments on the effectiveness of this procedure). Letting h denote the step size we have for fixed a and t

$$f(t) \approx f_h(t) = \frac{h\mathrm{e}^{at}}{\pi} \mathrm{Re}\big(\hat{f}(a)\big) + \frac{2h\mathrm{e}^{at}}{\pi} \sum_{k=1}^{\infty} \mathrm{Re}\big(\hat{f}(a + kh\,\mathrm{i})\big) \cos(kht).$$

Choosing $h = \pi/(2t)$, $a = A/(2t)$ (with A to be made precise later) and truncating the infinite series to the first j terms we are led to define

$$s_j(t) := \frac{\mathrm{e}^{A/2}}{t} \sum_{k=0}^{j} (-1)^k a_k(t), \tag{4.3}$$

where

$$a_0(t) := \hat{f}\left(A/2t\right)/2$$

and

$$a_k(t) := \mathrm{Re}\left(\hat{f}\left((A + 2k\pi\,\mathrm{i})/2t\right) \right), \quad k = 1, 2, \ldots, j.$$

It is possible to accelerate the convergence of the series in (4.3) by recognizing it is approximately alternating and using the Euler summation with binomial weights (see page 50 in [1] for more on this point). Hence, the proposed final approximation with parameters m, n, and A is

$$f(t) \approx E(m,n,t) := \sum_{k=0}^{m} \binom{m}{k} 2^{-m} s_{n+k}(t),$$

In the examples below we use, following [1], $m = 11$ and $n = 15$. The error associated with Euler summation can be estimated by considering the difference

$$E(m,n+1,t) - E(m,n,t).$$

It is advantageous from numerical computational point of view to invert, instead of the density function, the complementary distribution function. Therefore consider

$$\hat{F}^c(\lambda) := \int_0^\infty \mathrm{e}^{-\lambda t} (1 - F(t))\, dt,$$

where F is the distribution function associated with f. Firstly, the fact $|1 - F(t)| \le 1$ can be used to show (see [1]) that

$$|e_d| \le \frac{\mathrm{e}^{-A}}{1 - \mathrm{e}^{-A}},$$

where e_d stands for the discretization error when approximating the integral in (4.2) for $\hat{F}^c$ via the trapezoidal rule. For instance, $A = 18.4$ gives the upper bound 10^{-8}. Secondly, under some additional smoothness assumption, it can be proved (see [1] Remark 1) that for $\hat{F}^c$ we have $a_k(t) > 0$ when k/t is large enough motivating the use of the Euler summation (since the series in (4.3) is now alternating).

For the first hitting time of r for the diffusion Y the Laplace transform of the complementary distribution function is given by

$$\hat{F}^c(\lambda) = \frac{1}{\lambda}\left(1 - \mathbf{E}_x\left(\mathrm{e}^{-\lambda H_r(Y)}\right)\right) = \frac{1}{\lambda}\left(1 - \frac{\psi_\lambda(x)}{\psi_\lambda(r)}\right)$$

$$= \frac{1}{\lambda} \frac{\psi_\lambda(r) - \psi_\lambda(x)}{\psi_\lambda(r)}.$$

4.2 Numerical Solutions of PDEs

In this section we describe using [27, 28] two finite difference methods known as the *Crank-Nicolson* (C-N) scheme and the *Backward Euler* (BE) method. The C-N scheme is used with satisfactory results in [22] for calculation of transition probability densities of certain diffusions. The BE method can be applied in connection with the C-N scheme for the first time step to damp some numerical oscillations typical to the C-N scheme. Both methods are unconditionally stable: the numerical solutions are well behaved (do not blow up) for any choice of Δt. Because of the singularities appearing in the drift coefficients of our equations, methods that do not have this property (such as the explicit, forward Euler method) are practically unusable since they would require extremely small time steps.

To start with, let us introduce a uniformly spaced grid on the rectangle $[0,T] \times [l,r]$ with $(M+1)\times(N+1)$ nodes, that is, for

$$\Delta t = \frac{T}{M}, \quad \Delta x = \frac{r-l}{N},$$

we let $t_m = m\Delta t$ and $x_n = l + n\Delta x$ where $m = 0,1,...,M$, $n = 0,1,...,N$. For a real valued function f we use the following finite difference approximations, which can be justified with Taylor's expansion:

- the *forward difference* approximation of the derivative of f is

$$\frac{\partial f}{\partial x}(x_i) = \frac{f(x_{i+1}) - f(x_i)}{\Delta x} + \mathcal{O}(\Delta x).$$

- the *centralized difference* approximation of the derivative of f is

$$\frac{\partial f}{\partial x}(x_i) = \frac{f(x_{i+1}) - f(x_{i-1})}{2\Delta x} + \mathcal{O}((\Delta x)^2).$$

- the centralized difference approximation of the second order derivative of f is

$$\frac{\partial^2 f}{\partial x^2}(x_i) = \frac{f(x_{i+1}) - 2f(x_i) + f(x_{i-1})}{(\Delta x)^2} + \mathcal{O}((\Delta x)^2).$$

The C-N scheme approximates the left hand side in equation (3.2) with the forward difference and the right hand side with the average of the centralized differences at two consequtive times. Denoting $u_n^m := u(t_m, x_n)$ and dropping the truncation error terms, the discretized equation then reads

$$\frac{u_n^{m+1} - u_n^m}{\Delta t} = \frac{1}{2}\sigma^2(x_n)\frac{1}{2}\left(\frac{u_{n+1}^{m+1} - 2u_n^{m+1} + u_{n-1}^{m+1}}{(\Delta x)^2} + \frac{u_{n+1}^m - 2u_n^m + u_{n-1}^m}{(\Delta x)^2}\right)$$

$$+ b(x_n)\frac{1}{2}\left(\frac{u_{n+1}^{m+1} - u_{n-1}^{m+1}}{2\Delta x} + \frac{u_{n+1}^m - u_{n-1}^m}{2\Delta x}\right).$$

Multiply both sides with Δt, and define $r_1 = \frac{\Delta t}{2\Delta x}$, $r_2 = \frac{\Delta t}{2(\Delta x)^2}$. Rearranging the terms so that the values at time t_{m+1} appear on the left hand side and values at time t_m appear on the right hand side, we have

$$A_n u_{n-1}^{m+1} + B_n u_n^{m+1} - C_n u_{n+1}^{m+1} = -A_n u_{n-1}^m + D_n u_n^m + C_n u_{n+1}^m$$

where

$$A_n = \frac{1}{2}\big(b(x_n)r_1 - \sigma^2(x_n)r_2\big), \tag{4.4}$$

$$B_n = 1 + \sigma^2(x_n)r_2, \tag{4.5}$$

$$C_n = \frac{1}{2}\big(b(x_n)r_1 + \sigma^2(x_n)r_2\big), \tag{4.6}$$

$$D_n = 1 - \sigma^2(x_n)r_2. \tag{4.7}$$

Together with the boundary conditions, these form a set of $N+1$ linear equations which we then solve for each $m = 1, 2, ..., M+1$, using the initial condition for $m = 0$. The Neumann boundary condition (which is needed at a reflecting or an entrance boundary point) is implemented with the second order approximation. In other words, from

$$\frac{u_1^m - u_{-1}^m}{2\Delta x} = 0$$

we have $u_{-1}^m = u_1^m$ for the value of u at a "ghost" point beyond the boundary. Plugging this into the discretized equation at $n = 0$ gives

$$(1 + r_2\sigma^2(x_n))u_0^{m+1} - r_2\sigma^2(x_n)u_1^{m+1} = (1 - r_2\sigma^2(x_n))u_0^m + r_2\sigma^2(x_n)u_1^m.$$

Using a second order approximation for the boundary condition seems to be important in order to have good convergence.

For the backward Euler method, one similarly takes the central approximations for the spatial derivatives but now only at the time step $m+1$. This leads to the equation

$$A_n u_{n-1}^{m+1} + B_n u_n^{m+1} - C_n u_{n+1}^{m+1} = u_n^m$$

for $n = 1, ..., N$ and $m = 1, ..., M$, where A_n, B_n, C_n are given in (4.4)-(4.6) but now with $r_1 = \frac{\Delta t}{\Delta x}$, $r_2 = \frac{\Delta t}{(\Delta x)^2}$. The second order implementation of the Neumann boundary condition becomes

$$(1 + r_2\sigma^2(x_n))u_0^{m+1} - r_2\sigma^2(x_n)u_1^{m+1} = u_0^m.$$

Both methods described above are second order accurate in space, but only the C-N scheme is second order accurate in time. However, C-N is known to produce numerical oscillations around discontinuities and sharp gradients if the drift term is large compared to the diffusion coefficient (see for example [5,15]), while the BE method does not have this problem. In our examples this is seen as oscillations near the killing boundary. Although these oscillations were damped quite rapidly both in space and time, they still produced a small phase shift in the numerical solution of the hitting time distribution $t \mapsto u(x,t)$. For this reason the first step in the C-N scheme is divided into

10–100 substeps, as suggested in [5]. For the substeps we used the BE method. While small oscillations still remained for some of the examples, this procedure provided sufficient damping to give very accurate results in our test cases.

A problem more serious than the oscillations appearing in the C-N scheme is faced when the diffusing particle is pushed away from an exit boundary by a drift term tending to infinity in the vicinity of the boundary. In such cases convergence can be very slow or even nonattainable without huge computer capacity. We discuss this problem in more detail in the examples below.

5 Examples

In this section some perpetuities are examined numerically. If the Laplace transforms of the functionals are available we use the Euler algorithm for computing the density and/or the distribution functions. Applying in these cases also the numerical methods based on the associated PDE's obtained from the hitting time representations of the functionals and checking that the solutions resulting from the different methods coincide we are able to verify the correctness of the implementations.

Numerical methods for solving PDEs constitute a powerful tool for computing hitting time distributions of diffusions in general. However, there are diffusions as in Example 9 for which the methods presented here work unsatisfactorily. At least in some particular cases it is possible to transform the diffusion to a new one for which the methods seem to work better. This is discussed in Example 9. Due to these difficulties one may consider the numerical inversion of the Laplace transforms when available as the first choice for the kind of numerical computations studied in the paper.

Below the notation $\{B_t^{(\mu,\sigma)} : t \geq 0\}$ is used for a Brownian motion with (infinitesimal) drift μ and variance σ, i.e., $B_t^{(\mu,\sigma)} = \sigma\, B_t + \mu\, t$ with B a standard Brownian motion. We assume that $\sigma > 0$ and $\mu > 0$. In case $\sigma = 1$ we write $B^{(\mu)}$ for $B^{(\mu,1)}$. For a Bessel process with dimension parameter δ we use the notation of $\{R_t^{(\delta)} : t \geq 0\}$, and refer to [3] for their properties. If nothing else is written it is assumed that $B_0^{(\mu,\sigma)} = 0$ and $R_0^{(\delta)} = 0$.

Example 6. Consider the perpetual integral functional

$$I_1 := \int_0^\infty \cosh^{-2}(B_t^{(\mu)})\, dt.$$

The Laplace transform of this functional is computed in [4,29] for an arbitrary initial value x; taking therein $x = 0$ gives

$$\mathbf{E}_0(\exp(-\rho\, I_1)) = K\, {}_2F_1(\alpha, \beta, 1 + \mu; 1/2).$$

where

$$\alpha = \frac{1}{2} + \frac{1}{2}\sqrt{1-8\rho}, \quad \beta = \frac{1}{2} - \frac{1}{2}\sqrt{1-8\rho},$$

$$K = \frac{\Gamma(\mu+\alpha)\,\Gamma(\mu+\beta)}{\Gamma(\mu)\,\Gamma(\mu+1)},$$

and ${}_2F_1$ denotes Gauss's hypergeometric function given by

$$\begin{aligned} {}_2F_1(a,b,c;x) &:= \frac{\Gamma(c)}{\Gamma(a)\,\Gamma(b)} \sum_{k=0}^{\infty} \frac{\Gamma(a+k)\,\Gamma(b+k)}{\Gamma(c+k)} \frac{x^k}{k!} \\ &= 1 + \sum_{k=1}^{\infty} \frac{a(a+1)\ldots(a+k-1)\,b(b+1)\ldots(b+k-1)}{c(c+1)\ldots(c+k-1)} \frac{x^k}{k!}. \end{aligned}$$

Applying Proposition 1 with $g(x) := 2\arctan \mathrm{e}^x$ we obtain

$$I_1 = H_\mu(Z) \qquad a.s.,$$

where Z satisfies

$$dZ_t = dB_t + \left(\frac{1}{2}\operatorname{ctn} Z_t + \frac{\mu}{\sin Z_t}\right) dt, \qquad Z_0 = \pi/2. \tag{5.1}$$

For the drift term in (5.1), it holds

$$G(g^{-1}(x)) = \frac{1}{2}\operatorname{ctn} x + \frac{\mu}{\sin x} = \frac{1}{2}\left(\mu - \frac{1}{2}\right)\tan\frac{x}{2} + \frac{1}{2}\left(\mu + \frac{1}{2}\right)\operatorname{ctn}\frac{x}{2}.$$

Notice that for $0 < \mu < 1/2$ the drift of Z tends to $-\infty$ when Z is approaching π.

In Figures 1 and 6 we present the density and the distribution functions, respectively, of I_1 computed with the Euler algorithm. It has been checked that the PDE method yields the same results. However, the convergence in the case $\mu = 0.3$ seems to be slow. In fact, for $\mu = 0.1$ the convergence rate of the PDE method is so slow that we were unable to get satisfactory results with our limited computation capacity (RAM).

Example 7. In this example we compare the functionals

$$I_2 := \int_0^\infty \exp(-2B_s^{(\mu,\sigma)})ds \quad \text{and} \quad I_3 := \int_0^\infty (\exp(B_s^{(\mu,\sigma)}) + 1)^{-2} ds.$$

Notice that

$$I_3 = \int_0^\infty \exp(-2B_s^{(\mu,\sigma)}) \frac{1}{(1+\exp(-B_s^{(\mu,\sigma)}))^2} ds,$$

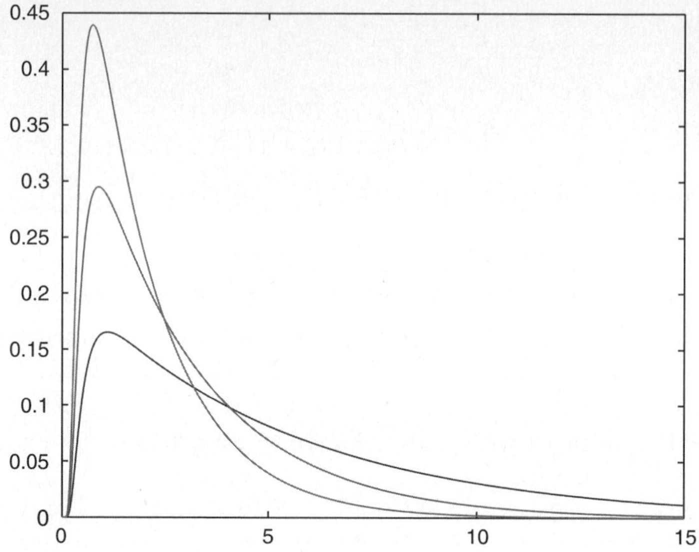

Fig. 1. Density of $\int_0^\infty \cosh^{-2}(B_t^{(\mu)})\,dt$ for $\mu = 0.3$ (lowest peak), $\mu = 0.5$ and $\mu = 0.7$ (highest peak)

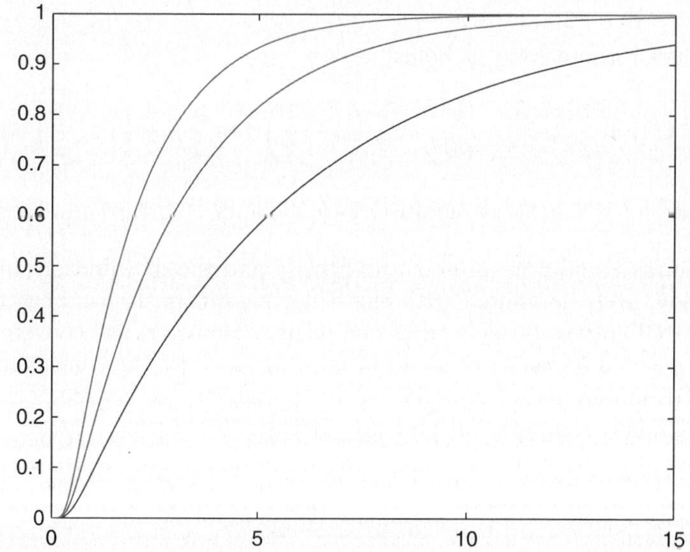

Fig. 2. Distribution function of $\int_0^\infty \cosh^{-2}(B_t^{(\mu)})\,dt$ for $\mu = 0.3$ (lowest curve), $\mu = 0.5$ and $\mu = 0.7$ (highest curve)

hence I_3 may be seen as a modification of I_2 which does not allow arbitrary large positive discounting. We remark also that I_3 has all moments which is not the case with I_2.

The Dufresne-Yor identity (cf. (1.2)) states that

$$I_2 = H_0(R^{(2-2\mu/\sigma^2)}) \qquad \text{a.s.},$$

where $R_0^{(2-2\mu/\sigma^2)} = 1/\sigma$. Consequently,

$$\mathbf{E}_0\left(\exp(-\rho\, I_2)\right) = \frac{\varphi_\rho(1/\sigma)}{\varphi_\rho(0)},$$

with (see [3] p. 133)

$$\varphi_\rho(x) = x^{-\nu} K_\nu(x\sqrt{2\rho}), \quad \text{and} \quad \varphi_\rho(0) = 2^{-(\nu+2)/2}\Gamma(-\nu)\,\rho^{\nu/2}$$

and $\nu = -\mu/\sigma^2$.

For I_3 we have the identity

$$I_3 = H_0(Z) \qquad \text{a.s.},$$

where Z is the diffusion associated with the SDE

$$dZ_t = dB_t + \left(\mu + (\mu - \frac{1}{2})\frac{\exp(Z_t)}{1-\exp(Z_t)}\right) dt, \qquad Z_0 = -\log 2.$$

Notice that here $g(x) := -\log(1+\exp(-x))$ (cf. Proposition 1) and that Z lives on $\mathbf{R}_-$. From [4] we recall the Laplace transform

$$\mathbf{E}_0\Big(\exp\Big(-\rho\, I_3\Big)\Big) = K\, 2^{\mu-\sqrt{\mu^2+2\rho}}\, {}_2F_1(\alpha,\beta,\alpha+\beta+2\mu\,;1/2),$$

where

$$\alpha = \frac{1}{2} - \mu + \sqrt{\mu^2+2\rho} + \sqrt{\frac{1}{4}+2\rho}, \quad \beta = \frac{1}{2} - \mu + \sqrt{\mu^2+2\rho} - \sqrt{\frac{1}{4}+2\rho},$$

and

$$K = \frac{\Gamma(2\mu+\alpha)\ \Gamma(2\mu+\beta)}{\Gamma(2\mu+\alpha+\beta)\ \Gamma(2\mu)}.$$

See Figures 3, 4, 5 and 6 for illustrations of the distributions of I_2 and I_3 computed with the Euler algorithm.

For both functionals in this example it was possible to solve the corresponding PDE numerically for $\mu \geq \frac{1}{2}$. For $\mu < 1/2$ the drift term tends to $-\infty$ as Z approaches the killing boundary 0. This again leads to very slow convergence. While it was still possible to achieve good results for some choices of $\mu < \frac{1}{2}$, for a small enough μ the results were bad even with the finest grid we could run on the computer. Notice that here we also need to truncate the semi-infinite domains into finite ones for numerical computations. This did not constitute a major problem, but with larger domains it is difficult to achieve (depending on the computer capacity) a grid which is spatially dense enough for accurate computations.

Example 8. We define the one-sided variants of I_2 and I_3 via

$$I_4 := \int_0^\infty \exp(-2B_s^{(\mu,\sigma)})\mathbf{1}_{\{B_s^{(\mu,\sigma)}>0\}} ds$$

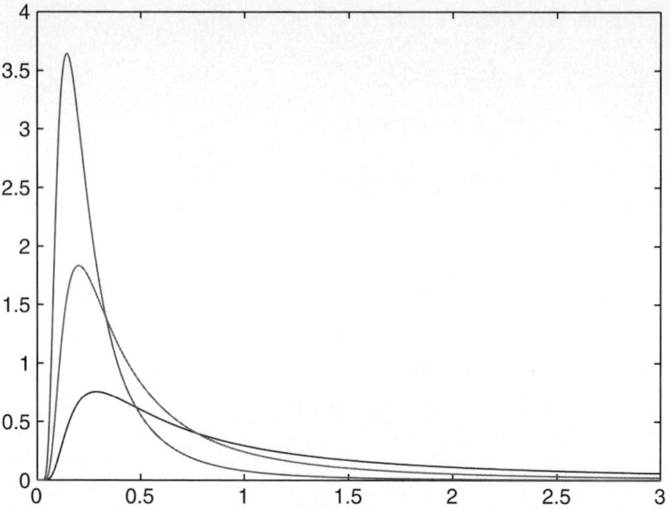

Fig. 3. Density of $\int_0^\infty \exp(-2\,B_t^{(\mu)})\,dt$ for $\mu = 0.75$ (lowest peak), $\mu = 1.50$ and $\mu = 2.50$ (highest peak)

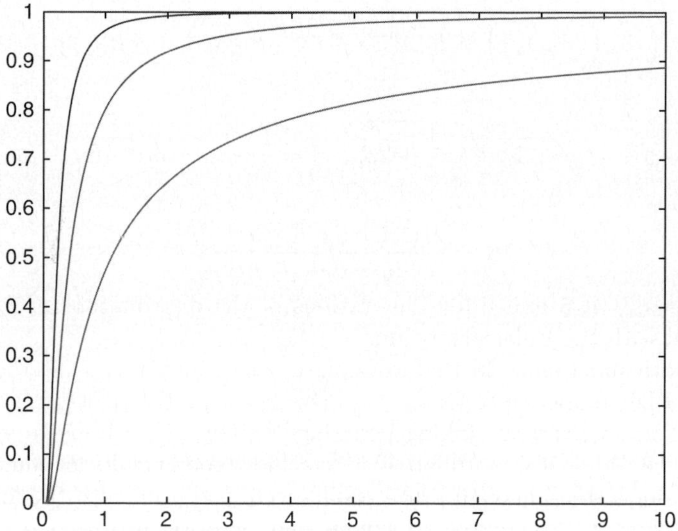

Fig. 4. Distribution function of $\int_0^\infty \exp(-2\,B_t^{(\mu)})\,dt$ for $\mu = 0.75$ (lowest curve), $\mu = 1.50$ and $\mu = 2.50$ (highest curve)

and

$$I_5 := \int_0^\infty (\exp(B_s^{(\mu,\sigma)}) + 1)^{-2} \mathbf{1}_{\{B_s^{(\mu,\sigma)} > 0\}} ds,$$

respectively.

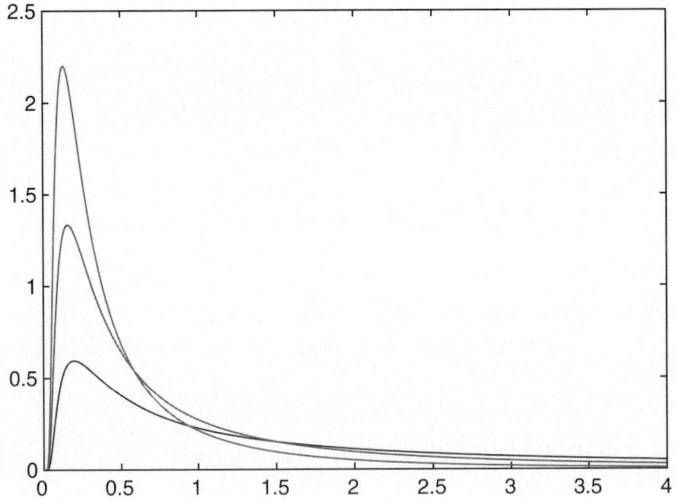

Fig. 5. Density function of $\int_0^\infty (\exp(B_t^{(\mu)}) + 1)^{-2}\, dt$ for $\mu = 0.25$ (lowest peak), $\mu = 0.50$ and $\mu = 0.75$ (highest peak)

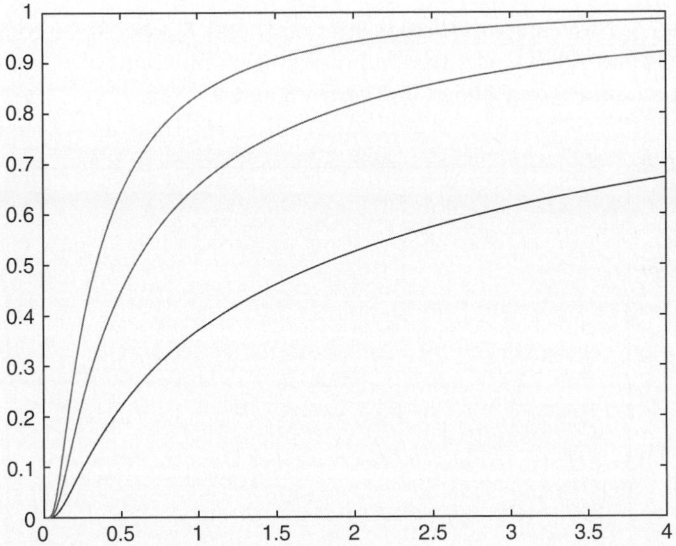

Fig. 6. Distribution function of $\int_0^\infty (\exp(B_t^{(\mu)})+1)^{-2}\, dt$ for $\mu = 0.25$ (lowest curve), $\mu = 0.50$ and $\mu = 0.75$ (highest curve)

In [24] it is shown that

$$I_4 = H_{1/\sigma}(R^{(2\mu/\sigma^2)}) \qquad \text{a.s.}$$

where $R_0^{(2\mu/\sigma^2)} = 0$. The Laplace transform of I_4 is hence given by

$$\mathbf{E}_0(\exp(-\rho I_4)) = \frac{\psi_\rho(0)}{\psi_\rho(1/\sigma)},$$

with (see [3] p. 133)

$$\psi_\rho(x) = x^{-\nu} I_\nu(x\sqrt{2\rho}) \quad \text{and} \quad \psi_\rho(0) = \frac{\rho^{\nu/2}}{2^{\nu/2}\,\Gamma(\nu+1)}$$

and $\nu = \mu/\sigma^2 - 1$.

The Laplace transform of the functional I_5 (in [24] this is called the one-sided translated Dufresne functional) is not known but the following identity (see [24]) holds

$$I_5 = H_0(Z) \qquad \text{a.s.},$$

where Z is a diffusion associated with the generator

$$\mathcal{G}f(x) = \frac{1}{2}\frac{d^2 f}{dx^2}(x) + \left(\frac{1}{2}\sigma + \frac{\mu - \frac{1}{2}\sigma^2}{\sigma\,(1 - \exp(\sigma x))}\right)\frac{df}{dx}(x)$$

living on $[-(\log 2)/\sigma, 0)$, having $-(\log 2)/\sigma$ as a reflecting barrier, and 0 as a killing barrier.

In Figure 7 we compare the densities of I_4 and I_5 (see [9] for comparisions between I_2 and I_4). The density and distribution functions of I_5 are displayed for different values of μ and σ in Figures 8 and 9.

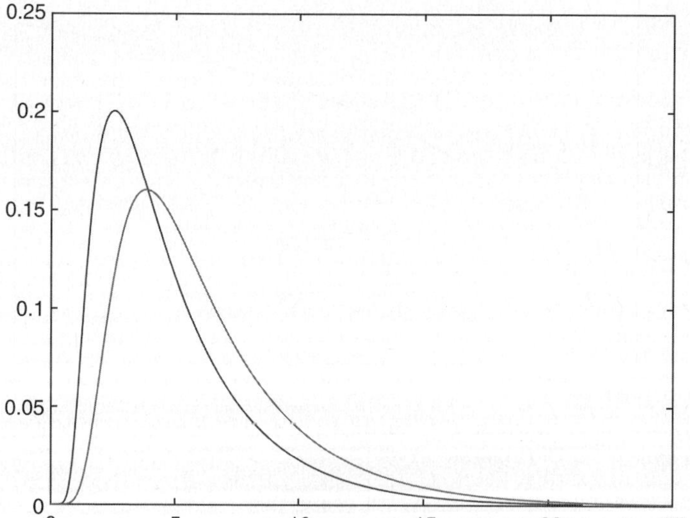

Fig. 7. Density function of $\int_0^\infty (\exp(B_t^{(\mu,\sigma)}) + 1)^{-2}\,\mathbf{1}_{\{B_t^{(\mu,\sigma)}>0\}}\,dt$ (upper peak) compared with the density function of $\int_0^\infty \exp(-2\,B_t^{(\mu,\sigma)})\,\mathbf{1}_{\{B_t^{(\mu,\sigma)}>0\}}\,dt$ for $\mu = 0.04$ and $\sigma = 0.20$

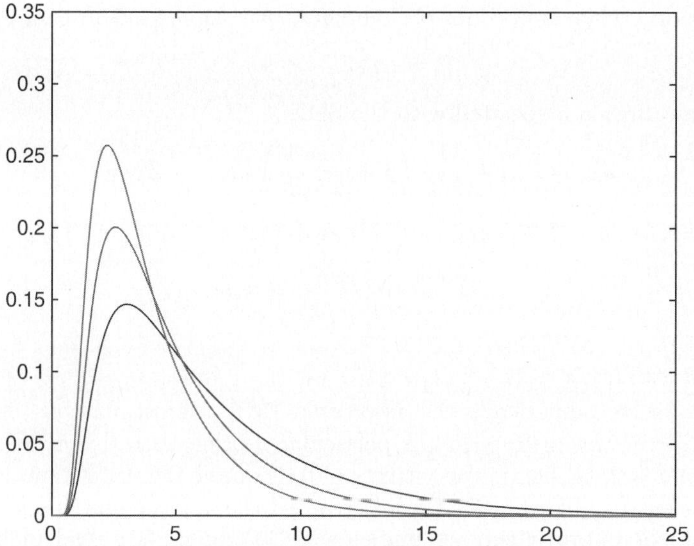

Fig. 8. Density function of $\int_0^\infty (\exp(B_t^{(\mu,\sigma)}) + 1)^{-2}\, \mathbf{1}_{\{B_t^{(\mu,\sigma)}>0\}}\, dt$ for $\sigma = 0.20$ and $\mu = 0.03$ (lowest peak), $\mu = 0.04$, and $\mu = 0.05$ (highest peak)

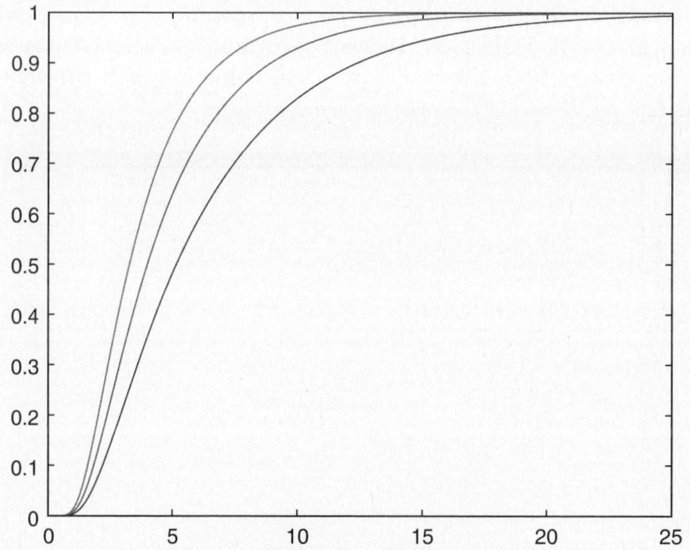

Fig. 9. Distribution function of $\int_0^\infty (\exp(B_t^{(\mu,\sigma)}) + 1)^{-2}\, \mathbf{1}_{\{B_t^{(\mu,\sigma)}>0\}}\, dt$ for $\sigma = 0.20$ and $\mu = 0.03$ (lowest curve), $\mu = 0.04$, and $\mu = 0.05$ (highest curve)

Example 9. In our final example we consider the functional

$$I_6^{(\delta)} := \int_0^\infty \exp(-2\, R_s^{(\delta)}) ds, \quad \delta \geq 2.$$

Proposition 1 when applied for $R^{(\delta)}$ and $g(x) := \exp(x)$ leads us to the identity

$$I_6 = H_0(Z) \qquad \text{a.s.} \tag{5.2}$$

with Z a diffusion associated with the SDE

$$dZ_t = dB_t + \frac{1}{2\,Z_t}\left(1 + \frac{\delta - 1}{\log Z_t}\right) dt, \qquad Z_0 = 1. \tag{5.3}$$

In the case $\delta = 3$ it is known (see Legall [14] and also [24]) that

$$I_6^{(3)} = H_1(R^{(2)}) \qquad \text{a.s.} \tag{5.4}$$

with $R_0^{(2)} = 0$.

Since we do not have an expression for the Laplace transform of $I_6^{(\delta)}$ for $\delta \neq 3$ we solve numerically the associated PDE. Unfortunately, due to the complexity of the drift term (in particular, notice that this tends, for all values on $\delta \geq 2$, to $+\infty$ in the vicinity of $0+$) simple finite difference schemes do not seem to give solutions converging to the correct one, see Figure 10. In search for improvement we implemented a nonuniform grid making the spatial discretization denser near the boundaries, and used a fourth-order implementation at the Neumann boundary. While this yielded better results than what is seen in Figure 10, full convergence still remained out of reach.

These difficulties can at least partly be overcome by transforming the diffusion Z given via SDE (5.3). Indeed, we study now the h-transform of Z with $h(x) = S(x) - S(0)$ where S is the scale function of Z Straightforward computations (cf. [3] p. 17) show that we may take

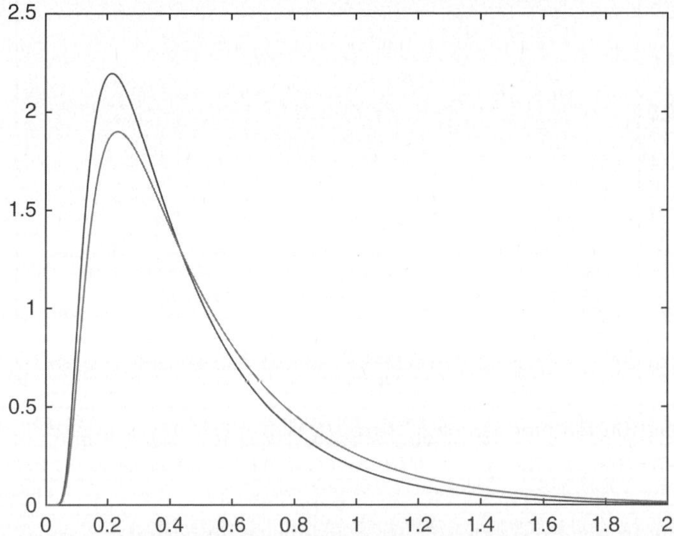

Fig. 10. Density function of $\int_0^\infty \exp(-2\,R_t^{(\delta)})\,dt$ for $\delta = 3.0$ (upper peak) obtained using the drift $\frac{1}{2x}(1 + \frac{\delta-1}{\log x})$ compared with the correct one (lower peak)

$$S(x) = \frac{1}{\delta - 2} \left|\log x\right|^{2-\delta}, \quad 0 < x < 1$$

(for simplicity we consider only the case $\delta > 2$). Then

$$\lim_{x \to 0} S(x) = 0 \quad \text{and} \quad S'(x) = x^{-1} |\log x|^{1-\delta}.$$

Consequently, the generator of the h-transform is given by

$$\begin{aligned} \mathcal{G}^{\uparrow} f &= \frac{1}{2} \frac{d^2 f}{dx^2} + \frac{1}{2x} \left(1 + \frac{\delta - 1}{\log x}\right) \frac{df}{dx} + \frac{S'(x)}{S(x)} \frac{df}{dx} \\ &= \frac{1}{2} \frac{d^2 f}{dx^2} + \frac{1}{2x} \left(1 + \frac{3 - \delta}{\log x}\right) \frac{df}{dx}, \qquad 0 < x < 1. \end{aligned}$$

Let $Z^{\uparrow}$ denote the h-transform, i.e., $Z^{\uparrow}$ is the diffusion associated with the generator $\mathcal{G}^{\uparrow}$. By Williams [30] time reversal result (see [3] p. 35, also for further references)

$$H_0(Z) \overset{(d)}{=} H_1(Z^{\uparrow}).$$

The PDE associated with $Z^{\uparrow}$ seems to be well suited for numerical computations. Notice, in particular, that if $\delta > 3$ the drift term of $Z^{\uparrow}$ tends to $+\infty$ as $x \to 1$ which fact is in strong contrast with the corresponding behaviour of the drift term of Z. Hereby it is also of interest to classify the boundaries of Z and $Z^{\uparrow}$. It holds for Z that the boundary point 0 is exit-not-entrance and 1 is entrance-not-exit. For the process $Z^{\uparrow}$ we have that 0 is entrance-not-exit and 1 is entrance-exit (regular) if $2 < \delta < 4$ and entrance-not-exit if $\delta \geq 4$. Figures 11, 12 show the density and distribution functions of I_6 for some choises of δ computed from the PDE associated with $Z^{\uparrow}$.

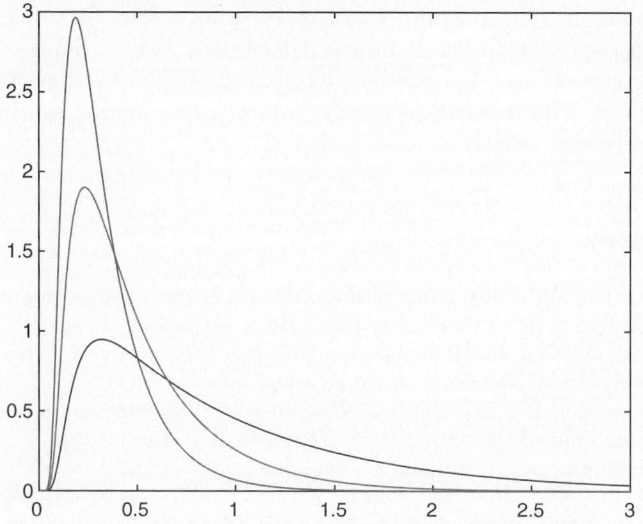

Fig. 11. Density function of $\int_0^\infty \exp(-2\,R_t^{(\delta)})\,dt$ for $\delta = 2.5$ (lowest peak), $\delta = 3.0$, $\delta = 3.5$ (highest peak)

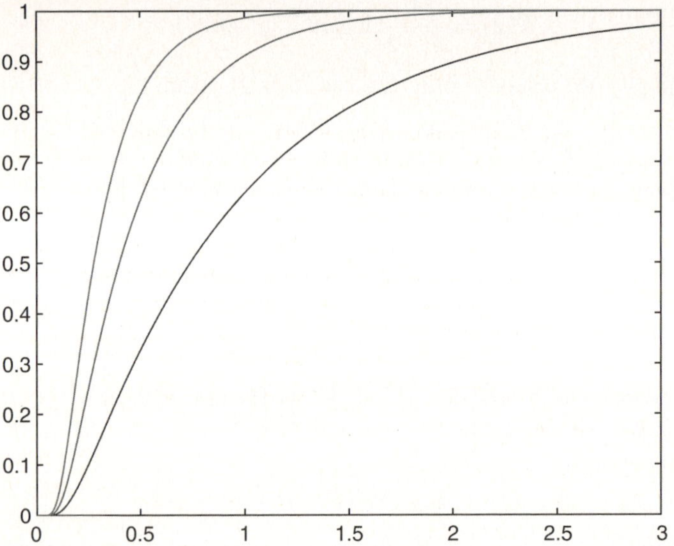

Fig. 12. Distribution function of $\int_0^\infty \exp(-2\,R_t^{(\delta)})\,dt$ for $\delta = 2.5$ (lowest curve), $\delta = 3.0$, $\delta = 3.5$ (highest curve)

As a final comment, and as an extra bonus from our transformation, we remark that when $\delta = 3$ then $\mathcal{G}^\uparrow$ is the generator of $R^{(2)}$, and we have recovered the identity (5.4).

Acknowledgements

Paavo Salminen thanks Vadim Linetsky for information on numerical inversion of Laplace transforms of hitting times. Olli Wallin thanks Siddhartha Mishra for several useful discussions on numerical solution of partial differential equations. The research of Paavo Salminen was funded by the Academy of Finland (grant 105849).

References

1. J. Abate and W. Whitt. Numerical inversion of Laplace transforms of probability distributions. *ORSA Journal of Computing*, 7(1):36–43, 1995.
2. R.M. Blumenthal and R.K. Getoor. *Markov Processes and Potential Theory.* Academic Press, New York, London, 1968.
3. A.N. Borodin and P. Salminen. *Handbook of Brownian Motion – Facts and Formulae, 2nd edition.* Birkhäuser, Basel, Boston, Berlin, 2002.
4. A.N. Borodin and P. Salminen. On some exponential integral functionals of BM(μ) and BES(3). *Zap. Nauchn. Semin. POMI*, 311:51–78, 2004. Preprint available in http://arxiv.org/abs/math.PR/0408367.
5. D. Britz, O. Østerby, and J. Strutwolf. Damping of Crank–Nicolson error oscillations. *Computational Biology and Chemistry*, 27:253–263, 2003.

6. P. Carr and V. Linetsky. The valuation of executive stock options in an intensity-based framework. *European Finance Review*, 4:211–230, 2000.
7. D. Davydov and V. Linetsky. Pricing and hedging path dependent options under the CEV process. *Management Science*, 47(7):949–965, 2001.
8. D. Davydov and V. Linetsky. Structuring, pricing and hedging double barrier step options. *J. Comput. Finance*, 5(2):55–87, 2001/02.
9. M. Decamps, A. De Schepper, M. Goovaerts, and W. Schoutens. A note on some new perpetuities. *Scand. Actuarial J.*, pages 261–270, 2005(4).
10. D. Dufresne. The distribution of a perpetuity, with applications to risk theory and pension funding. *Scand. Actuarial J.*, 1–2:39–79, 1990.
11. H.J. Engelbert and T. Senf. On functionals of Wiener process with drift and exponential local martingales. In M. Dozzi, H.J. Engelbert, and D. Nualart, editors, *Stochastic processes and related topics. Proc. Wintersch. Stochastic Processes, Optim. Control, Georgenthal/Ger. 1990*, number 61 in Math. Res., Academic Verlag, pages 45–58, Berlin, 1991.
12. W. Feller. The parabolic differential equations and the associated semi-groups of transformations. *Ann. Math.*, 55(3):468–519, 1952.
13. M. Fu, D. Madan, and T. Wang. Pricing asian options: a comparison of analytical and Monte Carlo methods. *J. Comput. Finance*, 2:49–74, 1997.
14. J.F. Le Gall. Sur la mesure de Hausdorff de la courbe brownienne. In J. Azéma and M. Yor, editors, *Séminaire de Probabilités XIX*, number 1123 in Springer Lecture Notes in Mathematics, pages 297–313, Berlin, Heidelberg, New York, 1985.
15. G. W. Harrison. Numerical solution of the Fokker Planck equation using moving finite elements. *Numerical methods for Partial differential Equations*, 4:219–232, 1988.
16. K. Itô and H.P. McKean. *Diffusion Processes and Their Sample Paths.* Springer Verlag, Berlin, Heidelberg, 1974.
17. J. Kent. Eigenvalue expansions for diffusion hitting times. *Z. Wahrscheinlichkeitstheorie verw. Gebiete*, 52:309–320, 1980.
18. D. Khoshnevisan, P. Salminen, and M. Yor. A note on a.s. finiteness of perpetual integral functionals of diffusions. *Electr. Comm. Prob.* 11:107–117, 2006.
19. H.P. McKean. Elementary solutions for certain parabolic differential equations. *TAMS*, 82:519–548, 1956.
20. H.P. McKean. *Stochastic Integrals.* Academic Press, New York, London, 1969.
21. M.A. Milevsky. The present value of stochastic perpetuity and the Gamma distribution. *Insur. Math. Econ.*, 20:243–250, 1997.
22. R. Poulsen. Approximate maximum likelihood estimation of discretely observed diffusion processes. *CAF working paper series*, 29, 1999.
23. P. Salminen and M. Yor. On Dufresne's perpetuity, translated and reflected. In J. Akahori, S. Ogawa, and S. Watanabe, editors, *Proceedings of Ritsumeikan International Symposium: Stochastic Processes and Applications to Mathematical Finance*, Singapore, 2004. World Scientific Publishing Co.
24. P. Salminen and M. Yor. Perpetual integral functionals as hitting times and occupation times. *Elect. J. Prob.*, 10:371–419, 2005.
25. P. Salminen and M. Yor. Properties of perpetual integral functionals of Brownian motion with drift. *Ann. I.H.P.*, 41(3):335–347, 2005.
26. A. De Schepper, M. Goovaerts, and F. Delbaen. The Laplace transform of annuities certain with exponential random time. *Insur. Math. Econ.*, 11(4):291–294, 1992.

27. J. Strikwerda. *Finite difference schemes and partial differential equations.* Wadsworth & Brooks/Cole, Pacific Grove, CA, 1989.
28. J.W. Thomas. *Numerical partial differential equations: finite difference methods.* Springer Verlag, New York, 1995.
29. I.V. Vagurina. On diffusion processes corresponding to hypergeometric equation. *Zap. Nauchn. Semin. POMI*, 311:79–91, 2004.
30. D. Williams. Path decompositions and continuity of local time for one-dimensional diffusions. *Proc. London Math. Soc.*, 28:738–768, 1974.
31. M. Yor. Sur certaines fonctionnelles exponentielles du mouvement brownien réel. *J. Appl. Probab.*, 29:202–208, 1992, (translated in English in [32].
32. M. Yor. *Exponential functionals of Brownian motion and related processes* in series Springer Finance. Springer Verlag, Berlin, Heidelberg, New York, 2001.

Chaos Expansions and Malliavin Calculus for Lévy Processes

Josep Lluís Solé, Frederic Utzet, and Josep Vives

Departament de Matemàtiques, Facultat de Ciències, Universitat Autónoma de Barcelona, 08193 Bellaterra (Barcelona), Spain. jllsole@mat.uab.cat, utzet@mat.uab.cat and vives@mat.uab.cat*

Summary. There are two different chaos expansions of a square integrable functional of a Lévy process: one proved by Itô [9] and the other by Nualart and Schoutens [17]. Related to each expansion a Malliavin type Calculus has been developed, being both quite different. In this paper we review the relationship between both approaches, and compare the corresponding Clark–Ocone–Haussmann representation formula.

Introduction

Kyoshi Itô, in his paper of 1956 [9], proved a chaotic representation of any square integrable functional of a Lévy process. However, in the Brownian motion case (Itô [8]) (and also in the Poisson process case, see, for example, [18]), the representation is expressed in terms of multiple integrals with respect to Brownian motion (respectively, the Poisson process), whereas in the general Lévy case it is necessary to introduce a two-parameter random measure associated with the Lévy process and the representation of a functional is written using multiple integrals with respect this random measure. On the other hand, Nualart and Schoutens [17], under some exponential integrability conditions on the Lévy measure, give a kind of chaotic representation expressed, roughly speaking, in terms of iterated integrals with respect the powers of the jumps of the Lévy processes. Both approaches enable the construction of a Malliavin calculus type for Lévy processes.

To be more specific, the Itô representation gives a Fock space structure to $L^2(\Omega)$, and then it is natural to define a two parameter (time and space) derivative operator $D_{t,x}$ as an annihilator operator, and a Skorohod integral δ of a two parameter process as a creation operator. These ideas have been developed by several authors: see Benth *et al.* [2], Øksendal and Proske [21], Di Nunno *et al.* [6,7], Løkka [13] (all these authors using a different random

* Supported by grants BFM2003-00261 Ministerio de Educación y Ciencia and 2001SGR-00174 Generalitat de Catalunya.

measure than Itô [9]); see also Lee and Shih [10, 11] for a pure white noise approach (with the Itô random measure), Privault [23], in the context of quantum probability, and Sole *et al.* [26].

In the Nualart and Schoutens approach, a family of strongly orthogonal martingales is introduced and the representation is expressed in terms of iterated integrals with respect these martingales; then, we can define a (one parameter) derivative operator for each n, $D_t^{(n)}$, as an annihilation operator of the n-th martingale, and a Skorohod integral $\delta^{(n)}$ as the adjoint of $\mathrm{Dom}D^{(n)}$ that turns out to be a creation operator for the martingale n: see Leon *et al.* [12] and Davis and Johansson [3].

In this paper we review some of the above cited papers, we extend some results proved only for purely jumps process or for simple Lévy processes to a general Lévy process with both continuous and jump part, and we study in deep the relationships between both approaches; in particular, we study the Clark–Ocone–Haussmann representation formula in both contexts.

1 Itô Chaotic Calculus for Lévy Processes

We will restrict ourselves to a square integrable Lévy processes because of the main results that we study need this hypothesis; for the same reason, we will only consider processes on a finite time interval $[0, T]$. This simplifies the definitions and notations; however, we should point out that Itô [9] proved the chaotic representation property for a completely general Lévy process with index on $\mathbb{R}$, and also the definition of the Malliavin derivative and the Skorohod integral can be done without the square integrability hypothesis (see Sole *et al.* [26]).

1.1 Itô Multiple Integrals

Consider a complete probability space $(\Omega, \mathcal{F}, \mathbb{P})$ and let $X = \{X_t,\, t \in [0, T]\}$ be a Lévy process (that means, X has stationary and independent increments, is continuous in probability and $X_0 = 0$), cadlag, centered, with $\mathbb{E}[X_1^2] < \infty$, and with Lévy measure ν. For all these concepts we refer to Sato [24]. Note that the square integrability of X_1 implies that $\int_{\mathbb{R}} x^2\, \nu(dx) < \infty$. Write $\mathbb{R}_0 = \mathbb{R} - \{0\}$. The process X can be represented as

$$X_t = \sigma W_t + \iint_{(0,t] \times \mathbb{R}_0} x\, d\widetilde{N}(t, x),$$

where $\{W_t,\, t \geq 0\}$ is a standard Brownian motion and $\sigma \geq 0$,

$$N(B) = \#\{t : (t, \Delta X_t) \in B\}, \quad B \in \mathcal{B}([0, T] \times \mathbb{R}_0),$$

is the jump measure of the process, where $\Delta X_t = X_t - X_{t-}$, and

$$d\widetilde{N}(t,x) = dN(t,x) - dt\, d\nu(x)$$

is the compensated jump measure.

Itô [9] extends the process X to an independent random measure M on $([0,T] \times \mathbb{R}, \mathcal{B}([0,T] \times \mathbb{R}))$: for a set $E \in \mathcal{B}([0,T] \times \mathbb{R})$, define

$$M(E) = \sigma \int_{E(0)} dW_t + \iint_{E'} x\, d\widetilde{N}(t,x), \tag{1.1}$$

where $E(0) = \{t \in \mathbb{R}_+ : (t,0) \in E\}$ and $E' = E - \{(t,0) \in E\}$. Then, for $E_1, E_2 \in \mathcal{B}([0,T] \times \mathbb{R})$

$$\mathbb{E}\left[M(E_1)M(E_2)\right] = \mu(E_1 \cap E_2),$$

where μ is the measure on $\mathcal{B}([0,T] \times \mathbb{R})$ given by

$$\mu(E) = \sigma^2 \int_{E(0)} dt + \iint_E x^2\, dt\, d\nu(x). \tag{1.2}$$

As in the Brownian setup, it is essential that this measure is continuous (Itô [9], page 256).

The following step is to construct multiple integrals with respect to M, and that is done in the same way as in the Brownian case: First consider an elementary function of the form

$$f = \mathbf{1}_{E_1 \times \cdots \times E_n},$$

where $E_1, \ldots, E_n \in \mathcal{B}(\mathbb{R}_+ \times \mathbb{R})$, are pairwise disjoints, with

$$\mu(E_1) < \infty, \ldots, \mu(E_n) < \infty,$$

and let

$$I_n(f) = M(E_1) \cdots M(E_n).$$

By linearity and continuity, I_n is extended to

$$L_n^2 =_{\text{def}} L^2\left(([0,T] \times \mathbb{R})^n, \mathcal{B}([0,T] \times \mathbb{R})^n, \mu^{\otimes n}\right).$$

This integral has the usual properties (Itô [9], Theorem 1):

1. $$I_n(f) = I_n(\widetilde{f}),$$
   where $\widetilde{f}$ is the symmetrization of f:
   $$\widetilde{f}(z_1, \ldots, z_n) = \frac{1}{n!} \sum_{\pi \in \mathfrak{G}_n} f\left(z_{\pi(1)}, \ldots, z_{\pi(n)}\right),$$
   $z_i = (t_i, x_i)$, $i = 1, \ldots, n$ and $\mathfrak{G}_n$ is the set of permutations of $\{1, 2, \ldots, n\}$.
2. $$I_n(af + bg) = aI_n(f) + bI_n(g).$$

3. $$\mathbb{E}[I_n(f)I_m(g)] = \delta_{n,m} n! \int_{(\mathbb{R}_+\times\mathbb{R})^n} \widetilde{f}\,\widetilde{g}\, d\mu^{\otimes n},$$

where $\delta_{n,m} = 1$, if $n = m$, and 0 otherwise.

Let $\{\mathcal{F}_t^X, t \in [0,T]\}$ be the natural filtration of X completed with the null sets of $\mathcal{F}$; it is well known that this filtration is right continuous. Write $L^2(\Omega) = L^2(\Omega, \mathcal{F}_T^X, \mathbb{P})$. Itô [9, Theorem 2], proves that every functional $F \in L^2(\Omega)$ can be represented as

$$F = \sum_{n=0}^{\infty} I_n(f_n), \quad f_n \in L_n^2, \tag{1.3}$$

and the representation is unique if we take every f_n symmetric. We will always assume the symmetry in the kernels f_n of such decompositions.

1.2 Derivative Operators

From the chaotic representation property, it is natural to apply all the machinery of the annihilation operators (Malliavin derivatives) and creation operators (Skorohod integrals) on Fock spaces as is exposed in Nualart and Vives [18, 19]. This point of view has been developed by Benth *et al.* [2], Øksendal and Proske [21], Di Nunno *et al.* [6, 7], Løkka [13], Yablonski [28], Lee and Shih [10, 11] and Solé *et al.* [26].

Let Dom$D$ be the set of functionals $F \in L^2(\Omega)$ with chaotic representation $F = \sum_{n=0}^{\infty} I_n(f_n)$, ($f_n$ symmetric) that satisfies

$$\sum_{n=1}^{\infty} n\, n! \|f_n\|_{L_n^2}^2 < \infty. \tag{1.4}$$

For $F \in \mathrm{Dom}D$, the Malliavin derivative is the stochastic process

$$DF : [0,T] \times \mathbb{R} \times \Omega \longrightarrow \mathbb{R}$$

defined by

$$D_z F = \sum_{n=1}^{\infty} n I_{n-1}\Big(f_n(z,\cdot)\Big), \quad z \in [0,T] \times \mathbb{R},$$

where the convergence of the series is in $L^2([0,T] \times \mathbb{R} \times \Omega, \mu \otimes \mathbb{P})$. The set Dom$D$ is a Hilbert space with the scalar product

$$\langle F, G \rangle = \mathbb{E}[F\,G] + \mathbb{E}\left[\iint_{[0,T]\times\mathbb{R}} D_z F\, D_z G\, d\mu(z)\right],$$

and D is a closed operator from Dom$D$ to $L^2\left([0,T] \times \mathbb{R} \times \Omega, \mu \otimes \mathbb{P}\right)$.

Given the form of the measure μ, for $f : ([0,T] \times \mathbb{R})^n \to \mathbb{R}$ measurable, positive or $\mu^{\otimes n}$ integrable, we have

$$\int_{(\mathbb{R}_+\times\mathbb{R})^n} f\,d\mu^{\otimes n}$$
$$= \int_{[0,T]\times([0,T]\times\mathbb{R})^{n-1}} f\big((t,0),z_1,\dots z_{n-1})\big)\,dt\,d\mu^{\otimes(n-1)}(z_1,\dots,z_{n-1})$$
$$+ \int_{[0,T]\times\mathbb{R}_0\times([0,T]\times\mathbb{R})^{n-1}} f(z_1,z_2,\dots z_n)\,d\mu^{\otimes(n)}(z_1,z_2,\dots,z_n),$$

As a consequence, when $\sigma > 0$ or $\nu \neq 0$, it is natural to consider two more spaces:

1. Let DomD^0 (if $\sigma > 0$) be the set of $F \in L^2(\Omega)$ with decomposition $F = \sum_{n=0}^{\infty} I_n(f_n)$ such that

$$\sum_{n=1}^{\infty} n\,n! \int_{[0,T]\times(R_+\times\mathbb{R})^{n-1}} f^2((t,0),z_1,\dots,z_{n-1})\,dt \times d\mu^{\otimes(n-1)}(z_1,\dots,z_{n-1}) < \infty, \tag{1.5}$$

   For such F we can define the square integrable stochastic process

$$D_{t,0}F = \sum_{n=1}^{\infty} n\,I_{n-1}\Big(f_n\left((t,0),\cdot\right)\Big),$$

   with convergence in $L^2([0,T]\times\Omega, dt\otimes\mathbb{P})$.
2. When $\nu \neq 0$, let DomD^J be the set of $F = \sum_{n=0}^{\infty} I_n(f_n)$ such that

$$\sum_{n=1}^{\infty} n\,n! \int_{R_+\times\mathbb{R}_0\times([0,T]\times\mathbb{R})^{n-1}} f_n^2\,d\mu^{\otimes n} < \infty,$$

   and for F satisfying this condition, define

$$D_z F = \sum_{n=1}^{\infty} n I_{n-1}\Big(f_n\left(z,\cdot\right)\Big),$$

   convergence in $L^2([0,T]\times\mathbb{R}_0\times\Omega, x^2\,dt\,d\nu(x)\otimes\mathbb{P}))$.

   When both $\sigma > 0$ and $\nu \neq 0$, then Dom$D$ = Dom$D^0 \cap$ DomD^J.

1.3 Alternative Definition of the Derivative Operator and Practical Rules

For practical purposes, the above definitions are not very useful since, in general, the chaos expansion of a functional is not known, and therefore some rules to compute the derivative are needed. In the case of Brownian motion, through the definition of the derivative as a week derivative on the canonical space, a chain rule is proved (see, for example, Nualart [16]). For the case of the Poisson process, it is proved that the Malliavin derivative coincides with a difference operator on the canonical space (Nualart and Vives [18]). In our context we can proceed as following: Since the Brownian part and the jumps part of X are independent, it can be constructed a canonical space of X as a product space $\Omega_W \times \Omega_J, \mathcal{F}_W \otimes \mathcal{F}_J, \mathbb{P}_W \otimes \mathbb{P}_J$, where:

- $(\Omega_W, \mathcal{F}_W, \mathbb{P}_W, \{W_t,\, t \in [0,T]\})$ is the canonical Brownian process; that is $\Omega_W = \mathcal{C}([0,T])$ is the space of continuous functions on $[0,T]$, null at the origin, with the topology of the uniform convergence, $\mathcal{F}_W$ the Borel σ-algebra and P_W the probability that makes the projections $W_t : \Omega_W \to \mathbb{R}$ a Brownian motion.
- $(\Omega_J, \mathcal{F}_J, \mathbb{P}_J, \{J_t, t \in [0,T]\})$ is the canonical pure jump Lévy process

$$J_t = \iint_{(0,t]\times\mathbb{R}_0} x\, d\widetilde{N}(t,x).$$

See [26] for a complete construction of this space, where we extend the ideas of Neveu [20] to a general Lévy process. Essentially, Ω_J is formed by infinite sequences $\omega = \big((t_1,x_1),(t_2,x_2),\ldots\big) \in \big([0,T]\times\mathbb{R}_0\big)^{\mathbb{N}}$, such that for every $\varepsilon > 0$, there is only a finite number of (t_i,x_i) with $|x_i| > \varepsilon$, where the t_i are the instants of jump and x_i the size of the corresponding jump.

Derivative $D_{t,0}$

In order to compute the derivative $D_{t,0}F$ for $F \in L^2(\Omega_W \times \Omega_J)$ we can use the the isometry

$$L^2(\Omega_W \times \Omega_J) \simeq L^2(\Omega_W; L^2(\Omega_J)),$$

and consider F as element of $L^2(\Omega_W; L^2(\Omega_J))$ and apply the theory of Malliavin derivative of a Hilbert space valued random variable following Nualart [16, page 61]. Under some restrictions, it is proved (see Solé *et al.* [26]) that both derivatives coincide. This is proved in the following way: By definition, a $L^2(\Omega_J)$-valued smooth random variable has the form

$$F = \sum_{i=1}^{n} G_i\, H_i,$$

where G_i are ordinary Brownian smooth random variables and $H_i \in L^2(\Omega_J)$. Define the Malliavin derivative of F as

$$D_t^{W^*} F = \sum_{i=1}^{n} D_t G_i \otimes H_i, \tag{1.6}$$

where D_t is the ordinary Malliavin derivative. This definition is extended to a subspace $\mathrm{Dom} D^{W^*}$.

Proposition 1. *$DomD^{W^*} \subset \mathrm{Dom}D^0$, and for $F \in DomD^{W^*}$,*

$$D_t^{W^*} F = \sigma\, D_{t,0}F. \tag{1.7}$$

Idea of the Proof.

First we consider the functionals of the form

$$F = N(B_1)\cdots N(B_m)W(C_1)\cdots W(C_k),$$

where $B_1,\dots,B_m \in \mathcal{B}(\mathbb{R}_+ \times \mathbb{R}_0)$ are pairwise disjoints, $0 \notin \overline{B}_i$, and $C_1,\dots,C_k \in \mathcal{B}(\mathbb{R}_+)$ pairwise disjoints, with $\int_{C_j} dt < \infty$. Itô [9, Lemma 2], shows that such F can be written as a sum of multiple integrals:

$$F = I_0(f_0) + \cdots + I_{m+k}(f_{m+k}),$$

and then the derivatives are easy to compute.

The final step is the extension of the the formula (1.7) to $\mathrm{Dom}D^{W^*}$, that is done by a density argument. □

Remark. Note that when there is no jump part, then $X_t = \sigma\, W_t$. Therefore, the derivative $D_{t,0}F$ coincides with the classical Malliavin derivative, except in a factor σ, due to the fact that in $D_{t,x}F$ we are differentiating with respect to $\sigma\, W$.

From the above Proposition, we deduce the following rule of differentiation:

Proposition 2 (Chain Rule). *Let $F = f(Z, Z') \in L^2(\Omega_W \times \Omega_J)$ with $Z \in DomD^W$ and $Z' \in L^2(\Omega_J)$, and $f(x,y)$ a continuously differentiable function with bounded partial derivatives in the variable x. Then $F \in DomD^0$ and*

$$D_{t,0}F = \frac{1}{\sigma}\frac{\partial f}{\partial x}(Z,Z')\, D_t^W Z,$$

where D^W is the Malliavin derivative in $(\Omega_W, \mathcal{F}_W, \mathbb{P}_W)$ and $DomD^W$ its domain.

Derivative $D_{t,x}$, $x \neq 0$

Consider $\omega = (\omega^W, \omega^J) \in \Omega_W \times \Omega_J$, $\omega^J = \big((t_1,x_1),(t_2,x_2),\dots\big) \in \big([0,T] \times \mathbb{R}_0\big)^{\mathbb{N}}$, with the restrictions pointed out above.

Given $z = (t,x) \in [0,T] \times \mathbb{R}_0$, we introduce in ω^J a jump of size x at instant t, and call the new element $\omega_z^J = \big((t_1,x_1),(t_2,x_2),\dots,(t,x),\dots\big)$, and write $\omega_z = (\omega^W, \omega_z^J)$. For a random variable F, we define the translation operator

$$\boldsymbol{\Psi}_{t,x}F = \frac{F(\omega_{t,x}) - F(\omega)}{x}.$$

This operator has the following property (Solé *et al.* [26]).

Proposition 3. *Let $F \in L^2(\Omega_W \times \Omega_J)$ such that*

$$E\left[\iint_{[0,T]\times\mathbb{R}_0} (\boldsymbol{\Psi}_z F)^2 \, \mu(dz)\right] < \infty.$$

Then $F \in DomD^J$ and

$$D_z F(\omega) = \boldsymbol{\Psi}_z F(\omega), \quad \mu \otimes \mathbb{P} \text{ a.e. } (z,\omega) \in [0,T] \times \mathbb{R}_0 \times \Omega.$$

Idea of the Proof.

We can restrict to a purely jump process. First, a compound Poisson process is considered; for this class of processes, as in the Poisson case, the multiple integral can be computed by each ω. Also, for a functional F with a finite chaos expansion, the translation operator can be computed explicitly. Then the equality of the derivative operator $D_{t,x}$ and the translation operator $\boldsymbol{\Psi}$ is checked on these functionals, and then extended to more general random variables. Finally, the result for a general purely jump Lévy process is obtained considering such process as limit a of compound processes. Of course, all the proof relies heavily on the structure of Ω_J that we have commented before. □

As an example of the power of the chain rule and the increment quotient formula, in Solé *et al.* [27] there is the computation of the minimal quadratic hedging of an Asian option for a general jump-diffusion process, where the main difficulty is that the functional depends on the entire trajectory of the process, and not only on the final value as in the European case.

1.4 The Skorohod Integral

Following the scheme of Nualart and Vives [18], we can define a creation operator (Skorohod integral) in the following way: Let $f \in L^2\big([0,T] \times \mathbb{R} \times \Omega, \mathcal{B}([0,T] \times \mathbb{R}) \otimes \mathcal{F}^X, \mu \otimes \mathbb{P}\big)$. There is a chaotic decomposition

$$f(z) = \sum_{n=0}^{\infty} I_n(f_n(z,\cdot)), \tag{1.8}$$

where $f \in L^2_{n+1}$ is symmetric in the n last variables. Now, denote by $\widehat{f_n}$ the symmetrization in all $n+1$ variables. If

$$\sum_{n=0}^{\infty} (n+1)! \, \|\widehat{f_n}\|^2_{L^2_{n+1}} < \infty, \tag{1.9}$$

define the Skorohod integral of f by

$$\delta(f) = \sum_{n=0}^{\infty} I_{n+1}(\widehat{f_n}),$$

convergence in $L^2(\Omega)$. Denote by Domδ the set of f that satisfy (1.9). The main properties of the operator δ are:

1. Duality formula: A process $f \in L^2([0,T]\times\mathbb{R}\times\Omega, \mu\times\mathbb{P})$ belongs to Domδ if and only if there is a constant C such that for all $F \in \mathrm{Dom}D$,
$$\left|\mathbb{E}\iint_{[0,T]\times R} f(z)\, D_z F\, d\mu(z)\right| \le C\left(\mathbb{E}[F^2]\right)^{1/2}.$$
If $f \in \mathrm{Dom}\delta$, then $\delta(f)$ is the element of $L^2(\Omega)$ characterized by
$$\mathbb{E}[\delta(f)\, F] = \mathbb{E}\iint_{[0,T]\times R} f(z)\, D_z F\, d\mu(z), \tag{1.10}$$
for any $F \in \mathrm{Dom}D$.
2. Isometry: Denote by $\mathbb{L}^{1,2}$ the set of elements $f \in L^2\big([0,T]\times\mathbb{R}\times\Omega, \mathcal{B}([0,T]\times\mathbb{R})\otimes\mathcal{F}^X, \mu\otimes\mathbb{P}\big)$ such that $f(z)\in\mathrm{Dom}D$, $\forall z$ μ – a.e and that $D_{\bullet}f(\cdot)\in L^2\big(([0,T]\times\mathbb{R})^2\times\Omega\big)$. In terms of the chaotic expression (1.8) of f, both conditions are equivalent to
$$\sum_{n=1}^{\infty} n\, n!\|\widehat{f}_n\|^2_{L^2_{n+1}} < \infty,$$
and, in particular, this implies $\mathbb{L}^{1,2}\subset\mathrm{Dom}\delta$. For $f, g\in\mathbb{L}^{1,2}$,
$$\begin{aligned} E[\delta(f)\delta(g)] &= \mathbb{E}\iint_{[0,T]\times\mathbb{R}} f(z)\, g(z) d\mu(z) \\ &\quad + \mathbb{E}\iint_{([0,T]\times R)^2} D_z f(z')\, D_{z'} g(z')\, d\mu(z)\, d\mu(z'). \end{aligned}$$
3. Differentiability of δ. Let $f\in\mathbb{L}^{1,2}$ such that $D_z f\in\mathrm{Dom}\delta, \forall z,\ \mu$ – a.e. Then $\delta(f)\in\mathrm{Dom}D$ and
$$D_z\delta(f) = f(z) + \delta(D_z f),\ \forall z,\ \mu - \text{a.e.}$$

1.5 Integral Respect to the Random Measure M and Skorohod Integral

The random measure M, with the filtration $\{\mathcal{F}_t^X, t\in[0,T]\}$, induces a martingale-valued measure, and it can be defined a stochastic integral $\iint_{[0,T]\times\mathbb{R}} f(z)\, dM_z$, of a predictable process f such that $E\iint_{[0,T]\times\mathbb{R}} f^2(z)\,\mu(dz) < \infty$. (See Applebaum [1, Chapter 4]). For f and g μ-square integrable predictable processes
$$\mathbb{E}\left[\iint_{[0,T]\times\mathbb{R}} f(z)\, dM_z\cdot\iint_{[0,T]\times\mathbb{R}} g(z)\, dM_z\right] = \mathbb{E}\left[\iint_{[0,T]\times\mathbb{R}} f(z)g(z)\, d\mu(z)\right].$$

An explicit expression for the integral $\iint_{[0,T]\times\mathbb{R}} f(z)\,dM_z$ is given in the following Proposition:

Proposition 4. *Let $f = \{f(z),\, z \in [0,T]\times\mathbb{R}\}$ be a predictable process such that $E\iint_{[0,T]\times\mathbb{R}} f^2(z)\,\mu(dz) < \infty$. Then*

$$\iint_{[0,T]\times\mathbb{R}} f(z)\,dM_z = \sigma\int_0^T f(t,0)\,dW_t + \iint_{[0,T]\times\mathbb{R}_0} xf(t,x)\,d\widetilde{N}(t,x). \quad (1.11)$$

As in the Brownian case, the Skorohod integral restricted to predictable processes coincides with the integral respect to the random measure M. The next theorem is a version of Privault [23, Proposition 11], Di Nunno *et al.* [6, Proposition 3.15], and Oksendal and Proske [21, Proposition 3.7]:

Theorem 5. *Let $f \in L^2([0,T]\times\mathbb{R}\times\Omega)$ be a predictable process. Then $f \in$ Domδ and*

$$\delta(f) = \iint_{[0,T]\times\mathbb{R}} f(z)\,dM_z.$$

Sketch of the Proof.

Consider first a simple predictable process f of the form:

$$f(z) = \beta\mathbf{1}_{(r,s]\times B}(z),$$

where $0 \le r < s \le T$, β is a bounded random variable $\mathcal{F}_r^X$ measurable, and $B \in \mathcal{B}(\mathbb{R})$. In order to prove that $f \in \mathrm{Dom}(\delta)$ and $\delta(f) = \iint_{[0,T]\times\mathbb{R}} f(z)\,dM_z$, by the duality relation (1.10) we need to check that for all $F \in \mathrm{Dom}D$, we have

$$\mathbb{E}\left[\iint_{[0,T]\times R} f(z)\,D_zF\,d\mu(z)\right] = \mathbb{E}\left[F\iint_{[0,T]\times\mathbb{R}} f(z)\,dM_z\right].$$

By a density argument, it is enough to consider the case $F = I_n(\widetilde{f_n})$, where $f_n = \mathbf{1}_{E_1\times\cdots\times E_n}$, with $E_1\ldots,E_n \in \mathcal{B}([0,T]\times\mathbb{R})$ pairwise disjoints with $\mu(E_i) < \infty$. □

1.6 The Lévy Process as a Normal Martingale: Multiple Integrals Respect to *X*

The Lévy process X is a square integrable martingale with predictable quadratic variation $\langle X,X\rangle_t = ct$, where $c = \sigma^2 + \int_{\mathbb{R}} x^2\,d\nu(x)$. Then, X is a normal martingale in the sense of Dellacherie and Meyer [15, page 199]. If $g = \{g(t),\, t \in [0,T]\}$ is a (unidimensional) predictable process such that $E\int_0^T g(t)^2dt < \infty$, then g will be integrable with respect to X. Moreover, $g(t)\,x$ is (two-parameter) predictable and it is also integrable with respect to $\widetilde{N}$, and

$$\int_0^T g(t)\,dX_t = \sigma \int_0^T g(t)\,dW_t + \iint_{[0,T]\times\mathbb{R}_0} g(t)\,x\,d\widetilde{N}(t,x).$$

From Proposition 4 we deduce

$$\int_0^T g(t)\,dX_t = \iint_{[0,T]\times\mathbb{R}} g(t)\,dM(t,x). \tag{1.12}$$

On the other hand, it can be defined a multiple integral respect to X: see Meyer [15]. For $f \in L^2\big([0,T]^n, (dt)^n\big)$, symmetric, we can consider $I_n^X(f)$, which coincides with $n!$ times the iterated integral:

$$I_n^X(f) = n! \int_0^T \int_0^{t_n^-} \cdots \int_0^{t_2^-} f(t_1,\ldots,t_n)\,dX_{t_n}\cdots dX_{t_2}\,dX_{t_1}. \tag{1.13}$$

The multiple integral with respect to X, I_n^X, coincides with the restriction of the multiple integral I_n to $L^2([0,T]^n)$: working first with elementary functions of the form $\mathbf{1}_{A_1\times\cdots\times A_n}$, $A_1,\ldots,A_n \in \mathcal{B}\big([0,T]\big)$ pairwise disjoints, and by a density argument, it is proved that for $f \in L^2\big([0,T]^n, (dt)^n\big)$, symmetric, we have $I_n^X(f) = I_n(f)$. This is a multivariate extension of (1.12).

2 Malliavin Calculus and Teugels Martingales

In the rest of the paper we will assume the condition of Nualart-Schoutens [17] (see also Schoutens [25]) about the exponential integrability of the Lévy measure, that is, there are $\varepsilon > 0$ and $\lambda > 0$ such that

$$\int_{(-\varepsilon,\varepsilon)^c} \exp\{\lambda\,|x|\}\,\nu(dx) < \infty.$$

This condition implies that the Lévy measure has moments of all order ≥ 2, that X_t has moments of all order, and that the polynomials are dense in $L^2(\mathbb{R}, P\circ X_1^{-1})$.

2.1 Teugels Martingales

Following Nualart and Schoutens [17], consider the square integrable martingales, called ***Teugel martingales***

$$Y_t^{(1)} = X_t,$$

$$Y_t^{(n)} = \sum_{0<s\leq t} \big(\Delta X_s\big)^n - m_n\,t, \quad n \geq 2,$$

where $m_n = \int_{-\infty}^{\infty} x^n \, \nu(dx)$, $n \geq 2$. Introduce also the martingales

$$H_t^{(n)} = \sum_{j=1}^{n} a_{nj} Y_t^{(j)}, \quad n \geq 1,$$

where the constants a_{nj} are chosen in such a way that $a_{nn} = 1$ and the martingales $H^{(n)}$, $n = 1, 2, \ldots$ are pairwise strongly orthogonal. Its predictable quadratic variation process is

$$\langle H^{(n)}, H^{(m)} \rangle_t = \delta_{nm} \, q_n \, t, \tag{2.1}$$

for some constants q_n; so they are normal martingales. The strong orthogonality of the martingales $H^{(n)}$ is equivalent to the existence of an orthogonal family of polynomials with respect to the measure

$$d\eta(x) = \sigma^2 \, d\delta_0(x) + x^2 \, d\nu(x),$$

Note that $d\mu(t, x) = dt \, d\eta(x)$. Specifically, the polynomials p_n defined by

$$p_n(x) = \sum_{j=1}^{n} a_{nj} x^{j-1}$$

are orthogonal with respect to de measure η:

$$\int_{\mathbb{R}} p_n(x) p_m(x) \, d\eta(x) = 0, \quad n \neq m.$$

and the q_n appearing in (2.1) is $q_n = \int_{\mathbb{R}} p_n^2(x) \, d\eta(x)$,

Proposition 6. *Let $g = \{g(t),\, t \in [0, T]\}$ be a predictable process such that $\mathbb{E}\left[\int_0^T g^2(t) \, dt\right] < \infty$. Then the process $f(t, x) = g(t) \, p_n(x)$ is integrable with respect to M and*

$$\int_0^T g(t) \, dH_t^{(n)} = \iint_{[0,T] \times \mathbb{R}} g(t) p_n(x) \, dM(t, x). \tag{2.2}$$

Proof.

For $n = 1$, we have $p_1(x) = 1$, $H^{(1)} = Y^{(1)} = X$, and then (2.2) is the formula (1.12).

For $n \geq 2$, the process $g(t) \, x^n$ is predictable and

$$\mathbb{E}\left[\iint_{[0,T] \times \mathbb{R}} g^2(t) \, x^{2n} \, dt d\nu(x)\right] < \infty.$$

Therefore it is integrable with respect to the random measure $\widetilde{N}(t,x)$ and

$$\int_0^T g(t)\,dY_t^{(n)} = \iint_{[0,T]\times\mathbb{R}_0} g(t)\,x^n\,d\widetilde{N}(t,x) = \iint_{[0,T]\times\mathbb{R}} g(t)\,x^{n-1}\,dM(t,x)$$

It follows that

$$\int_0^T g(t)\,dH_t^{(n)} = \sum_{j=1}^{n} a_{n,j} \int_0^T g(t)\,dY_t^{(j)} = \sum_{j=1}^{n} a_{n,j} \iint_{[0,T]\times\mathbb{R}} g(t)\,x^{j-1} dM(t,x)$$
$$= \iint_{[0,T]\times\mathbb{R}} g(t) p_n(x)\,dM(t,x). \;\square$$

Note that in particular,

$$H_t^{(n)} = \iint_{[0,t]\times\mathbb{R}} p_n(x)\,dM(s,x). \tag{2.3}$$

2.2 The Chaotic Representation of Nualart and Schoutens

The chaotic representation of Nualart and Schoutens uses iterated integrals of the form

$$J_n^{(i_1,\dots,i_n)}(g) = \int_0^T \Bigg(\int_0^{t_n-} \cdots \left(\int_0^{t_2-} g(t_1,\dots,t_n) dH^{(i_1)}(t_1) \right) \cdots dH^{(i_{n-1})}(t_{n-1}) \Bigg) dH^{(i_n)}(t_n),$$

for $g \in L^2([0,T]^n)$. For different indices, these integrals are orthogonal (see Leon *et al.* [12]): Given $f \in L^2([0,T]^n)$ and $g \in L^2([0,T]^m)$,

$$E\big[J_n^{(i_1,\dots,i_n)}(f)\, J_m^{(j_1,\dots,j_m)}(g)\big]$$
$$= \begin{cases} q_{i_1}\cdots q_{i_n} \int_{\Sigma_n} f(t_1,\dots,t_n)\, g(t_1,\dots,t_n)\, dt_1 \cdots dt_n, \\ \qquad\qquad \text{if } n=m \text{ and } (i_1,\dots,i_n) = (j_1,\dots,j_n), \\ 0, \qquad \text{otherwise,} \end{cases}$$

where $\Sigma_n = \{(t_1,\dots,t_n) \in \mathbb{R}_+^n : 0 < t_1 < t_2 < \cdots < t_n \le T\}$ is the positive simplex of $[0,T]^n$.

Nualart and Schoutens [17, Theorem 5] prove that every square integrable random variable F has a representation of the form

$$F = EF + \sum_{n=1}^{\infty} \sum_{i_1,\dots,i_n \ge 1} J_n^{(i_1,\dots,i_n)}(g_{i_1,\dots,i_n}), \tag{2.4}$$

where $g_{i_1,\dots,i_n} \in L^2([0,T]^n)$. We will assume that the functions $g_{i_1,\dots,i_n}$ in that representation are symmetric.

The next proposition gives the relationship between the iterated integral and the multiple integral and it is an extension of formulas (1.13) and (2.2):

Proposition 7. *Let $g \in L^2([0,T]^n)$. Then*

$$J_n^{(i_1,\dots,i_n)}(g) = I_n\big(g(t_1,\dots,t_n)\mathbf{1}_{\Sigma_n}(t_1,\dots,t_n)\,p_{i_1}(x_1)\cdots p_{i_n}(x_n)\big). \quad (2.5)$$

2.3 Malliavin Derivative in the k-th Direction

Let $g \in L^2([0,T]^n)$ and $k \geq 1$. Leon *et al.* [12] define the derivative of $J_n^{(i_1,\dots,i_n)}(g)$ in the k–th direction as the process

$$D_t^{(k)} J_n^{(i_1,\dots,i_n)}(g) = \sum_{\ell=1}^{n} \mathbf{1}_{\{i_\ell=k\}} J_{n-1}^{(i_1,\dots,\widehat{i_\ell},\dots,i_n)}\big(g(\underbrace{\cdots}_{\ell-1},t,\cdots)\mathbf{1}_{\Sigma_n^{(\ell)}(t)}(\cdot)\big), \quad (2.6)$$

where

$$\Sigma_n^{(\ell)}(t) = \{(t_1,\dots,\widehat{t_\ell},\dots,t_n) \in \Sigma_{n-1}\colon 0 < t_1 < \cdots < t_{\ell-1} < t \leq t_{\ell+1} < \cdots < t_n\}$$

and $\widehat{i}$ means that the i-th index is omitted. This definition is extended in a natural way to the subspace of $L^2(\Omega)$:

$$\mathrm{Dom}D^{(k)} = \Big\{F \in L^2(\Omega),\ F = EF + \sum_{n=1}^{\infty} \sum_{i_1,\dots,i_n \geq 1} J_n^{(i_1,\dots,i_n)}(g_{i_1,\dots,i_n}) :$$
$$\sum_{n=1}^{\infty} \frac{1}{n!} \sum_{i_1,\dots,i_n\geq 1} \sum_{\ell=1}^{n} \mathbf{1}_{\{i_\ell=k\}} q_{i_1}\cdots \widehat{q_{i_\ell}} \cdots q_{i_n} \|g_{i_1,\dots,i_n}\|^2_{L^2([0,T]^n)} < \infty\Big\}. \quad (2.7)$$

From the relationship between the kernels of the representations (1.3) and (2.4) (see Benth *et al.* [2, Formula 3.22]), we obtain that F with the representation

$$F = EF + \sum_{n=1}^{\infty} \sum_{i_1,\dots,i_n\geq 1} J_n^{(i_1,\dots,i_n)}(g_{i_1,\dots,i_n})$$

belongs to Dom$D$ if and only if

$$\sum_{n=1}^{\infty} \frac{1}{(n-1)!} \sum_{i_1,\dots,i_n\geq 1} q_{i_1}\cdots q_{i_n} \|g_{i_1,\dots,i_n}\|^2_{L^2([0,T]^n)} < \infty. \quad (2.8)$$

The following Proposition is based on Benth *et al.* [2, Proposition 3.8] and Di Nunno *et al.* [6, Remark 4.3]:

Proposition 8.

$F \in \mathrm{Dom}D$ if and only if $F \in \mathrm{Dom}D^{(k)}$,

$$\textit{for every } k \geq 1, \quad \textit{and} \sum_{n=1}^{\infty} q_n \|D^{(n)}F\|^2 < \infty.$$

In that case,

$$D_{t,x}F = \sum_{n=1}^{\infty} p_n(x)\, D_t^{(n)} F, \tag{2.9}$$

and

$$\begin{aligned} D_t^{(n)} F &= \frac{1}{\int_{\mathbb{R}} p_n^2(x)\, d\eta(x)} \int_{\mathbb{R}} p_n(x) D_{t,x}F\, d\eta(x) \\ &= \frac{1}{p_n^2(0)\,\sigma^2 + \int_{\mathbb{R}} x^2 p_n^2(x)\, d\nu(x)} \\ &\quad \left(p_n(0) D_{t,0}F + \int_{\mathbb{R}} p_n(x) D_{t,x}F\, x^2\, d\nu(x) \right). \end{aligned}$$

Proof. The equivalence between the domains of D and $D^{(k)}$ is a consequence of (2.7) and (2.8).

The relationship (2.9), is proved by checking the formula for an iterated integral $J_n^{(i_1,\dots,i_n)}(g)$, where $g(t_1,\dots,t_n) = g_1(t_1)\cdots g_n(t_n)$, with $\int_0^T g_i^2(t)\, dt < \infty$, and extending the equality by linearity and continuity to an arbitrary $F \in \mathrm{Dom}D$.

The second formula is deduced from the orthogonality of the polynomials $p_n(x)$. □

2.4 Skorohod Integrals

Following Davis and Johanson [3] we define the Skorohod integral in the direction k as the adjoint of the densely defined operator $D^{(k)}$: Let $f = \{f(t),\, t \in [0,T]\} \in L^2([0,T] \times \Omega, dt \otimes \mathbb{P})$ be a stochastic process such that there is a constant C such that for every $F \in \mathrm{Dom}D^{(k)}$,

$$\left| \mathbb{E}\left[\int_0^T \left(f(t)\, D_t^{(k)} F \right) \right] \right| \le C \left(\mathbb{E}\left[F^2 \right] \right)^{1/2}. \tag{2.10}$$

Then, we say that f is Skorohod integrable in the k direction, and define $\delta^{(k)}$ as the element of $L^2(\Omega)$ such that

$$\mathbb{E}\left[\int_0^T \left(f(t)\, D_t^{(k)} F \right) \right] = \mathbb{E}\left[\delta^{(k)}(f)\, F \right], \text{ for any } F \in \mathrm{Dom}D^{(k)}.$$

Denote by $Dom\delta^{(k)}$ the set of processes that satisfy (2.10).

The following Proposition is a consequence of Proposition 8 and the characterization (1.10) of $\mathrm{Dom}\delta$.

Proposition 9. *Let* $f \in \mathrm{Dom}\delta^{(k)}$. *Then* $f(t)p_k(x) \in \mathrm{Dom}\delta$ *and*

$$\delta^{(k)}\big(f(t)\big) = \delta\big(f(t)p_k(x)\big).$$

2.5 Clark–Ocone–Haussmann Formula

There are two Clark–Ocone–Haussmann formulas, one with the derivative $D_{t,x}F$ and the other one with $D_t^{(n)}F$. Thanks to Proposition 8 it is easy to go from one to the other. The next Theorem appears in Benth *et al.* [2], Di Nunno *et al.* [6, 7] and Oksendal and Proske [21]. We stated it in our context:

Theorem 10 (Clark–Ocone–Haussmann Formula, Two Parameter Derivative.). *Let $F \in L^2(\Omega)$ such that $F \in DomD$ and $\mathbb{E}\iint_{[0,T]\times\mathbb{R}} (D_z F)^2 d\mu(z) < \infty$. Then*

$$F = \mathbb{E}[F] + \sigma \int_0^T {}^{\mathrm{P}}(D_{t,0}F)\, dW_t + \iint_{[0,T]\times\mathbb{R}_0} {}^{\mathrm{P}}(DF_{t,x})x\, d\widetilde{N}(t,x),$$

where ${}^{\mathrm{P}}G$ means the predictable projection of a stochastic process G.

Idea of the Proof.

The proof is divided in two steps: First, we look for a process $h(x,t)$ such that

$$F = \mathbb{E}[F] + \delta(h).$$

From the experience with the Brownian motion (Nualart [16]) and with normal martingales (Ma, Protter and Sanmartin [14]), we check that $h(t,x) = \mathbb{E}[D_{t,x}F/\mathcal{F}_t]$ satisfies this condition.

In the second step, we look for a predictable process $h'(t,x)$ such that $\delta(h) = \delta(h')$. Also, by previous experience, we take $h'(t,x) = {}^{\mathrm{P}}D_{t,x}F$ and then we apply Theorem 5 and Proposition 4.

Remarks on the Predictable Projection.

1. The notion of *Predictable Projection* of a stochastic process indexed by $t \geq 0$ can be extended easily to our context for a two-parameter process. Let $Y = \{Y_{t,x},\ (t,x) \in [0,T] \times \mathbb{R}\}$ be a positive or bounded measurable process. There exists a predictable process $Z = \{Z_{t,x},\ (t,x) \in [0,T] \times \mathbb{R}\}$ such that for every predictable stopping time τ

$$Z_{\tau,x} = \mathbb{E}[Y_{\tau,x}/\mathcal{F}_{\tau-}] \quad \text{on} \quad \{\tau < \infty\}.$$

   This result is based in the same ideas as in the ordinary case (see, for example, Dellacherie et Meyer [4], pp. 114–115).
2. To compute the projective projection of a process is, in general, not easy. However, one can skip this task if one can find a predictable process h such that

$$h(t,x) = \mathbb{E}\left[D_{t,x}F/\mathcal{F}_t^X\right], \quad \forall(t,x,\omega) - \ \mu \otimes \mathbb{P} \text{ a.e.}$$

   Fortunately, this happens in many applications (see Solé *et al.* [27]).

From Proposition 8 and Theorem 10 we can get a new proof of the Clark-Ocone-Haussman formula using the derivatives in the direction k (Leon *et al.* [12, Theorem 1.8]):

Theorem 11 (Clark–Ocone–Haussmann Formula, Derivatives in the k-th Direction.). *Let $F \in L^2(\Omega)$ such that $F \in DomD$ and $\mathbb{E} \iint_{[0,T]\times\mathbb{R}} (D_z F)^2 d\mu(z) < \infty$. Then*

$$F = \mathbb{E}[F] + \sum_{k=1}^{\infty} \int_0^T {}^{\mathrm{p}}\big(D_t^{(k)} F\big)\, dH_t^{(k)}.$$

References

1. Applebaum, D., Lévy Processes and Stochastic Calculus. Cambridge University Press, Cambridge (2004).
2. Benth, F. E., Di Nunno, G., Løkka, A., Øksendal, B. and Proske, F., Explicit representations of the minimal variance protfolio in markets driven by Lévy processes. *Mathematical Finanace* **13** (2003) 55–72.
3. Davis, M. H. A. and Johanson, M. P., *Malliavin Monte Carlo Greeks for jump diffusions. Stoch. Proc. and their Appl.* **116** (2006) 101–129.
4. Dellacherie, C. et Meyer, P. A., Probabilités et Potentiel. Theorie des Martingales. Hermann, Paris (1982).
5. Dellacherie, C. et Meyer, P. A., Probabilités et Potentiel. Processus de Markov (fin). Compléments de calcul stochastique. Hermann, Paris (1992).
6. Di Nunno, G., Øksendal, B. and Proske, F., White Noise Analysis for Lévy processes. *Journal of Functional Analysis* **206** (2004) 109–148.
7. Di Nunno, G., Meyer-Brandis, Th., Øksendal, B. and Proske, F., Malliavin calculus and anticipative Itô formulae for Lévy processes. *Preprint series in Pure Mathematics, University of Oslo* (2004) **16**.
8. Itô, K., Multiple Wiener integral. *J. Math. Soc. Japan* **3** (1951) 157–169.
9. Itô, K., Spectral type of the shift transformation of differential processes with stationary increments. *Trans. Am. Math. Soc.* **81** (1956) 252–263
10. Lee, Y.-J. and Shih, H.-H., The product formula of multiple Lévy-Itô integrals. *Bull. Inst. Math. Acad. Sinica* **32** (2004) 71–95.
11. Lee, Y.-J. and Shih, H.-H., Analysis of generalized Lévy white noise functionals. *Journal of Functional Analysis* **211** (2004) 1–70.
12. Leon, J., Solé, J.L, Utzet, F. and Vives, J., On Lévy processes, malliavin calculus and market models with jumps. *Finance and Stochastics* **6** (2002) 197–225.
13. Løkka, A., Martingale representations and functionals of Lévy processes. *Preprint series in Pure Mathematics, University of Oslo* (2001) **21**.
14. Ma, J, Protter, P. and San Martin, J., Anticipating integrals for a class of martingales. *Bernoulli* **4** (1998) 81–114.
15. Meyer, P. A., Un cours sur les integrales stochastiques, *In: SÚminaire de ProbailitÚs X*, Lecture Notes in Mathematics, 511, 245–400, Springer, New York, 1976.

16. NUALART, D., The Malliavin Calculus and Related topics. Springer, Berlin (1995).
17. NUALART, D. AND SCHOUTENS, W., Chaotic and predictable representation for Lévy processes. *Stochastic Processes and their Applications* **90** (2000) 109–122.
18. NUALART, D. AND VIVES, J., Anticipative calculus for the Poisson Process based on the Fock space. *In: Séminaire de Probabilités* XXIV (Lecture Notes in Mathematics, 1426), Springer, Berlin Heidelberg New York: Springer 154–165 (1990).
19. NUALART, D. AND VIVES, J., A duality formula on the Poisson space and some applications. *In: Proceedings of the Acona Conference on Stochastic Analysis. (Progress in Probability).* Birkhäuser (1995).
20. NEVEU, J., Processus Pontuels. *In: École d'Eté de Probabilités de Saint Flour, VI,* (Lecture Notes in Mathematics, 598), Springer, Berlin (1977).
21. ØKSENDAL, B. AND PROSKE, F., White Noise of Poisson random measure. *Potential Analysis* **21** (2004) 375–403.
22. PICARD, J., On the existence of smooth densities for jump process, *Probab. Theory Relat. Fields* **105** (1996) 481–511.
23. PRIVAULT, N., An extension of stochastic calculus to certain non-Markovian processes Preprint 49, Universite d'Evry, (1997) `http://www.maths.unive-evry.fr/prepubli/49.ps`.
24. SATO, K., Lévy Processes and Infinitely Divisible Distributions. Cambridge University Press, Cambridge (1999).
25. SCHOUTENS, W., Stochastic Processes and Orthogonal Polynomial, Lecture Notes in Statistics, 146, Springer, New York, 2000.
26. SOLÉ, J., UTZET, F. AND VIVES, J., Canonic Lévy process and Malliavin Calculus. *Stoch. Proc. and their Appl.*, to appear (2006).
27. SOLÉ, J., UTZET, F. AND VIVES, J., Quadratic hedging of Asian options in jump diffusion models via Malliavin Calculus. *Preprint* (2006).
28. YABLONSKI, A., The calculus of variations for processes with independent increments, Rocky Mountain Journal of Mathematics, to appear (2006).

Study of Simple but Challenging Diffusion Equation

Daniel W. Stroock

M.I.T., 2-272, Cambridge, MA 02139-4307, USA, dwsamath.mit.edu

Summary. This note contains a summary of results, obtained in collaboration with David Williams, about a simple diffusion equation which does not fit the standard probabilistic model. In particular, the usual minimum principle does not apply, and its absence gives rise to some phenomena which are, at least to us, both unfamiliar and interesting.

Introduction

For the past two years David Williams and I have been devoting an embarrassing amount of effort[1] to understand solutions to the linear, constant coefficient partial differential equation

$$\dot{u} = \frac{1}{2}u'' + \mu u' \text{ in } I \times [0, \infty) \quad \text{with } \dot{u}(t, 0) = \sigma u'(t, 0) \text{ for } t \in I, \tag{1}$$

where $\dot{u} = \partial_t u$ and $u' = \partial_x u$, $I \subseteq \mathbb{R}$ is an open interval, and $(\mu, \sigma) \in \mathbb{R}^2$. It is essential for our analysis that a solution u be continuously differentiable at least once in t and twice in x in the whole of $I \times [0, \infty)$, including the spacial boundary $I \times \{0\}$.

Our initial result (cf. Theorem 1.1 in [3]) deals with the Cauchy initial value problem for (1). In its statement, the set F of initial values consists of bounded $f : [0, \infty) \longrightarrow \mathbb{R}$ which are continuous on $(0, \infty)$, but not necessarily at 0. The set U from which solutions come consists of $u \in C^{1,2}\big((0, \infty) \times [0, \infty); \mathbb{R}\big)$ which are bounded in $(0, 1) \times [0, \infty)$ and have the property that u, $\dot{u}$, and u'' are bounded on each vertical slice $[T_1, T_2] \times [0, \infty)$, where $0 < T_1 < T_2 < \infty$.

The author aknowledges support from NSF Grant DMS 0244991

[1] As explained in [3], our original reason for looking at these equations came from the study of certain Wiener–Hopf decompositions. However, our continued study of them has been motivated by pure intellectual curiosity.

Theorem 1. *Suppose that $u \in U$ satisfies (1) in $(0,\infty) \times [0,\infty)$ and that $f(x) \equiv \lim_{t\searrow 0} u(t,x)$ exists for each $x \in (0,\infty)$. Then $f(0) = \lim_{t\searrow 0} u(t,0)$ exists and the convergence of $u(t,\,\cdot\,)$ to f takes place uniformly on compact subsets of $(0,\infty)$. In particular, $f \in F$. Conversely, for each $f \in F$ there is a unique solution $u = u_f \in U$ to (1) such that $\lim_{t\searrow 0} u(t,x) = f(x)$ for each $x \in [0,\infty)$, and the convergence is uniform on compact subsets of $(0,\infty)$. In particular, if $\mathbf{Q}_t f \equiv u_f(t,\,\cdot\,)$ for $t > 0$ and $f \in F$, then, for each $t > 0$, $\mathbf{Q}_t$ maps F boundedly into $C_b([0,\infty);\mathbb{R})$ and $\{\mathbf{Q}_t : t \geq 0\}$ is a semigroup. Finally, if $\{f_n\}_1^\infty \subseteq F$ is a bounded sequence which tends to $f \in F$ in the sense that $f_n(0) \longrightarrow f(0)$ and $f_n \longrightarrow f$ uniformly on compact subsets of $(0,\infty)$, then $\mathbf{Q}_t f_n(x) \longrightarrow \mathbf{Q}_t f(x)$ uniformly for (t,x) in compact subsets of $(0,\infty) \times [0,\infty)$.*

When $\sigma \geq 0$, nothing in Theorem 1 is surprising, with the possible exception of the regularity of solutions at the spacial boundary. Moreover, the solutions when $\sigma \geq 0$ admit a familiar probabilistic interpretation. Namely, when $\sigma = 0$, the associated Markov process is simply Brownian motion with drift μ which is absorbed when it hits 0. When $\sigma > 0$, the associated Markov process is again Brownian motion with drift μ, only now 0 is a "sticky" reflection point. More precisely, let $\{B_t : t \geq 0\}$ be a standard, $\mathbb{R}$-valued Brownian motion, and set

$$\begin{aligned} L_t(x) &= \max\{(x + B_s + \mu s)^- : s \in [0,t]\} \text{ and} \\ X_t(x) &= x + B_t + \mu t + L_t(x). \end{aligned} \tag{2}$$

Then $\{X_t(x) : t \geq 0\}$ is Brownian motion with drift μ reflected at 0, and $\{L_t(x) : t \geq 0\}$ is its local time at 0. Finally, for any $\sigma \geq 0$, take

$$\tau_t(x) = \inf\{\tau : \tau + \sigma^{-1} L_\tau(x) \geq t\} \big(\equiv \inf\{\tau : X_\tau(x) = 0\} \text{ when } \sigma = 0\big).$$

Then $\{X_{\tau_t(x)}(x) : t \geq 0\}$ is, depending on whether $\sigma = 0$ or $\sigma > 0$, Brownian motion with drift μ which is either absorbed at 0 or has a "sticky" reflection at 0. In addition, an elementary application of Itô's calculus combined with Doob's stopping time theorem shows that

$$u_f(t,x) = \mathbb{E}\big[f\big(X_{\tau_t(x)}(x)\big)\big].$$

For a complete account of one-dimensional diffusion equations which are amenable to probabilistic interpretation, see Dynkin's classic interpretation in [1] of Feller's theory.

1 Preservation of Non-negativity when $\sigma < 0$

From the preceding it is clear that $\{\mathbf{Q}_t : t \geq 0\}$ is a conservative, Markov semigroup when $\sigma \geq 0$, a conclusion which can be drawn (with much less effort) via a purely analytic minimum principle argument. On the other hand, when $\sigma < 0$, the minimum principle is lost and $\{\mathbf{Q}_t : t \geq 0\}$ need not preserve non-negativity. In fact, we have (cf. Theorem 1.2 in [3]) the following.

Theorem 2. *Assume that $\sigma < 0$. For $f \in F$, u_f is non-negative if and only if f is non-negative and*

$$f(0) \geq 2|\sigma| \int_{(0,\infty)} e^{2\sigma\wedge\mu y} f(y)\, dy.$$

Moreover, if $F(\sigma,\mu)$ is the subset of $f \in F$ which satisfy

$$f(0) = 2|\sigma| \int_{(0,\infty)} e^{2\sigma\wedge\mu y} f(y)\, dy,$$

then $\{\mathbf{Q}_t : t \geq 0\}$ leaves $F(\sigma,\mu)$ invariant and its restriction to $F(\sigma,\mu)$ is Markov. Finally, $\mathbf{Q}_t\mathbf{1} \leq \mathbf{1}$ for all $t > 0$, and equality holds if and only if $\sigma \geq \mu$.

Even without going into the details, it is reasonably easy to understand why the function $J(y) \equiv 2|\sigma|e^{2\sigma\wedge\mu y}$ enters in the preceding. Namely, given a solution u, one can use integration by parts to see that

$$\frac{d}{dt}\big(u(t,0) - \langle J, u(t,\,\cdot\,)\rangle\big) = -2\sigma(\mu-\sigma)^+\big(u(t,0) - \langle J, u(t,\,\cdot\,)\rangle\big),$$

where $\langle \varphi, \Psi\rangle \equiv \int_{(0,\infty)} \varphi(y)\Psi(y)\, dy$. Hence,

$$u_f(t,0) - \langle J, u_f(t,\,\cdot\,) = e^{-2\sigma(\mu-\sigma)^+ t}\big(f(0) - \langle J, f\rangle\big), \tag{1.1}$$

which makes it clear why $F(\sigma,\mu)$ is $\{\mathbf{Q}_t : t \geq 0\}$ invariant. More generally, (1.1) shows that

$$f(0) \geq \langle J, f\rangle \implies u_f(t,0) \geq \langle J, u_f(t,\,\cdot\,)\rangle \quad \text{for all } t \geq 0.$$

Hence, if $f \geq 0$ and $f(0) \geq \langle J, f\rangle$, then an easy minimum principle argument shows that $u_f \geq 0$. Namely, choose $\eta \in C^\infty([0,\infty);(0,1))$ so that $\eta(0) > \langle J, \eta\rangle$, $\frac{1}{2}\eta'' + \mu\eta' > -1$, and $\lim_{x\to\infty} \eta(x) = \infty$. For $\epsilon > 0$, set $v_\epsilon(t,x) = u_f(t,x) + \epsilon(t + \eta(x))$, and note that $\dot{v}_\epsilon > \frac{1}{2}v_\epsilon'' + \mu v_e'$, $v_\epsilon(t,0) > \langle v_\epsilon(t,\,\cdot\,)\rangle$, $\lim_{t\to 0} v_\epsilon(t,\,\cdot\,) > 0$, and, for each $t > 0$, $\inf_{\tau\in(0,t]} v_\epsilon(\tau,x) \longrightarrow \infty$ as $x \to \infty$. From these it is easy to conclude that $v_\epsilon \geq 0$ everywhere for each $\epsilon > 0$.

The proof that $u_f \geq 0 \implies f(0) \geq \langle J, f\rangle$ is more challenging. One way to proceed is via probability theory. Namely, take (cf. 2) $X_t = X_t(0)$ and $L_t = L_t(0)$, and define $\Psi_t = |\sigma|^{-1}L_t - t$ and $\zeta = \inf\{t : \Psi_t = 0\}$. By Itô's calculus, one knows that $t \rightsquigarrow u_f\big(\Psi_t, X_t\big)$ is a local martingale on $[0,\zeta)$. Thus, if $u_f \geq 0$, then $t \rightsquigarrow u_f\big(\Psi_t, X_t\big)$ is a non-negative supermartingale on $[0,\zeta)$, and so

$$\begin{aligned} f(0) = u_f(0,0) &\geq \lim_{t\to\infty} \mathbb{E}\big[u_f\big(\Psi_{t\wedge\zeta}, X_{t\wedge\zeta}\big)\big] \\ &= \mathbb{E}\big[f(X_\zeta),\, \zeta < \infty\big] + \lim_{t\to\infty} \mathbb{E}\big[u_f\big(\Psi_t, X_t\big),\, \zeta = \infty\big]. \end{aligned}$$

Similarly, if u_f is bounded, then the inequality in the preceding can be replaced by equality. Hence

$$\begin{aligned} u_f \geq 0 &\implies f(0) \geq \mathbb{E}\big[f(X_\zeta),\, \zeta < \infty\big] \\ u_f \text{ bounded} &\implies f(0) = \mathbb{E}\big[f(X_\zeta),\, \zeta < \infty\big] + \lim_{t\to\infty} \mathbb{E}\big[u_f(\Psi_t, X_t),\, \zeta = \infty\big]. \end{aligned} \tag{1.2}$$

At this point, one has to check that, with probability one,

$$\begin{aligned} \mu > \sigma &\implies \lim_{t\to\infty} \Psi_t = -\infty \\ \mu = \sigma &\implies \overline{\lim_{t\to\infty}}\, \Psi_t = \infty = -\underline{\lim_{t\to\infty}}\, \Psi_t \\ \mu < \sigma &\implies \lim_{t\to\infty} \Psi_t = \infty. \end{aligned} \tag{1.3}$$

In particular, if $\mu \geq \sigma$, then $\mathbb{P}(\zeta < \infty) = 1$, and so the second line of (1.2) applied to $f \in F(\sigma, \mu)$ says that

$$\langle J, f\rangle = \mathbb{E}\big[f(X_\zeta),\, \zeta < \infty\big] \quad \text{when } \mu \geq \sigma.$$

That is, when $\mu \geq \sigma$, then $2|\sigma|e^{2\sigma y}$ is the distribution of X_ζ, and so the first line of (1.2) completes the proof that $u_f \geq 0 \implies f(0) \geq \langle J, f\rangle$ when $\mu \geq \sigma$. The case when $\sigma < \mu$ is a little trickier. To handle it, one must first observe that, by the last line of (1.3), the second term on the right in the second line of (1.2) can be replaced by

$$\mathbb{E}\Big[\lim_{t\to\infty} u_f(t, X_t),\, \zeta = \infty\Big]$$

when $\mu < \sigma$. Secondly, one has to show that

$$\mu < \sigma \;\&\; f \in F(\sigma, \mu) \implies \lim_{t\to\infty} u_f(t, x) = 0 \quad \text{for all } x \in [0, \infty).$$

Once this has been done, one can proceed as before to check that, when $\mu < \sigma$, $2|\sigma|e^{2\mu y}$ is the distribution of X_ζ on $\{\zeta < \infty\}$ and therefore $u_f \geq 0 \implies f(0) \geq \langle J, f\rangle$.

2 Bounded Solutions & Long Time Behavior

A closer examination of the probabilistic argument given in §1 reveals that, when $\sigma < 0$,

$$\begin{aligned} \mu \geq \sigma &\implies u_f \text{ is bounded} \iff f \in F(\sigma, \mu) \\ \mu < \sigma &\implies u_f \text{ is bounded for all } f \in F \text{ and } \lim_{t\to\infty} u_f(t, \cdot\,) = 0 \\ &\iff f \in F(\sigma, \mu). \end{aligned} \tag{2.1}$$

One gets more precise information from the following (cf. Theorem 3.2 in [4]).

Theorem 3. *Assume that $\sigma < 0$. Given $f \in F$, set (cf. the notation introduced following Theorem 2) $\delta f = f(0) - \langle J, f\rangle$. Then, as $t \nearrow \infty$,*

$$\mu < \sigma \Longrightarrow u_f(t,x) \longrightarrow \frac{\mu}{\mu-\sigma}\delta f$$

$$\mu = \sigma \Longrightarrow \frac{u_f(t,x)}{t} \longrightarrow 2\sigma^2\delta f$$

$$\mu > \sigma \Longrightarrow e^{2\sigma(\mu-\sigma)t}u_f(t,x) \longrightarrow \frac{\mu-2\sigma}{\mu-\sigma}e^{-2(\mu-\sigma)x}\delta f$$

uniformly on compacts. In particular,

$$\begin{aligned}\delta f \geq 0 &\Longleftrightarrow \overline{\lim_{t\to\infty}}\, u_f(t,x) \geq 0 \text{ for some } x \in [0,\infty)\\ &\Longleftrightarrow \underline{\lim_{t\to\infty}}\, u_f(t,x) \geq 0 \text{ for all } x \in [0,\infty).\end{aligned}$$

The proof of Theorem 3 involves no probability theory. The idea is to write $u_f = (\delta f)v + u_{\tilde f}$, where v is the solution with initial condition $\mathbf{1}_{\{0\}}$ and $\tilde f = f - (\delta f)\mathbf{1}_{\{0\}}$. Because $\tilde f \in F(\sigma,\mu)$, we know that $u_{\tilde f}$ is always bounded and, as $t \nearrow \infty$, tends to 0 if $\mu < \sigma$. Thus, everything comes down to the analysis of $v(t,\cdot)$ as $t \nearrow \infty$, and this analysis is carried out in Lemma 3.1 of [4].

3 Bounded, Ancient Solutions

Here we consider solutions to (1) which are *ancient* in the sense that they are solutions on $(-\infty,0)\times[0,\infty)$, and our goal is to classify all ancient solutions which are bounded. Our first result about such a solution is contained in the following regularity result (cf. Lemma 4.2 in [4]).

Lemma. *Assume that $u \in C^{1,2}\big((a,b)\times[0,\infty);\mathbb{R}\big)$ is a bounded solution to (1) for some $-\infty < a < b < \infty$. Then, $u \in C^\infty\big((a,b)\times[0,\infty);\mathbb{R}\big)$ and there is a K, which depends only of σ and μ, such that, for each $n \geq 1$ and $t \in (a,b)$,*

$$\|\partial_x^n u(t,\cdot)\|_{\mathrm{u}} \leq \left(\frac{Kn}{t-a}\right)^{\frac{n}{2}} e^{K(t-a)}\|u\|_{\mathrm{u}}.$$

In particular, if u is a bounded, ancient solution, then $u \in C_{\mathrm{b}}^\infty\big((-\infty,0)\times[0,\infty);\mathbb{R}\big)$ and there exists a $K' \in [1,\infty)$, depending only on σ and μ, such that

$$\|u\|_{C_{\mathrm{b}}^{n,2n}((-\infty,0)\times[0,\infty);\mathbb{R})} \leq (K')^n\|u\|_{\mathrm{u}}$$

for all $n \geq 0$. Hence, each bounded ancient solution admits a unique continuation as an entire, holomorphic function on $\mathbb{C}^2$.

Theorem 4. *If u is a bounded, ancient solution to (1), then*

$$u(t,x) = A + Be^{-2(\mu-\sigma)^+(\sigma t + x)}$$

for some $(A,B) \in \mathbb{R}^2$, where $B = 0$ if either $\sigma \geq \mu$ or $0 < \sigma < \mu$. Moreover, if u is a bounded solution to (1) in the whole of $\mathbb{R} \times [0,\infty)$, then u is constant.

Given the preceding lemma, the final assertion is an easy corollary of the initial statement. In order to prove the initial statement, one considers the function $w = \sigma u' - u$, which, by the lemma, is a bounded solution to $\dot{w} = \frac{1}{2}w'' + \mu w'$ in $(-\infty, 0) \times [0,\infty)$ which vanishes on the spacial boundary $(-\infty,0) \times \{0\}$. The desired result will follow once we show that such a w is constant in $t \in (0,\infty)$. To this end, let $Q(t,x,y)$ be the heat kernel for $\frac{1}{2}\partial^2 + \mu\partial$ in $(0,\infty)$ with boundary condition 0 at 0. Then

$$w(t_2,x) - w(t_1,x) = \int_{(0,\infty)} Q(T,x,y)\big(w(t_2 - T, y) - w(t_1 - T, y)\big)\,dy.$$

Since

$$\lim_{T\to\infty} \int_{(0,\infty)} Q(T,x,y)e^{-2\mu^+ y}\,dy = 0,$$

it suffices to show that

$$|w(t_2,y) - w(t_1,y)| \leq 2\|w\|_{\mathrm{u}} e^{-2\mu^+ y} \quad \text{for all } t_1 < t_2 < 0 \text{ and } y \geq 0. \qquad (*)$$

A proof of (*) can be based on a coupling argument. Namely, set $h = t_2 - t_1$, and let $\{B_t : t \geq 0\}$ and $\{B'_t : t \geq 0\}$ be a pair of mutually independent, $\mathbb{R}$-valued Brownian motions starting at 0. Next, define τ to be the first time at which the path $t \rightsquigarrow y + B_{t+h} + \mu(t+h)$ crosses the path $t \rightsquigarrow y + B'_t + \mu t$. Equivalently, $\tau = \inf\{t \geq 0 : B'_t - B_{t+h} = \mu h\}$. Then $\mathbb{P}(\tau < \infty) = 1$, and

$$t \rightsquigarrow U_t \equiv B'_{t\wedge\tau} + \big(B_{t+h} - B_{t\wedge\tau+h}\big)$$

is again a Brownian motion starting at 0. Because $y + B_{t+h} + \mu(t+h) = y + U_t + \mu t$ when $\tau \leq t$, $|w(t_2,y) - w(t_1,y)|$ is equal to

$$\begin{aligned}
&\lim_{t\to\infty}\Big|\mathbb{E}\big[w(t_1 - t, y + B_{t+h} + \mu(t+h)),\, \zeta^B_y > t+h\big] \\
&\qquad\qquad - \mathbb{E}\big[w(t_1 - t, y + U_t + \mu t),\, \zeta^U_y > t\big]\Big| \\
&= \lim_{t\to\infty}\Big|\mathbb{E}\big[w(t_1 - t, y + U_t + \mu t),\, \zeta^B_y > t+h\big] \\
&\qquad\qquad - \mathbb{E}\big[w(t_1 - t, y + U_t + \mu t),\, \zeta^U_y > t\big]\Big| \\
&\leq \|w\|_{\mathrm{u}} \lim_{t\to\infty}\Big(\mathbb{P}\big(\zeta^B_y > t+h \;\&\; \zeta^U_y \leq t\big) + \mathbb{P}\big(\zeta^B_y \leq t+h \;\&\; \zeta^U_y > t\big)\Big) \\
&\leq \|w\|_{\mathrm{u}}\Big(\mathbb{P}\big(\zeta^U_y < \infty\big) + \mathbb{P}\big(\zeta^B_y < \infty\big)\Big) = 2\|w\|_{\mathrm{u}}\mathbb{P}\big(\zeta^B_y < \infty\big),
\end{aligned}$$

where ζ^B_y and ζ^U_y are, respectively, the first time that $t \rightsquigarrow y + B_t + \mu t$ and $t \rightsquigarrow y + U_t + \mu t$ hit 0. Because $\mathbb{P}(\zeta^B_y < \infty) = e^{-2\mu^+ y}$, (*) follows.

4 A Harnack Principle

At first sight, (1) appears to be a parabolic equation. As such, one should expect that, if it satisfies a Harnack principle at all, that principle will be "one-sided." For example, when $\sigma = 0$,

$$w_\lambda(t,x) = e^{\lambda x + \frac{\mu^2+\lambda^2}{2}t}\sinh(\lambda x)$$

is a non-negative solution in $\mathbb{R} \times [0,\infty)$ for each $\lambda > 0$, and although

$$\sup_{\lambda>0} \frac{w_\lambda(s,x)}{w_\lambda(t,y)}$$

is bounded for each $0 < s < t$ and all (x,y) in compact subsets of $(0,\infty)$, it is infinite whenever $s > t$ or $s = t$ but $x \neq y$; and the same sort of one-sided Harnack principle holds whenever $\sigma \geq 0$. Thus, it is somewhat intriguing that a more robust Harnack principle holds as soon as $\sigma < 0$. To wit, one can prove (cf. the subsection following Theorem 5.6 in [4]) the following statement.

Theorem 5. *Assume that $\sigma < 0$. For each $0 < \ell < L$ and $0 < r < R$, there exists a $K < \infty$, depending only on σ, μ, ℓ, L, r, and R, such that $u(s,x) \leq Ku(t,y)$ whenever $(s,t) \in [0,\ell]^2$, $(x,y) \in [0,r]^2$, and u is a non-negative solution to (1) in $(-L,L) \times [0,R)$. Moreover, if $\mathcal{U}(L,R)$ is the set of all non-negative solutions to (1) in $(-L,L) \times [0,R)$, then, for each $(s,x) \in (-L,L) \times [0,R)$, $\{u \in \mathcal{U}(L,R) : u(s,x) \leq 1\}$ is compact in $C^\infty\big((-L,L) \times [0,R); [0,1]\big)$.*

In broad outline, the proof of this Harnack principle follows a line of reasoning which is familiar to experts in such matters. Namely, one shows that, for $(s,x) \in [-\ell,\ell] \times [0,r]$, $u(s,x)$ is bounded above and below by positive multiples of $u(\ell',0)$ plus the integral of u over $[-\ell',\ell'] \times [0,r']$, where $\ell < \ell' < L$ and $r < r' < R$. The most interesting part comes when one checks the lower bound, which comes from thinking about the behavior of Ψ_t. Namely, although Ψ_t does nothing but decrease while $X_t > 0$, it increases very rapidly whenever X_t visits 0, and it this increase which accounts for the "backdoor elliptic" nature of the resulting Harnack principle. In this connection, it should be observed that one cannot localize Harnack principle here to regions which do not include a healthy component of the spacial boundary.

5 Non-Negative, Ancient Solutions

When $\mu \neq \sigma$ and $\sigma \neq 0$,

$$e^{-\mu x + \frac{\lambda^2-\mu^2}{2}t}\left(\cosh(\lambda x)) + \frac{\lambda^2+\sigma^2-(\mu-\sigma)^2}{2\lambda\sigma}\sinh(\lambda x)\right)$$

is a non-constant, non-negative *global* (i.e., on $\mathbb{R}\times[0,\infty)$) solution to (1) for every $\lambda>0$ with the property that $\frac{\lambda^2+\mu^2-(\mu-\sigma)^2}{2\lambda\sigma}>-1$. When $\sigma=0$,

$$e^{-\mu x+\frac{\lambda^2-\mu^2}{2}t}\sinh(\lambda x)$$

is a non-constant, non-negative, global solution to (1) for each $\lambda>0$. Finally, in the *balanced case* (i.e., $\sigma=\mu$), if $\sigma\geq 0$, then

$$e^{-\mu x+\frac{\lambda^2-\mu^2}{2}t}\left(\sinh(\lambda x)+\frac{2\lambda\mu}{\lambda^2+\mu^2}\cosh(\lambda x)\right)$$

is a non-constant, non-negative, global to (1) for each $\lambda>0$. On the other hand, *in the balanced case, if $\sigma<0$, all non-negative, global solutions to* (1) *are constant.* An understanding of this last result can be found in (1.3). Namely, the middle line of (1.3) makes it reasonably easy to show that the process $\{(\Psi_t,X_t):t\geq 0\}$ is recurrent. Thus, since each non-negative, global solution is a non-negative supermartingale along this process, the constancy of such solutions follows from a standard argument based on Doob's martingale convergence theorem.

Of course, the situation is completely different when one considers non-negative ancient solutions. Indeed, even in the balanced case with $\sigma<0$, there are lots of non-constant, non-negative ancient solutions. The reason why such ancient solutions can exist even though global ones cannot is that the recurrence, alluded to above, disappears if the process is stopped when Ψ_t first visits 0, and it is the stopped process along which non-negative, ancient solutions will be non-negative supermartingales. In an attempt to understand what is the structure of the set of non-negative, ancient solutions in the balanced case, we proved the following.

Theorem 6. *When $\sigma=-1=\mu$, there is a one-to-one correspondence between non-negative ancient solutions u to* (1) *and triples (a,b,ν), $(a,b)\in[0,\infty)^2$ and ν a measure on $[0,\infty)$ with $\int e^{\lambda c}\,\nu(dc)<\infty$ for all $\lambda>0$, given by*

$$u(t,x)=a+b(x-t)+\int_0^\infty h_{1+c}(t,x)\,\nu(dc)$$

where

$$\begin{aligned}h_{1+c}(t,x)=4(c+1)\big[1-e^{(\frac{1}{2}c^2+c)t-cx}\big]\\+c^2e^{(\frac{1}{2}c^2+c)t}\big[e^{(2+c)x}-e^{-cx}\big].\end{aligned}$$

Given the requisite computations of the Green's function involved, our proof of this result is a straight-forward application of Martin's boundary theory. In order to interpret its conclusion as a statement about the associated Martin compactification, let Q denote the second quadrant $(-\infty,0)\times[0,\infty)$ in the Euclidean plane. Next, compactify Q by adjoining the line $\{0\}\times[0,\infty)$ and the quarter circle $\left[\frac{1}{2}\pi, \pi\right]$ at infinity. Now identify all points on $\{0\} \times [0, \infty)$ with a single point α and all points at infinity having "angle" in $\left[\frac{3}{4}\pi, \pi\right]$ with a (different) single point γ. That is, if $\{(t_n, x_n)\}_1^\infty \subseteq Q$ converges to a point on $\{0\} \times [0, \infty)$, we say that $(t_n, x_n) \longrightarrow \alpha$; and if it converges to infinity in such a way that $\arctan \frac{t_n}{x_n}$ converges to a point in $\left[\frac{3}{4}\pi, \pi\right]$, we say that (t_n, x_n) tends to γ. Finally, if $\{(t_n, x_n)\}_1^\infty$ tends to infinity so that $\arctan \frac{t_n}{x_n}$ converges to $\beta \in \left[\frac{1}{2}\pi, \frac{3}{4}\pi\right]$, we will say that $(t_n, x_n) \longrightarrow \beta$. With these conventions,

$$\{\alpha\} \cup \left[\frac{1}{2}\pi, \frac{3}{4}\pi\right] \cup \{\gamma\}$$

can be identified with the Martin boundary of Q in such a way that the associated Martin kernel with reference point $(t_0, x_0) \in Q$ is given on the boundary by

$$\kappa\big((t,x);\alpha\big) = \frac{x-t}{x_0 - t_0}, \quad \kappa\big((t,x);\gamma\big) = 1,$$
$$\kappa\big((t,x);\beta\big) = \frac{h_{-\tan\beta}(t,x)}{h_{-\tan\beta}(t_0,x_0)}.$$

References

1. Dynkin, E.B., *Markov Processes*, Grundlehren Series **122**, Springer–Verlag, 1965.
2. Feller, W., *Diffusion processes in one dimension*, TAMS (1954), 1–31.
3. Stroock, D. and Williams, D., *A simple PDE and Wiener-Hopf Riccati equations*, Comm. Pure & Appl. Math. **58 (8)** (2005), 1116–1148.
4. ——, *Further study of a simple PDE* (to appear in the volume of the Illinois J. of Math. dedicated to the memory of J.L. Doob).

Itô Calculus and Malliavin Calculus

Shinzo Watanabe

527-10, Chaya-cho, Higashiyama-ku, Kyoto, 605-0931, Japan

Dedicated to Professor Kiyosi Itô on his 90th birthday

1 Introduction

Since the *Wiener space* was established by N. Wiener as a mathematical model of Brownian motion in 1923, a rigorous theory of integrations on a function space started. In these almost eighty years, it has been providing us with important methods in stochastic analysis and its applications.

Around 1942, R. Feynman ([F 1, F 2]) had an epoch making idea of representing the propagators for Schrödinger equations by a path integral over trajectories of quantum mechanical particles. M. Kac noticed that its counterpart could be discussed rigorously on Wiener space and thus found the *Feynman-Kac formula.* He also applied probabilistic representations of heat kernels by Wiener functional expectations to study asymptotics of spectra of Schrödinger operators ([Ka 1, Ka 2, Ka 3]). This study was further developed in a fundamental paper by McKean and Singer [MS], which may be regarded as an origin of the *heat equation methods in the analysis of manifold.*

The approach by McKean and Singer is based on PDE theory, the method of parametrix for heat kernels, in particular. If we would give a similar probabilistic approach as Kac in the problems of McKean and Singer, we have several difficulties to overcome. In the case of Kac, the second order term of the Schrödinger operator is Laplacian in Euclidean space so that a use of Wiener process and pinned Wiener process is sufficient, which could be easily set up on a Wiener space. In the case of McKean and Singer, however, we need a Brownian motion on a curved Riemannian manifold; also the analysis of pinned Brownian motion requires some fine properties of heat kernels.

As we review in this expository article, a Brownian motion on a Riemannian manifold can be well set up on a Wiener space by appealing to the *Itô calculus*, and the pinned Brownian motion can be well handled by appealing to the *Malliavin calulus* on the Wiener space. We would study the conditional expectations of a class of Wiener functionals as integrations on a 'submanifold' embedded in the Wiener space so that we can develop a 'smooth' desintegration theory on Wiener space. In this study, an important role is played by the notion of *generalized Wiener functionals*, a notion similar to that of Schwartz distributions on Wiener space.

2 Wiener Space, Wiener Functionals and Wiener Maps

Let $(W_0(\mathbf{R}^d), P^W)$ be the d-dimensional classical Wiener space: $W_0(\mathbf{R}^d)$ is a path space $W_0(\mathbf{R^d}) := \{ w;\ [0,T] \ni t \mapsto w(t) \in \mathbf{R}^d,\ \text{continuous},\ w(0) = 0 \}$, which is a Banach space with the usual maximum norm, and P^W is the d-dimensional Wiener measure on it. Here T is a positive constant; sometimes, the time interval is taken to be $[0,\infty)$ and then $W_0(\mathbf{R}^d)$ is a Fréchet space with a family of maximum (semi)norms on subintervals. Let $H \subset W_0(\mathbf{R}^d)$ be the *Cameron-Martin subspace*, which is a real Hilbert space given by

$$H = \left\{ h \in \mathbf{W}_0(\mathbf{R^d}) \mid h(t) = \int_0^t \dot{h}(s)ds,\ \dot{h} \in L^2\left([0,T] \to \mathbf{R}^d\right) \right\},$$

$$||h||_H = ||\dot{h}||_{L^2}.$$

As we know well, the triple $(W_0(\mathbf{R}^d), H, P^W)$ is a typical example of more general notion of *abstract Wiener space*; that is, we may think of the Wiener space as a realization of *standard Gaussian measure on H*.

A P^W-measurable function $F : w \in W_0(\mathbf{R}^d) \mapsto F(w) \in S$, where S is a topological space endowed with the Borel σ-field $\mathcal{B}(S)$, is called an S-valued *Wiener functional* (or a *Wiener map* if we would regard F as a mapping). As usual, we identify two S-valued Wiener functionals F and F' if $P^W\{w; F(w) \neq F'(w)\} = 0$. Let B be any separable Banach space. Then $L_p(B) := L_p(W_0(\mathbf{R}^d) \to B)$, $1 \leq p \leq \infty$, is the usual L_p-space formed of B-valued Wiener functionals $F : W_0(\mathbf{R}^d) \to B$ such that $||F||_p := \left\{\int_{W_0(\mathbf{R}^d)} ||F(w)||_B^p P^W(dw)\right\}^{1/p} < \infty$. $L_p(\mathbf{R})$ is denoted simply by L_p. As usual, we denote the integral $\int_{W_0(\mathbf{R}^d)} F(w)P^W(dw)$ for $F \in L_1$ by $E(F)$ and call it the expectation of F.

If $F \in L_1$, then the conditional expectation $E_{0,0}^{T,x}(F) = E(F|w(T) = x)$ is defined, as usual, by a Radon-Nikodym density, so that, as a function of x, it is determined almost everywhere with an ambiguity of a set of Lebesgue measure 0. However, the Brownian bridge measure $P_{0,0}^{T,x}$ on $W_{0,0}^{T,x}(\mathbf{R}^d) := \{w;\ [0,T] \ni t \mapsto w(t) \in \mathbf{R}^d,\ \text{continuous},\ w(0) = 0,\ w(T) = x\}$, is well-defined for each x as the image measure of P^W under the map $w \in W_0(\mathbf{R}^d) \mapsto \hat{w} \in W_{0,0}^{T,x}(\mathbf{R}^d)$ defined by $\hat{w}(t) = w(t) + \frac{t}{T}(x - w(T))$, $0 \leq t \leq T$, and the conditional expectation $E_{0,0}^{T,x}(F)$ is defined without ambiguity if F is a Borel function on $W_{0,0}^{T,x}(\mathbf{R}^d)$ which is $P_{0,0}^{T,x}$-integrable. The expectation $E_{0,0}^{T,x}(F)$ may be symbolically written as $E[\delta_x(w(T))F(w)]/p(T,x)$, where $\delta_x(\cdot) = \delta_0(\cdot - x)$ is the Dirac delta function and

$$p(t,x) = (2\pi t)^{-d/2} \exp\left\{-\frac{|x|^2}{2t}\right\}, \quad t > 0, \quad x \in \mathbf{R}^d. \tag{2.1}$$

This kind of formal expressions will be rigorously and more generally defined in Section 3 below.

In the following, as a typical and important application of Wiener functional expectations, we would review probabilistic expressions of solution $u = (u(t,x))$ for initial value problem (IVP) of heat equations

$$\frac{\partial u}{\partial t} = Lu, \quad u|_{t=0} = f. \tag{2.2}$$

where L is a second-order semi-elliptic differential operator. In this section, we deal with the case of heat equations on $\mathbf{R}^d$ in which the principal second-order term of L is the half Laplacian: $\frac{1}{2}\Delta$. We introduce a usual notation $u = e^{tL}f$ or $u(t,x) = (e^{tL}f)(x)$ for the solution u of (2.2).

Solutions of IVP (2.2) by Wiener Functional Expectations. I

[1] The case $L = \frac{1}{2}\Delta$. Then $u = e^{tL}f$ is given by

$$u(t,x) = E\left[f(x + w(t))\right]. \tag{2.3}$$

[2] (Feynman-Kac formula) The case of a Schrödinger operator $L = \frac{1}{2}\Delta - V$ where the potential $V(x)$ is a Borel function bounded from below. Then $u = e^{tL}f$ is given by

$$u(t,x) = E\left[\exp\left\{-\int_0^t V(x + w(s))ds\right\} f(x + w(t))\right]. \tag{2.4}$$

[3] The case of operator $L = \frac{1}{2}\Delta + \sum_{i=1}^d b^i \frac{\partial}{\partial x^i} - V$ where the drift coefficients $b^i(x)$, $i = 1, \ldots, d$, are bounded Borel functions and the potential $V(x)$ is a Borel function bounded from below. Then, setting $w(t) = (w^1(t), \cdots, w^d(t))$, $u = e^{tL}f$ is given by

$$u(t,x) = E\left[\exp\left\{\sum_{i=1}^d \int_0^t b^i(x + w(s))dw^i(s) - \frac{1}{2}\sum_{i=1}^d \int_0^t b^i\left(x + w(s)\right)^2 ds \right.\right.$$
$$\left.\left. - \int_0^t V(x + w(s))ds\right\} f(x + w(t))\right]. \tag{2.5}$$

Here, the Wiener functionals $\int_0^t b^i(x + w(s))dw^i(s)$ are defined by Itô's stochastic integrals. Thus, we have encountered now a case of Wiener functional expectations in which a use of the Itô calculus is indispensable.

Now we can see that the solution of (2.2) has, in each case, the Lebesgue integral representation by the heat kernel:

$$u(t,x) = \int_{\mathbf{R^d}} \langle x|e^{tL}|y\rangle f(y)dy.$$

and the heat kernel $\langle x|e^{tL}|y\rangle$ is given as follows:

In the case [1], $\langle x|e^{tL}|y\rangle = p(t,y-x)$, where $p(t,x)$ is the Gauss kernel given by (2.1).

In the case [2],

$$\begin{aligned}\langle x|e^{tL}|y\rangle &= E_{0,0}^{t,y-x}\left[\exp\left\{-\int_0^t V(x+w(s))ds\right\}\right]p(t,y-x)\\ &= E\left[\exp\left\{-\int_0^t V(x+w(s))ds\right\}\delta_y(x+w(t))\right].\end{aligned}$$

In the case [3],

$$\begin{aligned}\langle x|e^{tL}|y\rangle = E_{0,0}^{t,y-x}\Bigg[&\exp\Bigg\{\sum_{i=1}^d\int_0^t b^i(x+w(s))dw^i(s)\\ &-\frac{1}{2}\sum_{i=1}^d\int_0^t b^i(x+w(s))^2ds - \int_0^t V(x+w(s))ds\Bigg\}\Bigg]p(t,y-x)\\ = E\Bigg[&\exp\Bigg\{\sum_{i=1}^d\int_0^t b^i(x+w(s))dw^i(s) - \frac{1}{2}\sum_{i=1}^d\int_0^t b^i(x+w(s))^2ds\\ &-\int_0^t V(x+w(s))ds\Bigg\}\delta_y(x+w(t))\Bigg].\end{aligned}$$

Strictly speaking, it is by no means obvious in this case that the Wiener functional under the expectation $E_{0,0}^{t,y-x}$ is $P_{0,0}^{t,y-x}$-measurable. Such a difficulty will be completely resolved by a general theory of desintegrations and quasi sure analysis in the Malliavin calculus, as we review in Section 3.

3 Itô Calculus on Wiener Space, Itô Functionals and Itô Maps

We would continue the same problem of probabilistic solutions of IVP (2.2) in which the second order differential operator L is of variable coefficients. If it is elliptic, then it is essentially the case of differential operators on a Riemannian manifold M in which the principal second order term is the half Laplacian $\frac{1}{2}\Delta_M$. Hence we need a Brownian motion, i.e., the diffusion on M with the infinitesimal generator $\frac{1}{2}\Delta_M$. As we would review now, this can be realized on a Wiener space by an application of the Itô calculus.

Let $(W_0(\mathbf{R}^d), P^W)$ be the d-dimensional Wiener space. Then, the coordinate $w(t) = (w^1(t), \cdots, w^d(t))$ of $w \in W_0(\mathbf{R}^d)$ is a realization of the d-dimensional Wiener process.

Let M be a smooth manifold of dimension n and let $A_0, A_1, \ldots, A_d$ be a smooth and complete vector fields on M. Consider the following stochastic differential equation (SDE) on M in which $\circ$ denotes the stochastic differential in the Stratonovich sense:

$$dX(t) = \sum_{i=1}^{d} A_i(X(t)) \circ dw^i(t) + A_0(X(t))dt, \quad X(0) = x \tag{3.1}$$

and we obtain the pathwise unique solution $X(t) = (X(t, x; w))$. In this note, we assume for simplicity that solutions exist globally; otherwise, we must consider solutions which may tend to the point at infinity of M in a finite time and many of definitions given below need some modifications. For the global existence, it is sufficient to assume that M is compact, or assume that M is embedded into a higher dimensional Euclidean space and, in the global Euclidean coordinates, the coefficients of vector fields have all the derivatives of order ≥ 1 bounded, cf. e.g. [IW] for details.

Let $\mathcal{C}([0,T] \to M)$ be the space of continuous paths $\xi : [0,T] \ni t \mapsto \xi(t) \in M$ endowed with the topology of uniform convergence and, for $x \in M$, $\mathcal{C}_x([0,T] \to M)$ be its subspace consisting of paths ξ such that $\xi(0) = x$. Then the solution defines the following Wiener map (called also an *Itô map*); for each $x \in M$,

$$X^x : w \in W_0(\mathbf{R}^d) \mapsto X^x(w) := [t \mapsto X(t, x; w)] \in \mathcal{C}_x([0,T] \to M).$$

If P_x is the image measure on $\mathcal{C}_x([0,T] \to M)$ of P^W under the Itô map X^x, then the system $\{P_x; x \in M\}$ defines a diffusion process on M with the infinitesimal generator $A = \frac{1}{2}\sum_{i=1}^{r}(A_i)^2 + A_0$. Thus, the Itô map provides us with the A-diffusion process. It actually provides us with something more; if we regard the solution as

$$w \in W_0(\mathbf{R}^d) \mapsto X(w) := [t \mapsto [x \mapsto X(t, x; w)]],$$

then $[x \mapsto X(t, x; w)] \in \mathrm{Diff}(M \to M)$, i.e. a diffeomorphism of M, so that we have a *stochastic flow* of diffeomorphisms (cf. [Ku]).

Wiener functionals which are defined by using the Itô calculus, particularly such functionals as are associated with an Itô map, are often called *Itô functionals.*

In order to discuss the Brownian motion on a Riemannian manifold, the following Itô map, called a *stochastic moving frame*, is very important and useful.

Stochastic Moving Frames

Let M be a Riemannian manifold of dimension d and $x \in M$. By a *frame* at x, we mean an orthonormal base (ONB) $\mathbf{e} = [\mathbf{e}_1, \ldots, \mathbf{e}_d]$ of the tangent space T_xM at x. A frame at x is denoted by $r = (x, \mathbf{e})$. We denote the totality of frames at all points of M by $O(M)$. It is given a natural structure of manifold and the projection $\pi : O(M) \to M$ is defined by $\pi(r) = x$ if $r = (x, \mathbf{e})$. The d-dimensional orthogonal group $O(d)$ acts on $O(M)$ from the right: $rg = (x, \mathbf{e}g = [(\mathbf{e}g)_1, \ldots, (\mathbf{e}g)_d])$, $g = (g^i_j) \in O(d)$, where $(\mathbf{e}g)_k = g^i_k \mathbf{e}_i$ (by the usual convention for summation), so that $O(M)$ forms a principal fibre bundle over M with the structure group $O(d)$, which we call the *bundle of orthonormal frames.*

We can identify $r \in O(M)$ with an isometric isomorphism $\tilde{r} : \mathbf{R}^d \to T_xM$, $x = \pi(r)$, defined by sending each of the canonical base δ_i, $\delta_i := (0, \cdots, 0, \overset{i-th}{1}, 0 \cdots 0)$, to $\mathbf{e}_i$, $i = 1, \ldots, d$, where $r = (x, \mathbf{e})$, $\mathbf{e} = [\mathbf{e}_1, \ldots, \mathbf{e}_d]$. This isomorphism $\tilde{r}$ is called the *canonical isomorphism* associated with the frame r. It holds that $\widetilde{rg} = \tilde{r} \circ g$, $g = (g^i_j) \in O(d)$; here g is identified with the orthogonal transformation $g : x = (x^i) \in \mathbf{R}^d \mapsto gx = (g^i_j x^j) \in \mathbf{R}^d$.

Before giving a formal definition of the stochastic moving frame in general, we explain the idea in a simple case of M being a two dimensional sphere $\mathbf{S}^2$. We take a plane $\mathbf{R}^2$ and consider a Brownian motion $w(t)$ on it canonically realized on the two dimensional Wiener space. We assign at each point $w(t) \in \mathbf{R}^2$ the canonical bases $\delta_1 = (1, 0)$ and $\delta_2 = (0, 1)$ so that $\delta = [\delta_1, \delta_2]$ forms an ONB in the tangent space $T_{w(t)}\mathbf{R}^2 \cong \mathbf{R}^2$. Then these bases at different points of the curve are parallel to each other. Given a sphere $\mathbf{S}^2$, choose a point x on it and an ONB $\mathbf{e} = [\mathbf{e}_1, \mathbf{e}_2]$ in the tangent space $T_x\mathbf{S}^2$. We put the sphere on the plane so that x touches at the origin of the plane and the ONB $\mathbf{e}$ coincides with the ONB δ. Now we roll the sphere on the plane along the Brownian curve $w(t)$ without slipping. Suppose that the Brownian curve is traced in ink. Then the trace of $w(t)$ together with the ONB δ at $w(t)$ is transferred into a curve $X(t)$ on $\mathbf{S}^2$ with an ONB $\mathbf{e}(t) = [\mathbf{e}_1(t), \mathbf{e}_1(t)]$ in $T_{X(t)}\mathbf{S}^2$. Thus, a random curve $r(t) = (X(t), \mathbf{e}(t))$ on the orthonormal frame bundle $O(\mathbf{S}^2)$ is obtained and this is precisely the stochastic moving frame we want. We can see that the random curve $X(t)$ thus obtained is a Brownian motion on the sphere.

Now we give a formal definition. There is a notion of the *system of canonical horizontal vector fields* $A_1, \ldots, A_d$ on $O(M)$: For each $i = 1, \cdots, d$, $A_i(r)$ is a smooth vector field on $O(M)$ uniquely determined by the property that the integral curve, i.e. the solution, of the following ordinary differential equation (ODE)

$$\frac{dr(t)}{dt} = A_i(r(t)), \quad r(0) = r, \quad r = (x, \mathbf{e}), \quad \mathbf{e} = [\mathbf{e}_1, \ldots, \mathbf{e}_d]$$

coincides with the curve $r(t) = (x(t), \mathbf{e}(t))$, $\mathbf{e}(t) = [\mathbf{e}_1(t), \ldots, \mathbf{e}_d(t)]$, where $x(t)$ is the geodesic with $x(0) = x$ and $\frac{dx}{dt}|_{t=0} = \mathbf{e}_i$, and $\mathbf{e}(t) = [\mathbf{e}_1(t), \ldots, \mathbf{e}_d(t)]$ is the parallel translate, in the sense of Levi-Civita, of $\mathbf{e} = [\mathbf{e}_1, \ldots, \mathbf{e}_d]$ along the curve $x(t)$.

Let $(W_0(\mathbf{R}^d), P^W)$ be the d-dimensional Wiener space. The stochastic moving frames on M starting at a frame r is, by definition, the solution

$$r(t) = (r(t,r;w)), \quad r(t,r;w) = (X(t,r;w), \mathbf{e}(t,r;w)) \tag{3.2}$$

of the following SDE on $O(M)$:

$$dr(t) = A_k(r(t)) \circ dw^k(t), \quad r(0) = r. \tag{3.3}$$

The assumption that solutions exist globally is equivalent to that the manifold is *stochastically complete*. We have the following important property of the stochastic moving frame under the right action of the structure group $O(d)$; for each $g \in O(d)$,

$$r(t,r;w)g = r(t,rg;g^{-1}w), \quad t \geq 0, \quad r \in O(M),$$

where $g^{-1}w \in W_0(\mathbf{R}^d)$ is defined by $(g^{-1}w)(t) = g^{-1}[w(t)]$. This implies, in particular, that

$$X(t,r;w) = X(t,rg;g^{-1}w), \quad t \geq 0, \quad r \in O(M), \quad g \in O(d).$$

By the rotation invariance of Wiener process, we have $g^{-1}w \stackrel{d}{=} w$, and hence

$$\{X(t,rg;w);\ t \geq 0\} \stackrel{d}{=} \{X(t,r;w);\ t \geq 0\}, \quad r \in O(M), \quad g \in O(d).$$

In other words, the law P_r on $C_x([0,T] \to M), x = \pi(r)$, of $[t \mapsto X(t,r;w)]$ satisfies $P_{rg} = P_r$ for all $g \in O(d)$. This implies that P_r depends only on $x = \pi(r)$ and we may write $P_r = P_x$. Then the family $\{P_x\}$ defines a diffusion process on M. If we note the identity: $\sum_{k=1}^d A_k^2 \tilde{f} = \widetilde{\Delta_M f}$, which holds for any smooth function f on M and $\tilde{f} := f \circ \pi$, we can see that its generator coincides with $\frac{1}{2}\Delta_M$ so that it is a Brownian motion on M. In this way, the Brownian motion on a Riemannian manifold can be obtained as the projection on the base manifold of the stochastic moving frame, cf. [IW] for details.

Solutions of IVP (2.2) by Wiener Functional Expectations. II

We consider the case of heat equations on a Riemannian manifold M of dimension d and we set up on d-dimensional Wiener space the stochastic moving frame $\{r(t,r;w) = (X(t,r;w), \mathbf{e}(t,r;w))\}$ as above.

[4] We consider the case $L = \frac{1}{2}\Delta_M$. Then $u = e^{tL}f$ is given by

$$u(t,x) = E[f(X(t,r;w))], \quad x = \pi(r). \tag{3.4}$$

As we explained above, the right-hand side (RHS) depends only on $x = \pi(r)$.

[5] (Feynman-Kac formula) We consider the case of a Schrödinger operator $L = \frac{1}{2}\Delta_M - V$ where the potential $V(x)$ is a real Borel function bounded from below. Then $u = e^{tL}f$ is given by

$$u(t,x) = E\left[\exp\left\{-\int_0^t V(X(s,r;w))ds\right\} f(X(t,r;w))\right], \quad x = \pi(r). \tag{3.5}$$

[6] (Schrödinger operators with magnetic fields) We consider the case of operator

$$Lu = \frac{1}{2}\left[\Delta_M u + 2\sqrt{-1}(du,\theta) - \left(\sqrt{-1}d^*\theta + \|\theta\|^2 + 2V\right)u\right] := H(\theta, V),$$

where θ is a real one-form (called a vector potential) and V is a real function (called a scalar potential). $(*,**)$ and $\| * \|$ are the Riemannian inner product and norm on the cotangent space $T^*_\cdot(M)$, respectively. d^* is the adjoint of exterior differentiation d so that $d^*\theta$ is a real function. Then $u = e^{tL}f$ is given by

$$u(t,x) = E\left[\exp\left\{\sqrt{-1}\int_0^t \bar{\theta}_i(X(s,r;w)) \circ dw^i(s) \right.\right.$$
$$\left.\left. - \int_0^t V(X(s,r;w))ds\right\} f(X(t,r;w))\right], \quad x = \pi(r). \qquad (3.6)$$

Here, $\bar{\theta}_i(r) = e_i^k\theta_k(x)$ if $r = (x,\mathbf{e})$, $\mathbf{e} = [\mathbf{e}_1,\ldots,\mathbf{e}_d]$ and $\theta(x) = \theta_i(x)dx^i$, $\mathbf{e}_i = e_i^k\frac{\partial}{\partial x^k}$ in a local coordinate $x = (x^1,\cdots,x^d)$. Obviously, $\bar{\theta}$ is defined independently of a particular choice of local coordinates.

[7] (Heat equations on vector bundles) We consider the case of the exterior product $\bigwedge T^*M$ of cotangent bundle T^*M, so that its section is a differential form on M, and the case of $L = \frac{1}{2}\Box$, where $\Box := -(d^*d + dd^*)$ is the de Rham-Hodge-Kodaira Laplacian acting on differential forms. We assume that M is compact and orientable.

The canonical isomorphism $\tilde{r} : \mathbf{R}^d \to T_xM$ associated with a frame $r = (x,\mathbf{e}) \in O(M)$ naturally induces an isomorphism $\tilde{r} : \mathbf{R}^d \to T_x^*M$ and an isomorphim $\tilde{r} : \bigwedge\mathbf{R}^d \to \bigwedge T_x^*M$, by sending bases δ_i and $\delta_{i_1}\wedge\cdots\wedge\delta_{i_p}$ to $\mathbf{f}^i$ and $\mathbf{f}^{i_1}\wedge\cdots\wedge\mathbf{f}^{i_p}$, for $i = 1,\cdots,d$ and $1 \le i_1 < \cdots < i_p \le d$, respectively, where $\mathbf{f} = [\mathbf{f}^1,\cdots,\mathbf{f}^d]$ is the ONB in T_x^*M dual to the ONB $\mathbf{e} = [\mathbf{e}_1,\ldots,\mathbf{e}_d]$ in T_xM. Here, we recall that the exterior product $\bigwedge\mathbf{R}^d = \sum_{p=0}^d \oplus \bigwedge^p\mathbf{R}^d$ is a 2^d-dimensional Euclidean space with the canonical base $\delta_{i_1}\wedge\cdots\wedge\delta_{i_p}$, forming an algebra under the exterior product $\wedge$. Let $\mathrm{End}(\bigwedge\mathbf{R}^d)$ be the algebra of linear tranformations on $\bigwedge\mathbf{R}^d$ and let $a_i^* \in \mathrm{End}(\bigwedge\mathbf{R}^d)$ be defined by

$$a_i^*(\lambda) = \delta_i\wedge\lambda, \quad \lambda \in \bigwedge\mathbf{R}^d, \quad i = 1,\ldots,d.$$

Let a_i be the dual of a_i^*. Then the system $a_{i_1}a_{i_2}\cdots a_{i_p}a_{j_1}^*a_{j_2}^*\cdots a_{j_q}^*$, where $1 \le i_1 < \cdots < i_p \le d$, $1 \le j_1 < \cdots < j_q \le d$, $p,q = 0,1,\ldots,d$, forms a basis in $\mathrm{End}(\bigwedge\mathbf{R}^d)$. Let $J^{ijkl}(r)$ be the scalarization (equivariant representation) of the Riemann curvature tensor; in a local coordinate, $J^{ijkl}(r) = R_{\alpha\beta\gamma\delta}(x)e_i^\alpha e_j^\beta e_k^\gamma e_l^\delta$, $r = (x,\mathbf{e})$. Let $D_2[J](r) \in \mathrm{End}(\bigwedge\mathbf{R}^d)$ be defined by $D_2[J](r) = J^{ijkl}(r)a_i^*a_ja_k^*a_l$. We define an $\mathrm{End}(\bigwedge\mathbf{R}^d)$-valued process $t \to M(t,r;w)$ by the solution to the following ODE on $\mathrm{End}(\bigwedge\mathbf{R}^d)$:

$$\frac{dM(t)}{dt} = \frac{1}{2} D_2[J](r(t,r;w)) \cdot M(t), \quad M(0) = I.$$

Then, $u = e^{tL}f$, $f \in \bigwedge(M), := \mathbf{e}^\infty(M \to \bigwedge T^*M)$, is given by

$$u(t,x) = E\left[\tilde{r} M(t,r;w) \widetilde{r(t,r;w)}^{-1} f(X(t,r;w)\right], \quad r = (x, \mathbf{e}). \tag{3.7}$$

Note that $\tilde{r} : \bigwedge \mathbf{R}^d \to \bigwedge T_x^*M$, $\widetilde{r(t,r;w)}^{-1} : \bigwedge T^*_{X(t,r;w)}M \to \bigwedge \mathbf{R}^d$, and $f(X(t,r;w)) \in \bigwedge T^*_{X(t,r;w)}M$, so that the Itô functional under the expectation takes values in $\bigwedge T_x^*M$. Hence the expectation is well-defined and takes its value in $\bigwedge T_x^*M$. Also, it does not depend on a particular choice of $r \in O(M)$ over the point x and so, we may write it as $u(t,x)$. Cf. [IW], for details.

The solutions $u = e^{tL}f$ obtained by Wiener functional expectations as above can also possess *heat kernel representations* in the form

$$u(t,x) = \int_M \langle x | e^{tL} | y \rangle f(y) dy$$

where dy is the Riemannian volume of M. The heat kernel $\langle x|e^{tL}|y\rangle$ is usually constructed by the method of parametrix in PDE theory. Here, we would apply our probabilistic approach by Wiener functional expectations also to this problem; this is indeed possible by appealing to the Malliavin calculus on Wiener space.

4 Malliavin Calculus on Wiener Space

The Malliavin calculus is a differential and integral calculus on an infinite dimensional vector space endowed with a Gaussian measure. Here, we restrict ourselves to the case of the r-dimensional Wiener space $(W_0(\mathbf{R}^r), P^W)$; the Malliavin calculus in this case is well suited to the analysis of Itô functionals as we shall see. We would develop the Malliavin calculus as a Sobolev differential calculus on Wiener space by introducing a family of *Sobolev spaces* of Wiener functionals.

For a real separable Hilbert space E, we denote by $L_p(E)$, $1 \le p < \infty$, the usual L^p-space of E-valued Wiener functionals. It is convenient to introduce the Fréchet space $L_{\infty-}(E) := \cap_{1<p<\infty} L_p(E)$ and its dual $L_{1+}(E) := \cup_{1<p<\infty} L_p(E)$, (the dual E^* being always identified with E by the Riesz theorem). When $E = \mathbf{R}$, $L_p(E)$, $(L_{\infty-}(E), L_{1+}(E))$ is denoted simply by L_p, (resp. $L_{\infty-}$, L_{1+}).

Typical differential operators are, the Gross-Malliavin-Shigekawa gradient operator D which sends an E-valued Wiener functional to an $H \otimes E$-valued Wiener functional, its dual operator or Skorohod operator D^* and the Ornstein-Uhrenbeck operator $L = -D^*D$. D is defined formally, for a Wiener functional $F = (F(w))$, by

$$\langle DF(w), h \otimes e \rangle_{H\otimes E} = \left\langle \lim_{\epsilon\to 0}(F(w+\epsilon h) - F(w))/\epsilon, e \right\rangle_E \qquad h \in H,\ e \in E.$$

These operators are defined, first, for polynomial functionals and also, the fractional power $(I-L)^\alpha$, $\alpha \in \mathbf{R}$, is defined for polynomial functionals by using the Wiener chaos expansion; a polynomial functional F is a finite sum $F = \oplus\sum_n F_n$ where F_n is a polynomial functional in the chaos subspace of order n, and then, $(I-L)^\alpha F$ is defined to be $\oplus\sum_n (1+n)^\alpha F_n$, which is also a polynomial functional. Let $\mathcal{P}(E)$ be the real vector space of all E-valued polynomial functionals. Noting that $\mathcal{P}(E) \subset L_{\infty-}(E)$, we define the norm $||*||_{p,\alpha}$ on $\mathcal{P}(E)$, $1<p<\infty$, $\alpha \in \mathbf{R}$, by $||F||_{p,\alpha} = ||(I-L)^{\alpha/2}F||_p$. Let $\mathcal{P}(E)^*$ be the algebraic dual of $\mathcal{P}(E)$, which is a real vector space formed of all $\mathbf{R}$-linear mappings $T : F \in \mathcal{P}(E) \mapsto T(F) \in \mathbf{R}$. For $G \in L_2(E)$, we define $T_G \in \mathcal{P}(E)^*$ by $T_G(F) = E(\langle G, F\rangle_E)$, and identify G with T_G. Then, $L_2(E) \subset \mathcal{P}(E)^*$ and hence, $\mathcal{P}(E) \subset L_2(E) \subset \mathcal{P}(E)^*$. Define the norm on $\mathcal{P}(E)^*$ by setting

$$||T||_{p,\alpha} = \sup\{T(F); F \in \mathcal{P}(E),\ ||F||_{q,-\alpha} \le 1\},$$
$$1 < p < \infty,\ \alpha \in \mathbf{R},\ \frac{1}{p} + \frac{1}{q} = 1.$$

This definition is compatible with the norm defined already on $\mathcal{P}(E)$, which is a subspace of $\mathcal{P}(E)^*$ as we saw above. We now define the family of Sobolev spaces:

$$\mathbf{D}_p^\alpha(E) = \{\, T \in \mathcal{P}(E)^*;\ \ ||T||_{p,\alpha} < \infty \,\}, \quad 1 < p < \infty, \quad \alpha \in \mathbf{R}.$$

Then $\mathbf{D}_p^\alpha(E)$, endowed with the norm $||*||_{p,\alpha}$, is a real separable Banach space in which the space $\mathcal{P}(E)$ of E-valued polynomial functionals is densely included. Our definition given here is of course equivalent to the usual one given by the completion of $\mathcal{P}(E)$ with respect to the norm $||*||_{p,\alpha}$ (cf. e.g. [IW]); this elegant idea of avoiding the use of an abstract notion like completion is due to Itô [It].

We have $\mathbf{D}_p^0(E) = L_p(E)$, $\mathbf{D}_p^\alpha(E) \subset \mathbf{D}_{p'}^{\alpha'}(E)$ if $p' \le p$, $\alpha' \le \alpha$, and $\mathbf{D}_p^\alpha(E)^* = \mathbf{D}_q^{-\alpha}(E)$ if $p^{-1} + q^{-1} = 1$. Again, $\mathbf{D}_p^\alpha(E)$ in the case of $E = \mathbf{R}$ is denoted simply by $\mathbf{D}_p^\alpha$.

We set $\mathbf{D}_p^\infty(E) = \bigcap_{\alpha>0}\mathbf{D}_p^\alpha(E)$, $\mathbf{D}_p^{-\infty}(E) = \bigcup_{\alpha>0}\mathbf{D}_p^{-\alpha}(E)$. We also denote $\mathbf{D}_{\infty-}^\infty(E) = \bigcap_{1<p<\infty}\mathbf{D}_p^\infty(E)$ and $\mathbf{D}_{1+}^{-\infty}(E) = \bigcup_{1<p<\infty}\mathbf{D}_p^{-\infty}(E)$. Again, we omit E in these notations when $E = \mathbf{R}$.

Now, the differential operators D and L can be extended uniquely to act on the space $\mathbf{D}_{1+}^{-\infty}(E)$ and D^* on $\mathbf{D}_{1+}^{-\infty}(H \otimes E)$, so that $D : \mathbf{D}_p^{\alpha+1}(E) \to \mathbf{D}_p^\alpha(H \otimes E)$, $D^* : \mathbf{D}_p^{\alpha+1}(H \otimes E) \to \mathbf{D}_p^\alpha(E)$ and $L : \mathbf{D}_p^{\alpha+2}(E) \to \mathbf{D}_p^\alpha(E)$ are continuous for every $1 < p < \infty$ and $\alpha \in \mathbf{R}$.

An element F in the space $\mathbf{D}_p^\infty(E)$ may be called *smooth* because it is infinitely differentiable. Typical examples of smooth functionals are give by Itô functionals; if $F(w) = f(X(t,x;w))$, $t > 0, x \in M$, where $X = (X(t,x;w))$ is the solution of SDE (3.1) when $d = r$ and f is a smooth function on M with a suitable growth condition at the point at infinity of M, then $F \in \mathbf{D}^{\infty}_{\infty-}$. It should be remarked, however, that smooth functionals are not continuous, in general. A typical example is Lévy's stochastic area $S(t,w) = \frac{1}{2}\int_0^t w^1(s)dw^2(s) - w^2(s)dw^1(s)$ on the two-dimensional Wiener space, which is an element in $\mathbf{D}^{\infty}_{\infty-}$ for each $t > 0$. However, there is no continuous function on the Wiener space which coincides with $S(t,w)$, P^W-a.e.; in fact, Sugita ([S 2]) showed more strongly that, on any separable Banach space continuously included in $W_0(\mathbf{R}^2)$ which has nonetheless P^W-measure 1, there exists no continuous function which coincides with $S(t,w)$, P^W-a.e. Thus, we see that Sobolev's embedding theorem no longer holds in our Sobolev differential calculus on Wiener space.

When $\alpha > 0$, some elements in $\mathbf{D}_p^{-\alpha}(E)$ are no more Wiener functionals in the sense of P^W-measurable functions; they are something like Schwartz distributions on Wiener space. We call them *generalized Wiener functionals.* The natural coupling between $F \in \mathbf{D}_p^{\alpha}$ and $G \in (\mathbf{D}_p^{\alpha})^* = \mathbf{D}_q^{-\alpha}$ is denoted by $E(FG)$; this notation is compatible with the usual one when $\alpha = 0$. In particular, the natural coupling of $F \in \mathbf{D}_{1+}^{-\infty}$ with $\mathbf{1} \in \mathbf{D}^{\infty}_{\infty-}$, $\mathbf{1}$ being the Wiener fuctional identically equal to 1, is denoted by $E(F)$ and is called the *generalized expectation* of F.

Typical examples of generalized Wiener functionals are obtained by the composite of Schwartz distributions on $\mathbf{R}^n$ (or on a manifold M) with a smooth $\mathbf{R}^n$-valued (resp. M-valued) Wiener functional which is *nondegenerate* in the sense given below. We mainly discuss the case of $\mathbf{R}^n$; the case of manifold can be discussed similarly and, indeed, can be reduced to the case of $\mathbf{R}^n$ by choosing a suitable local coordinate.

Let $F = (F^i(w)) \in \mathbf{D}^{\infty}_{\infty-}(\mathbf{R}^n)$ and define the *Malliavin covariance* $\sigma_F = (\sigma_F^{ij}(w))$ of F by

$$\sigma_F^{ij}(w) = \langle DF^i(w), DF^j(w)\rangle_H, \quad i,j = 1.\cdots,n.$$

It is nonnegative definite so that $\det\sigma_F \geq 0$, P^W-a.s.. We set $(\det\sigma_F)^{-1} = +\infty$ if $\det\sigma_F = 0$. For a domain U in $\mathbf{R}^n$, we say that F *is nondegenerate in* U if $1_U(F)\cdot(\det\sigma_F)^{-1} \in L_{\infty-} = \bigcap_{1<p<\infty} L_p$. Then, for every Schwartz distribution T on $\mathbf{R}^n$ with support contained in U, a generalized Wiener functional $T \circ F = T(F(w))$ can be defined uniquely as an element in $\mathbf{D}^{-\infty}_{\infty-} := \bigcap_{1<p<\infty} \mathbf{D}_p^{-\infty}$ so that this notion has the following two properties: (i) if T is given by a smooth function $f(x)$ on $\mathbf{R}^n$ with support contained in U, then $T\circ F = f(F(w))$, (ii) if $T_\nu \to T$ in the sense of a Sobolev norm with negative differentiability index, then $T_\nu \circ F \to T\circ F$ in $\bigcap_{1<p<\infty}\mathbf{D}_p^{-\alpha}$ for some $\alpha > 0$. $T\circ F$ is called the *composite* of the Schwartz distribution T on $\mathbf{R}^n$ and the Wiener functional F, or the *pull-back* of the Schwartz distribution T on $\mathbf{R}^n$ by the Wiener map $F: W_0(\mathbf{R}^d) \to \mathbf{R}^n$.

In particular, for Dirac δ-functions δ_x, $x \in U$, $\delta_x(F)$ is defined as an element in $\mathbf{D}^{-\infty}_{\infty-}$. By the continuity property (ii), we can deduce that $x \in U \mapsto$

$\delta_x(F) \in \mathbf{D}_{\infty-}^{-\alpha}$ is C^∞ and hence, $x \in U \mapsto E[\Phi \cdot \delta_x(F)]$ is a C^∞-function for $\Phi \in \mathbf{D}_{1+}^{\infty} =: \bigcup_{1<p<\infty} \mathbf{D}_p^{\infty}$. We can easily deduce that $p_F(x) := E(\delta_x(F))$, $x \in U$, is density in U, with respect to the Lebesgue measure, of the law of F and $E[\Phi \cdot \delta_x(F)] = p_F(x)E[\Phi | F = x]$, so that the conditional expectation of Φ given $F = x$ can be defined smoothly and pointwise on a set $\{x \in U | p_F(x) > 0\}$.

Let $r(t, r; w) = (X(t, r; w), \mathbf{e}(t, r, w))$ be the stochastic moving frame on a Riemannian manifold M as introduced in Section 2. Let δ_x, $x \in M$, be the Dirac delta function on M with pole at x defined with respect to the Riemannian volume dx. For each $t > 0$ and $r \in O(M)$, M-valued Wiener functional $X(t, r; w)$ is smooth and nondegenerate, so that the composite $\delta_y(X(t, r; w))$ is defined as an element in $\mathbf{D}_{\infty-}^{-\infty}$ for each $y \in M$. Using this notion, we can now give a probabilistic expression for heat kernels $\langle x | e^{tL} | y \rangle$ for heat equations studied in Section 2: here, heat kernels are defined with respect to the Riemannian volume dy on M, so that $u = e^{tL} f$ is given by $u(t, x) = \int_M \langle x | e^{tL} | y \rangle f(y) dy$.

In the case [4], i.e., $L = \frac{1}{2}\Delta_M$,

$$\langle x | e^{tL} | y \rangle = E[\delta_y(X(t, r; w))], \quad x = \pi(r).$$

By considering the right action of $O(d)$, we deduce as above that the expectation in the RHS does not depend on a particular choice of $r \in O(M)$ such that $x = \pi(r)$.

In the case of [5], i.e., $L = \frac{1}{2}\Delta_M - V$,

$$\langle x | e^{tL} | y \rangle = E\left[\exp\left\{ -\int_0^t V(X(s, r; w)) ds \right\} \cdot \delta_y(X(t, r; w)) \right], \quad x = \pi(r).$$

In the case of [6], i.e,, $L = H(\theta, V)$,

$$\langle x | e^{tL} | y \rangle = E\left[\exp\left\{ \sqrt{-1} \int_0^t \bar{\theta}_i(X(s, r; w)) \circ dw^i(s) \right.\right.$$
$$\left.\left. - \int_0^t V(X(s, r; w)) ds \right\} \cdot \delta_y(X(t, r; w)) \right], \quad x = \pi(r).$$

In the case of [7], i.e., $L = \frac{1}{2}\Box$ acting on $\bigwedge(M)$, we need a more careful consideration; the heat kernel $\langle x | e^{tL} | y \rangle$ takes its value in the vector space $\mathrm{Hom}(\bigwedge T_y^* M, T_x^* M)$ formed of all linear mappings from $\bigwedge T_y^* M$ to $\bigwedge T_x^* M$, and it should be given formally by

$$\langle x | e^{tL} | y \rangle = E\left[\tilde{r} M(t, r; w) \widetilde{r(t, r; w)}^{-1} \cdot \delta_y(X(t, r; w)) \right], \quad r = (x, \mathbf{e}).$$

However, the meaning of the generalized expectation in the RHS is not clear because the Wiener functional $\tilde{r} M(t, r; w) \widetilde{r(t, r; w)}^{-1}$ takes its value in the vector space $\mathrm{Hom}(\bigwedge T_{X(t,r;w)}^* M, \bigwedge T_x^* M)$, which is not a fixed vector space when w varies. We can overcome this difficulty by appealing to the *quasi-sure analysis* in the Malliavin calculus (cf. e.g., Malliavin [M], Lescot [Le], Itô [It]).

As we remarked above, a smooth Wiener functional cannot have a continuous modification, in general. It can possess however a modification, called *quasi-continuous modification* or *redefinition* of it. If $F \in \mathbf{D}^{\infty}_{\infty-}(\mathbf{R}^n)$ is non-degenerate in $U \subset \mathbf{R}^n$, then, as was shown by Airault-Malliavin ([AM]) and Sugita ([S 1]), there exists a finite Borel measure μ_x on the Wiener space $W_0(\mathbf{R}^d)$ associated uniquely with $x \in U$ such that, for every $\Phi \in \mathbf{D}^{\infty}_{1+}$, its quasi-continuous modification $\tilde{\Phi}$ is μ_x-integrable and the following identity holds:

$$\int_{W_0(\mathbf{R}^d)} \tilde{\Phi}(w)\mu_x(dw) = E[\Phi \cdot \delta_x(F)].$$

The measure μ_x has its full measure on the set $\mathcal{S}_x := \{w \in W_0(\mathbf{R}^d) \mid \tilde{F}(w) = x\}$. We may think of the measure μ_x as having the formal density $\delta_x(F)$, or we may think of it as the 'surface measure' on a 'hypersurface' $\mathcal{S}_x$ embedded in the Wiener space.

A similar theory can be developed in the case of $O(M)$ and we have a measure $\mu_y^{t,r}$ associated with $\delta_y(X(t,r;w))$, $y \in M$. If $\tilde{X}(t,r;w)$ is a quasi-continous modification of $X(t,r;w)$, then $\mu_y^{t,r}(\{w \mid \tilde{X}(t,r;w) \neq y\}) = 0$. Then, a quasi-continuous modification $\left[\tilde{r}M(t,r;w)\widetilde{r(t,r;w)}^{-1}\right]^{\sim}$ of $\tilde{r}M(t,r;w)\widetilde{r(t,r;w)}^{-1}$ takes values in $\mathrm{Hom}(\bigwedge T_y^*M, T_x^*M)$ quasi-surely and hence $\mu_y^{t,r}$-almost surely, so that it can be integrated by the measure $\mu_y^{t,r}$ to get an element in $\mathrm{Hom}(\bigwedge T_y^*M, T_x^*M)$. Now we have

$$\langle x|e^{tL}|y\rangle = \int_{W_0(\mathbf{R}^d)} \left[\tilde{r}M(t,r;w)\widetilde{r(t,r;w)}^{-1}\right]^{\sim} \mu_y^{t,r}(dw), \quad r = (x, \mathbf{e}).$$

5 Concluding Remarks

Probabilistic representations of heat kernels given above can be applied to study various properties of heat kernels; regularities, estimates, short time asymptotic expansions and so on. There are a huge amount of literatures and it is beyond the scope of this work to review of them. We would only refer to two survey articles [Ik, W] in which we can see several effective applications of our heat kernel representation by generalized Wiener functional expectations to the problems of McKean and Singer. Here we would content ourselves with giving a remark on quasi-sure analysis discussed above.

Quasi-Sure Analysis and the Theory of Rough Paths by T. Lyons

We saw in Section 3 an example of heat kernel representations in which some refinement of generalized expectations is necessary and we did it by appealing to the quasi-sure analysis on Wiener space. We would remark that another approach is possible to such a refinement by using the theory of rough paths due to T. Lyons. We first recall this theory: (cf. [Ly]).

Let $W_0(\mathbf{R^d}) := \{w;\ [0,T] \ni t \mapsto w(t) \in \mathbf{R}^d,\ \text{continuous},\ w(0)=0\}$ be the d-dimensional path space as above and H be its Cameron-Martin subspace. Let

$$\mathcal{H}_d \left(\cong \mathbf{R}^{d(d+1)/2} \cong \mathbf{R}^d \times so(d)\right) := \left\{x = (x^i, x^{(i,j)}) \mid 1 \le i < j \le d\right\}$$

endowed with the group multiplication $x \cdot y,\ x, y \in \mathcal{H}_d,$ defined by $x \cdot y := z = (z^i, z^{(i,j)})$ where $z^i = x^i + y^i,\ z^{(i,j)} = x^{(i,j)} + y^{(i,j)} + \frac{1}{2}(x^i y^j - x^j y^i)$. $\mathcal{H}_d$ is called the free nilpotent Lie group with step 2 and d generators.

Let $\triangle_T = \{(s,t) \mid 0 \le s \le t \le T\} (\subset [0,T]^2)$ and set

$$\begin{aligned} \Omega(\mathbf{R}^d) = \{\omega = (\omega(s,t)) : \triangle_T \ni (s,t) \mapsto \omega(s,t) \in \mathcal{H}_d,\ \ &\text{continuous}, \\ \omega(s,u) = \omega(s,t)\cdot\omega(t,u)\ \ \text{for every}\ \ 0 \le s \le t \le u \le T\}. \end{aligned}$$

Let $2 < p < 3$ and define a metric d_p on $\Omega(\mathbf{R}^d)$ by

$$d_p(\omega,\theta) = \sup_{0 \le s < t \le T} \left\{ \frac{|\omega^{(1)}(s,t) - \theta^{(1)}(s,t)|}{(t-s)^{1/p}} + \frac{|\omega^{(2)}(s,t) - \theta^{(2)}(s,t)|}{(t-s)^{2/p}} \right\}$$

where we denote $x^{(1)} = (x^i) \in \mathbf{R}^d$ and $x^{(2)} = (x^{(i,j)}) \in \mathbf{R}^{d(d-1)/2}$ for $x = (x^i, x^{(i,j)}) \in \mathcal{H}_d$.

Define a subspace $\Omega^{smooth}(\mathbf{R}^d)$ of $\Omega(\mathbf{R}^d)$ by setting $\Omega^{smooth}(\mathbf{R}^d) = \{\omega(h) \mid h \in H\}$, where $\omega(h) \in \Omega(\mathbf{R}^d)$ is defined by

$$\begin{aligned} \omega(h)(s,t)^i &= h^i(t) - h^i(s), \\ \omega(h)(s,t)^{(i,j)} &= \frac{1}{2}\int\int_{s \le t_1 \le t_2 \le t} (\dot{h}^i(t_1)\dot{h}^j(t_2) - \dot{h}^j(t_1)\dot{h}^i(t_2))dt_1 dt_2. \end{aligned}$$

Finally, we set

$$G\Omega_p(\mathbf{R}^d) = \overline{\Omega^{smooth}(\mathbf{R}^d))}^{d_p}$$

and call it the *space of geometric rouph paths.* It is a separable and complete metric space (i.e. a Polish space) under the metric d_p.

We can define a Wiener map $\rho : W_0(\mathbf{R}^d) \ni w \mapsto \rho[w] \in G\Omega_p(\mathbf{R}^d)$, by setting

$$\begin{aligned} \rho[w](s,t)^i &= w^i(t) - w^i(s), \\ \rho[w](s,t)^{(i,j)} &= \frac{1}{2}\int\int_{s \le t_1 \le t_2 \le t} (dw^i(t_1)dw^j(t_2) - dw^j(t_1)dw^i(t_2)) \\ &= \frac{1}{2}\int_s^t ([w^i(\tau) - w^i(s)]dw^j(\tau) - [w^j(\tau) - w^j(s)]dw^i(\tau)), \end{aligned}$$

the integral being in the sense of Itô's stochastic integrals. The image measure $\rho_*(P^W)$ on $G\Omega_p(\mathbf{R}^d)$ is denoted by $\bar{P}_p$ or simply by $\bar{P}$ when p is well understood.

Define the projection $\pi : G\Omega_p(\mathbf{R}^d) \ni \omega \mapsto \pi[\omega] \in W_0(\mathbf{R}^d)$ by setting $\pi[\omega](t)^i = \omega(0,t)^i$, $t \in [0,T]$, $i = 1, \ldots, d$. Then, $\pi : G\Omega_p(\mathbf{R}^d) \to W_0(\mathbf{R}^d)$ is $\bar{P}$-measurable and the image measure $\pi_*(\bar{P})$ coincides with the Wiener measure P^W on $W_0(\mathbf{R}^d)$. Then every Wiener functional (i.e. P^W-measurable function) F on $W_0(\mathbf{R}^d)$ can be lifted to a $\bar{P}$-measurable function $\bar{F}$ on $G\Omega_p(\mathbf{R}^d)$ by setting $\bar{F} = F \circ \pi (:= \pi_*(F))$. We call $\bar{F}$ the *lift of F on $G\Omega_p(\mathbf{R}^d)$*. We have $\rho \circ \pi = \mathrm{id}|_{G\Omega_p(\mathbf{R}^d)}$, $\bar{P}$-a.s., and $\pi \circ \rho = \mathrm{id}|_{W_0(\mathbf{R}^d)}$, P^W-a.s., so that the lifting is obviously an isomorphism between P^W-measurable functions and $\bar{P}$-measurable functions, (strictly speaking, an isomorphism between the equivalence classes of functions coinciding each other almost surely.) F can be recovered from $\bar{F}$ as $F = \bar{F} \circ \rho (:= \rho_*(\bar{F}))$, P^W-a.s.

By this isomorphism, every notion concerning Wiener functionals can be lifted to that concerning $\bar{P}$-measurable functions on $G\Omega_p(\mathbf{R}^d)$; for example, differential operators D, D^*, L are lifted to $\bar{D} := \pi_* D \rho_*$, $\bar{D}^* := \pi_* D^* \rho_*$, $\bar{L} := \pi_* L \rho_*$, and so on. So the lifting gives isomorphisms between Sobolev spaces $\mathbf{D}_p^\alpha(E)$ on the the Wiener space and Sobolev spaces $\bar{\mathbf{D}}_p^\alpha(E)$ on the space of geometric rough paths.

We consider a SDE like (3.1) on $\mathbf{R}^n$, which is set up on the Wiener space $W_0(\mathbf{R}^d)$:

$$dX(t) = \sum_{i=1}^{d} A_i(X(t)) \circ dw^i(t) + A_0(X(t))dt, \quad X(0) = x \tag{5.1}$$

and denote the solution by $X = (X(t,x;w))$. We assume that all coefficients of SDE are smooth with bounded derivatives of all orders. By the *skeleton* of X, we mean the solution $\Xi = (\xi(t,x;h))$ of ODE for given $h \in H$:

$$\frac{d\xi}{dt}(t) = \sum_{i=1}^{d} A_i(\xi(t)) \cdot \dot{h}^i(t) + A_0(\xi(t)), \quad \xi(0) = x. \tag{5.2}$$

Note that $\xi(t,x;h)$ is, for fixed $t > 0$ and $x \in \mathbf{R}^n$, a smooth functional of $h \in H$ and also the map $\mathbf{R}^n \times H \ni (x,h) \mapsto [t \mapsto \xi(t,x;h)] \in \mathcal{C}([0,T] \to \mathbf{R}^n)$ is continuous. A skeleton is something like a restriction of X on H; since $P^W(H) = 0$, however, the restriction is usually meaningless.

Now, one of the fundamental theorems of T. Lyons can be stated as follows: There exists a *continuous* map

$$\phi : (x,\omega) \in \mathbf{R}^n \times G\Omega_p(\mathbf{R}^d) \mapsto \phi(x,\omega) := [t \mapsto \phi(x,\omega)(t)] \in \mathcal{C}([0,T] \mapsto \mathbf{R}^n),$$

and hence a continuous map $\phi_{(x,t)} : \omega \in G\Omega_p(\mathbf{R}^d) \mapsto \phi_{(x,t)}(\omega) := \phi(x,\omega)(t) \in \mathbf{R}^n$, for each fixed x and t, such that the following hold:

(i) $\xi(t,x;h) = \phi(x,\omega[h])(t)$, for all $h \in H$, $t \in [0,T]$ and $x \in \mathbf{R}^n$,
(ii) If $\bar{X} = \big(\bar{X}(t,x;\omega)\big)$ is the lift of $X = (X(t,x;w))$ on $G\Omega_p(\mathbf{R}^d)$, then it holds that $\phi(x,\omega)(t) = \bar{X}(t,x;\omega)$, $\bar{P}$-a.s.

Hence, $\bar{X} = \big(\bar{X}(t,x;\omega)\big)$ has a modification *which is continuous in* ω. This continuous modification can be used in place of a quasi-continuous modification in quasi-sure analysis. (We should note that the original result of T. Lyons is much stonger than what we stated above: He introduced the notion of a differential equation driven by rough paths and constructed its solution which is given as rough paths in $\mathbf{R}^n$. The function ϕ above is obtained from the first component of the solution. This theory of Lyons, indeed, is a pure real analysis.)

Consider the stochastic moving frame $r = (r(t,r;w))$ realized on the Wiener space as above. Then its lift $\bar{r} = (\bar{r}(t,r;\omega))$ has a continuous modification on $G\Omega_p(\mathbf{R}^d)$ which we denote by the same notation $\bar{r}$. The measure $\mu_y^{t,r}(dw)$ on $W_0(\mathbf{R}^d)$ is now lifted to a measure $\bar{\mu}_y^{t,r}(d\omega)$ on $G\Omega_p(\mathbf{R}^d)$ and it is supported on the *closed set* $\{\omega \mid \bar{r}(t,r;\omega) = y\}$. Now, returning to the heat kernel $\langle x|e^{tL}|y\rangle$ in the case of de Rham-Hodge-Kodaira Laplacian $\frac{1}{2}\Box$, it is easy to see that the lift $\tilde{r}\bar{M}(t,r;\omega)\widetilde{\bar{r}(t,r;\omega)}^{-1}$ is continuous in ω which takes values in $\mathrm{Hom}(\bigwedge T_y^*M, T_x^*M)$ on the set $\{\ \omega \mid \bar{r}(t,r;\omega) = y\ \}$, where $x = \pi(r)$. Then we have

$$\langle x|e^{tL}|y\rangle = \int_{\{\omega|\bar{r}(t,r;\omega)=y\}} \left[\tilde{r}\bar{M}(t,r;\omega)\widetilde{\bar{r}(t,r;\omega)}^{-1}\right] \bar{\mu}_y^{t,r}(d\omega), \quad r = (x,\mathbf{e}).$$

References

[AM] H. Airault and P. Malliavin, Intégration géométrique sur l'espace de Wiener, *Bull. Sc. math.(2)*, **112**(1988), 35–2

[F 1] R. Feynman, A principle of least action in quantum mechanics, Thesis, Princeton Univ. 1942

[F 2] R. Feynman, Space-time approach to non-relativistic quantum mechanics, Rev. Mod. Phys. **20**(1948), 321–341

[Ik] N. Ikeda, Probabilistic methods in the studies of asymptotics, *École d'Été de Probabilités de Saint-Flour*, XVIII-1988, **LNM 1427**, Springer (1990), 197–325

[IW] N. Ikeda and S. Watanabe, *Stochastic Differential Equations and Diffusion Processes*, Second Edition, North-Holland/Kodansha, Amsterdam/ Tokyo, 1988

[It] K. Itô, A measure-theoretic approach to Malliavin calculus, in *New Trends in Stochastic Analysis, Proc. Taniguchi Workshop 1994, eds. Elworthy, Kusuoka and Shigekawa*, World Scientific 1997, 220–287

[IM] K. Itô and H. P. McKean, Jr., *Diffusion Processes and their Sample Paths*, Springer, Berlin, 1965, Second Printing 1974, in *Classics in Mathematics*, 1996

[Ka 1] M. Kac, On some connections between probability theory and differential and integral equations, Proc. Second Berkeley Symp. Univ. California Press (1951), 189–215

[Ka 2] M. Kac, *Probabity and Related Topics in Physical Sciences*, Interscience, New York, 1959

[Ka 3] M. Kac, Can one hear a shape of a drum?, Amer. Math. Monthly **73**(1966), 1–23

[Ku] H. Kunita, *Stochastic flows and stochastic differential equations*, Cambridge University Press, 1990

[Le] P. Lescot, Un théorème de desintégration en analyse quasi-sure, *Sém. Probab. XXVII*, **LNM 1557**, Springer (1991), 256-275

[Ly] T. Lyons, Differential equations driven by rough signals, Rev. Math. Iberoamer. **14**(1998), 215–310

[M] P. Malliavin, *Stochastic Analysis*, Springer, 1997

[MS] H. P. McKean, Jr. and I. M. Singer, Curvature and eigenvalues of Laplacian, J. Differential Geometry **1**(1967), 43–69

[S 1] H. Sugita, Positive generalized Wiener functions and potential theory over abstract Wiener spaces, Osaka J. Math. **25**(1988), 665–698

[S 2] H. Sugita, Hu-Meyer's multiple Stratonovich integral and essential continuity of multiple Wiener integral, *Bull. Sc. math.(2)*, **113**(1989), 463–474

[W] S. Watanabe, Short time asymptotic problems in Wiener functional integration theory, Applications to heat kernels and index theorems, *Stochastic Analysis and Related Topics, Proc. Silivri 1988, eds. Korezliogru and Ustunel*, **LNM 1444**, Springer (1990), 1–62

The Malliavin Calculus for Processes with Conditionally Independent Increments

Aleh L. Yablonski

Department of Functional Analysis, Belarusian State University, F.Skaryna av., 4, 220050, Minsk, Belarus, `yablonski@bsu.by`. Supported by INTAS grant 03-55-1861

Summary. The purpose of this paper is to construct the analog of Malliavin derivative D and Skorohod integral δ for some class of processes which include, in particular, processes with conditionally independent increments. We introduce the family of orthogonal polynomials. By using these polynomials it is proved the chaos decomposition theorem of $L^2(\Omega)$. The definition of Malliavin derivative and Skorohod integral for a certain class of stochastic processes is given and it is shown that they are equal respectively to the annihilation and the creation operators on the Fock space representation of $L^2(\Omega)$. The analogue of Clark–Haussmann–Ocone formula for processes with conditionally independent increments is also established.

Keywords and Phrases: processes with conditionally independent increments, Malliavin calculus, Skorohod integral, multiple integral, orthogonal polynomials, chaos expansion, Clark–Haussmann–Ocone formula

1 Introduction

It was shown by Karatzas and Ocone [14] how the stochastic calculus of variations developed by Malliavin [20] can be used in mathematical finance. This discovery led to an increase in the interest in the Malliavin calculus.

In the Brownian setup the calculus of variations has a complete form (see the elegant presentation of Nualart [22]). It is based on the operators D and δ which are called Malliavin derivative and Skorohod integral, respectively. There are two equivalent approaches to definition of the operator D: as a variational derivative and through the chaos decomposition.

For discontinuous processes it is possible to develop the Malliavin-type calculus by using some "generalized" or "weak" derivatives (see, e.g., [1, 4, 15] and references therein). Nevertheless, it was shown in [24] that in the Poisson case small perturbations of the trajectories lead to a certain difference operator. This idea was extended for Lévy processes in [26, 28, 31, 33].

Alternatively, the operator D can be defined by its action on the chaos representation of L^2-functionals. The case of normal martingales with chaotic representation property was considered in [19]. But, in general, a Lévy process has no chaotic representation property in the sense that Brownian motion and Poisson process do. There are two different chaotic expansions introduced in [11] and [23]. By using these expansions two types of Malliavin operators for some classes of Lévy processes have been studied in the papers [2,7,8,16,18,25, 31]. The relationship between them has been shown in [2,31]. The connection of such derivative to the difference operator from [26,28] was studied in [18,31, 33]. For random Lévy measures the Skorohod integral and Malliavin derivative were considered in [5,6].

The purpose of the paper is to construct the Malliavin calculus for some class of processes which includes, in particular, the processes with conditionally independent increments. It is also proved the chaos representation theorem for such type of processes.

In general the processes with conditionally independent increments can be described in terms of their triplets of characteristics (B, μ, ν), where B represents the "drift" part, μ is connected with continuous martingale part (Gaussian part for Lévy processes) and ν is a compensator of measure associated to the jumps of the original process. Since B is a bounded variation process then for our purposes we can set $B = 0$ without loss of generality. Therefore we start with two random measures μ and ν on certain measurable spaces which describe, respectively, the continuous and discontinuous parts of the process. In Section 2 we define the appropriate Hilbert space H connected to these measures and stochastic process indexed by elements of H. This construction allows us to consider aa rather general class of processes. The system of generalized orthogonal polynomials defined in [33] is used in the proof of the chaos decomposition of L^2 functionals.

Section 3 deals with multiple integrals with respect to the L^2-valued measure generated by the considering process. Their connection to generalized orthogonal polynomials and chaos expansion is also proved.

In Sections 4 and 5 we define the operator D and its adjoint operator δ. Then we show that they are generalizations of the Malliavin derivative and Skorohod integral. It is also proved that their action on the Fock space representation of L^2-functionals coincides with annihilation and creation operators. In the end of the last section we prove the analogue of Clark-Haussmann-Ocone formula for processes with conditionally independent increments.

2 The Chaos Decomposition

Let $(\Omega, \mathcal{F}, \mathsf{P})$ be a complete probability space. Suppose that μ and ν are random measures defined on the measurable spaces $(T, \mathcal{A})$ and $(T \times X_0, \mathcal{B})$ respectively, such that the following conditions are satisfied:

1. $\mu(A,\cdot)$ and $\nu(B,\cdot)$ are $\mathcal{F}$-measurable for all $A \in \mathcal{A}$ and $B \in \mathcal{B}$,
2. $\mu(\cdot,\omega)$ and $\nu(\cdot,\omega)$ are σ-finite measures without atoms for all $\omega \in \Omega$.

Consider $\Delta \notin X_0$ and denote $X = X_0 \cup \{\Delta\}$, $\mathcal{G} = \sigma(\mathcal{A} \times \{\Delta\}, \mathcal{B})$. Define a new measure $\pi(dtdx) = \mu(dt)\delta_\Delta(dx) + \nu(dtdx \cap (T \times X_0))$ on the σ-algebra $\mathcal{G}$. Here $\delta_\Delta(dx)$ is the measure which gives mass one to the point Δ. Let $\mathcal{H}$ be a σ-algebra generated by measure π and the collection $\mathcal{N}$ of P-null events of $\mathcal{F}$, i.e.,

$$\mathcal{H} = \sigma\{\pi(A) : A \in \mathcal{G}\} \vee \mathcal{N}.$$

Define a new measure M_π on σ-algebra $\mathcal{G} \otimes \mathcal{F}$ in the following way. For any $A \in \mathcal{G}$ and $B \in \mathcal{F}$ we set $M_\pi(A \times B) = \mathsf{E}[\mathbf{1}_B \pi(A)]$. Then extension of it on the σ-algebra $\mathcal{G} \otimes \mathcal{F}$ can be done as usual (see e.g., [27, Ch. 4]). Suppose that measure M_π is σ-finite. Then there exists the sequence of the sets $U_n \in \mathcal{G} \otimes \mathcal{F}$ such that $\bigcup_{n=1}^\infty U_n = T \times X \times \Omega$ and $M_\pi(U_n) < \infty$. We can choose $U_n = A_n \times B_n$, where $A_n \in \mathcal{G}$ and $B_n \in \mathcal{F}$. Indeed for any $\epsilon > 0$ and each U_n we can find $A_n^k \in \mathcal{G}$ and $B_n^k \in \mathcal{F}$ such that $U_n \subset \bigcup_{k=1}^\infty A_n^k \times B_n^k$ and $M_\pi(U_n) \le \sum_{k=1}^\infty M_\pi(A_n^k \times B_n^k) \le M_\pi(U_n) + \epsilon$. Hence $M_\pi(A_n^k \times B_n^k) < \infty$ and $T \times X \times \Omega = \bigcup_{k,n=1}^\infty A_n^k \times B_n^k$. Renumeration of the sets $A_n^k \times B_n^k$ yields the desired result.

If we consider the restriction of the measure M_π on the σ-algebra $\mathcal{G} \otimes \mathcal{H}$ then it is possible to show that it will be σ-finite. Indeed, let $U_n = A_n \times B_n$, $A_n \in \mathcal{G}$ and $B_n \in \mathcal{F}$ be as above then $\bigcup_{n=1}^\infty A_n = T \times X$ and $\bigcup_{n=1}^\infty B_n = \Omega$. Denote by C_n^m the following sets: $C_n^m = \{\omega \in \Omega : \pi(A_n) \le m\}$, $m = 1, 2, \ldots$. Then $C_n^m \in \mathcal{H}$ and $\bigcup_{m=1}^\infty C_n^m = \{\pi(A_n) < \infty\} \supset (B_n \setminus N_n)$, where $\mathsf{P}(N_n) = 0$. Hence $\bigcup_{n,m=1}^\infty A_n \times (C_n^m \cup N_n) = T \times X \times \Omega$ and $M_\pi(A_n \times (C_n^m \cup N_n)) \le m < \infty$. In fact we have a stronger property: $\pi(A_n, \omega) < \infty$ for all $\omega \in \bigcup_{m=1}^\infty C_n^m$. This property implies the sigma-finiteness of the π in the sense [27, Ch. 4, Def. 21].

Consider the Hilbert space $H = L^2(T \times X \times \Omega, \mathcal{G} \otimes \mathcal{H}, M_\pi)$ and assume that it is separable. Denote by $\pi(f)$ the integral of f with respect to measure π:

$$\pi(f) = \int_{T \times X} f(t,x)\pi(dtdx).$$

It was shown in [27, Ch. 4] that if $\mathsf{E}[\pi(|f|)] < \infty$ then $\pi(f)$ is $\mathcal{H}$-measurable or $\mathcal{F}$-measurable whenever f is $\mathcal{G} \otimes \mathcal{H}$-measurable or $\mathcal{G} \otimes \mathcal{F}$-measurable respectively. The scalar product and the norm in H will be denoted by $\langle \cdot ; \cdot \rangle_H$ and $||\cdot||_H$ respectively, i.e. for any $f,\ g \in H$

$$\langle f; g \rangle_H = \mathsf{E}(\pi(hg)) = \mathsf{E} \int_{T \times X} h(t,x) g(t,x) \pi(dtdx), \ ||f||_H^2 = \mathsf{E}(\pi(h^2)).$$

Definition 1. *We say that a stochastic process $L = \{L(h), h \in H\}$ is a conditional additive process on H if the following conditions are satisfied.*

1. *For all $h,\ g \in H$ and $\alpha, \beta \in L^\infty(\Omega, \mathcal{H}, \mathsf{P})$ we have P-a.s.*

$$L(\alpha h + \beta g) = \alpha L(h) + \beta L(g),$$

2. For all $z \in \mathbb{R}$ and $h \in H$

$$\mathsf{E}[e^{izL(h)}|\mathcal{H}] = \exp\left(-\frac{1}{2}z^2\int_T h^2(t,\Delta)\mu(dt) + \int_{T\times X_0}\left(e^{izh(t,x)} - 1 - izh(t,x)\right)\nu(dtdx)\right). \quad (2.1)$$

Remark 2.

1. In this definition we suppose that the process $L(h)$ can be defined on the original probability space $(\Omega, \mathcal{F}, \mathsf{P})$. If it is not the case then it is possible to define μ, ν, and $L(h)$ verifying the above conditions on some extension $(\Omega', \mathcal{F}', \mathsf{P}')$ of the original probability space. So we can always assume that original probability space $(\Omega, \mathcal{F}, \mathsf{P})$ is rich enough for defining all necessary objects.
2. The definition 1 shows that the random variable $L(h)$ has a conditionally infinitely divisible distribution.
3. If measure ν is zero and measure μ is deterministic then L is an isonormal Gaussian process (see, e.g., [22, Def. 1.1.1, p. 4]).
4. If measures μ and ν are deterministic then L is an isonormal Lévy process (see, e.g., [33]).

Example 3. Let $L_t, t \geq 0$ be a càdlàg real-valued process with $\mathcal{H}$-conditionally independent increments on complete probability space $(\Omega, \mathcal{F}, \mathsf{P})$, where $\mathcal{H} \subset \mathcal{F}$. Suppose that L_t is a quasi-left-continuous semimartingale with respect to filtration $\mathcal{F}_t$, $t \geq 0$ generated by the natural filtration and σ-algebra $\mathcal{H}$, i.e. $\mathcal{F}_t = \bigcap_{s>t}(\mathcal{F}_t^0 \vee \mathcal{H})$, where $\mathcal{F}_t^0 = \sigma\{L_s : s \leq t\}$. In this case there exists a version of the characteristics (B, μ, ν) of L_t (see, e.g., [12, 17]) such that:

1. $B_t, t \geq 0$ is a continuous process of locally bounded variation with $B_0 = 0$;
2. μ_t, $t \geq 0$ is a continuous nondecreasing process with $\mu_0 = 0$;
3. $\nu(dtdx, \omega)$ is a predictable random measure defined on the Borel σ-algebra of $\mathbb{R}_+ \times \mathbb{R}_0$, where $\mathbb{R}_0 = \mathbb{R} \setminus \{0\}$ such that $\nu(\{t\} \times \mathbb{R}_0) = 0$, $\int_0^t \int_{\mathbb{R}_0}(|x|^2 \wedge 1)\nu(dsdx) < \infty$ for all $t \geq 0$ P a.s.

Moreover, B, μ and ν are $\mathcal{H}$-measurable and we have for all $z \in \mathbb{R}$ and $s \leq t$

$$\mathsf{E}[\exp(iz(L_t - L_s))|\mathcal{H}] = \exp\left[iz(B_t - B_s) - \frac{1}{2}z^2(\mu_t - \mu_s) + \int_s^t \int_{\mathbb{R}_0}(e^{izx} - 1 - izx\mathbf{1}_{|x|\leq 1})\nu(dtdx)\right].$$

Here $\mu_t = \langle L^c; L^c\rangle_t$ is a quadratic variation of the continuous parts of L, ν is a compensator of the random measure $N(dtdx)$ associated to the jumps of L. Hence the following canonical representation holds:

$$L_t = L_0 + L_t^c + \int_0^t \int_{|x|\le 1} x(N(dsdx) - \nu(dsdx)) + \int_0^t \int_{|x|>1} xN(dsdx) + B_t. \tag{2.2}$$

We can define L^2-valued measure $L(dtdx)$ with conditionally independent values on the disjoint sets by

$$L(A) = \int_{A(0)} dL_t^c + \iint_{A\setminus A(0)} (N(dtdx) - \nu(dtdx)),$$

where $A \in \mathcal{G}$, $M_\pi(A) < \infty$, $A(0) = A \cap (T \times \{0\})$.

Suppose that $T = \mathbb{R}_+$, $X_0 = \mathbb{R}_0$, $\Delta = 0$. Let $\mathcal{A}$ and $\mathcal{B}$ be the Borel σ-algebras of T and $T \times X_0$ respectively. Let $\mu(dt)$ be the measure on $\mathcal{A}$ generated by process μ_t. Suppose that σ-algebra $\mathcal{H}$ is generated by the measures μ and ν. Construct measure π and Hilbert space $H = L^2(\mathbb{R}_+ \times \mathbb{R} \times \Omega, \mathcal{G} \otimes \mathcal{H}, M_\pi)$ as above. Then it is easy to show that for any $h \in H$ the random variable

$$\begin{aligned} L(h) &= \int_0^\infty h(s,0)dL_s^c + \int_0^\infty \int_{\mathbb{R}_0} h(s,x)(N(dsdx) - \nu(dsdx)) \\ &= \iint_{\mathbb{R}_+\times\mathbb{R}} h(t,x)L(dtds) \end{aligned} \tag{2.3}$$

is well defined and $L(h)$ is a conditional additive process on H.

On the other hand if we have a conditional additive process $L(h)$ on H then the L^2-valued measure $L(dtdx)$ which is given by $L(A) = L(\mathbf{1}_A)$ for all $A \in \mathcal{G}$ with $M_\pi(A) < \infty$ has conditionally independent values on the disjoint sets. In order to express the process L_t in terms of process $L(h)$ we can not write $L_t = L(h_t)$, where $h_t(s,x) = \mathbf{1}_{[0;t]}(s)\mathbf{1}_{\{0\}}(x) + x\mathbf{1}_{[0;t]}(s)$ because, in general, $h_t \notin H$. Therefore we define for any $n \ge 1$ the random variable $\tau_n = \inf\{t > 0 : \mu_t \le n\}$. Obviously τ_n is an increasing sequence and $h_{t\wedge\tau_n}\mathbf{1}_{\{x=0\}} \in H$ for all $t \ge 0$ and $n \ge 1$. Moreover, the process $L(h_{t\wedge\tau_n}\mathbf{1}_{\{x=0\}})$, $t \ge 0$ has a version with continuous sample paths and $L(h_{t\wedge\tau_n}\mathbf{1}_{\{x=0\}}) = L(h_{t\wedge\tau_m}\mathbf{1}_{\{x=0\}})$ if $t \le \tau_n$ and $n < m$. Therefore $L^c(t) = \lim_{n\to\infty} L(h_{t\wedge\tau_n}\mathbf{1}_{\{x=0\}})$ well defined continuous process. Furthermore for any set $A \in \mathcal{B}$ such that $M_\pi(A) = \mathsf{E}[\nu(A)] < \infty$ the random variable $N(A) = L(\mathbf{1}_A) + \nu(A)$ is an integer valued random measure with compensator measure ν. And we can define

$$\bar{L}_t = L^c(t) + \int_0^t \int_{|x|\le 1} x(N(dsdx) - \nu(dsdx)) + \int_0^t \int_{|x|>1} xN(dsdx). \tag{2.4}$$

Comparing equalities (2.2) and (2.4) we deduce that the only characteristics which cannot be determined from $L(h)$ is the drift process B and initial value $L(0)$.

Denote by K the following subset of H:

$$K = \{h \in H : h\mathbf{1}_{X_0} \in L^\infty(T \times X \times \Omega, \mathcal{G} \otimes \mathcal{H}, M_\pi),\ \pi(h^2) \in L^\infty(\Omega, \mathcal{H}, \mathsf{P})\} \tag{2.5}$$

The elements of K satisfy the following properties.

Lemma 4. *Suppose that $h \in K$ then*

1. $|h| \in K$ *and* $zh \in K$ *for all* $z \in \mathbb{R}$.
2. $\mathsf{E}[\exp(\pi(h^2)/2)] < \infty$, $\mathsf{E}[\pi(|h|^k \mathbf{1}_{X_0})] < \infty$ *and* $\mathsf{E}[\pi(h^2)^k] < \infty$ *for all* $k \geq 2$.
3. $\mathsf{E}\left[\exp\left(\frac{1}{2}\int_T h^2(t,\Delta)\mu(dt) + \int_{T\times X_0}\left(e^{h(t,x)} - 1 - h(t,x)\right)\nu(dtdx)\right)\right] < \infty.$

Proof. The first two properties are evident. The proof of the last statement is based on the boundedness of $h\mathbf{1}_{X_0}$ and inequality $|e^x - 1 - x| \leq e^{|x|}x^2/2$, which can be proved by using Taylor's formula. We omit the details.

Lemma 5. *The set K is dense in H.*

Proof. Choose a $h \in H$. Then $\pi(h^2) < \infty$ a.s. Consider two sequences of sets $B_m = \{\pi(h^2) \leq m\} \in \mathcal{H}$ and $C_k = \{(t,x,\omega) : |h(t,x,\omega)| \leq k\} \in \mathcal{G} \otimes \mathcal{H}$. Denote $h_{k,m} = h\mathbf{1}_{C_k}\mathbf{1}_{B_m}$ It is evident that $h_{k,m} \in K$ for all integers $k,\ m \geq 1$. By dominated convergence theorem we have $||h_{k,m} - h\mathbf{1}_{B_m}||_H \to 0$ as $k \to \infty$ and $||h\mathbf{1}_{B_m} - h||_H = \mathsf{E}[\pi(h^2)\mathbf{1}_{\Omega\setminus B_m}] \to 0$ as $m \to \infty$ which completes the proof of the lemma.

The following lemma describes some properties of $L(h)$.

Lemma 6.

1. *If $h \in H$ then $L(h) \in L^2(\Omega, \mathcal{F}, \mathsf{P})$. Furthermore $\mathsf{E}[L(h)|\mathcal{H}] = 0$ and $\mathsf{E}[L(h)L(g)|\mathcal{H}] = \pi(hg)$.*
2. *Let $h_n \in H$ be a sequence such that $||h_n - h||_H \to 0$ as $n \to \infty$ for some $h \in H$. Then $\mathsf{E}[(L(h_n) - L(h))^2|\mathcal{H}] \to 0$ as $n \to \infty$ in $L^1(\Omega, \mathcal{F}, \mathsf{P})$.*
3. *If $h \in K$ then $L(h) \in L^p(\Omega, \mathcal{F}, \mathsf{P})$ for all $p \geq 1$, $\mathsf{E}[\exp(c|L(h)|)] < \infty$ for all $c \in \mathbb{R}$, and*

$$\mathsf{E}[\exp(L(h))|\mathcal{H}] = \exp\left(\frac{1}{2}\int_T h^2(t,\Delta)\mu(dt) + \int_{T\times X_0}\left(e^{h(t,x)} - 1 - h(t,x)\right)\nu(dtdx)\right).$$

4. *Let $h_1, \ldots h_n \in K$ be such that $h_i h_j = 0$ M_π-a.s. if $i \neq j$. Then for any integers $p_1, \ldots, p_n \geq 1$ we have*

$$\mathsf{E}[L(h_1)^{p_1} \cdots L(h_n)^{p_n}|\mathcal{H}] = \mathsf{E}[L(h_1)^{p_1}|\mathcal{H}] \cdots \mathsf{E}[L(h_n)^{p_n}|\mathcal{H}].$$

Proof. 1. Choose a $B \in \mathcal{H}$, $\mathsf{P}(B) > 0$. Denote by $f_B(z)$ the characteristic function of the random variable $L(h)$ with respect to restriction of probability on the set B, i.e., $f_B(z) = \mathsf{E}[\mathbf{1}_B e^{izL(h)}]$. Equality (1) implies that $f_B(z)$ can be written in the following form:

$$f_B(z) = \mathsf{E}\left[\mathbf{1}_B \exp\left(-\frac{z^2}{2}\int_T h^2(t,\Delta)\mu(dt)\right.\right.$$
$$\left.\left.+\int_{T\times X_0}\left(e^{izh(t,x)} - 1 - izh(t,x)\right)\nu(dtdx)\right)\right].$$

By using this equality it is easy to show that $f_B(z)$ two times differentiable function. Hence $L(h) \in L^2(\Omega, \mathcal{F}, \mathsf{P})$. Taking the first and second derivatives at $z = 0$ in both sides of the equality above yields $\mathsf{E}[L(h)\mathbf{1}_B] = 0$ and $\mathsf{E}[L(h)^2\mathbf{1}_B] = \mathsf{E}[\pi(h^2)\mathbf{1}_B]$. Then $\mathsf{E}[L(h)|\mathcal{H}] = 0$, $\mathsf{E}[L(h)^2|\mathcal{H}] = \pi(h^2)$ and $\mathsf{E}[L(h)L(g)|\mathcal{H}] = \mathsf{E}[L(h+g)^2 - L(h-g)^2|\mathcal{H}]/4 = \pi(hg)$ which completes the proof of the first statement of the lemma.

2. Suppose that $h_n \to h$ as $n \to \infty$ in H. It means that $\mathsf{E}(\pi((h_n - h)^2)) \to 0$ as $n \to \infty$. Then from the first part of the lemma we have $\mathsf{E}[(L(h_n) - L(h))^2|\mathcal{H}] = \pi((h_n - h)^2)$ which implies the proof of second statement of the lemma.

3. Denote by $u(z)$ the following expression:

$$u(z) = \mathsf{E}\left[\exp\left(\frac{z^2}{2}\int_T h^2(t,\Delta)\mu(dt)\right.\right.$$
$$\left.\left.+\int_{T\times X_0}\left(e^{zh(t,x)} - 1 - zh(t,x)\right)\nu(dtdx)\right)\right].$$

Since $h \in K$ then Lemma 4 implies that $u(z)$ is finite for all $z \in \mathbb{R}$. The right hand side in the equality above is meaningful even z is complex. Indeed if $z = a + ib$ then $\operatorname{Re}(e^{zh} - 1 - zh) = e^{ah}\cos(bh) - 1 - ah = (e^{ah} - 1 - ah) + e^{ah}(\cos(bh) - 1) \le (e^{ah} - 1 - ah)$. Hence from Lemma 4 we have

$$\left|\mathsf{E}\left[\exp\left(\frac{z^2}{2}\int_T h^2(t,\Delta)\mu(dt) + \int_{T\times X_0}\left(e^{zh(t,x)} - 1 - zh(t,x)\right)\nu(dtdx)\right)\right]\right|$$
$$\le \mathsf{E}\left[\exp\left(\frac{a^2}{2}\int_T h^2(t,\Delta)\mu(dt) + \int_{T\times X_0}\left(e^{ah(t,x)} - 1 - ah(t,x)\right)\nu(dtdx)\right)\right] < \infty.$$

The function $u(z)$ is analytic function for all $z \in \mathbb{C}$. If $z = it$, $t \in \mathbb{R}$ then $u(z) = f(t) = \mathsf{E}[e^{itL(h)}]$ coincides with characteristic function of $L(h)$. Hence characteristic function $f(t)$ infinitely differentiable for all $t \in \mathbb{R}$ and $L(h)$ has finite moments of all orders, i.e. $L(h) \in L^p(\Omega, \mathcal{F}, \mathsf{P})$ for all $p \ge 1$. Moreover

$$u(z) = \sum_{k=0}^{\infty} \frac{1}{k!} z^k \mathsf{E}[L(h)^k],$$

where the radius of convergence of the series being infinite. It follows that

$$\mathsf{E}[\exp(c|L(h)|)] = \sum_{k=0}^{\infty} \frac{c^k}{k!} \mathsf{E}[|L(h)|^k] < \infty$$

for all $c \in \mathbb{R}$. Hence $f(z) = \mathsf{E}[\exp(zL(h))]$ is analytic for all complex z. The uniqueness theorem yields $u(z) = f(z)$ which implies the third statement of the lemma.

4. From the previous parts of the lemma we have $L(h_k) \in L^p(\Omega, \mathcal{F}, \mathsf{P})$ for all $p \geq 1$ and conditional characteristic function of random variables $L(h_k)$ is infinitely differentiable if $h_1, \dots h_n \in K$. Since $h_i h_j = 0$ M_π-a.s. if $i \neq j$ then we have the following equality:

$$\begin{aligned}
&\mathsf{E}\left[\exp\left(i\sum_{k=1}^{n} z_k L(h_k)\right)\middle|\mathcal{H}\right] \\
&= \exp\left(\frac{1}{2}\sum_{k=1}^{n} z_k^2 \int_T h_k^2(t,\Delta)\mu(dt)\right. \\
&\qquad \left. + \sum_{k=1}^{n}\int_{T\times X_0}\left(e^{iz_k h_k(t,x)} - 1 - iz_k h_k(t,x)\right)\nu(dtdx)\right) \\
&= \mathsf{E}\left[\exp\left(iz_1 L(h_1)\right)\middle|\mathcal{H}\right]\cdots\mathsf{E}\left[\exp\left(iz_n L(h_n)\right)\middle|\mathcal{H}\right].
\end{aligned}$$

Taking the $(p_1 + p_2 + \cdots + p_n)$th partial derivative $\frac{\partial^{p_1+p_2+\cdots+p_n}}{\partial z_1^{p_1}\cdots\partial z_n^{p_n}}$ at $z_1 = z_2 = \cdots = 0$ in both sides of the above equality yields

$$\mathsf{E}[L(h_1)^{p_1}\cdots L(h_n)^{p_n}|\mathcal{H}] = \mathsf{E}[L(h_1)^{p_1}|\mathcal{H}]\cdots\mathsf{E}[L(h_n)^{p_n}|\mathcal{H}].$$

Lemma 7. *Let $\mathcal{N}$ be a collection of* P*-null events of $\mathcal{F}$. Then*

$$\mathcal{H} \subset \mathcal{F}_L = \sigma\{L(h) : h \in H\} \vee \mathcal{N}.$$

Proof. It is suffice to show that $\pi(C)$ is $\mathcal{F}_L$-measurable for all $C \in \mathcal{G}$.

Since the measure M_π is σ-finite then there exists a sequence of pairwise-disjoint sets $A_n \times B_n$, $A_n \in \mathcal{G}$ and $B_n \in \mathcal{H}$ such that $\bigcup_{n=1}^\infty A_n \times B_n = T \times X \times \Omega$ and $M_\pi(A_n \times B_n) < \infty$. Then we have $M_\pi(C \times \Omega) = \sum_{n=1}^\infty M_\pi((C \cap A_n) \times B_n)$ and $M_\pi((C \cap A_n) \times B_n) < \infty$. Therefore $\pi(C \cap A_n)\mathbf{1}_{B_n} < \infty$ P-a.s. for all $n = 1, 2, \dots$ and $\sum_{k=1}^n \pi(C \cap A_k)\mathbf{1}_{B_k} \to \pi(C)$ P-a.s. as $n \to \infty$. Hence it is suffice to show that $\pi(C \cap A_k)\mathbf{1}_{B_k}$ is $\mathcal{F}_L$-measurable.

Denote by C_k^m the following sets $C_k^m = \{\pi(C \cap A_k) \leq m\}$. Then $C_k^m \in \mathcal{H}$, $\pi(C \cap A_k)\mathbf{1}_{C_k^m}\mathbf{1}_{B_k} \leq m$ and $\pi(C \cap A_k)\mathbf{1}_{C_k^m}\mathbf{1}_{B_k} \to \pi(C \cap A_k)\mathbf{1}_{B_k}$ P-a.s. as $m \to \infty$. Therefore the proof will be complete if we show that $\pi(U)\mathbf{1}_D$ is $\mathcal{F}_L$-measurable for all $U \in \mathcal{G}$ and $D \in \mathcal{H}$ such that $\pi(U)\mathbf{1}_D$ is bounded.

Let $U \in \mathcal{G}$ and $D \in \mathcal{H}$ be arbitrary sets such that $\pi(U)\mathbf{1}_D \leq Q$ a.s. For any $\omega \in D$ define measure $\pi^{\otimes 2}(dt_1 dx_1 dt_2 dx_2, \omega) = \pi(dt_1 dx_1, \omega)\pi(dt_2 dx_2, \omega)$ on $\mathcal{G}^{\otimes 2}$. Since measures μ and ν without atoms then the measure π has no atoms and $\pi^{\otimes 2}(\Delta_2^U, \omega) = 0$, where $\Delta_2^U = \{(u,u) : u \in U\}$. Therefore if we define measure $M_\pi^2(dz d\omega) = \pi^{\otimes 2}(dz, \omega)\mathsf{P}(d\omega)$ on $\mathcal{G}^{\otimes 2} \otimes \mathcal{H}$ then $M_\pi^2(\Delta_2^U \times D) = \mathsf{E}(\pi^{\otimes 2}(\Delta_2^U)\mathbf{1}_D) = 0$. Hence for any $m = 1, 2, \dots$ there exists a system of sets $U_k^m \in \mathcal{G}$ and $D_k^m \in \mathcal{H}$, $k = 1, 2, \dots$ such that $\bigcup_{k=1}^\infty U_k^m \times U_k^m \times D_k^m \supset \Delta_2^U \times D$, $\bigcup_{k=1}^\infty U_k^m = U$, $\bigcup_{k=1}^\infty D_k^m = D$ and $\sum_{k=1}^\infty \mathsf{E}(\pi(U_k^m)^2\mathbf{1}_{D_k^m}) \leq 1/m$.

For any $n \geq 1$ we can find a system of pairwise-disjoint sets $\{V_1^n, \dots, V_{p_n}^n\} \subset \mathcal{G}$ and $\{E_1^n, \dots, E_{q_n}^n\} \subset \mathcal{H}$, such that each U_k^m and D_k^m $k = 1, \dots, n$ can be expressed as a disjoint union of some V_j^n or E_j^n respectively. Then we have

$$\mathbf{1}_{\bigcup_{k=1}^n U_k^m \times U_k^m \times D_k^m} = \sum_{k=1}^{q_n} \sum_{j_1,j_2=1}^{p_n} \epsilon_{j_1,j_2,k}^n \mathbf{1}_{V_{j_1}^n \times V_{j_2}^n \times E_k^n}, \tag{2.6}$$

where ϵ_{j_1,j_2}^n is equal to 0 or 1. Set

$$\mathbf{1}_{Z_n} = \sum_{k=1}^{q_n} \sum_{j=1}^{p_n} \epsilon_{j,j,k}^n \mathbf{1}_{V_j^n \times E_k^n}.$$

It is evident that $Z_n \subset U \times D$. Furthermore $\mathbf{1}_{Z_n}(s,\omega) \to \mathbf{1}_{U \times D}(s,\omega)$ as $n \to \infty$. Indeed, if $s \in U$ and $\omega \in D$ then $(s,s,\omega) \in \Delta_2^U \times D$ and there exists $n_0 \geq 1$ such that $(s,s,\omega) \in \bigcup_{k=1}^n U_k^m \times U_k^m \times D_k^m$ for all $n \geq n_0$. Hence one can find j_n and k_n for all $n \geq n_0$ such that $\epsilon_{j_n,j_n,k_n}^n = 1$. Therefore $Z_n(s,\omega) = 1$ for all $n \geq n_0$.

It follows from dominated convergence theorem that

$$\mathsf{E}\left(\sum_{k=1}^{q_n} \sum_{j=1}^{p_n} \epsilon_{j,j,k}^n \pi(V_j^n) \mathbf{1}_{E_k^n} - \pi(U)\mathbf{1}_D\right)^2 = M_\pi(Z_n) - M_\pi(U \times D) \to 0. \tag{2.7}$$

Set $V_{0j}^n = V_j^n \cap (T \times X_0)$ and let us calculate the following expectation:

$$S_n = \mathsf{E}\left(\sum_{k=1}^{q_n} \sum_{j=1}^{p_n} \epsilon_{j,j,k}^n (L(\mathbf{1}_{V_j^n}\mathbf{1}_{E_k^n})^2 - L(\mathbf{1}_{V_{0j}^n}\mathbf{1}_{E_k^n}) - \pi(V_j^n))\mathbf{1}_{E_k^n}\right)^2.$$

Since $\pi(V_{0j}^n)\mathbf{1}_{E_k^n} \leq \pi(V_j^n)\mathbf{1}_{E_k^n} \leq \pi(U)\mathbf{1}_D \leq Q$ then $\mathbf{1}_{V_j^n}\mathbf{1}_{E_k^n} \in K$ and from Lemma 6 we get

$$\begin{aligned}
S_n &= \sum_{k=1}^{q_n} \sum_{i,j=1}^{p_n} \epsilon_{j,j,k}^n \epsilon_{i,i,k}^n \mathsf{E}\Big((L(\mathbf{1}_{V_i^n}\mathbf{1}_{E_k^n})^2 - L(\mathbf{1}_{V_{0i}^n}\mathbf{1}_{E_k^n}) - \pi(V_i^n)) \\
&\quad \times (L(\mathbf{1}_{V_j^n}\mathbf{1}_{E_k^n})^2 - L(\mathbf{1}_{V_{0j}^n}\mathbf{1}_{E_k^n}) - \pi(V_j^n))\mathbf{1}_{E_k^n}\Big) \\
&= \sum_{k=1}^{q_n} \sum_{i \neq j} \epsilon_{j,j,k}^n \epsilon_{i,i,k}^n \mathsf{E}\Big((\pi(V_i^n)\pi(V_j^n) - \pi(V_i^n)\pi(V_j^n) - \pi(V_i^n)\pi(V_j^n) \\
&\quad + \pi(V_i^n)\pi(V_j^n))\mathbf{1}_{E_k^n}\Big) + \sum_{k=1}^{q_n} \sum_{j=1}^{p_n} \epsilon_{j,j,k}^n \mathsf{E}\Big((L(\mathbf{1}_{V_j^n}\mathbf{1}_{E_k^n})^4 + L(\mathbf{1}_{V_{0j}^n}\mathbf{1}_{E_k^n})^2 \\
&\quad + \pi(V_j^n)^2 - 2L(\mathbf{1}_{V_j^n}\mathbf{1}_{E_k^n})^2 L(\mathbf{1}_{V_{0j}^n}\mathbf{1}_{E_k^n}) - 2L(\mathbf{1}_{V_j^n}\mathbf{1}_{E_k^n})^2 \pi(V_j^n))\mathbf{1}_{E_k^n}\Big).
\end{aligned}$$

Taking respective derivatives at zero of conditional characteristic function of L given by formula (1) yields $\mathsf{E}[L(\mathbf{1}_{V_j^n}\mathbf{1}_{E_k^n})^4|\mathcal{H}] = (3\pi(V_j^n)^2 + \pi(V_{0j}^n))\mathbf{1}_{E_k^n}$ and $\mathsf{E}[L(\mathbf{1}_{V_j^n}\mathbf{1}_{E_k^n})^2 L(\mathbf{1}_{V_{0j}^n}\mathbf{1}_{E_k^n})|\mathcal{H}] = \pi(V_{0j}^n)\mathbf{1}_{E_k^n}$. Therefore we have

$$S_n = \sum_{k=1}^{q_n}\sum_{j=1}^{p_n} \epsilon_{j,j,k}^n \mathsf{E}\Big((3\pi(V_j^n)^2 + \pi(V_{0j}^n) + \pi(V_{0j}^n) + \pi(V_j^n)^2$$

$$-2\pi(V_{0j}^n) - 2\pi(V_j^n)^2)\mathbf{1}_{E_k^n}\Big)$$

$$= 2\mathsf{E}\left(\sum_{k=1}^{q_n}\sum_{j=1}^{p_n} \epsilon_{j,j,k}^n \pi(V_j^n)^2 \mathbf{1}_{E_k^n}\right).$$

The last equality and formula (2.6) imply

$$S_n \le 2\mathsf{E}\left(\sum_{k=1}^{q_n}\sum_{i,j=1}^{p_n} \epsilon_{i,j,k}^n \pi(V_i^n)\pi(V_j^n)\mathbf{1}_{E_k^n}\right) \le 2\sum_{k=1}^{\infty} \mathsf{E}(\pi(U_k^m)^2\mathbf{1}_{D_k^m}) \le 2/m. \tag{2.8}$$

Therefore from expressions (2.7) and (2.8) we deduce that

$$\sum_{k=1}^{q_n}\sum_{j=1}^{p_n} \epsilon_{j,j,k}^n (L(\mathbf{1}_{V_j^n}\mathbf{1}_{E_k^n})^2 - L(\mathbf{1}_{V_{0j}^n}\mathbf{1}_{E_k^n}))\mathbf{1}_{E_k^n} \to \pi(U)\mathbf{1}_D$$

as $n,\ m \to \infty$. Hence $\pi(U)\mathbf{1}_D$ is $\mathcal{F}_L$ measurable, which completes the proof of the lemma.

In what follows we will always assume that $\mathcal{F}$ is a completion of $\mathcal{F}_L = \sigma\{L(h), h \in H\}$.

Now we will introduce the generalized orthogonal polynomials P_n (see, e.g. [33]). Denote by $\overline{x} = (x_1, x_2, \dots, x_n, \dots)$ a sequence of real numbers.

Define a function $F(z, \overline{x})$ by

$$F(z, \overline{x}) = \exp\left(\sum_{k=1}^{\infty} (-1)^{k+1} \frac{z^k}{k} x_k\right). \tag{2.9}$$

If $R(\overline{x}) = (\limsup |x_k|^{1/k})^{-1} > 0$ then the series in (2.9) converges for all $|z| < R(\overline{x})$. So the function $F(z, \overline{x})$ is analytic for $|z| < R(\overline{x})$.

Consider an expansion in powers of z of the function $F(z, \overline{x})$

$$F(z, \overline{x}) = \sum_{n=0}^{\infty} z^n P_n(\overline{x}).$$

Using this development, one can easily show the following equalities:

$$(n+1)P_{n+1}(\overline{x}) = \sum_{k=0}^{n} (-1)^k x_{k+1} P_{n-k}(\overline{x}), \quad n \ge 0, \tag{2.10}$$

$$\frac{\partial}{\partial x_l} P_n(\overline{x}) = \begin{cases} 0, & \text{if } l > n, \\ (-1)^{l+1}\frac{1}{l} P_{n-l}(\overline{x}), & \text{if } l \le n. \end{cases} \tag{2.11}$$

Indeed, (2.10) and (2.11) follow from $\frac{\partial F}{\partial z} = \sum_{k=0}^{\infty}(-1)^k z^k x_{k+1} F$, respectively, and $\frac{\partial F}{\partial x_l} = (-1)^{l+1}\frac{F}{l}z^l$. From (2.11) it follows that P_n depends only on finite number of variables, namely $x_1, x_2, \ldots, x_n$. Since $P_0 \equiv 1$, then (2.10) implies that $P_n(x_1, x_2, \ldots, x_n)$ is a polynomial with the highest order term $\frac{x_1^n}{n!}$. The first polynomials are $P_1(x_1) = x_1$ and $P_2(x_1, x_2) = \frac{1}{2}(x_1^2 - x_2)$.

Using the equality $F(z, \overline{x} + \overline{y}) = F(z, \overline{x})F(z, \overline{y})$, where $\overline{y} = (y_1, y_2, \ldots, y_n, \ldots)$ and $\overline{x} + \overline{y} = (x_1 + y_1, x_2 + y_2, \ldots, x_n + y_n, \ldots)$ it is easy to show that

$$P_n(\overline{x} + \overline{y}) = \sum_{k=0}^{n} P_k(\overline{x})P_{n-k}(\overline{y}). \tag{2.12}$$

If $\overline{u}(y) = (y, y^2, y^3, \ldots, y^n, \ldots)$ then $F(z, \overline{u}(y)) = 1 + zy$ for $|zy| < 1$. Hence $P_1(\overline{u}(y)) = y$ and $P_n(\overline{u}(y)) = 0$ for all $n \geq 2$. Furthermore, equation (2.12) implies that

$$P_n(\overline{x} + \overline{u}(y)) - P_n(\overline{x}) = yP_{n-1}(\overline{x}). \tag{2.13}$$

It is possible to find the explicit formula for polynomials P_n. Indeed, P_n can be written in the following form:

$$P_n(x_1, x_2, \ldots x_n) = \sum_{i_1+i_2+\cdots+i_n \leq n} a_{i_1,i_2,\ldots i_n} x_1^{i_1} x_2^{i_2} \cdots x_n^{i_n}.$$

It is easy to see that

$$\frac{\partial^{i_1+i_2+\cdots+i_n} P_n}{\partial x_1^{i_1} \partial x_2^{i_2} \cdots \partial x_n^{i_n}}(0, \ldots, 0) = i_1! i_2! \cdots i_n! a_{i_1,i_2,\ldots i_n}.$$

It follows from the equality (2.11) that

$$\frac{\partial^{i_1+i_2+\cdots+i_n} P_n}{\partial x_1^{i_1} \partial x_2^{i_2} \cdots \partial x_n^{i_n}}(0, \ldots, 0) =$$

$$\begin{cases} 0, & \text{if } i_1 + 2i_2 + 3i_3 \cdots + ni_n \neq n, \\ (-1)^{n+i_1+i_2+\cdots+i_n} 2^{-i_2} 3^{-i_3} \cdots n^{-i_n}, & \text{if } i_1 + 2i_2 + 3i_3 \cdots + ni_n = n. \end{cases}$$

Hence

$$P_n(x_1, x_2, \ldots x_n) = \sum_{i_1+2i_2+3i_3\cdots+ni_n=n} (-1)^{n+i_1+i_2+\cdots+i_n} \frac{x_1^{i_1} x_2^{i_2} \cdots x_n^{i_n}}{i_1! i_2! \cdots i_n! 2^{i_2} 3^{i_3} \cdots n^{i_n}}. \tag{2.14}$$

For $h \in K$ let $\overline{x}(h) = (x_1(h), x_2(h), \ldots x_n(h), \ldots)$ denote the sequence of the random variables, such that $x_1(h) = L(h)$, $x_2(h) = L(h^2 \mathbf{1}_{X_0}) + \int_{T \times X} h^k(t,x)\pi(dtdx) = L(h^2\mathbf{1}_{X_0}) + \pi(h^2)$, $x_k(h) = L(h^k \mathbf{1}_{X_0}) + \int_{T \times X_0} h^k(t,x)$ $\nu(dtdx) = L(h^k\mathbf{1}_{X_0}) + \pi(h^k\mathbf{1}_{X_0})$, $k = 3, 4, \ldots$.

The relationship between generalized orthogonal polynomials and conditional additive processes on H is given by the following result.

Lemma 8. *Let h and $g \in K$. Then for all $n, m \geq 0$ we have $P_n(\overline{x}(h))$ and $P_m(\overline{x}(g)) \in L^2(\Omega)$, and*

$$\mathsf{E}[P_n(\overline{x}(h))P_m(\overline{x}(g))|\mathcal{H}] = \begin{cases} 0, & \text{if } n \neq m, \\ \frac{1}{n!}\left(\mathsf{E}[L(h)L(g)|\mathcal{H}]\right)^n, & \text{if } n = m. \end{cases}$$

Proof. Since $h, g \in K$ and P_n, P_m are the polynomials, then by Lemma 6 we have $P_n(\overline{x}(h))$ and $P_m(\overline{x}(g)) \in L^2(\Omega, \mathcal{F}, \mathsf{P})$.

Denote by $\phi(z, \overline{x})$ the power of the exponent in the formula (2.9):

$$\phi(z, \overline{x}) = \sum_{k=1}^{\infty} (-1)^{k+1} \frac{z^k}{k} x_k.$$

Since

$$\begin{aligned} \frac{1}{R} &= \limsup_{k\to\infty} ||x_k(h)||_{L^2(\Omega)}^{1/k} = \lim_{k\to\infty} \left[\mathsf{E}(L(h^k \mathbf{1}_{X_0})^2) + \mathsf{E}(\pi(h^k \mathbf{1}_{X_0})^2)\right]^{1/2k} \\ &\leq \lim_{k\to\infty} \left(||h\mathbf{1}_{X_0}||_{L^\infty}^{2k-2} \mathsf{E}(\pi(h^2\mathbf{1}_{X_0})) + ||h\mathbf{1}_{X_0}||_{L^\infty}^{2k-4} \mathsf{E}(\pi(h^2\mathbf{1}_{X_0})^2)\right)^{1/2k} \\ &\leq ||h\mathbf{1}_{X_0}||_{L^\infty}. \end{aligned}$$

Then the series

$$\sum_{k=1}^{\infty} \frac{|z|^k}{k} ||x_k(h)||_{L^2(\Omega)}$$

converges if $|z| < 1/\,||h\mathbf{1}_{X_0}||_{L^\infty} \leq R$, which implies that $\phi(z, \overline{x}(h)) \in L^2(\Omega)$ for all $|z| < 1/\,||h\mathbf{1}_{X_0}||_{L^\infty}$.

Let's note that for all $|z| < 1/\,||h\mathbf{1}_{X_0}||_{L^\infty}$ we have $\ln(1 + zh\mathbf{1}_{X_0}) \in H$. Indeed, by using Taylor's formula, we get

$$(\ln(1 + zh\mathbf{1}_{X_0}))^2 \leq \frac{z^2 h^2 \mathbf{1}_{X_0}}{(1 - |z|\,||h\mathbf{1}_{X_0}||_{L^\infty})^2}.$$

In the same way one can obtain the following inequality

$$|\ln(1 + zh\mathbf{1}_{X_0}) - zh\mathbf{1}_{X_0}| \leq \frac{z^2 h^2 \mathbf{1}_{X_0}}{2(1 - |z|\,||h\mathbf{1}_{X_0}||_{L^\infty})^2},$$

which implies that $\ln(1 + zh(t,x)\mathbf{1}_{X_0}) - zh(t,x)\mathbf{1}_{X_0}$ is integrable with respect to measure M_π for all $|z| < 1/\,||h\mathbf{1}_{X_0}||_{L^\infty}$.

So by using the linearity and the continuity of the mapping $h \to L(h)$ we have for all $|z| < 1/\,||h\mathbf{1}_{X_0}||_{L^\infty}$

$$\begin{aligned} \phi(z, \overline{x}(h)) &= \sum_{k=2}^{\infty} (-1)^{k+1} \frac{z^k}{k} \left(L(h^k \mathbf{1}_{X_0}) + \pi(h^k \mathbf{1}_{X_0})\right) + zL(h) - \frac{z^2}{2}\pi(h\mathbf{1}_\Delta) \\ &= L\left(\ln(1 + zh\mathbf{1}_{X_0}) + zh\mathbf{1}_\Delta\right) + \int_{T\times X_0} (\ln(1 + zh(t,x)) \\ &\quad - zh(t,x))\nu(dtdx) - \frac{z^2}{2}\pi(h\mathbf{1}_\Delta). \qquad (2.15) \end{aligned}$$

We claim that $w = \ln(1 + zh\mathbf{1}_{X_0}) + zh\mathbf{1}_\Delta \in K$ for all $|z| < 1/\|h\mathbf{1}_{X_0}\|_{L^\infty}$. Indeed

$$|w\mathbf{1}_{X_0}| = |\ln(1 + zh\mathbf{1}_{X_0})| \le \frac{|z|\,\|h\mathbf{1}_{X_0}\|_{L^\infty}}{|1 - |z|\,\|h\mathbf{1}_{X_0}\|_{L^\infty}|}$$

and since $h \in K$ then we have for some constant $C_1 > 0$

$$\begin{aligned}\pi(w^2) &= z^2\pi(h^2\mathbf{1}_\Delta) + \pi(\ln^2(1 + zh\mathbf{1}_{X_0}))\\ &\le z^2\pi(h^2\mathbf{1}_\Delta) + \frac{z^2\pi(h^2\mathbf{1}_{X_0})}{(1 - |z|\,\|h\mathbf{1}_{X_0}\|_{L^\infty})^2} \le C_1.\end{aligned}$$

Hence from inequality $\ln(1+x) \le x$, equality (2.15) and Lemma 6 we have for all $|z| < 1/\|h\mathbf{1}_{X_0}\|_{L^\infty}$

$$\begin{aligned}\mathsf{E}(F(z, \overline{x}(h))^2) &= \mathsf{E}[\exp(2\phi(z, \overline{x}(h)))]\\ &\le \mathsf{E}[\exp(2L(\ln(1 + zh\mathbf{1}_{X_0}) + zh\mathbf{1}_\Delta))] < \infty.\end{aligned}$$

So $F(z, \overline{x}(h)) \in L^2(\Omega)$ if $|z| < 1/\|h\mathbf{1}_{X_0}\|_{L^\infty}$.

Consequently for $|z| < 1/\|h\mathbf{1}_{X_0}\|_{L^\infty}$ and $|y| < 1/\|g\mathbf{1}_{X_0}\|_{L^\infty}$ we get from (2.15)

$$\begin{aligned}&\mathsf{E}[F(z, \overline{x}(h))F(y, \overline{x}(g))|\mathcal{H}] = \mathsf{E}[\exp(\phi(z, \overline{x}(h)) + \phi(y, \overline{x}(g)))|\mathcal{H}]\\ &= \mathsf{E}[\exp(L(\ln[(1 + zh\mathbf{1}_{X_0})(1 + yg\mathbf{1}_{X_0})])\\ &+ \int_{T\times X_0}(\ln[(1 + zh(t,x))(1 + yg(t,x))] - zh(t,x) - yg(t,x))\nu(dtdx)\\ &+ L(zh\mathbf{1}_\Delta + yg\mathbf{1}_\Delta) - \frac{1}{2}\int_T(z^2h^2(t,\Delta) + y^2g^2(t,\Delta))\mu(dt))|\mathcal{H}]\\ &= \exp\Bigg(\int_{T\times X_0}(e^{\ln[(1+zh(t,x))(1+yg(t,x))]} - 1\\ &- \ln[(1 + zh(t,x))(1 + yg(t,x))])\nu(dtdx)\\ &+ \int_{T\times X_0}(\ln[(1 + zh(t,x))(1 + yg(t,x))] - zh(t,x) - yg(t,x))\nu(dtdx)\\ &+ \frac{1}{2}\int_T((zh(t,\Delta) + yg(t,\Delta))^2 - z^2h^2(t,\Delta) - y^2g^2(t,\Delta))\mu(dt)\Bigg)\\ &= \exp(zy\int_{T\times X}h(t,x)g(t,x)\pi(dtdx)) = \exp(zy\mathsf{E}[L(h)L(g)|\mathcal{H}]),\end{aligned}$$

where we have used Lemma 6 to calculate the conditional expectation.

Taking the $(n+m)$-th partial derivative $\frac{\partial^{n+m}}{\partial z^n \partial y^m}$ at $z = y = 0$ in both sides of the above equality yields

$$\mathsf{E}[n!m!P_n(\overline{x}(h))P_m(\overline{x}(g))|\mathcal{H}] = \begin{cases}0, & \text{if } n \ne m,\\ n!\,(\mathsf{E}[L(h)L(g)|\mathcal{H}])^n, & \text{if } n = m.\end{cases}$$

Lemma 9. *The random variables* $\{e^{L(h)},\ h \in K\}$ *form a total subset of* $L^2(\Omega, \mathcal{F}, P)$.

Proof. It follows from Lemma 6 that $e^{L(h)} \in L^2(\Omega)$ if $h \in K$.

Let $\xi \in L^2(\Omega)$ be such that $\mathsf{E}(\xi e^{L(h)}) = 0$ for all $h \in K$. The linearity of the mapping $h \to L(h)$ implies

$$\mathsf{E}\left(\xi \exp \sum_{k=1}^{n} z_k L(h_k)\right) = 0 \tag{2.16}$$

for any $z_1, \ldots, z_n \in \mathbb{R}$, $h_1, \ldots, h_n \in K$, $n \geq 1$. Suppose that $n \geq 1$ and $h_1, \ldots, h_n \in K$ are fixed. Then (2.16) says that Laplace transform of the signed measure

$$\tau(B) = \mathsf{E}(\xi \mathbf{1}_B(L(h_1), \ldots, L(h_n))),$$

where B is a Borel subset of $\mathbb{R}^n$, is identically zero on $\mathbb{R}^n$. Consequently, this measure is zero, which implies $\mathsf{E}(\xi \mathbf{1}_G) = 0$ for any $G \in \mathcal{F}$. So $\xi = 0$, completing the proof of the lemma.

For each $n \geq 0$ we will denote by $\mathcal{P}_n$ the closed linear subspace of $L^2(\Omega, \mathcal{F}, P)$ generated by the random variables $\{\xi P_n(\overline{x}(h)) : h \in K, \xi \in L^\infty(\Omega, \mathcal{H}, \mathsf{P})\}$. $\mathcal{P}_0$ will be the set $L^2(\Omega, \mathcal{H}, P)$ of $\mathcal{H}$-measurable square integrable random variables. For $n = 1$, $\mathcal{P}_1$ coincides with the set of random variables $\{L(h) : h \in H\}$. From Lemma 8 we obtain that $\mathcal{P}_n$ and $\mathcal{P}_m$ are orthogonal whenever $n \neq m$. We will call the space $\mathcal{P}_n$ chaos of order n.

Theorem 10. *The space $L^2(\Omega, \mathcal{F}, P)$ can be decomposed into the infinite orthogonal sum of the subspaces $\mathcal{P}_n$:*

$$L^2(\Omega, \mathcal{F}, P) = \bigoplus_{n=0}^{\infty} \mathcal{P}_n.$$

Proof. Let $\xi \in L^2(\Omega, \mathcal{F}, P)$ such that ξ is orthogonal to all $\mathcal{P}_n$, $n \geq 0$. We have to show that $\xi = 0$. For all $h \in K$ and $\eta \in L^\infty(\Omega, \mathcal{H}, P)$ we get $\mathsf{E}(\xi \eta P_n(\overline{x}(h))) = 0$. Hence $\mathsf{E}[\xi P_n(\overline{x}(h)) | \mathcal{H}] = 0$. Since from the proof of Lemma 8 we have that $F(z, \overline{x}(h)) \in L^2(\Omega)$ for all $|z| < 1/\, ||h\mathbf{1}_{X_0}||_{L^\infty}$, then $\mathsf{E}[\xi F(z, \overline{x}(h)) | \mathcal{H}] = 0$ for $|z| < 1/\, ||h\mathbf{1}_{X_0}||_{L^\infty}$. Using equality (2.15) we obtain

$$\begin{aligned} 0 = \mathsf{E}[\xi F(z, \overline{x}(h)) | \mathcal{H}] &= \mathsf{E}[\xi e^{\phi(z, \overline{x}(h))} | \mathcal{H}] = \mathsf{E}[\xi \exp(L(\ln(1 + zh\mathbf{1}_{X_0})) \\ &+ \int_{T \times X_0} (\ln(1 + zh(t,x)) - zh(t,x)) \nu(dtdx) + L(zh\mathbf{1}_\Delta) \\ &- \frac{1}{2} \int_T z^2 h^2(t, \Delta) \mu(dt)) | \mathcal{H}]. \end{aligned}$$

Thus for any $|z| < 1/\, ||h\mathbf{1}_{X_0}||_{L^\infty}$

$$\mathsf{E}[\xi \exp(L(\ln(1 + zh\mathbf{1}_{X_0})) + L(zh\mathbf{1}_\Delta))] = 0. \tag{2.17}$$

We claim that if $h \in K$ such that $h \geq \varepsilon - 1$ M_π-a.e. for some $1 \geq \varepsilon > 0$ then equality (2.17) holds for $z = 1$. Indeed the right-hand side of the expression (2.17) meaningful in this case for all $z \in [0;1]$. The extension to the complex numbers $\mathrm{Re}\, z \in [0;1]$ is evident. Denote this function by $\Phi(z)$. Since $|h\mathbf{1}_{X_0}/(1+zh\mathbf{1}_{X_0})| \le ||h\mathbf{1}_{X_0}||_{L^\infty}/\varepsilon$ and $\pi(h^2\mathbf{1}_{X_0}/(1+zh\mathbf{1}_{X_0})^2) \le \pi(h^2)/\varepsilon^2$ then $h\mathbf{1}_{X_0}/(1+zh\mathbf{1}_{X_0}) \in K$ and the straightforward calculation shows that $\Phi(z)$ is differentiable and

$$\Phi'(z) = \mathsf{E}[\xi L(h\mathbf{1}_{X_0}/(1+zh\mathbf{1}_{X_0}) + h\mathbf{1}_\Delta)\exp(L(\ln(1+zh\mathbf{1}_{X_0})) + L(zh\mathbf{1}_\Delta))].$$

Hence $\Phi(z)$ is an analytical function for $\mathrm{Re} z \in [0;1]$. Consequently the uniqueness theorem yields the desired statement.

For any $g \in K$ we have $(e^g - 1) \in K$ and $(e^g - 1)\mathbf{1}_{X_0} > -1 + \varepsilon$ M_π-a.e. for some $1 \ge \varepsilon > 0$. Putting in (2.17) $h = (e^g - 1)\mathbf{1}_{X_0} + g\mathbf{1}_\Delta$ and $z = 1$ we deduce that $\mathsf{E}(\xi e^{L(g)}) = 0$ for all $g \in K$. By Lemma 9 we get $\xi = 0$, which completes the proof of the theorem.

3 Multiple Integrals

The purpose of the section is to define multiple stochastic integrals with respect to L and to show that the nth chaos $\mathcal{P}_n$ is generated by these multiple stochastic integrals. The construction of multiple stochastic integrals for processes with independent increments provided by Itô in [11]. For its generalization to other classes of processes the reader referred to [9, 13, 21, 32].

Recall that random measure $\pi(dtdx,\omega)$ has no atoms for all $\omega \in \Omega$. It means that neither measure μ nor measure ν has no atoms for all $\omega \in \Omega$. Since a separable Hilbert space H has the form $H = L^2(T \times X \times \Omega, \mathcal{G}\otimes\mathcal{H}, M_\pi)$, where $M_\pi(dtdxd\omega) = \pi(dtdx,\omega)P(d\omega)$ is a σ-finite measure, then the process L is characterized by the family of random variables $\{L(A), A \in \mathcal{G}\otimes\mathcal{H}, M_\pi(A) < \infty\}$, where $L(A) = L(\mathbf{1}_A)$. We can consider $L(A)$ as a $L^2(\Omega,\mathcal{F},P)$-valued measure on the parametric space $(T\times X\times\Omega, \mathcal{G}\otimes\mathcal{H})$, which takes conditionally independent values on any family of disjoint subsets of $T \times X \times \Omega$.

Fix $m \ge 1$. Denote by M_π^m the following measure

$$M_\pi^m(dt_1dx_1\cdots dt_mdx_md\omega) = \pi(dt_1dx_1,\omega)\cdots\pi(dt_mdx_m,\omega)\mathsf{P}(d\omega),$$

defined on the σ-algebra $\mathcal{G}^{\otimes m}\otimes\mathcal{F}$. In this section will consider only the restriction of this measure on the σ-algebra $\mathcal{G}^{\otimes m}\otimes\mathcal{H}$. Since the measure M_π is σ-finite then M_π^m will be σ-finite. Indeed if the sequence of pairwise-disjoint sets $A_n \times B_n$, $A_n \in \mathcal{G}$ and $B_n \in \mathcal{H}$ such that $\bigcup_{n=1}^\infty A_n \times B_n = T \times X \times \Omega$ and $M_\pi(A_n \times B_n) < \infty$, then setting $B_n^k = \{k-1 \le \pi(A_n \times B_n) < k\}$ we have $\bigcup_{k=1}^\infty A_n \times B_n^k = A_n \times B_n$. Denote $B_{n_1n_2\ldots n_m}^{k_1k_2\ldots k_m} = \bigcap_{j=1}^m B_{n_j}^{k_j}$ then $M_\pi^m(A_{n_1}\times A_{n_2}\times\cdots A_{n_m}\times B_{n_1n_2\ldots n_m}^{k_1k_2\ldots k_m}) \le k_1k_2\cdots k_m < \infty$ and $(T\times X)^m\times\Omega = \bigcup_{n_1,n_2,\ldots,n_m=1}^\infty \bigcup_{k_1,k_2,\ldots,k_m=1}^\infty A_{n_1}\times A_{n_2}\times\cdots A_{n_m}\times B_{n_1n_2\ldots n_m}^{k_1k_2\ldots k_m}$.

For any $\omega \in \Omega$ we can define measure $\pi^{\otimes m}(dt_1dx_1\cdots dt_mdx_m,\omega)$ on the σ-algebra $\mathcal{G}^{\otimes m}$ as a m-th power of the measure π. Since measure π is σ-finite and without atoms then measure $\pi^{\otimes m}$ is σ-finite and without atoms. Moreover, $\pi^{\otimes m}(\Delta_m, \omega) = 0$ for all $\omega \in \Omega$, where $\Delta_m = \{(t_1, \ldots, t_m) : \exists t_i = t_j, i \neq j\}$ is a 'diagonal' set. Indeed, for fixed $\omega \in \Omega$ σ-finiteness of π implies that $T \times X = \bigcup_{i=1}^{\infty} T_i$, where $T_1, T_2, \ldots$ are pairwise-disjoint sets in $\mathcal{G}$ and $\pi(T_i) < \infty$. Then $(T \times X)^m = \bigcup_{i_1,\ldots,i_m=1}^{\infty} T_{i_1} \times \cdots \times T_{i_m}$ and $\pi^{\otimes m}(T_{i_1} \times \cdots \times T_{i_m}) < \infty$. Define $C_{i_1,\ldots,i_m} = (T_{i_1} \times \cdots \times T_{i_m}) \bigcap \Delta_m$. Then $\Delta_m = \bigcup_{i_1,\ldots,i_m=1}^{\infty} C_{i_1,\ldots,i_m}$. It is easy to see that $C_{i_1,\ldots,i_m} = \emptyset$ if all the indices $i_1, \ldots, i_m$ are different. Hence it is enough to prove that $\pi^{\otimes m}(C_{i_1,\ldots,i_m}) = 0$ if some of indices $i_1, \ldots, i_m$ are equal. Suppose that $i_1 = i_2$. Using the nonexistence of atoms for the measure π for any $n \in \mathbb{N}$ we can determine a system of pairwise-disjoints sets $\{V_1, \ldots, V_n\} \subset \mathcal{G}$, such that $\bigcup_{i=1}^{n} V_i = T_{i_1}$ and $\pi(V_i) = \pi(T_{i_1})/n$ for every $i = 1, \ldots, n$. Then $C_{i_1,\ldots,i_m} \subset \bigcup_{i=1}^{n} V_i \times V_i \times T_{i_3} \times \cdots \times T_{i_m}$. Hence $\pi^{\otimes m}(C_{i_1,\ldots,i_m}) \leq \sum_{i=1}^{n} \pi(V_i)^2 \pi(T_{i_3}) \cdots \pi(T_{i_m}) = \pi(T_{i_1})^2 \pi(T_{i_2}) \pi(T_{i_3}) \cdots \pi(T_{i_m})/n$. Letting n tend to ∞ we obtain the desired result.

Precisely speaking the set Δ_m may not be an element of the σ-algebra $\mathcal{G}^{\otimes m}$ but the calculations above show that Δ_m belongs to completion $\overline{\mathcal{G}^{\otimes m}}^{\omega}$ of $\mathcal{G}^{\otimes m}$ with respect to the measure $\pi^{\otimes m}(\cdot, \omega)$ for all $\omega \in \Omega$. Since $\pi^{\otimes m}(\cdot, \omega)$ can be extended to the σ-algebra $\overline{\mathcal{G}^{\otimes m}} = \bigcap_{\omega \in \Omega} \overline{\mathcal{G}^{\otimes m}}^{\omega}$ and $\Delta_m \in \overline{\mathcal{G}^{\otimes m}}$ then measure M_π^m can be extended on the σ-algebra $\overline{\mathcal{G}^{\otimes m}} \otimes \mathcal{H}$ and $M_\pi^m(\Delta_m \times \Omega) = 0$. This fact is very important for definition of the multiple stochastic integral.

Set $\mathcal{G}_0 = \{A \in \mathcal{G} \otimes \mathcal{H} : \ \mathbf{1}_A \in K\}$. We will define the multiple stochastic integral $I_m(f)$ of a function $f \in L^2((T \times X)^m \times \Omega, \mathcal{G}^{\otimes m} \otimes \mathcal{H}, M_\pi^m)$. Denote by $\mathcal{E}_m$ the set of elementary functions of the form

$$\begin{aligned} & f(t_1, x_1, \ldots, t_m, x_m, \omega) \\ &= \sum_{i_1,\ldots,i_m=1}^{n} a_{i_1,\ldots,i_m}(\omega) \mathbf{1}_{A_{i_1}}(t_1, x_1, \omega) \cdots \mathbf{1}_{A_{i_m}}(t_m, x_m, \omega), \end{aligned} \tag{3.1}$$

where $A_1, \ldots, A_n$ are pairwise-disjoint sets in $\mathcal{G}_0$, and the coefficients $a_{i_1,\ldots,i_m} \in L^\infty(\Omega, \mathcal{H}, \mathsf{P})$ are zero if any two of indices $i_1, \ldots, i_m$ are equal.

For a function of the form (3.1) we define the multiple integral of the m-th order

$$I_m(f) = \sum_{i_1,\ldots,i_m=1}^{n} a_{i_1,\ldots,i_m} L(A_{i_1}) \cdots L(A_{i_m}).$$

The definition does not depend on particular representation of f, and the following properties hold:

(i) $I_m(\alpha f + \beta g) = \alpha I_m(f) + \beta I_m(g)$ for all α and $\beta \in L^\infty(\Omega, \mathcal{H}, \mathsf{P})$, f and g in $\mathcal{E}_m$.

(ii) $I_m(f) = I_m(\widetilde{f})$, where $\widetilde{f}$ denotes the symmetrization of f with respect to pairs of nonrandom variables, which is defined by

$$\widetilde{f}(t_1, x_1, \ldots, t_m, x_m, \omega) = \frac{1}{m!} \sum_{\sigma} f(t_{\sigma(1)}, x_{\sigma(1)}, \ldots, t_{\sigma(m)}, x_{\sigma(m)}, \omega),$$

σ running over all permutations of $\{1, \ldots, m\}$.

(iii)

$$\mathsf{E}[I_m(f)I_p(g)|\mathcal{H}] = \begin{cases} 0, & \text{if } p \neq m, \\ m!\pi(\widetilde{f}\widetilde{g}), & \text{if } p = m. \end{cases}$$

The properties can be proved using Lemma 6 and exactly the same arguments as those used, for example, in [22, p. 8-9].

In order to extend the multiple stochastic integral to the space $L^2(M_\pi^m)$ we have to prove the following lemma.

Lemma 11. *The space $\mathcal{E}_m$ is dense in $L^2((T\times X)^m \times \Omega, \mathcal{G}^{\otimes m} \otimes \mathcal{H}, M_\pi^m)$.*

Proof. In order to show that $\mathcal{E}_m$ is dense in $L^2(M_\pi^m)$ it is suffices to show that the indicator function of any set $A \times B = A_1 \times A_2 \times \cdots \times A_m \times B$, where $A_1, \ldots, A_m \in \mathcal{G}$, $B \in \mathcal{H}$ and $M_\pi^m(A\times B) < \infty$ can be approximated by elementary functions in $\mathcal{E}_m$.

Denote by B'_k the following set

$$B'_k = \{\omega \in \Omega : \pi(A_1,\omega)\pi(A_2,\omega)\cdots\pi(A_m,\omega) \le k\}.$$

Since $M_\pi^m(A\times B) < \infty$ then $\mathbf{1}_{A_1\times\cdots\times A_m\times B_k} \to \mathbf{1}_{A\times B}$ as $k\to\infty$. Hence it is possible to assume that $\pi(A_1,\omega)\pi(A_2,\omega)\cdots\pi(A_m,\omega)\mathbf{1}_B \le C$ a.s. for some positive constant C. This implies that $A_i\times B \in \mathcal{G}_0$, $i=1,\ldots,m$. Furthermore, it is possible to suppose that any sets A_i and A_j either equal or disjoint. Indeed, there exists a finite system of pairwise disjoint sets $\{A'_1,\ldots,A'_n\} \subset \mathcal{G}$ such that each A_i can be expressed as a disjoint union of some of A'_j. Then the indicator function of the set $A\times B$ can be represent as a finite sum of the indicator functions of the sets $A'_{i_1}\times A'_{i_2}\times\cdots\times A'_{i_m}\times B$. If all indices $i_1,\ldots,i_m$ are different then it is an element of $\mathcal{E}_m$. For other indices some of the sets A'_{i_k} are equal.

Since $M_\pi^m(\Delta_m\times\Omega) = 0$ then $M_\pi^m((A\times B)\bigcap(\Delta_m\times\Omega)) = 0$ and for any $\epsilon > 0$ there exists a system of sets $U_k^\epsilon = A_1^k\times\cdots A_m^k\times B_k$, $k=1,2,\ldots$ such that $\bigcup_{k=1}^\infty U_k^\epsilon \supset ((A\times B)\bigcap(\Delta_m\times\Omega))$, $\sum_{k=1}^\infty M_\pi^m(U_k^\epsilon) < \epsilon$, $\bigcup_{k=1}^\infty A_i^k = A_i$, $i=1,\ldots,m$ and $\bigcup_{k=1}^\infty B_k = B$. Therefore $A_i^k\times B_k \in \mathcal{G}_0$, $i=1,\ldots,m$, $k=1,2,\ldots$ and $\bigcup_{k=1}^\infty U_k^\epsilon \subset A\times B$.

For any $n\ge 1$ we can find a system of pairwise disjoint sets $\{C_1^n, C_2^n,\ldots,C_{p_n}^n\}\subset\mathcal{G}$ and $\{B_1^n,B_2^n,\ldots,B_{q_n}^n\}\subset\mathcal{G}$, such that each A_i^k and B_k, $i=1,\ldots,m$, $k=1,\ldots,n$ can be expressed as a disjoint union of some of C_j^n or B_j^n respectively. Notice that $C_j^n\times B_i^n\in\mathcal{G}_0$ for all $j=1,\ldots,p_n$, $i=1,\ldots,q_n$. We have

$$\mathbf{1}_{\bigcup_{k=1}^n U_k^\epsilon} = \sum_{k=1}^{q_n}\sum_{j_1,\ldots,j_m=1}^{p_n} \epsilon^n_{j_1,\ldots,j_m,k}\mathbf{1}_{C_{j_1}^n\times C_{j_2}^n\times\cdots C_{j_m}^n\times B_k^n},$$

where $\epsilon^n_{j_1,\ldots,j_m,k}$ is 0 or 1. Let J_n be the set of m-tuples $(j_1,\ldots,j_m)$, $j_i\in\{1,2,\ldots,p_n\}$, $i=1,\ldots,m$, where all the indices are different. We set

$$\mathbf{1}_{V_n^\varepsilon} = \sum_{k=1}^{q_n} \sum_{(j_1,\dots,j_m)\in J_n} (1-\epsilon^n_{j_1,\dots,j_m,k})\mathbf{1}_{C^n_{j_1}\times C^n_{j_2}\times\cdots C^n_{j_m}\times B^n_k}.$$

Then $\mathbf{1}_{V_n^\varepsilon}$ belongs to $\mathcal{E}_m$ and for M_π^m-a.e. $(t,\omega)\in(A\setminus\Delta_m)\times B$ we have $\mathbf{1}_{V_n^\varepsilon}(t,\omega)=1$ for all $\varepsilon\geq\varepsilon_0$ and $n\geq n_0$. Hence $\mathbf{1}_{V_n^\varepsilon}\to\mathbf{1}_{(A\setminus\Delta_m)\times B}$ M_π^m-a.e. and in $L^2(M_\pi^m)$ as $n\to\infty$ and $\epsilon\to 0$. Finally the fact $M_\pi^m(\Delta_m\times\Omega)=0$ implies the proof of the lemma.

Letting $f=g$ in property (iii) obtains

$$\mathsf{E}(I_n(f)^2) = m!\|\tilde{f}\|^2_{L^2(M_\pi^m)} \leq m!\,\|f\|^2_{L^2(M_\pi^m)}.$$

Therefore, the operator I_m can be extended to a linear and continuous operator from $L^2(M_\pi^m)$ to $L^2(\Omega,\mathcal{F},\mathsf{P})$, which satisfies properties (i), (ii) and (iii).

If $f\in L^2(M_\pi^p)$ is symmetric function and $g\in K$ the contraction of one index of f and g is denoted by $f\otimes_1 g$ and is defined by

$$(f\otimes_1 g)(t_1,x_1,\dots,t_{p-1},x_{p-1},\omega) = \int_{T\times X} f(t_1,x_1,\dots,t_{p-1},x_{p-1},s,z,\omega)g(s,z,\omega)\pi(dsdz,\omega).$$

The tensor product $f\otimes g$ will be understood as tensor product with respect to nonrandom variables, i.e.

$$(f\otimes g)(t_1,x_1,\dots,t_{p+1},x_{p+1},\omega) = f(t_1,x_1,\dots,t_p,x_p,\omega)g(t_{p+1},x_{p+1},\omega).$$

Notice that $f\otimes_1 g\in L^2(M_\pi^{p-1})$ and $f\otimes g\in L^2(M_\pi^{p+1})$ if $g\in K$.

The tensor product $f\otimes g$ and the contractions $f\otimes_1 g$ are not necessarily symmetric. We will denote their symmetrization by $f\tilde{\otimes}g$ and $f\tilde{\otimes}_1 g$ respectively.

The following, so called product formula, will be useful in the sequel. It was initially derived by Itô [10] for Gaussian case and by Kabanov [13] for Poisson case, then extended by Russo and Vallois [29] to products of two multiple stochastic integrals with respect to a normal martingale.

Proposition 12. *Let $f\in L^2(M_\pi^p)$ be a symmetric function and let $g\in K$. Then*

$$I_p(f)I_1(g) = I_{p+1}(f\otimes g) + pI_{p-1}(f\otimes_1 g) + pI_p(fg\mathbf{1}_{X_0}). \tag{3.2}$$

Proof. Since $g\in K$ then $fg\mathbf{1}_{X_0}\in L^2(M_\pi^p)$ and the right-hand side is correctly defined.

By the density of elementary functions in $L^2(M_\pi^p)$ and by properties (i) and (ii) we can assume that f is the symmetrization of the function $\mathbf{1}_{A_1}(t_1,x_1,\omega)\mathbf{1}_{A_2}(t_2,x_2,\omega)\cdots\mathbf{1}_{A_p}(t_p,x_p,\omega)$, where the A_i are pairwise-disjoint sets of $\mathcal{G}_0$, and $g=\mathbf{1}_{A_1}$ or $\mathbf{1}_{A_0}$, where $A_0\in\mathcal{G}_0$ is disjoint with $A_1,\dots,A_p$. The case $g=\mathbf{1}_{A_0}$ is immediate because $f\otimes_1 g=fg=0$ and $f\otimes g\in\mathcal{E}_{p+1}$.

So, we assume $g = \mathbf{1}_{A_1}$. Since $A_i \in \mathcal{G}_0$ then $\pi(A_1)\pi(A_2)\cdots\pi(A_p) \le C$ for some real constant $C > 0$. Given $\epsilon > 0$, according to Lemma 11 we can find the function $v'_\epsilon \in \mathcal{E}_2$ such that $||v'_\epsilon - \mathbf{1}_{A_1} \otimes \mathbf{1}_{A_1}||_{L^2(M_\pi^2)} \le \epsilon$. It is possible to write v'_ϵ in the following form

$$v'_\epsilon(t_1, x_1, t_2, x_2, \omega) = \sum_{i,j=1}^{k_\epsilon} a_{ij}^\epsilon(\omega)\mathbf{1}_{C_i}(t_1, x_1, \omega)\mathbf{1}_{C_j}(t_2, x_2, \omega), \tag{3.3}$$

where $C_1, \ldots, C_{k_\epsilon}$ are pairwise disjoint subsets of A_1 in $\mathcal{G}_0$, $a_{ij}^\epsilon \in L^\infty(\Omega, \mathcal{H}, \mathsf{P})$ and $a_{ii}^\epsilon = 0$, $i, j = 1, \ldots, k_\epsilon$. Set $A_1^0 = A_1 \cap (T \times X_0 \times \Omega)$ and

$$\begin{aligned} &v_\epsilon(t_1, x_1, \ldots, t_{p+1}, x_{p+1}, \omega) \\ &\quad = v'_\epsilon(t_1, x_1, t_2, x_2, \omega)\mathbf{1}_{A_2}(t_3, x_3, \omega)\cdots\mathbf{1}_{A_p}(t_{p+1}, x_{p+1}, \omega). \end{aligned}$$

Then v_ϵ is an elementary function and we have

$$\begin{aligned} I_p(f)I_1(g) &= L(A_1)^2 L(A_2)\cdots L(A_p) \\ &= I_{p+1}(v_\epsilon) + \pi(A_1)L(A_2)\cdots L(A_p) + L(A_1^0)L(A_2)\cdots L(A_p) \\ &\quad + [(L(A_1)^2 - \pi(A_1) - L(A_1^0))L(A_2)\cdots L(A_p) - I_{p+1}(v_\epsilon)] \\ &= I_{p+1}(v_\epsilon) + pI_{p-1}(f \otimes_1 g) + pI_p(fg\mathbf{1}_{X_0}) + R_\epsilon. \end{aligned} \tag{3.4}$$

Indeed

$$f \otimes_1 g = \frac{1}{p}\pi(A_1)\,\mathrm{symm}(\mathbf{1}_{A_2} \otimes \cdots \otimes \mathbf{1}_{A_p}),$$

and

$$fg\mathbf{1}_{X_0} = \frac{1}{p}\mathbf{1}_{A_1^0} \otimes \mathrm{symm}(\mathbf{1}_{A_2} \otimes \cdots \otimes \mathbf{1}_{A_p}),$$

where $\mathrm{symm}(\cdot)$ denotes the symmetrization with respect to the pairs of non-random variables of the function in parentheses. We have

$$\begin{aligned} &\left|\left|\tilde{v}_\epsilon - f\tilde{\otimes}g\right|\right|^2_{L^2(M_\pi^{p+1})} \\ &\quad = \left|\left|\tilde{v}_\epsilon - \mathrm{symm}(\mathbf{1}_{A_1} \otimes \mathbf{1}_{A_1} \otimes \mathbf{1}_{A_2} \otimes \cdots \otimes \mathbf{1}_{A_p})\right|\right|^2_{L^2(M_\pi^{p+1})} \\ &\quad \le \left|\left|v_\epsilon - \mathbf{1}_{A_1} \otimes \mathbf{1}_{A_1} \otimes \mathbf{1}_{A_2} \otimes \cdots \otimes \mathbf{1}_{A_p}\right|\right|^2_{L^2(M_\pi^{p+1})} \\ &\quad = \mathsf{E}\Big[\pi(A_2)\cdots\pi(A_p)\int_{(T\times X)^2}(v'_\epsilon(t_1, x_1, t_2, x_2) \\ &\qquad\quad - \mathbf{1}_{A_1}(t_1, x_1)\mathbf{1}_{A_1}(t_2, x_2))^2\,\pi(dt_1dx_1)\pi(dt_2dx_2)\Big] \\ &\quad \le C\,||v'_\epsilon - \mathbf{1}_{A_1} \otimes \mathbf{1}_{A_1}||^2_{L^2(M_\pi^2)} \le C\epsilon^2, \end{aligned} \tag{3.5}$$

and

$$\mathsf{E}(R_\epsilon^2) = \mathsf{E}\Big([L(A_1)^2 - \pi(A_1) - L(A_1^0) - I_2(v'_\epsilon)]^2 L(A_2)^2 L(A_3)^2 \cdots L(A_p)^2\Big).$$

Lemma 6 and properties of multiple integral imply

$$\begin{aligned}\mathsf{E}(R_\epsilon^2) = \mathsf{E}\Big(\big[&L(A_1)^4 + \pi(A_1)^2 + \pi(A_1^0) + 2\pi^{\otimes 2}(\widetilde{v'}_\epsilon^2)\\ &-2\pi(A_1)^2 - 2L(A_1)^2 L(A_1^0) - 2L(A_1)^2 I_2(v'_\epsilon)\big]\pi(A_2)\cdots\pi(A_p)\Big). \quad (3.6)\end{aligned}$$

Taking fourth derivative at $z = 0$ of conditional characteristic function of L given by formula (3.1) yields

$$\mathsf{E}[L(A_1)^4|\mathcal{H}] = 3\pi(A_1)^2 + \pi(A_1^0). \quad (3.7)$$

By using the same arguments since $A_1 = A_1^0 \cup (A_1 \setminus A_1^0)$ we get

$$\begin{aligned}&\mathsf{E}[L(A_1)^2 L(A_1^0)|\mathcal{H}]\\ &\quad = \mathsf{E}[L(A_1^0)^3 + 2L(A_1 \setminus A_1^0)L(A_1^0)^2 + L(A_1 \setminus A_1^0)^2 L(A_1^0)|\mathcal{H}] = \pi(A_1^0). \quad (3.8)\end{aligned}$$

It follows from equality (3.3) and Lemma 6 that

$$\begin{aligned}\mathsf{E}[L(A_1)^2 I_2(v'_\epsilon)|\mathcal{H}] &= \sum_{i,j=1}^{k_\epsilon} a_{ij}^\epsilon \mathsf{E}[L(A_1)^2 L(C_i)L(C_j)|\mathcal{H}]\\ &= \sum_{i,j=1}^{k_\epsilon} a_{ij}^\epsilon \mathsf{E}[2L(C_i)^2 L(C_j)^2|\mathcal{H}]\\ &= 2\sum_{i,j=1}^{k_\epsilon} a_{ij}^\epsilon \pi(C_i)\pi(C_j)\\ &= 2\int_{(T\times X)^2} \widetilde{v'}_\epsilon d\pi^{\otimes 2} = 2\pi^{\otimes 2}(\widetilde{v'}_\epsilon). \quad (3.9)\end{aligned}$$

Substituting expressions (3.7), (3.8) and (3.9) into (3.6) we have

$$\begin{aligned}\mathsf{E}(R_\epsilon^2) &= \mathsf{E}[(2\pi(A_1)^2 + 2\pi^{\otimes 2}(\widetilde{v'}_\epsilon^2) - 4\pi^{\otimes 2}(\widetilde{v'}_\epsilon))\pi(A_2)\cdots\pi(A_p)]\\ &= 2\mathsf{E}[\pi^{\otimes 2}((\widetilde{v'}_\epsilon - \mathbf{1}_{A_1} \otimes \mathbf{1}_{A_1})^2)\pi(A_2)\cdots\pi(A_p)]\\ &\le 2C\,||v'_\epsilon - \mathbf{1}_{A_1} \otimes \mathbf{1}_{A_1}||^2_{L^2(M_\pi^2)} \le 2C\epsilon^2. \quad (3.10)\end{aligned}$$

From formulae (3.4), (3.5) and (3.10) we obtain the desired result.

The next result gives the relationship between generalized orthogonal polynomials and multiple stochastic integrals.

Theorem 13. *Let P_n be the nth generalized orthogonal polynomial, and $\overline{x}(h) = (x_k(h))_{k=1}^{\infty}$, where $x_1(h) = L(h)$, $x_2(h) = L(h^2\mathbf{1}_{X_0}) + ||h||_H^2$, $x_k(h) = L(h^k\mathbf{1}_{X_0}) + \int_{T\times X_0} h^k(t,x)\nu(dtdx)$, $k = 3,4,\dots$ and $h \in K$. Then it holds that*

$$n!P_n(\overline{x}(h)) = I_n(h^{\otimes n}), \tag{3.11}$$

where $h^{\otimes n}(t_1,x_1,\dots,t_n,x_n,\omega) = h(t_1,x_1,\omega)\cdots h(t_n,x_n,\omega)$.

Proof. We will prove the theorem by induction on n. For $n = 1$ it is immediate. Assume it holds for $1,2,\dots,n$. Using the product formula (3.2) and recursive relation for generalized orthogonal polynomials (2.10), we have

$$\begin{aligned}
&I_{n+1}(h^{\otimes(n+1)})\\
&= I_n(h^{\otimes n})I_1(h) - nI_{n-1}\left(h^{\otimes(n-1)}\pi(h^2)\right) - nI_n(h^{\otimes(n-1)}\otimes(h^2\mathbf{1}_{X_0}))\\
&= n!P_n(\overline{x}(h))L(h) - n!\pi(h^2)P_{n-1}(\overline{x}(h)) - nI_{n-1}(h^{\otimes(n-1)})I_1(h^2\mathbf{1}_{X_0})\\
&\quad + n(n-1)I_{n-2}(h^{\otimes(n-2)})\pi(h^3\mathbf{1}_{X_0}) + n(n-1)I_{n-1}(h^{\otimes(n-2)}\otimes(h^3\mathbf{1}_{X_0}))\\
&= n!\sum_{k=0}^{1}(-1)^{k+1}x_{k+1}(h)P_{n-k}(\overline{x}(h)) + n!P_{n-2}(\overline{x}(h))\pi(h^3\mathbf{1}_{X_0})\\
&\quad + n(n-1)I_{n-1}(h^{\otimes(n-2)}\otimes(h^3\mathbf{1}_{X_0})) = \dots\\
&= n!\sum_{k=0}^{n-1}(-1)^{k+1}x_{k+1}(h)P_{n-k}(\overline{x}(h)) + n!(-1)^n P_0(\overline{x}(h))\pi(h^{n+1}\mathbf{1}_{X_0})\\
&\quad + n!(-1)^n I_1(h^{n+1}) = n!\sum_{k=0}^{n}(-1)^{k+1}x_{k+1}(h)P_{n-k}(\overline{x}(h))\\
&= (n+1)!P_{n+1}(\overline{x}(h)),
\end{aligned}$$

which completes the proof of the theorem.

From this theorem and Theorem 10 we deduce the following result.

Corollary 14. *Any square integrable random variable $\xi \in L^2(\Omega,\mathcal{F},\mathsf{P})$ can be expanded into a series of multiple stochastic integrals:*

$$\xi = \sum_{k=0}^{\infty} I_k(f_k).$$

Here $f_0 = \mathsf{E}[\xi|\mathcal{H}]$, and I_0 is the identity mapping on the $L^2(\Omega,\mathcal{H},\mathsf{P})$. Furthermore, this representation is unique provided the functions $f_k \in L^2(M_\pi^k)$ are symmetric with respect to the pairs of nonrandom variables.

Proof. The proof uses the same arguments as those used, for example, in [22, Th. 1.1.2], so we omit it.

The following technical lemma will be needed in the sequel.

Lemma 15. *Let $f_k \in L^2((T\times X)^k\times\Omega,\mathcal{G}^{\otimes k}\otimes\mathcal{H},M_\pi^k)$ and $g_m \in L^2((T\times X)^m\times\Omega,\mathcal{G}^{\otimes m}\otimes\mathcal{H},M_\pi^m)$ be a symmetric with respect to pairs of nonrandom variables functions and $p\le k\wedge m$. Then there exist $\mathcal{G}^{\otimes p}\otimes\mathcal{F}$ measurable versions of the processes $I_{k-p}(f_k(\cdot,t_1,x_1,\dots,t_p,x_p))$ and $I_{m-p}(g_m(\cdot,t_1,x_1,\dots,t_p,x_p))$ which belong to $L^2((T\times X)^p\times\Omega,\mathcal{G}^{\otimes p}\otimes\mathcal{F},M_\pi^p)$ and the following equality holds*

$$\mathsf{E}\Bigg[\int_{(T\times X)^p} I_{k-p}(f_k(\cdot,t_1,x_1,\dots,t_p,x_p))$$

$$\times I_{m-p}(g_m(\cdot,t_1,x_1,\dots,t_p,x_p))\pi(dt_1dx_1)\cdots\pi(dt_pdx_p)\Bigg]$$

$$=\begin{cases}0, & \text{if } m\neq k,,\\ (m-p)!\mathsf{E}\left[\int_{(T\times X)^m}(g_mf_m)(t_1,x_1,\dots,t_m,x_m)\right. & \\ \qquad\left.\pi(dt_1dx_1)\cdots\pi(dt_mdx_m)\right], & \text{if } m=k.\end{cases}$$

Proof. It is easy to verify that statement of the lemma is valid for elementary functions from $\mathcal{E}_k$ and $\mathcal{E}_m$. The general case will follow by the limit argument.

Now let $T=\mathbb{R}_+$, $X_0=\mathbb{R}_0$, $\Delta=0$ and $L(dtdx)$ be as in Example 3. Denote by Σ_n the 'increasing simplex' of $(\mathbb{R}_+\times\mathbb{R})^n$:

$$\Sigma_n=\{(t_1,x_1,\dots,t_n,x_n)\in(\mathbb{R}_+\times\mathbb{R})^n : 0<t_1<\cdots<t_n\},$$

and we extend a function f_n defined on $\Sigma_n\times\Omega$ by making f_n symmetric with respect to pairs of nonrandom variables. If the function f_n square integrable with respect to measure M_π^n then we have

$$I_n(f_n)=n!\int_{\Sigma_n} f_n(t_1,x_1,\dots,t_n,x_n)L(dt_1dx_1)\cdots L(dt_ndx_n).$$

Indeed, this equality is clear if f_n is an elementary function of the form (3.1), and in the the general case equality will follow by the density argument, taking in account that iterated integral verifies the same properties as the multiple integral. In particular Lemma 15 holds for iterated integral. Note that the domain Σ_n and its symmetrization do not cover $(\mathbb{R}_+\times\mathbb{R})^n$: we are ignoring the 'diagonal set'. Since in the beginning of this section was proved that the 'diagonal set' has M_π^n measure zero and we consider the functions as an elements of L^2 which are the the equivalence classes, then we will always choose the representative that vanishes on the 'diagonal set'.

4 The Derivative Operator

In this section we introduce the operator D. Then we will show that it is equal to the Malliavin derivatives in the Gaussian case (see, e.g., [22]) and to the difference operator defined in [24, 26] in the Poisson case. We will also proof that the derivatives operators defined via the chaos decomposition in [2,3,18,19,25,28,31] for certain Lévy processes coincide with the operator D.

We denote by $C_b^\infty(\mathbb{R}^n)$ the set of all infinitely continuously differentiable functions $f : \mathbb{R}^n \to \mathbb{R}$ such that f and all of its partial derivatives are bounded.

Let $\mathcal{S}$ denote the class of smooth random variables such that a random variable $\xi \in \mathcal{S}$ has the form

$$\xi = f(L(h_1), \ldots, L(h_n)), \tag{4.1}$$

where f belongs to $C_b^\infty(\mathbb{R}^n)$, $h_1, \ldots, h_n$ are in K, and $n \geq 1$.

Lemma 16.

1. The set $\mathcal{S}$ is dense in $L^p(\Omega, \mathcal{F}, \mathsf{P})$, for any $p \geq 1$.
2. The set $\{\xi h : \xi \in \mathcal{S}, h \in K\}$ is dense in $L^2(T \times X \times \Omega, \mathcal{G} \otimes \mathcal{F}, M_\pi)$.
3. The set $\{ue^{L(v)} : u, v \in K\}$ is a total set of $L^2(T \times X \times \Omega, \mathcal{G} \otimes \mathcal{F}, M_\pi)$.

Proof.

1. Let $\{h_k\}_{k=1}^\infty \subset K$ be a dense subset of H. Define $\mathcal{F}_n = \sigma(L(h_1), \ldots, L(h_n))$. Then $\mathcal{F}_n \subset \mathcal{F}_{n+1}$ and $\mathcal{F}$ is the smallest σ-algebra containing all the $\mathcal{F}_n$'s. Choose a $g \in L^p(\Omega)$. Then

$$g = \mathsf{E}(g|\mathcal{F}) = \lim_{n\to\infty} \mathsf{E}(g|\mathcal{F}_n).$$

By the Doob-Dynkin Lemma we have that for each n, there exist a Borel measurable function $g_n : \mathbb{R}^n \to \mathbb{R}$ such that

$$\mathsf{E}(g|\mathcal{F}_n) = g_n(L(h_1), \ldots, L(h_n)).$$

Each such g_n can be approximated by functions $f_m^{(n)}$ where $f_m^{(n)} \in C_b^\infty(\mathbb{R}^n)$ such that $||f_m^{(n)}(L(h_1), \ldots, L(h_n)) - g_n(L(h_1), \ldots, L(h_n))||_{L^p(\Omega)}$ converges to zero as $m \to \infty$. Since $f_m^{(n)}(L(h_1), \ldots, L(h_n)) \in \mathcal{S}$ we have the first statement of the lemma.
2. It is enough to show that indicator function $\mathbf{1}_{A\times B}$, where $A \in \mathcal{G}$, $B \in \mathcal{F}$ and $M_\pi(A \times B) < \infty$ can be approximated by the processes of the form ξh, where $\xi \in \mathcal{S}$ and $h \in K$. It follows from the previous part of the lemma that there exists the sequence ξ_n in $\mathcal{S}$ such that $\xi_n \to \mathbf{1}_B$ as $n \to \infty$ in $L^2(\Omega)$. Set $C_m = \{\pi(A) \leq m\}$. Then $C_m \in \mathcal{H}$ and $\bigcup_{m\geq 1} C_m = \{\pi(A) < \infty\} \supset B$ a.s. The processes $\mathbf{1}_{A\times C_m}\xi_n$ have required form and letting $m \to \infty$ and then $n \to \infty$ we obtain the desired result.
3. Lemma 9 implies that finite linear combinations of the random variables $e^{L(v)}$, $v \in K$ are dense in $L^2(\Omega)$. The same arguments as in previous part of the lemma yield the density of the set of the linear combinations of the processes $ue^{L(v)}$, $u, v \in K$, which completes the proof of the lemma.

Definition 17. *The stochastic derivative of a smooth random variable ξ of the form (4.1) is the stochastic process $D\xi = \{D_{t,x}\xi,\ (t,x) \in T \times X\}$ indexed by the parameter space $T \times X$ given by*

$$\begin{aligned} D_{t,x}\xi = &\sum_{k=1}^{n} \frac{\partial f}{\partial y_k}(L(h_1), \dots, L(h_n))h_k(t,x)\mathbf{1}_{\Delta}(x) \\ &+\Big(f(L(h_1) + h_1(t,x), \dots, L(h_n) + h_n(t,x)) \\ &-f(L(h_1), \dots, L(h_n))\Big)\mathbf{1}_{X_0}(x). \end{aligned} \tag{4.2}$$

Remark 18.

1. If the measure ν is zero and the measure μ is deterministic then $D\xi$ coincides with the Malliavin derivative (see, for example, [22, Def. 1.2.1, p. 24]).
2. If the measure μ is zero and the measure ν is deterministic then D coincides with the difference operator defined in [26].
3. If $T = \mathbb{R}_+$, the measure μ is the Lebesgue measure and X is a metric space, and the measure ν is the product of the Lebesgue measure times the measure β satisfying $\int_M (|x|^2 \wedge 1)\beta(dx)$, then D is the operator ∇^- from [28].
4. If measures μ and ν are both deterministic then D coincides with operator defined in [33], see also [31].

Lemma 19. *Suppose that ξ is smooth functional of the form (4.1) and $h \in H$. Then*

$$\mathsf{E}\left[\int_{T\times X} D_{t,x}\xi h(t,x)\pi(dtdx)\bigg|\mathcal{H}\right] = \mathsf{E}[\xi L(h)|\mathcal{H}]. \tag{4.3}$$

Proof. The proof will be done in three steps.

Step 1. Suppose first that

$$\xi = e^{iz_1 L(h_1)} \dots e^{iz_n L(h_n)}.$$

Then $\mathrm{Re}\xi \in \mathcal{S}$ and $\mathrm{Im}\xi \in \mathcal{S}$ and

$$\begin{aligned} \mathsf{E}[\xi L(h)|\mathcal{H}] &= \frac{1}{i}\frac{d}{dz}\left(\mathsf{E}\left[\exp\left(i\sum_{k=1}^{n} z_k L(h_k) + izL(h)\right)\bigg|\mathcal{H}\right]\right)\bigg|_{z=0} \\ &= \frac{1}{i}\frac{d}{dz}\exp\left(-\frac{1}{2}\int_T\left(\sum_{k=1}^{n} z_k h_k(t,\Delta) + zh(t,\Delta)\right)^2\mu(dt)\right. \\ &\quad + \int_{T\times X_0}\left(\exp\left(i\sum_{k=1}^{n} z_k h_k(t,x) + izh(t,x)\right) - 1\right. \end{aligned}$$

$$
-i\left(\sum_{k=1}^{n} z_k h_k(t,x) + z h(t,x)\right)\Bigg)\,\nu(dtdx)\Bigg)\Bigg|_{z=0}
$$

$$
\begin{aligned}
&= \left(\int_{T\times X_0} h(t,x)\left(\exp\left(i\sum_{k=1}^{n} z_k h_k(t,x)\right) - 1\right)\nu(dtdx)\right. \\
&\quad \left. + i\int_T h(t,\Delta)\sum_{k=1}^{n} z_k h_k(t,\Delta)\mu(dt)\right) \\
&\quad \times \exp\left(-\frac{1}{2}\int_T\left(\sum_{k=1}^{n} z_k h_k(t,\Delta)\right)^2 \mu(dt)\right. \\
&\qquad \left. + \int_{T\times X_0}\left(\exp\left(i\sum_{k=1}^{n} z_k h_k(t,x)\right) - 1 - i\sum_{k=1}^{n} z_k h_k(t,x)\right)\nu(dtdx)\right) \\
&= \mathsf{E}[\xi|\mathcal{H}]\left(\int_{T\times X_0} h(t,x)\left(\exp\Big(i\sum_{k=1}^{n} z_k h_k(t,x)\Big) - 1\right)\nu(dtdx)\right. \\
&\qquad \left. + i\int_T h(t,\Delta)\sum_{k=1}^{n} z_k h_k(t,\Delta)\mu(dt)\right).
\end{aligned}
$$

On the other hand

$$
\begin{aligned}
&\mathsf{E}\left[\int_{T\times X} D_{t,x}\xi h(t,x)\pi(dtdx)\,\middle|\,\mathcal{H}\right] \\
&= \mathsf{E}\left[\int_{T\times X_0}\left(\exp\Big(i\sum_{k=1}^{n} z_k(L(h_k)+h_k(t,x))\Big) - \exp\Big(i\sum_{k=1}^{n} z_k L(h_k)\Big)\right)\right. \\
&\quad \left. h(t,x)\nu(dtdx)\,\middle|\,\mathcal{H}\right] + \mathsf{E}\left[\int_T i\sum_{j=1}^{n} z_j \exp(i\sum_{k=1}^{n} z_k L(h_k))h_j(t,\Delta)\mu(dt)\,\middle|\,\mathcal{H}\right] \\
&= \mathsf{E}[\xi|\mathcal{H}]\left(\int_{T\times X_0} h(t,x)\left(\exp\Big(i\sum_{k=1}^{n} z_k h_k(t,x)\Big) - 1\right)\nu(dtdx)\right. \\
&\qquad \left. + i\int_T h(t,\Delta)\sum_{k=1}^{n} z_k h_k(t,\Delta)\mu(dt)\right).
\end{aligned}
$$

Hence we have (4.3). By linearity we deduce that (4.3) also holds for smooth variables of the form (4.1), where the function f is a trigonometric polynomial.

Step 2. Assume that ξ of the form (4.1) such that $f \in C_b^\infty(\mathbb{R}^n)$ is periodic on every variable function. Then there is a sequence of trigonometric polynomials g_m such that $g_m \to f$ and $\partial g_m/\partial x_k \to \partial f/\partial x_k$ for every $k = 1, \ldots n$ uniformly on $\mathbb{R}^n$ as $m \to \infty$. Denote $\eta_m = g_m(L(h_1), \ldots, L(h_n))$. Then $\eta_m \in \mathcal{S}$, and by Step 1 we get

$$\mathsf{E}[\eta_m L(h)|\mathcal{H}] = \mathsf{E}\left[\int_{T\times X} D_{t,x}\eta_m h(t,x)\pi(dtdx)\middle|\mathcal{H}\right]. \tag{4.4}$$

Since $\eta_m \to \xi$ in $L^2(\Omega)$ and $D\eta_m \to D\xi$ in $L^2(T\times X\times\Omega, \mathcal{G}\otimes\mathcal{F}, M_\pi)$ then letting $m\to\infty$ in (4.4) we obtain (4.3).

Step 3. Assume that ξ of the form (4.1). Consider the sequence $\{\chi_m, m = 1,2,\dots\}$ of functions, such that $\chi_m \in C^\infty(\mathbb{R}^n)$, $0\le\chi_m\le 1$, $\chi_m(x) = 1$ if $|x|\le m$, $\chi(x) = 0$, if $|x| > m+1$ and $|\nabla\chi_m|\le 2$. Define g_m as a periodic extension on all variables of the function $f\chi_m$. Then $\zeta_m = g_m(L(h_1),\dots,L(h_n))$ is smooth variable such that $|\zeta_m|\le ||f||_{L^\infty}$ and $|D\zeta_m|\le ||\nabla f||_{L^\infty}\sum_{i=1}^n |h_i|$. Hence by the dominated convergence theorem $\zeta_m\to\xi$ in $L^2(\Omega)$ and $D\zeta_m\to D\xi$ in $L^2(T\times X\times\Omega,\mathcal{G}\otimes\mathcal{F}, M_\pi)$ as $m\to\infty$. Since by Step 2 formula (4.3) is true for ζ_m, then letting $m\to\infty$ completes the proof of the lemma.

Applying this lemma to the product of two smooth functionals we obtain the "integration by parts" formula.

Lemma 20. *Suppose ξ and η are the smooth functionals and $h\in H$, then*

$$\mathsf{E}[\xi\eta L(h)|\mathcal{H}] =$$
$$\mathsf{E}\left[\int_{T\times X}(\xi D_{t,x}\eta + \eta D_{t,x}\xi + \mathbf{1}_{X_0}D_{t,x}\xi D_{t,x}\eta)h(t,x)\pi(dtdx)\middle|\mathcal{H}\right]. \tag{4.5}$$

As a consequence of the above lemma it can be shown in the same way as in [33] that the expression of the derivative $D\xi$ given in (4.2) does not depend on the particular representation of ξ in (4.1).

For $p\ge 1$ define a norm for $\mathcal{G}\otimes\mathcal{F}$ measurable function by the following expression

$$||u||_{2,p} = \left(\mathsf{E}\left[\left(\int_{T\times X} u(t,x)^2\pi(dtdx)\right)^{p/2}\right]\right)^{1/p}.$$

Let $L^{2,p}(M_\pi) = L^{2,p}(T\times X\times\Omega,\mathcal{G}\otimes\mathcal{F}, M_\pi)$ be the set of all (equivalent classes of) functions $u(t,x,\omega)$ on $T\times X\times\Omega$ such that $||u||_{2,p} < \infty$.

Lemma 21. *The operator D is closable as an operator from $L^p(\Omega,\mathcal{F},\mathsf{P})$ to $L^{2,p}(M_\pi)$, for any $p\ge 1$.*

Proof. Let $\{\xi_n, n\ge 1\}$ be a sequence of smooth random variables such that $\mathsf{E}|\xi_n|^p\to 0$ and $D\xi_n$ converges to ζ in $L^{2,p}(M_\pi)$. Then from Lemma 20 it follows that for any $h\in K$ and $\eta\in\mathcal{S}$ we have

$$\mathsf{E}(\xi_n\eta L(h)) = \langle D\eta;\xi_n h\rangle_{L^2(M_\pi)} + \langle D\xi_n;\eta h\rangle_{L^2(M_\pi)} + \langle D\xi_n;\mathbf{1}_{X_0}hD\eta\rangle_{L^2(M_\pi)}.$$

Taking the limit as $n\to\infty$, since η, $D\eta$ are bounded, and $h\in K$ we obtain

$$\langle\zeta;\eta h\rangle_{L^2(M_\pi)} + \langle\zeta;\mathbf{1}_{X_0}hD\eta\rangle_{L^2(M_\pi)} = 0. \tag{4.6}$$

If $h(t,x)=0$ for $x\neq\Delta$, then (4.6) implies, that

$$\langle\zeta;\eta h\rangle_{L^2(M_\pi)}=0.$$

Thus from Lemma 16 we deduce $\zeta_{t,\Delta}=0$ for M_π-almost all $(t,\Delta,\omega)\in T\times\{\Delta\}\times\Omega$. Substituting this expression into (4.6) we have for any $h\in H$

$$\langle\zeta;hD\eta\rangle_{L^2(M_\pi)}=0. \tag{4.7}$$

Let $\phi_n\in C_b^\infty(\mathbb{R})$ such that $0\le\phi_n(x)\le e^x$ and $\phi_n(x)\to e^x$ for all $x\in\mathbb{R}$. Putting in (4.7) $\eta=\phi_n(L(g))$ and $h(t,x)=u(t,x)e^{-g(t,x)}$, where $u,\ g\in K$ and then letting $n\to\infty$ we get

$$\langle\zeta;ue^{L(g)}\rangle_{L^2(M_\pi)}=0.$$

It follows from Lemma 16 that $\zeta_{t,x}=0$ for M_π-a.a. $(t,x,\omega)\in T\times X\times\Omega$ completing the proof of the lemma.

We will denote the closure of D again D and its domain in $L^p(\Omega)$ by $\mathbb{D}^{1,p}$. Now we will state the chain rule.

Proposition 22. *Suppose $p\ge1$ is fixed and $\xi=(\xi^1,\dots,\xi^m)$ is a random vector whose components belong to the space $\mathbb{D}^{1,p}$. Let $\phi\in C^1(\mathbb{R}^m)$ be a function with bounded partial derivatives. Then $\phi(\xi)\in\mathbb{D}^{1,p}$ and*

$$D_{t,x}\phi(\xi)=\begin{cases}\sum_{k=1}^m\frac{\partial\phi}{\partial x_k}(\xi)D_{t,\Delta}\xi^k, & \text{if } x=\Delta,\\ \phi(\xi^1+D_{t,x}\xi^1,\dots,\xi^m+D_{t,x}\xi^m)-\phi(\xi^1,\dots,\xi^m), & \text{if } x\neq\Delta.\end{cases} \tag{4.8}$$

Proof. The proof can be easily obtain by approximation ξ by smooth random variables and the function ϕ by smooth functions with compact support.

Applying the above proposition we obtain, that $L(h)\in\mathbb{D}^{1,2}$ for all $h\in H$ and $D_{t,x}L(h)=h(t,x)$.

Lemma 23. *It holds that $P_n(\overline{x}(h))\in\mathbb{D}^{1,p}$ for all $p\ge1$, $h\in K$, $n=1,2,\dots$ and*

$$D_{t,x}P_n(\overline{x}(h))=P_{n-1}(\overline{x}(h))h(t,x). \tag{4.9}$$

Proof. As in the proof of Proposition 22 one can obtain that $P_n(\overline{x}(h))\in\mathbb{D}^{1,p}$ for all $p\ge1$, $h\in K$, $n=1,2,\dots$ and (4.8) holds. Then the definition of $\overline{x}(h)$ and equality (2.11) imply

$$D_{t,\Delta}P_n(\overline{x}(h))=\frac{\partial P_n}{\partial x_1}(\overline{x}(h))h(t,\Delta)=P_{n-1}(\overline{x}(h))h(t,\Delta).$$

It follows from the relationships (4.8) and (2.13) that for $x\neq\Delta$ we have

$$D_{t,x}P_n(\overline{x}(h))=P_n(\overline{x}(h)+\overline{u}(h(t,x)))-P_n(\overline{x}(h))=h(t,x)P_{n-1}(\overline{x}(h)),$$

where $\overline{u}(y)=(y,y^2,\dots,y^k,\dots)$. The proof is complete.

The product rule can be proved in the same manner.

Proposition 24. *Let $\xi \in \mathbb{D}^{1,p}$, $p \geq 1$ and η is a smooth variable from $\mathcal{S}$. Then $\xi\eta \in \mathbb{D}^{1,p}$ and*

$$D(\xi\eta) = \xi D\eta + \eta D\xi + D\xi D\eta \mathbf{1}_{X_0}. \tag{4.10}$$

Proof. The equation (4.10) holds if ξ and η are smooth variables. Then, the general case follows by a limit argument, using the fact that D is closed.

The following proposition is more or less evident.

Proposition 25. *Let ξ be $\mathcal{H}$-measurable random variable such that $\xi \in L^p(\Omega, \mathcal{H}, \mathsf{P})$ for some $p \geq 1$. Then $\xi \in \mathbb{D}^{1,p}$ and $D\xi = 0$ M_π-a.e.*

Proof. By the density arguments we can assume that $\xi = \mathbf{1}_U$, where $U \in \mathcal{H}$. Then for any $h \in K$ as in the proof of Lemma 7 we have $\xi \mathbf{1}_{\{L(h)\neq 0\}} = L(\xi h)^2/L(h)^2 \mathbf{1}_{\{L(h)\neq 0\}}$. Since $h \in K$ it easy to show that $\eta_\varepsilon(h) = L(\xi h)^2/(L(h)^2 + \varepsilon) = \xi L(h)^2/(L(h)^2+\varepsilon) \to \xi \mathbf{1}_{\{L(h)\neq 0\}}$ as $\varepsilon \to 0$ in $L^p(\Omega)$. If we will show that $\eta_\varepsilon(h) \in \mathbb{D}^{1,p}$ and $D\eta_\varepsilon(h) \to 0$ in $L^{2,p}(M_\pi)$ for any $h \in K$ then $\xi \mathbf{1}_{\{L(h)\neq 0\}}$ will be in $\mathbb{D}^{1,p}$ and $D(\xi \mathbf{1}_{\{L(h)\neq 0\}}) = 0$ for any $h \in K$ implying as in the proof of Lemma 7 that $\xi \in \mathbb{D}^{1,p}$ and $D\xi = 0$.

Let us show that $\eta_\varepsilon(h) \in \mathbb{D}^{1,p}$ and $D\eta_\varepsilon(h) \to 0$. Set $f(x,y) = x^2/(y^2+\varepsilon)$. Then $f(x,y)e^{-(x^2+y^2)/n} = f_n(x,y) \in C_b^\infty(\mathbb{R}^2)$ and by dominated convergence theorem we have $f_n(L(\xi h), L(h)) \to f(L(\xi h), L(h)) = \eta_\varepsilon(h)$ as $n \to \infty$ in $L^p(\Omega)$. In the same way we obtain that the derivative $Df_n(L(\xi h), L(h))$ converges in $L^{2,p}(M_\pi)$ which implies $\eta_\varepsilon(h) \in \mathbb{D}^{1,p}$ and the limit $D(\eta_\varepsilon(h))$ is given by

$$\begin{aligned} D(\eta_\varepsilon(h)) &= \left(\frac{2L(\xi h)\xi h}{L(h)^2+\varepsilon} - \frac{2L(h)L(\xi h)^2 h}{(L(h)^2+\varepsilon)^2} \right) \mathbf{1}_\Delta \\ &\quad + \left(\frac{(L(\xi h)+\xi h)^2}{(L(h)+h)^2+\varepsilon} - \frac{L(\xi h)^2}{L(h)^2+\varepsilon} \right) \mathbf{1}_{X_0} \\ &= \left(\frac{2\xi^2 L(h)h}{(L(h)^2+\varepsilon)^2} \mathbf{1}_\Delta + \xi^2 \frac{(L(h)+h)^2 - L(h)^2}{((L(h)+h)^2+\varepsilon)(L(h)^2+\varepsilon)} \mathbf{1}_{X_0} \right) \varepsilon. \end{aligned}$$

Letting $\varepsilon \to 0$ in the equality above we obtain the desired result.

The following lemma shows the action of the operator D via the chaos decomposition.

Proposition 26. *Let $\xi \in L^2(\Omega)$ with a development*

$$\xi = \sum_{k=0}^{\infty} I_k(f_k), \tag{4.11}$$

where $f_k \in L^2(M_\pi^k)$ are symmetric with respect to pairs of nonrandom variables. Then $\xi \in \mathbb{D}^{1,2}$ if and only if

$$\sum_{k=1}^{\infty} kk!\,||f_k||^2_{L^2(M_\pi^k)} < \infty \tag{4.12}$$

and in this case we have

$$D_{t,x}\xi = \sum_{k=1}^{\infty} kI_{k-1}(f_k(\cdot,t,x)). \tag{4.13}$$

Moreover

$$\mathsf{E}\left[\int_{T\times X}(D_{t,x}\xi)^2\pi(dtdx)\bigg|\mathcal{H}\right] =$$

$$\sum_{k=1}^{\infty} kk!\int_{(T\times X)^k} f_k(t_1,x_1,\ldots,t_k,x_k)^2\pi(dt_1dx_1)\cdots\pi(dt_kdx_k).$$

Proof. The proof will be done in three steps.

Step 1. Suppose first that $k \geq 1$ and

$$\xi = P_k(\overline{x}(h)) = \frac{1}{k!}I_k(h^{\otimes k}) = I_k(f_k), \tag{4.14}$$

with $h \in K$. Then by Lemma 23 $\xi \in \mathbb{D}^{1,2}$ and by equality (4.9) we get

$$D_{t,x}P_k(\overline{x}(h)) = P_{k-1}(\overline{x}(h))h(t,x).$$

Hence

$$D_{t,x}\xi = kI_{k-1}(f_k(\cdot,t,x)). \tag{4.15}$$

¿From Proposition 25 and formula (4.10) we deduce that equality (4.15) holds for any linear combination of random variables of the form $\eta P_k(\overline{x}(h))$, where η is $\mathcal{H}$-measurable bounded random variable. Since formula (4.15) implies that $||D\xi||^2_{L^2(M_\pi)} = k\mathsf{E}\xi^2$ then it follows that $\mathcal{P}_k$, $k \geq 1$ is included in $\mathbb{D}^{1,2}$.

If $k = 0$ then Proposition 25 implies that $\mathcal{P}_0 = L^2(\Omega,\mathcal{H},\mathsf{P}) \subset \mathbb{D}^{1,2}$.

Step 2. Let $\xi \in L^2(\Omega)$ has an expansion (4.11). Suppose that (4.12) holds. Define

$$\xi_n = \sum_{k=0}^{n} I_k(f_k).$$

Then the sequence ξ_n converges to ξ in $L^2(\Omega)$, and by Step 1 we have $\xi_n \in \mathbb{D}^{1,2}$ and $D_{t,x}\xi = \sum_{k=1}^{n} kI_{k-1}(f_k(\cdot,t,x))$. It follows from Lemma 15 and equality (4.12) that $D_{t,x}\xi_n$ converges in $L^2(M_\pi)$ to the right-hand side of (4.13). Therefore $\xi \in \mathbb{D}^{1,2}$ and (4.13) holds.

Step 3. Suppose $\xi \in \mathbb{D}^{1,2}$. Note that formula (4.5) holds for $\xi \in \mathbb{D}^{1,2}$ and $\eta \in \mathbb{D}^{1,p}$ for some $p > 2$ if $h \in K$. Since by Proposition 23 $\eta = P_m(\overline{x}(g)) \in \mathbb{D}^{1,p}$ for all $p \geq 1$ and $g \in K$, then we have

$$\begin{aligned}\lim_{n\to\infty}(\langle D\xi_n;\eta h\rangle_{L^2(M_\pi)} + \langle D\xi_n; D\eta h\mathbf{1}_{X_0}\rangle_{L^2(M_\pi)})\\ &= \lim_{n\to\infty}(\mathsf{E}(\xi_n\eta L(h)) - \langle D\eta;\xi_n h\rangle_{L^2(M_\pi)})\\ &= \mathsf{E}(\xi\eta L(h)) - \langle D\eta;\xi h\rangle_{L^2(M_\pi)}\\ &= \langle D\xi;\eta h\rangle_{L^2(M_\pi)} + \langle D\xi; D\eta h\mathbf{1}_{X_0}\rangle_{L^2(M_\pi)}.\end{aligned}$$

It follows from equation (23) that

$$\eta + \mathbf{1}_{X_0}D\eta = P_m(\overline{x}(g)) + \mathbf{1}_{X_0}gP_{m-1}(\overline{x}(g)).$$

Then for all $m = 1, 2, \ldots$ we obtain

$$\begin{aligned}\lim_{n\to\infty}(\langle D\xi_n; P_m(\overline{x}(g))h\rangle_{L^2(M_\pi)} + \langle D\xi_n; P_{m-1}(\overline{x}(g))gh\mathbf{1}_{X_0}\rangle_{L^2(M_\pi)})\\ = \langle D\xi; P_m(\overline{x}(g))h\rangle_{L^2(M_\pi)} + \langle D\xi; P_{m-1}(\overline{x}(g))gh\mathbf{1}_{X_0}\rangle_{L^2(M_\pi)}.\end{aligned}$$

Since $P_0 = 1$ and $\lim_{n\to\infty}\langle D\xi_n; P_0(\overline{x}(g))h\rangle_{L^2(M_\pi)} = \langle D\xi; P_0(\overline{x}(g))h\rangle_{L^2(M_\pi)}$ for all $h \in L^2(M_\pi)$, then we deduce by induction that

$$\lim_{n\to\infty}\langle D\xi_n; P_m(\overline{x}(g))h\rangle_{L^2(M_\pi)} = \langle D\xi; P_m(\overline{x}(g))h\rangle_{L^2(M_\pi)}.$$

From Lemma 15 we deduce that for $n > m$ the expression

$$\langle D\xi_n; P_m(\overline{x}(g))h\rangle_{L^2(M_\pi)}$$

is equal to

$$\mathsf{E}\left(\int_{T\times X}(m+1)I_m\left(f_{m+1}(\cdot,t,x)\right)h(t,x)\pi(dtdx)P_m(\overline{x}(g))\right).$$

Hence the projection of $\int_{T\times X} D_{t,x}\xi h(t,x)\pi(dtdx)$ on the m-th chaos is equal to

$$\int_{T\times X}(m+1)I_m\left(f_{m+1}(\cdot,t,x)\right)h(t,x)\pi(dtdx).$$

Thus for any $\zeta \in L^2(\Omega,\mathcal{F},\mathsf{P})$ we have

$$\begin{aligned}\langle D\xi; h\zeta\rangle_{L^2(M_\pi)} &= \mathsf{E}\left(\int_{T\times X} D_{t,x}\xi h(t,x)\pi(dtdx)\zeta\right)\\ &= \mathsf{E}\left(\sum_{m=0}^{\infty}\int_{T\times X}(m+1)I_m\left(f_{m+1}(\cdot,t,x)\right)h(t,x)\pi(dtdx)\zeta\right)\\ &= \left\langle\sum_{m=0}^{\infty}(m+1)I_m\left(f_{m+1}(\cdot,t,x)\right)h\zeta\right\rangle_{L^2(M_\pi)}.\end{aligned}$$

Since the set $\{h\zeta : h \in K, \zeta \in L^2(\Omega,\mathcal{F},\mathsf{P})\}$ is dense in $L^2(T\times X\times\Omega, \mathcal{G}\otimes\mathcal{F}, M_\pi)$ then

$$D_{t,x}\xi = \sum_{m=0}^{\infty} (m+1) I_m \left(f_{m+1}(\cdot, t, x) \right),$$

which completes the proof of the proposition.

Remark 27. This proposition implies that the operator D is an annihilation operator on the Fock space on Hilbert space H.

The equations (4.13) can be considered as a definition of the operator D. This approach was developed for pure jump Lévy process, the particular case of Poisson processes, the case of general Lévy process with no drift and the case of certain class of martingales in [2, 3, 18, 19, 25, 28].

Let $A \in \mathcal{G} \otimes \mathcal{H}$. We will denote by $\mathcal{F}_A^0$ the σ-algebra generated by the random variables $\{L(B), B \subset A, B \in \mathcal{G}_0\}$. Set $\mathcal{F}_A = \mathcal{F}_A^0 \vee \mathcal{H}$. The following results are modification of Proposition 1.2.5 from [22, p. 32] and it shows how to compute the derivative of a conditional expectation with respect to a σ-algebra generated by stochastic process.

Lemma 28. *Suppose that $\xi \in L^2(\Omega, \mathcal{F}, \mathsf{P})$ with the expansion (4.11). Let $A \in \mathcal{G} \otimes \mathcal{H}$. Then*

$$\mathsf{E}[\xi | \mathcal{F}_A] = \sum_{k=0}^{\infty} I_k(f_k \mathbf{1}_A^{\otimes k}). \tag{4.16}$$

Proof. By the density of elementary functions in $L^2(M_\pi^k)$ and by linearity we can assume that $\xi = I_k(f_k)$, where $f_k = \eta \mathbf{1}_{A_1} \otimes \cdots \otimes \mathbf{1}_{A_k}$ with pairwise-disjoint sets $A_1, \ldots, A_k \in \mathcal{G}_0$ and $\eta \in L^\infty(\Omega, \mathcal{H}, \mathsf{P})$. Then we have

$$\mathsf{E}[\xi|\mathcal{F}_A] = \mathsf{E}[\eta L(A_1) \cdots L(A_k)|\mathcal{F}_A] = \eta \mathsf{E}\left[\prod_{i=1}^{k} (L(A_i \cap A) + L(A_i \setminus A)) \middle| \mathcal{F}_A \right]$$
$$= \eta \mathsf{E}[L(A_1 \cap A) \cdots L(A_m \cap A)|\mathcal{F}_A] =$$
$$= I_k(\eta \mathbf{1}_{A_1 \cap A} \otimes \cdots \otimes \mathbf{1}_{A_k \cap A}) = I_k(f_k \mathbf{1}_A^{\otimes k}).$$

Proposition 29. *Suppose that $\xi \in \mathbb{D}^{1,2}$, and $A \in \mathcal{G} \otimes \mathcal{H}$. Then $\mathsf{E}(\xi|\mathcal{F}_A) \in \mathbb{D}^{1,2}$ and we have*

$$D_{t,x}(\mathsf{E}(\xi|\mathcal{F}_A)) = \mathsf{E}(D_{t,x}\xi|\mathcal{F}_A)\mathbf{1}_A(t,x)$$

M_π-a.e. in $T \times X \times \Omega$.

Proof. By Lemma 28 and Proposition 26 we obtain

$$\mathsf{E}(D_{t,x}\xi|\mathcal{F}_A)\mathbf{1}_A(t,x) = \sum_{k=1}^{\infty} k I_{k-1}(f_k(\cdot, t, x)\mathbf{1}_A^{\otimes(k-1)})\mathbf{1}_A(t,x) = D_{t,x}(\mathsf{E}(\xi|\mathcal{F}_A)).$$

Remark 30. In particular, if ξ is $\mathcal{F}_A$-measurable and belongs to $\mathbb{D}^{1,2}$, then $D_{t,x}\xi = 0$ M_π-a.e. in A^c.

5 The Skorohod Integral

In this section we consider the adjoint of the operator D, and we will show that it coincides with the Skorohod integral [30] in the Gaussian case and with the extended stochastic integral introduced by Kabanov [13] in the pure jump Lévy case. See also [2, 3, 18, 28]. So it can be considered as a generalization of the stochastic integral. We will call it Skorohod integral and will establish the expression of it in terms of the chaos expansion as well as prove some of its properties.

We recall that the derivative operator D is a closed and unbounded operator defined on the dense subset $\mathbb{D}^{1,2}$ of $L^2(\Omega)$ with values in $L^2(T \times X \times \Omega, \mathcal{G} \otimes \mathcal{F}, M_\pi)$.

Definition 31. *We denote by δ the adjoint of the operator D and will call it Skorohod integral.*

The operator δ is closed unbounded operator on $L^2(T \times X \times \Omega, \mathcal{G} \otimes \mathcal{F}, M_\pi)$ with values in $L^2(\Omega)$ defined on $\mathrm{Dom}\delta$, where $\mathrm{Dom}\delta$ is the set of processes $u \in L^2(M_\pi)$ such that

$$\left| \mathsf{E} \int_{T\times X} D_{t,x}\xi u(t,x)\pi(dtdx) \right| \leq c\,||\xi||_{L^2(\Omega)}$$

for all $\xi \in \mathbb{D}^{1,2}$, where c is some constant depending on u.

If $u \in \mathrm{Dom}\delta$, then $\delta(u)$ is the element of $L^2(\Omega)$ such that

$$\mathsf{E}(\xi\delta(u)) = \mathsf{E} \int_{T\times X} D_{t,x}\xi u(t,x)\pi(dtdx) \qquad (5.1)$$

for any $\xi \in \mathbb{D}^{1,2}$.

The following proposition shows the behavior of δ in terms of the chaos expansion.

Proposition 32. *Let $u \in L^2(T \times X \times \Omega, \mathcal{G} \otimes \mathcal{F}, M_\pi)$ with the expansion*

$$u(t,x) = \sum_{k=0}^{\infty} I_k(f_k(\cdot,t,x)). \qquad (5.2)$$

Then $u \in \mathrm{Dom}\delta$ if and only if the series

$$\delta(u) = \sum_{k=0}^{\infty} I_{k+1}(\tilde{f}_k) \qquad (5.3)$$

converges in $L^2(\Omega)$.

Recall that $\tilde{f}_k$ is a symmetrization of f_k in all its pairs of nonrandom variables is given by

$$\tilde{f}_k(t_1, x_1, \ldots, t_k, x_k, t, x, \omega) = \frac{1}{k+1}\Bigg(f_k(t_1, x_1, \ldots, t_k, x_k, t, x, \omega)$$

$$+ \sum_{i=1}^{k} f_k(t_1, x_1, \ldots, t_{i-1}, x_{i-1}, t, x, t_{i+1}, x_{i+1}, \ldots, t_i, x_i, \omega) \Bigg).$$

Proof. The proof is the same as in the Gaussian case (see, e.g., [22, Prop. 1.3.1, p. 36]).

Remark 33. It follows from Proposition 32 that the operator δ coincides with Skorohod integral in the Gaussian case and with extended stochastic integral introduced by Kabanov for pure jump Lévy processes (see, e.g., [2, 3, 13, 18, 22, 28, 30]).

It follows from proposition above that Domδ is the subspace of $L^2(M_\pi)$ formed by the processes that satisfy the following condition:

$$\sum_{k=1}^{\infty} (k+1)! ||\tilde{f}_k||^2_{L^2(M_\pi^{k+1})} < \infty. \tag{5.4}$$

If $u \in$ Domδ, then the sum of the series (5.4) is equal to $\mathsf{E}\delta(u)^2$.

Note that the Skorohod integral is a linear operator and has zero mean, e.g., $\mathsf{E}(\delta(u)) = 0$ if $u \in$ Domδ. The following statements prove some properties of δ.

Proposition 34. *Let $u, v \in$ Domδ be arbitrary stochastic process. Then for all α and β in $L^\infty(\Omega, \mathcal{H}, \mathsf{P})$ we have $\alpha u + \beta v \in$ Domδ and*

$$\delta(\alpha u + \beta v) = \alpha\delta(u) + \beta\delta(v).$$

Moreover $\mathsf{E}[\delta(u)|\mathcal{H}] = 0$.

Proof. The proof follows from the properties (i) and (iii) of the multiple integral.

Proposition 35. *Suppose that u is a Skorohod integrable process. Let $\xi \in \mathbb{D}^{1,2}$ such that $\mathsf{E}(\int_{T\times X}(\xi^2 + (D_{t,x}\xi)^2 \mathbf{1}_{X_0})u(t,x)^2\pi(dtdx)) < \infty$. Then it holds that*

$$\delta((\xi + \mathbf{1}_{X_0} D\xi)u) = \xi\delta(u) - \int_{T\times X} (D_{t,x}\xi)u(t,x)\pi(dtdx), \tag{5.5}$$

provided that one of the two sides of the equality (5.5) exists.

Proof. Let $\eta \in \mathcal{S}$ be a smooth random variables. Then by the product rule (4.10) and by the duality relation (5.1), we get

$$\begin{aligned} &\mathsf{E}\left(\int_{T\times X}(D_{t,x}\eta)(\xi+\mathbf{1}_{X_0}(x)D_{t,x}\xi)u(t,x)\pi(dtdx)\right)\\ &=\int_{T\times X}\mathsf{E}(u(t,x)(D_{t,x}(\xi\eta)-\eta D_{t,x}\xi))\pi(dtdx)\\ &=\mathsf{E}\left(\eta(\xi\delta(u)-\int_{T\times X}(D_{t,x}\xi)u(t,x)\pi(dtdx))\right), \end{aligned}$$

and the result follows.

As in the Gaussian case or in the case of processes with independent increments in order to prove some other properties of Skorohod integral we will define a class of processes contained in Domδ (see [22, 33]).

Definition 36. *Let $\mathbb{L}^{1,2}$ denote the class of processes $u \in L^2(T\times X\times\Omega, \mathcal{G}\otimes\mathcal{F}, M_\pi)$ such that $u(t,x)\in\mathbb{D}^{1,2}$ for all $(t,x)\notin R$, where $R\subset T\times X$ and $M_\pi(R\times\Omega)=0$, and there exists a measurable version of the multiparametrical process $D_{t,x}u(s,y)$ satisfying $\mathsf{E}\int_{T\times X}\int_{T\times X}(D_{t,x}u(s,y))^2\pi(dtdx)\pi(dsdy)<\infty$.*

If the process u has the expansion (5.2), then $u\in\mathbb{L}^{1,2}$ if and only if the series

$$\begin{aligned} &\int_{T\times X}\int_{T\times X}\mathsf{E}\left(\sum_{k=1}^{\infty}kI_{k-1}(f_k(\cdot,t,x,s,y))\right)^2\pi(dtdx)\pi(dsdy)\\ &=\sum_{k=1}^{\infty}kk!\,||f_k||^2_{L^2(M_\pi^{k+1})} \end{aligned}$$

converges.

Since $||\tilde{f}_k||_{L^2(M_\pi^{k+1})}\le||f_k||_{L^2(M_\pi^{k+1})}$ then from (5.4) we deduce that $\mathbb{L}^{1,2}\subset$ Domδ.

The proofs of the following propositions use the chaos expansion therefore they can be done as in the Gaussian case (see, for instance [22, pp. 38–40]).

Proposition 37. *Suppose that $u\in\mathbb{L}^{1,2}$ and for all $(t,x)\notin R$, where $R\subset T\times X$ and $M_\pi(R\times\Omega)=0$ the two-parameter process $\{D_{t,x}u(s,y),(s,y)\in T\times X\}$ is Skorohod integrable, and there exists a version of the process $\{\delta(D_{t,x}u(\cdot,\cdot)),(t,x)\in T\times X\}$ which belongs to $L^2(M_\pi)$. Then $\delta(u)\in\mathbb{D}^{1,2}$ and we have*

$$D_{t,x}\delta(u)=u(t,x)+\delta(D_{t,x}u(\cdot,\cdot)). \tag{5.6}$$

Proposition 38. *Suppose that $u\in\mathbb{L}^{1,2}$ and $v\in\mathbb{L}^{1,2}$. Then we have*

$$\begin{aligned} \mathsf{E}[\delta(u)\delta(v)|\mathcal{H}]&=\mathsf{E}\left[\int_{T\times X}u(t,x)v(t,x)\pi(dtdx)\middle|\mathcal{H}\right]\\ &+\mathsf{E}\left[\int_{T\times X}\int_{T\times X}D_{s,y}u(t,x)D_{t,x}v(s,y)\pi(dtdx)\pi(dsdy)\middle|\mathcal{H}\right]. \end{aligned}\tag{5.7}$$

Now we will show that the operator δ is an extension of the Itô integral. Let L_t, $t \in [0;1]$ be a processes with $\mathcal{H}$-conditionally independent increments. Assume that the canonical triplet of its characteristics (B, μ, ν) such that $B = 0$. Then as in Example 3 we have random measure $N(dtdx)$ associated to jumps of L with compensator measure ν, the measure μ connected with continuous part of L and conditional additive process $L(h)$ on H. We denote by L_p^2 the subset of $L^2(M_\pi)$ formed by $\mathcal{F}_t$-predictable processes.

The following technical lemma will be needed.

Lemma 39. *Let $A \in \mathcal{G} \otimes \mathcal{H}$ be a set with finite M_π measure, and let ξ be a square integrable random variable that is measurable with respect to the σ-algebra $\mathcal{F}_{A^c}$. Then the process $\xi \mathbf{1}_A$ is Skorohod integrable and*

$$\delta(\xi \mathbf{1}_A) = \xi L(A).$$

Proof. Suppose first that $\xi \in \mathbb{D}^{1,2}$ and $\mathbf{1}_A \in K$. By using Proposition 35 and Remark 30 we have

$$\delta(\xi \mathbf{1}_A) = \delta((\xi + \mathbf{1}_{X_0} D\xi)\mathbf{1}_A) = \xi\delta(\mathbf{1}_A) - \int_{T\times X} (D_{t,x}\xi)\mathbf{1}_A(t,x)\pi(dtdx) = \xi L(A).$$

The general case follows by a limit argument, using the facts that $\mathbb{D}^{1,2}$ and K are dense and δ is closed.

Proposition 40. $L_p^2 \subset \mathrm{Dom}\delta$, *and the restriction of the operator δ to the space L_p^2 coincides with the usual stochastic integral, that is*

$$\delta(u) = \int_0^1 u(t,0)dL^c(t) + \int_0^1 \int_{\mathbb{R}_0} u(t,x)(N(dtdx) - \nu(dtdx)).$$

Proof. Suppose that u is an elementary adapted processes of the form

$$u_{t,x} = \sum_{i=1}^{n} \xi_i \mathbf{1}_{A_i} \mathbf{1}_{(t_i;t_{i+1}]}(t)\mathbf{1}_{B_i}(x),$$

where $0 \le t_1 < \cdots < t_{n+1} \le 1$, B_i is a borel set of $\mathbb{R}$, $A_i \in \mathcal{H}$ such that $\mathbf{1}_{(t_i;t_{i+1}]\times B_i \times A_i}(t,x) \in K$ and ξ_i is square integrable and $\mathcal{F}_{t_i}$ measurable random variable. Then from the Lemma 39 we obtain $u \in \mathrm{Dom}\delta$ and

$$\delta(u) = \sum_{i=1}^{\infty} \xi_i((L_{t_{i+1}} - L_{t_i})\mathbf{1}_{B_i}(0) + \int_{t_i}^{t_{i+1}} \int_{B_i} (N(dtdx) - \nu(dtdx)).$$

The general case follows by monotone class argument since δ is closed.

The predictable projection of a stochastic process indexed by $t \ge 0$ and $x \in \mathbb{R}$ can be defined similarly as in a one parametrical case (see, i.g, [12, 17, 31]). Let $Y = \{Y(t,x), t \ge 0, x \in \mathbb{R}\}$ be a measurable integrable process.

There exists a predictable process $Z = \{Z(t,x), t \geq 0, x \in \mathbb{R}\}$ such that for every predictable stopping time τ

$$Z(\tau, x)\mathbf{1}_{\{\tau<\infty\}} = \mathsf{E}[Y(\tau, x)\mathbf{1}_{\{\tau<\infty\}}|\mathcal{F}_{\tau-}].$$

We will call the process Z predictable projection of process Y and will denote it by ${}_p(Y(t,x))$. The following result is so-called Clark-Haussmann-Ocone formula.

Proposition 41. *Let $\xi \in \mathbb{D}^{1,2}$, and suppose that a process with conditionally independent increments L_t, $t \in [0;1]$ has the form*

$$L_t = L_t^c + \int_0^t \int_{|x|\leq 1} x(N(dsdx) - \nu(dsdx)) + \int_0^t \int_{|x|>1} xN(dsdx).$$

Then

$$\xi = \mathsf{E}[\xi|\mathcal{H}] + \int_0^1 {}^p(D_{t,0}\xi)dL_t^c + \int_0^1 \int_{\mathbb{R}_0} {}^p(D_{t,x}\xi)(N(dtdx) - \nu(dtdx)). \quad (5.8)$$

Proof. Let $\xi \in \mathbb{D}^{1,2}$ have an expansion $\xi = \sum_{n=0}^{\infty} I_n(f_n)$. Using (4.13) and (4.16) we have

$$\mathsf{E}[D_{t,x}\xi|\mathcal{F}_t] = \sum_{n=1}^{\infty} n\mathsf{E}[I_{n-1}(f_n(\cdot,t,x))|\mathcal{F}_t] = \sum_{n=1}^{\infty} nI_{n-1}(f_n(\cdot,t,x)\mathbf{1}_{[0;t]}^{\otimes(n-1)}).$$

It follows from the arguments in the beginning of the Section 3, that

$$f_n(\cdot,t,x)\mathbf{1}_{[0;t]}^{\otimes(n-1)} = f_n(\cdot,t,x)\mathbf{1}_{[0;t)}^{\otimes(n-1)}$$

as elements of $L^2(M_\pi^{n-1})$. Thus

$$\mathsf{E}[D_{t,x}\xi|\mathcal{F}_t] = \sum_{n=1}^{\infty} nI_{n-1}(f_n(\cdot,t,x)\mathbf{1}_{[0;t]}^{\otimes(n-1)}) = \sum_{n=1}^{\infty} nI_{n-1}(f_n(\cdot,t,x)\mathbf{1}_{[0;t)}^{\otimes(n-1)}).$$

Since $I_{n-1}(f_n(\cdot,t,x)\mathbf{1}_{[0;t)}^{\otimes(n-1)})$ is predictable then it is easy to show that ${}^p\big(I_{n-1}(f_n(\cdot,t,x))\big) = I_{n-1}(f_n(\cdot,t,x)\mathbf{1}_{[0;t)}^{\otimes(n-1)})$. Hence

$${}^p(D_{t,x}\xi) = \sum_{n=1}^{\infty} n\, {}^p(I_{n-1}(f_n(\cdot,t,x))) = \sum_{n=1}^{\infty} nI_{n-1}(f_n(\cdot,t,x)\mathbf{1}_{[0;t)}^{\otimes(n-1)}).$$

Therefore ${}^p(D_{t,x}\xi) = \mathsf{E}[D_{t,x}\xi|\mathcal{F}_t]$ as an $L^2(M_\pi)$ processes.

Set $\phi(t,x) = \mathsf{E}[D_{t,x}|\mathcal{F}_t]$ and $\psi(t,x) = {}^p(D_{t,x}\xi)$. Then from equality (5.3) we deduce

$$\delta(\psi) = \delta(\phi) = \sum_{n=1}^{\infty} I_n(f_n) = \xi - \mathsf{E}[\xi|\mathcal{H}],$$

which shows the desired result because Proposition 40 implies that

$$\delta(\psi) = \int_0^1 {}^p(D_{t,0}\xi)dL_t^c + \int_0^1 \int_{\mathbb{R}_0} {}^p(D_{t,x}\xi)(N(dtdx) - \nu(dtdx)).$$

Acknowledgments

I would like to thank Bernt Øksendal for his encouragement and interest, Paul Kettler for the attentive reading and the Department of Mathematics, University of Oslo, for its warm hospitality. This work was supported by INTAS grant 03-55-1861.

References

1. Bass, R.F., Cranston, M.: The Malliavin calculus for pure jump Lévy processes and applications to local time. Ann. Probab. 14 (1986), pp. 490–532.
2. Benth, F.E., Di Nunno, G., Løkka, A., Øksendal, B., Proske, F.: Explicit representation of the minimal variance portfolio in markets driven by Lévy processes. Math. Finance 13 (2003), pp. 54–72.
3. Benth, F.E., Løkka, A.: Anticipative calculus for Lévy processes and stochastic differential equations. Stoch. Stoch. Rep. 76 (2004), no. 3, 191–211.
4. Bichteler, K., Gravereaux, J.B., Jacod, J.: Malliavin Calculus for Processes with Jumps. Gordon and reach Science Publisher, New York 1987.
5. Di Nunno, G.: On orthogonal polynomials and the Malliavin derivative for Lévy stochastic measures. Preprint series in Pure Mathematics, University of Oslo, 10, 2004.
6. Di Nunno, G.: Random Fields: Skorohod integral and Malliavin derivative. Preprint series in Pure Mathematics, University of Oslo, 36, 2004.
7. Di Nunno, G., Meyer-Brandis, T., Øksendal, B., Proske, F.: Malliavin calculus and anticipative Itô formulae for Lévy processes. Infin. Dimens. Anal. Quantum Probab. Relat. Top. 8 (2005), no. 2, 235–258.
8. Di Nunno, G., Øksendal, B., Proske, F.: White noise analysis for Lévy processes. J. Funct. Anal. 206 (2004), no. 1, 109–148.
9. Engel, D.D.: The multiple stochastic integral. Mem. Amer. Math. Soc. 265 (1982).
10. Itô, K.: Multiple Wiener integral. J. Math. Soc. Japan. 3 (1951), pp. 157–169.
11. Itô, K.: Spectral type of the shift transformation of differential processes with stationary increments. Trans. Am. Math. Soc. 81 (1956), pp. 253–263.
12. Jacod, J., Shiryaev, A.N.: Limit theorems for stochastic processes. 2nd edition. Grundlehren der Mathematischen Wissenschaften [Fundamental Principles of Mathematical Sciences], 288. Springer-Verlag, Berlin, 2003.
13. Kabanov, Yu. M.: On extended stochastic integrals. Th. Probab. Appl. 20 (1975), pp. 710–722.
14. Karatzas, I., Ocone, D.L.; A generalized Clark representation formula, with application to optimal portfolios. Stochastics Stochastics Rep. 34 (1991), no. 3-4, 187–220.
15. Kulik, A.M.: Malliavin calculus for functionals with generalized derivatives and some applications to stable processes. Ukrainian Math. J. 54 (2002), no. 2, 266–279.
16. Léon, J.A., Solé, J.L., Utzet, F., Vives, J.: On Lévy processes, Malliavin calculus and market models with jumps. Finance Stochast. 6 (2002), pp. 197–225.

17. Liptser, R. Sh., Shiryayev, A.N.: Theory of martingales. Mathematics and its Applications (Soviet Series), 49. Kluwer Academic Publishers Group, Dordrecht, 1989.
18. Løkka, A.: Martingale representation and functionals of Lévy processes. Stochastic Anal. Appl. 22 (2004), no. 4, 867–892.
19. Ma, J., Protter, P., San Martin, J.: Anticipating integrals for a class of martingales. Bernoulli 4 (1998), no. 1, pp. 81–114.
20. Malliavin, P.: Stochastic Analysis. Springer-Verlag, New York 1997.
21. Meyer, P.A. Un cours sur les inteégrales stochastiques. Séminare de Probabilités X, Lect. Math. Notes, Vol. 511, Springer Verlag 1976, pp. 245–400.
22. Nualart, D.: The Malliavin Calculus and Related Topics. Springer, Berlin 1995.
23. Nualart, D., Schoutens, W.: Chaotic and predictable representations for Lévy processes. Stochastic Process. Appl. 90 (2000), pp. 109–122.
24. Nualart, D., Vives, J.: Anticipating calculus for the Poisson process based on the Fock space. Séminare de Probabilités XXIV, Lect. Math. Notes, Vol. 1426, Springer Verlag 1990, pp. 154–165.
25. Øksendal, B., Proske, F.: White noise of Poisson random measures. Potential Anal. 21 (2004), no. 4, 375–403.
26. Picard, J.: On the existence of smooth densities for jump processes. Probab. Theory Rel. Fields 105 (1996), pp. 481–511.
27. Pollard, P: A user's guide to measure theoretic probability. Cambridge Series in Statistical and Probabilistic Mathematics, 8. Cambridge University Press, Cambridge, 2002.
28. Privault, N.: An extension of stochastic calculus to certain non-Markovian processes. Preprint. (1997).
29. Russo, F., Valois, P.: Product of two multiple stochastic integrals with respect to a normal martingale. Stoch. Proc. and Appl. 73 (1998), pp. 47–68.
30. Skorohod, A.V.: On generalization of a stochastic integral. Theory Probab. Appl. 20 (1975), pp. 219–233.
31. Solé, J.L., Utzet, F. and Vives, J.: Canonical Lévy process and Malliavin calculus. Preprint. (2005).
32. Surgailis, D.: On L^2 and non-L^2 multiple stochastic integration. Lecture Notes in Control and Information Sci., Vol. 36, Springer, Berlin-New York, 1981, pp. 212–226.
33. Yablonski, A.: The calculus of variations for processes with independent increments. Preprint series in Pure Mathematics, University of Oslo, 15, 2004.

Printing: Krips bv, Meppel
Binding: Stürtz, Würzburg